J. Sündermann (Ed.)

Circulation and Contaminant Fluxes in the North Sea

With 391 Figures

Springer-Verlag
Berlin Heidelberg New York London Paris
Tokyo Hong Kong Barcelona Budapest

Professor Dr. JÜRGEN SÜNDERMANN

Zentrum für Meeres- und Klimaforschung
Universität Hamburg
Institut für Meereskunde
Troplowitzstraße 7
D-22529 Hamburg
Germany

Cover picture: Suspended matter distribution in the southern North Sea in May 1986, derived from satellite data.

ISBN 3-540-56825-5 Springer-Verlag Berlin Heidelberg New York
ISBN 0-387-56825-5 Springer-Verlag New York Berlin Heidelberg

Library of Congress Cataloging-in-Publication Data. Circulation and contaminant fluxes in the North Sea / J. Sündermann (ed.). p. cm. Includes bibliographical references and index. ISBN 3-540-56825-5 (Berlin : acid-free). - ISBN 0-387-56825-5 (New York : acid-free) 1. Ocean circulation - North Sea. 2. Marine pollution - North Sea. I. Sündermann, Jürgen. GC228.6N69C57 1994 551.46'136-dc20 93-38496

© Springer-Verlag Berlin Heidelberg 1994
Printed in Germany

Cover design: Struve & Partner, Heidelberg
Typesetting: Camera ready by author
32/3145-5 4 3 2 1 0 - Printed on acid-free paper

Contents

3 Model Experiments

4 Interdisciplinary Evaluation of Field and Model Data

5 A Composite View of the North Sea Ecosystem and Future Research Needs

List of Contributors

L. Aletsee
RWTH-Aachen, Abteilung Systematik
Worringer Weg
D-52074 Aachen

P.W. Balls
Scottish Office Agriculture and Fisheries Department
Marine Laboratory,
Victoria Road, PO Box 101
Aberdeen AB9 8DB
United Kingdom

G. Becker
Bundesamt für Seeschifffahrt und Hydrographie
Bernhard-Nocht-Str. 78
D-20359 Hamburg

M. Bohle-Carbonell
CEC DG XII/E, SQDM
200 Rue de Loi
B-1049 Bruxelles

U.H. Brockmann
Institut für Biogeochemie und Meereschemie
Martin-Luther-King-Platz 6
20146 Hamburg

P. Damm
Institut für Meereskunde
Troplowitzstraße 7
D-22529 Hamburg

W. DANNECKER
Institut für Anorganische und Angewandte Chemie
Martin-Luther-King-Platz 6
D-20146 Hamburg

R. DOERFFER
GKSS-Forschungszentrum
Max-Planck-Straße
D-21502 Geesthacht

C.-D. DÜRSELEN
RWTH-Aachen
Abteilung Systematik
Worringer Weg
D-52074 Aachen

M. ENGEL
Institut für Meereskunde
Troplowitzstraße 7
D-22529 Hamburg

H.-H. ESSEN
Institut für Meereskunde
Troplowitzstraße 7
D-22529 Hamburg

A. FAUBEL
Institut für Hydrobiologie
 und Fischereiwissenschaft
Zeiseweg 9
D-22765 Hamburg

A. FROHSE
Bundesamt für Seeschifffahrt und Hydrographie
Bernhard-Nocht-Straße 78
D-20359 Hamburg

H. GRAßL
Meteorologisches Institut
Bundesstraße 55
D-20146 Hamburg

N. GREEN
Norwegian Institute for Water Research - NIVA
P.O. Box 69 Korsvoll
N-0808 Oslo

K.-W. GURGEL
Institut für Meereskunde
Troplowitzstraße 7
D-22529 Hamburg

M. HAARICH
Bundesforschungsanstalt für Fischerei
Wüstland 2
D-22589 Hamburg

K. HESSNER
Institut für Meereskunde
Troplowitzstraße 7
D-22529 Hamburg

K. HEYER
Institut für Hydrobiologie und Fischereiwissenschaft
Zeiseweg 9
D-22765 Hamburg

H. HINZPETER
Meteorologisches Institut
Bundesstraße 55
D-20146 Hamburg

H. HÜHNERFUSS
FB 13, Abteilung für Organomeereschemie
Martin-Luther-King-Platz 6
D-20146 Hamburg

U. KAMMANN
Institut für Biochemie und Lebensmittelchemie
Martin-Luther-King-Platz 6
D-20146 Hamburg

L. KARBE
Institut für Hydrobiologie
 und Fischereiwissenschaft
Zeiseweg 9
D-22765 Hamburg

M. KERSTEN
TU HH-Harburg
FB Umweltschutztechnik
Eißendorfer Straße 38
D-21073 Hamburg

W. KIENZ
TU HH-Harburg
FB Umweltschutztechnik
Eißendorfer Straße 38
D-21073 Hamburg

H.-J. KIRZEL
Buchenstraße 52
D-25421 Pinneberg

H. KLEIN
Bundesamt für Seeschifffahrt und Hydrographie
Bernhard-Nocht-Straße 78
D-20359 Hamburg

R. KNICKMEYER
Beim Schlump 23
D-20146 Hamburg

P. KÖNIG
Bundesamt für Seeschifffahrt und Hydrographie
Bernhard-Nocht-Straße 78
D-20359 Hamburg

K.J.M. KRAMER
TNO Laboatorium
Ambachtsweg 8
1785AJ Den Helder
The Netherlands

M. KRAUSE
Institut für Allgemeine Botanik
Ohnhorststraße 18
D-22609 Hamburg

U. KRELL
Hofweg 1
D-78244 Gottmadingen

M. KRIEWS
Alfred-Wegener-Institut für Polar- und Meeresforschung
Columbusstraße
D-28217 Bremerhaven

W. KÜHN
Institut für Meereskunde
Troplowitzstraße 7
D-22529 Hamburg

O. LANDGRAFF
Institut für Biochemie und Lebensmittelchemie
Martin-Luther-King-Platz 6
D-20146 Hamburg

H. LUTHARDT
Deutsches Klimarechenzentrum
Bundesstraße 55
D-20146 Hamburg

P. MARTENS
Biologische Anstalt Helgoland
Hafenstraße 43
D-25992 List/Sylt

A. MOLL
Institut für Meereskunde
Troplowitzstraße 7
D-22529 Hamburg

F. MONTENY
Vrije Universiteit Brussel
Laboratorium voor Analytische Scheikunde & Geochemie
Pleinlaan 2
B-1050 Brussel

K. NAUMANN
Institut für Anorganische und Angewandte Chemie
Martin-Luther-King-Platz 6
D-20146 Hamburg

D. PAN
Second Institute of Oceanography
Hang Zhou
P.R. China

T. POHLMANN
Institut für Meereskunde
Troplowitzstraße 7
D-22529 Hamburg

W. PULS
GKSS-Forschungszentrum
Max-Planck-Straße
D-21502 Geesthacht

G. RADACH
Institut für Meereskunde
Troplowitzstraße 7
D-22529 Hamburg

S. REGIER
Papenstraße 130
D-22089 Hamburg

H.-J. RICK
Universität Oldenburg
ICBM
Ammerländer Heerstraße 114-118
D-26129 Oldenburg

F. SCHIRMER
Institut für Meereskunde
Troplowitzstraße 7
D-22529 Hamburg

T. SCHLICK
Institut für Meereskunde
Troplowitzstraße 7
D-22529 Hamburg

K.H. SCHLÜNZEN
Meteorologisches Institut
Bundesstraße 55
D-20146 Hamburg

D. SCHMIDT
Bundesamt für Seeschifffahrt und Hydrographie
Wüstland 2
D-22589 Hamburg

M. Schulz
Laboratoire de Modélisation du Climat et de l'Environment
C.E. Saclay 709
F-191 Gif-Sur-Yvette CEDEX

M. Schwikowski-Gigar
Paul-Scherrer-Institut
CH-5232 Villingen

M. Steiger
Institut für Anorganische und Angewandte Chemie
Martin-Luther-King-Platz 6
D-20146 Hamburg

H. Steinhart
Institut für Biochemie und Lebensmittelchemie
Martin-Luther-King-Platz 6
D-20146 Hamburg

J. Sündermann
Institut für Meereskunde
Troplowitzstraße 7
D-22529 Hamburg

U. Terzenbach
Reyes Weg 24
D-22081 Hamburg

R.J. van Enk
Geochem-research BV
Piet Heinlaan 5
3941 VE Doorn
The Netherlands

J.J.G. Zwolsman
Waterloopkundig Laboratorium
Postbus 177
2600 MH Delft
The Netherlands

1 Introduction

This book presents in an integrated approach results from five years of interdisciplinary research in the project *Zirkulation und Schadstoffumsatz in der Nordsee (ZISCH)* (Circulation and Contaminant Fluxes in the North Sea). From 1984 to 1989 over 30 scientists and technicians worked in the project, which was supported by the Bundesminister für Forschung und Technologie (German Minister for Research and Technology). Together with the Universität Hamburg, which initiated ZISCH, the following institutions participated: Technische Universität Hamburg-Harburg, Rheinisch-Westfälische Technische Hochschule Aachen, Deutsches Hydrographisches Institut (now Bundesamt für Seeschiffahrt und Hydrographie), GKSS Forschungszentrum Geesthacht, Alfred-Wegener-Institut Bremerhaven, Biologische Anstalt Helgoland, Universität Paderborn. More than other national and international projects, the ZISCH project aimed at a holistic view of the marine ecosystem. For this reason, the different disciplines working together were not limited to purely marine sciences (oceanography, marine biology and chemistry), but atmospheric sciences (meteorology, atmospheric chemistry) and food chemistry also participated. Previous to ZISCH, no North Sea wide investigation of comparable complexity had ever been carried out. The area under investigation is shown in Fig. 1.1.

The ZISCH research project was established in view of the obvious threat to the North Sea environment generated by human activities. The primary causes are considered to be industrial, agricultural and municipal discharges which reach the sea over different pathways. Table 1.1 summarizes the sources and amounts of various contaminants entering the North Sea.

As the ZISCH project began, the first International North Sea Conference had just taken place (November/December 1984 in Bremen). Policy-makers from the North Sea countries had recognized that it was necessary

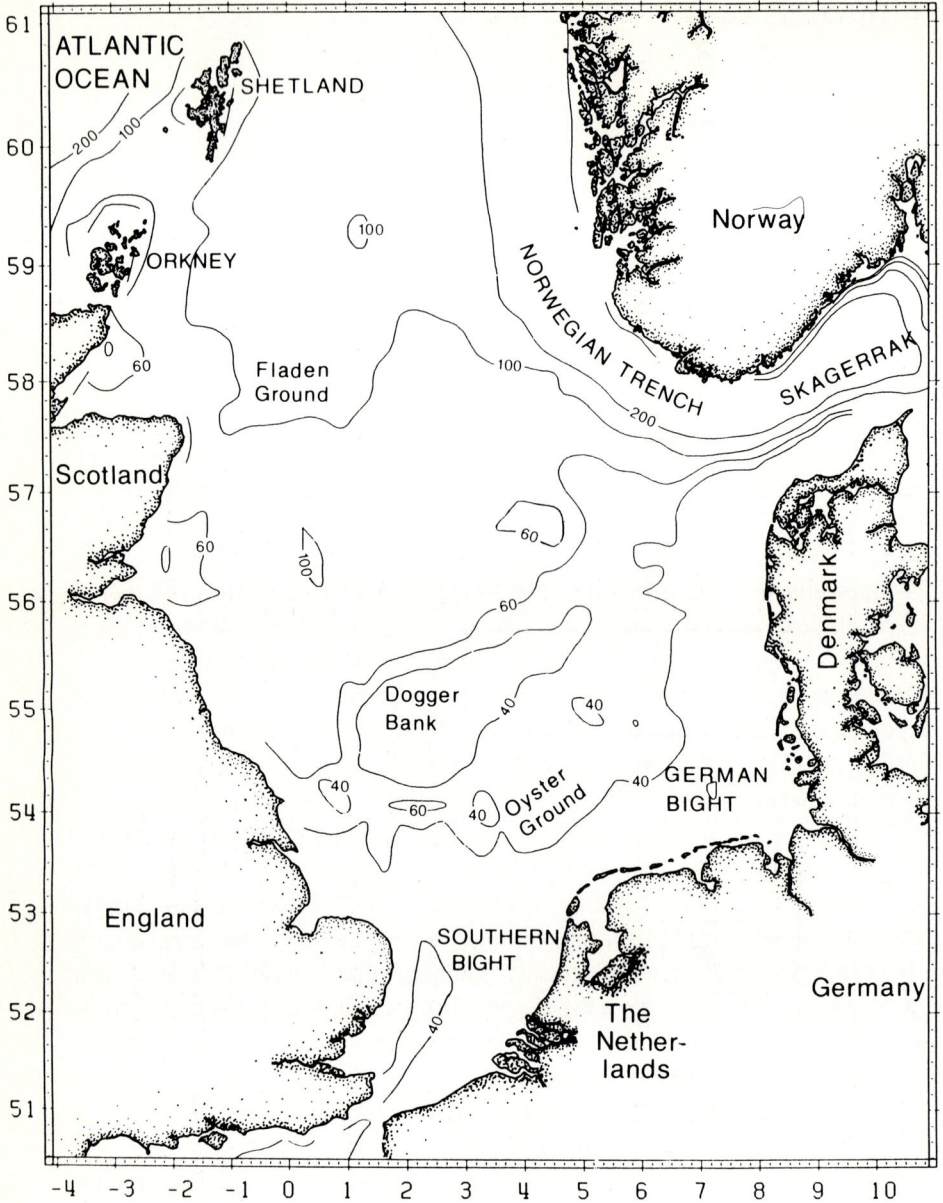

Fig. 1.1. Topography of the North Sea

Table 1.1.   Estimates of inputs (tonnes) via the various pathways to the North Sea in 1990

Pathway	Cd	Hg	As	Cr	Cu	Ni	Pb	Zn	PCB*	HCH**	N	P
Riverine inputs[a]	55	25	NI	NI	1 400	NI	1 000	6 800	2.7	1.1	920 000	48 000
Direct inputs[a]	17	2.1	NI	NI	330	NI	160	1 300	0.2	0.2	120 000	7 100
Atmosphere[a,b]	74	6.9	220	[180]	740	400	1 700	5 500	NI	[8.1]	520 000	NI
Disposal at sea												
Incineration[a]	0.1	0.1	0.1	4.9	4.6	5.0	4.9	5.7	NI	NI	NI	NI
Industrial waste[c]	0.3	0.2	2.4	24	180	64	220	440	NI	NI	NI	NI
Sewage sludge	1.2	0.7	0.1	21	76	11	77	160	NI	NI	6 300	570
Dredged material[d]	71	19	720	2 800	1 300	1 200	2 700	7 900	0.6	NI	NI	NI

NI = No information;   [ ] = not reliable;   * = IUPAC Nos. 28, 52, 101, 118, 153, 138,180 ;   ** = γ-HCH
Source: Oslo and Paris Commissions (Reports on Monitoring and Assessment & Dumping and Incineration at Sea, July 1992)

[a]Maximum estimate.
[b]Based on deposition measurements at coastal stations (calculated for an area of 525 000 km^2).
[c]Chemical waste, fly ash, rock, tailings, sediments. Dumping in internal waters is included.
[d]Dredged material from harbours, estuaries and navigation channels. Dumping in internal waters is included. Metal load is overestimated because it contains an unknown share of metals of natural origin. In addition, dumping of dredged material often means merely relocating material without any new input of constituents to the North Sea.

to act in concert to alleviate the environmental problems of this shelf sea. But what was the basis for their decision making? It turned out that despite considerable data acquisition in the North Sea over the years, making it one of the most intensively studied marine areas worldwide, no comprehensive, sufficiently representative observational material was available for assessing the ecosystem, not to mention for understanding the relationships between anthropogenic effects, climatological factors and ecological state. Previous investigations had not been carried out on a holistic, integrated basis and had only covered a small spectrum of ecological parameters (for example only heavy metals without regarding phytoplankton at the same time). Moreover, the diversity of the relevant spatial and temporal scales caused by the high degree of stochastic variability in the North Sea had not been taken adequately into consideration, so that the many data sets did not result in a representative picture. It was not possible to develop a clear protection strategy for the North Sea on this scientific basis. It was necessary to create a new, integrated and interdisciplinary concept for ecological North Sea research which would make allowance for the complex interrelationships between the physical, chemical and biological processes. ZISCH realized this approach.

The research program had the objective of quantifying fluxes of major contaminants in the North Sea. This entailed:

1.   Acquisition of consistent, comprehensive data sets for the North

Sea ecosystem on the whole together with the distribution of key contaminants within it.

2.   Calculation of transport paths and fate of critical contaminants as well as mass budgets, considering given anthropogenic sources. Hereby, present conditions were analyzed, and predictions were attempted on the basis of various scenarios.

The conservation of mass permits formulation of the concentration of a particular substance in water according to a transport equation. Simply stated, the evolution of the concentration at a fixed location (for example, an increase in contamination of the Dogger Bank) results from the sum of local sources and sinks (for example, input from the atmosphere) and advection and diffusion in the flow field (for example, as carried by currents from the polluted plume of the River Tyne). In order to evaluate contamination, knowledge is thus necessary regarding the three-dimensional circulation and turbulent diffusion of the water masses (including suspended matter) as well as the contaminant fluxes resulting from the interactions of the hydrosphere with atmosphere, biosphere and lithosphere.

While advection and diffusion can be characterized by meteorological and hydrographical parameters, the sources and sinks of the system also include biological, chemical and sedimentological processes. A research program for investigation of contaminant fluxes must thus include physical, chemical and biological elements, closely interwoven. The individual components of the system together with the interaction mechanisms are shown in Fig. 1.2. Horizontally, it represents the transport equation, vertically, the transport media air, water, suspended matter and sediment. The arrows refer to the dominant processes.

First, in the purely hydrodynamic subsystem, the circulation of the air forces the circulation of the water. This causes dissolved substances, suspended matter and planktonic organisms to be transported, these, in turn, interacting with each other and with the sediment. Contaminants entering this system are then carried by air and water, in the latter in dissolved or particulate form. The input of contaminants occurs by means of anthropogenic "external" sources via the atmosphere, the rivers, the entrances to the North Sea and dumping. There are fluxes in both directions between the transport media air, water and sediment. Finally, the subsystem "contaminant transport" also contains "internal" sources and sinks. These include chemical transformation in the air and in the water as well as accumulation in organisms and in suspended matter and exchange processes between water and sediment. A comprehensive programme *Circulation and Contaminant Fluxes in the North Sea* must encompass all of the constituents shown in Fig. 1.2.

The ZISCH investigations represented a qualitatively and quantitatively new approach to North Sea research in several respects. Among these:

- the first simultaneous blanket coverage of all important biological, chemical and physical parameters in the entire North Sea ecosystem;
- the first simultaneous measurements of major contaminants (metals and organohaline compounds) in the different ecosystem compartments;
- simultaneous determinations of atmospheric inputs of momentum, energy and matter as important ecosystem boundary conditions;
- performance of the complex measurement program during two seasons, namely the spring plankton bloom and the subsequent winter period of minimal biological activity;
- support of data analysis and interpretation by oceanographic and meteorological numerical models on the same scales.

The present volume addresses colleagues in the marine sciences but also scientists in related fields and scientific policymakers as well as the inter-

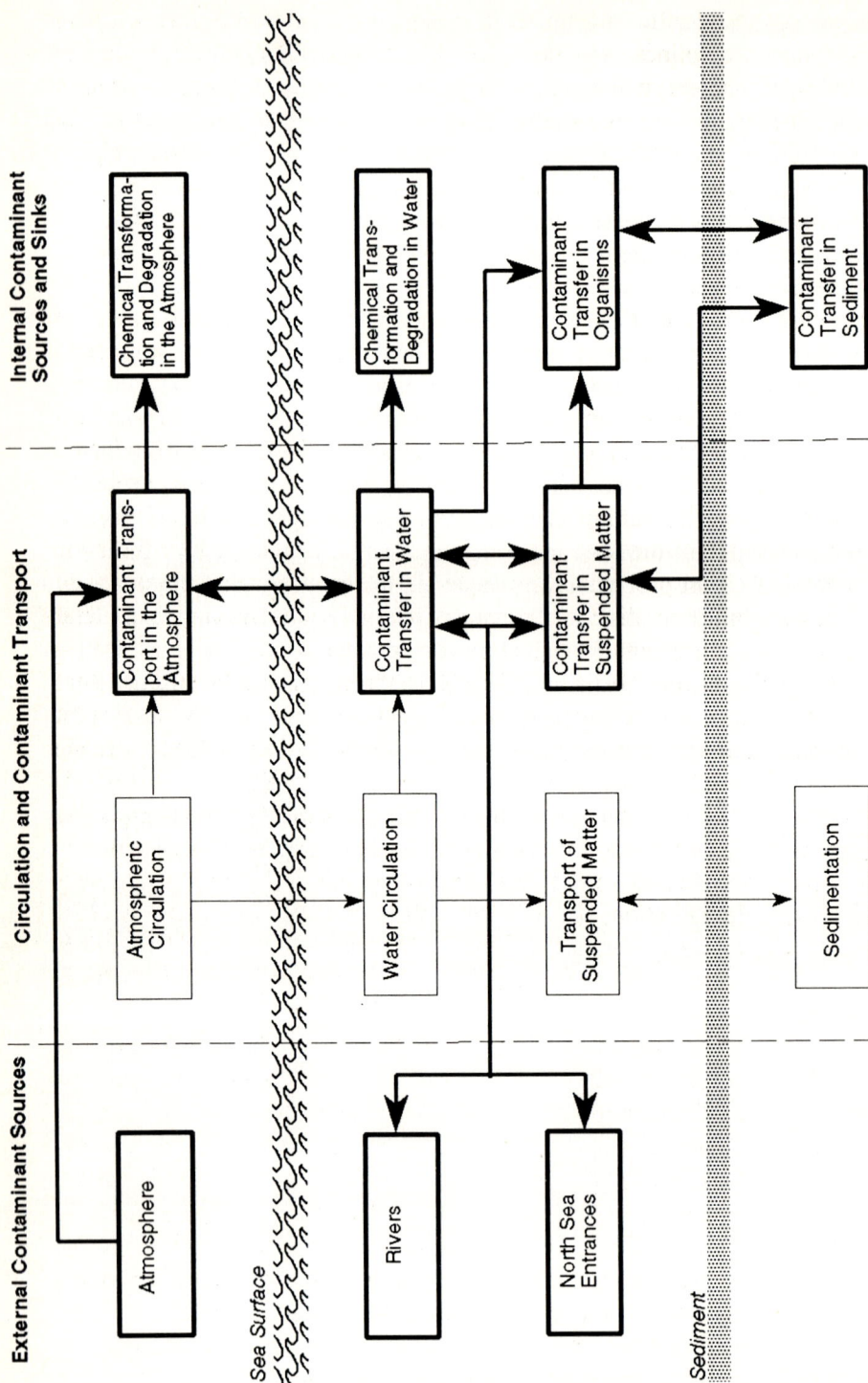

Fig. 1.2. Schematic representation of contaminant fluxes in the marine ecosystem. The thin arrows show the hydrodynamic interactions, the thick ones the transport of matter. Only the dominant components are considered.

ested general public. Rather than presenting the results according to individual disciplines, we have tried to consolidate the diverse data in summary chapters which strive for a holistic understanding of causal relationships. A first step in this approach was taken by us previously in avery successful publication for the general public, *The North Sea - Water Exchange and Pollution* (Sündermann and Degens 1989).

For comparison, some other recently published books with reference to the North Sea ecosystem shall be mentioned:

Biogeochemistry and Distribution of Suspended Matter in the North Sea and Implications to Fisheries Biology (Kempe et al. 1988) is based on results of a German project primarily concerned with the role of suspended matter/sediment as vehicles and as long term accumulators for contaminants. In this book it is emphasized that the long term cycles of sediment fluxes (several decades) and the "ultimate rubbish dump" of contaminated sediment in the Norwegian Trench could be a threat to the ecosystem for generations, even if all inputs of contaminants were to be stopped immediately. Changes in the spectrum of species (including extraordinary algal blooms) can most probably be attributed to eutrophication and the increase in susceptibility to diseases in fish, birds and mammals to a general weakening of immunity provoked by contaminants.

Pollution of the North Sea - an Assessment (Salomons et al. 1988) attempts to take a more comprehensive, systematic approach. Authors from different countries bordering the North Sea and from different marine disciplines were invited to contribute articles describing the ecosystem, contaminant fluxes and, particularly, the effects of contaminants on organisms. These form a mosaic picture of the pollutant load and the transport mechanisms in the North Sea. The editors point out apparent ecosystem deterioration, but they only speak of indications. The inherent complexity of the system has prevented formulation of causal relationships and, thus, of prognoses. A precautionary reduction in inputs of potential pollutants is called for.

The final report *Stickstoff, Phosphor, Plankton und Sauerstoffmangel in der Deutschen Bucht und in der Kieler Bucht* (Gerlach 1990) is based on results from a German project on eutrophication of the North Sea and the Baltic. It includes a comprehensive description of the ecosystems and the influence of excessive nutrient inputs. It establishes that phytoplankton biomass has increased considerably during the past 30 years, paralleling increased concentrations of phosphorus and nitrogen compounds. Interannual fluctuations (which can coincide with extreme algal blooms and oxygen deficits) seem to be governed by atmospheric processes.

Warnsignale aus der Nordsee (Lozan et al. 1990) stresses contaminant concentrations in water, sediment and organisms as well as the changes in flora and fauna which have occurred in the past decades. It has been written from an environmentally active point of view and, accordingly, includes a critical review of current protective measures and a list of necessary actions for the future. Demands for drastic reductions in the inputs of contaminants are made and justified by the lack of understanding of the complex interrelationships in the marine ecosystem.

The most integrated presentation of the current environmental situation in the North Sea and the associated risks is given in *Nordsee - ein Lebensraum ohne Zukunft?* (Buchwald 1990). A conscientious evaluation of the environmental state of the North Sea is undertaken, written in almost popular form on the basis of a thorough study of the literature. The author warns that the inputs into the North Sea are irreversible and that a creeping decline will eventually lead to an acute deterioration of the environment.

Each of these books has its own specific value, highlights and objectives. The basic assertions are supported and supplemented by the present volume. Through continued research, a more complete and more reliable picture of the North Sea ecosystem thus emerges.

ZISCH consisted of field research and of model simulations. Accordingly, we have presented our observational and model results in Chapters 2 and 3, respectively. In view of the long-term stability, depiction of isobaths is considered permissable. We carried out long debates about the permissability and usefulness of depicting highly variable North Sea parameters such as temperature or nitrate, for example, as isolines, particularly when they had only been measured quasi-synoptically. Such isolines are optical and didactic supports, which only reflect nature to a certain degree. Still, they can offer valuable information to the experienced, cautious scientist, who remains conscious of the variability of the system. In view of this, we settled on isoline portrayals of the results - even when the measurements were not carried out synoptically.

Chapter 4 is a first holistic, interdisciplinary analysis of the contributions of the individual working groups. Chapter 5 is a summary of the ZISCH results and characterizes the present state of North Sea research from the point of view of the ZISCH group.

Stated briefly, it can be said that the current problems to be addressed by marine research are interdisciplinary in nature and, accordingly, require an interdisciplinary research concept. Clarification of the highly complex interrelationships in the ecosystem cannot succeed using a traditionally monodisciplinary approach; new programs are warranted and indispensable. The present volume should be seen in this light, since it is one of several contributions to North Sea research. It differs in its aims and in its specific combination of disciplines from the others, complementing them and at the same time carrying them further.

British *North Sea Project*, influenced several projects in the Marine Science and Technology Program (MAST) of the European Community.

The substance of this volume was produced by a large number of com-mitted scientists and technicians from various disciplines. As director of the ZISCH Project and editor of this book, I would like to sincerely thank my colleagues for their contributions and their identification with the project and with the object of the research. They did a great job of keeping up a truly interdisciplinary cooperation over the years. This is true especially for Susan Beddig, who coordinated the project and revised the manuscript of this book. Thanks are also due to the institutions which actively supported the University of Hamburg in carrying out ZISCH. Without their logistical aid, the results could not have been obtained. I also thank the Bundes-minister für Forschung und Technologie for providing the necessary funding and for the administrators who identified themselves personally with the work. In conclusion, I would like to express my gratitude to our colleagues from the Netherlands Institute of Sea Research (NIOZ) in Texel for their helpful critical reviewing of the manuscript.

Hamburg, in April 1993   J. SÜNDERMANN

References

Buchwald K (1990) Nordsee - ein Lebensraum ohne Zukunft? Die Werkstatt, Göttingen, 552 pp

Gerlach S (1990) Stickstoff, Phosphor, Plankton und Sauerstoffmangel in der Deutschen Bucht und in der Kieler Bucht. Forschungsbericht 10204215, Umweltbundesamt, Schmidt, Berlin, 357 pp

Kempe S, Liebezeit G, Dethlefsen V, Harms U (eds) (1988) Biogeochemistry and distribution of suspended matter in the North Sea and implications to fisheries biology. Mitt Geol-Paläont Inst Univ Hamburg 65, 547 pp

Lozan JL, Lenz W, Rachor E, Watermann B, von Westernhagen H (eds) (1990) Warnsignale aus der Nordsee. Parey, Hamburg, 428 pp

Salomons W, Bayne BL, Duursma EK, Förstner U (eds) (1988) Pollution of the North Sea - an Assessment. Springer, Berlin Heidelberg New York, 687 pp

Sündermann J, Degens ET (1989) The North Sea - Water Exchange and Pollution. Univ Hamburg, 49 pp

2 Field Data

2.1 Seasonal and Interannual Variability in the Atmosphere and in the Sea

P. DAMM, H. HINZPETER, H. LUTHARDT and U. TERZENBACH

2.1.1 Wind and Air Temperature in the North Sea Region

During the ZISCH project, several cruises were carried out in the period 1986-1989 to investigate concentration and transport of contaminants in the North Sea. Since this transport is also affected by atmospheric fields, wind and temperature fields were determined for a region encompassing the North Sea and the adjoining parts of the Atlantic for the years mentioned above. In order to evaluate the representativity of the cruise results it is necessary to know the major parameters of the meteorological field and their deviation from long-term means over the North Sea for the years in question. A publication on the North Sea climate (Korevaar 1990) is available; the extensive climate data and maps of the German Marine Weather Office have not yet been published. Since variances of the meteorological field are unknown, it is only possible to detect sizeable deviations from the long-term mean state. Such is the case for the winters of 1987/88 and 1988/89, which were particularly mild and which - together with the third mild winter of 1989/90 - have gained the attention of climatologists.

Results of the analyses of monthly mean meteorological parameters for the years 1986 to 1989 are shown, e.g. from January to June in Figs. 2.1.2a,b-2.1.7a,b. They are computed for every month of the year and based on data collected at coastal sites, on oil rigs and on ships, and distributed by the Global Telecommunication System (GTS). Using the

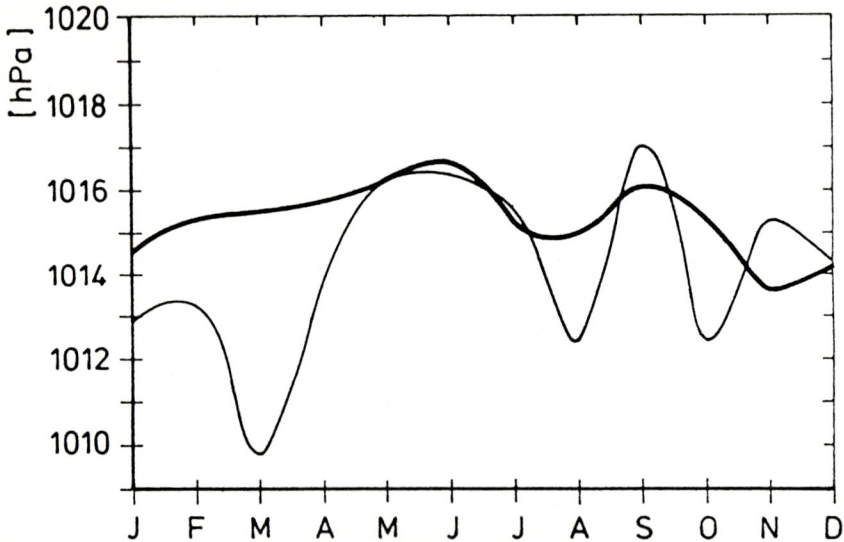

Fig. 2.1.1. Seasonal development of the air pressure. *Thick line* over London (1931-1960); *thin line* site 4 in Figs. 2.1.2a-2.1.7a (1986-1989). The seasonal development of the mean for the entire North Sea in the years 1986-1989 did not differ significantly from that of site 4

method of Luthardt (1987), pressure, wind, air temperature and sea-surface temperature were computed at the points in a 42 x 42 km grid covering the North Sea. The displayed frequency distributions of the air-sea temperature difference refer to the positions marked by enclosed numbers in Figs. 2.1.2b-2.1.7b. For technical reasons, the values are printed at a distance from these numbers. This difference is a characteristic parameter for the climate, but it is also indicative for the stability of the boundary layer, which influences vertical mixing and thus contaminant dissipation from the atmosphere into the ocean.

Figure 2.1.1 implies strikingly the anomaly of the winters 1987/88 and 1988/89. The development of the mean air pressure in London over the years 1930-1960, which is commonly regarded as a normal climate period, is characterized by a - be it ever so small - minimum in November. The mean pressures of the years 1986-1989 over the North Sea as well as the pressure at site 4 (Figs. 2.1.2a,b-2.1.7a,b) show almost unanimously a distinct minimum in March. For the years 1988 and 1989 this would be even more extreme, since the rather normal pressure developments in March 1987 tempers the extreme in the four years' mean.

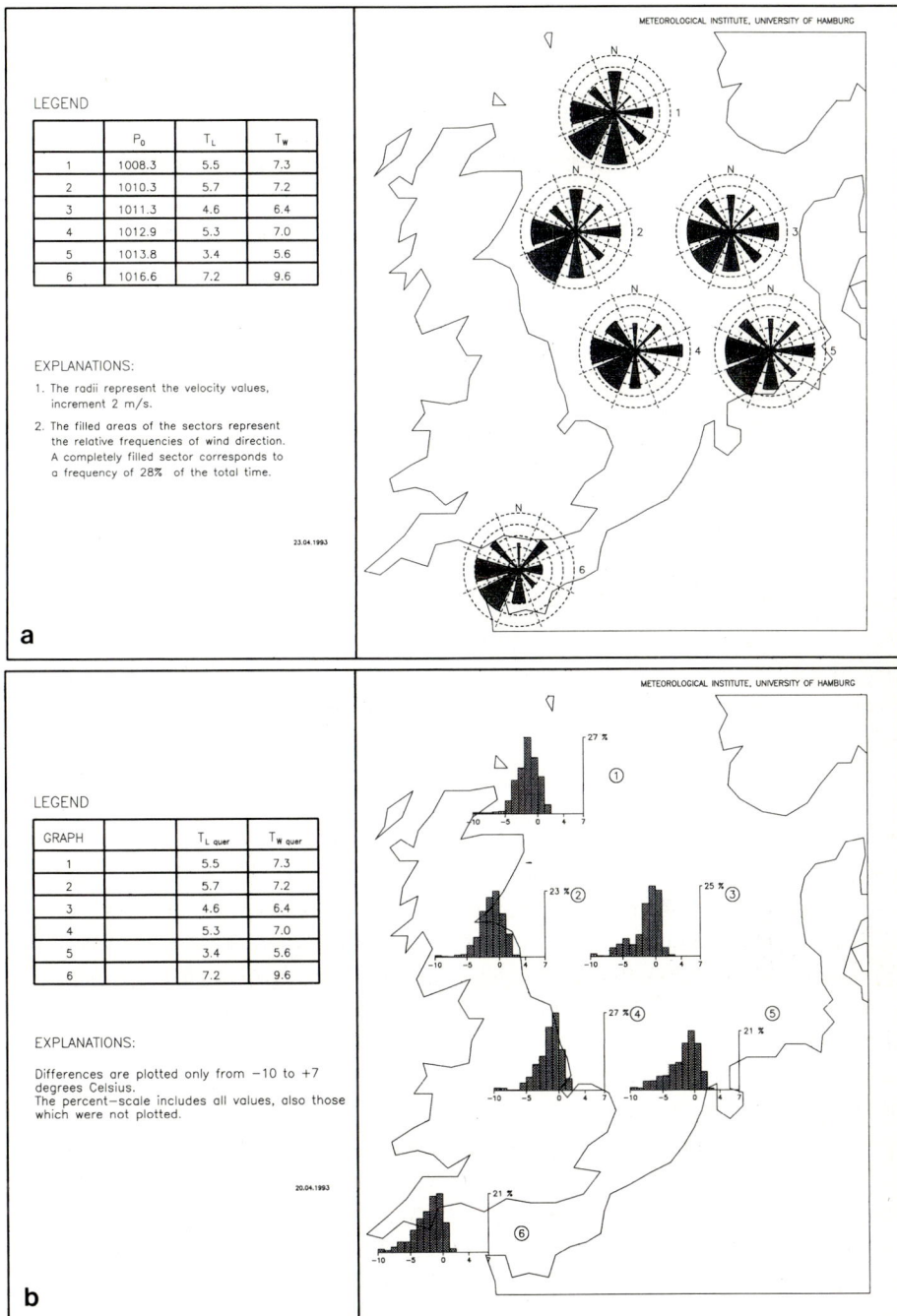

LEGEND

	P_0	T_L	T_W
1	1008.3	5.5	7.3
2	1010.3	5.7	7.2
3	1011.3	4.6	6.4
4	1012.9	5.3	7.0
5	1013.8	3.4	5.6
6	1016.6	7.2	9.6

EXPLANATIONS:

1. The radii represent the velocity values,
   increment 2 m/s.

2. The filled areas of the sectors represent
   the relative frequencies of wind direction.
   A completely filled sector corresponds to
   a frequency of 28% of the total time.

23.04.1993

a

LEGEND

GRAPH		$T_{L\,quer}$	$T_{W\,quer}$
1		5.5	7.3
2		5.7	7.2
3		4.6	6.4
4		5.3	7.0
5		3.4	5.6
6		7.2	9.6

EXPLANATIONS:

Differences are plotted only from −10 to +7
degrees Celsius.
The percent−scale includes all values, also those
which were not plotted.

20.04.1993

b

Fig. 2.1.2. Mean wind velocity and direction *(a)* and mean frequency distribution of the difference between air and water temperature *(b)* for January of the years 1986-1989

The North Sea, a marginal sea, is affected less by meridional heat transport of the Atlantic than by the continental influences. The water temperature in the north (site l) increases only by 7 °C from March (Fig. 2.1.4b) to August (without fig.) - from 6.2 °C to 13.5 °C - while in the German Bight (site 5) it increases from 3.8 °C to 16.2 °C during the same period. Only from May to June is the air temperature higher than the water temperature.

January
The mean air pressure over the North Sea region (Fig. 2.1.2a) increases steadily from around 1008 hPa in the northern part (site l) to around 1017 hPa in the English channel (site 6). Corresponding to the course of the isobars, over 50% of the time the entire North Sea area is characterized by southwest and west winds, with speeds of 9 m/s in the northwestern North Sea and 8 m/s in the southeastern parts. Aside from the English Channel, the lack of northeasterly winds may indicate mild winters within these four years mean. In January, the North Sea is colder than the air 20-30% of the time. Air temperatures 5 °C lower than water temperatures hardly occur in the northern parts; in the south, however, they occur 10-15% of the time (Fig. 2.1.2b).

February
The pressure difference between the northernmost and the southernmost areas is on the average 6 hPa within the four years' mean (Fig. 2.1.3a). The wind field has hardly changed compared to January. Southwesterly winds are dominant; average wind speeds have increased by about 1 m/s. The distribution of the air-water temperature difference corresponds roughly to that of January, although air temperatures more than 5 °C lower than those of the water practically no longer occur (Fig. 2.1.3b).

March
The mean pressure reaches the lowest value of all months in March and changes by about 9 hPa from the northernmost to the southernmost regions (Fig. 2.1.4a). The frequency of easterly winds decreases clearly, in the southern part of the North Sea southwesterly and westerly winds dominate. The very low frequency of northeasterly winds remains noteworthy, and the easterly winds, as well, remain extremely rare in the northern North Sea. The wind speeds in March have decreased in the southern and eastern parts compared with the month before. In the North Sea the mean air temperature is about 1 °C lower than that of the water (Fig. 2.1.4b). Only in the far north (site l) is the difference somewhat greater, in the central

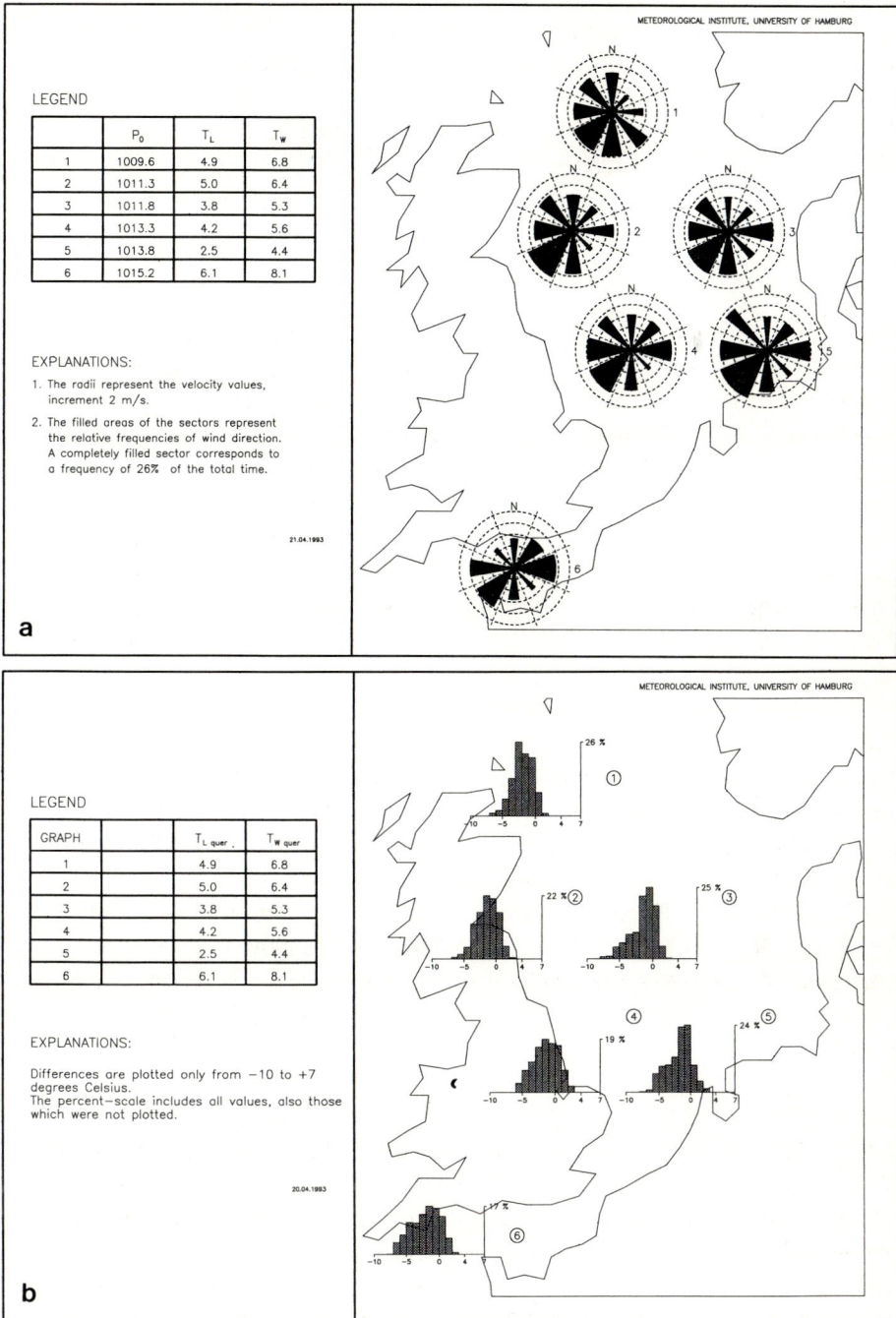

LEGEND

	P_0	T_L	T_W
1	1009.6	4.9	6.8
2	1011.3	5.0	6.4
3	1011.8	3.8	5.3
4	1013.3	4.2	5.6
5	1013.8	2.5	4.4
6	1015.2	6.1	8.1

EXPLANATIONS:

1. The radii represent the velocity values, increment 2 m/s.

2. The filled areas of the sectors represent the relative frequencies of wind direction. A completely filled sector corresponds to a frequency of 26% of the total time.

METEOROLOGICAL INSTITUTE, UNIVERSITY OF HAMBURG

21.04.1993

a

LEGEND

GRAPH		$T_{L\,quer}$	$T_{W\,quer}$
1		4.9	6.8
2		5.0	6.4
3		3.8	5.3
4		4.2	5.6
5		2.5	4.4
6		6.1	8.1

EXPLANATIONS:

Differences are plotted only from −10 to +7 degrees Celsius.
The percent−scale includes all values, also those which were not plotted.

METEOROLOGICAL INSTITUTE, UNIVERSITY OF HAMBURG

20.04.1993

b

Fig. 2.1.3. Mean wind velocity and direction *(a)* and mean frequency distribution of the difference between air and water temperature *(b)* for February of the years 1986-1989

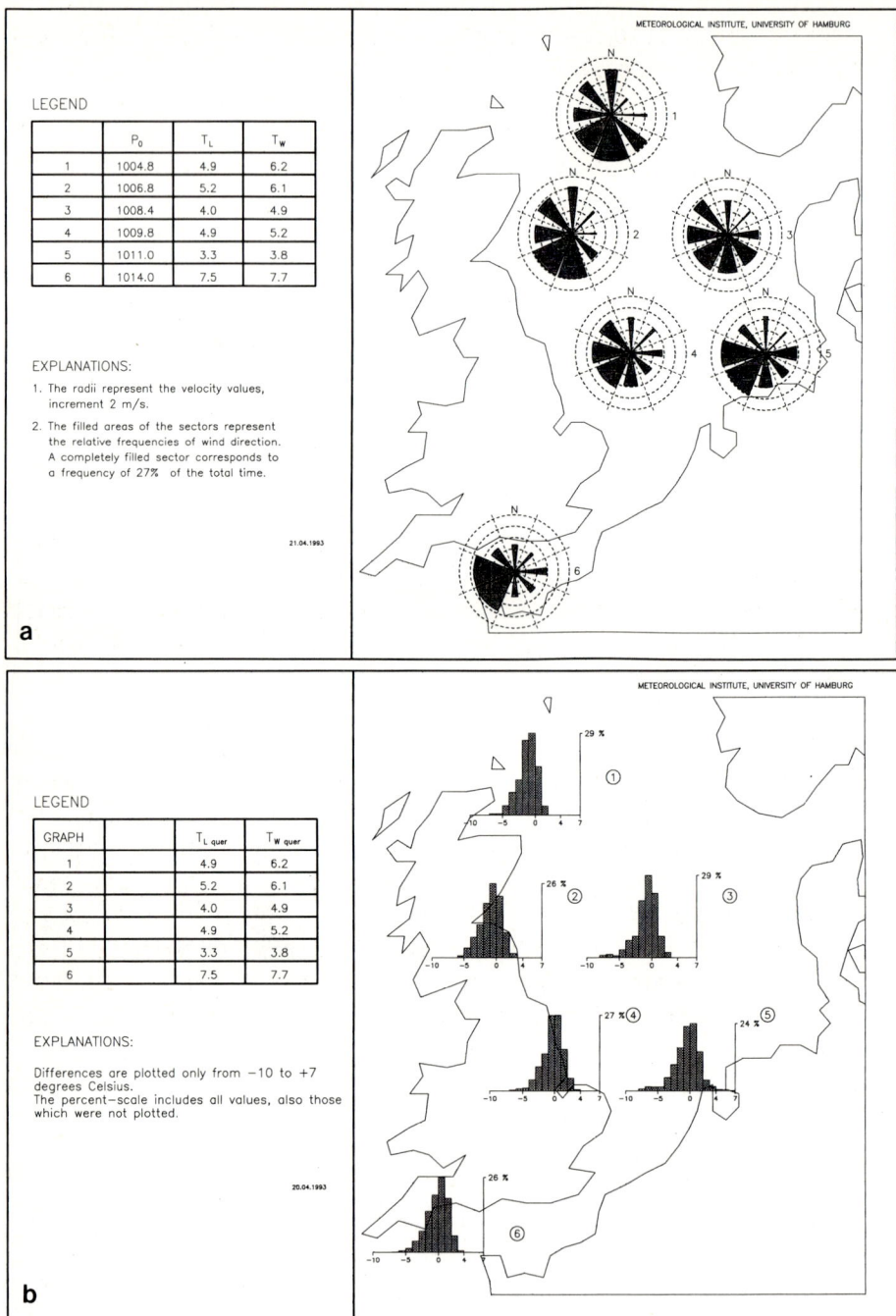

Fig. 2.1.4. Mean wind velocity and direction *(a)* and mean frequency distribution of the difference between air and water temperature *(b)* for March of the years 1986-1989

western North Sea somewhat smaller, and in the Channel the air and water have the same temperature.

April

In April the pressure field has changed distinctly compared with March (Fig. 2.1.5a). The mean pressure has increased again by 5 hPa and is widely uniform over the North Sea - only towards the Channel does it decrease by about 1 hPa. Accordingly, the wind conditions have changed completely. Wind speeds have decreased to an average of 4-5 m/s. In the central North Sea easterly winds prevail about 20% of the time with speeds of 6-7 m/s, and only in the northernmost area is no direction predominant. In the Channel, the picture has changed the same way as in the central and southern North Sea. Instead of westerly and southwesterly winds, in April easterly and northeasterly winds prevail 40% of the time. This frequency of east winds is anomalous compared to the long term climatological mean. In April the temperature differences between air and water have disappeared in the southern half (Fig. 2.1.5b). Towards the north light differences occur. In the northernmost area (site 1) the air is an average of 0.5 °C cooler than the water; temperature differences of more than 5 °C hardly occur here as well.

May

In this month the mean pressure over the North Sea increases again by about 1-3 hPa, reaching 1014 hPa in the north and 1017 hPa in the south (Fig. 2.1.6a). All wind directions are nearly equally frequent, with wind speeds of 4-6 m/s. Only in the Channel do easterly and northeasterly winds dominate around 40% of the time. In the rest of the North Sea the lack of a dominant east component is typical for these 4 years, although not for the climatological mean. The air is now warmer than the water over the entire North Sea, in the north only a few tenths of a degree, in the eastern and central North Sea almost 1 °C. Everywhere else the difference is around 0.5 °C (Fig. 2.1.6b).

June

In June the mean pressure over the North Sea is nearly equal to that of May (Fig. 2.1.7a). Only the air pressure distribution has changed. The pressure at sites 1 and 2 increases by 1 hPa while it decreases at sites 3 and 5 about 1 hPa. In consequence, the horizontal pressure gradients have become smaller and the wind force decreases down to about 3-5 m/s; the frequency of easterly winds, however, is strongly diminished, northerly and northwesterly winds being the most frequent only in the north. The air and

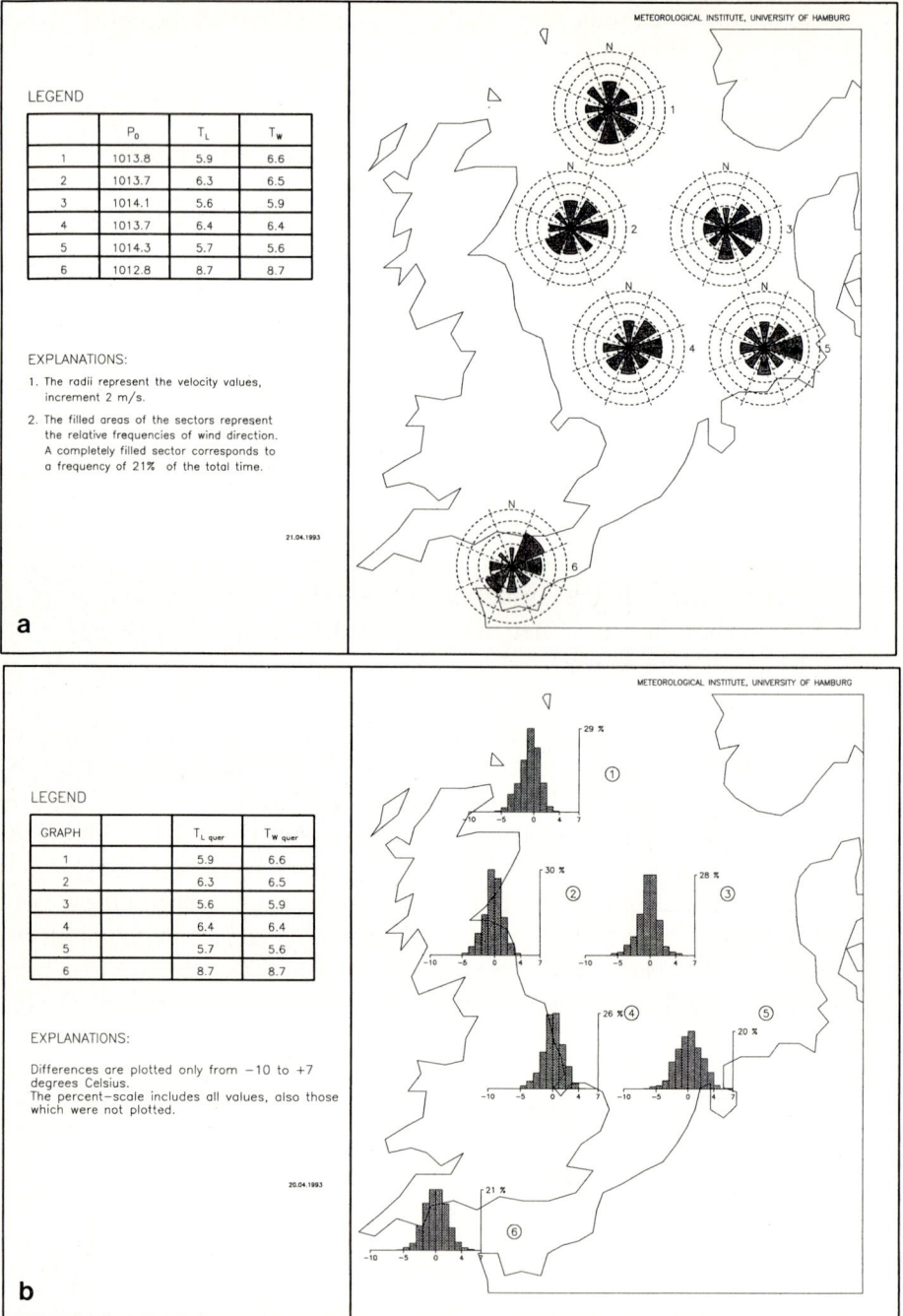

Fig. 2.1.5. Mean wind velocity and direction *(a)* and mean frequency distribution of the difference between air and water temperature *(b)* for April of the years 1986-1989

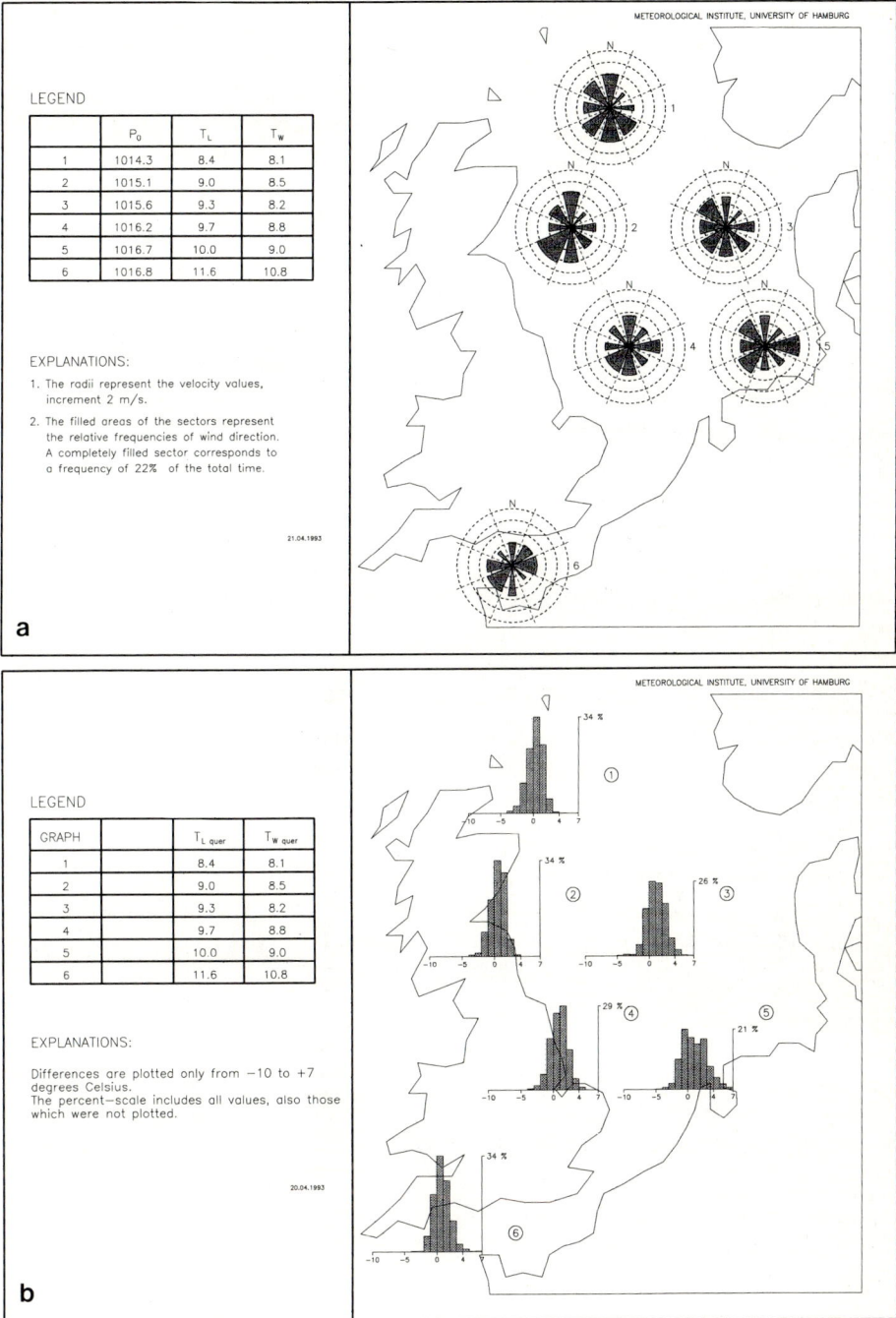

LEGEND

	P_0	T_L	T_W
1	1014.3	8.4	8.1
2	1015.1	9.0	8.5
3	1015.6	9.3	8.2
4	1016.2	9.7	8.8
5	1016.7	10.0	9.0
6	1016.8	11.6	10.8

EXPLANATIONS:

1. The radii represent the velocity values, increment 2 m/s.

2. The filled areas of the sectors represent the relative frequencies of wind direction. A completely filled sector corresponds to a frequency of 22% of the total time.

21.04.1993

METEOROLOGICAL INSTITUTE, UNIVERSITY OF HAMBURG

a

LEGEND

GRAPH		$T_{L\ quer}$	$T_{W\ quer}$
1		8.4	8.1
2		9.0	8.5
3		9.3	8.2
4		9.7	8.8
5		10.0	9.0
6		11.6	10.8

EXPLANATIONS:

Differences are plotted only from −10 to +7 degrees Celsius.
The percent—scale includes all values, also those which were not plotted.

20.04.1993

METEOROLOGICAL INSTITUTE, UNIVERSITY OF HAMBURG

b

Fig. 2.1.6. Mean wind velocity and direction *(a)* and mean frequency distribution of the difference between air and water temperature *(b)* for May of the years 1986-1989

water temperatures are now 3 °C higher than in May, but the picture of the differences has hardly changed in comparison to May (Fig. 2.1.7b). In the northernmost region, this difference lies around a few tenths of a degree, and this difference increases going southwards to about 0.5 °C. Only in the area of the German Bight do the differences disappear. For the summer months, it can be generalized that extreme, positive deviations of the air temperature from that of the water are more seldom and smaller than is the case for the negative ones in the winter half year.

July

The air pressure distribution in July is quite similar to that in June. However, the pressure in the northernmost North Sea has decreased by about 2 hPa. The mean wind speeds are slightly greater than in June, and winds with westerly components occur most frequently. The air and water temperatures have increased again by 2-3 °C and are practically equal. Positive as well as negative deviations from the monthly mean occur with about equal magnitude and frequency. This is also the case for the German Bight, while in the southwestern North Sea the air temperature is almost 0.5 °C higher than the water temperature.

August

In August the mean pressure decreases to about 1012 hPa. There is a difference however between the decrease in the north and in the south. As it amounts to 3-4 hPa in the north and 1-2 hPa in the south, the pressure gradient increases further. Consequently, with the exception of the English Channel, the wind speed has increased again to 4-6 m/s. Contrary to July, winds with easterly components hardly occur. Water and air temperatures show almost the same picture as in July, except that in the German Bight area the air is about 0.5 °C cooler than the water. The temperature has hardly changed in comparison to July.

September

In September the pressure increases again by about 4 hPa, and the pressure difference continues to increase. Between the northern and southern North Sea it is about 3 hPa, increasing further by 2-3 hPa towards the English Channel. Accordingly, the mean wind speed has increased to 5-7 m/s. In the southern North Sea westerly and southwesterly directions are now most frequent; in the eastern North Sea the frequency of northwesterly winds has decreased; only in the northernmost area are southerly winds also frequently observed. The air temperature has decreased by about 1 °C compared to August. In the Channel and in the German Bight the air is about 1 °C cooler than the water. In the rest of the North Sea the differ-

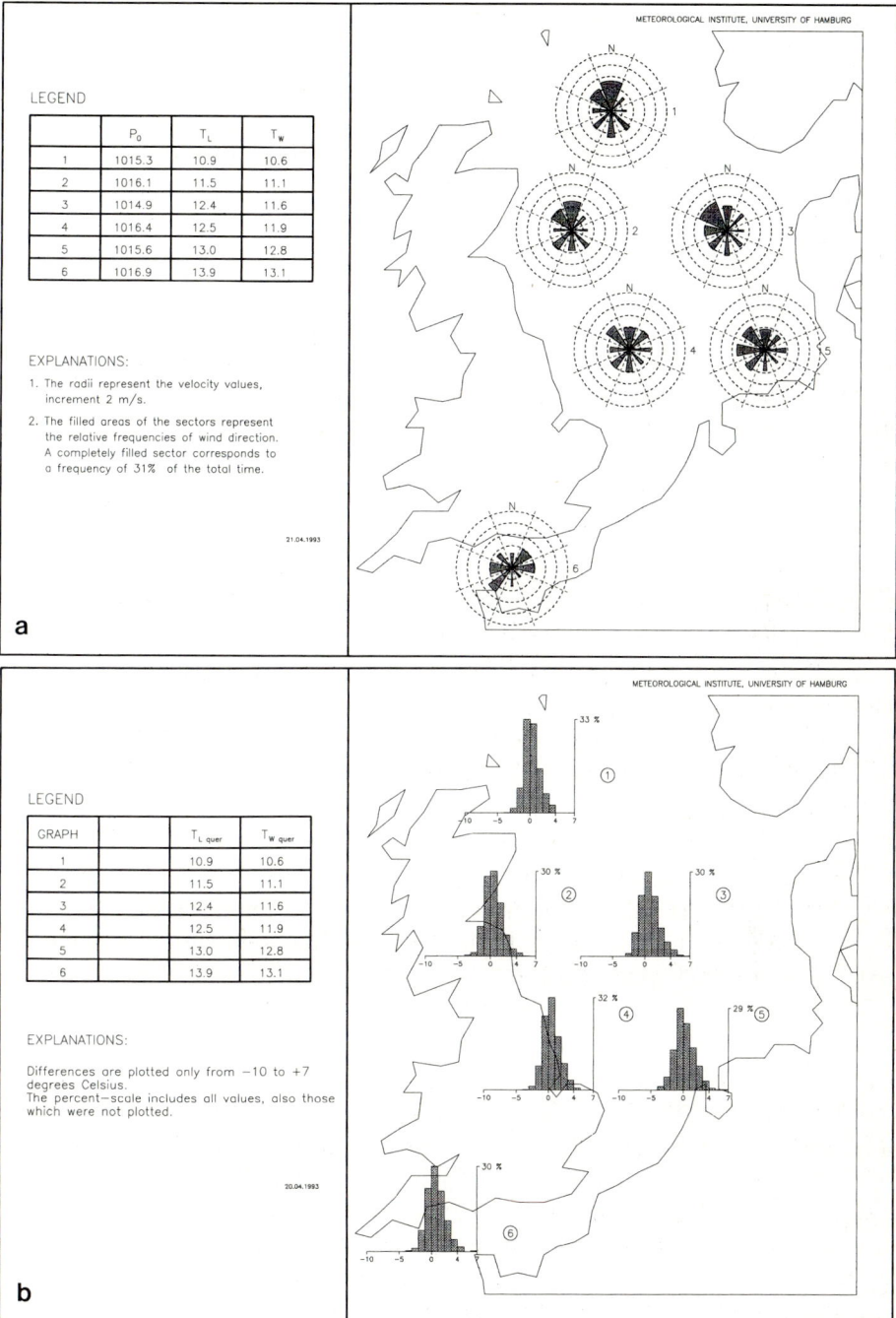

Fig. 2.1.7. Mean wind velocity and direction *(a)* and mean frequency distribution of the difference between air and water temperature *(b)* for June of the years 1986-1989

ence is around 1 °C and only seldom is the air more than 5 °C cooler than the water.

October

In this month the pressure decreases again by an average of 2 hPa, even 5 hPa in the English Channel. Corresponding to the pressure difference of 3 hPa between the northernmost area and the continental North Sea coast, southwesterly winds predominate. The mean wind speed has increased again by about 1 m/s compared with September. About 60% of the time, however, the wind has now a southerly component. In particular in the southeastern North Sea these winds occur more frequently than in the preceding months. The air temperature has decreased further by 2 to 10-12 °C. In the Channel and in the German Bight it is 1.5-2 °C lower than that of the water. In the rest of the North Sea this difference is about 1 °C.

November

In November the pressure is 2-3 hPa greater than in October, having a value of about 1015 hPa. The pressure difference between the northernmost area and the English Channel is 4 hPa. In this month, the distribution of wind direction shows again the old picture with the low frequency of easterly winds, which have particularly decreased in the eastern North Sea; only in the Channel, with 30% of all cases, do easterly winds still occur frequently. November is a further 2-3 °C cooler. The frequency distributions of the temperature difference are now strongly left-skewed. Again, the difference between the air and water temperature is greatest in the Channel and the German Bight with 3 °C. In the central North Sea it is about 1.5-2 °C. In the German Bight the air is more than 5-10 °C cooler than the water 25% of the time.

December

The pressure hardly differs in December from January, having a mean value of about 1014 hPa and increasing quite uniformly from the northernmost area to the Channel by 7 hPa. The windspeed is 7-8 m/s; westerly winds prevail. Only in the Channel do southwesterly and easterly winds occur equally frequently, i.e. 25% of the time each. With barely 6 °C, the air temperatures reach their lowest values in the German Bight. Except for the English Channel they are 7-8 °C elsewhere. The frequency distribution of the temperature differences is extremely left-skewed, particularly in the German Bight. Here again, the difference is 5-10 °C 25% of the time.

LEGEND

	P_0	T_L	T_W
1	1007.4	8.2	7.5
2	1009.0	8.7	7.5
3	1012.8	9.0	7.3
4	1013.5	9.6	7.8
5	1015.9	10.0	7.5
6	1016.1	10.6	9.6

EXPLANATIONS:

1. The radii represent the velocity values, increment 3 m/s.

2. The filled areas of the sectors represent the relative frequencies of wind direction. A completely filled sector corresponds to a frequency of 49% of the total time.

21.04.1993

METEOROLOGICAL INSTITUTE, UNIVERSITY OF HAMBURG

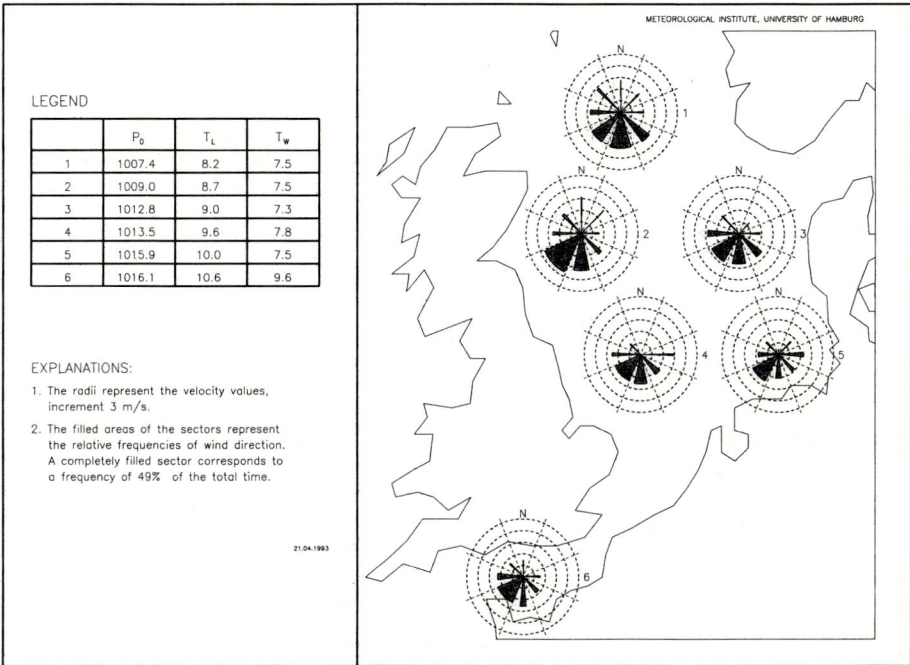

Fig. 2.1.8. Mean wind velocity and direction for May 1986

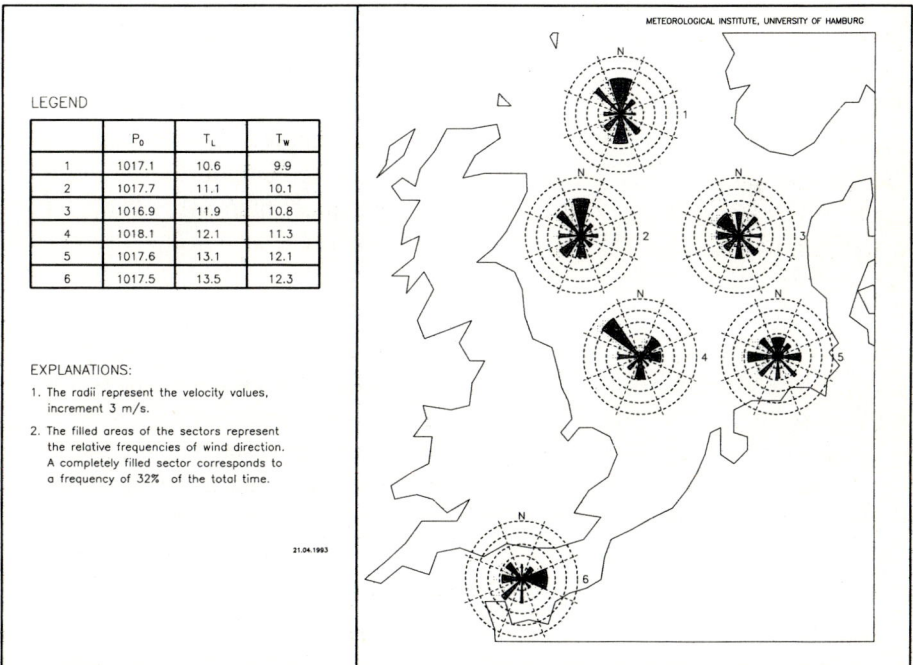

LEGEND

	P_0	T_L	T_W
1	1017.1	10.6	9.9
2	1017.7	11.1	10.1
3	1016.9	11.9	10.8
4	1018.1	12.1	11.3
5	1017.6	13.1	12.1
6	1017.5	13.5	12.3

EXPLANATIONS:

1. The radii represent the velocity values, increment 3 m/s.

2. The filled areas of the sectors represent the relative frequencies of wind direction. A completely filled sector corresponds to a frequency of 32% of the total time.

21.04.1993

METEOROLOGICAL INSTITUTE, UNIVERSITY OF HAMBURG

Fig. 2.1.9. Mean wind velocity and direction for June 1986

2.1.1.1 Distinctive Features During the Months of the VALDIVIA Cruises

During the 4 years considered, the VALDIVIA cruises were conducted from 2 May to 13 June 1986 and from 26 January to 9 March 1987. For this period the wind field over the North Sea will be examined more closely.

May and June 1986 (Figs. 2.1.8 and 2.1.9) are distinctly different from the wind fields of the 4-year mean. While the mean distribution of wind directions is quite uniform - only off the Scottish coast do southwest winds prevail more frequently (20%) - May 1986 is characterized by an almost complete lack of winds from the northwest over north to east. Almost everywhere only southwesterly and southerly winds occur. On the other hand, the picture in June agrees with the 4-year mean; only the speeds of the northerly and northwesterly winds are slightly greater in June 1986.

February 1987 had distinctly lower wind speeds - about 2-3 m/s below the 4-year mean and characterized by an almost complete lack of easterly and northeasterly winds (Fig. 2.1.10). However, winter 86/87 does not belong to the mild winters, which were characteristic for the years 1987/88

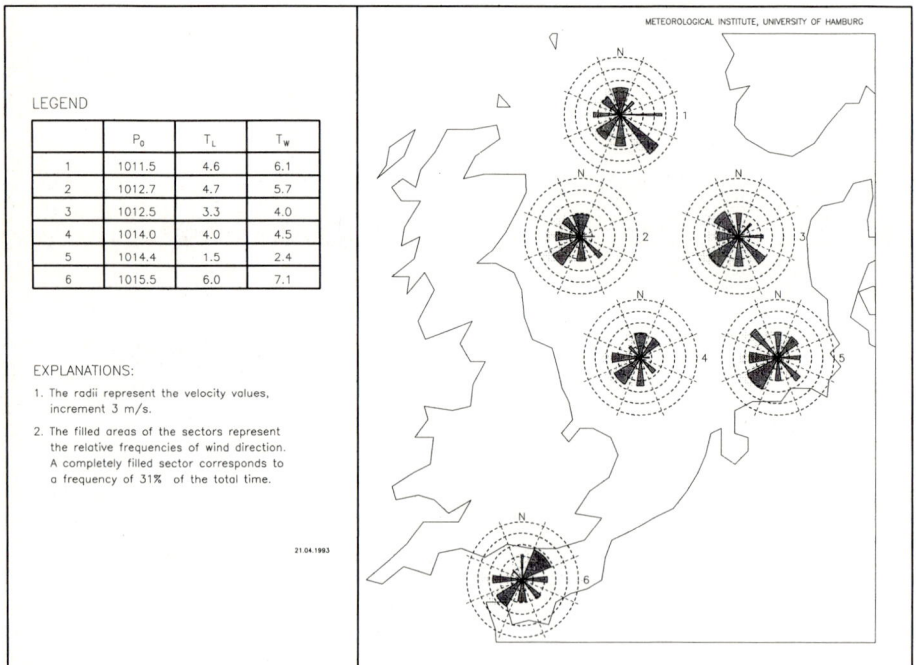

METEOROLOGICAL INSTITUTE, UNIVERSITY OF HAMBURG

LEGEND

	P_0	T_L	T_W
1	1011.5	4.6	6.1
2	1012.7	4.7	5.7
3	1012.5	3.3	4.0
4	1014.0	4.0	4.5
5	1014.4	1.5	2.4
6	1015.5	6.0	7.1

EXPLANATIONS:

1. The radii represent the velocity values, increment 3 m/s.

2. The filled areas of the sectors represent the relative frequencies of wind direction. A completely filled sector corresponds to a frequency of 31% of the total time.

21.04.1993

Fig. 2.1.10. Mean wind velocity and direction for February 1987

and 1988/89 and which were also characterized by an almost complete lack of easterly winds in February. February 1987 has been only a warmer period between the cold months January and March 1987. In the 4-year mean this is not so distinct, since February 1986 - a cold winter - stabilizes the average with almost exclusively easterly winds (Fig. 2.1.11). This example shows that the fields represented in Figs. 2.1.2-2.1.7 should be seen only as representative for the ZISCH activities.

Comparing mean values of air and water temperatures during the VALDIVIA cruises with the 8-years mean from 1982 to 1989 of the same period, significant differences are found. For the experimental period, both air and water show lower temperatures over the North Sea region. Looking at a horizontal cross section from the northwest to the southeast part of the North Sea, air temperature is about 0.5 °C and water temperature about 1.0 °C lower than the 8-year mean (Figs. 2.1.12a,b-2.1.13a,b). For the air temperature values over Northern Germany and Great Britain there are no such differences at all. As a consequence of this fact, it is concluded that the lower air temperatures in May and June 1986 over the North Sea were caused by the significantly lower sea-surface temperatures at this period. Because the air-sea temperature differences are essential for the density

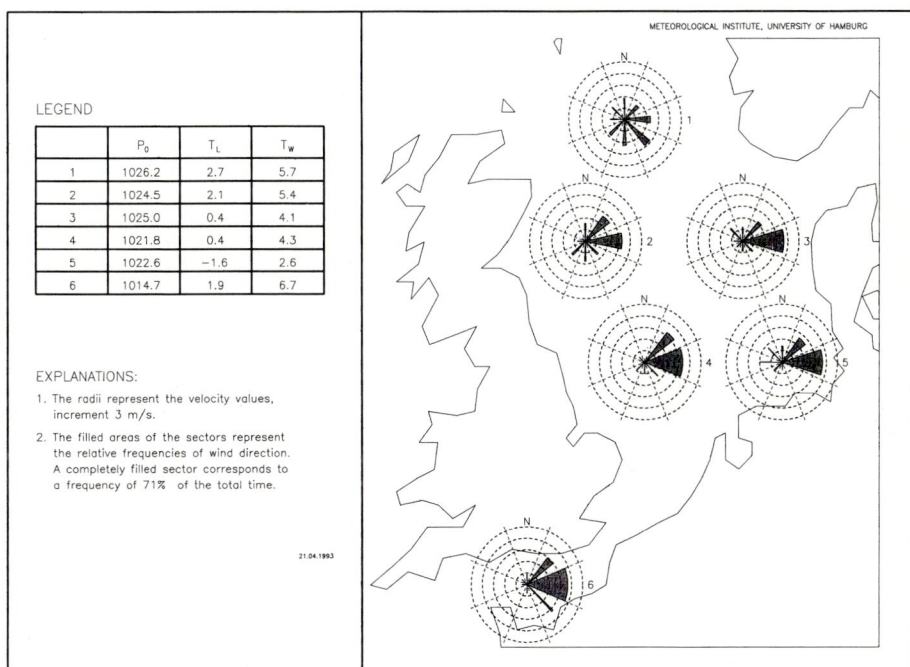

METEOROLOGICAL INSTITUTE, UNIVERSITY OF HAMBURG

LEGEND

	P_0	T_L	T_W
1	1026.2	2.7	5.7
2	1024.5	2.1	5.4
3	1025.0	0.4	4.1
4	1021.8	0.4	4.3
5	1022.6	−1.6	2.6
6	1014.7	1.9	6.7

EXPLANATIONS:

1. The radii represent the velocity values, increment 3 m/s.

2. The filled areas of the sectors represent the relative frequencies of wind direction. A completely filled sector corresponds to a frequency of 71% of the total time.

21.04.1993

Fig. 2.1.11. Mean wind velocity and direction for February 1986

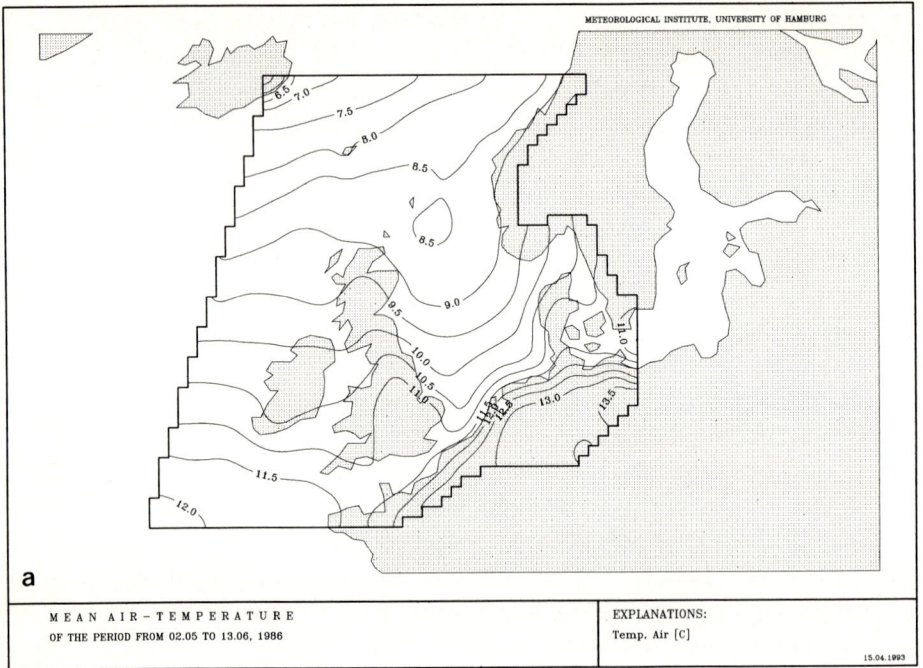

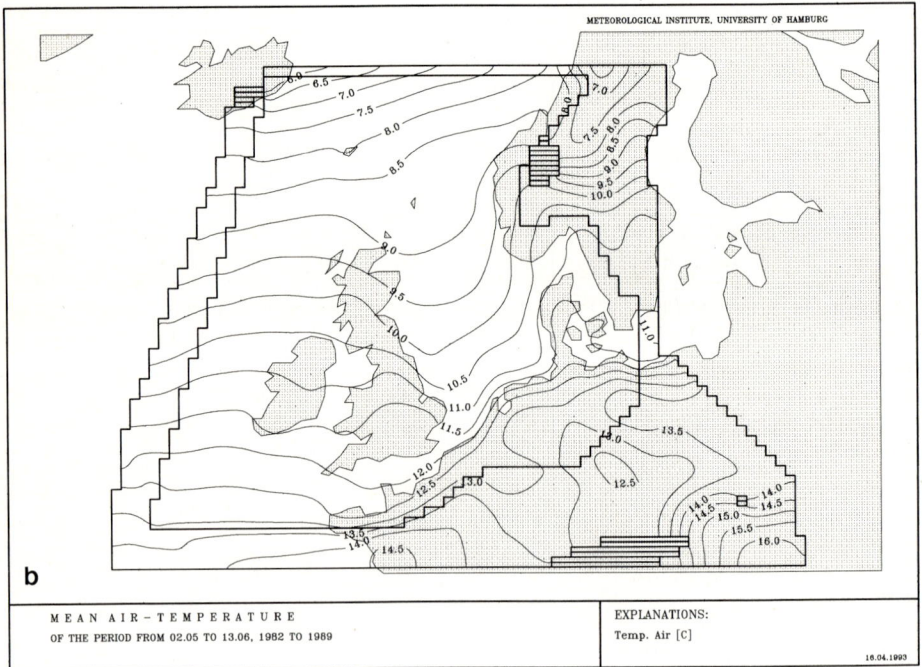

Fig. 2.1.12. Mean air temperature from 2.5.-13.6.1986 *(a)* and from 2.5.-13.6. for the years 1982-1989 *(b)*

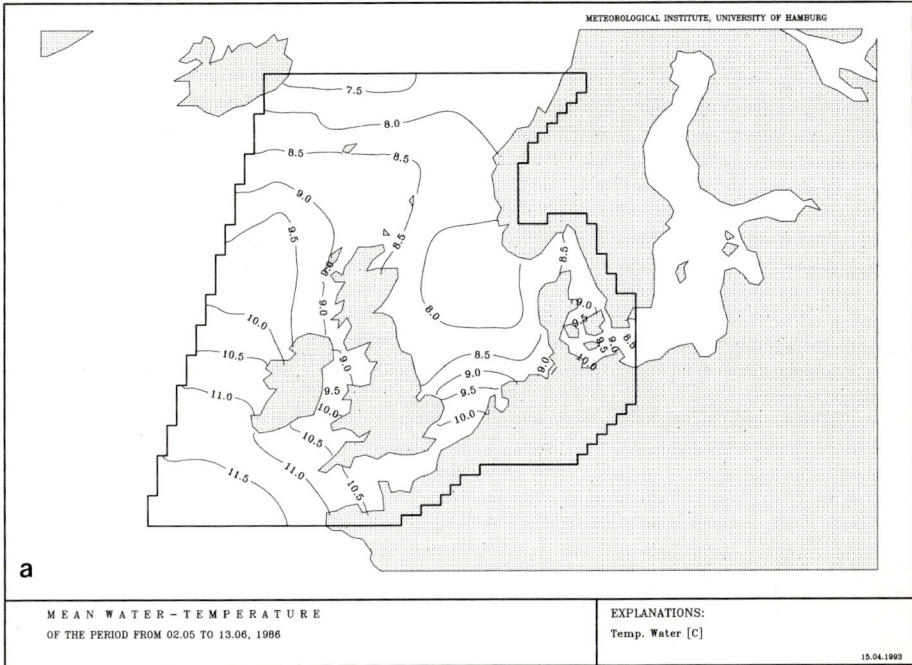

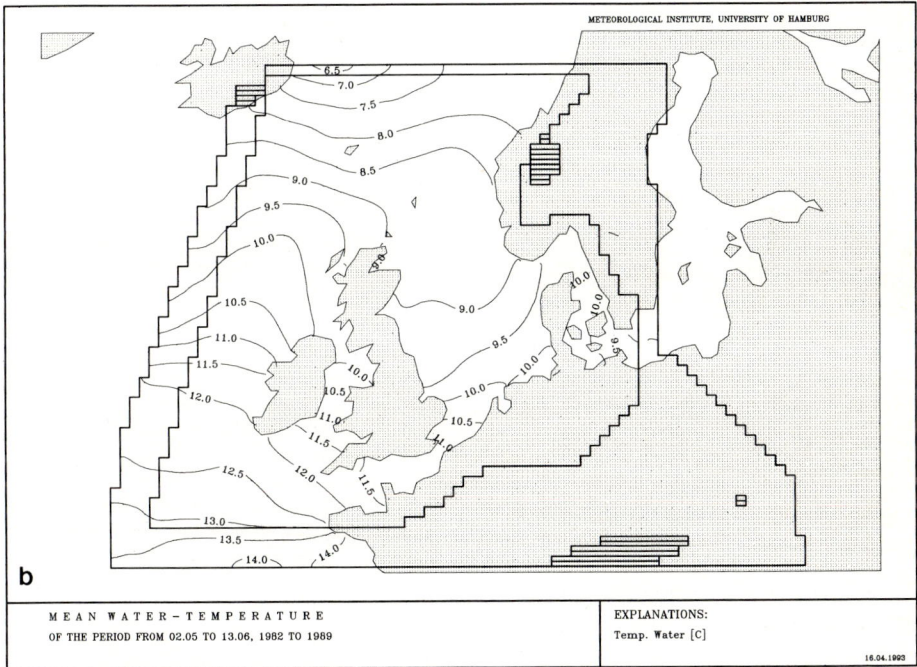

Fig. 2.1.13. Mean water temperature from 2.5.-13.6.1986 *(a)* and from 2.5.-13.6. for the years 1982-1989 *(b)*

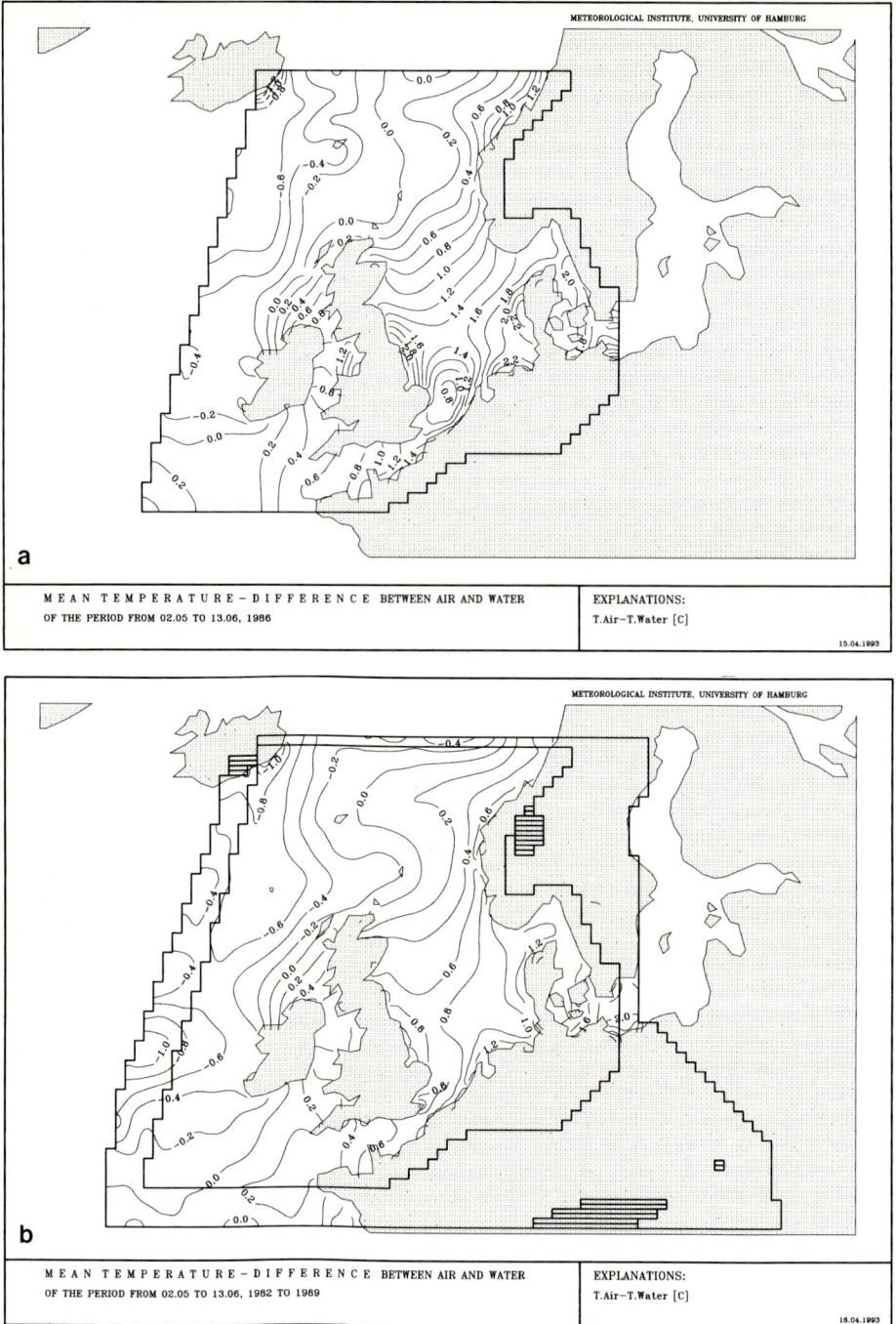

a

MEAN TEMPERATURE–DIFFERENCE BETWEEN AIR AND WATER
OF THE PERIOD FROM 02.05 TO 13.06, 1986

EXPLANATIONS:
T.Air–T.Water [C]

15.04.1993

b

MEAN TEMPERATURE–DIFFERENCE BETWEEN AIR AND WATER
OF THE PERIOD FROM 02.05 TO 13.06, 1982 TO 1989

EXPLANATIONS:
T.Air–T.Water [C]

16.04.1993

Fig. 2.1.14. Mean temperature difference between air and water from 2.5.-13.6.1986 *(a)* and from 2.5.-13.6. for the years 1982-1989 *(b)*

a

MEAN VERTICAL HEAT−FLUX
OF THE PERIOD FROM 02.05 TO 13.06, 1986

EXPLANATIONS:
T'w' [m K/s]

METEOROLOGICAL INSTITUTE, UNIVERSITY OF HAMBURG

19.04.1993

b

MEAN VERTICAL HEAT−FLUX
OF THE PERIOD FROM 02.05 TO 13.06, 1982 TO 1989

EXPLANATIONS:
T'w' [m K/s]

METEOROLOGICAL INSTITUTE, UNIVERSITY OF HAMBURG

19.04.1993

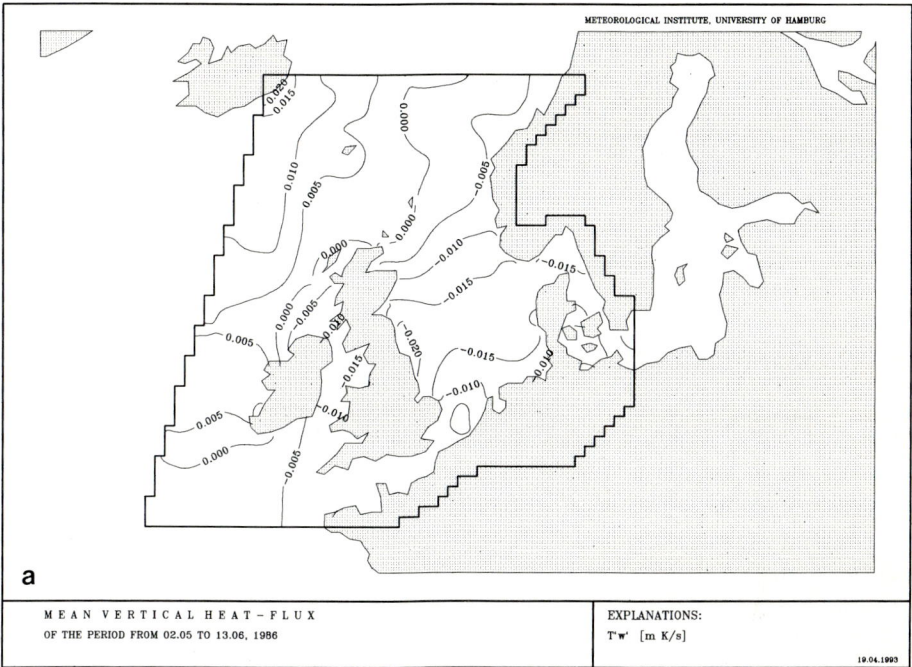

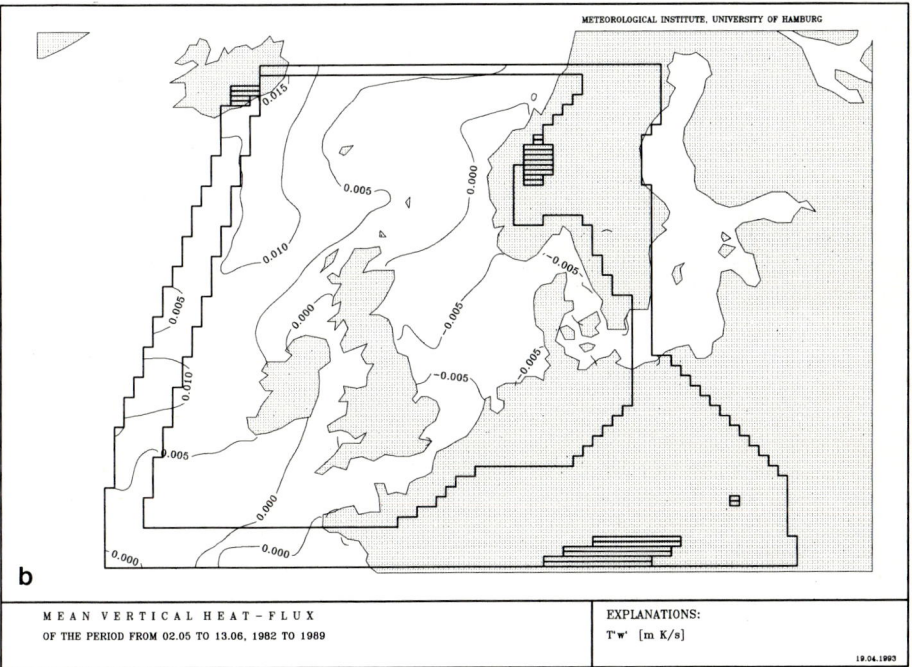

Fig. 2.1.15. Mean vertical heat flux from 2.5.-13.6.1986 *(a)* and from 2.5.-13.6. for the years 1982-1989 *(b)*

stratification of the atmosphere, the analyzed differences in Fig. 2.1.14a suggest a more stable stratified boundary layer during the experimental period compared to the averaged conditions for the years 1982 to 1989 in Fig. 2.1.14b. It must be expected that the vertical turbulent fluxes of the atmosphere caused an efficient warming of the North Sea. While the momentum transport does not differ much, the vertical turbulent heat flux from the atmosphere to the sea-surface was found to be larger by a factor of 2.5 for the experimental period (Fig. 2.1.15a,b).

2.1.2 Marine Variability

The North Sea, a semi-enclosed basin within the northwest European shelf sea, is one of the most productive regions of the world ocean. Obviously the environmental conditions, if not disturbed by human activity, are ideal for marine life. In a shallow shelf sea temperature and salinity, turbulence and currents, almost directly reflect the activity of the atmosphere. This is in contrast to the deep ocean, which reacts to atmospheric disturbances (climate variability and/or changes) with a time-lag of some decades, or even centuries.

Important factors for the marine variability of the North Sea are the inflowing water masses from the Atlantic and the continental freshwater runoff. The saline Atlantic water and the freshwater drained via a number of rivers and via the Baltic Sea from the huge hinterland of western Europe are merged and mixed by the action of the tides and of the atmospherically induced turbulence of waves and currents. This permanently ongoing mixing process maintains a lateral stratification, which has a considerable influence on the dynamics of the North Sea. The haline stratification is present throughout the year but it shows an annual variation due to changes in the precipitation over the land and in the Atlantic inflow. The role of the precipitation over the North Sea itself has hardly been investigated; usually it is assumed that it is balanced by evaporation.

The rapid absorption of radiative solar input into the sea accounts for a thermal stratification, which is most pronounced in the summer. Wide areas of the North Sea form a two-layer system with a sharp interface, the thermocline, which separates heated surface water from the bottom water. The latter tends to keep the temperature of the previous winter season, because it is thermally isolated from the surface heat input by the pycnocline. In autumn and winter, this stratification breaks down due to net cooling at the sea surface and as a consequence of mixing induced by the tides, wind and waves. Then the two-layer system reduces to one

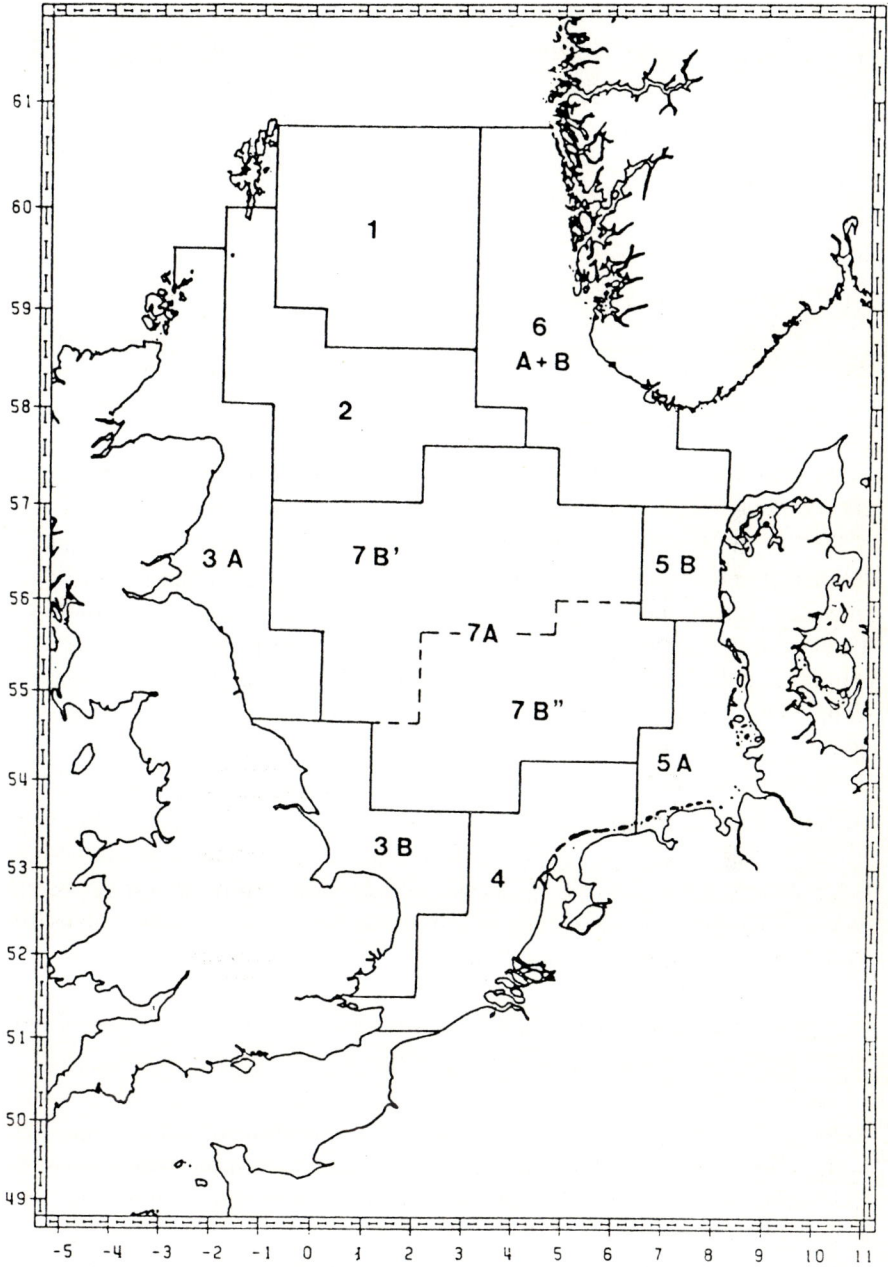

Fig. 2.1.16. ICES boxes: hydrographic regions of the North Sea in accordance with the ICES Study Group (1977)

vertically well mixed layer of uniform temperature. In addition to the vertical thermal stratification during summer, a lateral thermal stratification is present throughout the year. This is caused by the fact that the North Sea is a transition region from coastal to oceanic water masses. The temperature gradient between these water masses changes sign between summer and winter.

Both the atmospheric and the marine variability are characterized by the transition between the oceanic and the continental mid-latitude climate regimes. The temporal and spatial changes in the respective dominance of the regimes account for a considerable annual, sub-annual and inter-annual variability in the atmosphere and in the sea.

2.1.3 Selected Areas of the North Sea

Several attempts have been made to divide the complex system "North Sea" into characteristic areas.

Giving main emphasis to the hydrographical, biological and chemical characteristics; in 1977 the North Sea was divided into boxes, by the *Study Group on Flushing Times of the North Sea ICES* (International Council for the Exploration of the Sea): the so-called ICES boxes (Fig. 2.1.16). Of course, this division is a compromise with respect to the different points of view, but many scientists use this system, so it is useful for interdisciplinary work.

The characteristics of the North Sea boxes, given by the above mentioned study group are (Becker 1981):

ICES Box 1: water masses of oceanic origin with relatively low transport rates. During the summer thermally stratified. Mixed water plankton community. Spawning area of the haddock.

ICES Box 2: coastal waters and ocean waters mixed. Water transport to the North Sea. Partial stratification in summer.

ICES Box 3 A,B: relatively slow southward transport rates. Transition zone between mixed and stratified water. Salinity decreases southward with increasing freshwater inflow. Neritic plankton community. Spawning area of herring and sprat.

ICES Box 4: Channel water inflow, mixed with coastal waters. Strong horizontal salinity gradients. Haline stratification only right offshore. Neritic-plankton community. Spawning area of flatfish.

ICES Box 5 A,B: generally northward water transport. Coastal waters, partly haline, occasional thermal stratification. Neritic-plankton community.

ICES Box 6A: surface layer: Northward flow of the Norwegian coastal waters and the Baltic outflow. Permanent density stratification. Neritic-plankton community.

ICES Box 6B: Deep layer: Horizontally inhomogenous with southward directed oceanic water transport in the west, and northward directed mixed water transport in the east.

ICES Box 7A: covers ICES boxes 7B' and 7B''. Water transport into several areas with low speeds. A well-defined summer thermocline. Mixed-plankton community with neritic component. Spawning area of the cod.

ICES Box 7B'B'': bottom water layer. During the summer north and south of the Dogger Bank two separated water masses.

2.1.4 Water Mass Transport

Through the Shetland Passage and the Fair Isle-Shetland Passage, water masses from the Atlantic flow into the northern North Sea. They partly spread into the central North Sea and recirculate through the Norwegian Trench back into the Atlantic. The south receives Atlantic water that flows in through the English channel and spreads into the southern North Sea. Then it flows into the German Bight and moves north to leave the North Sea via the Norwegian Trench. A part of the Baltic Sea water flows through the Skagerrak into the North Sea.

The circulation in the North Sea is affected by the tides. The net water mass transport is determined by the residual flow regime. A definition of the residual flow is given by Backhaus (1980). For a first approximation, a one day low-pass filter of the flow is sufficient.

In contrast to other ocean areas, a large number of velocity time series exist. Nevertheless, it is still not possible to obtain a circulation field of the

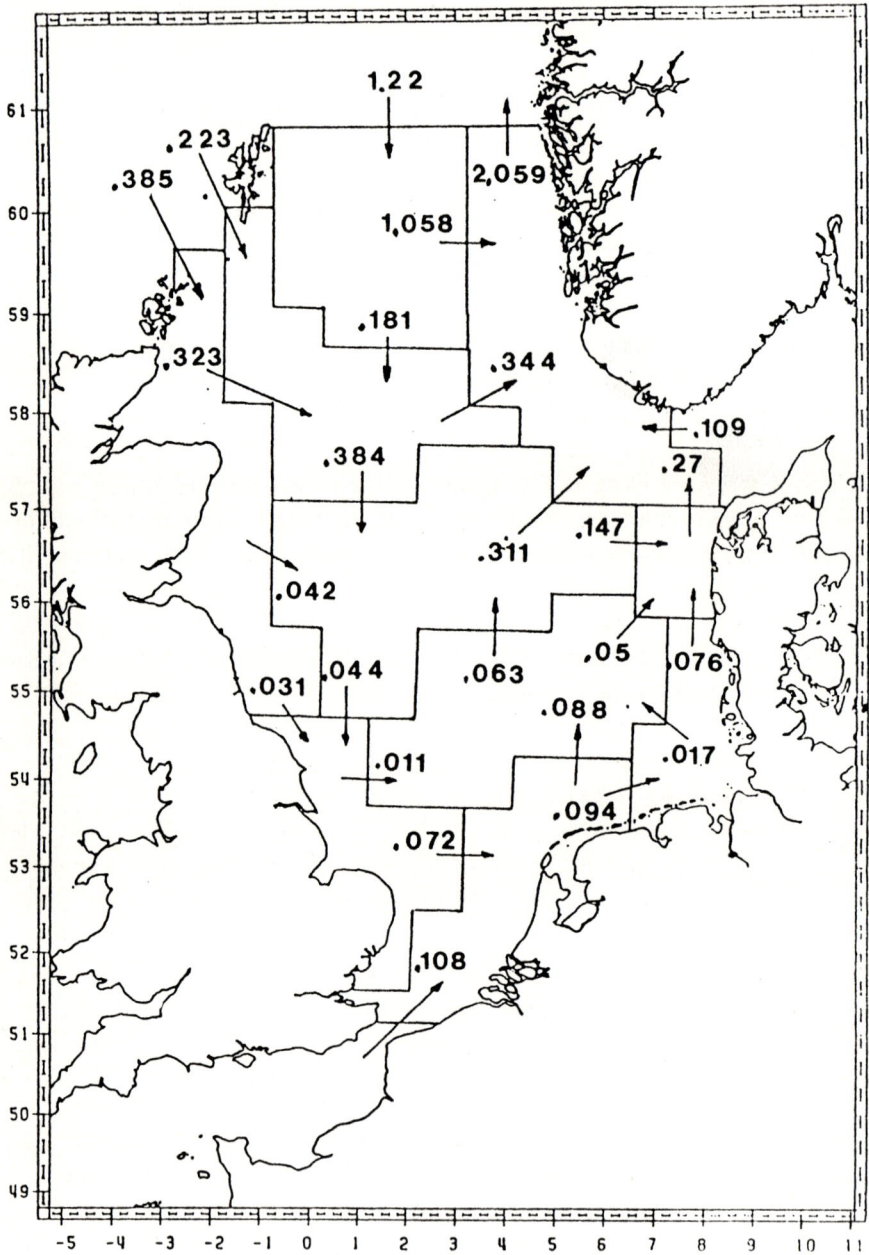

Fig. 2.1.17. Vertically integrated net water mass transport (1970-1981) through the ICES boxes in Sv (1 Sv $= 10^6 \text{m}^3/\text{s}$). (After Radach et al. 1990)

whole North Sea by measured data, because the time series were taken at different time. Often the observation period is not even long enough to filter out the tidal current, i.e. to determine the residual flow.

As part of ZISCH, three-dimensional circulation fields were estimated for the period between 1969 and 1981 (Hainbucher et al. 1986). They are based on a numerical simulation driven by wind fields, a mean winter and summer salinity and temperature distribution and a M_2 tidal forcing at the open boundary. The data gained from the model were stored as daily mean values. Based on these model results, Radach et al. (1990) and Lenhart (1990) calculated water mass balances and water exchange times for each ICES box; Radach for successive winters and Lenhart for annual means. Hainbucher et al. (1986) gives the water mass transport and its standard deviation through selected sections in the North Sea.

Regarding the net transport rates between the ICES boxes (Fig. 2.1.17), calculated by Lenhart (1990), 24% of the incoming Atlantic water masses through the northern boundaries reach the central North Sea (ICES Box 7B') north of the Dogger Bank, and only about 5% circulate south of it. The recirculation of water masses in the north of North Sea is clearly visible, looking at the net transport rates of the ICES Boxes 1 and 2.

Also Atlantic water masses moves through the the English Channel towards the continental coast into the German Bight and the ICES Box 5B.

The water masses leave the North Sea through the ICES Box 6, running parallel to the Norwegian coast.

The described mass transports between the ICES boxes are based on the mean circulation over one decade and deliver a quasi-static picture. An idea of the variability of the circulation can be gained with the help of mass transports through selected sections in the North Sea in Fig. 2.1.18, where arrows point out the positive direction of calculated mass transport.

For the different sections the transport time series (Figs. 2.1.19, 2.1.20, 2.1.21, 2.1.22 , 2.1.23) from 1969-1981 are described. The horizontal solid line is the mean transport rate for the observed period of time, and the varying solid line represents 90 day low-pass filtered transport rates. The shaded areas of the demonstrated time series show band pass filtered transports (6.5 - 90 days), and give an idea of existing fluctuations. For the 90 day low pass filtered transports a high variation is still visible: The transport rates in the outflow area (Fig. 2.1.19) of the North Sea vary at a range of 55% with a standard deviation of about 15%, related to the mean transport. A similar variability is found at the inflow areas (Figs. 2.1.20, 2.1.21, 2.1.22) and also through the southwest boundaries of the German Bight (Fig. 2.1.23).

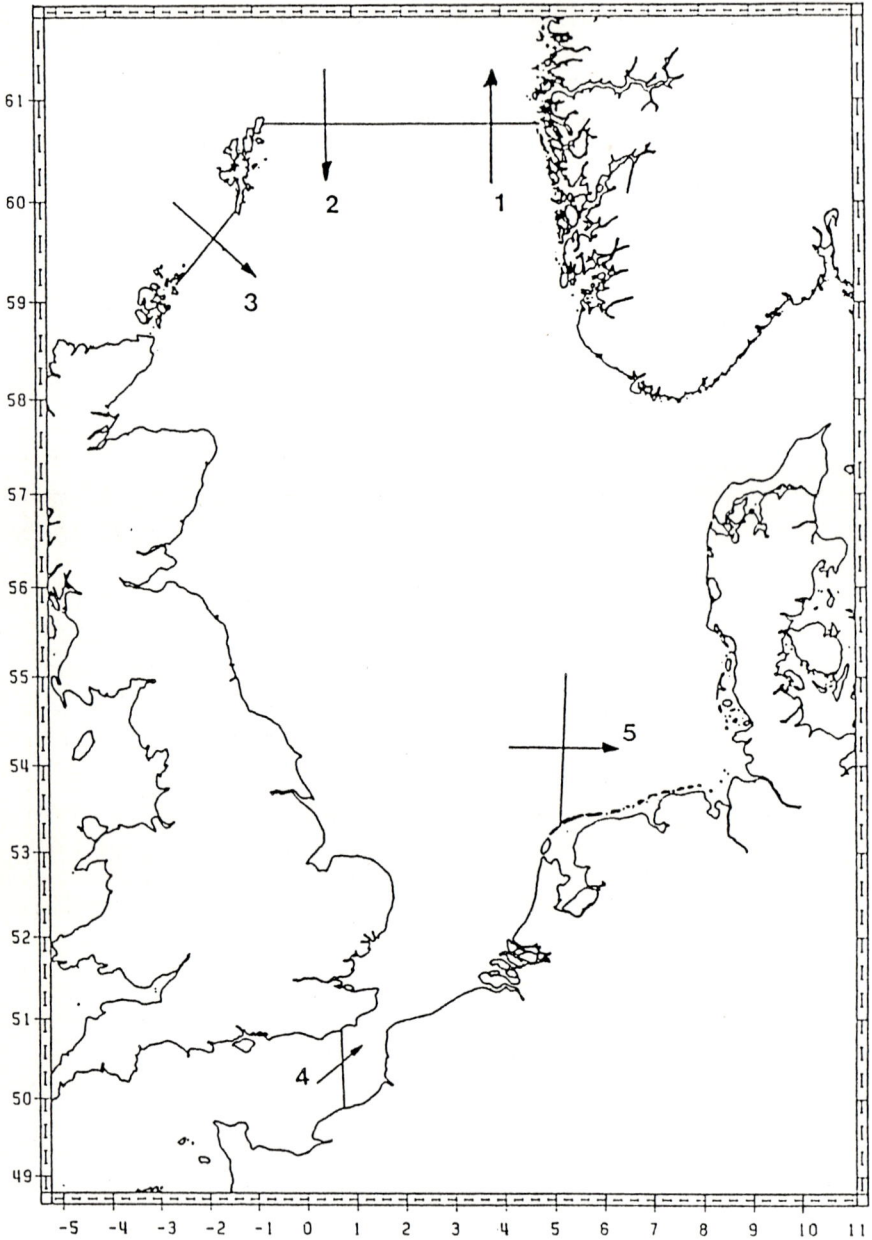

Fig. 2.1.18. Location of transport sections. *1* Norw. Trench -N-; *2* Shetland Trench; *3* Fair Isle; *4* Dover Strait; *5* German Bight; *arrows* point out the direction of positive transport

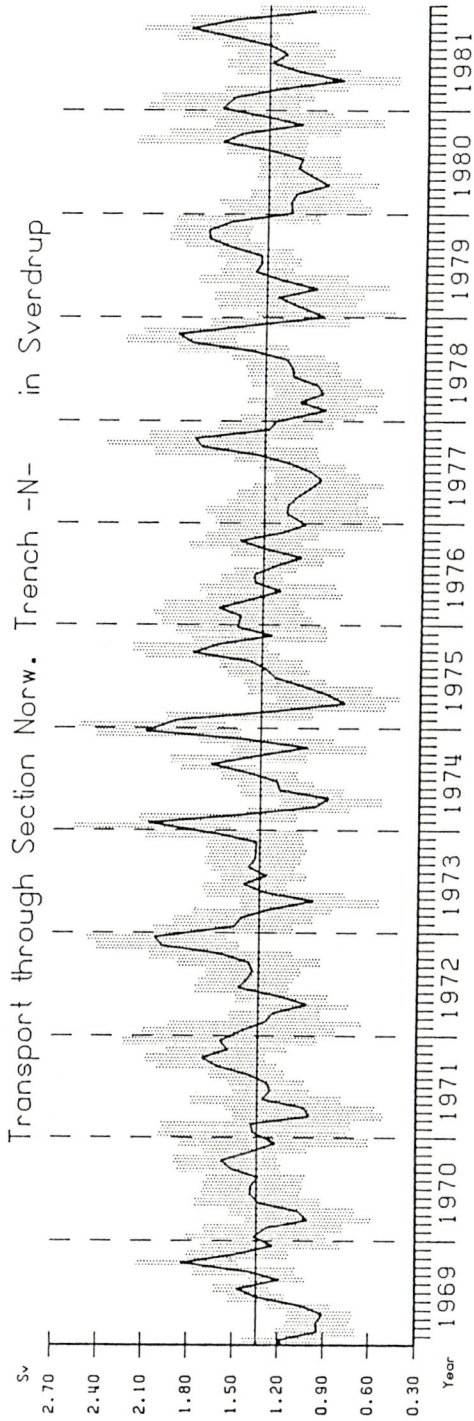

Fig. 2.1.19. Water mass transport through section 1 (see Fig. 2.1.18 for direction of transport): Norwegian Trench -N- (in Sverdrup = 10^6 m^3/s). *Shaded area* standard deviation. (After Hainbucher et al. 1986)

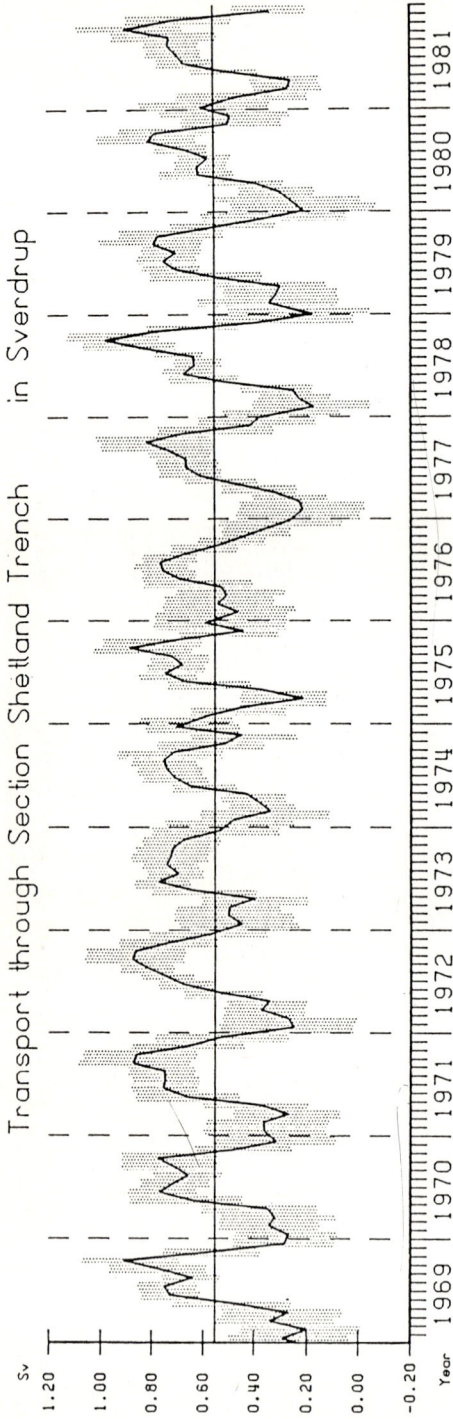

Fig. 2.1.20. Water mass transport through section 2 (see Fig. 2.1.18 for direction of transport), Shetland Trench (in Sverdrup $= 10^6 \text{m}^3/\text{s}$). *Shaded area* standard deviation. (After Hainbucher et al. 1986)

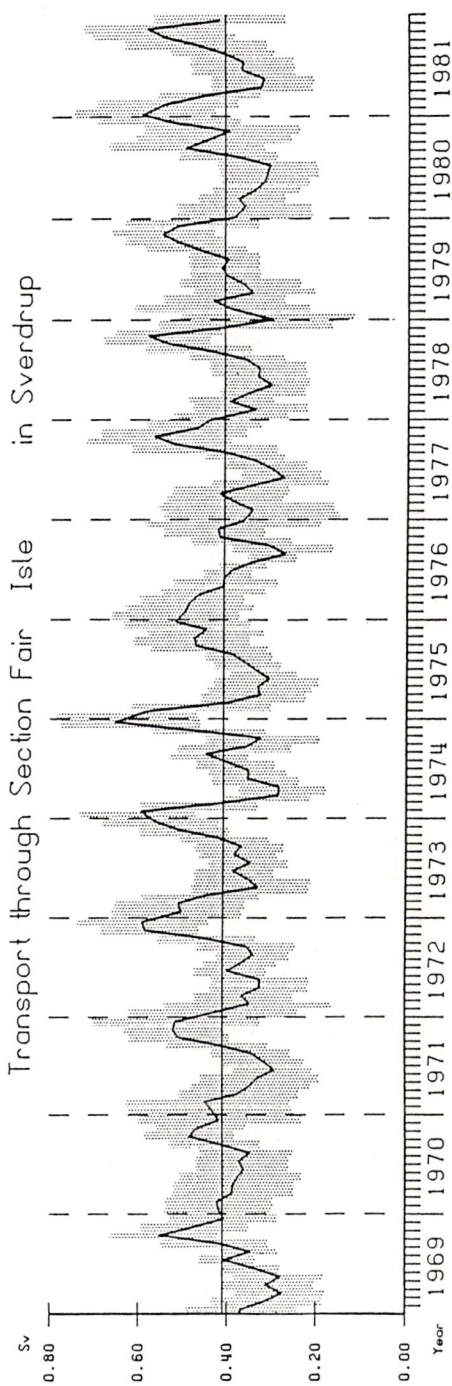

Fig. 2.1.21. Water mass transport through section 3 (see Fig. 2.1.18 for direction of transport), Fair Isle (in Sverdrup = $10^6 m^3/s$). *Shaded area* standard deviation. (After Hainbucher et al. 1986)

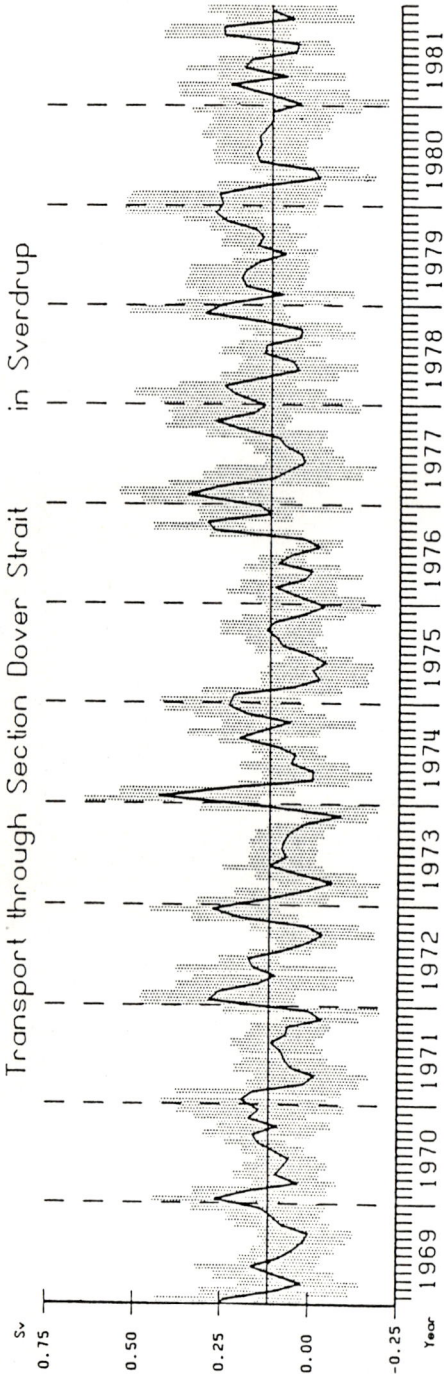

Fig. 2.1.22. Water mass transport through section 4 (see Fig. 2.1.18 for direction of transport), Dover Strait (in Sverdrup $= 10^6 \text{m}^3/\text{s}$). *Shaded area* standard deviation. (After Hainbucher et al. 1986)

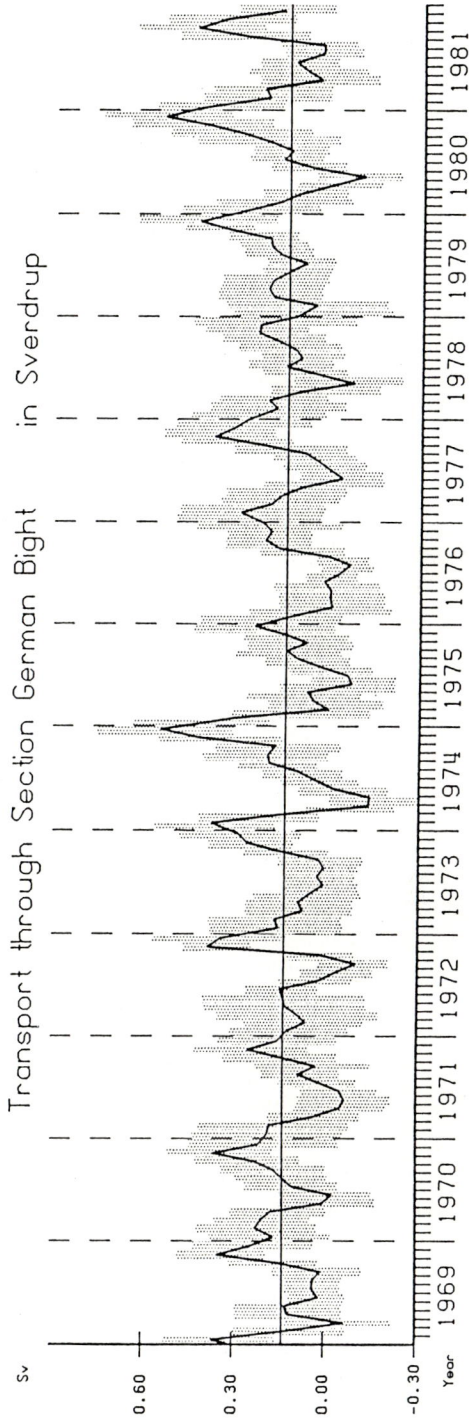

Fig. 2.1.23. Water mass transport through section 5 (see Fig. 2.1.18 for direction of transport), German Bight (in Sverdrup $= 10^6\,\mathrm{m}^3/\mathrm{s}$). *Shaded area* standard deviation. (After Hainbucher et al. 1986)

The high fluctuations of the transport here demonstrated are typical for the North Sea and have to be considered carefully when dealing with averaged values or when interpreting observed data for this area.

There are different possibilities for defining the water mass exchange for a certain area, the calculation method and the time definition change accordingly: Maier-Reimer (1977), for example, takes the exchange time range for the mass as the time the water needs to leave the North Sea (flushing-time). In contrast, Davis (1980), Backhaus (1984), Lenhart (1990) and Radach et al. (1990) take the water mass exchange time as the time elapsed until a volume equivalence A flows into a certain area V (turnover time).

The first definition results in short flushing times in the outflow areas. The second definition leaves recirculation unconsidered and the turnover times are dependent on the volumes of the sea area observed. Therefore, Radach and Lenhart propose to relate them to a certain volume (pers. comm.). In this way, a comparable dimension for the active mass exchange would be given.

All of the stated definitions are examples for one physical fact being looked at from different points of view. The decision regarding which method is to be preferred depends on the question at hand.

Especially for the calculation of dispersion of matter, it is important to know the time range during which the contaminated water masses of a certain area are exchanged (turnover time). The turnover times calculated by Lenhart give a long term mean (12 years), a 1-year mean (1986) and the actual exchange times for the time range of the two large scale ZISCH surveys experiments (1986,1987) for the ICES boxes (Table 2.1.1). The different turnover times of the same ICES boxes for different averaging time intervals (1986; ZISCH 1986; ZISCH 1987) are in the range of the standard deviations of the period 1970-1981.

During the field measurements from 2 May to 13 June 1986, the turnover times for the water masses of the individual ICES boxes were longer than the climatological mean (1970-1981). In these months (May and June) the wind fields over the North Sea differed significantly from the 4-year mean (see Sect. 2.1.1). In May 1986 the winds blow with an anomalous prevalence from the southwest and south, while in June 1986 exceedingly high wind speeds from the northerly and northwesterly directions occur. Water mass exchange is generally weakened by these anomalous wind fields, the flushing times become longer.

For the field experiments during February 1987, the flushing times are again larger than those of the climatological mean. They even exceeded the mean turnover times (1970-1981) increased by their standard deviations.

Table 2.1.1. Ranges of turnover time calculated by Lenhart (1990)

min: minimum of exchange time (days); max: maximum of exchange time (days); mean: average of exchange time (days); σ: standard deviation of exchange time (days); A/V: exchange time per volume (minutes); ZISCH 1986: May to June 1986 (days); ZISCH 1987: February 1987 (days); -: exchange time > observation time.

| ICES box | Observation time | | | | | | | | | | |
| | 1970-1981 | | | | | 1986 | | | | ZISCH 1986 | ZISCH 1987 |
No.	Min	Mean	Max	σ	A/V	Min	Mean	Max	A/V	Mean	Mean
1	26	42	55	5.2	9.5	26	39	49	8.9	36	-
2	16	29	39	4.7	7.4	16	34	42	8.7	32	38
3A	18	33	53	8.1	15.0	22	42	56	19.0	38	-
3B	9	23	44	6.9	29.1	16	29	47	36.7	23	36
4	7	20	41	6.8	21.8	11	25	48	27.2	20	34
5A	8	27	55	10.1	64.6	15	36	56	86.1	-	-
5B	3	11	27	4.5	39.2	2	11	26	39.2	12	17
6AB	33	48	63	7.1	5.4	32	48	58	5.4	-	-
7B'	23	40	68	10.1	9.3	37	56	71	13.0	-	-
7B''	15	32	56	9.3	16.6	22	39	60	20.3	39	-

This is an effect of the anomalously weak winds during this period (see Sect. 2.1.1) on the circulation system of the North Sea.

In areas of continental river input, as the Rhine estuary into the Hoofden (ICES Box 4) and the Elbe-Weser estuary into the German Bight (ICES Box 5), the standard deviation of the turnover time is 35%, extremely high related to the mean value. The mean residence time of inflowing matter is very variable in those shallow areas; in unfavourable situations a correspondingly high accumulation of matter will occur.

Through the ICES Box 6, off the Norwegian coast, water masses leave the North Sea: on the average every 48 days a water amount equal to the

volume of the box. Relative to the size of this region, every 5.4 min 1 km^3 water leaves this area and will be replaced by new North Sea water. These numbers are related to a vertical integrated transport through the open boundaries. The area itself has a distinct northward directed flow at the surface, indicating that the surface water exchange is a lot faster. The shallow German Bight has a different behaviour. In the mean the water will be replaced every 27 days. Every 65 min 1 km^3 of water will be exchanged. Here a distinction of the exchange rates in a vertical dimension, as in the Norwegian Trench, is not necessary. Not counting the high variability of the turnover time, the German Bight needs by far the longest time to exchange water with the surroundings.

2.1.5 The Representativity of Observed Data

As part of ZISCH, aside from a few local experiments, two large field surveys were carried out, each of them covering the whole North Sea. One took place during summer (May/June 1986) and the other during winter (February 1987). In the following, the representativity of these data will be investigated by comparing them with climatological data.

For the sake of simplicity, it was decided to discuss here only one typical data subset for each experiment. Surface temperature was chosen for the summer survey. With this selection the higher variability of temperature and response to changes in the seasonal thermocline is emphasized. As a typical example for low variability, the salinity of the bottom water is examined more closely for the winter survey.

2.1.5.1 ZISCH Surface Water Temperature During Summer

During May and June, the surface of the North Sea warms up and a thermocline forms. The surface temperature distribution found during the summer survey is shown in Fig. 2.1.24. It shows relatively warm temperatures (10-12 °C) in off shore waters off the continent and southern Norway. Towards the Shetland Islands the temperatures drop to 7 °C.

In the climatological average (1968-1985) of the surface temperatures of the North Sea for the month of May (Damm 1989) is in the central and northern North Sea about 0.5-1 °C higher than at the time of the summer campaign in 1986 (Fig. 2.1.25). The measured temperatures at the continental coast are slightly higher than the climatological average values. Temperature deviations similar to the field observations are seen in the climatological average (1902-1954) by Tomczak and Goedecke (1962), which is based on much fewer observation data than Damm's (about 10 %).

The deviations between the actual measurements and the climatological average of the surface temperature are in the order of magnitude of the variability (Fig. 2.1.26). The latter is derived from the standard deviation of the measured data based on the climatological map (Damm 1989).

The climatological weekly temperature distribution (Becker 1981) (Figs. 2.1.27, 2.1.28) demonstrates to which extent the surface temperature of the North Sea change in average during the ZISCH-survey period. On the climatological average the surface temperature of the North Sea increased in general about 3 °C from the beginning (19th week) until the end (23rd week).

This strong increase in surface temperature compared with the climatological mean is still manifested during the measurement period May/June 1986. During this period the mean difference between air and water temperature is particularly large, and the heat flux from the atmosphere to the hydrosphere is larger than the climatological mean by a factor of 2.5 (see Sect. 2.1.1). From this it can be concluded that the surface water temperature was significantly increased on a large scale during the field experiment and that the temperature data not accounting for this trend cannot be regarded as a synoptic surface temperature map.

2.1.5.2 ZISCH Salinity in the Bottom Water During Winter

The water masses in the North Sea are well mixed in winter. The temporal variability of the temperature and salinity distribution is small compared to the other seasons. The salinity distribution at the bottom (Fig. 2.1.29),

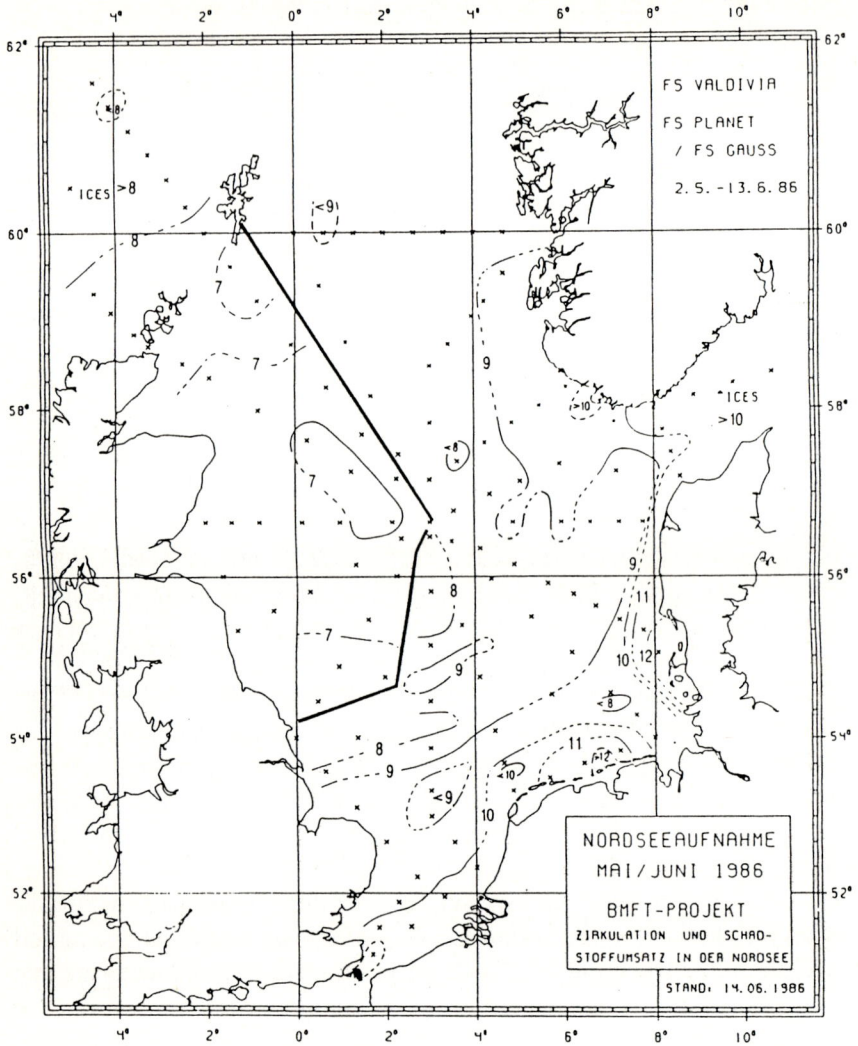

Fig. 2.1.24. ZISCH field experiment: May/June 1986. Surface temperature (0-5 m) of the North Sea (in °C)

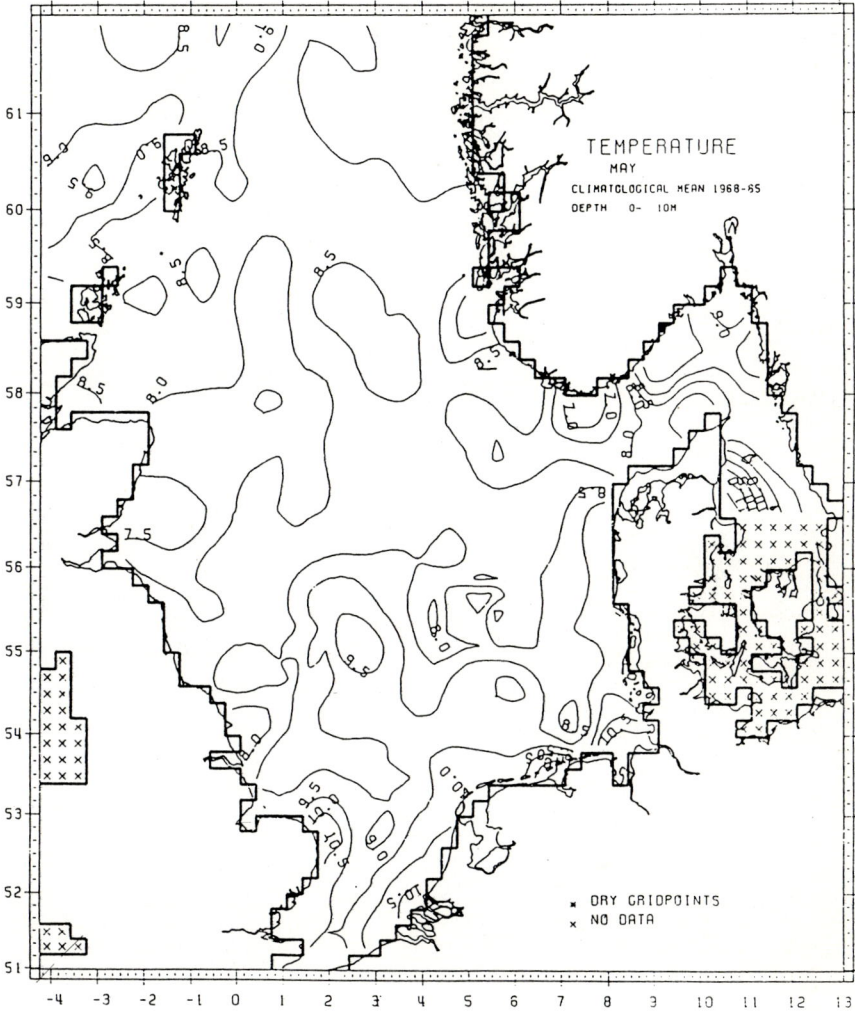

Fig. 2.1.25. Climatological mean (1968-1985) in May. Surface temperature (0-10 m) of the North Sea (in °C)

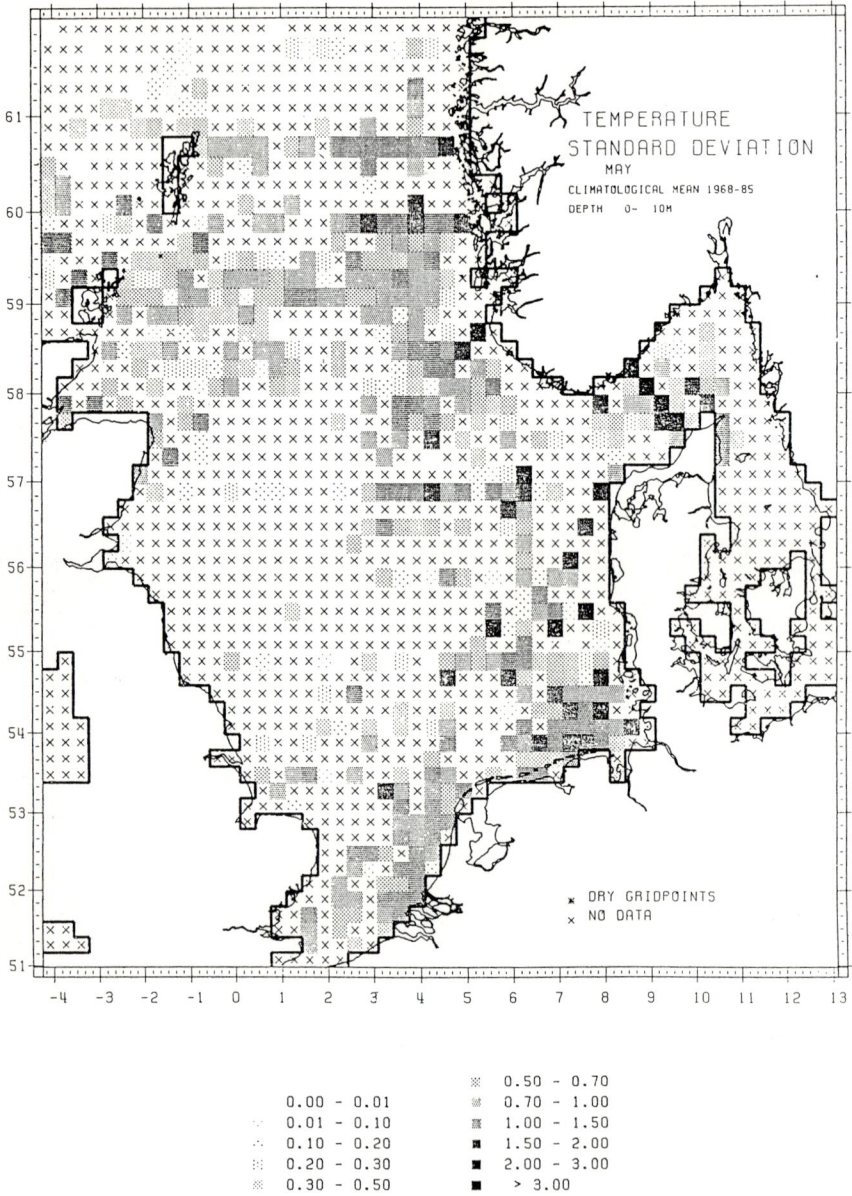

TEMPERATURE
STANDARD DEVIATION
MAY
CLIMATOLOGICAL MEAN 1968-85
DEPTH 0- 10M

DRY GRIDPOINTS
NO DATA

	0.50 - 0.70
0.00 - 0.01	0.70 - 1.00
0.01 - 0.10	1.00 - 1.50
0.10 - 0.20	1.50 - 2.00
0.20 - 0.30	2.00 - 3.00
0.30 - 0.50	> 3.00

Fig. 2.1.26. Climatological mean (1968-1985) in May. Standard deviation of surface temperature (0-10 m) of the North Sea (in °C)

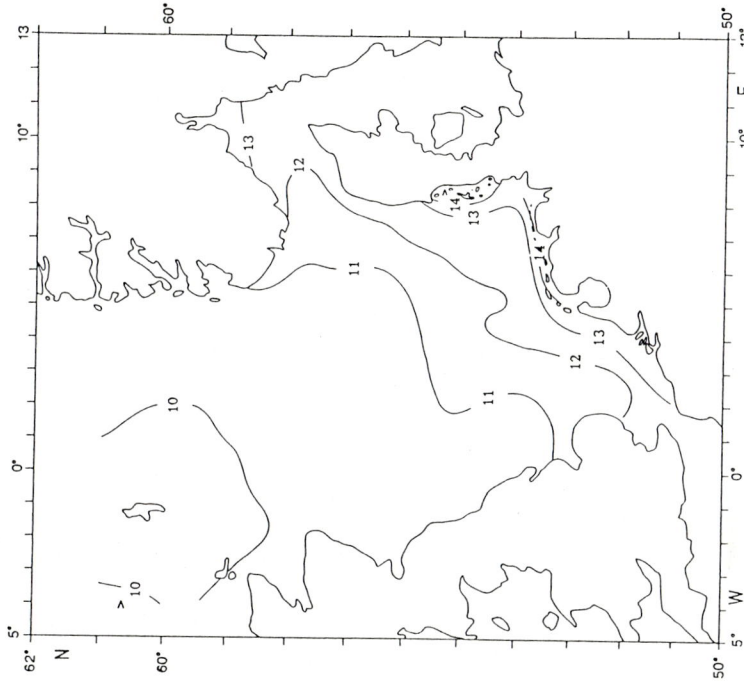

Fig. 2.1.28. Climatological mean values of the North Sea surface temperature (in °C) in the 23rd week (4th June to 10th June). (After Becker 1981)

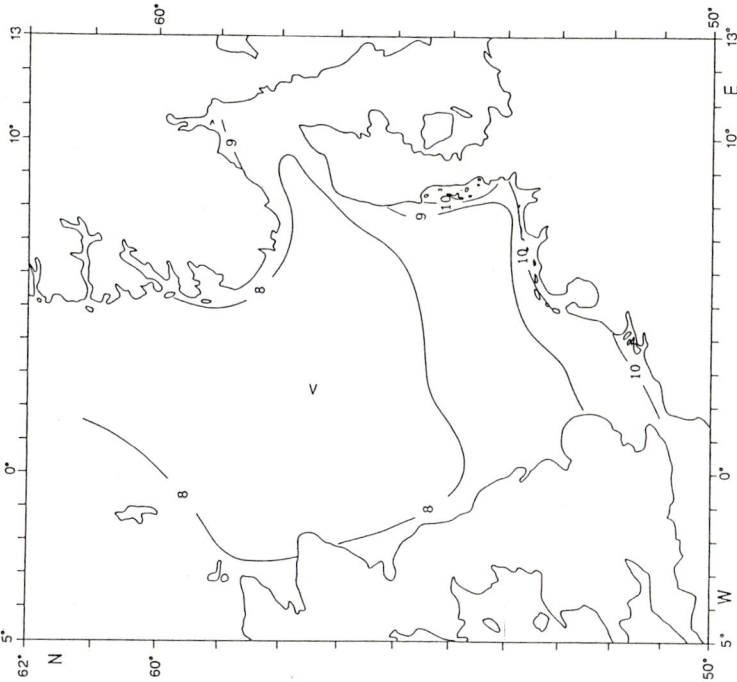

Fig. 2.1.27. Climatological mean values of the North Sea surface temperature (in °C) in the 19th week (7th May to 13th May). (After Becker 1981)

obtained during the winter survey has the same structure as the one from the climatological average (Fig. 2.1.30). In the latter, the 35 isohaline reaches further south to the Dogger Bank.

The difference between the actual data and the climatological map is within on standard deviation (Fig. 2.1.31). The spatial variability of the salinity at the bottom of the central North Sea is shown by van de Kamp (1990) in the area from 0° to 4° E. Over the last 20 years (1970-1989) regions with salinity between 34.9 and 35 are indicated by the shading in the pictures (Figs. 2.1.32, 2.1.33, 2.1.34). They demonstrate the variability of the salinity more distinctly than a presentation of the standard deviation could do.

A change of salinity in the bottom water in the central North Sea during the experimental period in February 1987 cannot be derived from the data, since measurements were not carried out more than once at any positions, and comparable data are not available. However, from other experiments (Soetje and Huber 1980), it is known that the temporal variability of the bottom salinity occurs on time scales which are longer than the experiment described here. That would indicate that this data set for bottom salinity can be regarded as synoptic.

2.1.6 Summary

The mean transport rates between the ICES boxes represent an averaged circulation. The magnitude of the actual variability of this circulation, however, could be shown in time series for the water mass transport through selected sections. The long flushing times of water masses found for the period of the experiment, by a numerical model, could be attributed to anomalous meteorological conditions. Similarly, the enhanced warming of the surface water in May/June 1986 could also be ascribed to anomalous meteorological conditions. Nevertheless, the surface temperatures measured during this period under mean warming conditions could not be regarded as a synoptic data set.

The bottom salinity conditions in the North Sea are different. Here, it could only be shown that salinity fluctuations in the bottom water were measured for several Februaries. It cannot be determined to what degree the data set can be regarded as synoptic for lack of comparable measurements.

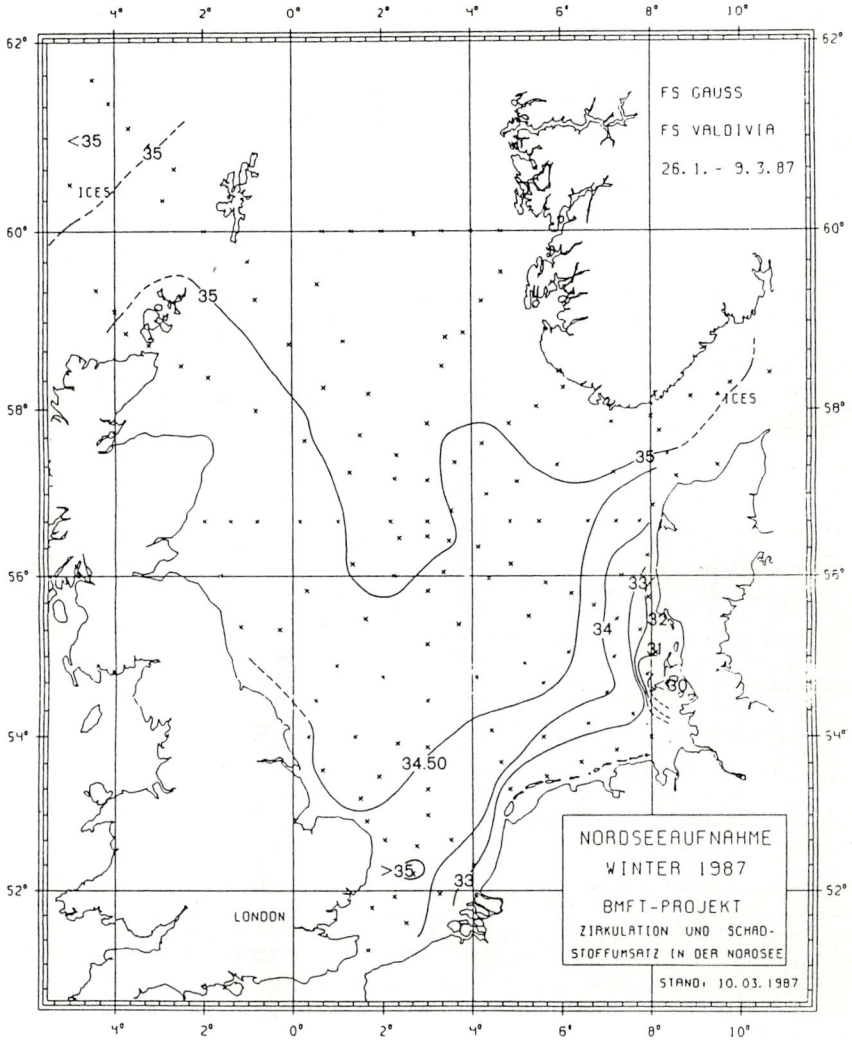

Fig. 2.1.29. ZISCH field experiment: February 1987. Bottom salinity of the North Sea (in psu)

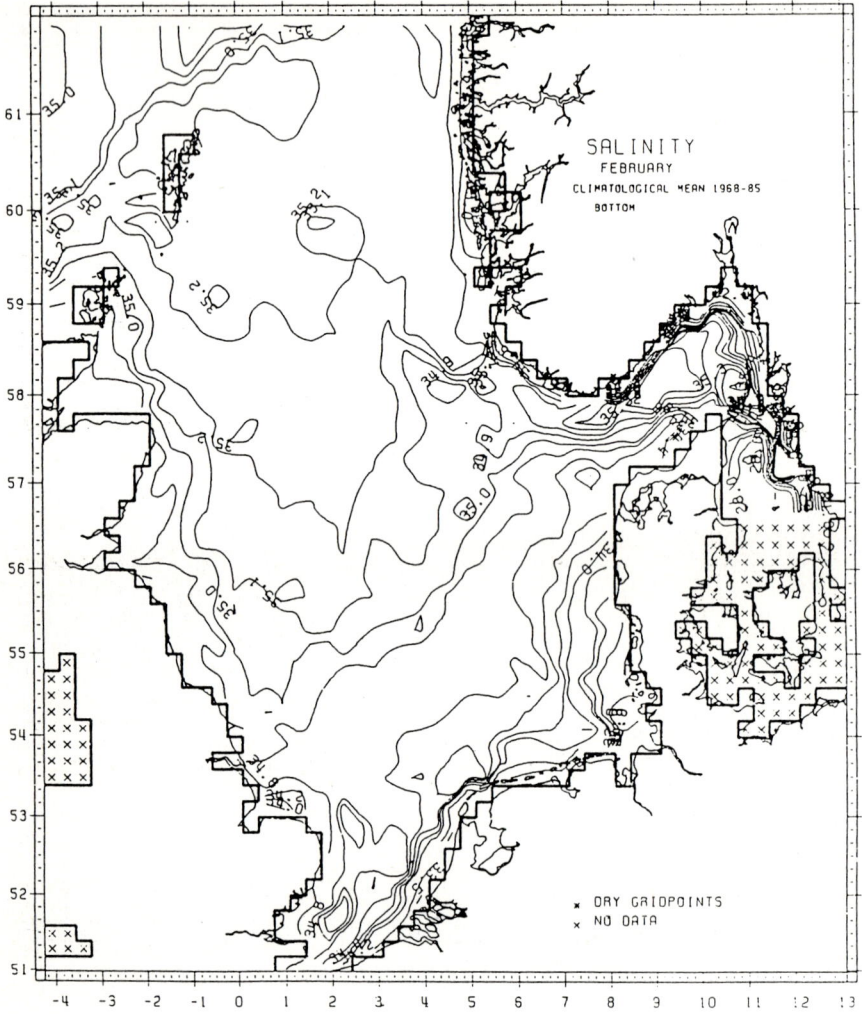

Fig. 2.1.30. Climatological mean (1968-1985) in February. Bottom salinity of the North Sea (in psu)

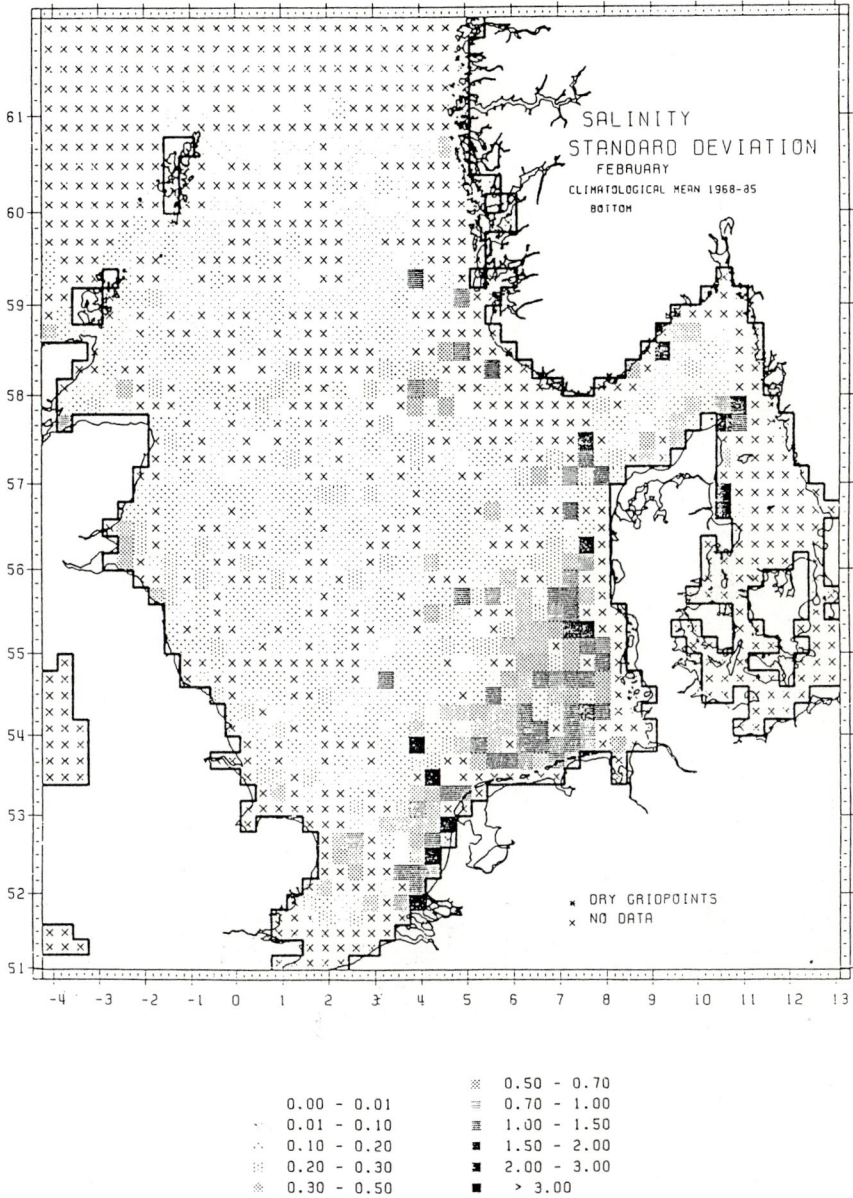

Fig. 2.1.31. Climatological mean (1968-1985) in February. Standard deviation of bottom salinity of the North Sea (in psu)

Fig. 2.1.32. Bottom salinity charts in February 1978 (in psu). Index areas mentioned in the text are *shaded*. (After van de Kamp 1990)

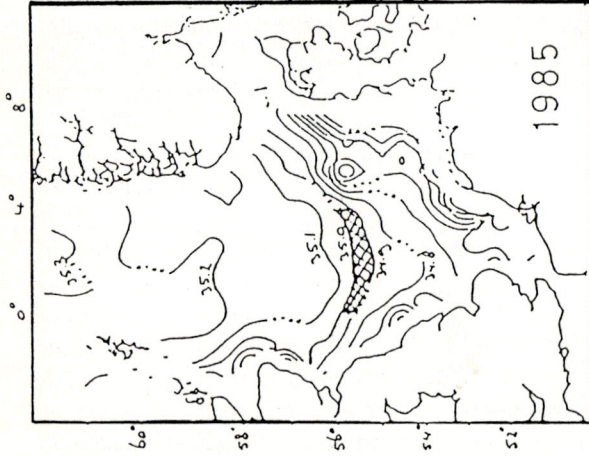

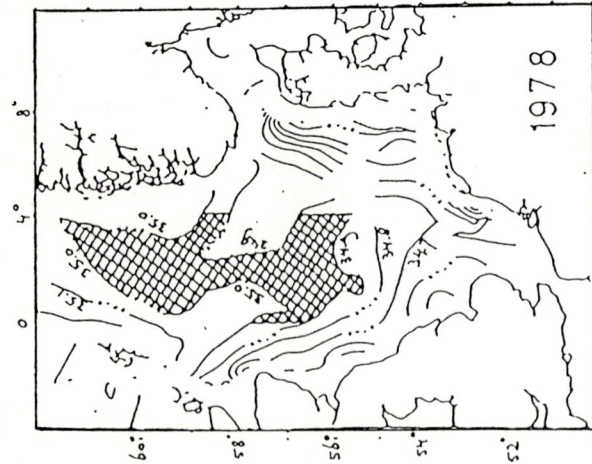

Fig. 2.1.33. Bottom salinity charts in February 1985 (in psu). Index areas mentioned in the text are *shaded*. (After van de Kamp 1990)

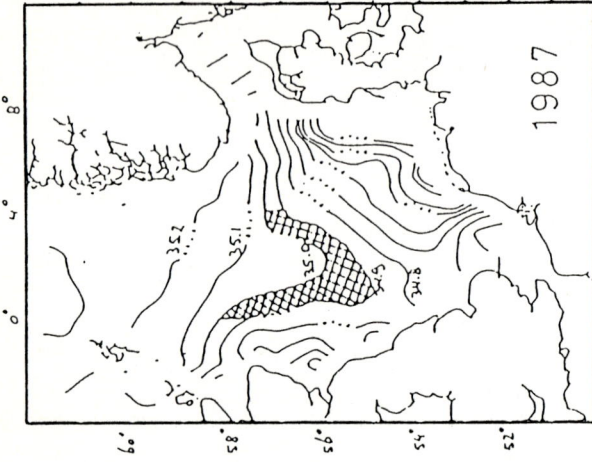

Fig. 2.1.34. Bottom salinity charts in February 1987 (in psu). Index areas mentioned in the text are *shaded*. (After van de Kamp 1990)

References

Backhaus JO (1980) Simulation von Bewegungsvorgängen in der Deutschen Bucht. Dtsch Hydrogr Z Ergänzungsh B 15:7-56

Backhaus JO (1984) Estimates of the variability of low frequency currents and flushing times of the North Sea. ICES Hydrography Committee, Copenhagen, C.M. 1984 / C:24

Becker GA (1981) Beiträge zur Hydrographie und Wärmebilanz der Nordsee Dtsch Hydrogr Z 34:167-262

Damm P (1989) Klimatologischer Atlas des Salzgehaltes, der Temperatur und der Dichte, 1968-1985. Mitt Inst Meeresk Univ Hamb Techn Ber, pp 6-89

Davis AM (1980) Flushing times of the North Sea (Section 17: Outcome of mathematical models, First Draft Flushing Times Group). Int Council for the Exploration of the Sea, Copenhagen

Hainbucher D, Backhaus JO, Pohlmann T (1986) Atlas of climatological and actual seasonal circulation patterns in the North Sea and adjacent shelf regions: 1968-1981. Mitt Inst Meeresk Univ Hamb Techn Ber, pp 1-86

Korevaar CG (1990) North Sea climate based on observations from ships and light vessels. Kluwer Academic, Dortrecht

Lenhart HJ (1990) Phosphatbilanz der Nordsee - eine Abschätzung auf der Grundlage der ICES-Boxen. Dipl Univ Hamburg

Luthardt H (1987) Analyse der wassernahen Druck- und Windfelder über der Nordsee aus Routinebeobachtungen. Hamburger Geophys Einzelschriften, Reihe A, Heft 83, Wittenborn, Hamburg

Maier-Reimer E (1977) Residual circulation in the North Sea due to the M2-tide and mean annual wind stress. Dtsch Hydrogr Z 30:69-80

Radach G, Schönfeld W, Lenhart H (1990) Nährstoffe in der Nordsee-Eutrophierung, Hypertrophierung und deren Auswirkungen. In: Lozan JL, Lenz W, Rachor E, Watermann B, von Westernhagen H (Hrsg) Warnsignale aus der Nordsee. Parey, Hamburg, pp 48-65

Soetje KC, Huber K (1980) A compilation of data on the thermal stratification at the central station in the northern North Sea during FLEX '76. Meteor-Forschungsergeb (A) 22:69-77

Tomczak G, Goedecke E (1962) Monatskarten der Temperatur der Nordsee, dargestellt für verschiedene Tiefenhorizonte. Dtsch Hydrogr Z Ergänzungsh Reihe B (4°)

van de Kamp G (1990) Long-term changes in salinity distribution in the central North Sea. ICES Hydrography Committee, Copenhagen, C.M. 1990 / C:22

2.2 Ecological Situation in the North Sea During Spring and Winter 1986/87

U.H. BROCKMANN, T. POHLMANN, G. BECKER, P. KÖNIG,
L. ALETSEE, H.-J. RICK, M. KRAUSE, P. MARTENS,
R. KNICKMEYER and K. HEYER

2.2.1 Introduction

On the North Sea shelf, the ecological processes are controlled by physical forces which define the boundary conditions for biological development. The most important factors are light, turbulence and stratification, which determine the depth of the euphotic zone and the residence time of the phytoplankton in it. The light regime is directly affected by the turbulence in shallow regions via erosion. The suspended load can cause light limitation. Winter storms generate maximum turbulence, further reducing the already minimal light available. Thus, biological growth in the North Sea follows a seasonal cycle in which the growth period is governed by the local light climate, the temperature regime and the wind forces. During spring, increasing temperature generally enhances biological processes, but is more important for the stratification of the water column, which prevents the downward mixing of phytoplankton cells out of the euphotic zone, triggering the phytoplankton bloom.

Thermal stratification prevails in the central and northern North Sea during summer. Wind stress and strong tidal currents prevent stratification in the shallower southern parts (Pingree et al. 1968). In the stratified areas, nutrients are soon diminished in the mixed layer by phytoplankton growth and enriched in the bottom layer by mineralization of sedimented biomass, consisting of phytoplankton cells as well as faecal pellets and organic detritus. The primary production in the euphotic zone becomes nutrient-limited. Nutrient recycling due to mineralization processes in the upper layer is a dominant process, but bursts of new production follow upwelling events and mixing of the water column during storms.

During autumn, surface-cooling and, more often, storms break up the thermocline, which leads to a general entrainment of nutrients and may - after transient stabilization of the water column - cause an autumn bloom. During winter, phytoplankton growth in part of the North Sea is light-limited, caused by the decrease of incident light as well as by the short residence time of the phytoplankton in the euphotic zone due to extensive vertical mixing. During this time of the year remineralization processes dominate, as they already did in the bottom layer of the stratified areas during summer. In winter, significant primary production occurs only above shallow banks where the "critical depth" (Parsons et al. 1984) is equal to or greater than the depth of the water column, or in areas of stable haline stratification, i.e. in coastal regions with freshwater discharges. At the critical depths the assimilation energy equals the loss by dissimilation.

Production processes are enhanced by the influx of nutrient-rich Atlantic water and also affected by inputs from river discharge and shore-lines. The inflow of Atlantic water as well as the extent of river plumes can be detected by salinity gradients, but at certain times even better by nutrient gradients. Another important source for nutrients is the atmosphere, probably most effective during nutrient-limitation in the surface layer in summer.

In spring 1986, following the phytoplankton spring bloom, and in late winter 1987, at minimum primary production activity, the North Sea ecosystem was investigated on a station net covering the whole North Sea. The station net was shaped like a star (Fig. 2.2.1). Sampling started in the centre, followed by the northwest section and moving counterclockwise around the North Sea following the residual currents. By this strategy, a time series was measured in the central North Sea and more synoptic data sets were obtained in the individual sections. Generally, however, advection processes have to be considered when comparing the data from different stations. Therefore, the results presented in the following figures by lines of equal concentrations (isolines) are indicative of the local processes occurring only at that point in time during sampling. The entire sampling period lasted for more than 6 weeks in each case. Thus, a time-lag should be considered especially when comparing the data from the eastern and western part of the central and northern North Sea, where samples were taken at the beginning and at the end of the campaign.

The translocation of sampled waters has been calculated by a transport model (Hainbucher et al. 1987), showing a theoretical shift of water masses from west to east according to the residual currents modulated by the actual wind fields (Fig. 2.2.2 a,b). Due to stratification during summer, the trajec-

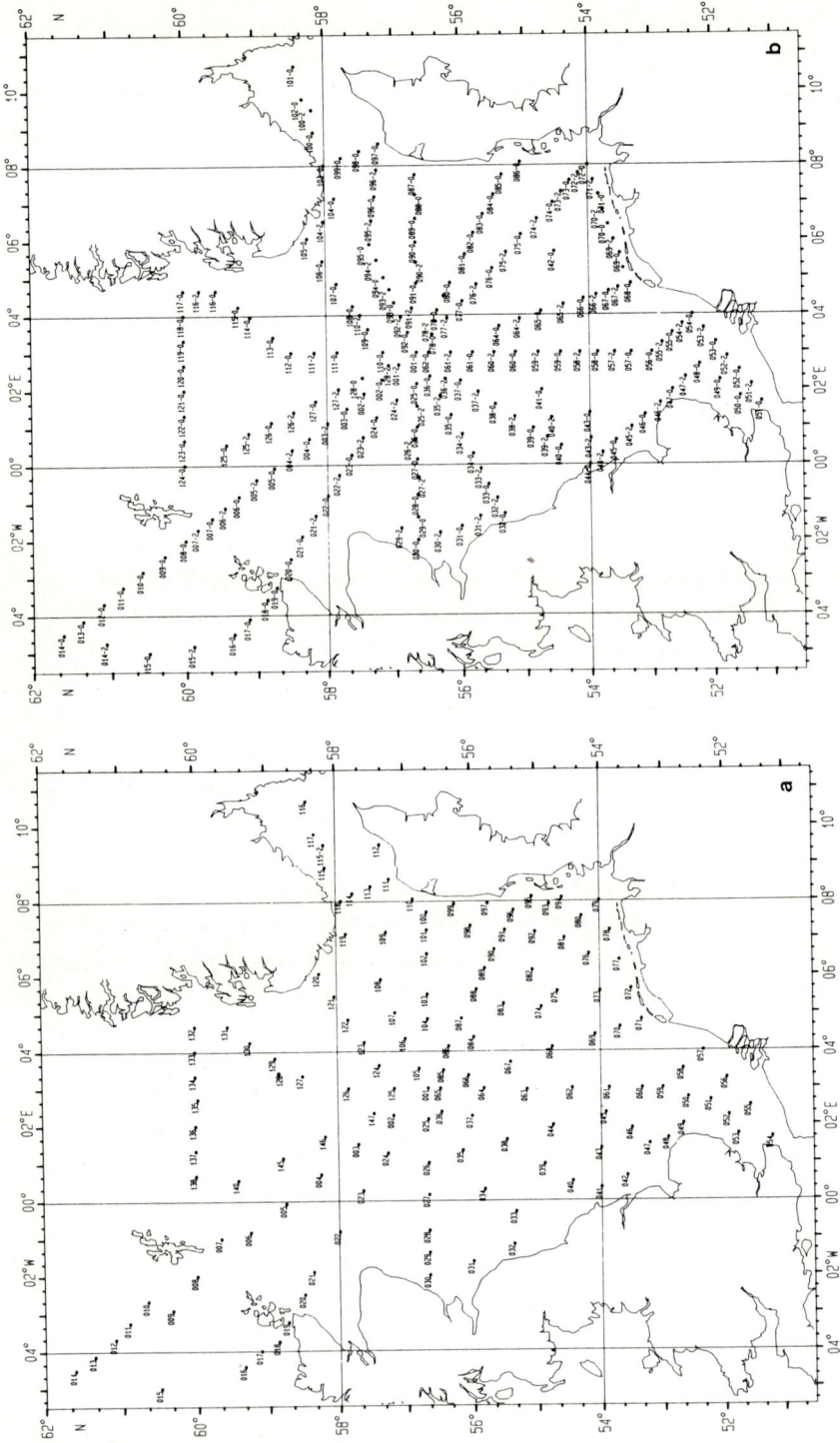

Fig. 2.2.1. Station nets of winter 1987 (a) and spring 1986 (b) cruises in the North Sea. Sampling dates were 28 January to 6 March 1987 and 3 May to 12 June 1986. Cruises started in the centre of the North Sea: towards the northwest investigating section by section anticlockwise. Each pair of tracks lasted for 5–10 days

tories for the surface layer are much longer than during winter. In addition, during winter an easterly wind situation of a longer duration caused a stagnation of water masses in the German Bight and off the Dutch coast, contrary to the usually counterclockwise residual current system. All trajectories were calculated for the time of the last sampling at a station east of the centre of the star. For this reason, the trajectories are longer for the first stations than for stations sampled later on during the cruises. The effect of the shelf edge current resulting in a jet towards the northeast is clearly documented by spring and winter calculations.

2.2.2 Salinity

Different hydrographic regimes in the North Sea have been distinguished according to their salinities (see Chap. 2.1). The Atlantic inflow was seen in the pattern of the surface salinity, especially at the northern boundary, by the 35 PSU-isoline (Fig. 2.2.3 a,b) and - more extended -in the bottom water layer (not depicted). On the other hand, freshwater discharges by the continental rivers and from the Baltic formed lenses of less saline water in the continental coastal water and in the Skagerrak. The spreading of the different water masses depended on the circulation pattern of the residual currents and on the discharge rates and local wind fields; wind forces spread river plumes towards the open sea or can block freshwater discharge when blowing onshore. In winter 1987 the freshwater discharge was higher than in spring 1986. For instance, the discharge of the Elbe river ranged between 1200 and 2800 m^3/s in winter (ARGE Elbe 1987, 1988), and in spring 1986 between 500 and 1400 m^3/s. Therefore, in winter 1987, water with a salinity of < 34 PSU was spread in a broad belt along the continental coast continuing northward with the Norwegian coastal current. In comparison, during spring 1986 the discharge was higher for the Baltic, resulting in a low salinity water mass (< 29 PSU) around Cap Lindesnes. The prevailing westerly winds at that time of the year gave rise to a frontal system at the outlet of the Skagerrak. In spring 1986, the inflow of the Humber/Thames river plume could be detected, while in winter 1987 the low salinity estuarine water was probably mixed faster due to higher turbulence.

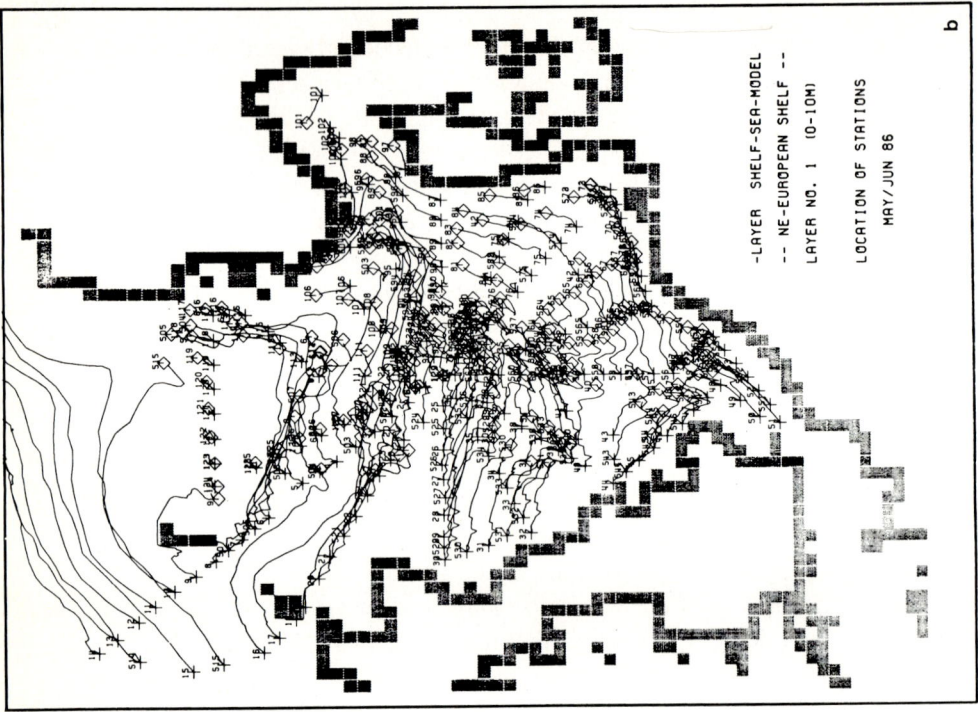

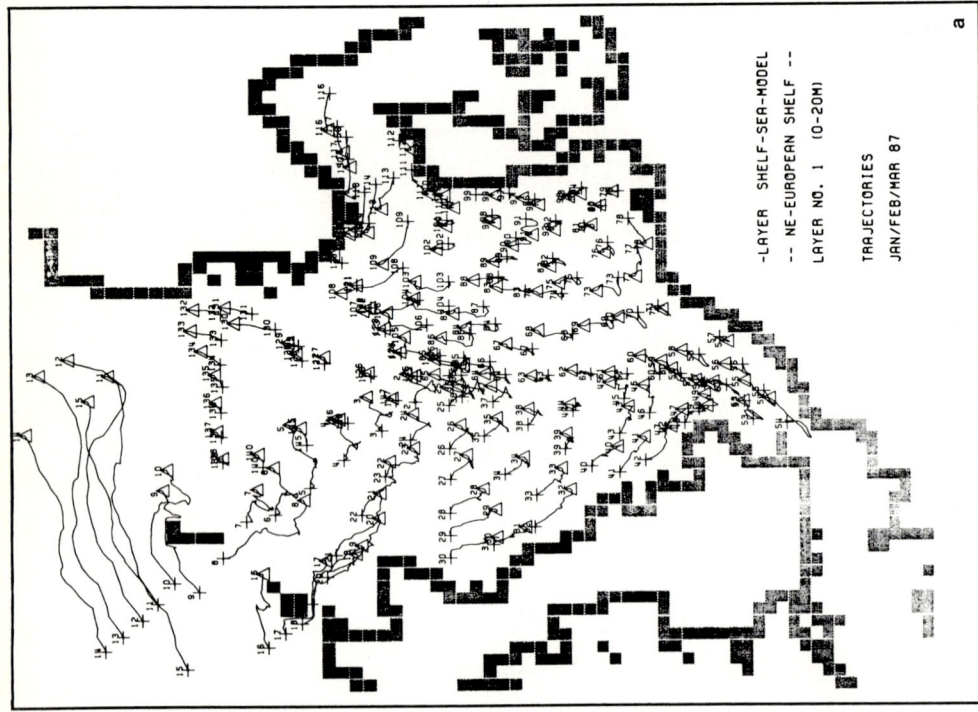

Fig. 2.2.2. Trajectories calculated for the transport of water masses at the surface during survey period. Winter 1987 (*a*) and spring 1986 *water masses are indicated by arrows. Calculated positions which were reached at the time of the last sampling are marked by triangles*

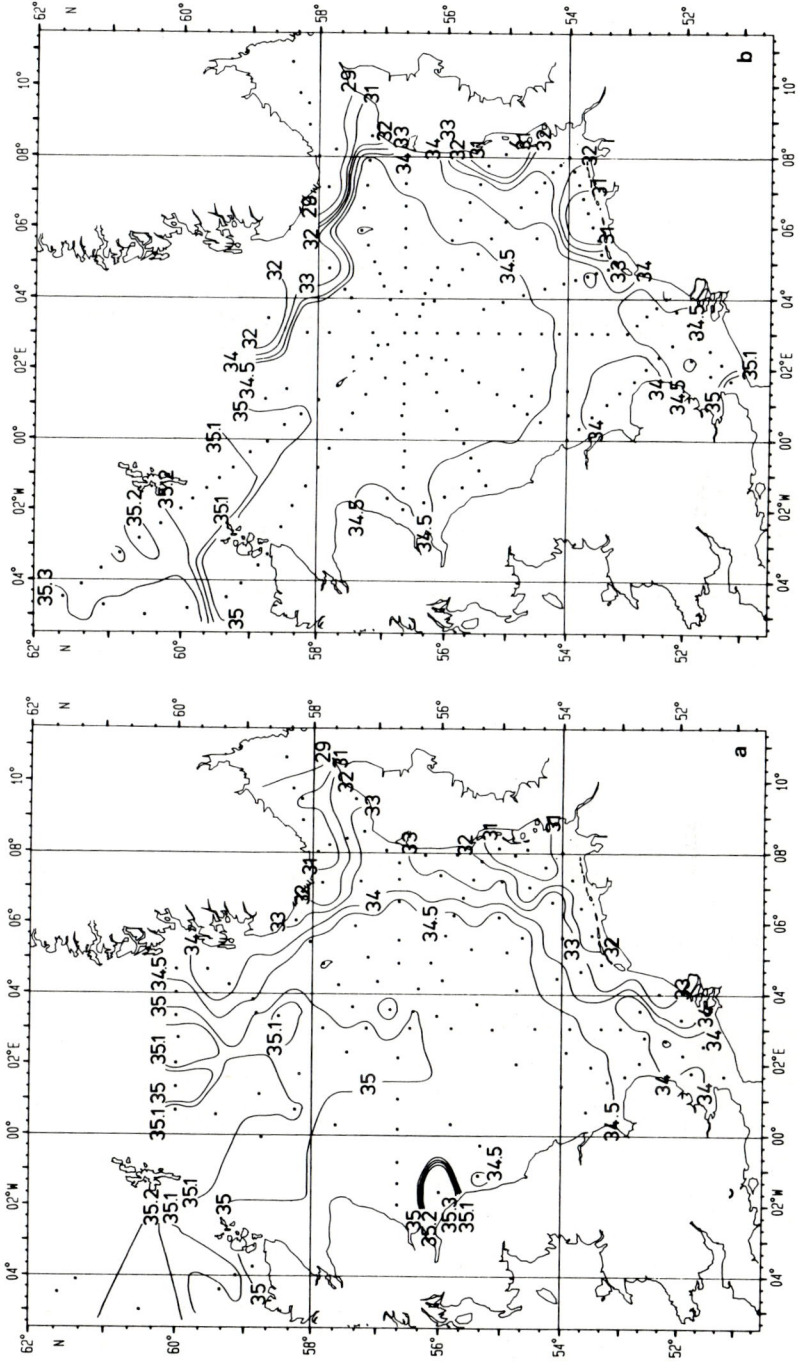

Fig. 2.2.3. Surface salinities during winter 1987 (*a*) and spring 1986 (*b*)

2.2.3 Temperature

Surface temperature conditions in the North Sea during the investigation periods have been compiled by Wegner (1987). During winter 1987, in the cold surface waters the inflow of the warm Atlantic water from the north could be traced all the way southward to the central North Sea, whereas the small inflow from the Channel was cooled very quickly in the shallow Southern Bight (Fig. 2.2.4a). The cooling effect during winter was especially evident in the continental coastal water and in the Skagerrak, areas which received cold freshwater discharges. The direct cooling on the shallow shelf contributed to this effect as well. Cooling was distinctly detected in the Dogger Bank area, where water temperature decreased to < 4 °C (Fig. 2.2.4a). During spring 1986, warming up was faster in the shallow areas than in the central part of the North Sea (Fig. 2.2.4b). In the Skagerrak, the low salinity Baltic outflow caused haline stratification, thus the warming of the stratified water there proceeded similarly to that of the continental coastal zone.

Especially the horizontal gradients of the surface temperature cannot be depicted continuously, due to the time lag of non-synoptic measurements in the central North Sea, whereas the bottom temperatures were not influenced by this effect.

The vertical temperature distribution along a section from south to north running through the central North Sea (Fig. 2.2.5) shows that in the central and northern North Sea a stable thermocline had been formed by the end of May, separating an upper mixed layer and a bottom layer at about 50 m depth. At the shelf edge the temperature of the bottom layer was higher, caused by inflow of warmer Atlantic water.

2.2.4 Light Regime

The light regime in temperate waters is dominated by the annual cycle of incident light (irradiance). Stratification increases the residence time of phytoplankton cells in the euphotic zone, thus phytoplankton growth is enhanced. On the other hand, at times of extensive blooming, high cell densities lead to self-shading and can cause a limitation of primary production below the surface layer. In shallow areas of turbid coastal waters, particulate material is kept in suspension throughout the water column. Therefore, light is significantly reduced and, especially in winter, light limitation may occur.

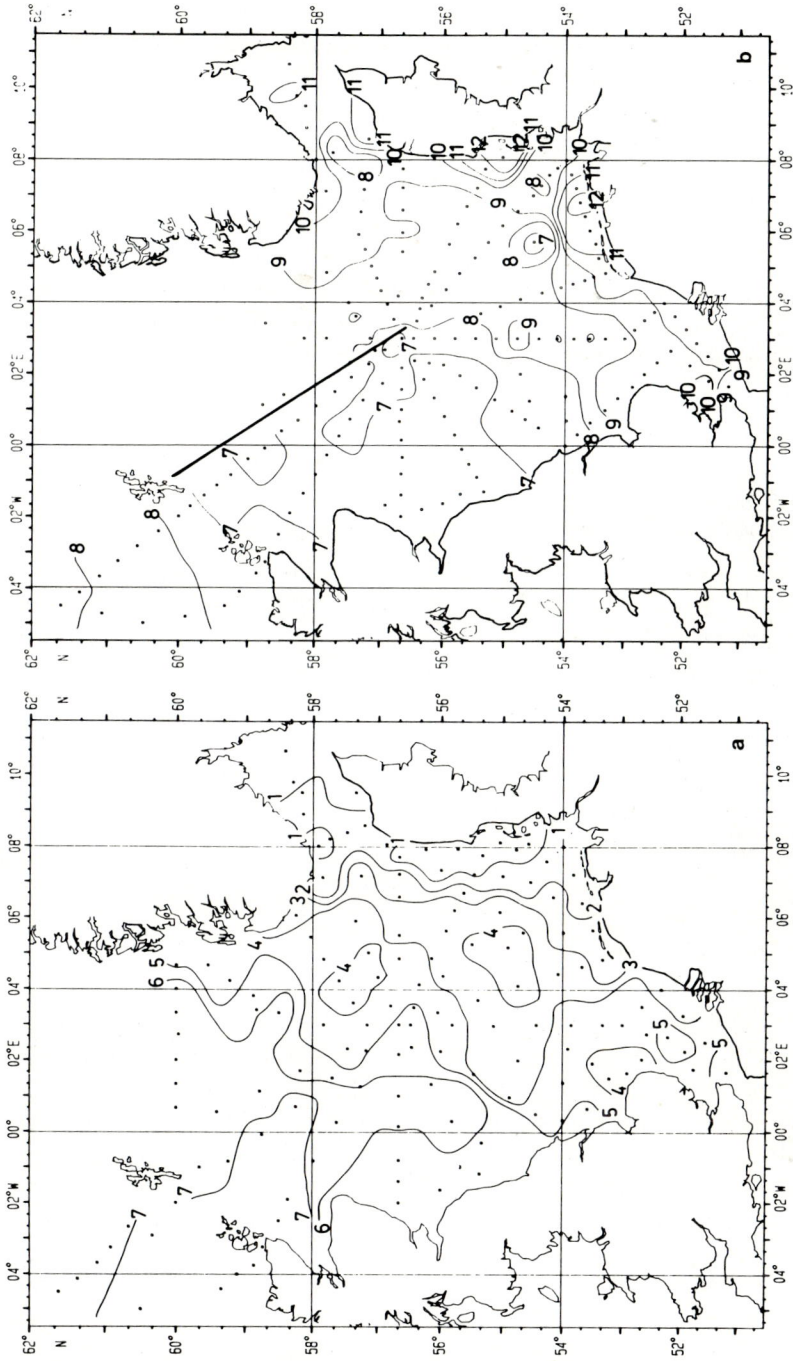

Fig. 2.2.4. Sea surface temperatures during winter 1987 (*a*) and spring 1986 (*b*). Due to the permanent warming during the spring cruise 1986, neighbouring stations sampled at a greater time-interval are separated by a *line*

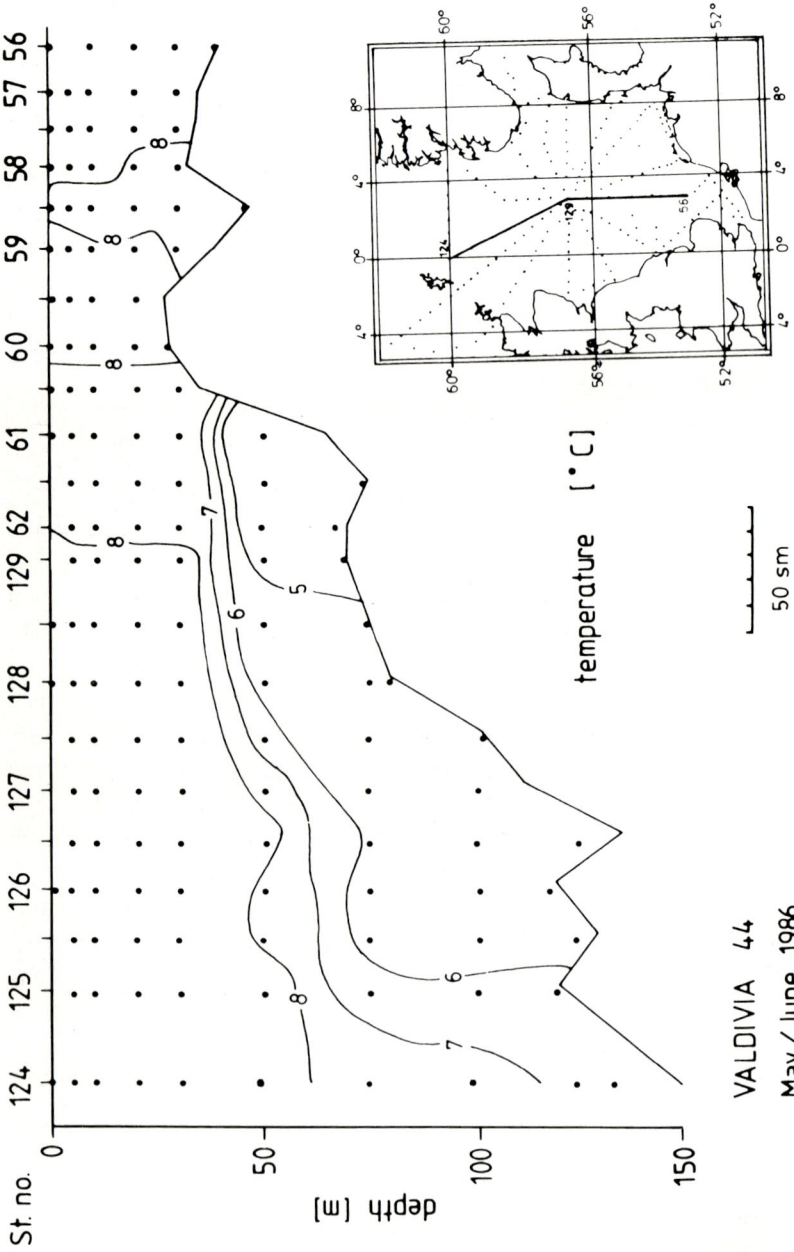

Fig. 2.2.5. Temperature cross-section through the North Sea from the south (*right*) to the middle and northwest (*left*) during spring 1986. Sampling dates: station nos. *56-62*, 26+27 May, station nos. *124-129*, 4-12 June

In the belt of continental coastal water, the turbidity during winter 1987 was increased by a factor of 2.5 in comparison to spring 1987 (Fig. 2.2.6 a,b). Maximum turbidity was detected at the margin of the Dutch Wadden Sea, probably caused by local erosion events. The plume of high concentrations of suspended matter off Ringköbing Fjord was due to easterly wind forces. These structures were confirmed by particle counts as well as by chemical measurements of particulate organic material. The area of less turbid water in the middle of the Southern Bight is generally the region where early diatom blooms develop (Postma 1981). This area of low turbidity was also present in spring 1986 and was caused by the Atlantic inflow passing through the Channel. A belt of turbid water originating from the coast of Norfolk is fed by permanent erosion and was first described by Joseph (1953). This feature is conspicuous on many satellite images (Holligan et al. 1989; Chap. 4.1, this Vol.).

2.2.5 Nutrient Distribution

The phosphate concentrations in winter 1987 (Fig. 2.2.7a) in the continental coastal water and, especially in the Elbe river plume spreading into the shallow German Bight, reached values between 0.7 and 2 μM, respectively (Brockmann et al. 1990). Other estuarine and coastal sources were indicated by concentrations of more than 0.7 μM in British estuaries. The inflow of the nutrient-rich Atlantic water led to concentrations of more than 0.7 μM at the surface west of Shetland.

In the central North Sea, a large area starting from the Dogger Bank and spreading to north-northeast was characterized by concentrations of less than 0.4 μM. These low values indicated that primary production occurred here during winter, thus diminishing nutrient concentrations significantly. This phenomenon was observed previously in February 1984 (Brockmann and Wegner 1985). The large extension of water low in phosphate concentration could be a consequence of the reduced mineralization rates during the cold winter temperatures. In addition, small water lenses of lower nutrient concentrations in the Skagerrak were indicative of areas where primary production occurred during winter due to haline stratification in the Norwegian coastal current.

The surface concentrations of nitrate and nitrite showed a broad belt of nutrient-rich continental coastal water connecting the British coast with the continental coastal water during winter 1987 (Fig. 2.2.8a). The highest concentrations of >50 μM were again found in the Elbe river plume. Reduced surface concentrations of nitrate during winter were observed in

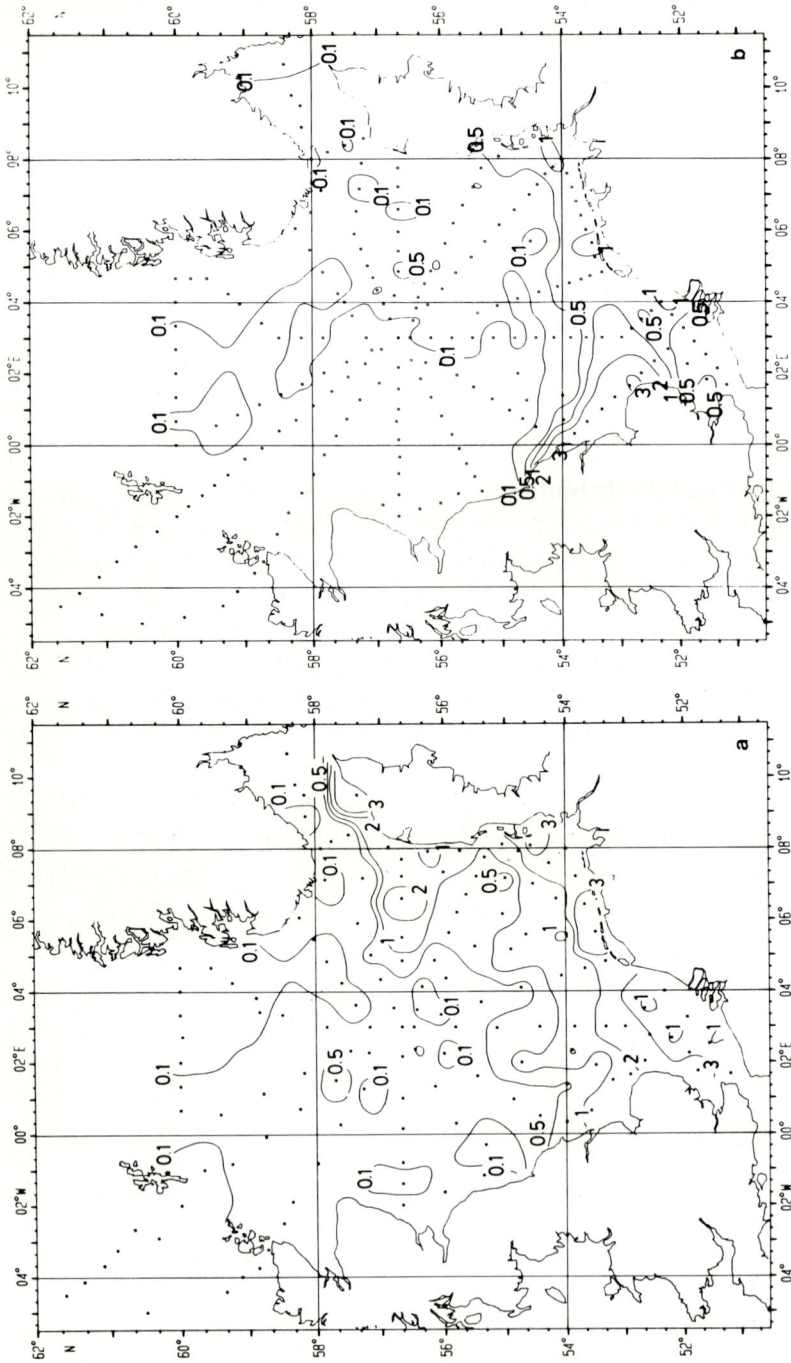

Fig. 2.2.6. Suspended material at a depth of 5 m in winter 1987 (*a*) and spring 1986 (*b*). Measurements were performed with a Nephelometer. Units are given as NTU (nephelometric turbidity units)

the vicinity of the Dogger Bank (less than 4 µM) and in some lenses in the Norwegian coastal current (less than 6 µM), confirming significant nutrient uptake by phytoplankton growth, as already discussed above for phosphate.

An area of reduced silicate concentrations was found at and near the Dogger Bank which was much more extended than the area of low nitrogen and phosphate concentrations. From that it could be concluded (1) that the phytoplankton bloom occurring in the Dogger Bank was dominated by diatoms (taking up silicate) and (2) that silicate, which was taken up at nearly the same rate as nitrogen, was remineralized more slowly than the other nutrients.

In late spring 1986 the surface concentrations of phosphate (Fig. 2.2.7b) were diminished in most regions of the North Sea, values as low as 0.2 µM and less than 0.1 µM were observed in the southeastern part of the North Sea. The discharge of the Elbe river was less than 500 m^3/s in May, 1986 resulting in only a small detected increase of phosphate concentrations in the plume area at the stations sampled. Due to the high productivity during spring, the fact that the sampling was not synoptic was particularly evident in the strong differences in phosphate concentrations at the sections from the central North Sea to the northwest and north. These were sampled at the beginning and close to the end of the campaign, respectively. High nutrient concentrations of the Atlantic water were present at the surface due to upwelling and turbulence at the shelf edge. The latter process was also indicated by the weak thermocline in this area (Fig. 2.2.5).

When the surface nutrient concentrations were diminished, actual discharges could be detected more easily than during winter due to steeper gradients, as for instance in the Humber and Thames plumes in summer 1986, where concentrations of more than 0.5 µM phosphate were found.

In spring 1986, the discharge of nitrate from the rivers caused high concentrations in the English coastal water as well as in the continental coastal water, reaching values of more than 25 µM in the plumes, whereas in the main parts of the North Sea surface the nitrate concentrations had already dropped below 1 µM (Fig. 2.2.8b). Again, the effect of the survey not being synoptic was seen by the still higher concentrations of nitrate in surface waters along the first sampling legs. At this time, the Atlantic waters in the northwest still had concentrations of more than 10 µM.

In spring 1986 the surface concentrations of silicate were reduced to less than 0.5 µM in vast regions of the North Sea except some areas in the central part (Fig. 2.2.9b). Here values were still high at the first stations sampled during the still progressing phytoplankton spring bloom, but also at stations at the end of the campaign, indicating that the spring diatom bloom was not consistent all over the area (see Fig. 2.2.11b).

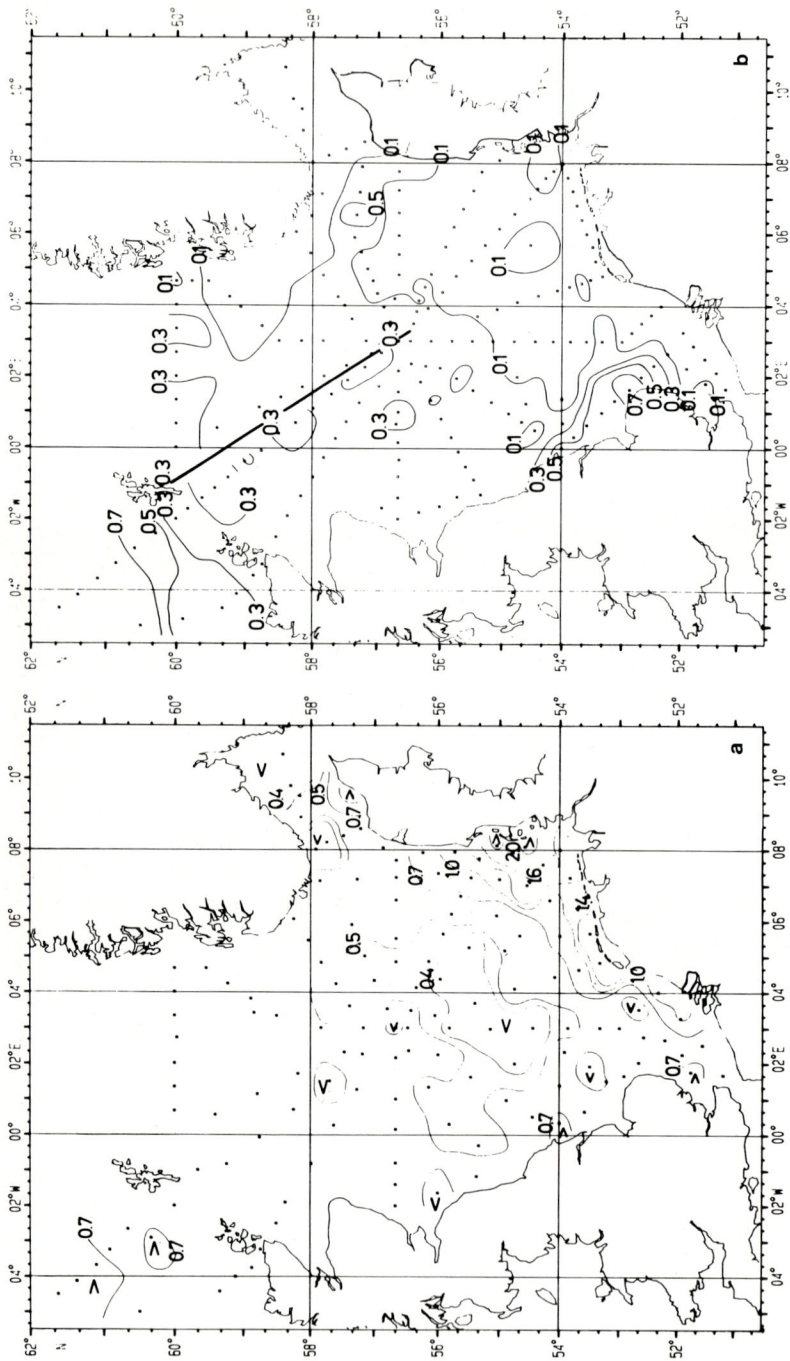

Fig. 2.2.7. Phosphate concentrations at the surface in winter 1987 *(a)* and spring 1986 *(b)* (μmol/l = μM)

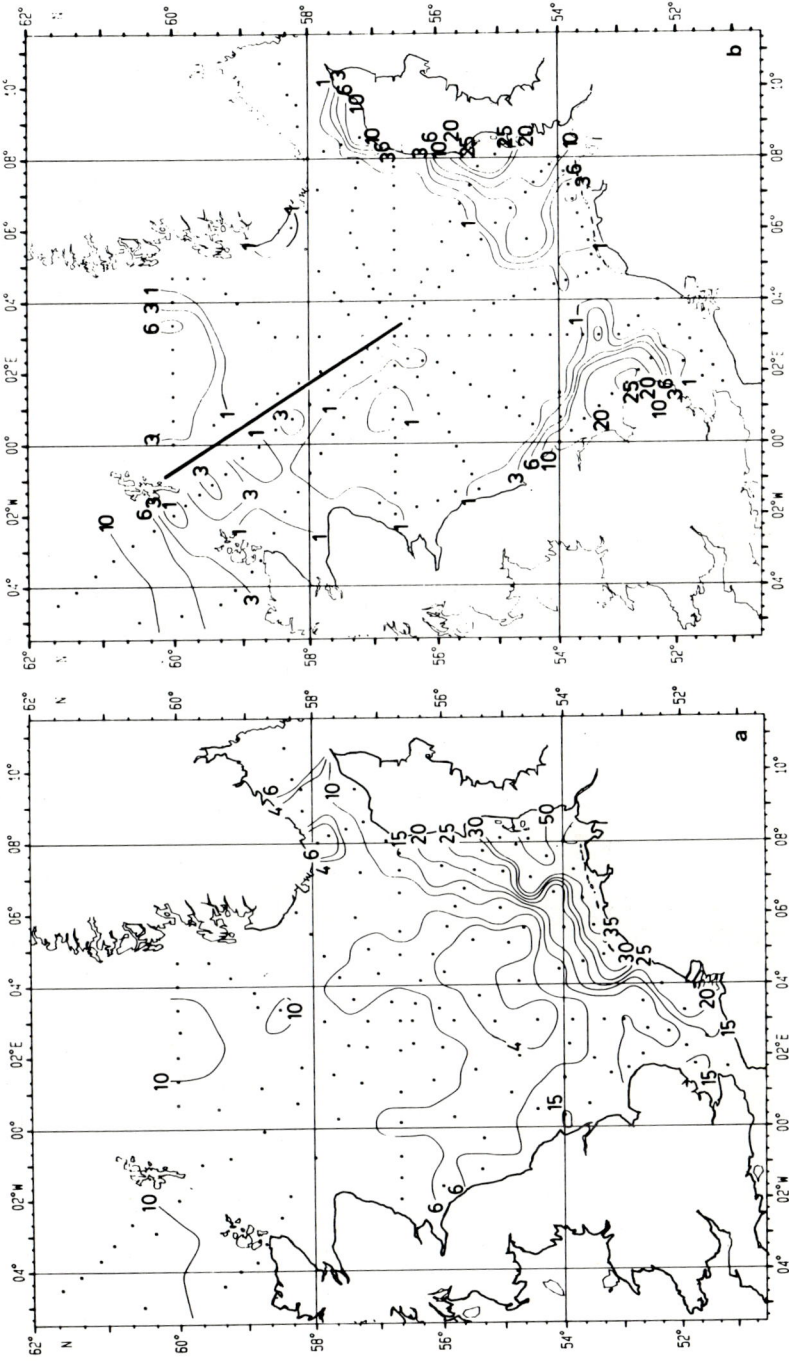

Fig. 2.2.8. Nitrate and nitrite concentrations at the surface in winter 1987 (*a*) and spring 1986 (*b*) (μmol/l)

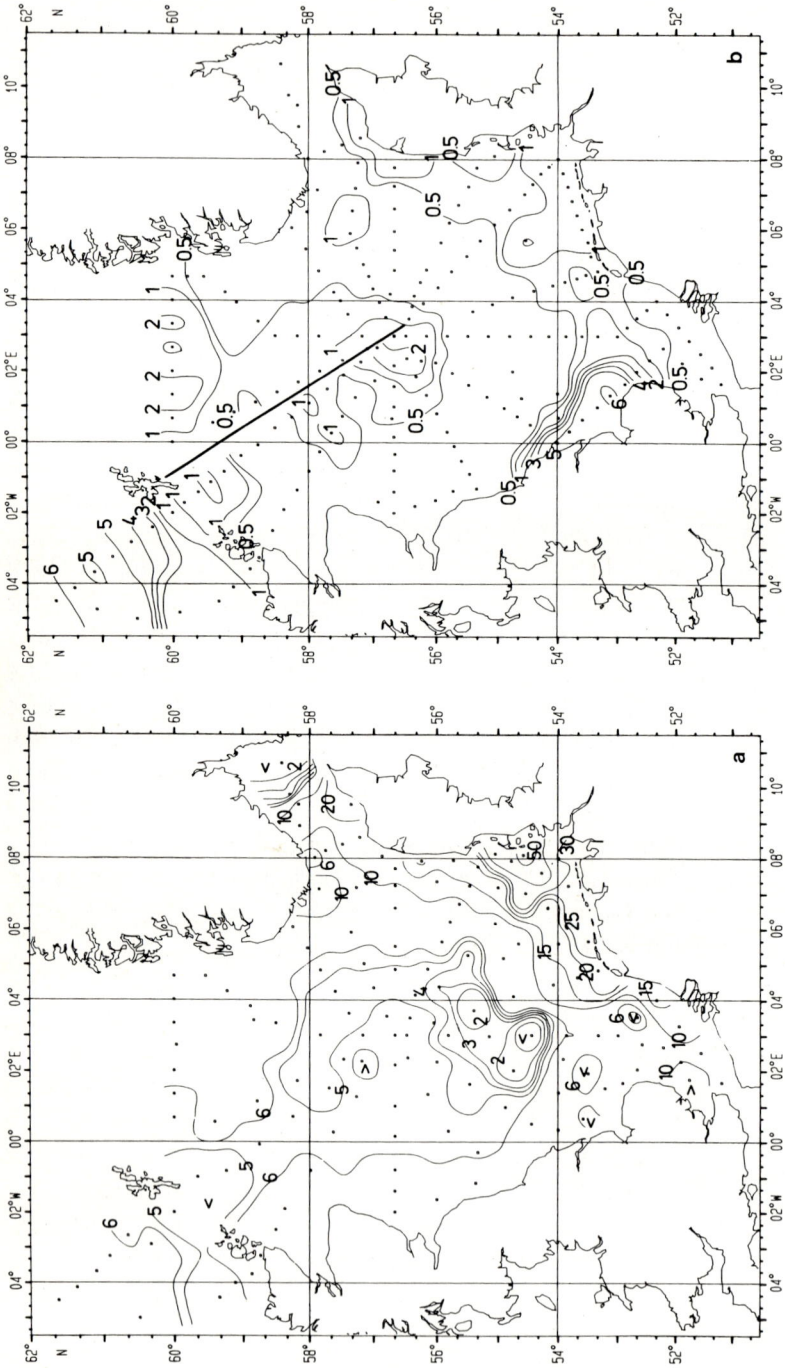

Fig. 2.2.9. Silicate concentrations at the surface in winter 1987 (*a*) and spring 1986 (*b*) (μmol/l)

Ammonium measurements are presented in connection with secondary production and decomposition (see below).

2.2.6 Primary Production and Phytoplankton Abundance

Primary production was calculated from uptake rates of ^{14}C-labelled carbonate. During winter 1987 the primary production was high in the Dogger Bank area and in the Skagerrak (Fig. 2.2.10a). Here, more than 100 mg C/m^{2}h were fixed in the upper 20 m of the water column (Rick 1990). These areas matched the nutrient-poor regions in the Skagerrak during winter and the nutrient minimum around the Dogger Bank area (Fig. 2.2.9a). Also, in the Southern Bight at a station in the zone of less turbid water (Fig. 2.2.6a), production reached a rate of more than 100 mg C/m^{2}h.

However, mean primary production in winter 1987 was very low, 25 mg C/m^{2}h (225 mg C/m^2 day). During May/June 1986, production rates were generally much higher (Fig. 2.2.10b), averaging 130 mg C/m^{2}h (2050 mg C/m^2 day). There were marked differences in production between areas in spring 1986 as well. In the eastern part of the central North Sea, where nutrients were already depleted when these stations were visited later in the cruise, production rates were generally below 50 mg C/m^{2}h. In the western central North Sea, where the nutrients were not yet depleted during the first tracks, production was found to be higher than 100 mg C/m^{2}h. Between the Shetland and Orkney Islands, due to permanent upwelling of nutrient-rich water, the production was high as well. A station with high production near Flamborough Head was probably associated with an area of persistent frontal zones, where nutrient-rich water was mixed into the upper layer (Pingree et al. 1968). Some very productive stations were encountered at the eastern side of the turbid, nutrient-rich waters along Norfolk. This reflects the rapid development of phytoplankton after sedimentation of the suspended matter once these waters enter an area with smaller tidal current velocities. At the station with a primary production above 1000 mg C/m^{2}h, a large amount of *Phaeocystis globosa* was observed (Fig. 2.2.11a) and diatom biomass was also higher than at neighbouring stations (Fig. 2.2.11b).

The high primary production which was detected off the Frisian Islands was also due to a massive bloom formed by *Phaeocystis globosa* (Fig. 2.2.11a); each year it follows the early diatom spring bloom in the shallow coastal water. The nutrients for this extended productivity were supplied by river discharges, in-situ remineralization and fast remobilization in the shal-

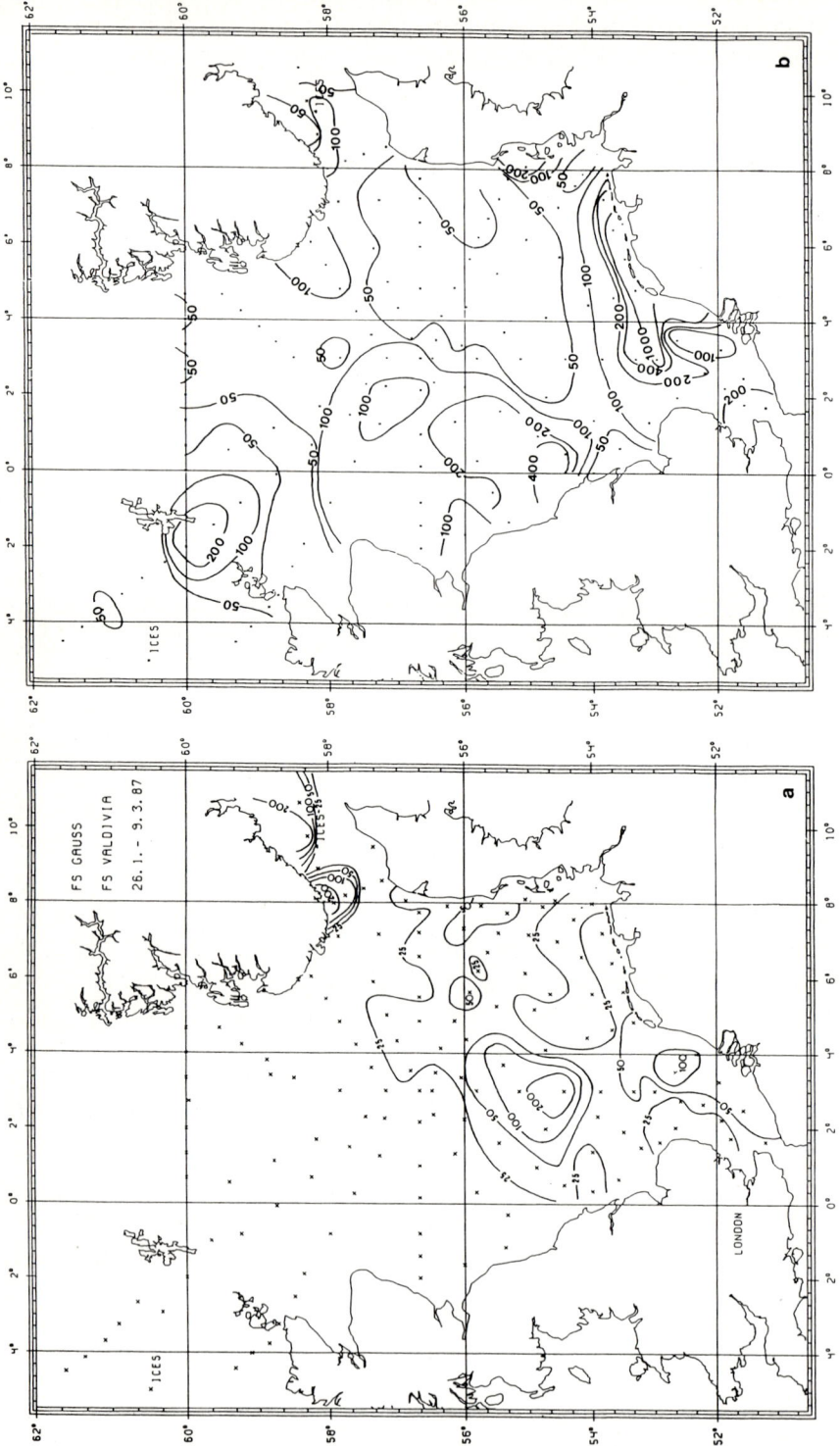

Fig. 2.2.10. Primary production in the upper layer (0-20 m, integrated, measured in mg C/m³h. (*a*) Winter 1987. (*b*) Spring 1986

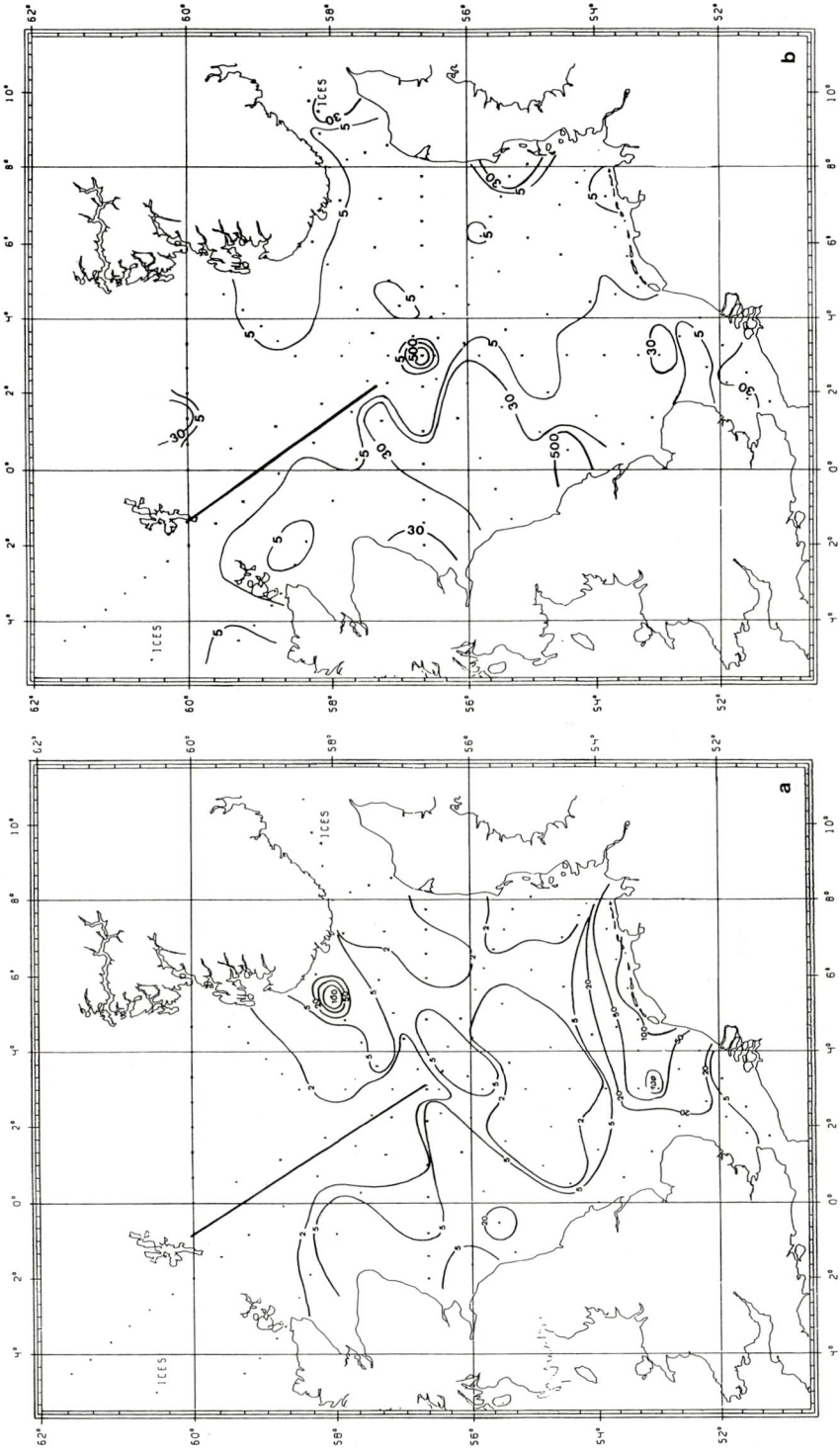

Fig. 2.2.11. Spring 1986 distribution in surface waters (0-20 m) of (*a*) Haptophyta (in 10^{-2} mg C/l, (*b*) diatoms (10^{-3} mg C/l)

low Wadden Sea. The production maximum in the north Frisian coastal water coincided with a nutrient maximum (especially nitrate; Fig. 2.2.8b), supplied by the Elbe river plume.

During the spring cruise, chlorophyll concentrations (Fig.2.2.12) matched the primary production, i.e. in areas of high primary production the chlorophyll concentrations were mostly high as well. This was especially true in the region between the Shetland and Orkney Islands as well as in the continental coastal water.

The horizontal distribution of diatoms (Fig. 2.2.11b) showed that the highest concentrations were generally found in the western part of the North Sea during the first half of the spring cruise in 1986. At the dates on which the eastern station legs were visited, spring diatom blooms already would have been succeeded by other phytoplankton species. Regions with high diatom concentrations, i.e. more than 30 µg C/l in the surface layer, were associated with productions of above 200 mg/m^{2}h: Shetland, central British coast, between Norfolk and Weel, near the Channel and northern German Bight.

During May/June 1986 the concentrations of dinoflagellates were already in the same order of magnitude as diatoms (Fig. 2.2.13), reaching values of up to 100 µg C/l. This high production was also found north of the German Bight, indicating here a succession of phytoplankton species in the spreading Elbe river plume where silicate was limiting diatom growth, but other nutrients were still available, favoring the dinoflagellates (without a silica-shell). Minimum cell numbers of dinoflagellates were found around the Shetland and Orkney Islands, and in the Atlantic inflow between Shetland and Norway. In parts of the central North Sea and in the non-stratified waters of the southern North Sea, dinoflagellates were generally a minor phytoplankton component as well.

2.2.7 Consumers and Decomposition

The different topographic areas of the North Sea are characterized by the distribution of the dominant copepods *Calanus* sp. (mainly *C. finmarchicus*) and *Temora longicornis* (Krause and Martens 1990). The most abundant species were *C. finmarchicus* in the northern part and *T. longicornis* in the south (Figs. 2.2.14, 2.2.15). The same distribution pattern was found during spring and winter. During winter, the largest concentrations of *T. longicornis* were spread more towards the north in the central North Sea. The distribution patterns of the prevailing zooplankton, with life cycles of a few weeks (during spring), are influenced by the different hydrographic

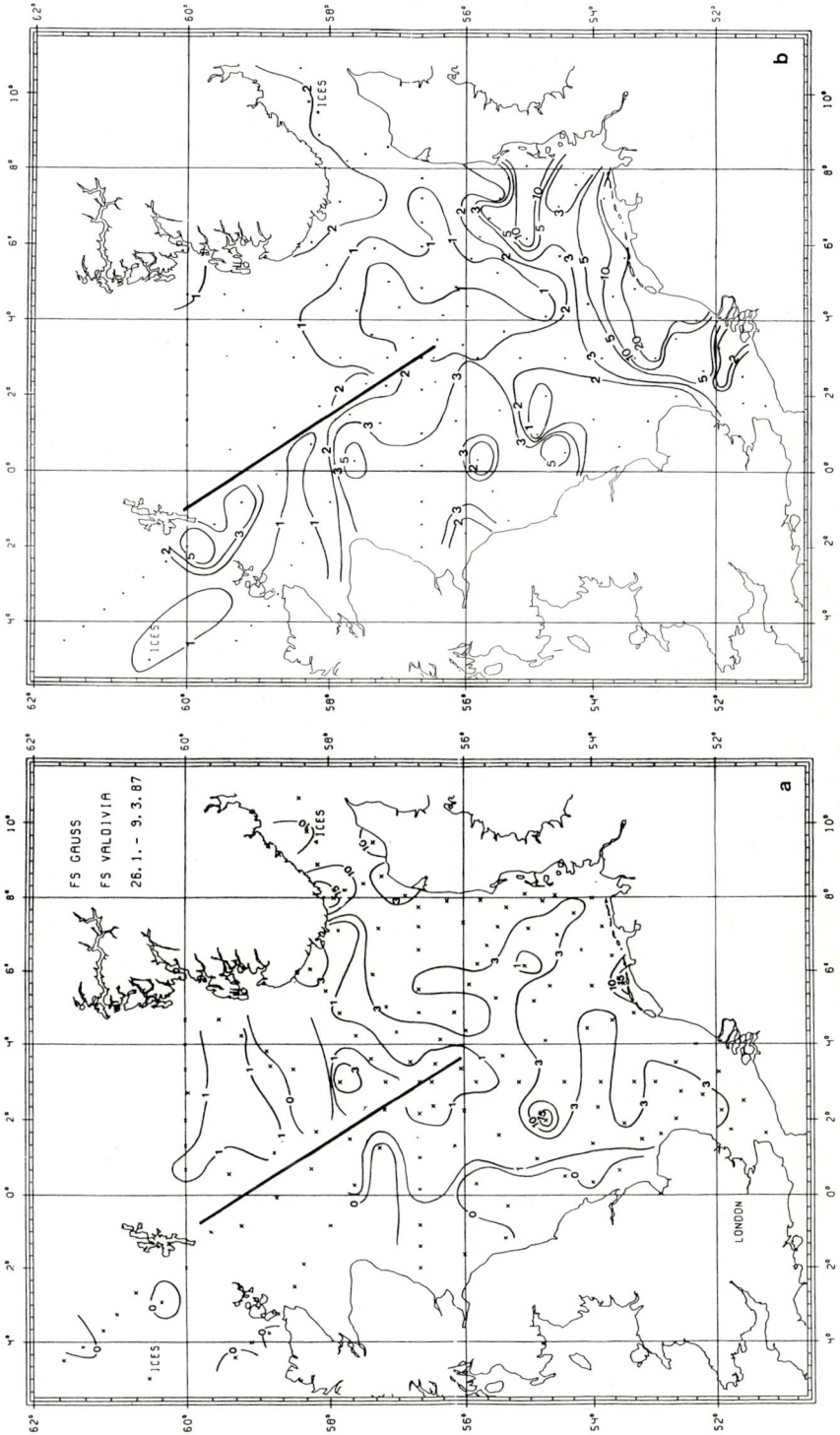

Fig. 2.2.12. Chlorophyll distribution (μg/l) for winter (*a*) and spring (*b*)

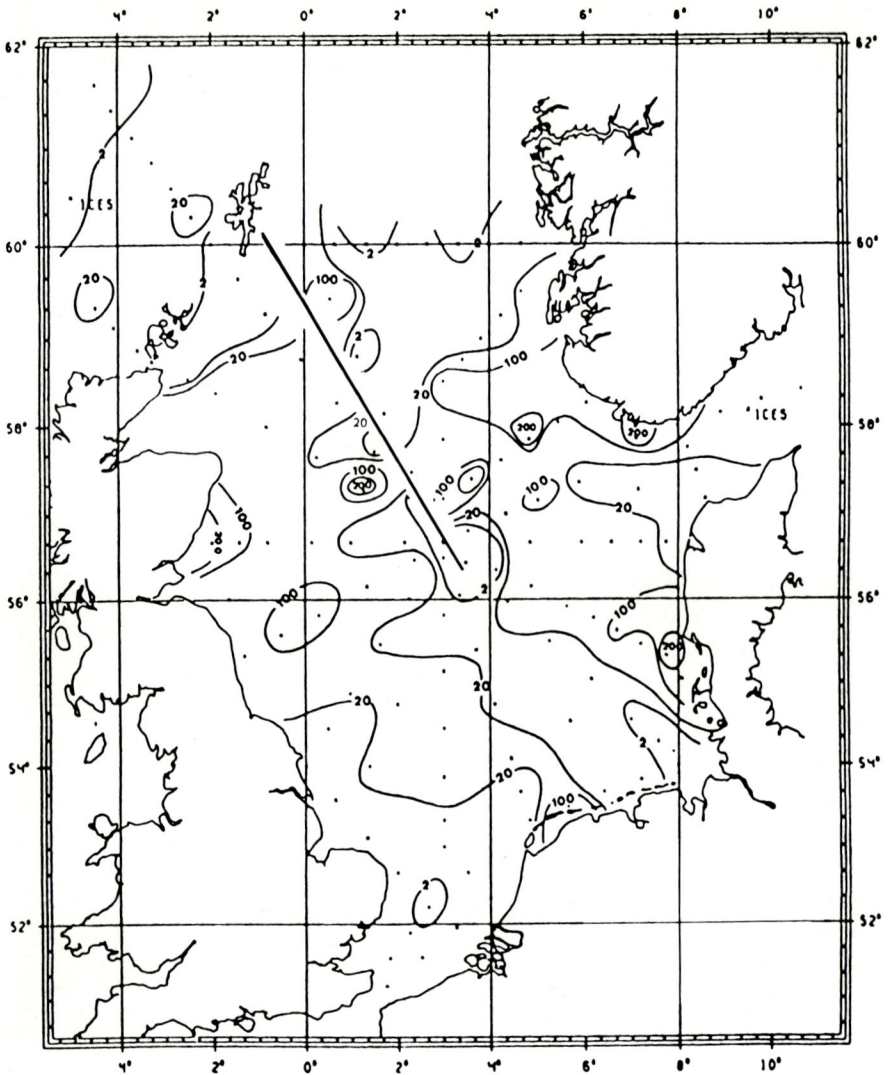

Fig. 2.2.13. Abundance of dinoflagellates, calculated from biomass measurements $(10^{-2}$ mg C/l) near the surface (0-20 m) in spring 1986

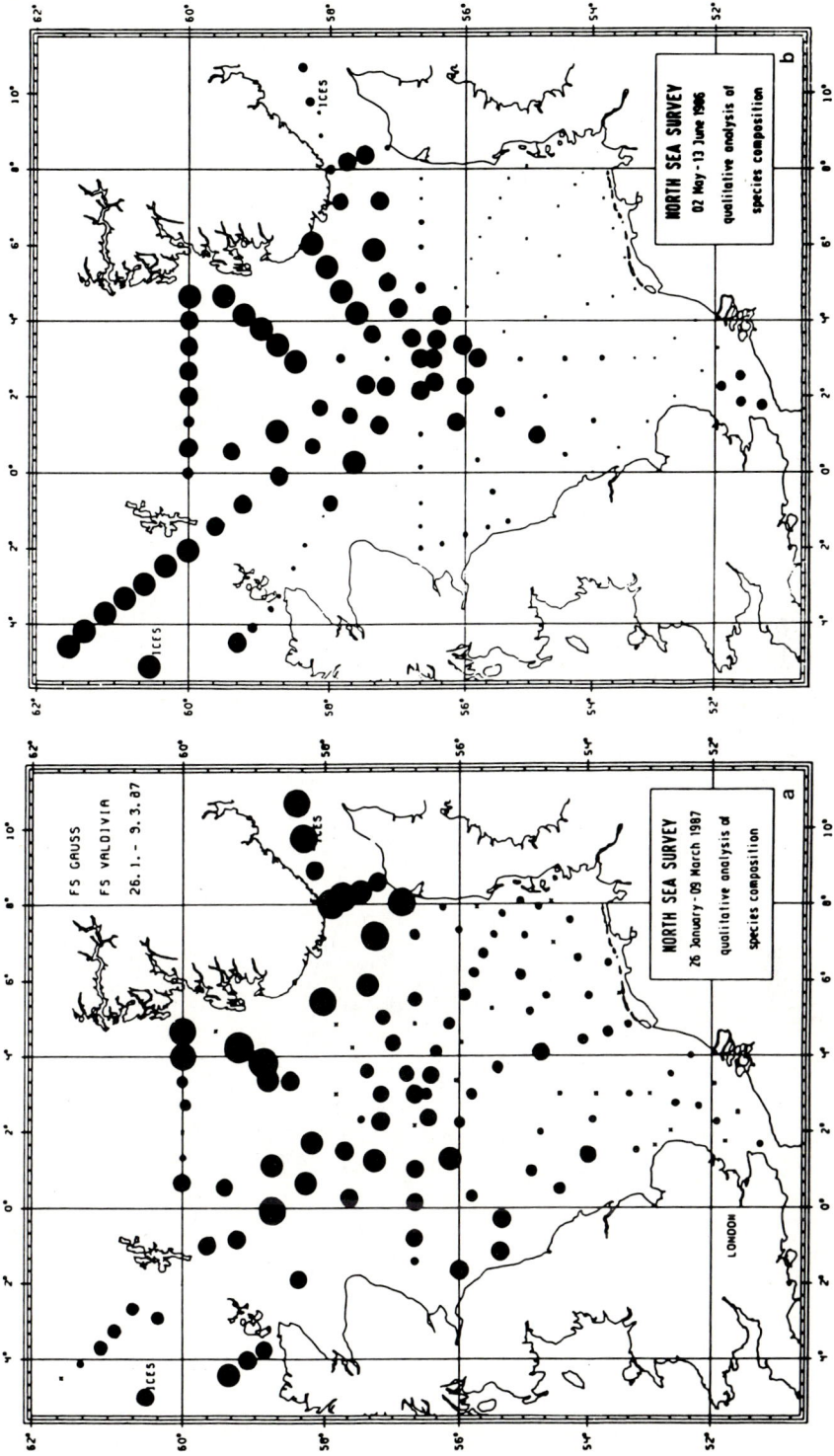

Fig. 2.2.14. Qualitative distribution of *Calanus* sp. (all stages) in the North Sea winter 1987 (*a*) and spring 1986 (*b*). The *sizes of the dots* indicate the relative abundance. *Smallest dots* less than 10 %; *largest circles* more than 90 % of total mesozooplankton

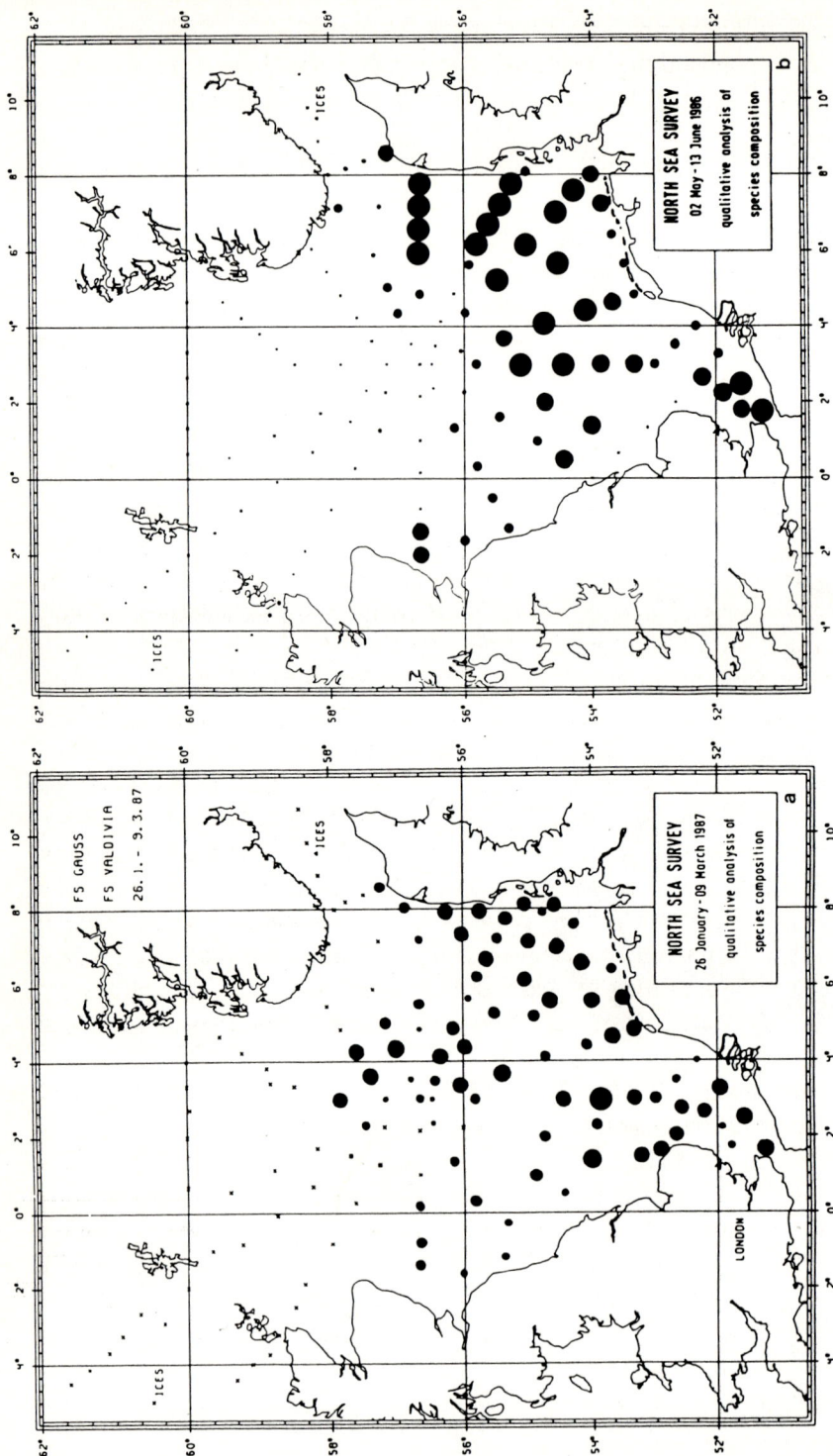

Fig. 2.2.15. Qualitative distribution of *Temora longicornis* (all stages) in (*a*) winter 1987 and (*b*) spring 1986. *Dots* see Fig. 14

regimes in the North Sea. In the deeper northern parts, *C. finmarchicus,* the largest copepod in the North Sea, overwinters in the warmer deep layer of Atlantic water, colonizing the whole water column in spring. A minor population of *C. helgolandicus,* found in the southern Atlantic inflow, was probably advected through the Channel.   These specific horizontal gradients were also observed for other zooplankton species. Some, for instance the larvae of some benthic animals as well as the copepod *Acartia,* were restricted to numerous coastal sites.

The grazing influence was estimated by the production of faecal pellets. During May/June densities of more than 30×10^{-6} m^3/l were observed. The presence of pellets mainly in the upper part of the water column indicated that disintegration was occurring already here and that fast sedimentation took place in the lower layer (Martens and Krause 1990). The first assumption was supported by observations that the length of the faecal pellets was negatively correlated with water depth. Since the faecal pellet concentrations were not only significantly correlated with the abundance of zooplankton, but also with chlorophyll-*a* values, a fast turnover within the food chain was evident. The second hypothesis was confirmed by early remineralization of sedimented biomass.

Ammonium measurements in spring 1986 showed that in the deep water concentrations had increased to more than 4 μM, due to remineralization of sedimented biomass from the spring bloom.

Ammonium, as a primary release product, is also excreted by the herbivorous zooplankton during grazing, causing small spots or even layers of ammonium-enriched water in the vicinity of maximum grazing activity. In the mixed layer, released ammonium is preferably taken up by the phytoplankton. Another dominant process of remineralization detectable by increasing ammonium concentrations occurs near the sediment surface, where the down-raining biomass, dying cells as well as faecal pellets and detritus, accumulate. Here, remineralization is controlled by benthic animals and bacteria.

Indeed, during May/June in the bottom layer of the central eastern North Sea (not at the stations of the first legs) ammonium concentrations increased to more than 4 μM (Fig. 2.2.16 a,b). This is many times the concentrations found at any other location and season. Even in the western central North Sea, a region characterized by Atlantic water inflow, the ammonium concentrations were higher than in winter, indicating that here also sedimenting biomass was already remineralized at that time of the year. A cross-section (Fig. 2.2.17) revealed that the ammonium enrichment was not only restricted to the very deep bottom water, but was spreading within the entire bottom layer of 50 m thickness. Here, the main ammonium sink was oxidation via nitrite to nitrate by nitrifying bacteria.

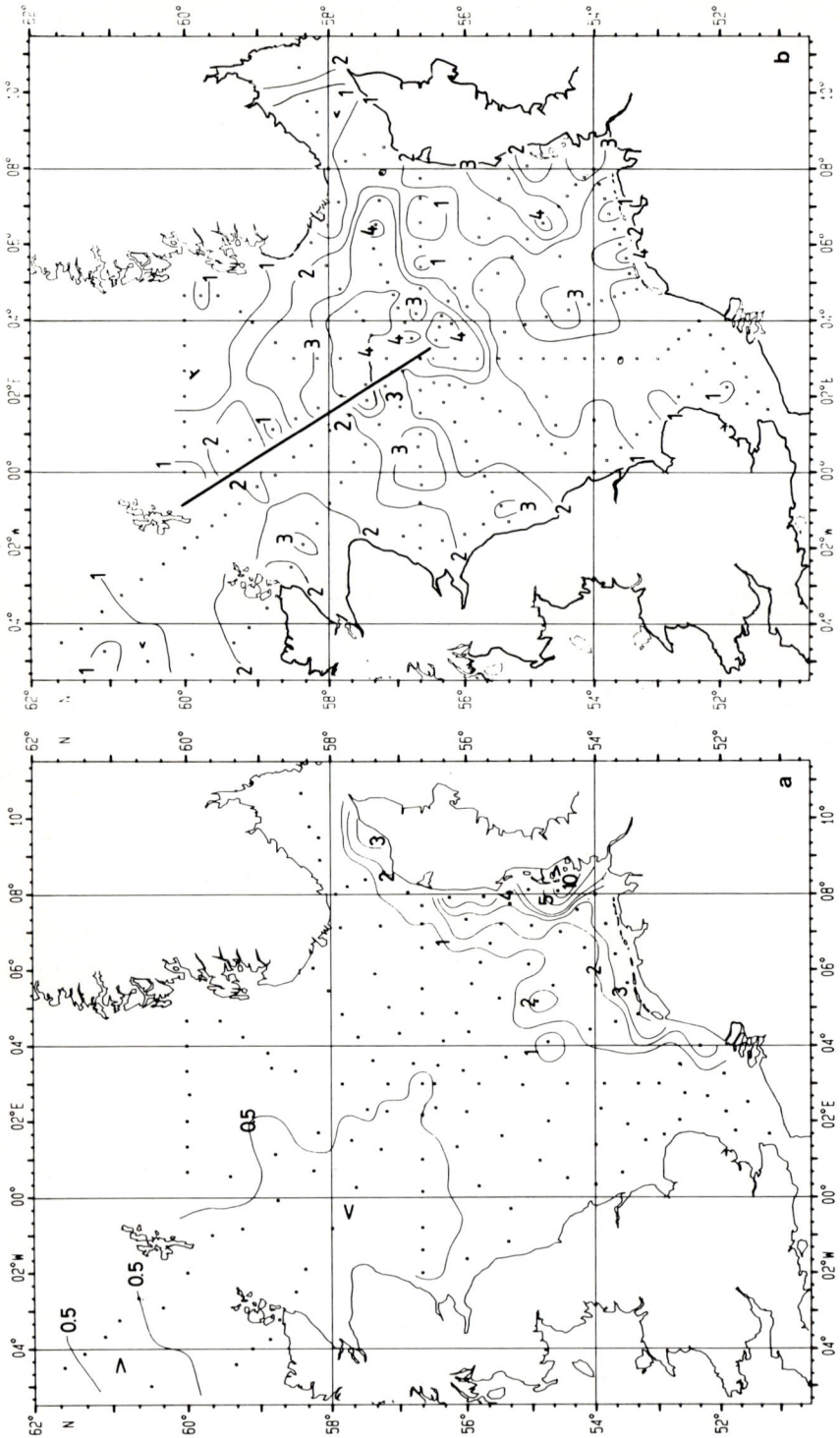

Fig. 2.2.16. Ammonium concentrations near the bottom in winter 1987 (*a*) and spring 1986 (*b*) (µmol/l)

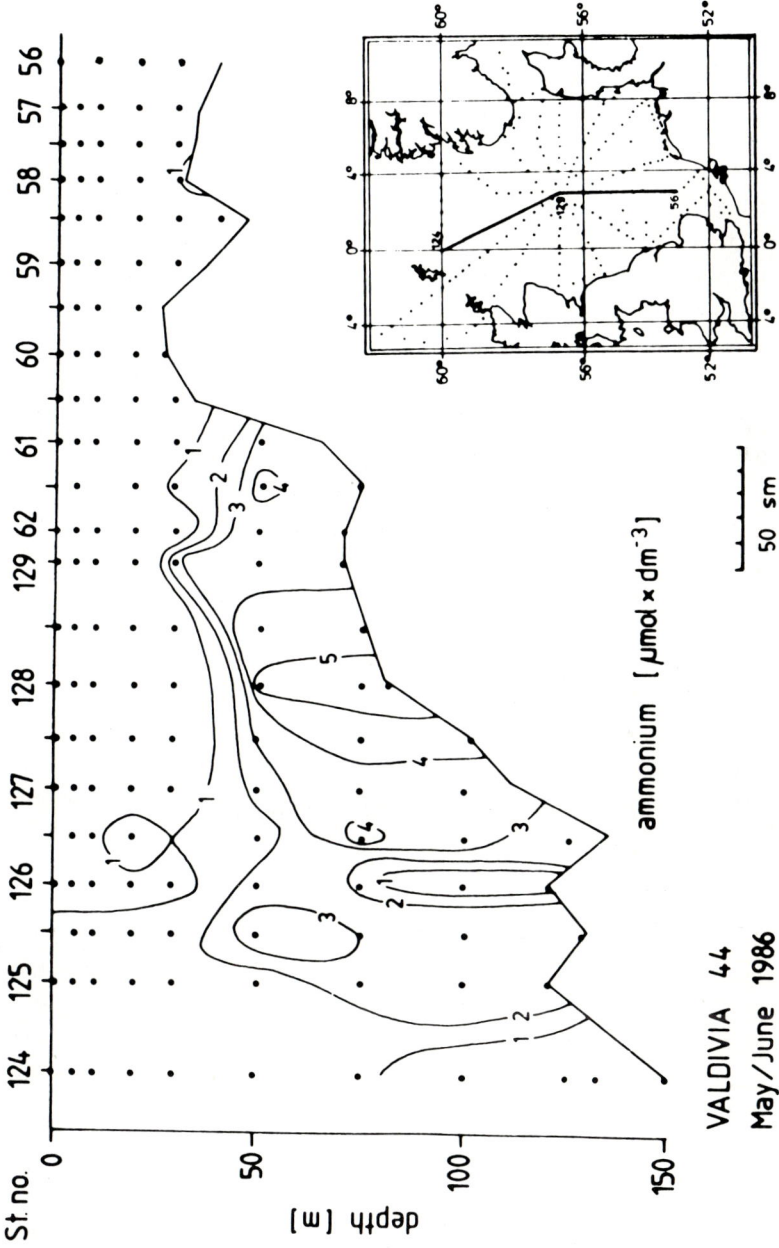

Fig. 2.2.17. Cross-section through the North Sea from south (St. 56) to north (St. 124) (see map) showing the ammonium concentrations in the different depths during spring 1986

Ammonium, as well as all the other nutrients which are enriched in the bottom layer, can be brought back into the euphotic zone following storm events, when the thermocline is eroded partially or broken up totally.

Dissolved organic nitrogen (DON) generally shows similar concentrations during spring and winter, indicating that the major part of DON is refractory. However, DON can also be released by healthy cells in small amounts. Mainly, DON will be formed by decomposition. In May/June 1986 for instance, the concentration of dissolved organic nitrogen was found to be highest in the continental coastal water with its permanent production and decomposition cycles (Fig. 2.2.18b). In particular, the tidal flats of the Wadden Sea can be identified as a source for dissolved organic material (Eberlein et al. 1985). This DON originates from particulate biomass produced in the nutrient-rich river plumes, mainly of Rhine and Elbe and has been transported to these flats prior to decomposition. Direct discharges from the shore can also contribute to the concentration of dissolved organic nitrogen. The concentrations of DON detected during winter in the continental coastal water (Fig. 2.2.18a), were much smaller than during spring, indicating that spring production and decomposition add a significant contribution to the dissolved organic carbon pool. These substances are normally degraded by heterotrophic bacteria but can - when nutrients are limiting - also be used by some mixotrophic phytoplankton species directly, for example in the nutrient-depleted mixed layer during the summer season (Brockmann et al. 1990).

2.2.8 Sediment and Epibenthos

The sediment composition in the North Sea is mainly controlled by hydrodynamic forces, which allow a permanent sedimentation only in a few restricted areas, such as the Norwegian Trough (Eisma 1981; Puls 1987). The investigation in spring 1986 showed that the highest concentrations of fine mud as well as particulate matter were found here (Knickmeyer and Steinhart 1989). Besides this, in the central part of the North Sea the organic content, estimated by weight loss during combustion as well as the percentage of the small size (< 20 μm) fraction, was higher in areas where rapid massive remineralization occurred, as indicated by the ammonium load (Fig. 2.2.19 a,b).

The distribution of the epibenthos reflects the hydrographic conditions in the North Sea (Frauenheim et al. 1989). With a cluster analysis (Jaccard-Index) which compares all stations based on the species composition (only presence and absence data), similar stations are assigned to one cluster. In

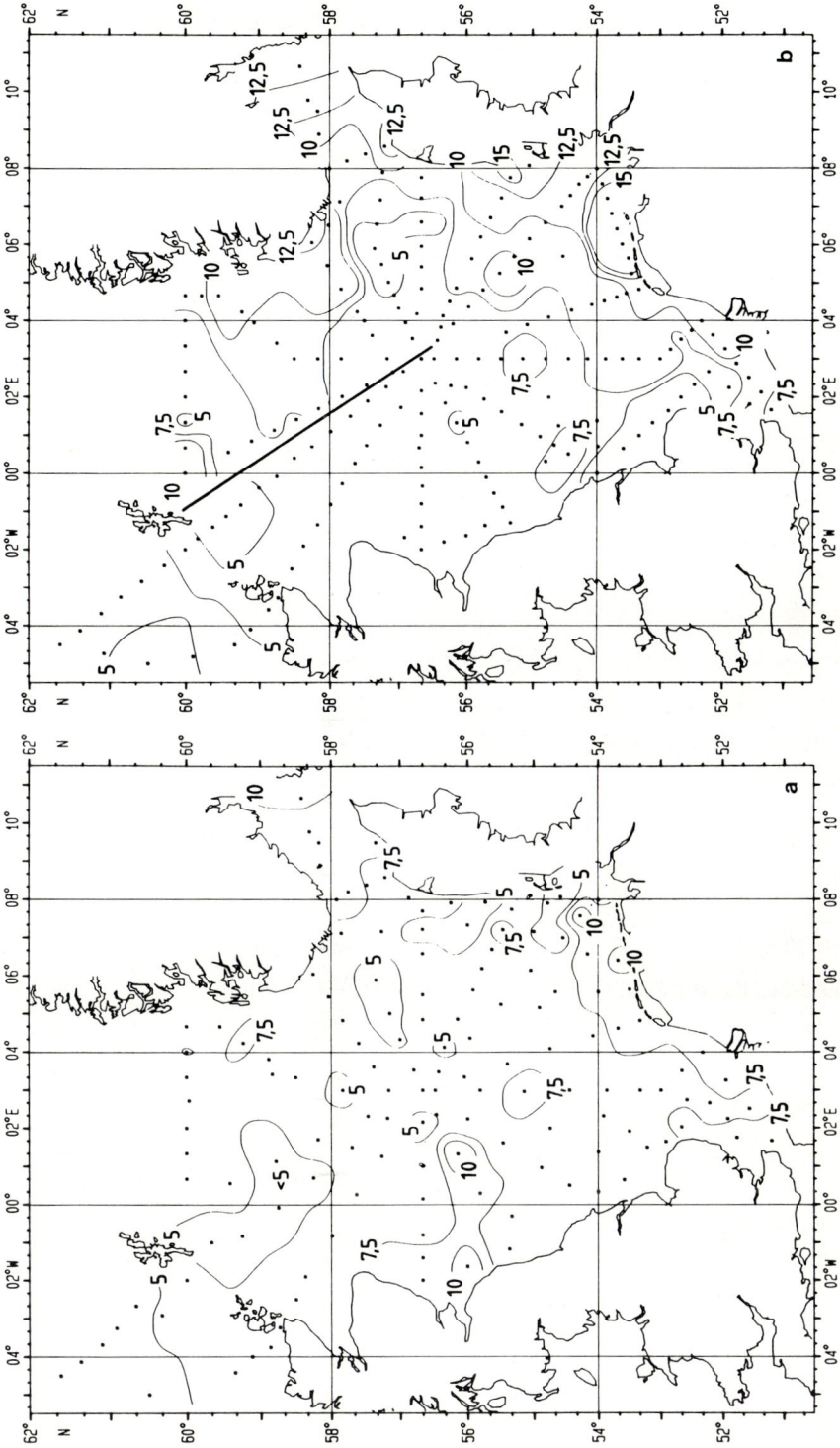

Fig. 2.2.18. Concentration of dissolved organic nitrogen in μg at N/l at the surface in winter 1987 (*a*) and spring 1986 (*b*)

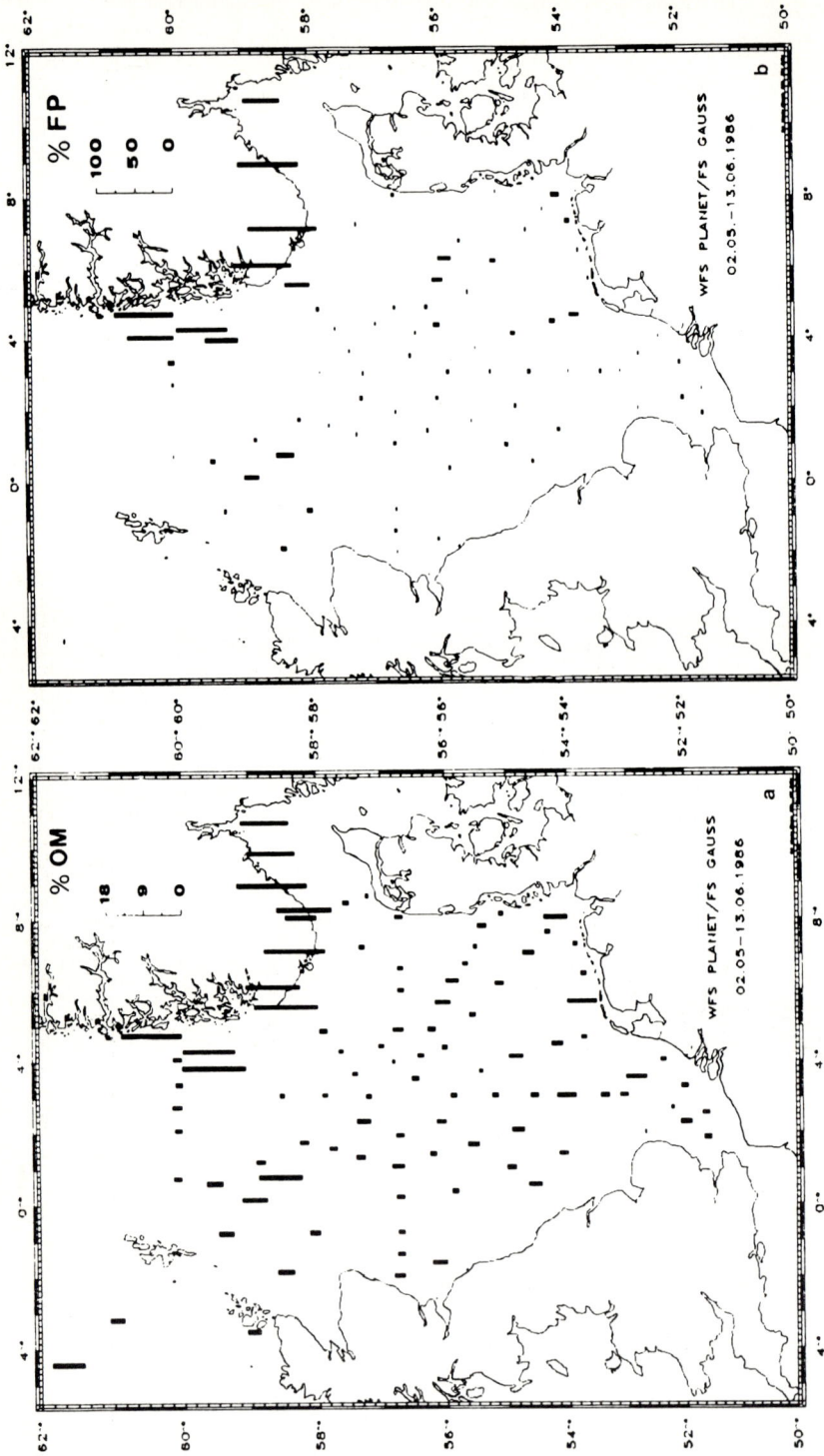

Fig. 2.2.19. (a) Distribution of organic material in the upper 2 cm of the sediment during spring 1986, estimated by combustion. (b) Fraction of fine particles <20 um (as % of the total particle volume)

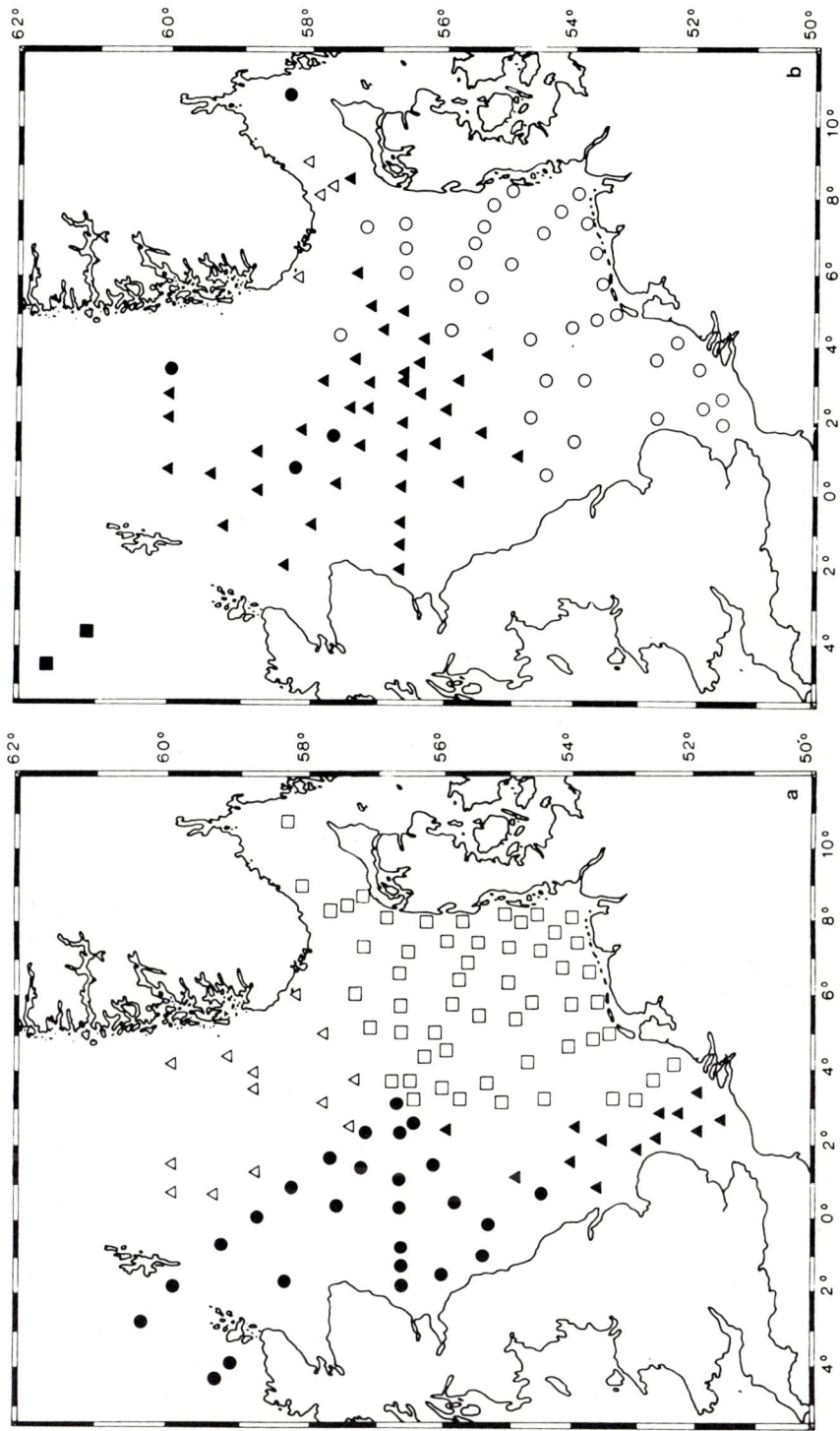

Fig. 2.2.20. Distribution of the main clusters of epibenthic animals in winter 1987 (a) and spring 1986 (b)

spring 1986 two clusters can be recognized (Fig. 2.2.20b): one includes the stations of the southern part of the North Sea, the other combines the stations of the northern part. This finding corresponds with previous results for the epibenthos (Dyer et al. 1982) and for the macro-zoobenthos (Künitzer 1990).

In winter two main clusters can be separated as well (Fig. 2.2.20a), but these divide the North Sea into a western and an eastern part. The differences between the winter and spring surveys will be due to the seasonal differences in the hydrographic situation. In winter the inflow from the North Atlantic reached far to the south along the British coast (Hainbucher et al. 1987). Therefore, northern species could appear together with species which were introduced from south through the Strait of Dover. The differences between the clusters in winter and spring are not caused by different distributions of abundant species since this distribution is constant in both seasons, but rather by the presence and absence of some rare species. The number of species caught in winter was higher than in spring.

Looking at single species of the epibenthos, one can distinguish species which occur in the entire North Sea (e.g. *Aphrodite aculeata, Buccinum undatum, Pagurus bernhardus, Asterias rubens*), others live only north (e.g. *Beringius turtoni, Pagurus pubescens, Henrica sanguinolenta*) or south (e.g. *Corystes cassivelaunus, Ophiura albida*) of the Dogger Bank, while others correlate with the bottom substrate or/and the depth. The hermit crab, *Pagurus bernhardus*, is abundant in the entire North Sea and can therefore be used for comparative analysis of pollutants.

2.2.9 Conclusions

The topographical characteristics of the North Sea exert a strong influence on the ecosystem. In the deeper northern part, following spring warming, the thermocline reduces recycling of dissolved compounds by separating surface and bottom layer for a longer period of time, whereas in shallow areas production and decomposition are coupled in the entire water column. The stratification causes an enrichment, e.g. of ammonium, in the bottom layer beneath productive areas. The two regions are furthermore characterized, for example, by the dominance of different zooplankton species, such as *Calanus finmarchicus* (northern area) and *Temora longicornis* (southeastern part). This difference was found during both seasons investigated. This was also true for the species of benthic fauna, which were distributed in regionally separated clusters.

During winter, the continental coastal water was characterized by high nutrient concentrations, whereas on the Dogger Bank and in some areas in the Skagerrak primary production continued at low irradiances. Especially the winter production in the Dogger Bank area has been observed previously (Brockmann and Wegner 1985). Also during spring local discharges by rivers were clearly detectable due to steep gradients in nutrient-exhausted areas following the phytoplankton spring bloom.

The extent, timing and species composition of the investigated spring bloom is not representative for other years. They reflect only the topical situation in the different regions during sampling time. Because of the variability of physical processes controlling the biological development as well as biological variability itself, the extent of nutrient consumption and plankton species abundance varies during late spring from year to year. The only common features are: nutrient depletion in the mixed layer, nutrient enrichment (especially ammonium) in the deep bottom layer and regional dominance of *Phaeocystis globosa* in continental coastal waters.

Areas of high phytoplankton productivity are patchily distributed within the extended regions of nutrient depletion, indicating the inhomogeneous development with temporary nutrient limitation. Especially in stratified areas, the time lag is evident between production in the euphotic zone and remineralization in the deeper layer, uncoupling these processes.

During spring, phytoplankton biomass was calculated to be as high as 20-50 mg C/m^3. The zooplankton biomass reached values up to 100 mg C/m^3 during spring in the central North Sea as well as in the northern continental coastal water (starting in the German Bight). This relation reflects the high productivity of the North Sea and the fast turnover of biomass. The standing crop of phytoplankton represents less than 1/5 of the biomass which is necessary to feed 100 mg C/m^3 of zooplankton (Steele 1974). However, the mean seasonal production figures for primary production in the 0-20 m layer in spring 1986 of 70 mg/m^{2}h (corresponding to 50 mg/m^3 day) match the high zooplankton biomass of 100 mg C/m^3, assuming that this will probably consume 25% of its carbon body weight per day. Additionally, it has to be considered that at spots of high zooplankton abundance, phytoplankton stocks and production numbers may have been larger in the prior weeks, during which the zooplankton stock developed.

The differences in phosphate concentrations between the surface and the bottom layer during late spring of 0.5 μM correspond to a high biomass production of 650 mg C/m^3 during spring [assuming C:P = 108:1 (Redfield et al. 1963)].

Acknowledgements

We thank Dr. M.A. Baars for his helpful comments on the manuscript.

References

Anonymous (1990) Interim report on the quality status of the North Sea. North Sea Conf. The Hague, 7-8 March, 1990, 51 pp

ARGE Elbe (1987) Wassergütedaten der Elbe von Schnackenburg bis zur See. Arbeitsgemeinschaft für die Reinhaltung der Elbe, Wassergütestelle Elbe

ARGE Elbe (1988) Wassergütedaten der Elbe von Schnackenburg bis zur See. Arbeitsgemeinschaft für die Reinhaltung der Elbe, Wassergütestelle Elbe

Brockmann UH, Wegner G (1985) Hydrography, nutrient and chlorophyll distribution in the North Sea in February, 1984. Arch Fisch Wass 36:27-45

Brockmann UH, Laane RWPM, Postma H (1990) Cycling of nutrient elements in the North Sea. Neth J Sea Res 26:239-264

Dyer MF, Fry WG, Fry PD, Cranmer GJ (1982) A series of North Sea benthos surveys with trawl and headline camera. J Mar Biol Assoc UK 62:297-313

Eberlein K, Leal MT, Hammer KD, Hickel W (1985) Dissolved organic substances during a *Phaeocystis pouchetii* bloom in the German Bight (North Sea) Mar Biol 89:311-316

Eisma D (1981) Supply and deposition of suspended matter in the North Sea. Spec Publ Int Assoc Sediment 5:415-428

Frauenheim K, Neumann V, Thiel H, Türkay M (1989) The distribution of the larger epifauna during summer and winter in the North Sea and its suitability for environmental monitoring. Senckenb Marit 20:101-118

Hainbucher D, Pohlmann T, Backhaus J (1987) Transport of conversative passive tracers in the North Sea: first results of a circulation transport model. Continent Shelf Res 7:1161-1179

Holligan PM, Aarup T, Groom SB (1989) The North Sea: satellite colour atlas. Continent Shelf Res 9:667-765

Knickmeyer R, Steinhart H (1989) Cyclic organochlorines in North Sea sediments. Relation with size and organic matter. Dtsch Hydrogr Z 42:43-59

Krause M, Martens P (1990) Distribution patterns of mesozooplankton biomass in the North Sea. Helgol Meeresunters 44:295-327

Künitzer A (1990) The benthic infauna of the North Sea: species distribution and assemblages. ICES, Copenhagen, CM 1990/Mini:2

Martens P, Krause M (1990) The fate of faecal pellets in the North Sea. Helgol Meeresunters 44:9-19

Parsons TR, Takahashi M, Hargrave B (1984) Biological oceanographic processes. Pergamon Press, 330 pp

Pingree RD, Holligan PM, Mardell GT (1968) The effects of vertical stability on phytoplankton distributions in the summer on the northwest European Shelf. Deep-Sea Res 25:1011-1028

Postma H (1981) Exchange of materials between the North Sea and the Wadden Sea. Mar Geol 40:199-213

Puls W (1987) Simulation of suspended sediment dispersion in the North Sea. ICES, Copenhagen, CM 1987/C:37, 24 pp

Redfield AC, Ketchum BH, Richards FA (1963) The influence of organisms on the composition of sea water. In: Hill MN (ed) The sea. Wiley-Interscience, London, pp 26-77

Rick H-J (1990) Ein Beitrag zur Abschätzung der Wechselbeziehung zwischen den planktischen Primärproduzenten des Nordseegebietes und den Schwermetallen Kupfer, Zink, Cadmium und Blei auf Grundlage von Untersuchungen an natürlichen Planktongemeinschaften und Laborexperimenten mit bestands-bildenden Arten. Diss RWTH Aachen, 94 pp

Steele JH (1974) The structure of marine ecosystems. Harvard Univ Press, Cambridge, Ma

Wegner G (1987) Map of the sea-surface temperature. DHI Rep, Hamburg

2.3 Large-Scale Distribution of Contaminants in Spring 1986 and Winter 1986/87

U. Kammann, M. Haarich, K. Heyer, H. Hühnerfuss,
L. Karbe, M. Kersten, D. Schmidt and H. Steinhart

Contaminants, like the nutrients discussed in the preceding chapter, are permanently involved in the dynamics of the ecosystem. In spite of the water exchange between North Sea and Atlantic, large amounts of contaminants are held back by ad- and absorptive processes on the North Sea shelf. The contaminants may be transported in the waterway adsorbed onto suspended matter. When these reach sedimentation areas, the particles settle on the bottom and form sediment layers. Resuspension or bioturbation may enable the contaminants in shallow regions to return to the water column. Describing the situation of contaminant distribution during two seasons and comparing contamination levels in the compartments, an attempt will be made to illuminate interrelationships and transport processes in the marine ecosystem. For this approach, the amounts of several organochlorine compounds and heavy metals were determined by different working groups for the compartments "water", "suspended matter", "sediment" and "benthic organisms" observed in the entire North Sea during two seasons.

2.3.1 Cyclic Organochlorines and Heavy Metals in Sea Water

The state of the North Sea with respect to the contamination of sea water by organochlorine compounds and heavy metals during the two surveys in early summer 1986 and in late winter 1987 will be demonstrated by the distribution of α-HCH, γ-HCH (Lindane), the sum of five PCB congeners (PCBs 52, 101, 138, 153 and 180), Hg (mercury), Cd (cadmium) and Pb (lead). The data of the 10 m depth horizon have been selected since they cover the entire area completely. Samples for PCB analysis were only taken during the summer cruise.

2.3.1.1 Cyclic Organochlorines

The distribution of γ-HCH, which is shown in Fig. 2.3.1, revealed maximum values in surface water along the Dutch and German coast, west of Jylland and along the Norwegian trench, while α-HCH (Fig. 2.3.2) showed elevated concentrations in surface water southwest of the Norwegian coast and in the Baltic outstream (Weber and Hühnerfuss, unpubl. results). In general, the qualitative distribution of the two HCH-isomers in winter 1987 was comparable to the results found during the summer campaign (Figs. 2.3.3 and 2.3.4).

Comparing the two data sets for summer 1986 and winter 1987, it can be concluded that significantly larger concentrations of γ-HCH were found in coastal areas during the summer 1986 campaign. A similar distribution could be reported for the hermit crab (Fig. 2.3.26). This is particularly the case in the estuaries of the rivers Elbe, Weser and Humber, the Britsh Channel and in the Skagerrak. In the central and northern North Sea, however, the concentrations of γ-HCH remained relatively constant within the error limits of the analysis method and appeared to be independent of the season. Therefore, it can be assumed that the seasonal variation of biological activity is not reflected in a comparable variability of γ-HCH concentration. The observed significant increase of γ-HCH concentrations in coastal areas during summer 1986 expresses the seasonal agricultural application of Lindane, which is transported into the North Sea via rivers in considerable amounts (Hühnerfuss and Weber, unpubl. results).

The main sources of α-HCH in the North Sea have been determined to be at the Norwegian coast (Figs. 2.3.2 and 2.3.4). In this region maximum α-HCH concentrations were observed in summer 1986, while other distant areas remained unaffected by this source and by seasonal variations of agricultural activity.

Based upon rough approximations, an overall load of 79.5 tons γ-HCH and 80 tons γ-HCH can be estimated for the North Sea during the summer campaign. In comparison to investigations in 1979, an increase in γ-HCH levels and a decrease in α-HCH contamination were observed in the southern North Sea. In the central and northern parts of the North Sea, however, no significant changes were found. Comparisons between the total

γ- HCH in 10m depth

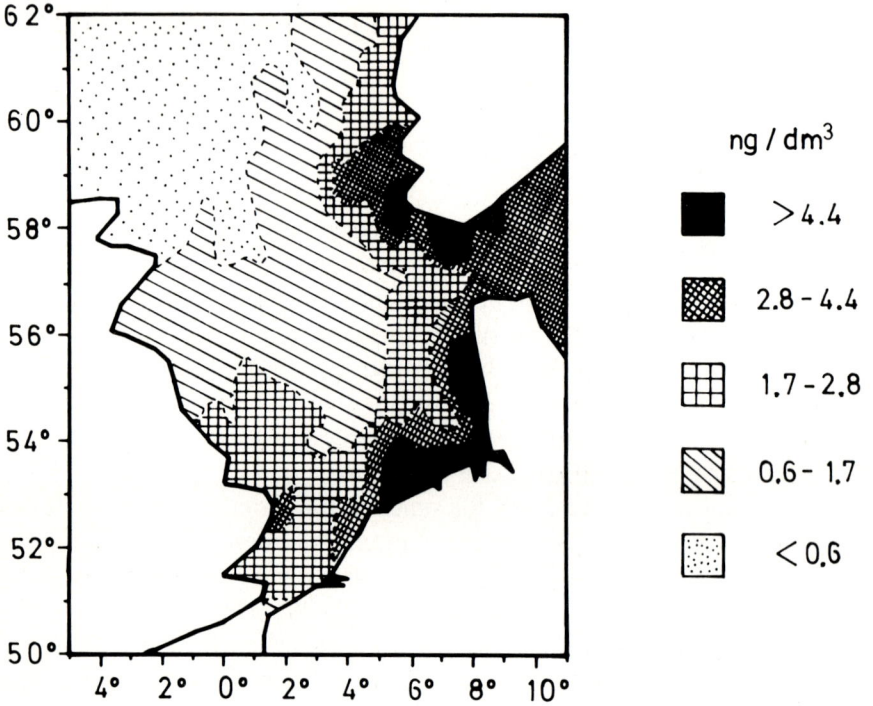

Fig. 2.3.1. Distribution of γ-HCH in sea water (ng/dm³ water) of the North Sea (10 m horizon), summer 1986

α - HCH in 10m depth

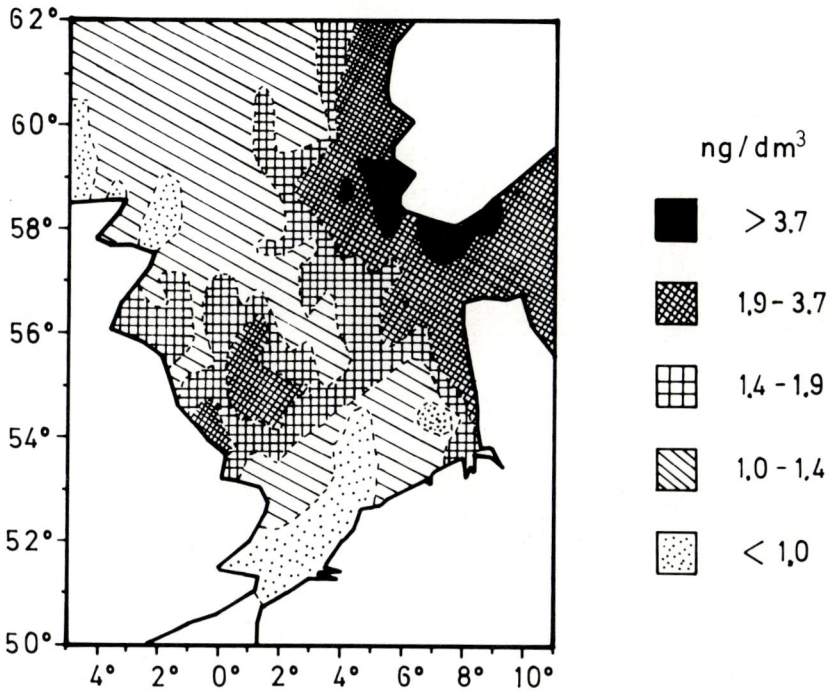

Fig. 2.3.2. Distribution of α-HCH in sea water (ng/dm³ water) of the North Sea (10 m horizon), summer 1986

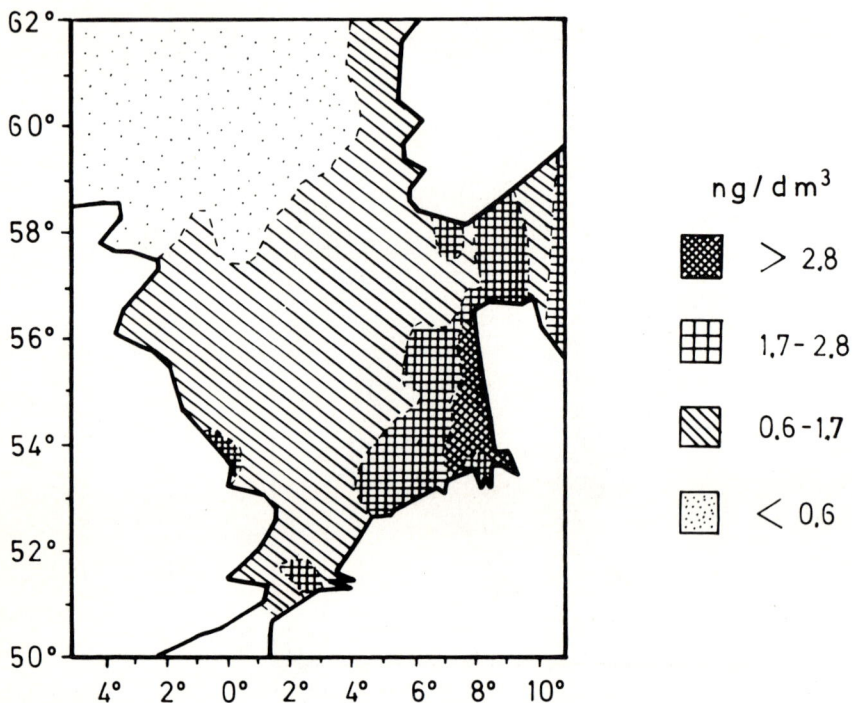

Fig. 2.3.3. Distribution of γ-HCH in sea water (ng/dm^3 water) of the North Sea (10 m horizon), winter 1987

α - HCH in 10m depth

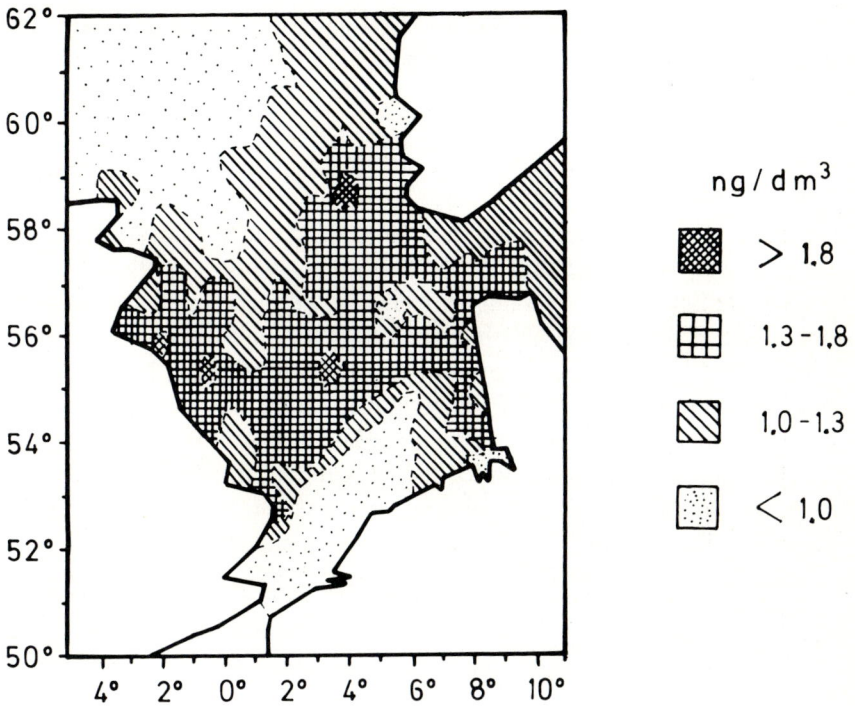

Fig. 2.3.4. Distribution of α-HCH in sea water (ng/dm^3 water) of the North Sea (10 m horizon), winter 1987

annual inflow of the two HCH-isomers into the North Sea and the corresponding outflow values reveals that the outflow of α-HCH is balanced by the inflow, while the outflow of γ-HCH is significantly lower than its inflow. This indicates that γ-HCH is less stable than α-HCH (Hühnerfuss and Weber, unpubl. results) (cf. Chap. 2.3.5). The overall load during the period of the winter campaign 1987 was calculated to be 57.5 tons α-HCH and 50 tons γ-HCH - a significant reduction of both HCH isomers compared with the summer 1986 campaign.

In contrast to expectation, the horizontal distribution of the PCB congeners (Fig. 2.3.5) revealed maximum values in the northwestern part of the North Sea and along the British coast, while significantly lower levels were determined in samples from the German Bight. In addition, a shift in the PCB pattern was observed for different areas of the North Sea (Weber and Hühnerfuss, unpubl. results). It should be noted that concentrations of individual PCB congeners are lower than the measured concentrations of HCH isomers by an order of magnitude.

An unequivocal explanation for the unexpected distribution of the PCBs in the North Sea is not yet possible, because our knowledge about the different transport processes for various pollutants is still incomplete.

Only tentative explanations can be offered:

1.   Transport of PCBs from the highly polluted Irish Sea along the northern Scottish and eastern British coasts.

or

2.   Atmospheric input into the ocean.

In order to gain further insight into the contribution of these two possible transport processes additional water samples were taken along the west coast of Great Britain and from the Irish Sea to the North Sea area north of Scotland in September/October 1990. It turned out that the inflow from the Irish Sea into the North Sea is not sufficient to explain the high PCB values found in the northwestern part of the North Sea during the summer 1986 campaign.

It cannot be excluded that atmospheric fluxes may have contributed to the strikingly high concentrations as suggested by Fig. 2.3.5. However, further work, including analysis of wind trajectories during the experimental period, is required in order to verify this hypothesis. Atmospheric transport is an important and well-known pathway for a number of other contaminants into the North Sea. Certain heavy metals, e.g. lead, reach the North Sea mainly via atmospheric transport (see Chap. 3.1).

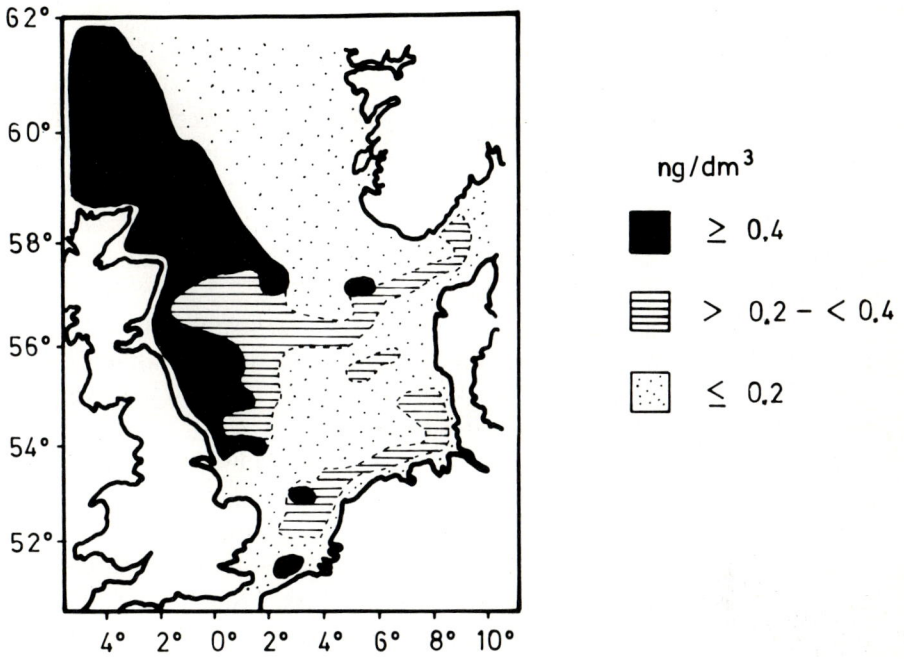

Fig. 2.3.5. Distribution of the sum of PCB 52, PCB 101, PCB 138, PCB 153 and PCB 180 in sea water (ng/dm^3 water) of the North Sea (10 m horizon), winter 1987

2.3.1.2 Heavy Metals

The analytical methods which were used for the determination of trace metals in unfiltered sea water are briefly presented and their differences discussed in Chapter 4.3. For comparison of concentration levels from condensed data, the median value has been favoured instead of arithmetic means, which are influenced more by extreme values.

In summer 1986 the concentrations of "reactive" Hg show low levels in most parts of the central and northern North Sea (Fig. 2.3.6). The median value for these less contaminated regions has been calculated to be 1.5 ng/l. In contrast, elevated values were found near the coast of southeast England, in the German Bight and in the Skagerrak, but also in the northwestern part east of Scotland, and near the Shetland Islands.

The central and northern North Sea are influenced by major rivers, by shallow seas or the outflow of the Baltic Sea. Since Hg is mainly transported by suspended matter, higher concentrations are often related to higher particle concentrations, as found in the estuaries of the Wash and the Humber river in southeast England or in the Wadden Sea.

To explain the strikingly high concentrations of Hg in the northwestern North Sea around Devils Hole and Witch Ground as well as in the Fair Isle passage, more detailed investigations are necessary.

In late winter 1987 the distribution of Hg shows a substantially changed pattern. Except in the northwestern part, where maximum concentrations occur in the same area as in summer, nearly the entire North Sea shows low levels of Hg of less than 1 ng/l (Fig. 2.3.7).

Since particle concentrations are still elevated in coastal regions of the southern part of the North Sea, the difference between summer 1986 and winter 1987 may be caused by the particle chemistry, especially the organic content of the suspended particulate matter (SPM).

The Cd distribution of the summer cruise shows a more uniform pattern than Hg or Pb. Whereas the majority of Cd concentrations are very low in the central and northern North Sea, elevated values are found again in the southern part near the coasts and in the Skagerrak (Fig. 2.3.9). Median for the 10 m depth horizon of the entire North Sea is about 10 ng/l for ASV (difference pulse anodic stripping voltametry) and 14 ng/l for CVAAS (cold vapour atomic absorption spectrometry) data. In comparison to low level regions in the northern North Sea, median values of sites with elevated concentrations are higher by a factor of 2.

Although in winter 1987 the distribution pattern of Cd looks quite similar to the summer distribution, particularly in the central part of the North Sea, the accumulation of elevated concentrations is more evident for the coastal zones from the Channel to North Denmark than in summer 1986 (Fig.

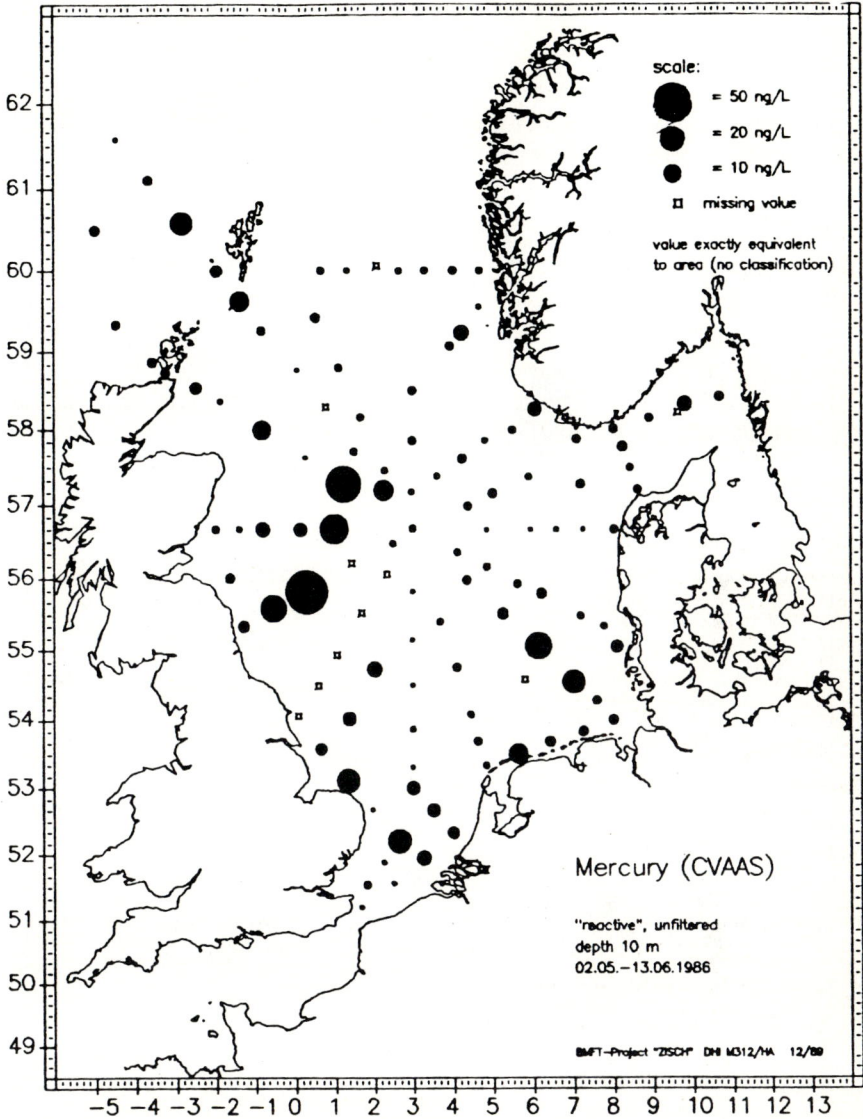

Fig. 2.3.6. Distribution of Hg in unfiltered sea water (10 m horizon) of the North Sea, summer 1986; method: cold vapour atomic absorption spectrometry (CVAAS)

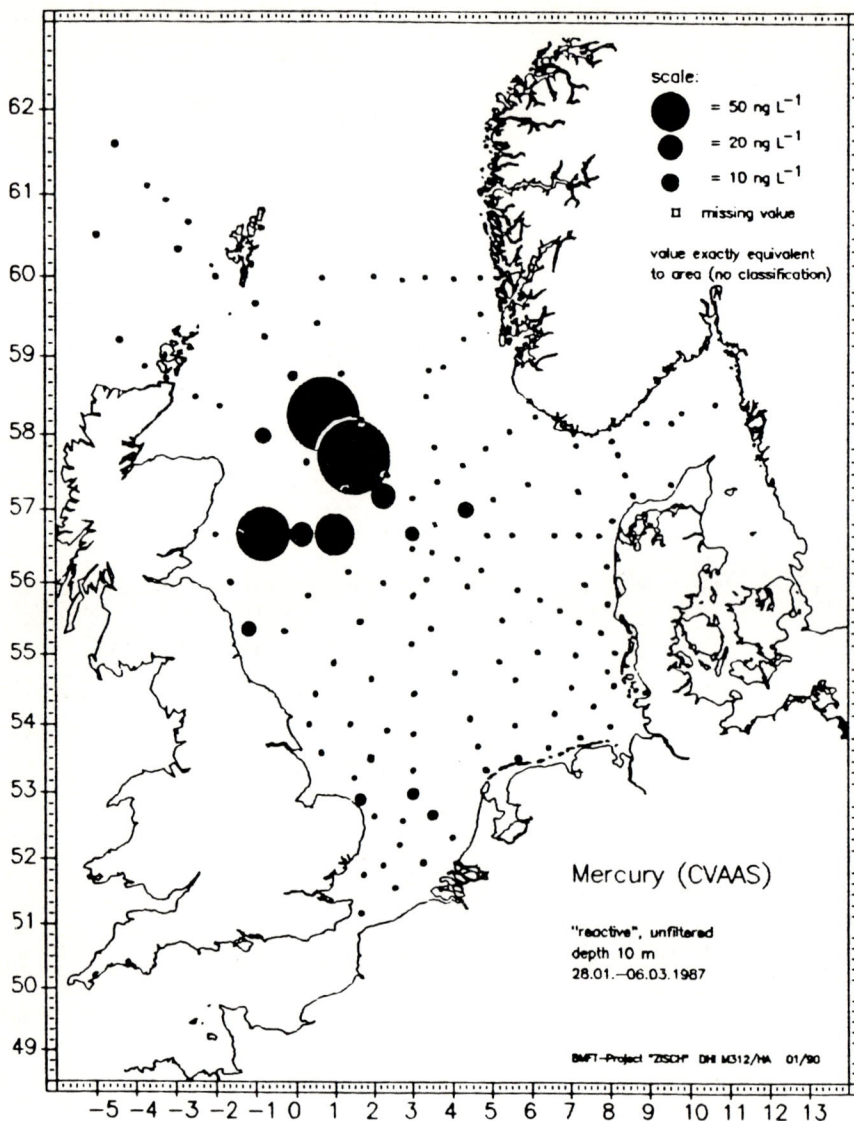

Fig. 2.3.7. Distribution of Hg in unfiltered sea water (10 m horizon) of the North Sea, winter 1987; method: cold vapour atomic absorption spectrometry (CVAAS)

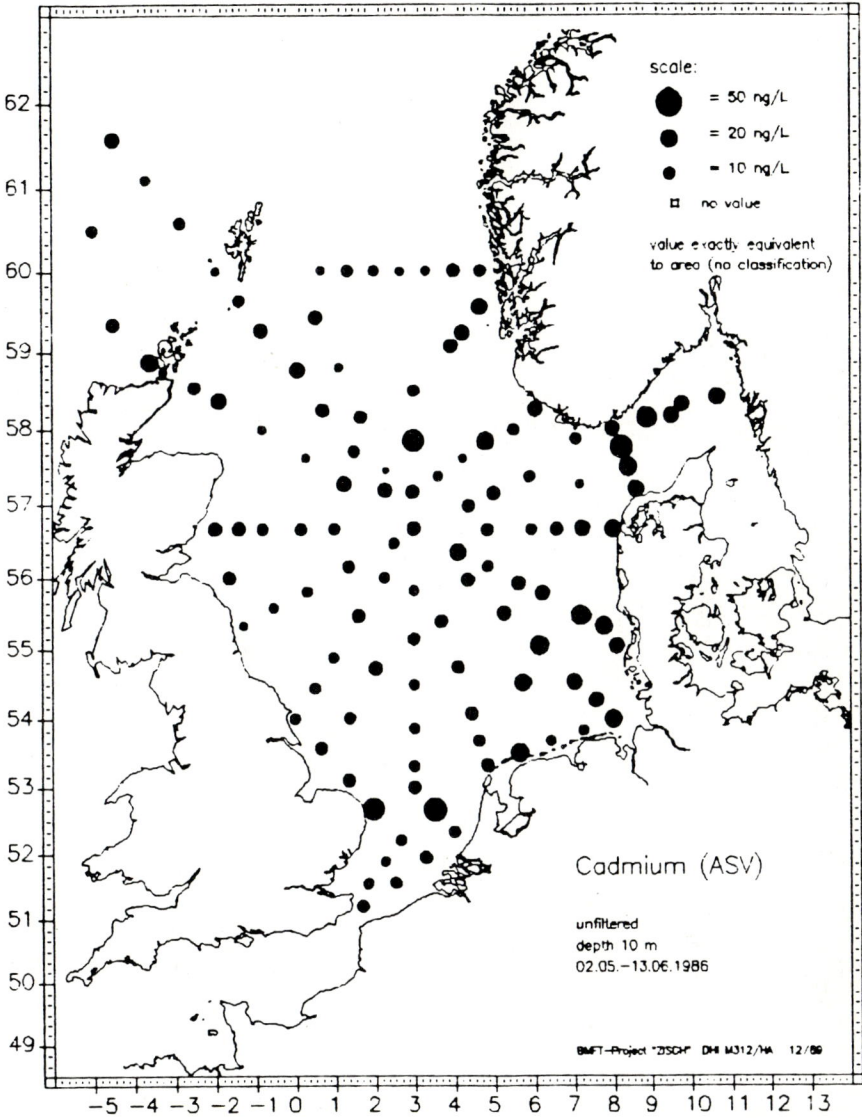

Fig. 2.3.8. Distribution of Cd in unfiltered sea water (10 m horizon) of the North Sea, summer 1986; method: difference pulse anoding stripping voltametry (ASV)

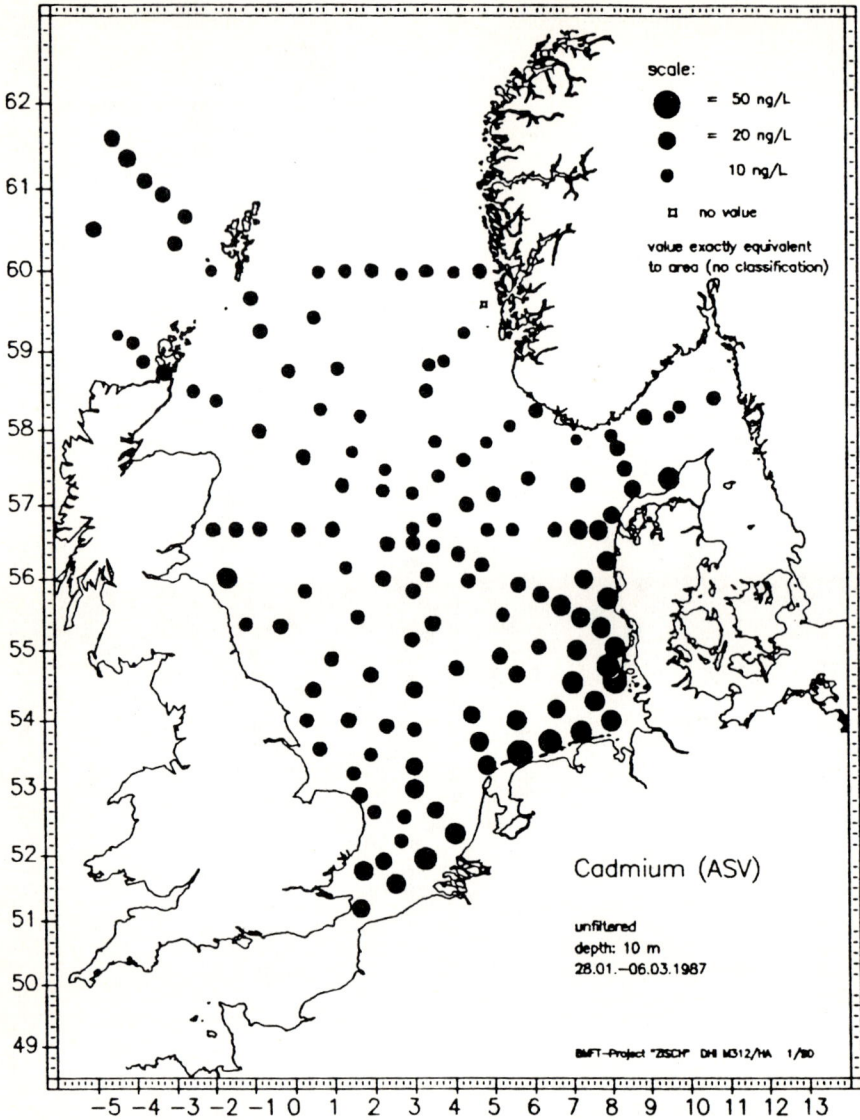

Fig. 2.3.9. Distribution of Cd in unfiltered sea water (10 m horizon) of the North Sea, winter 1987; method: difference pulse anoding stripping voltametry (ASV)

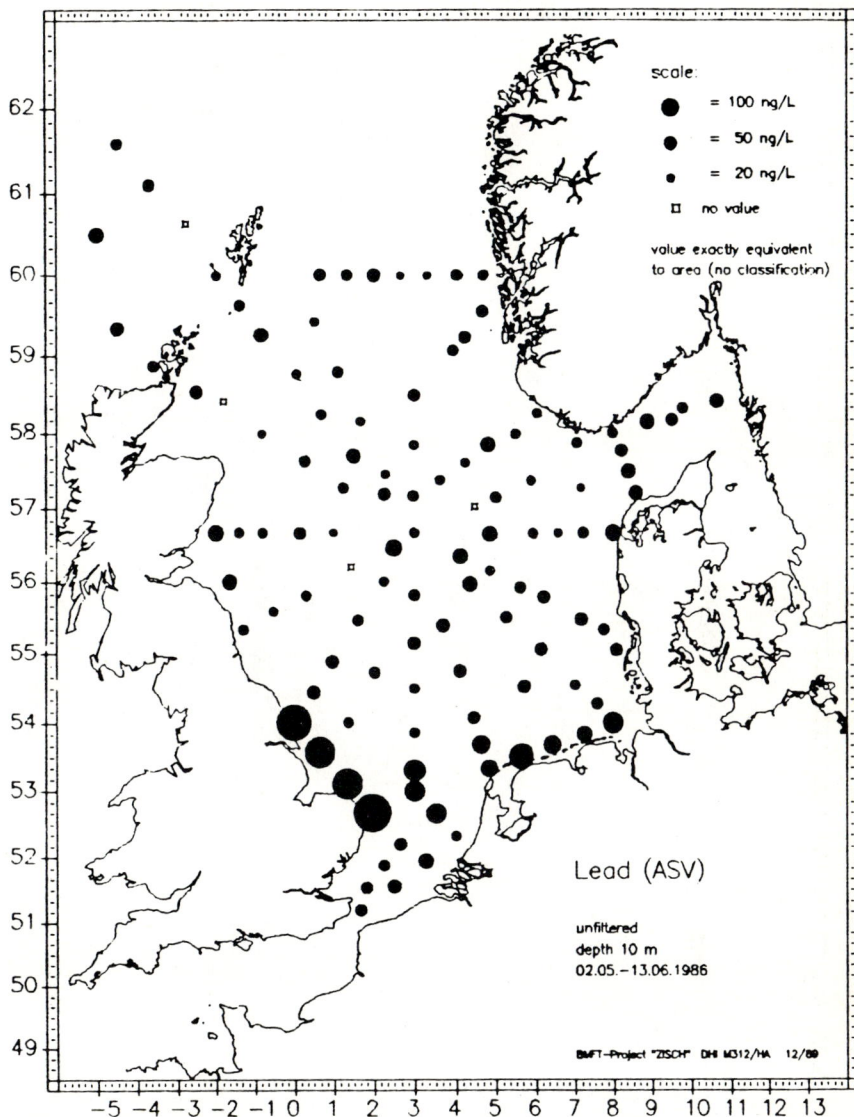

Fig. 2.3.10. Distribution of Pb in unfiltered sea water (10 m horizon) of the North Sea, summer 1986; method: difference pulse anoding stripping voltametry (ASV)

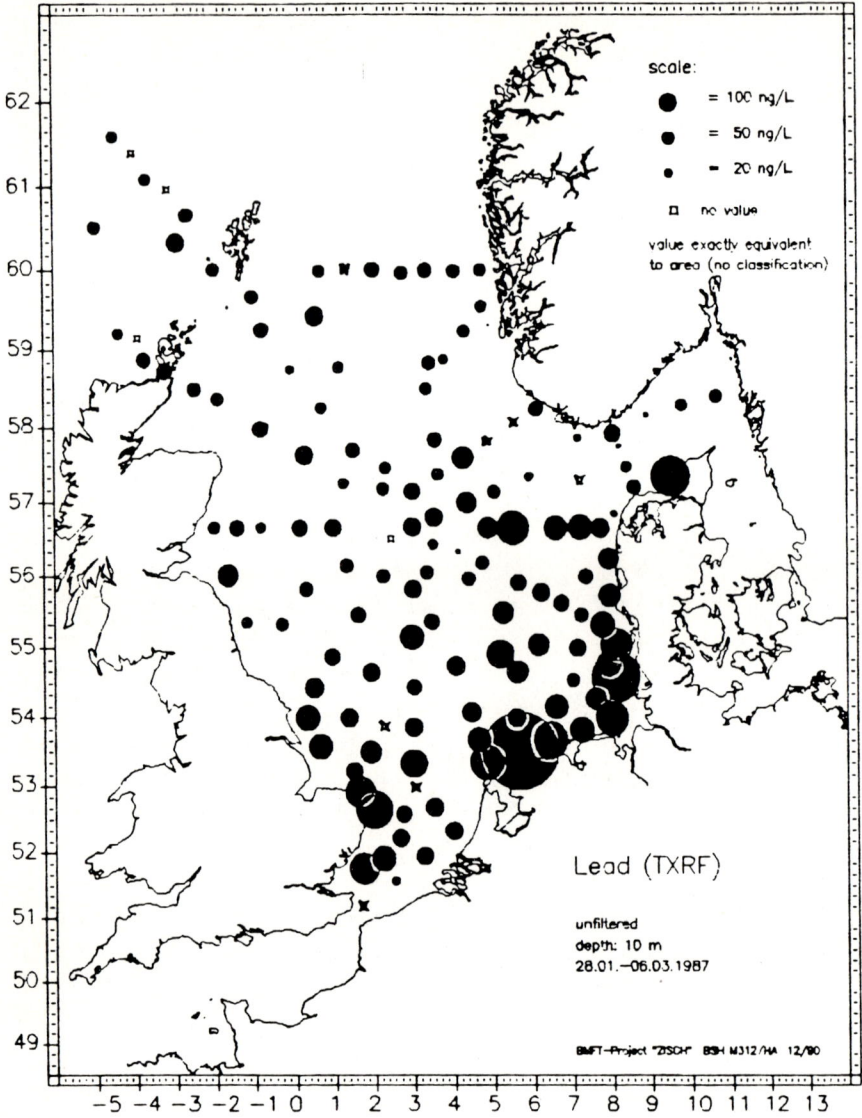

Fig. 2.3.11. Distribution of Pb in unfiltered sea water (10 m horizon) of the North Sea, winter 1987; method: total reflectance X-ray fluorescence (TXRF)

2.3.8). Apparently, resuspension and remobilization processes increase Cd concentrations in the water in winter, especially in regions influenced by the Wadden Sea. In the German Bight, low values for the Cd uptake by phytoplankton have been calculated at stations near the Wadden Sea, where Cd concentrations in water are elevated.

Generally, in comparison to the summer cruise, the median concentration for the entire North Sea has increased by a factor of 1.5. Whereas the level in the German Bight is two times higher than in summer, and in the coastal zone of the German Bight even three times higher, no significant changes have taken place in the Skagerrak.

For Pb, the major part of the North Sea including the central and the northern areas shows a median of about 30 ng/l. Slightly elevated Pb concentrations of about 35 ng/l were measured in the Skagerrak and around the Orkney and Shetland Islands (Fig. 2.3.10 and Fig. 2.3.11).

In contrast, for all coastal regions off Scotland and England, except the Thames estuary, and from the Scheldt estuary to the northern border of the Wadden Sea at Esbjerg, the median is 80 - 90 ng/l. The highest concentrations of up to 840 ng/l and a median of nearly 400 ng/l were found off the English southeastern coast, including the Humber and Wash estuaries. For this area, a good agreement between the Pb content calculated from analysis of unfiltered water and SPM has been achieved. This indicates, at least for this small region, that Pb is nearly totally transported by particles. For all other regions, however, no significant correlation between Pb in unfiltered sea water and suspended matter was found for the summer cruise.

Opposite to the summer cruise, where elevated Pb concentrations were restricted to the southeastern and southern coast of the North Sea, in winter higher levels occurred in nearly the entire southern part of the North Sea. However, the values obtained from the coastal regions of southeastern England and of West to North Frisia again stand out (Fig. 2.3.6). The level in the central North Sea was twice as high as in the northern North Sea, where the median was about 30 ng/l, as in summer. Again, the patch of maximum Pb concentrations was found off the coast of southeastern England, resulting in a regional median of 290 ng/l, about 100 ng/l lower than in summer 1986.

However, Pb concentrations were also elevated in the areas off the Frisian coasts in winter 1987. Although the median was "only" about 210 ng/l and therefore lower than off the English coast, it had increased by a factor of 2 in comparison to summer 1986, and extremely high values up to nearly 3 mg/l have been found. For the coastal region of the German Bight, Pb in water and SPM are correlated rather well, if data pairs with SPM values > 3 mg/l are excluded (linear, r = 0.88). From the data obtained, we can assume that the high concentrations of Pb (in unfiltered water samples) are mainly caused by particles originating from the Wadden Sea.

Finally, attention should be drawn to some striking features at certain stations, suggesting local sources or other reasons not yet resolved. For Hg, Cd, and Pb, elevated concentrations were found in summer and in winter near the islands Ameland and Terschelling off the Dutch coast. Cd and Pb, and also other heavy metals, showed high concentrations near the island Amrum off the North Frisian coast. Both stations are located close to the Wadden Sea, which should be taken into account for explaining these maxima.

Off the northern Danish coast, elevated levels for several heavy metals, including Cd and Pb, were found in the Jammer Bight in winter 1987. At the east-west transect that starts from the Limfjord, concentrations of Cd and Pb (as well as nickel, copper, manganese and iron) decreased towards the open sea in summer 1986. In winter 1987, however, Pb (as well as manganese and iron) increased from the coast towards the central North Sea. Since in winter the sampling at those stations was carried out after a heavy storm, the differences between summer and winter may not be typical for seasonal effects but caused by this event.

A discussion of medians and concentration ranges of selected metals from the ZISCH data in comparison with published data for different parts of the North Sea is reported in the literature (Kersten et al. 1988; Anonymous 1991; Haarich et al. 1993). In addition to condensed presentations of heavy metal data of the ZISCH cruises (Dicke et al. 1987; Schmidt and Dicke 1987, 1990), the entire data sets will be available in Haarich and Schmidt (1993a, b).

2.3.2 Heavy Metals in Suspended Matter

Geographical distributions of particulate Cd and Pb in the North Sea are given in Fig. 2.3.12. Sampling and analysis procedures, together with distribution patterns of 10 other elements as well as seston and POC concentrations, have been given elsewhere (Kersten et al. 1990). The total concentrations of these two elements in the suspended particulate matter samples for the cruise in winter 1987 show the following distribution in the surface water:

- Cd (Fig. 2.3.12a). The Cd content was high in the central and northern North Sea and in the Skagerrak (mostly > 2 µg/g maximum 14.2 µg/g). In the southern North Sea the Cd content was generally < 1 µg/g,

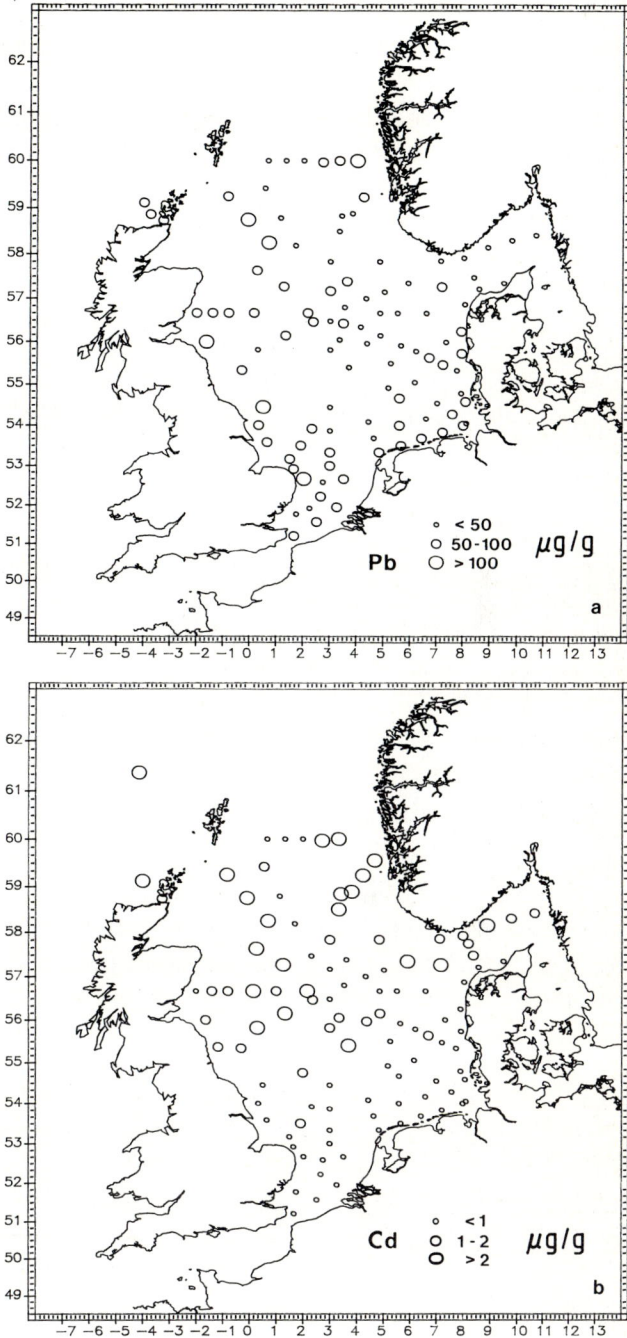

Fig. 2.3.12. Concentrations (µg/g) distribution maps for particulate Cd *(a)* and Pb *(b)* in the North Sea based on 115 SPM samples from surficial waters(10 m depths) collected in winter 1987

reaching minimum values of 0.12 µg/g. A significantly positive correlation was found between Cd and POC, while a negative was found between Cd and Al (Kersten et al. 1990).
- Pb (Fig. 2.3.12b). The Pb content was lowest in the central North Sea and in the Skagerrak (< 50 µg/g). In coastal areas around the North Sea, especially off Britain, it was generally higher (> 50 µg/g, reaching nearly 200 µg/g). A significantly positive correlation was found between Pb and Mn.

Before going on to discuss briefly the distributions of these two heavy metals, it seems necessary to obtain some background information on their SPM carrier. To decide whether the elements were associated with the inorganic or organic fraction, they were correlated with Al and POC, respectively. Moreover, this distinction can also be obtained from the relation between the element contents and total SPM concentrations in the water column.

When plotting the particulate metal concentrations C_{PM} (on a mass basis) versus the SPM concentrations C_{SPM}, distinct curves are generally observed which fit a hyperbolic relationship

$$C_{PM} = A + B \cdot C_{SPM}^{-1} \tag{1}$$

where A and B are hyperbolic regression coefficients. The hyperbolic relationship reflects the fact that the higher the SPM concentration in natural water is, the more the particulate metal concentration will tend to that of resuspended bottom sediment ($C_{PM} = A$). A is a measure of the bottom sediment composition. This is supported by the element/Al ratios in the SPM, which are very similar to that of bottom sediment at high SPM concentrations (typically in the order of several mg/l). The resuspended fraction has higher contents of Al and Pb. The hyperbolic relationship for these elements is thus inverse with a negative correlation of these elements with TOC. On the other hand, at low resuspended sediment values, the SPM composition is predominated by a "permanently suspended fraction" (Duinker et al. 1983). The latter fraction has higher concentrations of TOC and Cd and tends to be predominated by organic matter in composition. The strong positive correlation between TOC and this element might be used to suggest their association with the organic carbon component of SPM. The regression coefficient B may thus be understood as a measure of the scavenging efficiency of the organic matter, which is most important at low dilution by resuspended mineral particles. This non-linear regression model can be used taking the above mentioned sediment background concentrations

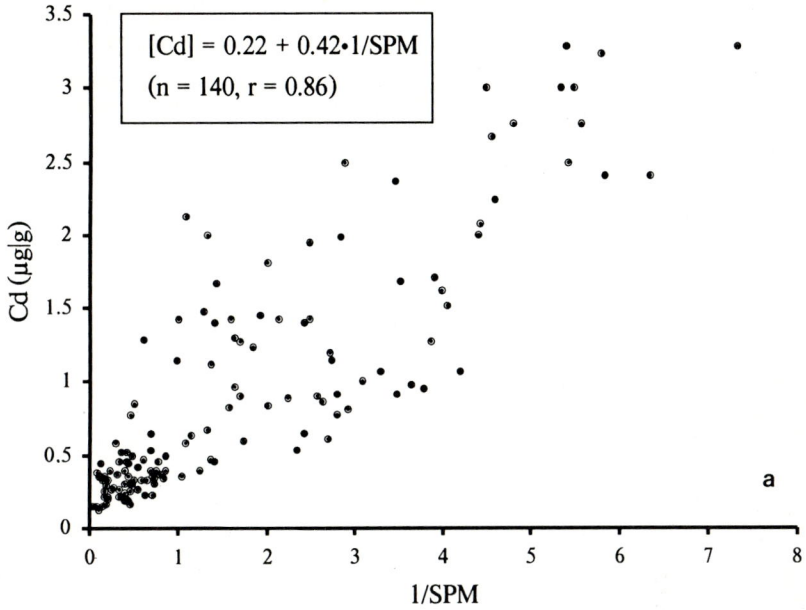

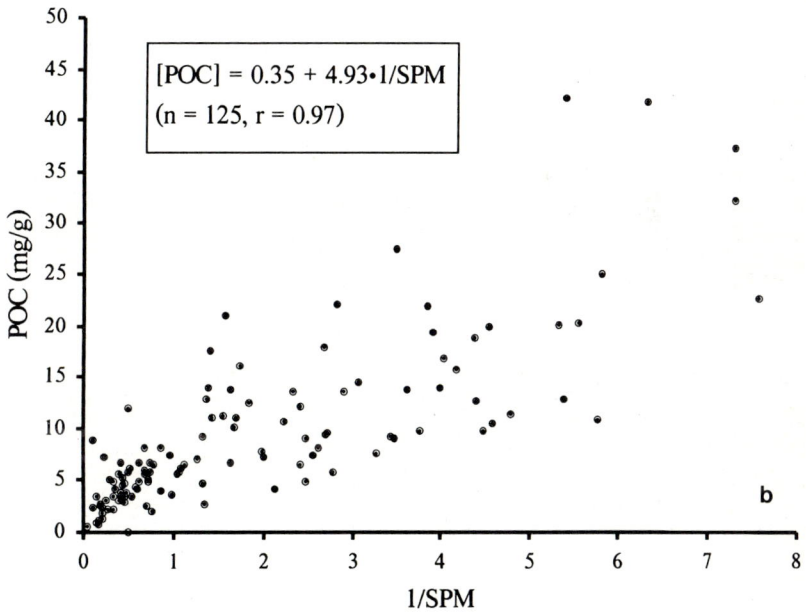

Fig. 2.3.13. Relationship of Cd *(a)* and POC *(b)*(mass/mass basis: µg/g and mg/g, respectively) vs. the inverse suspended particulate matter (SPM) concentration (mass/volume basis: mg/l) in the North Sea. Data from Kersten et al. (1990)

for the regression constant A in order to assess the degree of anthropogenic contamination of both Cd and Pb, but also other trace metals.

This threshold value was 53 µg/g for Pb and 0.22 µg/g for Cd. It is now possible to make allowances for varying SPM concentrations of the samples. Readings which are considerably above the metal vs. SPM concentration curve can be termed anomalous. Anthropogenic impact by an increase in contaminated SPM input will be readily detected in areas with high SPM loading for both elements. Pb concentrations higher than the threshold value are also indicative of anomalous element enrichment at a low SPM loading. For Cd, however, it seems to be difficult to differentiate natural and anthropogenic enrichments simply from its total particulate element concentration at a low SPM loading.

The brief considerations above give a sufficient background for interpretation of the distributions of particulate Cd and Pb concentrations in the surface North Sea waters as shown in Fig. 2.3.12. In spite of the low density of sampling sites, certain areas of increased trace element contents in the SPM samples indicate that the distributions are not even and that mixing of suspended matter in the North Sea is far from being complete. A clear division is obvious in the composition of suspended matter between a northern and a southern part including the coastal areas. The reason for that pattern must be seen mainly in the different SPM concentrations found in these areas. SPM concentrations vary considerably between North Sea areas, showing a drastic change from the shallow (15-30 m) southern and eastern bights to the deeper (70-100 m) central North Sea.

In the shallow southern areas, wind and tidal current-induced stresses on the floor resuspend bottom sediments into the water column (Eisma and Kalf 1987). This resuspended matter significantly determines the composition of SPM. In the deeper central and northern North Sea, resuspension of bottom sediments affects the composition of the suspended matter much less in the surface water. Here, organic matter derived from biological productivity is the major component of SPM (Eisma and Kalf 1987). The Skagerrak, which receives suspended matter from the Baltic Sea in its surface waters, takes a separate position. The different distribution patterns of Cd and Pb can thus be understood from their different affinities for scavenging substrates as discussed above.

Though atmospheric inputs are suspected to be the major source of Cd in the North Sea, recent studies indicated that Cd from contaminated aerosols is leached nearly 100 % upon contact with sea water (cf. Chap. 4.6). This process of rapid leaching of Cd from particles makes the most likely pathway from waste combustion as a potential contamination source (Kersten et al. 1988). Hence, enhanced particulate Cd concentrations in those areas may be the result of low particulate loads, biological scavenging processes, and remineralization of biogenic detritus at the sediment-water interface rather than simply being input of contaminated particles.

Enhanced particulate Pb concentrations were found along coastal areas, but especially in the western parts of the North Sea. The large inputs delivered by the estuaries seem not to reach very far beyond the coastal zone. Atmospheric input is suspected to be the major source of anthropogenic Pb in the North Sea as well. From atmospheric deposition modelling (cf. Chap. 3.1), a significant decrease of total Pb deposition is found from land to sea, which is still in good agreement with the air masses loaded with metals over Britain striking mostly the western North Sea (cf. Chap. 3.1). Although about half of the Pb load is leached from aerosols upon contact with sea water (cf. Chap.4.6), in the shallower southern part of the open North Sea, this Pb fraction is scavenged by the higher particulate load which increases the removal of anthropogenic Pb by relatively short particle residence times in the water column.

2.3.3 Cyclic Organochlorines and Heavy Metals in North Sea Sediment

A considerable amount of research during the last 20 years has proven that organochlorine compounds as well as heavy metals have only a short residence time in the water column. The seabed is the ultimate sink for virtually every class of pollution in the marine environment.

The results from the surveys show that sediments from the Kattegat/Skagerrak/Norwegian Trench and the inner German Bight are highly polluted with PCBs (Fig. 2.3.14). The Kattegat/Skagerrak is the largest sedimentation area for suspended matter in the North Sea. In the last 10 000 years a 10-20 m large sediment layer has been formed which is highly contaminated in its upper centimeters.

The contamination of North Sea sediments with PCB (as the sum of concentrations of 24 individual congeneres), HCB, Lindane, α-HCH and heavy metals (i.e. Cd and Pb) was determined in samples taken on the summer cruise 1986.

Many contaminants enter the sea from very localized (point) sources such as a river or a discharge point of industrial waste. Our results suggest that the "mud area" southeast of Helgoland is a point source for PCBs (Fig. 2.3.14). The patterns of individual PCB congeners in the sediments differ between depositional areas and offshore stations: samples from the Kattegat/Skagerrak/Norwegian Trench as well as the inner German Bight

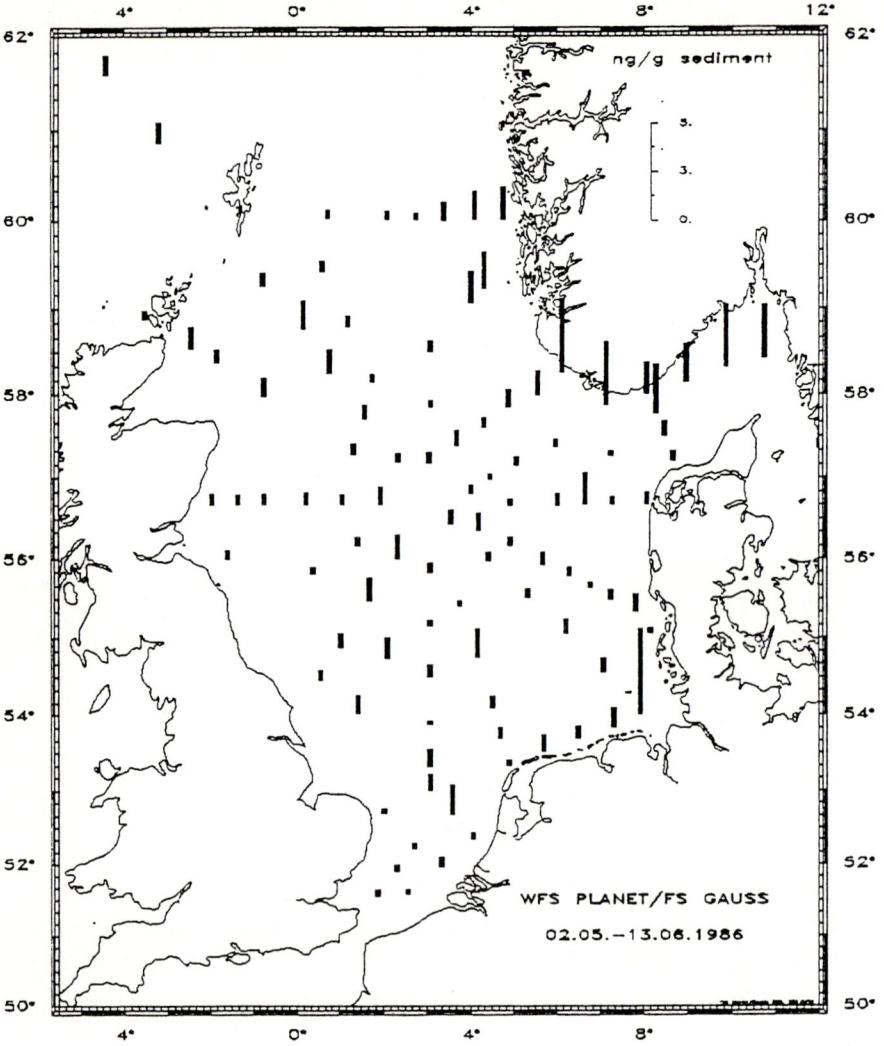

Fig. 2.3.14. Concentrations of Σ-PCB (as the sum of 24 components) in sediments throughout the North Sea. Concentration ng PCBs/g sediment, summer 1986

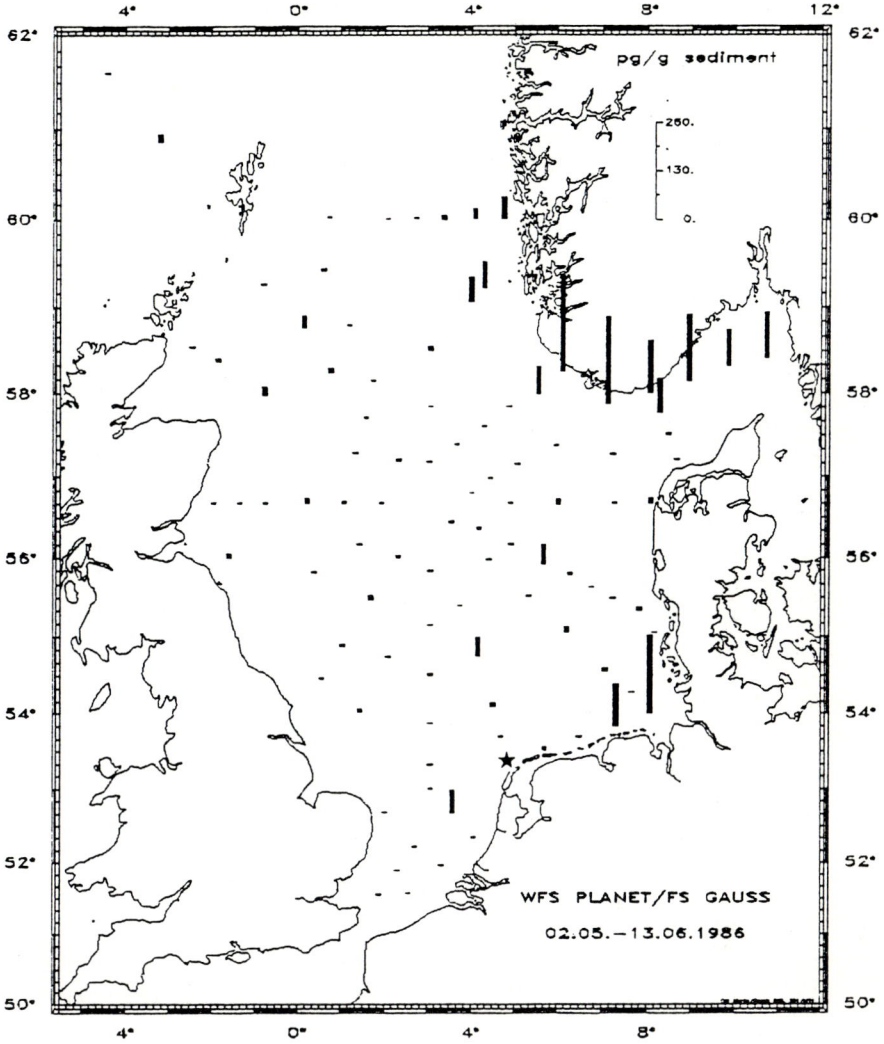

Fig. 2.3.15. Concentrations of HCB in sediments throughout the North Sea. Concentration pg HCB/g sediment, summer 1986

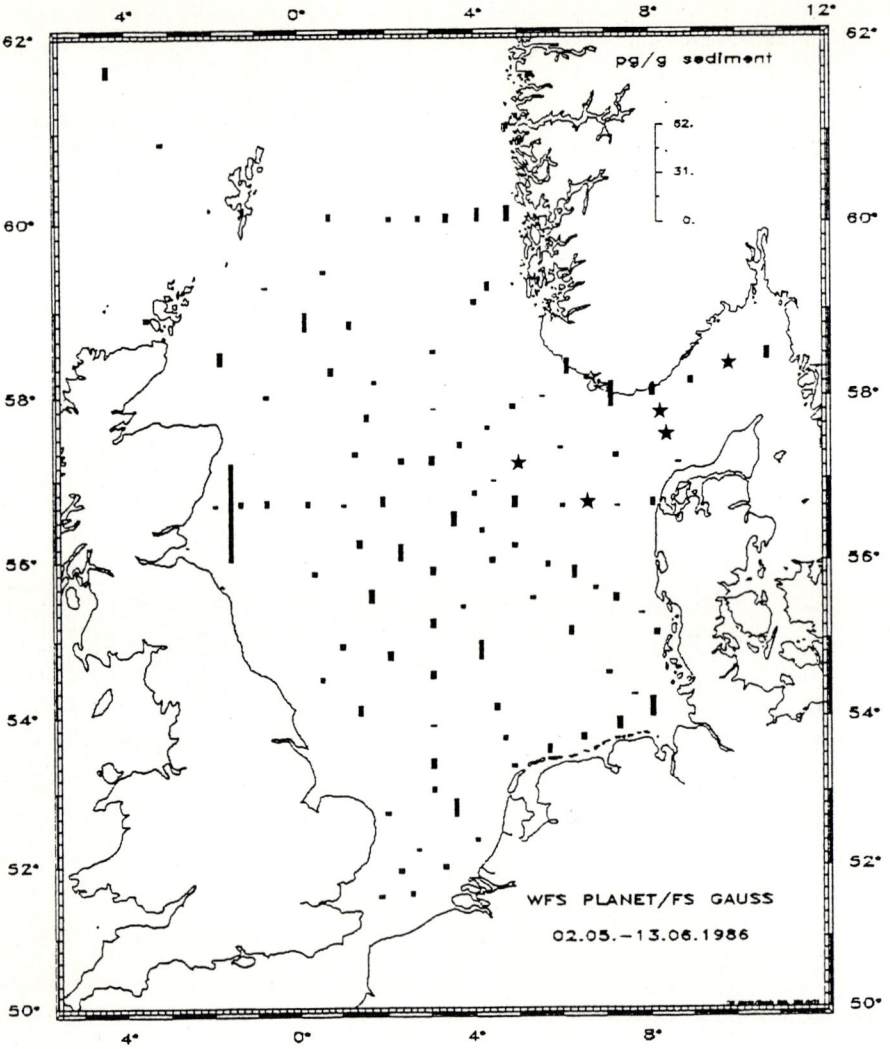

Fig. 2.3.16. Concentrations of Lindane in sediments throughout the North Sea. Concentration pg Lindane/g sediment, summer 1986

show higher amounts of highly chlorinated congeners than samples from the central North Sea (Knickmeyer and Steinhart 1988a; Landgraff et al. 1992).

Lindane does not show a distinct distribution pattern in the sediments investigated with the exception of a high amount of Lindane at station 31 near the Scottish coast (Fig. 2.3.16).

The amounts of HCB found in sediments at stations 71 and 72 (station net : Fig. 2.2.1, Chap. 2.2) in the inner German Bight correspond well to known HCB inputs from the dumping of harbour sludge and from the rivers Elbe and Ems (Duinker et al. 1983, 1984; Eder et al. 1987; Knickmeyer and Steinhart 1988a). Samples extremely contaminated with HCB were also obtained at nearshore stations next to the southern Norwegian coast (Fig. 2.3.15). In addition to HCB, also octachlorostyrene (OCS) and decachlorobiphenyl (DCB) could be detected in all of these samples. In Frierfjorden and Kristiansandsfjorden in southern Norway, serious local contamination with fully chlorinated aromatic compounds of industrial origin has been detected in different marine compartments (Lunde and Baumann Ofstad 1976; Bjerk and Brevik 1980; Underdal et al. 1981; Knutzen et al. 1984). These cyclic organochlorines are formed during production of magnesium and nickel through the reaction between chlorine and Sönderberg (graphite) electrodes.

Knutzen (1986) reported that seven Norwegian smelters between Stavanger and Oslo produce light metals, silicium, silicium carbide, ferrosilicium and ferromanganese. These smelters are probably sources for HCB contamination in the Skaggerrak/Norwegian Trench.

2.3.3.1 Heavy Metals in North Sea Sediment

The question arose, whether the SPM results are comparable with the contaminant distribution in North Sea sediments. Clearly, total contaminant concentrations in sediments of such a variety as found in the North Sea are by no means useful for comparison purposes. Physical grain size fractionation and subsequent chemical analyses can be used for both spatial and temporal trend identification and accurate determination of trace element enrichment factors in sediments (Förstner et al. 1982). One procedure entails normalization of bulk chemical data to the percent of a selected size range, and the other requires an actual physical separation of a selected size range prior to chemical analysis, typically < 63 or 20 µm. If these techniques are not used, intersample comparisons can be improper because intersample grain-size differences can superimpose metal concentration gradients which do not reflect actual conditions.

A quantitative measure of sediment pollution of the entire North Sea was recently achieved using chemical analyses of the < 20 µm fraction (Kersten and Klatt 1988). These data are used to calculate the "Geo-Index" as a sediment quality criterion (Müller 1979):

$$I_{geo} = \log_2 \left(\frac{C}{1.5 B} \right) \tag{2}$$

where C is the measured concentration of an element in the < 20 µm fraction and B is the geochemical background value in fossil argillaceous sediment ("average schale") or a regional sediment reference (e.g. the SPM background values discussed above). The factor 1.5 is used because of possible variations of the background levels due to lithogenic effects. The results rounded to the next integer from 0 to 6 give the seven grades of the I_{geo}-classification, whereby the highest grade (=6) reflects already a 100-fold enrichment above background levels (2^6 x 1.5 = 64 x 1.5 = nearly 100).

This classification is exemplified in Fig. 2.3.17 for the distribution of Cd and Pb contents in the sediment samples from the entire North Sea. Unlike the Cd distribution in SPM, no distinction in concentrations can be found between the southern and northern part of the North Sea. The I_{geo}-map of Cd looks rather patchy with some "hot" spots between the Thames and Rhine mouths, and the eastern North Sea.

An I_{geo}-class of 2-3 found for Pb in some areas represents a moderate to strong pollution. While the Pb contamination "hot" spots identified in the southwest- and southeast coastal zones agree well with that found in the SPM pattern, the significant sediment contamination identified in the central North Sea is not reflected by the SPM pattern.

There is neither an explanation for this disagreement nor a plausible reason for Pb contamination of the central North Sea sediments. Long term contamination processes seem to be responsible for that pattern rather than any short term transport, which should be mirrored in the SPM pattern. However, Kersten and Kröncke (1990) have evidenced that there is a significant positive correlation between Pb and total organic carbon (TOC) contents in North Sea sediments. This relationship is not sufficiently considered by the I_{geo} or Al normalization concept.

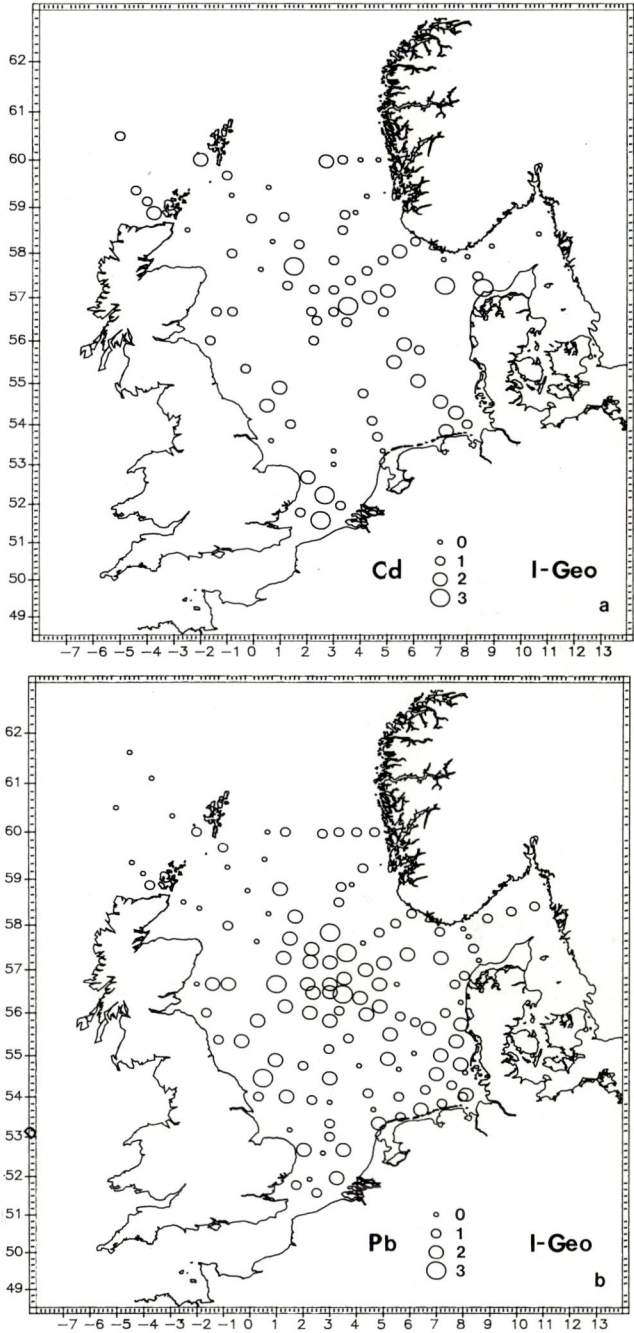

Fig. 2.3.17. I_{geo}-classification of the Cd *(a)* and Pb *(b)* concentration distribution maps for North Sea sediment (<20 µm fraction) as calculated from Eq. (2) and the SPM background values

2.3.4 Cyclic Organochlorines and Heavy Metals in Benthic Organisms and Bottom Dwelling Fish

There is a lot of information on heavy metals and cyclic organochlorine contamination of benthic organisms and bottom dwelling fish of the North Sea. However, most studies are related to estuarine and coastal waters or focussed on restricted areas such as the Dogger Bank (Claussen 1988; Kröncke 1988). Fewer studies have been carried out covering the total North Sea. To fill this gap, special attention has to be drawn to euryeceous species widely distributed within the North Sea. For this purpose, there is evidence that the dab (*Limanda limanda*) may be an adequate species of fish, the hermit crab (*Pagurus bernhardus*) a useful species of Crustacea. Both can be sampled in reasonable numbers in most areas of the North Sea from the coastal zones to deep waters down to about 150 m (Frauenheim et al. 1989).

The contents of several organochlorine compounds and heavy metals were determined in dab livers and the soft abdominal part of hermit crabs taken from the North Sea during two seasons. The data obtained show that regions with similar contaminant levels can be distinguished (cf. Chap. 2.3.5). Regions with high contamination were often those directly influenced by the known riverine and other land-based sources. However, high concentrations of Hg and Cd were also measured in extended areas north of the Dogger Bank and in the northernmost and northwestern North Sea.

2.3.4.1 Cyclic Organochlorines

Fig. 2.3.19 shows that the Rhine is an important source of PCB pollution in the southern North Sea. Following the general routes of water and suspended particulate matter, the concentration of PCB in hermit crabs decreases with increasing distance from the coast. Besides the Rhine, the Ems-Dollart and some extent the Weser and Elbe are PCB sources to the German Bight.

The Thames estuary, which is separated from the Rhine-Meuse estuary by the Atlantic residual currents through the Straits of Dover, gives rise to further PCB contamination of the southern Bight. Samples from the British east coast did not differ from the relatively less polluted central North Sea. The results reported here correspond well with the input pattern of rivers in the North Sea, which were calculated by model simulation studies (Müller-Navarra and Mittelstaedt 1985).

Several offshore stations showed higher concentrations of PCB in the hermit crab in summer 1986 than in winter 1987 (Figs. 2.3.18 and 2.3.19). Higher concentrations of PCB in the crab in summer correlate well with higher concentrations of ammonia in the bottom waters, which indicates the remineralization of organic nitrogen (cf. Chap. 2.2). The explanation may be that during the summer cruise the primary plankton bloom in most parts of the North Sea had already collapsed and was followed by the cyclic settlement of high amounts of particulate carbon and nitrogen to the sediments. The mass sedimentation of cells and fecal pellets causes a vertical transport of low chlorinated biphenyls associated with the organic matter to the sediments (Knickmeyer and Steinhart 1989a).

The seasonal changes in the pattern of the 24 individual PCB congeners are influenced by the uptake of food from the spring plankton bloom in summer and organic material from sediment and suspended matter in winter. A comparison of PCB patterns in abdomen and eggs of the hermit crab indicates that in summer pollution of lower chlorinated biphenyls is transferred from the food to the eggs of the hermit crab. In winter, stored lipids and pollutants are transferred from the hepatopancreas to the gonads (Knickmeyer and Steinhart 1988b). In accordance with these findings it could be shown that HCB and PCB transfer from female dab livers to ovaries causes significant alterations in organochlorine contamination and PCB patterns in dab (Kammann et al. 1990; Knickmeyer and Steinhart 1990).

The amounts of HCB in the hermit crab are higher in the central and northern parts of the North Sea than in the southern parts (Figs. 2.3.20 and 2.3.21). The highest concentration of HCB in whelks (*Buccinum undatum*) was found in the same area (Knickmeyer and Steinhart 1989b). The high amounts of HCB in the Danish coastal sea may be caused by the river Elbe, which is highly polluted, especially in spring (Gaul and Ziehbarth 1983; Köhler et al. 1986). HCB possesses the highest vapour pressure of all organic substances under investigation. The results suggest therefore that atmospheric transport is a further HCB input path for the North Sea. The amounts of Lindane in the hermit crab are presented in Figs. 2.3.22 and 2.3.23.

2.3.4.2 Heavy Metals

Different patterns of large scale distribution have been observed for Hg, Cd, Pb, Ag, Zn, Cu, Fe and Mn in dabs and hermit crabs sampled from various locations in the southern central and northern North Sea (Karbe et al. 1988, 1989). The results for Hg and Cd in hermit crabs are shown in

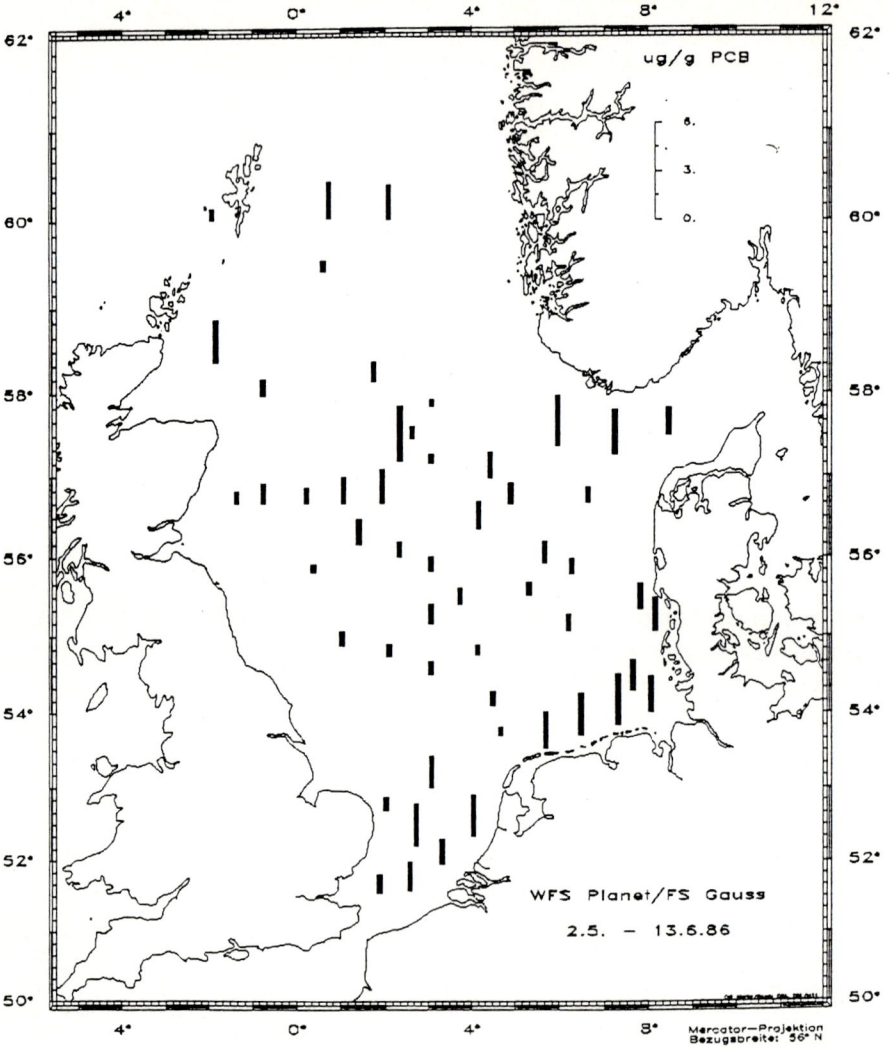

Fig. 2.3.18. Concentrations of Σ-PCB (as the sum of 24 components) of *Pagurus* ssp. in early summer 1986 throughout the North Sea. Concentrations in mg PCB/g n-hexane extractable lipid

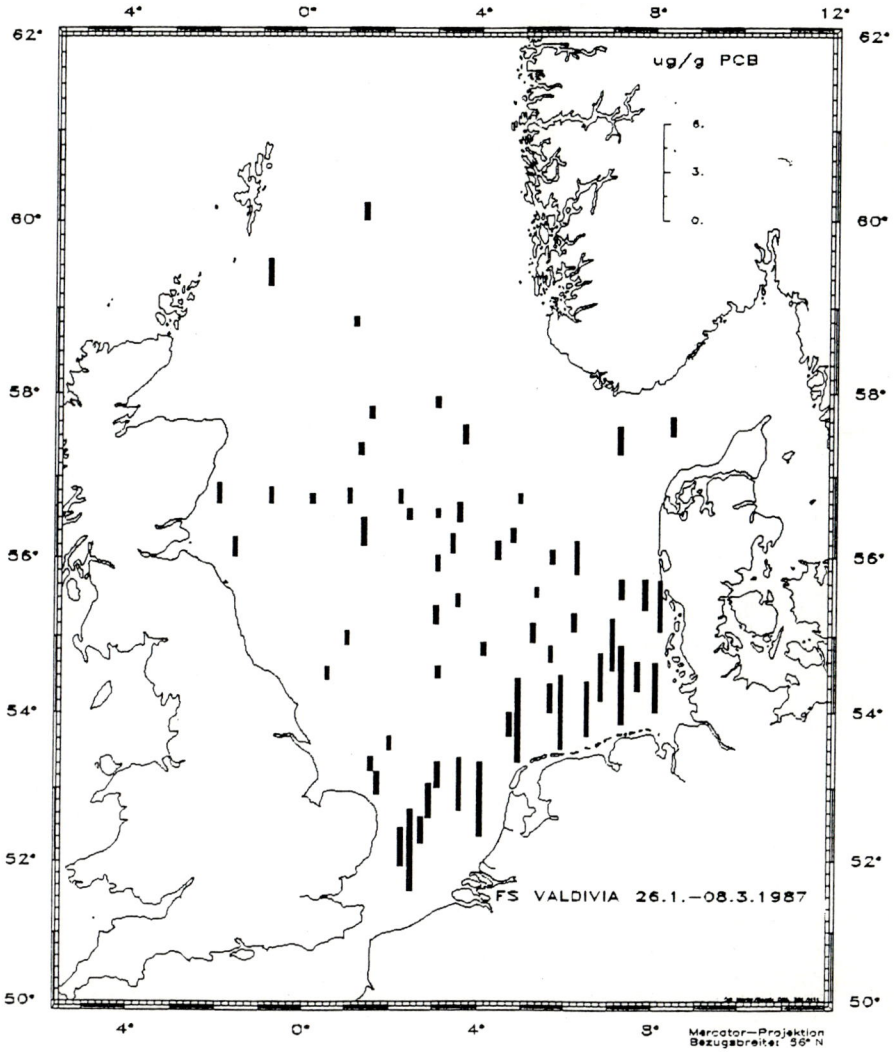

Fig. 2.3.19. Concentrations of Σ-PCB (as the sum of 24 components) of *Pagurus* ssp. in winter 1987 throughout the North Sea. Concentrations in mg PCB/g n-hexane extractable lipid

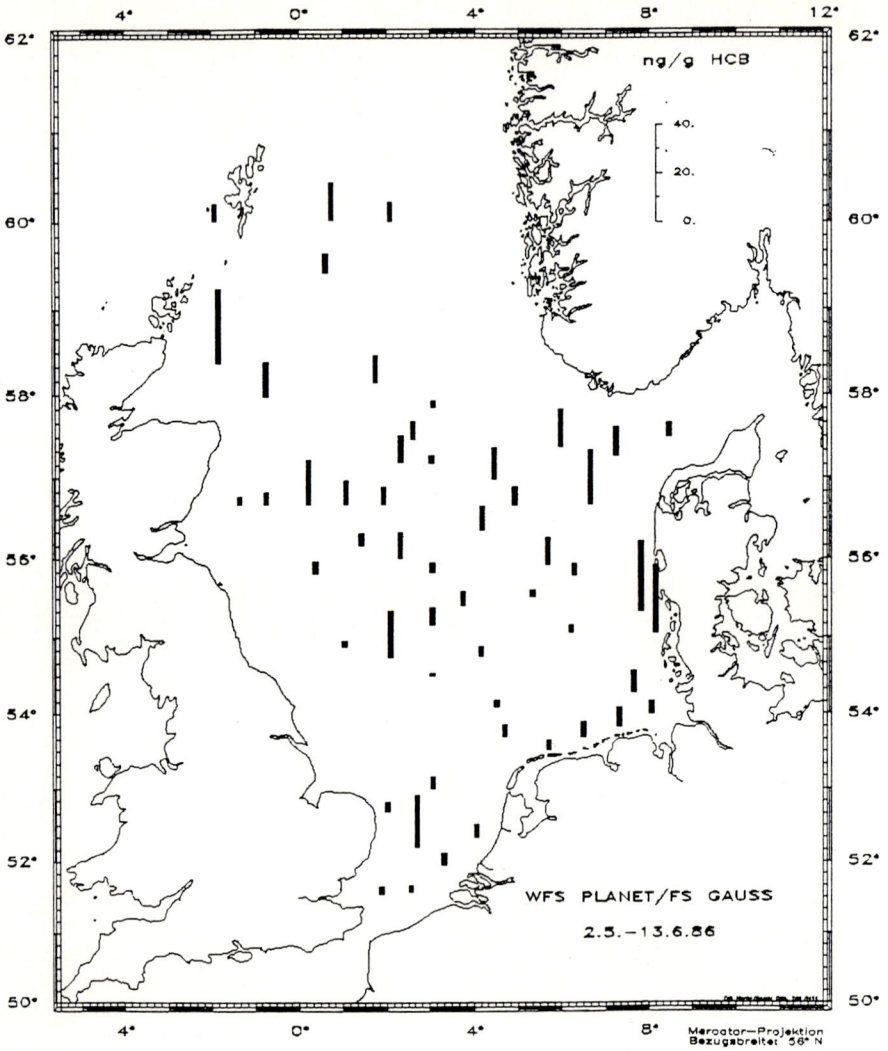

Fig. 2.3.20. Concentrations of HCB in *Pagurus* ssp. in early summer 1986 throughout the North Sea. Concentrations in ng HCB/g n-hexane extractable lipid

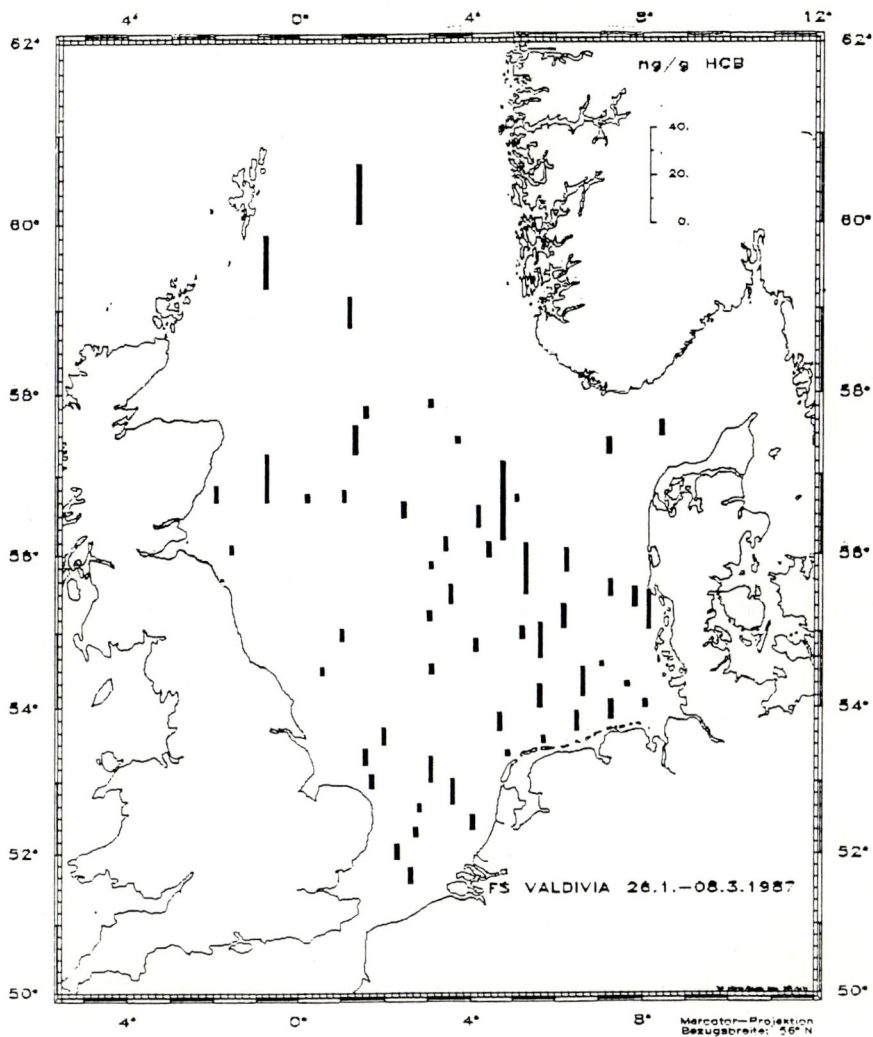

Fig. 2.3.21. Concentrations of HCB in *Pagurus* ssp. in winter 1987 throughout the North Sea. Concentrations in ng HCB/g n-hexane extractable lipid

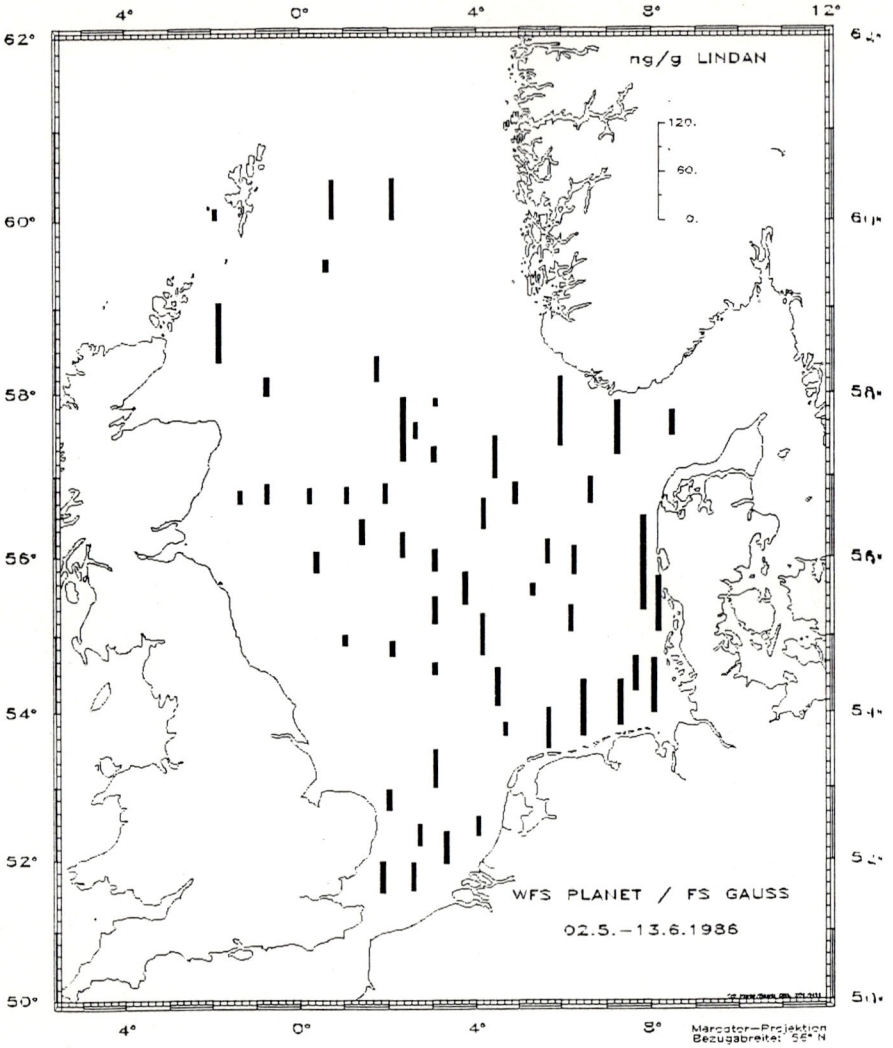

Fig. 2.3.22. Concentrations of Lindane in *Pagurus* ssp. in early summer 1986 throughout the North Sea. Concentrations in ng Lindane/g n-hexane extractable lipid

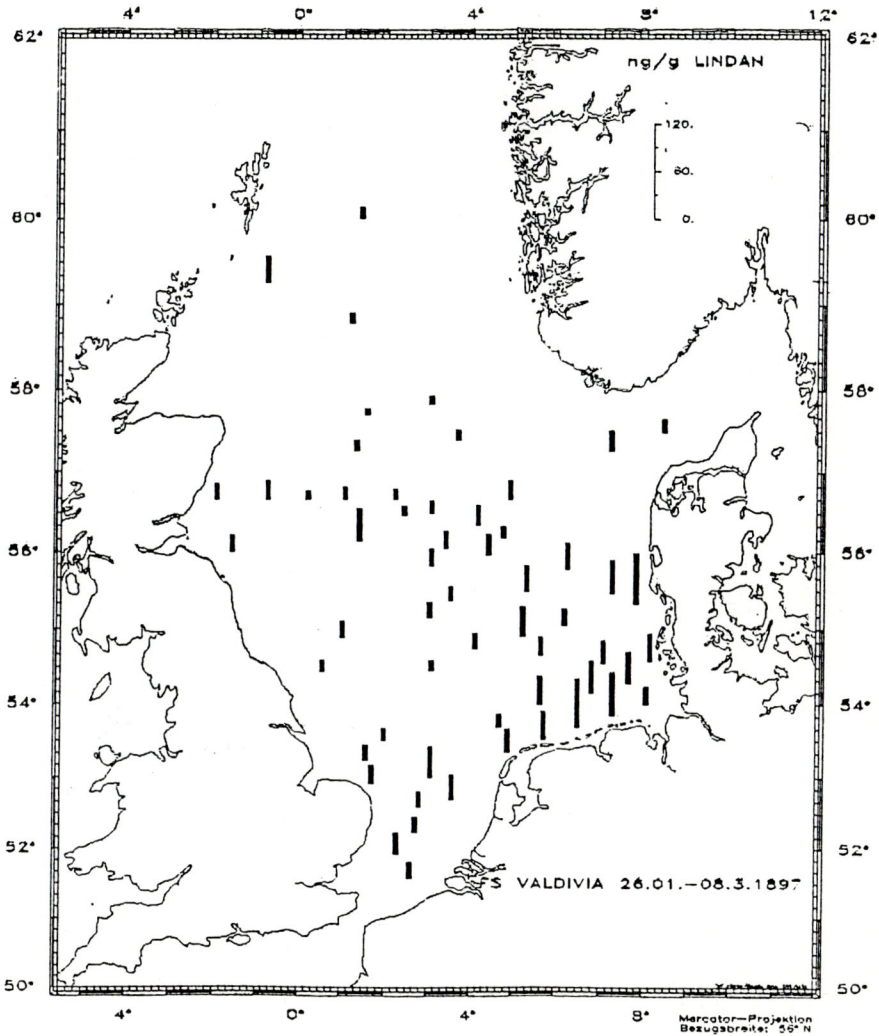

Fig. 2.3.23. Concentrations of Lindane in *Pagurus* ssp. in winter 1987 throughout the North Sea. Concentrations in ng Lindane/g n-hexane extractable lipid

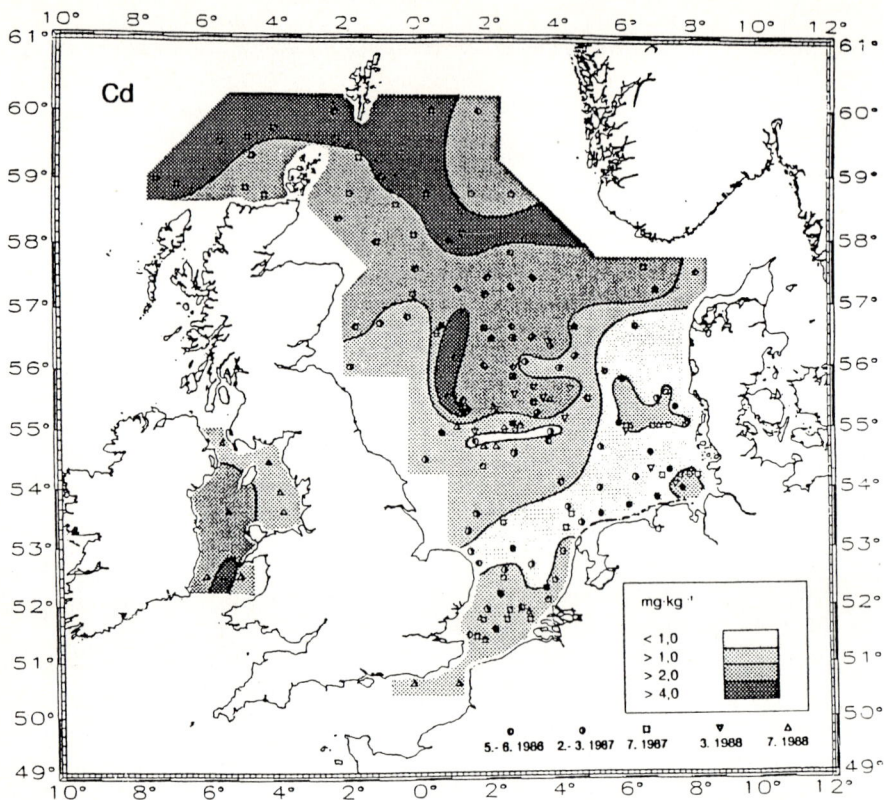

Fig. 2.3.24. Mercury: isolines for concentrations in the soft abdominal parts of hermit crabs (mg/kg dry weight). (After Karbe et al. 1989, supplemented by data from a cruise in late summer 1990)

Figs. 2.3.24 and 2.3.25. Because of little seasonal and year-to-year differences, the data of all samplings (1986 to 1990) are presented in one figure. Cd and Hg were selected, since these metals revealed the clearest distribution patterns.

The results for both metals show that the North Sea can be divided into a northern and a southern region. In the southern region, the highest concentrations of both Hg and Cd were measured within the areas directly influenced by the discharges from the Rhine-Meuse and Weser-Elbe estuaries. However, there are significant differences between these two metals in the offshore areas of the southern North Sea.

For a south to north transect crossing the Dogger Bank concentrations of Hg, Cd and Lindane in the hermit crab are given in Fig 2.3.26. Variations of temperature with respect to water depth are presented for this transect in Chapter 2.2. (Fig. 2.2.8). The concentration of Hg in the crab decreased

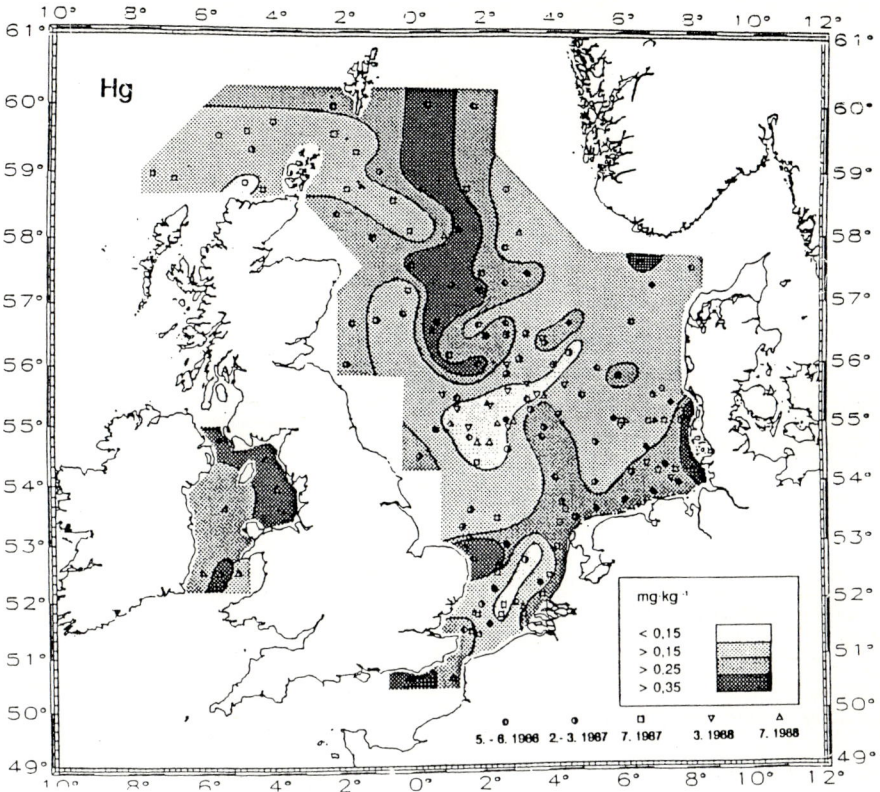

Fig. 2.3.25. Cadmium: isolines for concentrations in the soft abdominal parts of hermit crabs (mg/kg dry weight). (After Karbe et al.1989 supplemented by data from a cruise in late summer 1990)

with distance from the coast to the more offshore locations and revealed minimum values at the Dogger Bank. Similar observations could be made for Lindane in the crab. However, Lindane tended to decrease constantly towards the open sea, while Hg concentrations increased again towards the northern North Sea. In contrast, concentrations of Cd were relatively low within the near-shore German Bight and increased toward the more offshore areas of the southern North Sea and the Dogger Bank (Fig. 2.3.26). There are relations to changes in salinity and depth as well as to the type of sediment and faunal assemblages (cf. Chap. 2.2).

Similar results have been found by Claussen (1988) and Kröncke (1988) comparing Cd concentrations in livers of dab and several benthic organisms (*Nephthys* spp., *Echinocardium cordatum, Venus striatula*) sampled at the

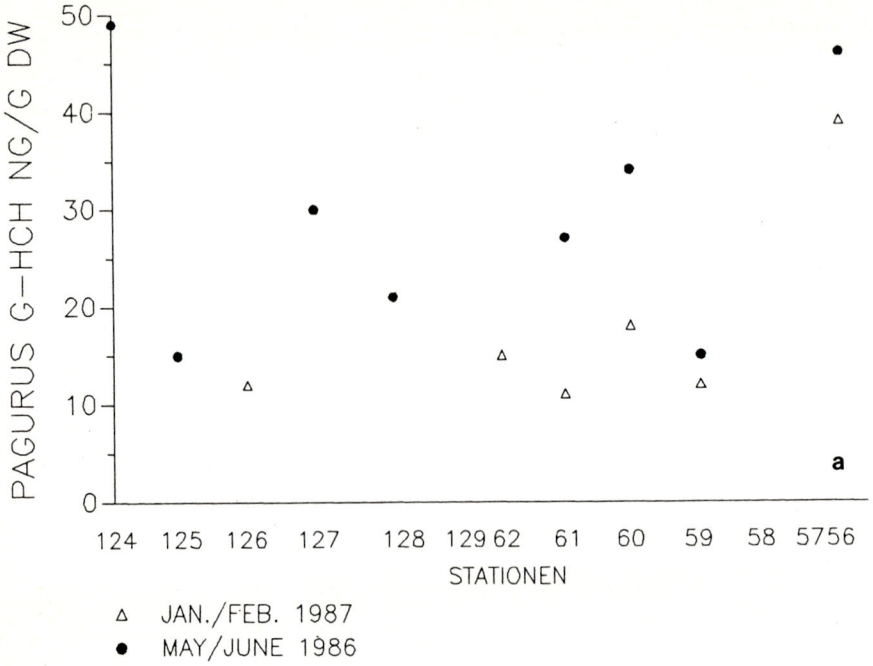

a

PAGURUS G–HCH NG/G DW

STATIONEN

△ JAN./FEB. 1987
● MAY/JUNE 1986

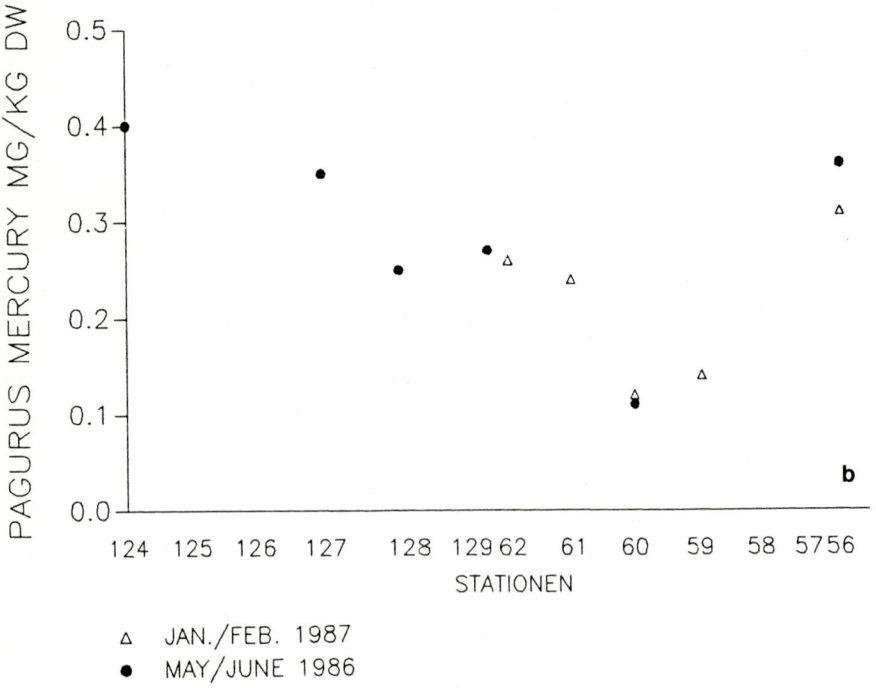

b

PAGURUS MERCURY MG/KG DW

STATIONEN

△ JAN./FEB. 1987
● MAY/JUNE 1986

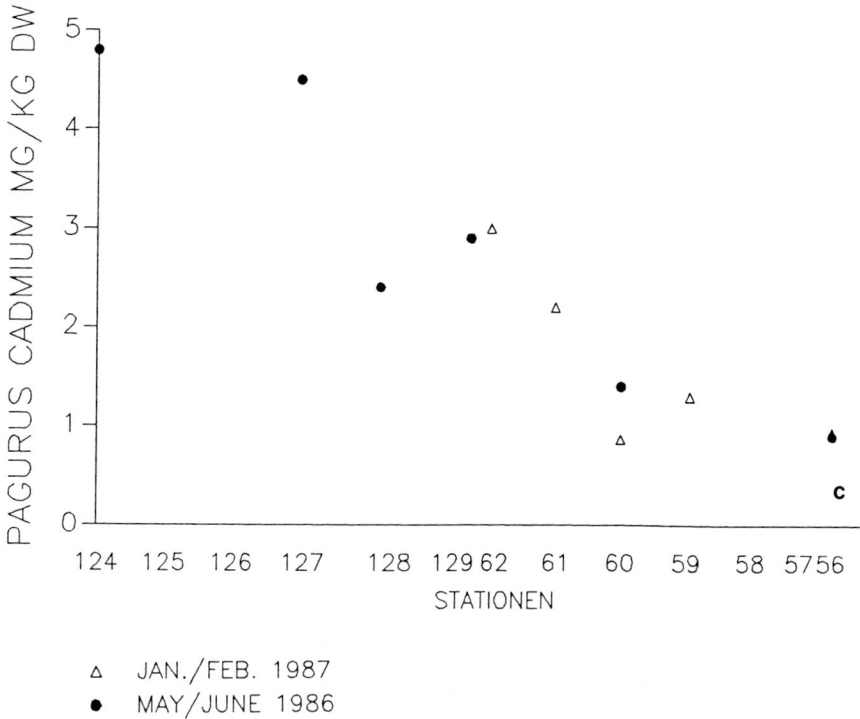

Fig. 2.3.26. Mercury *(a)*, cadmium *(b)* and Lindane *(c)* in hermit crabs: contaminant concentrations of locations crabs are sampled at a south to north transect from the continental coast crossing the Dogger Bank to northern most regions of the study area, summer/winter 1986/87

Dogger Bank with those from the German Bight, and by Borchardt et al. (1988) who compared the regional trends for Hg and Cd in offshore mussels (*Mytilus edulis*) within the German Bight.

In the northern North Sea, the most conspicuous results are unexpectedly high concentrations of both Cd and Hg within the deeper northern waters. Hg as well as Cd concentrations are at the highest level within an extended area originating from the North and reaching over the central North Sea southward to the northern slope of the Dogger Bank. Areas of highly contaminated hermit crabs correspond to the depth contours and sediment type and metal contamination determined by Basford and Eleftheriou (1988) for the sediment as a total (no single grain size fraction).

2.3.5 Correlations Between Contaminants and Additional Parameters in the Different Marine Compartments

The contents and spatial distribution of the organochlorines PCB, HCB, Lindane and α-HCH in the hermit crab, water and sediment samples obtained during the two surveys have been evaluated. The ICES boxes of the North Sea (Fig. 2.1.28 in Chap. 2.1), which divide the North Sea into areas of comparable hydrographic conditions, have been partitioned into a north-group (box 1, 2, 3.1 and 7.1) and a south-group (box 3.2, 4, 5 and 7.2).

Table 2.3.1. Spearman Rank Correlations of organochlorines between the different compartments in northern and southern North Sea. "+"/"−" positive/negative correlation with level of significance < 0.05. "++"/"−−" : positive/negative correlation with level of significance < 0.005. P hermit crab; W water; S sediment

SOUTH ICES boxes 3.2/4/5/7.2

W. depth	W. g/a HCH	W. Lindane	P. g/a HCH	P. PCB	P. HCB	P. Lindane	P. lipid	P. wetweight	S. g/a HCH	S. PCB	S. HCB	S. Lindane	S. <20 u	S. TOC	S. water cont.	
			−					−	−	+	+	+	++	++		S. water cont.
								−		+	+	+	+		+	S. TOC
								−	++	++	+			+	+	S. <20 u
										+	++					S. Lindane
									+	+						S. HCB
									+			−				S. PCB
−	++		++					+						+		S. g/a HCH
																P. wetweight
−	++	+	+	+												P. lipids
				++	+		−−									P. Lindane
				+												P. HCB
−		+				+		−						+		P. PCB
−	++	++						−				+				P. g/a HCH
−								−								W. Lindane
−																W. g/a HCH
	−	−														W. depth

NORTH ICES boxes 1/2/3.1/7.1

Both regions are separated by means of a line originating from the Flamborough Head crossing the Dogger Bank to the northeast. This line corresponds to a frontal system dividing the northern North Sea, influenced by North Atlantic inflow, and the southern North Sea, dominated by continental waters and Atlantic inflow via the English Channel (Backhaus and Hainbucher 1987). The separation of the North Sea into a northern and a southern region is also shown for the distribution of epibenthic fauna (Dyer et al. 1983, Frauenheim et al. 1989) and for macrobenthic endofauna (Künitzer et al. 1992).

The interrelationships between contaminants in the hermit crab and other compartments, sediment character, water depth, salinity and the body condi-

Table 2.3.2. Spearman Rank Correlations of heavy metals between the different compartments in northern and southern North Sea. "+"/"-" positive/negative correlation with level of significance < 0.01. "++"/"--" : positive/negative correlation with level of significance < 0.05. P hermit crab); W water; S sediment; C_I body condition index of the hermit crab

	P. Hg	P. Cd	P. Ag	P. Zn	P. Cu	P. Pb	P. Fe	P. Mn	P. cond. Index	W. depth	W. salinity	W. Hg	W. Cd	SPM Cd	S. Pb	S. water cont.	S. TOC	S. < 20 u
P. Hg	\	++								+								
P. Cd		\					++			++	++			+				
P. Ag	+	-	\	++							--		--					
P. Zn				\														
P. Cu	+				\													
P. Pb						\												
P. Fe					-	+	\	++								++		
P. Mn							++	\	++									
P. cond. Index				--	++			+	\									
W. depth		+								\	++				+			
W. salinity	-	++	-								\							
W. Hg							-		+			\	-					
W. Cd		--							--	--			\	--				
SPM Cd	+	-					+		-			+		\				
S. Pb	-								-	-				+	\	--	--	--
S. water cont.							+									\	++	+
S. TOC							+									++	\	+
S. < 20 u	+		+					+								++	++	\

NORTH ICES boxes 1/2/3.1/7.1

SOUTH ICES boxes 3.2/4/5/7.2

tion index (C_I) of the hermit crab (weight of abdomen/carapax length) were calculated for rank correlations according to Spearman, handling the data from the northern and the southern subregions separately (Tables 2.3.1 and 2.3.2).

In the southern part of the North Sea, more significant correlations could be observed than in the northern part. The lipid content of the crab decreased with increasing water depth in the southern North Sea (Table 2.3.1). Furthermore, negative correlations were also found between water depth and PCB contamination of the hermit crab and the γ/α-HCH ratio in the animal, water and sediment. In accordance to this, Fig. 2.3.19 shows the PCB gradient in the southern North Sea with elevated values near the German-Danish coast.

In the southern North Sea, a positive correlation between the contamination levels of Lindane, PCB and HCB in the hermit crab could be observed. Therefore similar near-shore sources could be assumed for these contaminants during the summer cruise in the southern North Sea. The only positive correlation between contaminants for the data from the northern North Sea was found between Lindane and PCB contamination of the crab.

A clear division between the northern and southern part is also obvious in the correlation between sediment parameters and cyclic organochlorines in sediment. In the southern North Sea these data are all connected which each other and show positive correlations (cf. Chap 4.5). In the northern North Sea, however, the sediment parameters are correlated which each other but, in contrast to the southern part, show no correlation with cyclic organochlorines except with HCB. The area of highest HCB contamination in the northern North Sea is the Norwegian Trench (Fig. 2.3.15). Samples from this area are extremely rich in fine particles but not in organic matter, which explains the correlation of HCB and sediment parameters.

The γ/α-HCH ratio in hermit crabs and sediment samples became smaller from south to northwest (negative correlation with water depth). The same phenomenon was found in water and zooplankton samples collected during the same cruises (Knickmeyer and Steinhart 1989a). It is possible that Lindane is degraded or isomerized to α-HCH by microorganisms or photochemically (Malaiyandi and Shah 1984; Macholz and Kujawa 1985) and that these reactions are more distinct in the northwestern part of the North Sea.

In the southern part of the North Sea, the concentrations of some metals decreased with increasing distance from the coast; this is true for Pb in sediment, Cd in water, as well as Hg and Zn in the hermit crab, discernible by significant correlations with water depth or salinity. Only the Cd concentrations in the crab increased with increasing salinity or water depth in the southern region, whereas in the northern part of the North Sea the Cd and also Hg and Pb concentrations of the crab correlated positively with water depth. In the northern part Ag and Cu correlated negatively with water depth. In both regions, Fe and Mn in the hermit crab were correlated positively at a high level of significance ($p < 0,01$), within the southern North Sea also Fe, Mn and Pb as well as Hg with Zn and Cu (Table 2.3.2).

For some metals, relations to sediment properties could be found. In the southern part of the North Sea Hg, Ag and Mn as well as Fe in the northern part determined in the hermit crab correlated positively with the fraction < 20 µm but also with POC and the sediment water content. In areas with a lower median of grain size distribution (areas of seston sedimentation) the concentrations of these metals were enhanced in the hermit crab.

The body condition index (C_I) may be used as a covariate to which metal concentrations can be related. Within the southern North Sea regions, concentrations of Cu and Fe proved to be negatively correlated and that of Zn positively correlated with the body condition of the crab. Larger specimens of the hermit crab revealed higher Cu and Fe contents than slimmer ones, whereas in slender organisms higher concentrations of Zn could be determined. Within the northern regions no correlations to the body condition could be calculated for any metals under investigation. For further interpretation of these south to north differences the reader is referred to Chapter 4.5.

Since Cd was determined in sea water as well as in seston and in the hermit crab it is possible to compare the relations between these different compartments of the ecosystem. In the southern region the concentrations of Cd in crabs increased with distance from the coast, whereas Cd concentrations in water decreased. In the northern North Sea, Cd in the hermit crab and in seston correlated positively but Cd in seston and in water negatively. The Cd concentration in sediment decreased in both regions with increasing distance from the coast. As described above for Cd, no significant correlation was found between Hg in the hermit crab and in water. Furthermore, no correlation was found between Pb in crabs and sediment samples.

The different correlations between metals in the southern and northern part of the North Sea and the different compartments of the ecosystem indicate differing causes for the distribution patterns of heavy metals. The most obvious differences are exhibited by the distributions of Hg and Cd in water, sediment, seston and a benthic organism (hermit crab) in the northern and southern North Sea. In the southern part of the North Sea, Cd concentrations were high in water and in sediment but low in seston and the hermit crab. These results suggest that the hermit crab is not able to accumulate Cd from water or sediment. In the northern part, Cd concentrations were low in water as well as in sediment, but high in seston and the hermit crab. Following the most likely pathway, the hermit crab may take up Cd from seston in the northen North Sea since the organic content of seston is high in this region (Nolting and Eisma 1988; Kersten et al. 1990), and it may therefore serve as food source for the crab.

In the souhern North Sea, Hg was high in water as well as in sediment and crab samples. Therefore, it may be possible that the hermit crab accumulates Hg from sediment or water. In the northern part of the North Sea, distinct areas could be found with enhanced Hg concentrations in water, sediment and crabs. The different distribution patterns of Cd and Hg can be explained with respect to different sources of these metals and a differing bioavailiability for the hermit crab in both regions of the North Sea. The interrelationships between water, suspended matter, sediments and biota are discussed in more detail in Chapter 4.5.

References

Anonymous (1991) A review of measurements of trace metals in coastal and shelf sea water samples collected by ICES and JMP Laboratories during 1985-1987. ICES Coop Research Report 178, Copenhagen 1991

Backhaus JO, Hainbucher D (1987) A finite difference circulation model for shelf seas and its application to low frequency variability on the European Shelf. In: Nilhout JCJ, Jamart BM (eds) Three-dimensional models of marine and estuarine dynamics. Elsevier, Amsterdam pp 221-244

Basford DJ, Eleftheriou A (1988) The benthic environment of the North Sea (56° to 61° N). J Mar Biol Assoc UK 68:125-141

Bjerk JE, Brevic EM (1980) Organochlorine compounds in aquatic environments. Arch Environ Contam Toxicol 9:743-750

Borchart T, Burchert S, Hablizel H, Karbe L, Zeitner R (1988) Trace metal concentrations in mussels - comparison between offshore regions in the southeastern North Sea from 1983 to 1986. Mar Ecol Prog Ser 42:17-31

Claussen T (1988) Levels and spatial distribution of trace metals in dabs (*Limanda limanda*) of the southern North Sea. In: Kempe S, Liebezeit G, Dethlefsen V, Harms U (eds) Biogeochemistry and distribution of suspended matter in the North Sea and implications to fisheries biology. Mitt Geol Paläontol Inst Univ Hamb 65:467-496

Dicke M, Schmidt D, Michel A (1987) Trace metal distribution in the North Sea. In: Lindberg SE, Hutchinson TC (eds) Proc Int Conf Heavy Metals in the Environment vol 2, New Orleans 1987. CEP Consultants, Edinburgh, pp 312-314

Duinker JC, Hillebrand MTJ, Boon JP (1983) Organochlorines in benthic invertebrates and sediments from the Dutch Wadden Sea: identification of individual PCB components. Neth J Sea Res 17:19-38

Duinker JC, Boon JP, Hillebrand MTJ (1984) Organochlorines in the Dutch Wadden Sea. Neth Inst Sea Res Publ Ser 10:211-228

Dyer MF, Fry WG, Fry PD, Cranmer GJ (1983) Benthic regions within the North Sea. J Mar Biol Ass UK 63:683-693

Eder G, Sturm R, Ernst W (1987) Chlorinated hydrocarbons in sediments of the Elbe river and the Elbe estuary. Chemosphere 16:2487-2496

Eisma D, Kalf J (1987) Distribution, organic content and particle size of suspended matter in the North Sea. Neth J Sea Res 21:265-285

Förstner U, Calmano W, Schoer J (1982) Heavy metals in bottom sediment and suspended material from the Elbe, Weser, and Ems Estuaries and from the German Bight. Thalassia Jugosl 18:97-122

Frauenheim K, Neumann V, Thiel H, Türkay M (1989) The distribution of the larger epifauna during summer and winter in the North Sea and its suitability for environmental monitoring. Senckenb Marit 20:101-118

Gaul H, Ziehbarth U (1983) Method for the analysis of lipophilic compounds in water and results about the distribution of different organochlorine compounds in the North Sea. Dtsch Hydrogr Z 36:191-212

Haarich M, Schmidt D (1993a) Schwermetalle in der Nordsee, Sommer-Aufnahme der gesamten Nordsee im Mai/Juni 1986 im Projekt ZISCH. Dtsch Hydrogr Z Ergänzungsh Hamb (in prep)

Haarich M, Schmidt D (1993b) Schwermetalle in der Nordsee, Winter-Aufnahme der gesamten Nordsee im Januar-März 1987 im Projekt ZISCH. Dtsch Hydrogr Z Ergänzungsh Hamb (in prep)

Haarich M, Schmidt D, Freimann P, Jacobsen, C (1993) North Sea research projects ZISCH and PRISMA - application of total-reflection X-ray spectrometry in sea-water analysis. Spectrochim Acta 48B:183-192

Kammann U, Knickmeyer R, Steinhart H (1990) Distribution of polychlorobiphenyls and hexachlorobenzene in different tissues of the dab (Limanda limanda) in relation to lipid polarity. Bull Environ Contam Toxicol 45:552-559

Karbe L, Gonzalez-Valero J, Borchardt T, Dembinski M, Duch A, Hablizel H, Zeitner R (1988) Heavy metals in fish and benthic organisms from the northwestern, central and southern North Sea: Regional patterns comparing dab, blue mussel and hermit crab. (Limanda limanda, Mytilus edulis, Pagurus bernhardus). ICES, Copenhagen, C.M.1988/E:22

Karbe L, Dembinski M, Gonzalez-Valero J, Müller M, Zeitner R (1989) Regionale Verteilungsmuster von Schwermetallen in Benthosorganismen der Nordsee und angrenzender Seegebiete. Arb Dtsch Fischereiverb 48:95-105

Kersten M, Klatt V (1988) Trace metal inventory and geochemistry of the North Sea shelf sediments. Mitt Geol Paläontol Inst Univ Hamb 65:289-311

Kersten M, Kröncke I (1990) Bioavailability of lead in North Sea Sediments. Helgol Meeresunters 45: 403-409

Kersten M, Dicke M, Kriews M, Naumann K, Schmidt D, Schulz M, Schwikowski M, Steiger M (1988) Distribution and fate of heavy metals in the North Sea. In: Salomons W, Bayne BL, Duursma EK Förstner U (eds) Pollution of the North Sea, an assessment. Springer, Berlin, Heidelberg, New York, pp 300-347

Kersten M, Kienz W, Koelling S, Schröder M, Förstner U (1990) Schwermetallbelastung in Schwebstoffen und Sedimenten der Nordsee. Vom Wasser 75:245-272

Knickmeyer R, Steinhart H (1988a) The distribution of cyclic organochlorines in North Sea sediments. Dtsch Hydrogr Z 41:1-21

Knickmeyer R, Steinhart H (1988b) Seasonal differences of cyclic organochlorines in eggs of the hermit crab *Pagurus bernhardus* L. from the North Sea. Sarsia 73:291-297

Knickmeyer R, Steinhart H (1989a) Cyclic organochlorines in plankton from the North Sea in spring. Estuarine Coastal Shelf Sci 28:117-127

Knickmeyer R, Steinhart H (1989b) Cyclic organochlorines in the whelks *Buccinum undatum* and *Neptunea antiqua* in the North Sea and Irish Sea. Dtsch Hydrogr Z 42:43-59

Knickmeyer R, Steinhart H (1990) On the distribution of polychlorinated biphenyl congeners and hexachlorobenzene in different tissues of the dab (*Limanda limanda*) from the North Sea. Chemosphere 19:1309-1320

Knutzen J (1986) Utslipp av polysykliske aromatiske hydrokarboner (PAH) fra norske smelteverk. Vann 20:133-138

Knutzen J, Martisen K, Naes K (1984) Om observasjoner av klororganiske stoffer i organismer og sedimenter fra Kristiansandsfjorden. Vann 18:392-400

Köhler A, Harms U, Luckas B (1986) Accumulation of organochlorines and mercury in flounder - an approach to pollution assessments. Helgol Wiss Meeresunters 40:431-440

Kröncke I (1988) Heavy metals in macrofauna of the North Sea. In: Kempe S, Liebezeit G, Dethlefsen V, Harms U (eds) Biogeochemistry and distribution of suspended matter in the North Sea and implications to fishery biology. Mitt Geol Paläontol Inst Univ Hamb 65:455-465

Künitzer A, Basford D, Craejmeersch JA, Dewarumez J-M, Dörjes J, Dueneveld GCA, Eleftheriou A, Heip C, Herman P, Kingston P, Niermann U, Rachor E, Rumohr H, De Wilde PAJ (1992) The benthic infauna of the North Sea: distribution and assemblages. J Mar Sci 49:127-143

Landgraff O, Kammann U, Steinhart H (1992) Distribution of cyclic organochlorines in sediments from the North Sea and the German Bight - an overview. Analysis 2016:M74-77

Lunde G, Baumann Ofstad E (1976) Determination of fat-soluble chlorinated compounds in fish. Z Anal Chem 282:395-399

Macholz RM, Kujawa M (1985) Recent state of Lindane metabolism. Part III, Residue Rev 94:119-149

Malaiyandi M, Shah SM (1984) Evidence of photoisomerization of hexachlorocyclohexane isomers in the ecosphere. J Environ Sci Health Part A A19:+887-910

Müller G (1979) Schwermetalle in den Sedimenten des Rheins - Veränderungen seit 1971. Umschau 79:778-783

Müller-Navarra S, Mittelstaedt E (1985) Schadstoffausbreitung und Schadstoffbelastung in der Nordsee. Dtsch Hydrogr Inst, Hamburg, pp 1-50

Nolting RF, Eisma D (1988) Elementary composition of suspended particulate matter in the North Sea. Neth J Sea Res 22/3:219-236

Rick H-J (1990) Ein Beitrag zur Abschätzung der Wechselbeziehung zwischen den planktischen Primärproduzenten des Nordseegebietes und den Schwermetallen Kupfer, Zink, Cadmium und Blei auf Grundlage von Untersuchungen an natürlichen Planktongemeinschaften und Laborexperimenten mit bestandsbildenden Arten. Diss RWTH Aachen

Schmidt D, Dicke M (1987) Trace determination of mercury in the water column: two surveys covering the entire North Sea. In: Lindberg SE,Hutchinson TC (eds) Proc Int Conf Heavy Metals in the Environment, vol 2, New Orleans 1987. CEP Consultants, Edinburgh, pp 315-317

Schmidt D, Dicke M (1990) Schwermetalle im Wasser. In: Lozan JL, Lenz W, Rachor E, Watermann B, von Westernhagen H (eds) (1990) Warnsignale aus der Nordsee. Parey, Hamburg, pp 30-41

Underdal B, Norheim G, Hoff H, Hastein T (1981) Kvikksolv og klorerte hydrokarboner i fisk fra Skiensvassdraget og fjordene i Grenlandsomradet. Rep National Veterinary Institute, 29 pp

2.4 Atmospheric Transport of Contaminants, Their Ambient Concentration and Input into the North Sea

W. Dannecker, H. Hinzpeter, H.-J. Kirzel, H. Luthardt, M. Kriews, K. Naumann, M. Schulz, M. Schwikowski, M. Steiger and U. Terzenbach

2.4.1 Introduction

Only recently has the importance of atmospheric input into marginal seas like the North Sea been emphasized for several heavy metals and nitrogen components. The few data for the open sea, however, have left large uncertainties in quantitative estimates (Cambray et al. 1975). Modelling indicated that there should be a steep decrease in deposition depending on meteorological conditions, and that coastal stations overpredict the input substantially (van Aalst et al. 1983; van Jaarsveld et al. 1986; Krell and Roeckner 1988, see also Chapt. 3.1). The presented studies therefore focussed on measurements at sea in comparison with coastal sites. On the island of Helgoland and at the research platform *Nordsee* as well as on cruises in the North Sea, concentrations of components of interest and their deposition were monitored over a period of years and compared with a reference station at the Westerhever lighthouse. Wind-directed sampling at the island of Helgoland was carried out to characterize the air masses reaching the German Bight from different source regions. The interpretation of these measurements and intensive field measurements together with calculated backward trajectories and long-term wind direction frequencies permitted quantification of the contributions from different source regions to the atmospheric input into the German Bight.

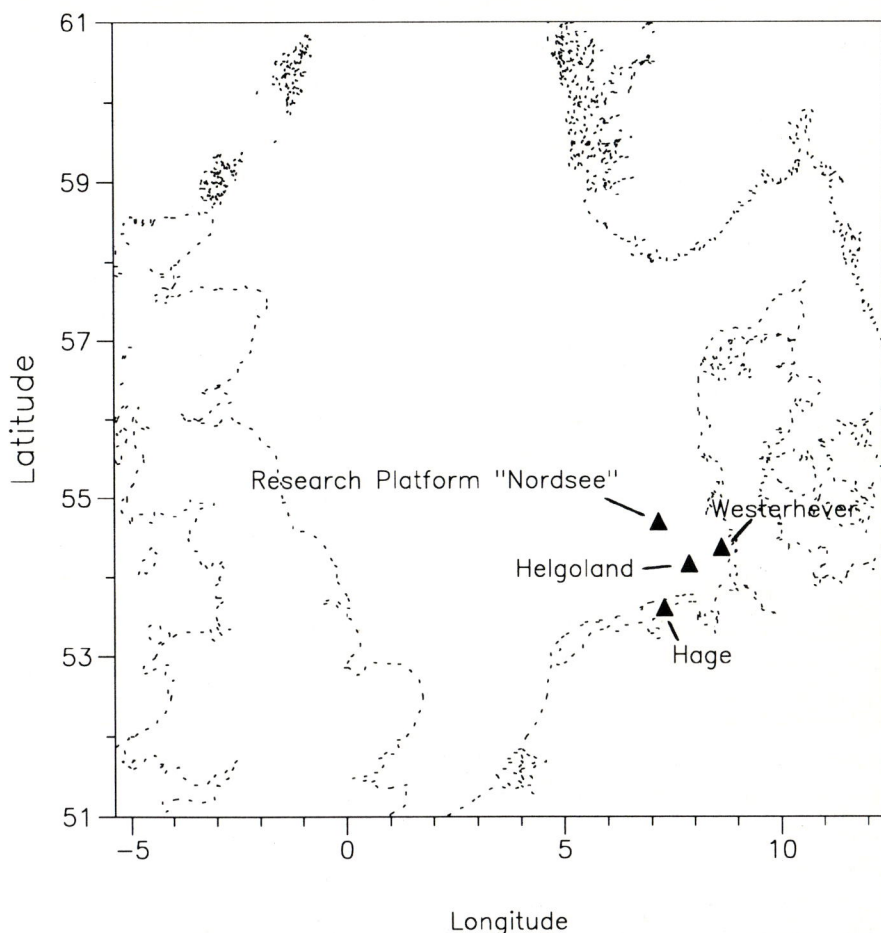

Fig. 2.4.1. Sites of aerosol measurements in the southeastern North Sea

Measurements of the characteristics of the planetary boundary layer during aircraft missions complemented the data set by giving insight into the transport conditions over sea and the distribution of atmospheric pollution in the southern North Sea region.

Data on elemental concentrations in aerosols from several measurement campaigns at the research platform *Nordsee* (FPN), the island of Helgoland, Hage (Ostfriesland) and Westerhever (Eiderstedt Peninsula), combined with the results from the long-term aerosol monitoring program at the island of Helgoland and the Westerhever lighthouse should give a first impression of concentration distribution patterns of some pollutants in the southeastern North Sea region (see Fig. 2.4.1).

2.4.2   Wind Structure and Turbulence in the Boundary Layer at Nordholz

To obtain information about the wind profile and turbulence in the marine boundary layer, a SODAR system as described by Peters (1991) was installed at the research platform *Nordsee* (FPN). These measurements were often disturbed by artificial sound sources, particularly due to mechanical work at the platform. Therefore, continuous measurements were nearly impossible and the measurements were continued at Nordholz air base (Lower Saxony). Between the estuaries of the rivers Elbe and Weser the direction normal to the coast is WNW. The distance of the SODAR-system from the coast was about 6 km, varying with the tide. From June 1986 to December 1988 SODAR measurements were performed on 269 days, and vertical profiles of the wind as well as of the standard deviation of the vertical wind component were determined. This was carried out for a height range between 30 and 450 m with a vertical resolution of 20 m.

In a first evaluation step, measurements were scanned for the occurrence of low level jets (LLJ) which may be important for air pollution transport. In a second step, mean values of wind speed and the standard deviation of its vertical component were determined for all measurement phases.

2.4.2.1 The Low Level Jet (LLJ)

Westerly and northerly winds were observed on 202 days, on 67 days easterly and southerly winds prevailed, and only for wind directions between 45 and 225 degrees, corresponding with winds from land, was the LLJ observed on 11 days. Thus, no LLJ was observed on days with winds from sea. The wind maximum occurred at heights between 100 and 300 m, the LLJ developed during nighttime and occasionally continued until the morning hours. The wind speed exceeded the geostrophic wind by at most 7-8 m/s, with a maximum absolute wind speed of 15 m/s.

As an example, Fig. 2.4.2a shows the developing profile of horizontal wind speed for the night 12/13th of October 1986. The LLJ developed continuously from 20.00 GMT in the evening to its maximum around 04.00 GMT but then collapsed in quite a short time.

The development of the wind vector at a height of 190 m is shown in Fig. 2.4.2b. The hodograph roughly approximates a circle, which would be expected in case of stationarity and constant Coriolis term. At Hanover, 176 km SE of Nordholz, a stable boundary layer with a temperature inversion of 5 °C strength and 250 m height was observed at midnight.

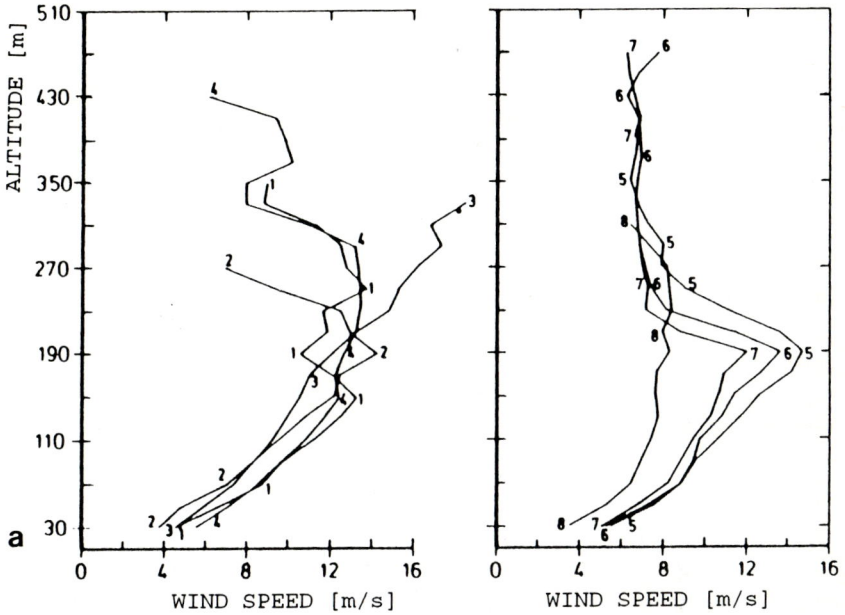

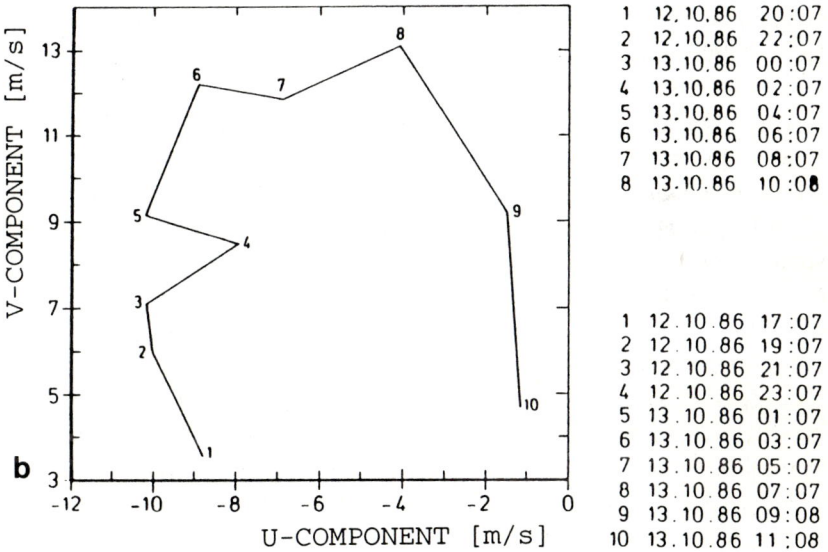

Fig. 2.4.2. *(a)* Series of vertical profiles of wind speed (1 hourly means) in hourly time intervals measured at Nordholz with Doppler Sodar. *(b)* Hodograph of the horizontal wind component at 190 m height. Time hourly means and hourly intervals as in Fig. 2.4.2a

Between the 1st and 3rd of October 1988 LLJ were also observed with southeasterly winds together with an inversion with 250 m height and 5 °C strength at Hanover. From the 19th to 21st of October 1988, the synoptic situation was similiar, but due to the cloudy and overcast daytime weather, partly with fog, and to the lack of an inversion at Hanover, no LLJ was observed. Therefore, we may assume that on cloudless days with SE winds having a continental fetch of about 200 km, the LLJ is a common appearance. The observations agree with the results of a simple analytical model by Thorpe and Trevor (1977). According to these results, the LLJ needs a well developed daily variation of the surface temperature and, thus, a strong daily variation of the boundary layer height.

Therefore, due to the nearly constant sea surface temperature no LLJ develops with sea breezes.

2.4.2.2 The Wind Profiles

Figures 2.4.3 and 2.4.4 show the daily mean wind speed profiles for different wind directions and for all seasons. These are defined by astronomic date (21.3., 21.6., 23.9., 22.12.). All figures contain the daily mean profiles. For the directions 135°-180° daytime as well as nighttime profiles are shown (Figs. 2.4.2a,b). For all seasons the daytime is defined from 06.00 to 18.00 h.

Winds from NNE are lowest in summer with about 4 m/s at 50 m height, reaching the highest value of 8.5 m/s in winter and exhibiting intermediate speeds in fall (5.8 m/s) and spring (6.2 m/s). In most seasons, the increase with height is small (2 m/s up to 450 m), except only in fall (4 m/s).

During daytime, the ENE direction generally exhibits lower wind speed: at 50 m height 3.8 m/s in summer, 4 m/s in fall, 5.5 m/s in winter and 6 m/s in spring, only slightly increasing with height. For the winter season the sample is too small, resulting in irregular profile variations.

At nighttime, the picture is similiar, but the increase with height is considerably larger, especially during spring, with an increase from 5-12 m/s. In summer and fall (21.6., 23.12.) there is no monotonic increase, but a small maximum between 100 and 150 m can be seen.

For winds from the SE direction in the daytime, a monotonic increase from about 4 m/s to 6-8 m/s in 450 m height is observed, with the greatest increase in autumn and winter.

During nighttime, we find nearly the same values at 50 m, but the increase with height is more pronounced and the wind reaches 8-12 m/s at 450 m. Again, autumn and winter show the largest values at that height.

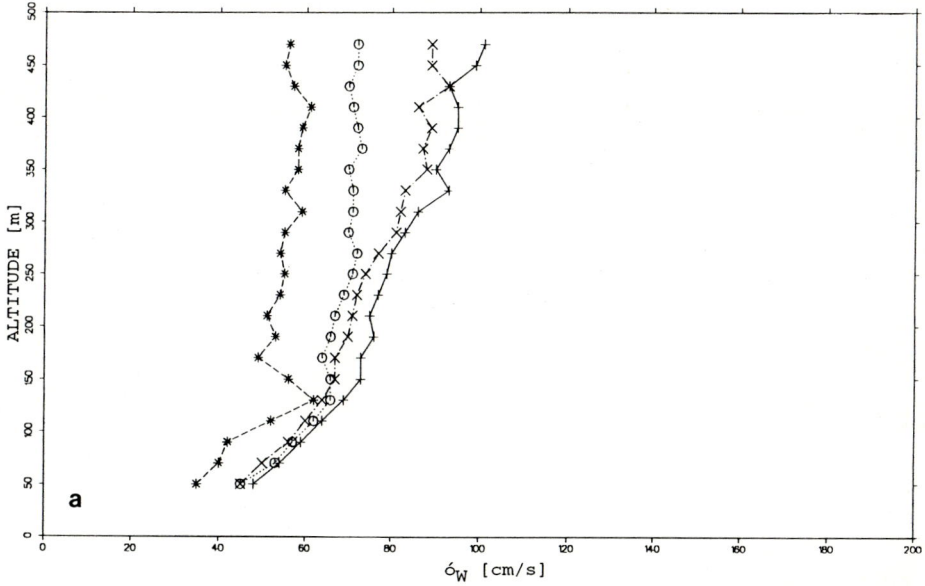

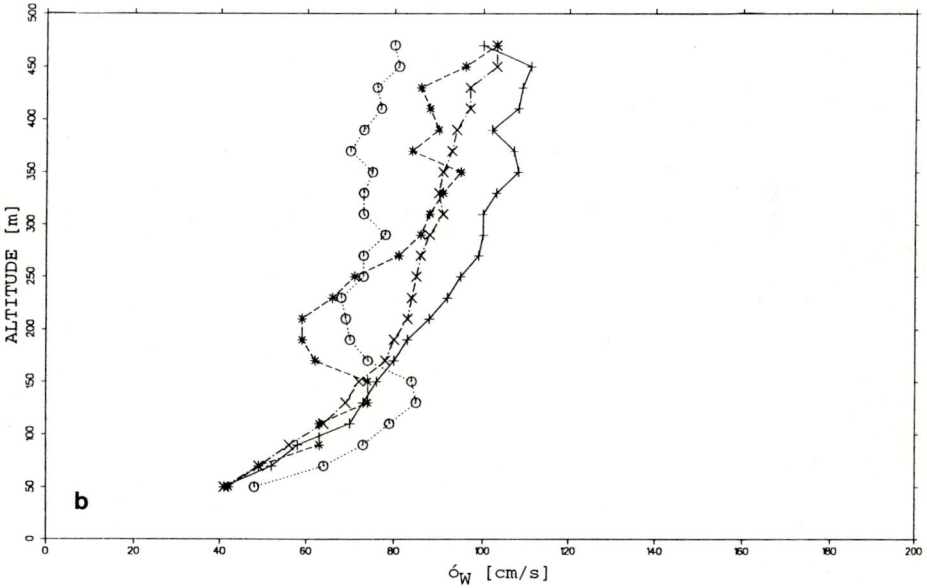

Fig. 2.4.3. *(a)* Wind profile (daytime mean) for sector 135-179° (x winter; + spring; * summer; o fall). *(b)* Wind profile (nightime mean) for sector 135°-179° (legend as in *a*)

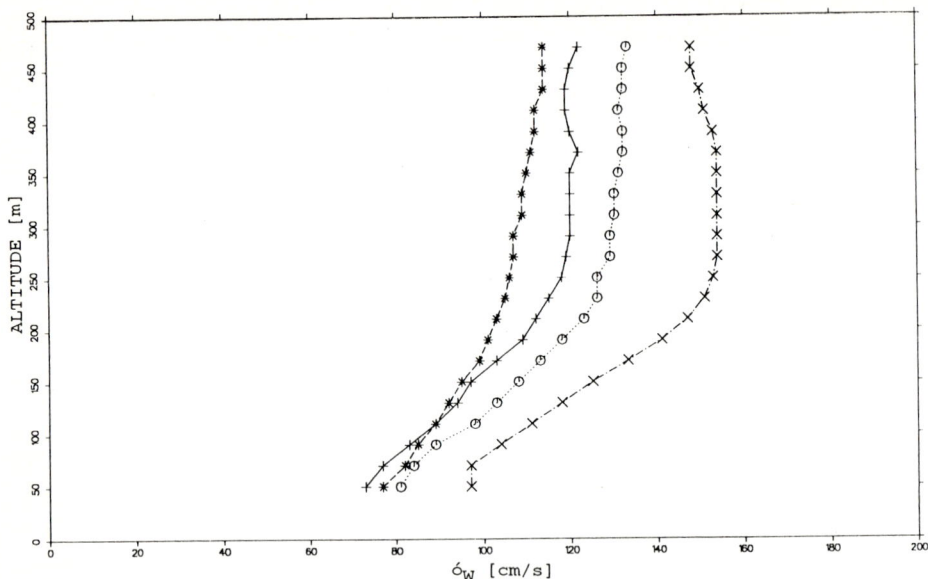

Fig. 2.4.4. Wind profile (daily mean) for sector 270°-314° (legend as in Fig. 2.4.3a)

More pronounced is the wind maximum in summer and fall at 100-150 m height.

Winds from SSE show about 4-5 m/s during daytime at 50 m, the greatest increase with height in winter and spring, and only a small increase in summer. For summer and fall even during daytime a small maximum can be seen, which is most pronounced in the night.

For the SSW direction, wind speed at 50 m is 5-6 m/s with a strong increase up to 13-14 m/s at 450 m height. Winds from WSW range from 6.5 to 8.5 m/s at 50 m with the strongest winds in winter and weakest in summer. The winter value increases to 15 m/s at 450 m, spring and autumn values to 13 m/s and summer values to 11 m/s.

Winds from WNW reach about 10 m/s for the winter and 8 m/s for the other seasons. In all seasons there is an increase up to 250 m and constant wind above that level. At 450 m height values reach 15 m/s in winter, 13.4 m/s in fall, 12 m/s in spring and 10.5 m/s in summer.

The NNW direction at 50 m shows greater differences: 4.5 m/s in summer, 6 m/s in spring and autumn and 9 m/s in winter. At the 450 m level the wind speed was 8 m/s in summer, 10-11 m/s in spring and fall, and 13.5 m/s in winter.

If not explicitly mentioned or evident from the figures, the difference between day and night profiles is small. Only for the SSW direction do

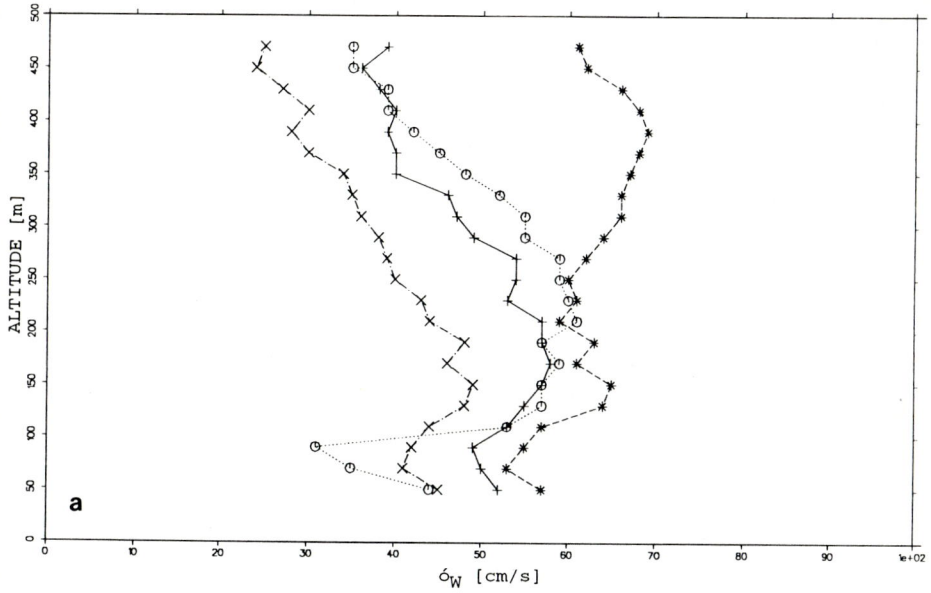

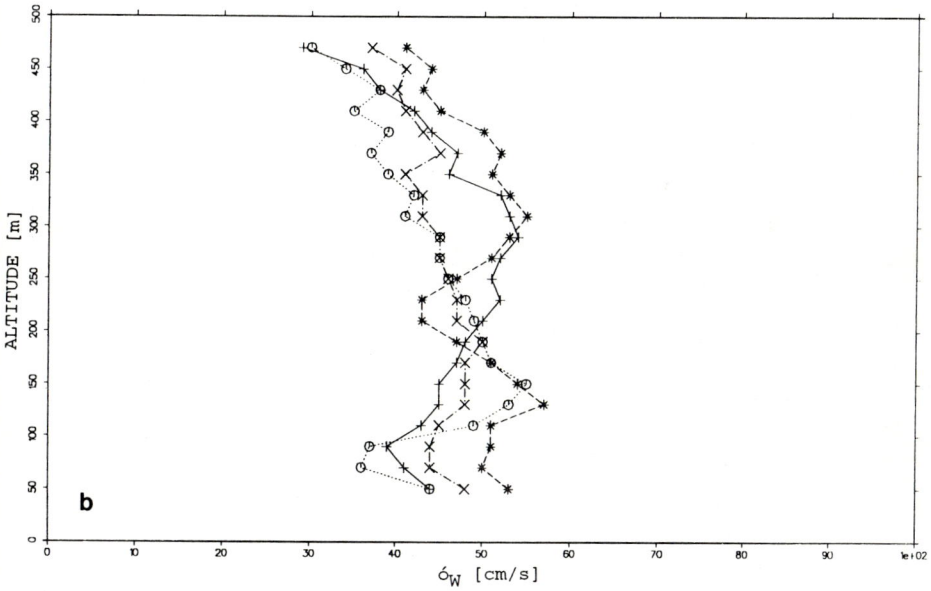

Fig. 2.4.5. *(a)* Standard deviation of the wind (daytime mean) for sector 135°-179° (legend as in Fig. 2.4.3a). *(b)* Standard deviation of the wind (nightime mean) for sector 135°-179° (legend as in Fig. 2.4.3a)

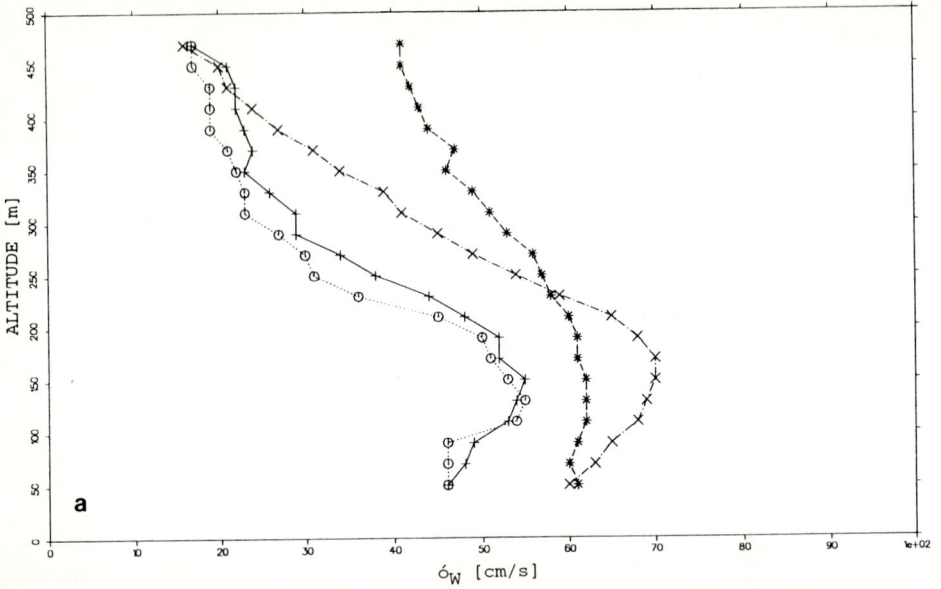

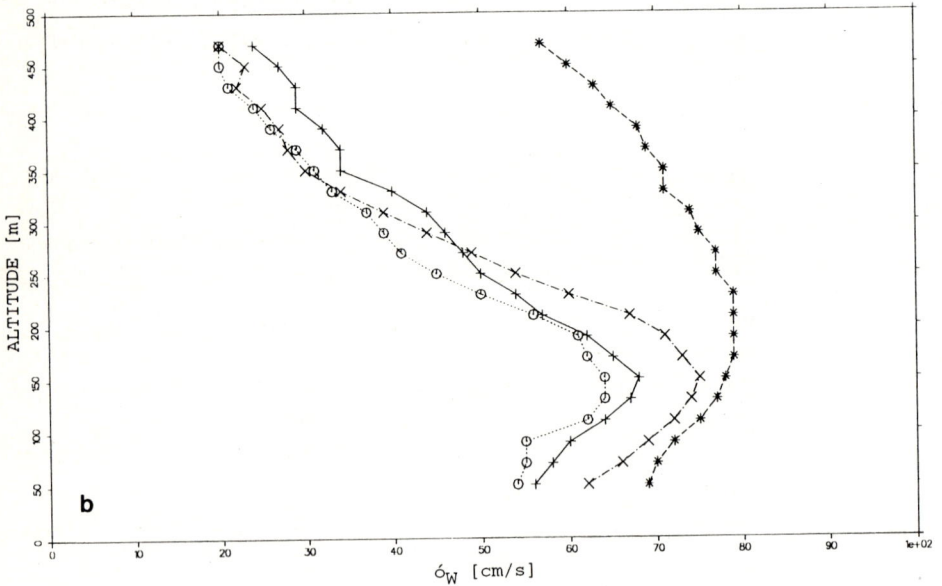

Fig. 2.4.6. *(a)* Standard deviation of the wind (daytime mean) for sector 270°-314° (legend as in Fig. 2.4.3a). *(b)* Standard deviation of the wind (nightime mean) for sector 270°-314° (legend as in Fig. 2.4.3a)

considerable differences exist, generally of 1-2 m/s increasing to 3 m/s in summer.

Since the wind profiles presented here are simple arithmetic means, the observed maxima during night in summer and fall for land breeze directions (Fig. 4.2.2b) are remarkable. They occur in the height range for which LLJ was found at nighttime during land breeze cases. Therefore, we may assume that LLJ occurs so frequently, at least in our sample, that even the otherwise monotonically increasing mean wind profile is modified.

2.4.2.3 The Standard Deviation of the Vertical Wind

The standard deviation of the vertical wind and its variation with height, wind speed, and direction for different seasons has been chosen as a measure for turbulence in the atmospheric boundary layer.

Figures 2.4.5 and 2.4.6 present the dependence of σ on the wind direction. Generally, there is a well expressed maximum at a height of about 150 m. In summer and fall maxima often occur at greater heights. During daytime, when convection and larger values of turbulent heat flux prevail, this would be expected; but since these mean values are samples over quite different weather conditions, further interpretation is insignificant at the present state of the analysis. An intensive analysis of the data with respect to similar scaling has to be done in the future (see Sorbjan et al. 1991).

2.4.3 Trajectories of Selected Sources over the North Sea

In order to gain insight regarding the influence of specific sources of air pollution, the wind field above the North European Shelf was analyzed for the years 1987, 1988, 1989 (Luthardt 1987). Three-hourly data were available and then linearly interpolated. With these analyzed and interpolated fields, trajectories of selected sources have been determined. To obtain an impression of the influence of the sources on different parts of the North Sea, six different regions were defined (see Figs. 2.4.7 and 2.4.8).

For each quarter of the years 1987, 1988, 1989 all trajectories (about 65,000) starting from a source point were calculated at hourly intervals. The trajectories were followed for 24, 36 and 48 h. For each full hour, the area was determined into which an air parcel originating from a source was guided by a trajectory. For each area, the frequency of trajectory passes

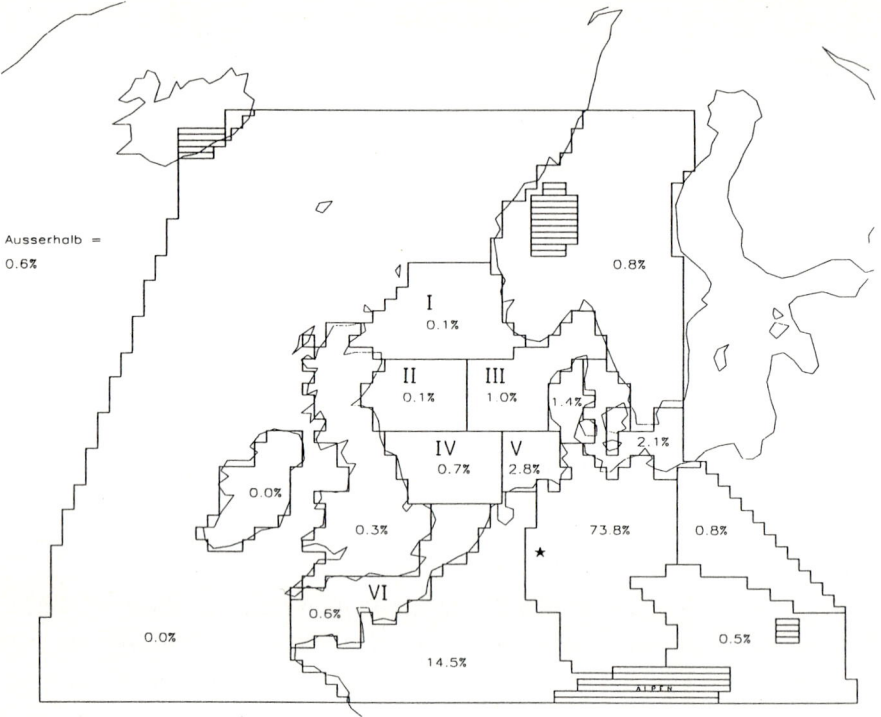

Fig. 2.4.7. Probability of residence integrated over 24 h for surface wind trajectories during spring (1987, 1988, 1989). Starting point of trajectories: Essen (n=6495)

was estimated by normalizing the number of trajectories which had reached the area with the total number for all areas. This percentage represents that part of the total travel time 24, 36 or 48 h from the source which the air parcel has spent in the considered area. This quantity is called the "probability of residence" in an area, and it is a measure of the potential atmospheric input into the sea, though it does not consider the reduced contaminant concentration along trajectory due to deposition on its way to the considered area.

As the height of the source and, therefore, the influence of boundary layer stability was not considered, the trajectories based upon the surface wind field as well as those based upon the geostrophic wind field have been calculated. The difference between the geostrophic and the surface wind probability of residence (TG and TS, respectively) indicates the influence of the source height. The statistics for the 3 years show only minor differences in most cases.

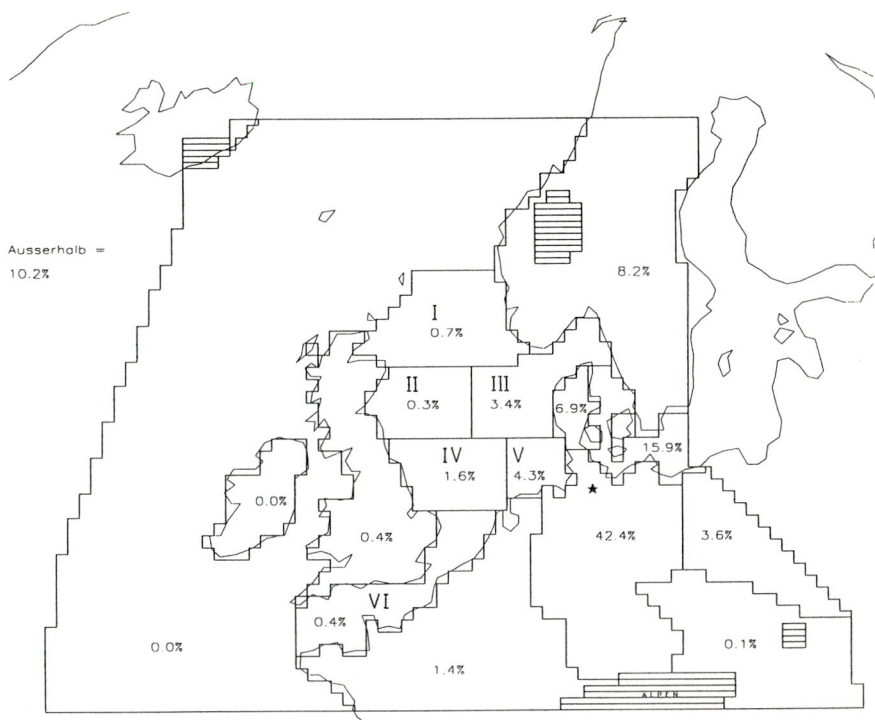

Fig. 2.4.8. Probability of residence integrated over 24 h for surface wind trajectories during spring (1987, 1988, 1989). Starting point of trajectories: Hamburg (n=6495)

Six sources were selected for these calculations: Ekofisk, Essen, Hamburg, Manchester and Rotterdam. Ekofisk, in the centre of the North Sea, is, of course, of eminent importance. Essen, at a distance of about 230 km from the North Sea coast may represent the Ruhr region. Hamburg, with a distance of only about 90 km from the sea, could be important for the southeastern area of the North Sea. Manchester was chosen to represent an important British industrial region, in spite of the large distance from the British North Sea coastline of about 180 km. Rotterdam, situated at the southern coastline of the Channel area, should be most important for that region.

Table 2.4.1 gives the distribution of the probability of residence for the six areas of the North Sea. As examples, Figs. 2.4.7 and 2.4.8 show the distribution for the first quarter of the year (TS, 24 h) for Essen as well as for Hamburg.

Due to its central position, Ekofisk has the greatest atmospheric potential for the North Sea. Hamburg and Essen have a much smaller importance,

Table 2.4.1. Distribution of the probability of residence for five source regions and six receptor areas in North Sea. All values are given in per cent.

Hamburg

Quarter h	I	II	III	IV	V	VI
I TS 24	0.7	0.3	3.4	1.6	4.3	0.5
48	1.9	1.0	3.6	1.8	2.8	0.6
TG 24	1.9	0.7	3.4	1.2	2.8	0.6
48	1.5	0.5	2.3	0.8	1.5	0.6
II TS 24	0.2	0.3	1.0	2.7	6.2	0.8
48	0.7	0.7	1.5	2.5	4.6	1.8
TG 24	1.2	1.4	3.5	3.0	6.3	1.1
48	1.7	1.1	3.0	2.3	3.5	2.5
III TS 24	0.1	0.1	1.6	0.7	4.8	0.1
48	0.9	0.5	3.0	1.2	3.4	0.4
TG 24	0.7	0.4	2.6	1.3	3.4	0.9
48	1.0	0.5	2.3	1.2	2.2	1.1
IV TS 24	1.5	1.1	4.2	1.8	6.3	0.1
48	2.6	1.6	3.9	2.2	3.6	0.4
TG 24	1.7	0.8	4.2	1.4	3.0	0.2
48	1.7	0.9	2.7	1.0	1.6	0.3

Rotterdam

Quarter h	I	II	III	IV	V	VI
I TS 24	1.6	1.5	4.8	7.7	7.3	11.3
48	2.6	1.7	4.7	5.3	4.7	6.5
TG 24	1.4	0.8	3.3	4.4	4.2	6.6
48	1.3	0.7	2.3	2.6	2.3	3.9
II TS 24	0.3	0.5	1.5	6.1	4.3	19.8
48	1.3	1.4	2.2	5.2	3.8	13.4
TG 24	0.8	1.2	1.7	5.5	3.1	19.6
48	1.5	1.5	1.7	3.8	2.1	12.5
III TS 24	0.4	0.6	2.4	5.9	7.1	11.1
48	1.7	1.3	3.5	4.6	5.3	7.4
TG 24	0.8	1.0	2.5	5.1	3.8	8.8
48	1.3	1.3	2.3	3.5	2.4	5.7
V TS 24	1.9	1.9	4.1	9.5	4.8	16.6
48	2.8	1.9	3.5	7.1	3.4	10.3
TG 24	1.7	1.5	2.8	7.5	3.3	10.2
48	1.6	1.3	2.2	4.8	2.1	5.9

Ekofisk

Quarter h	I	II	III	IV	V	VI
I TS 24	15.9	9.0	23.5	4.3	2.6	0.8
48	9.6	5.0	13.3	2.9	1.6	1.3
II 24	12.9	18.7	29.1	10.5	4.3	2.2
48	9.1	10.8	18.1	7.3	3.7	3.1
III 24	16.3	11.9	28.2	4.5	4.8	0.7
48	10.5	6.8	16.4	3.2	3.1	0.9
IV 24	15.5	12.1	24.8	5.2	3.7	0.8
48	9.4	6.3	14.2	4.0	2.4	1.0

Manchester

Quarter h	I	II	III	IV	V	VI
I TS 24	3.3	6.3	3.9	10.8	1.3	2.9
48	4.2	5.1	4.5	7.6	1.4	3.1
TG 24	3.5	4.5	3.7	9.8	1.9	5.4
48	2.3	2.6	2.5	5.4	1.3	3.5
II TS 24	0.5	2.5	0.5	6.3	0.2	1.5
48	1.8	3.3	1.5	7.1	0.8	4.1
TG 24	1.8	3.6	1.8	7.2	0.8	5.1
48	2.0	2.7	1.9	5.0	1.0	5.0
III TS 24	1.8	5.5	1.6	12.9	0.7	2.2
48	3.5	5.6	3.3	10.9	1.5	5.3
TG 24	3.2	4.8	3.5	11.6	1.7	8.3
48	3.0	3.2	3.1	7.3	1.5	6.9
IV TS 24	2.0	5.2	2.8	8.1	1.0	1.9
48	3.2	4.1	3.3	5.2	1.0	2.1
TG 24	3.1	4.4	2.3	7.7	1.6	3.9

Essen

Quarter h	I	II	III	IV	V	VI
I TS 24	0.1		1.0	0.7	2.8	0.6
48	0.8	0.5	2.7	1.3	2.9	0.9
TG 24	0.9	0.5	1.7	1.7	2.0	2.5
48	1.1		1.6	1.3	1.3	1.9
II TS 24	0.0	0.0	0.3	0.4	1.0	1.2
48	0.3	0.3	0.8	0.8	1.5	1.9
TG 24	0.3	0.5	0.7	2.7	1.4	5.9
48	0.9	0.9	1.1	2.3	1.2	5.0
III TS 24	0.0	0.0	0.3	0.1	1.1	0.4
48	0.5	0.1	1.0	0.7	1.7	1.0
TG 24	0.2	0.3	0.9	2.0	1.9	2.5
48	0.7	0.7	1.2	1.7	1.5	2.4
IV TS 24	0.2	0.2	0.9	1.6	2.2	1.5
48	1.6	0.9	2.2	2.3	2.2	3.0
TG 24	1.1	1.0	2.1	4.3	2.5	3.2

Hamburg due to its eastern position, Essen due to the greater distance from the sea. Although Manchester is relatively far from the British east coast, it has a greater influence than Hamburg and Essen due to the prevailing westerly winds. For the Channel region, which includes the sea off southeast England, Rotterdam is of major importance.

2.4.4 Aerosol Measurement in the Southern North Sea Region

In June 1986, a monitoring program was started on the island of Helgoland ($54°10'$N, $7°53'$E) to study the ambient air elemental concentrations associated with the aerosol above the southern North Sea region. This program was supplemented in 1987 by two other stations in this area (Westerhever and the German research platform *Nordsee*).

This aerosol monitoring program implied weekly aerosol sampling of atmospheric particulate matter on dried preweighed quartz microfiber filters (Munktell MK 360) using a high volume sampler with a flow rate of 100 m^3/h. After digestion of the aerosol-laden filters with a mixture of HF, HNO_3 and $HClO_4$ the samples were analyzed by atomic spectrometric techniques (ICP-OES, F-AAS and GF-AAS) for a set of 24 elements known as tracers for anthropogenic, crustal and marine aerosol sources (Kriews et al. 1989).

Table 2.4.2 shows the mean elemental concentrations and their standard deviations at the sites Helgoland and Westerhever for the presumably anthropogenic elements As, Cd, Cu, Ni, Pb, Sb, Se and Zn from June 1986 to May 1990.

This long-term study revealed a strong variability of elemental concentrations, as is demonstrated in Figs. 2.4.9-2.4.14 for the elements As, Cd, Cu, Ni, Pb and Zn. The elements Sb and Se are not shown as a graph, because their time series were similar to that of As. The observed elemental concentrations were strongly influenced by the meteorological conditions during the sampling period (Kriews et al. 1989).

It is interesting to note that the element concentrations at Helgoland and Westerhever showed a similar time-dependent distribution pattern. The highest concentrations for the anthropogenic elements were observed in winter, while in summer the concentrations were lower. The reasons for this are probably the influence of seasonal characteristics in the meteorological conditions. In winter, situations with stable stratification often occur in the boundary layer. These are associated with southeasterly air masses, while in summer, westerly flow and good mixing conditions dominate.

Table 2.4.2. Ambient elemental mean concentrations in air measured by several authors in the North Sea region (ng/m^3)

	Helgoland (1) 1986/90	Westerhever (1) 1987/89	Ostende (2) 1977	Westhinder (3) 1980/85	Tange (4) 1983	Pellworm (5) 1984/85
As	1.7	2.5	--	--	--	2.7
Cd	1.6	0.52	5	3.9	0.6	0.67
Cu	3.5	4.8	17	16.8	3.0	3.3
Ni	2.6	2.8	11	--	3.0	4.4
Pb	20.2	26.9	241	147	40.9	38.8
Sb	0.96	1.3	--	--	0.9	1.1
Se	0.98	1.3	--	--	0.7	0.9
Zn	38.2	51.6	250	150	29.7	40

(1) This chapter; (2) Kretzschmar and Cosemans (1979); (3) Dedeurwaerder and Artaxo (1987); (4) Kemp (1984) and Stössel (1987).

It can also be seen that the absolute concentrations were higher in Westerhever than at Helgoland with the exception of the element Cd. The outliers in Cd concentration at Helgoland are possibly due to unidentified local sources on the island.

Concentration differences at Westerhever and Helgoland are especially striking in samples from the winter and spring while they are lower in summer. This may be due to higher deposition fluxes associated with the meteorological conditions in winter.

The data sets can also be used to calculate the input into the North Sea on a long-term basis. For As, Cd, Cu, Ni, Pb Sb, Se and Zn concentrations we estimated dry deposition fluxes from the atmosphere to the sea surface (Table 2.4.6) based on the average mean concentrations and we calculated average dry deposition velocities using the dry deposition model of Slinn and Slinn (1980). The wet deposition flux in the same table is based on the average mean concentration, the yearly precipitation height and summarized washout-factors from the last GESAMP report (Duce et al. 1990).

These calculated total deposition fluxes agree well with data (see below) from a long-term study of deposition samples at the research platform *Nordsee*, which are shown in Table 2.4.9. However, large uncertainties have to be considered for both methods. These regard the use of scavenging ratios and dry deposition models and the problems in collecting representa-

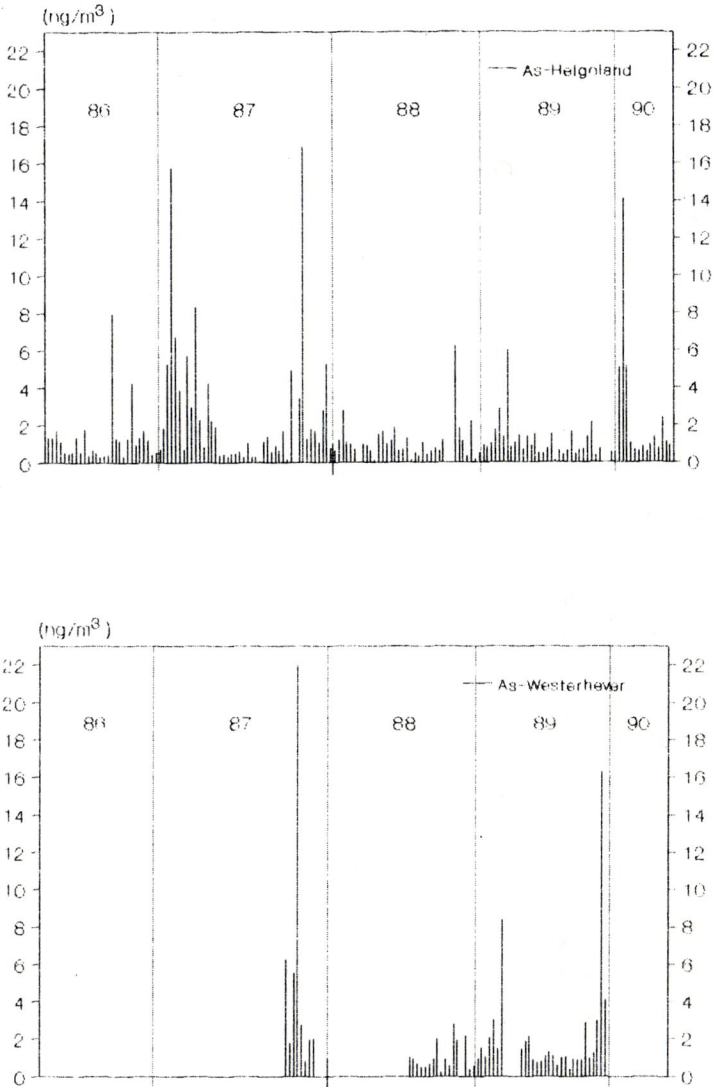

Fig. 2.4.9. Time series of ambient As-concentration during a long term monitoring program at Helgoland and Westerhever based on weekly aerosol sampling.

tive wet-only and total-deposition samples in the field.

It is interesting to note that the reduction of lead emissions, resulting from the German law banning leaded gasoline as of 1.1.1986, shows up in our long-term study at Helgoland. Our measurements indicate a reduction of Pb concentration in ambient air from the beginning of the monitoring

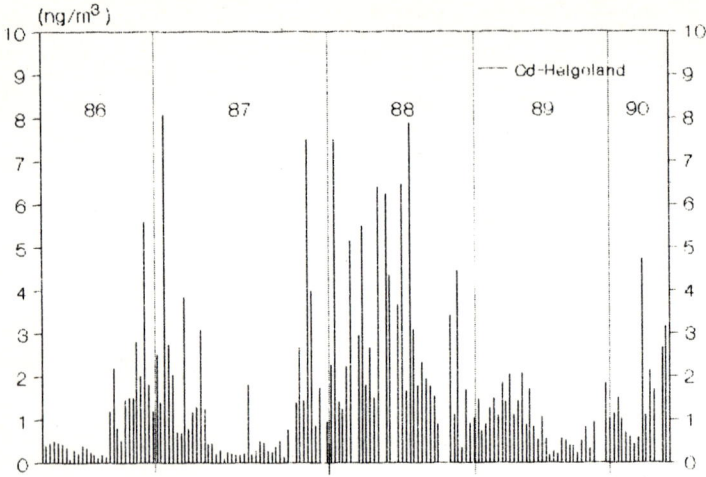

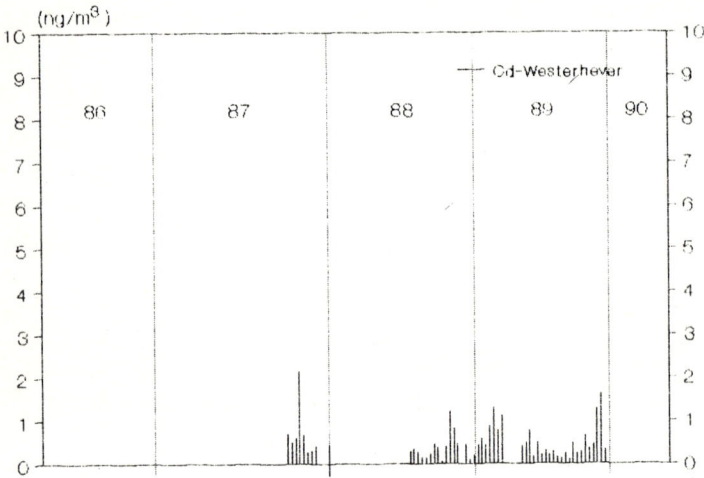

Fig. 2.4.10. Time series of ambient Cd concentration

program in 1986 until 1990.

This finding is demonstrated in Fig. 2.4.15, where the Pb concentration in air is normalized to the Al concentration. Al is a tracer element for crustal weathering. It is assumed that the aerosol sources for Al and Pb on the continent are similar. They both stem from widespread ground level sources and arrive in the German Bight in a well mixed state. The calcu-

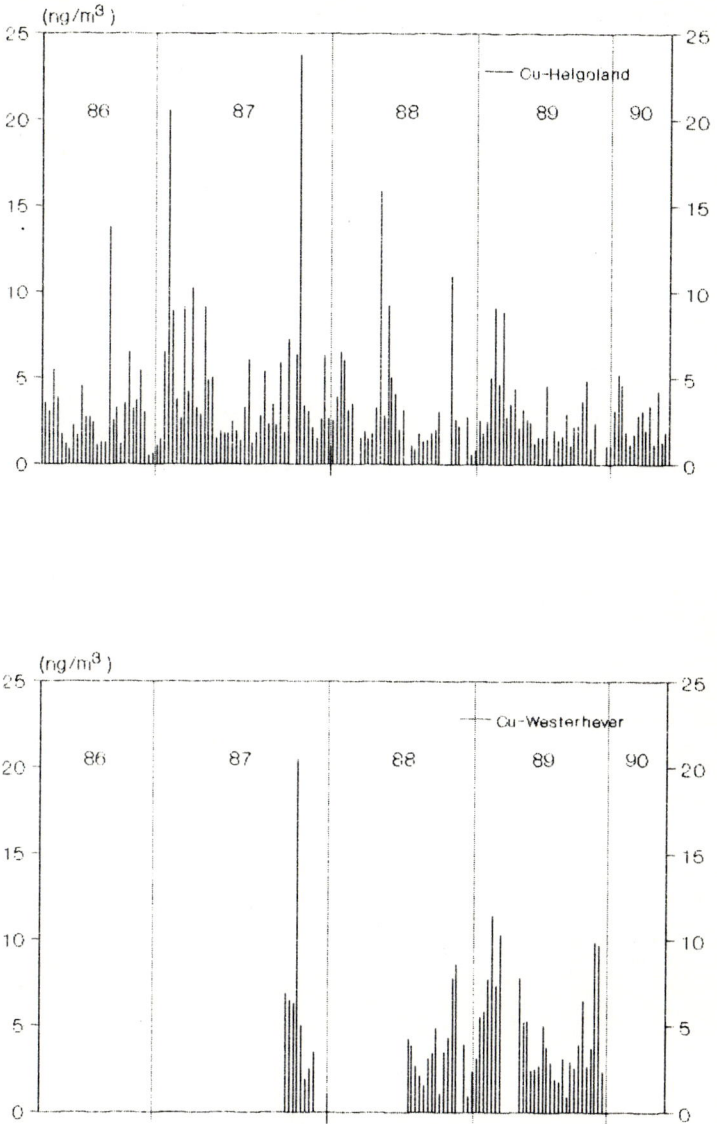

Fig. 2.4.11. Time series of ambient Cu concentration

lated enrichment factor (EF_{crust}) from the equation

$$EF_{(crust)} = \frac{Pb_{(ambient\,air)}/Al_{(ambient\,air)}}{Pb_{(earth\,crust)}/Al_{(earth\,crust)}}$$

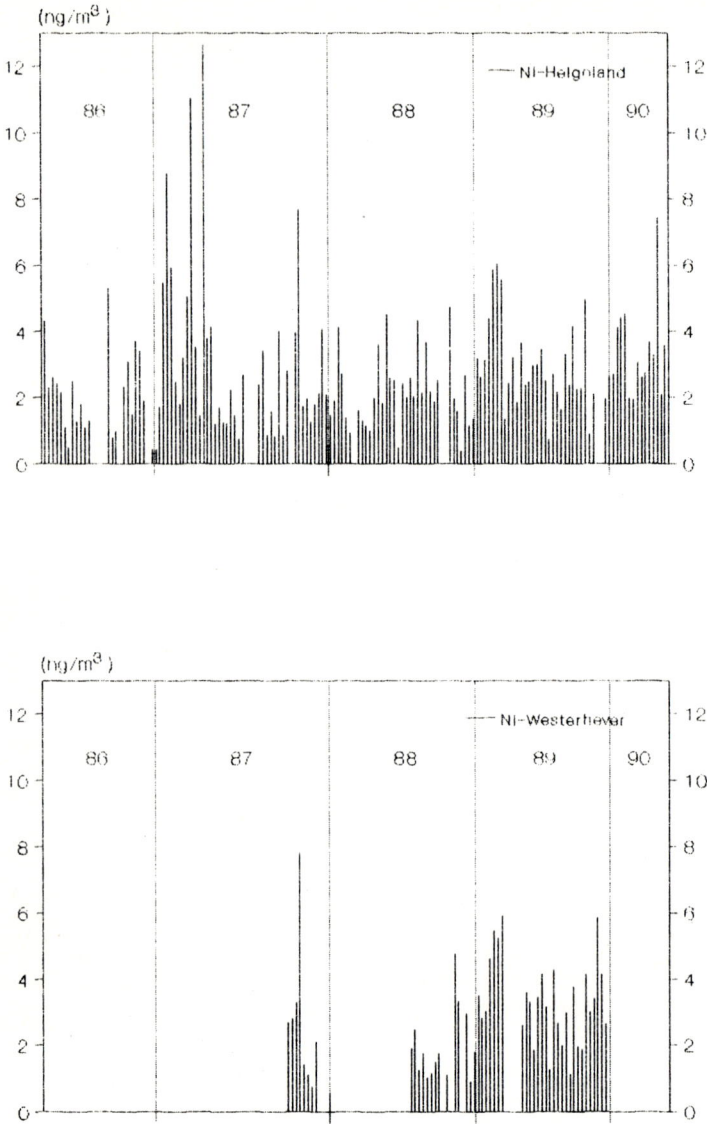

Fig. 2.4.12. Time series of ambient Ni concentration

eliminates the influence of meteorological conditions during the sampling period, for instance higher frequencies of less polluted westerly winds in some years.

To see whether the variable meteorological conditions were not respon-

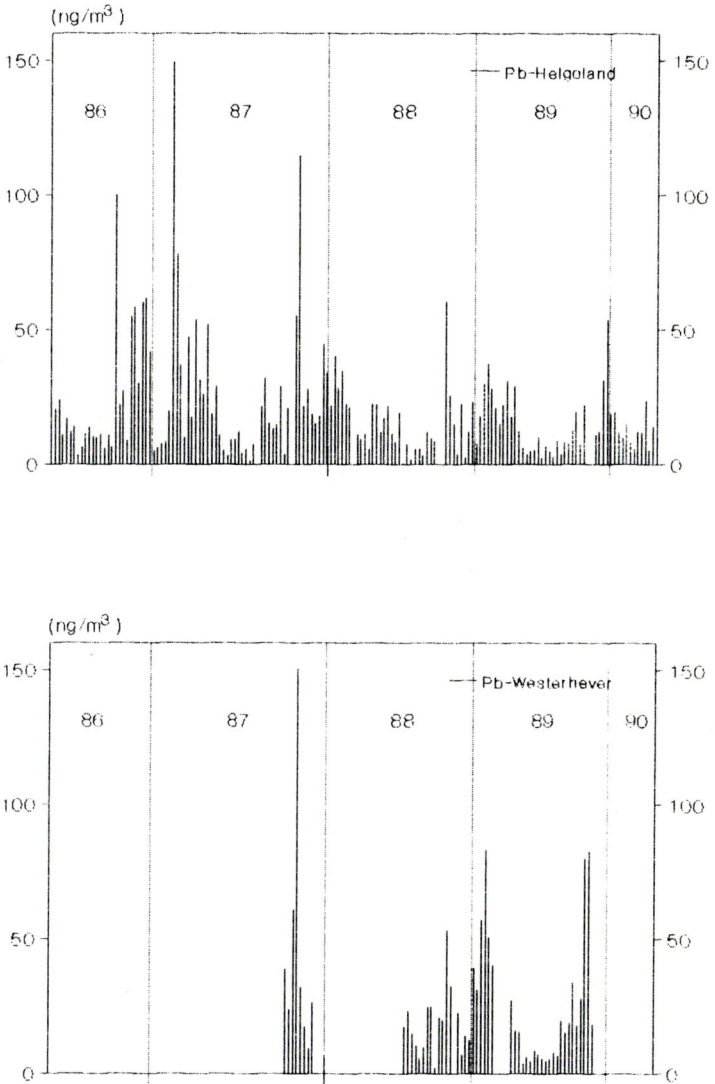

Fig. 2.4.13. Time series of ambient Pb concentration

sible for the decrease in Pb concentration, further calculations were
made. Taking into account the different wind direction frequencies in the
sampling period (1986-1990) yearly mean concentrations are predicted on
the basis of wind-directed sampling (see below in this Chapter). These

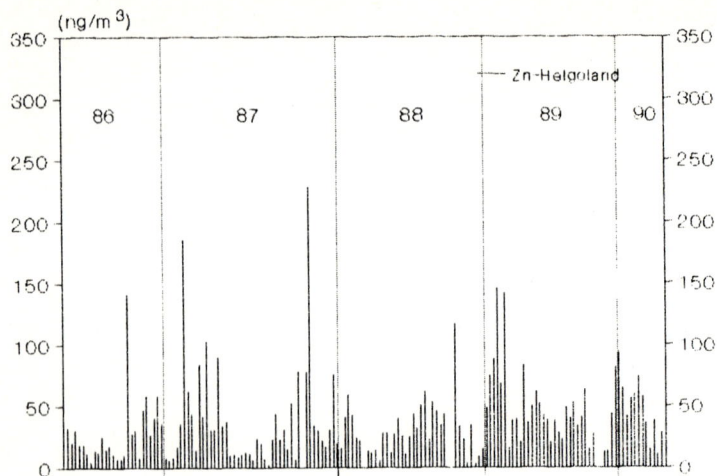

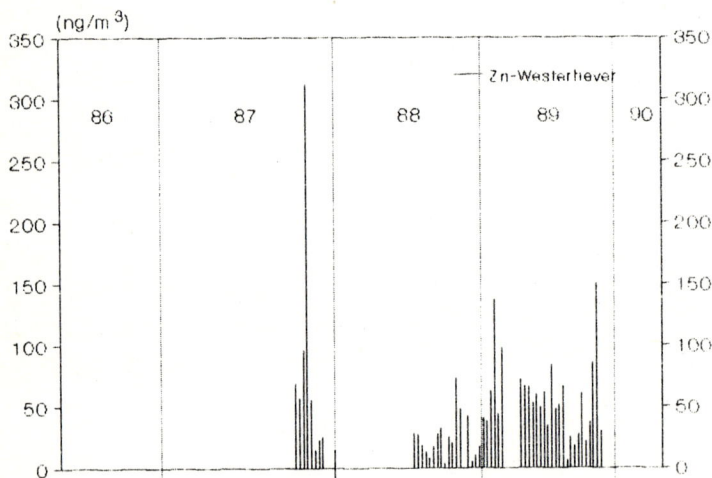

Fig. 2.4.14. Time series of ambient Zn concentration

results showed that the measured Pb concentration should not have decreased as much as was actually observed during the period of investigation. Predicted concentrations of Pb are 24 ng/m^3 for 1987, 22.2 ng/m^3 for 1988 and 22.3 ng/m^3 for 1989.

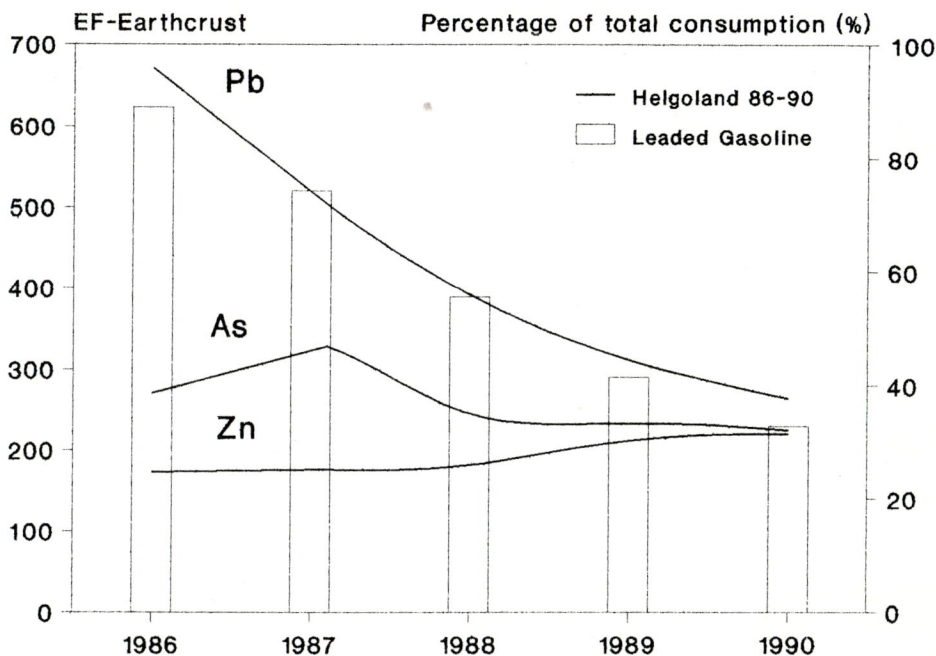

Fig. 2.4.15. Enrichment factor earth crust for three selected anthropogenic elements in the aerosol in comparison with percentage of leaded fuel consumption

2.4.5   Sources of Atmospheric Pollutants

2.4.5.1 Wind-directed Aerosol Sampling

The main purpose of the strategy described here (Kriews et al. 1988) is to obtain more information about the origin and spatial distribution patterns of trace elements polluting the atmosphere above the North Sea by using a sampling system taking into account direction and speed of the local horizontal wind.

One or more 30°-sectors can be chosen as sampling sectors on four channels. The system ensures that each filter is only charged with aerosol from a selected sampling sector. Situations with low wind speeds (< 3 m/s) and variable wind directions are sampled separately on two further units.

Figure 2.4.16 shows the position of the sampling site on Helgoland and the selected sectors. Channel 5 is not evaluated because of a local source

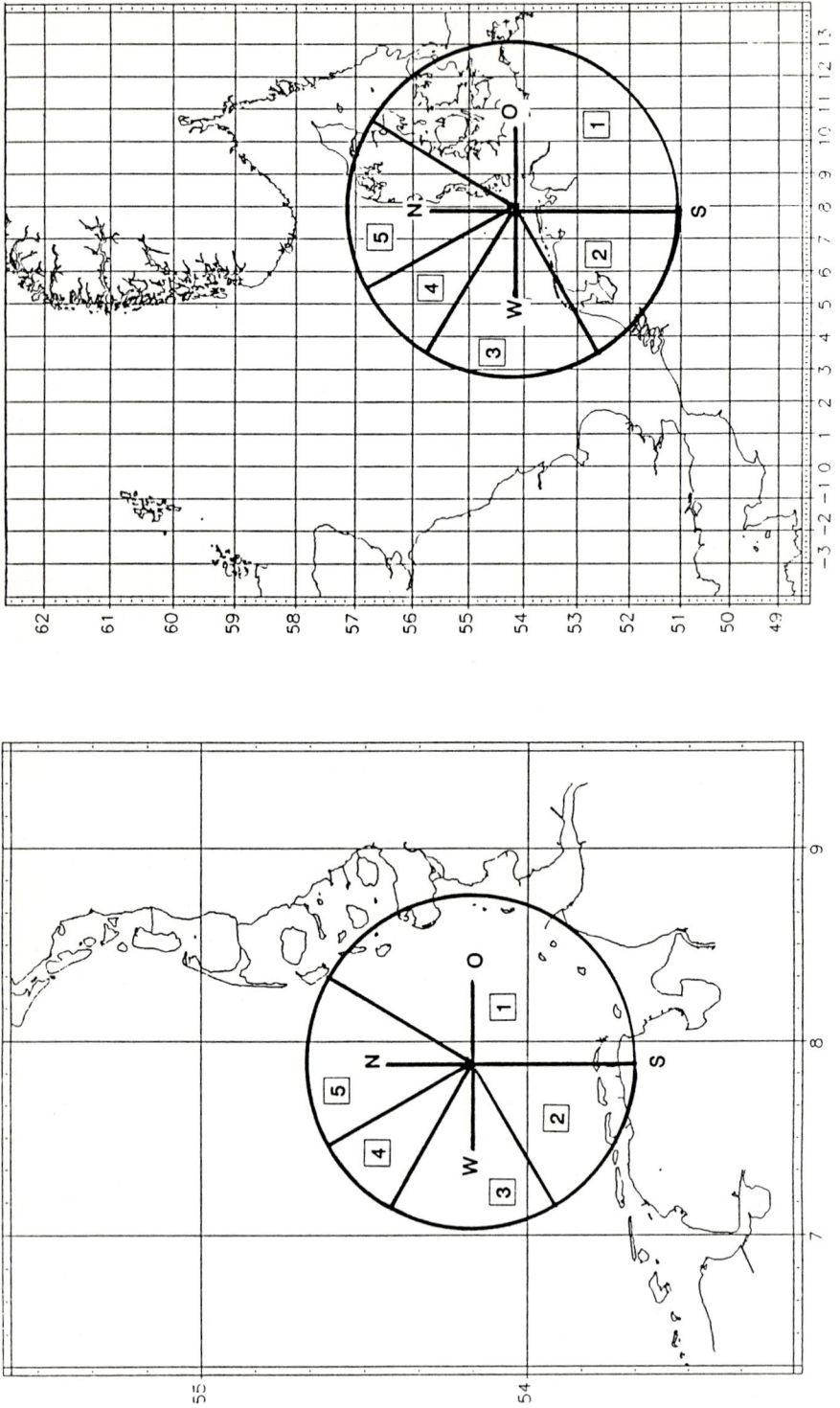

Fig. 2.4.16. Position of the sampling site at Helgoland in the German Bight and chosen sampling sectors for wind-directed aerosol sampling

(oil power plant) influencing this direction. The drifting winds are also sampled on the channel 5 filter.

First results from a sampling period between April 1987 and September 1988 (16 samples) are presented in Fig. 2.4.17a-g, together with an integral calculated, elemental concentrations as well as the weekly (n=58 samples) concentrations (Kriews 1992) during the same sampling period.

Since the filters for low wind speeds (< 3 m/s) and variable winds (sampled on channel 5 together with wind from the direction 330°-30°) do not show significant deviations from the integral value, they were not plotted in Fig. 2.4.17a-g.

The easterly and the southwesterly winds (sector 1 and 2) showed the highest concentrations of "anthropogenic elements". This clearly underlines the influence of the industrialized areas (northern and eastern part of FRG, Eastern Europe, the Ruhr region and the Benelux countries), which are closest to the German Bight with respect to air pollution.

By normalizing the measured elemental concentrations in each transport sector to the long-term wind frequency distribution reported by the German weather service for Helgoland (1971-1980 and 1987-1989), and assuming that these wind statistics are representative for the air masses arriving at Helgoland, yearly elemental concentrations and percentages for each transport sector are calculated. Our estimates for the elements which are shown in Fig. 2.4.17a-g are presented in Table 2.4.3.

Our estimates conform well to the yearly measured mean concentrations in 1986/90 and to samples from the same sampling period from weekly integral aerosol sampling.

These results and the results from Steiger (1991), with high time resolution aerosol sampling (12 h) during several measuring campaigns in 1987 evaluated in combination with backward trajectories agree well. The results show that wind-directed sampling gives representative aerosol samples and, consequently, distribution patterns for the trace elements investigated in this study.

2.4.5.2 Source Apportionment Based on Backward Trajectories

In order to estimate the contributions of different European source regions to the elemental concentrations in the German Bight, a total of 265 12-hourly aerosol samples collected in 1987 at the four sampling sites Hage, Westerhever, Helgoland and research platform *Nordsee* were classified according to transport sectors. The five sectors chosen (see Fig. 2.4.18) represent former East Germany, Poland and the northern part of former West Germany (sector 1), Belgium, the Netherlands, the northern part of

France as well as the Ruhr district (sector 2) and the UK (sector 3). The remaining two sectors represent the transport of less polluted air masses from the North Atlantic (sector 4) and Scandinavia (sector 5).

The samples were classified according to 72 h back trajectories of surface wind (see Chap. 3.1). Only those events were considered where the origin of the air masses arriving at the receptor region could be clearly associated with one of the five sectors. Therefore, 75 cases remained unclassified.

For correction of natural sources, a chemical element balance receptor model (CEB model) was applied to calculate the contributions from

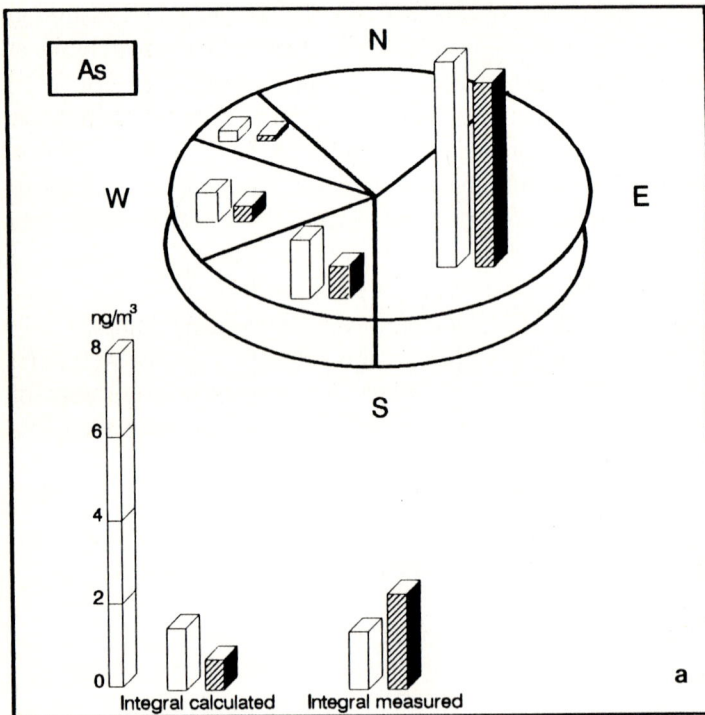

Fig. 2.4.17a-g. Ambient elemental concentration distribution for the transport sectors during the sampling period from 14.4.87-15.9.88. *White bars* indicate the average, *shaded bars* the standard deviation

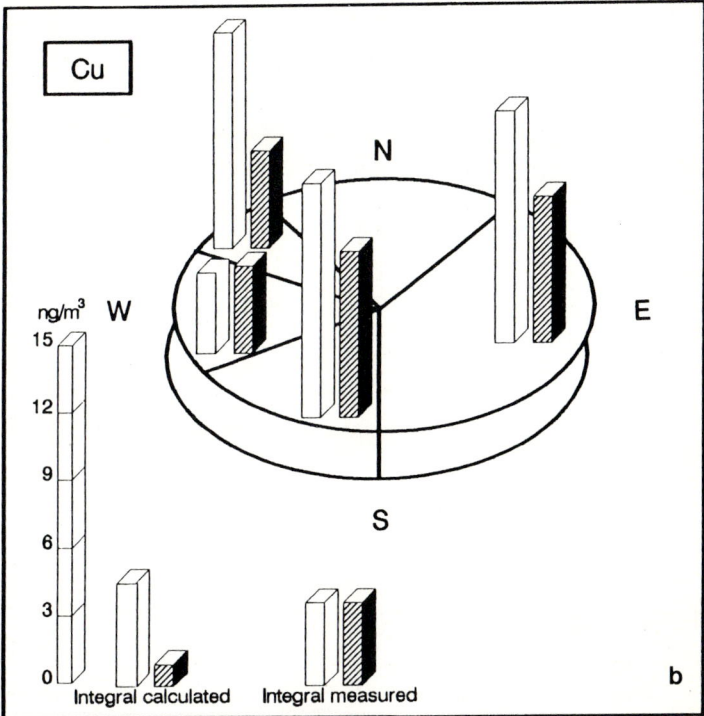

b

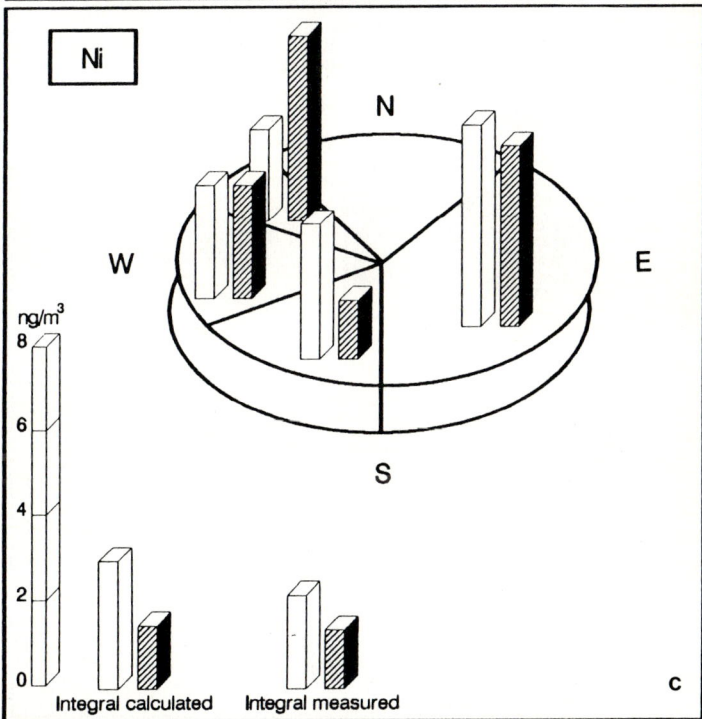

c

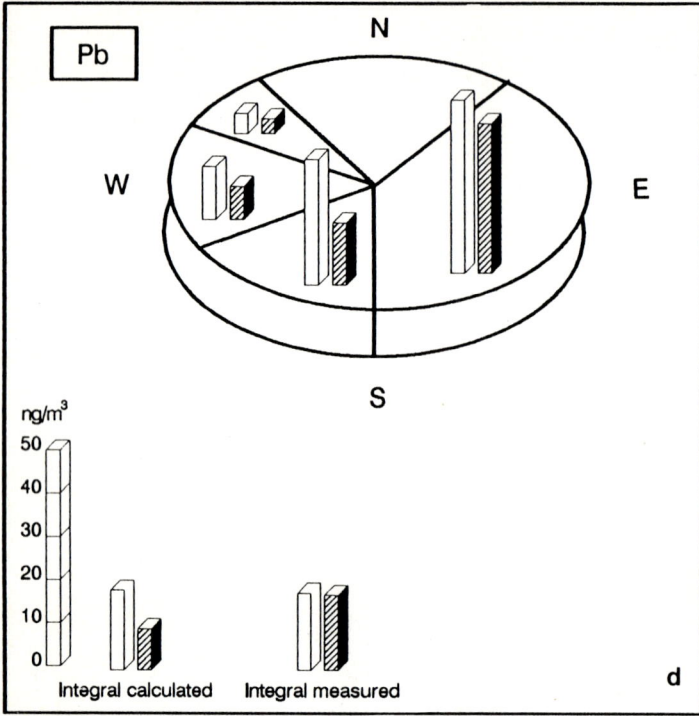

d

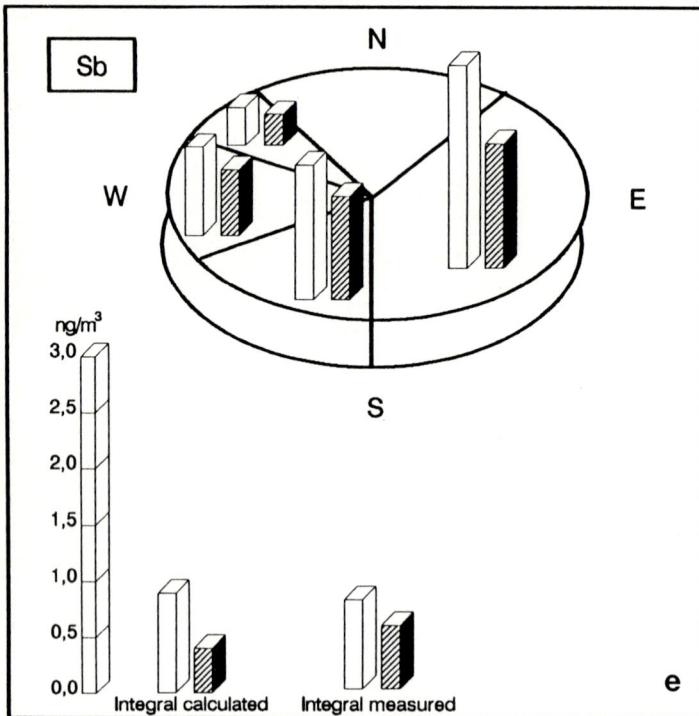

e

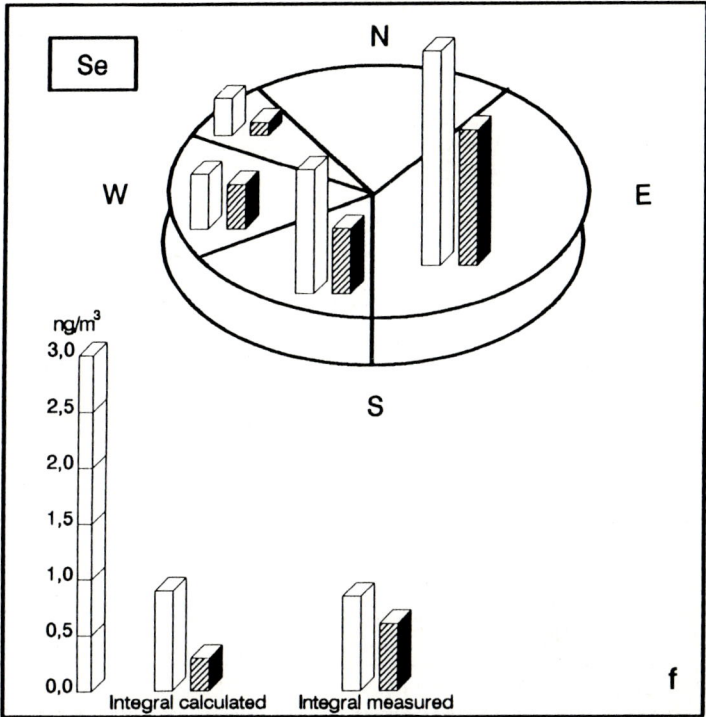

Se

ng/m³

N

W

E

S

Integral calculated   Integral measured

f

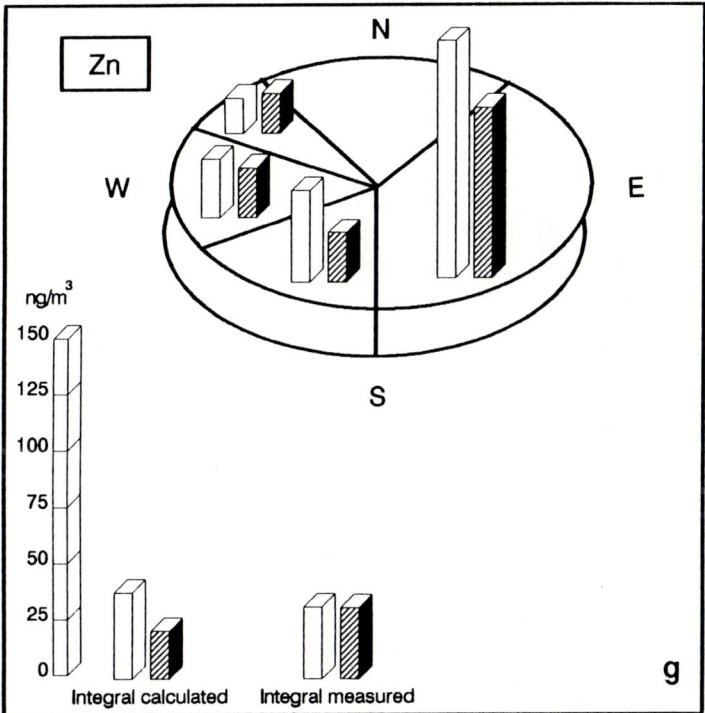

Zn

ng/m³

N

W

E

S

Integral calculated   Integral measured

g

Table 2.4.3. Wind direction frequency distribution (1971-1980 and 1987-1989) and percentage contribution of source regions in different wind sectors related to long-term average mean concentrations

Percentage contribution of wind sectors to integral concentration					Predicted means	Measured	
Sector[a]	30°- 180° (%)	180°- 240° (%)	240°- 300° (%)	300°- 330° (%)	(ng/m^3)	86/90[b] (ng/m^3)	87/88[c] (ng/m^3)
Wind frequency	33.7	17.0	22.2	11.1			
As	79.4	11.5	7.6	1.4	1.7	1.7	1.4
Cu	49.2	24.7	11.3	14.8	6.0	3.5	3.7
Ni	54.4	17.9	19.6	8.1	2.5	2.6	2.2
Pb	62.3	22.7	12.5	2.5	18.0	20.2	17.8
Sb	56.3	19.9	20.1	3.7	0.86	0.96	0.80
Se	66.2	19.1	10.9	3.7	0.82	0.98	0.84
Zn	70.8	13.9	11.7	3.6	42.3	38.2	32.3

[a]Wind-directed aerosol sampling from 14.4.87-15.9.88 (Kriews 1992)
[b]Yearly mean concentrations from weekly aerosol sampling
[c]Mean concentration for the period from 14.4.87-15.9.88 from weekly aerosol sampling (Kriews et al. 1989)

different sources based on the measured concentrations of the elements Al, Ti, Ca, Mg, Sr and Na. Details of this procedure are given in Steiger (1991). The contributions of natural sources to the concentrations of the elements discussed here are negligible (< 2 % for As, Cd, Pb, Sb, Se and Zn; < 5 % for Cu).

Assuming that the long-term frequency distribution of the local wind direction in the receptor area is a convenient approximation for the frequency of the arrival of air masses from the five sectors, it was possible to use this information to weight the elemental concentrations given in Table 2.4.4. Meteorological data obtained at Helgoland over a period of 9 years (1981-1989) were used to calculate the frequency distribution of wind direction according to the five-sector classification.

The results of this approach are summarized in Table 2.4.5. The same approach could not be used for the elements V and Ni, as it was found that

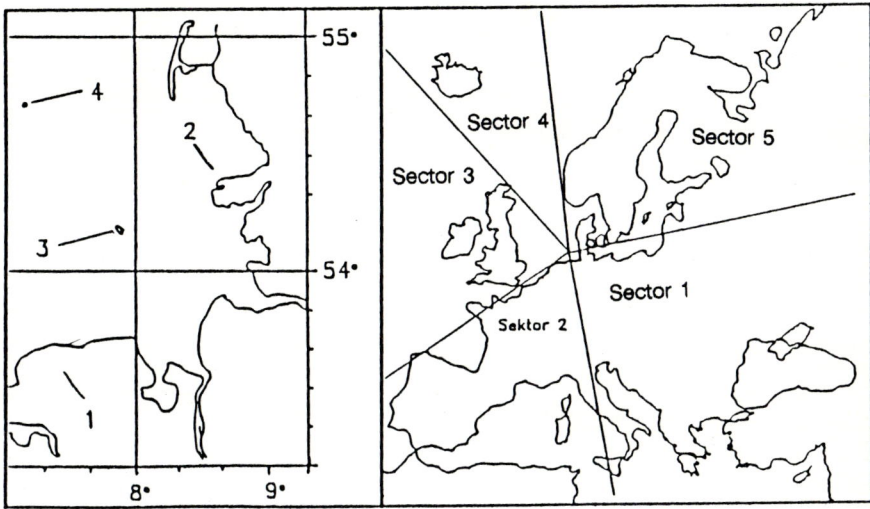

Fig. 2.4.18. Transport sectors chosen for classification of 12 hourly ambient concentrations with backward trajectories

both elements are strongly affected by local emissions from ship engines burning residual fuel oil (Schulz et al. 1988).

Steiger (1991) estimated the contribution of local ship emissions to account for about 27-52 % of the ambient V concentration on the average. Similar results were obtained for Ni.

For the other trace elements, an excellent agreement was found between calculated long-term average concentrations and those measured at the Helgoland site between 1986 and 1990, indicating the model assumptions to be reasonable. As could be expected, the contribution of air masses originating in sectors 4 and 5 is of minor importance. The most relevant source regions are located in sector 1, which is probably a result of the prevailing dispersion characteristics. Stable atmospheric conditions leading to high pollutant concentrations in the boundary layer are often associated with southeasterly flows. At present however, sources in the northern part of Germany, like the city of Hamburg, and more distant but possibly stronger sources in former East Germany and Poland cannot be distinguished from each other. For the reasons mentioned, the contributions of source regions located in sectors 2 and 3 are somewhat lower when compared to sector 1, but the results clearly indicate that they cannot be neglected.

Table 2.4.4. Averages and standard deviations of elemental concentrations for the five transport sectors (ng/m^3). N is the number of samples. The values are corrected by subtracting contributions from natural sources such as crustal shale and limestone as well as resuspended sea salt particles

	Sector 1 N=85	Sector 2 N=25	Sector 3 N=12	Sector 4 N=46	Sector 5 N=22
As	8.0 ±10.6	0.9 ±0.6	0.9 ±0.6	0.3 ±0.5	0.5 ±0.5
Cd	1.6 ±1.5	1.5 ±2.7	0.7 ±1.2	0.8 ±1.1	0.6 ±0.8
Cu	8.5 ±7.8	3.7 ±2.2	2.2 ±1.6	1.4 ±2.4	1.9 ±1.8
Pb	63.5 ±41.5	34.7 ±17.2	15.2 ±11.6	9.1 ±13.9	15.5 ±17.5
Sb	1.8 ±1.4	1.4 ±1.1	0.6 ±0.7	0.5 ±1.3	0.4 ±0.6
Se	2.4 ±1.9	0.9 ±0.4	0.8 ±0.5	0.3 ±0.3	0.3 ±0.3
Zn	98.9 ±85.0	31.9 ±18.1	15.3 ±12.7	9.7 ±13.1	15.6 ±13.4

75 cases not classified.

2.4.6 Measurements of the Atmospheric Input

Airborne pollutants reach the marine environment through dry and wet deposition, making a systematic division necessary according to the methods of measurement. Different approaches have been utilized and these are briefly discussed to promote understanding of the differences found in the flux measurements. The actual spatial and temporal variation of the atmospheric input has important consequences for the impact on the marine environment. In order to obtain information on the horizontal distribution of deposition, the most seaward permanent station available, the research platform *Nordsee* and a coastal station on the Eiderstedt Peninsula, 100 km away, were chosen as measurement sites.

2.4.6.1 Dry Deposition

The process of dry deposition is often considered to be controlled by diffusion and absorption processes in the marine boundary layer and at the sea surface. While the dry deposition of gaseous components like nitrogen oxides can be treated by Henry's law and the diffusivity in the atmosphere,

Table 2.4.5. Wind direction frequency distribution and percentage contributions of source regions in sectors 1 to 5 to long-term average elemental concentrations (corrected as in Table 2.4.4)

	Percentage contributions of source regions					Concentration	
Sector	1 (%)	2 (%)	3 (%)	4 (%)	5 (%)	Calculated (ng/m^3)	Helgoland (ng/m^3)
Wind frequency	20.3	19.3	34.8	8.6	16.9		
As	73.1	8.1	13.6	1.1	4.1	2.2	1.7
Cd	32.0	27.6	24.4	6.8	9.2	1.0	1.6
Cu	47.4	19.5	21.2	3.2	8.7	3.7	3.5
Pb	45.6	23.7	18.6	2.8	9.3	28.3	20.2
Sb	37.0	27.7	22.8	4.8	7.8	1.0	0.96
Se	47.9	17.4	27.2	2.2	5.3	1.0	0.98
Zn	57.3	17.6	15.2	2.4	7.5	35.1	38.2

further processes have to be considered for the deposition of airborne particulate matter (Hicks et al. 1980). Gravitational settling, impaction due to the inertia of the aerosol particles and the absorption properties of the sea surface act on a population of particles distributed in size from 0.01 to 100 µm. The different size classes are specifically affected by these deposition processes. The marine surface adds another complication by the fact that the high humidity close to the sea surface might alter the size distribution. In effect the growth of hygroscopic particles will increase the overall deposition velocity.

With a double impactor system, optimized to resolve the bimodal size distribution of many elements in the aerosol, 15 sets of measurements were carried out at the research platform *Nordsee* and during North Sea cruises

Table 2.4.6. Calculated mean dry deposition velocities (n = 52) and standard deviations from impactor measurements on Helgoland, the FPN and the RV *Poseidon* according to the calculation scheme given by Slinn and Slinn (1980), assuming 10 m/s wind speed and hydrophobic particles with a density of 2.5 g/cm^3. Average fluxes calculated with the concentrations measured at the three stations

	Overall dry deposition velocity (cm/s)	Dry deposition flux (μg/m^2/d)		
		FPN	Helgoland	Westerhever
As	0.18 ± 0.08	0.42	0.27	0.39
Cd	0.09 ± 0.06	0.07		0.04
Cu	0.26 ± 0.09	0.89	0.80	1.09
Ni	0.26 ± 0.13	0.53	0.58	0.62
Pb	0.15 ± 0.07	3.74	2.65	3.53
Zn	0.28 ± 0.11	10.30	9.28	12.33

with the research vessel *Poseidon* in order to obtain a more accurate calculation of the dry deposition flux. One result was that estimates strongly depend on the correct determination of the coarse particle mode (Steiger et al. 1989). Finally, another set of wind-directed impactor measurements at Helgoland was included to calculate overall average dry deposition velocities as found in Table 2.4.6. The contributions of the different size classes have been integrated for this purpose. These velocity values were multiplied with the mean ambient concentration found at the research platform, on Helgoland and at Westerhever to calculate a dry deposition flux of the selected heavy metals.

Interestingly, the elements with a more salient appearance in the coarse mode of the aerosol, having higher deposition velocities, show the greater variation when inferring the mean deposition velocity from impactor measurements. This can be explained by the higher variability of the coarse mode due to the rapid sedimentation process and by the relatively high uncertainty in determining the coarse particle size distribution. This finding has important consequences for measuring deposition with surrogate sur-

Table 2.4.7. Estimated mean dry deposition and standard deviations for each transport sector ($\mu g/m^2/d$) (16 samples)

	30°-180°	180°-240°	240°-300°	300°-330°
As	0.68	0.21	0.17	0.07
	±0.6	±0.21	±0.10	±0.04
Cu	1.70	2.33	0.98	2.56
	±1.07	±1.66	±1.04	±1.15
Ni	1.72	0.68	0.57	0.55
	±1.53	±0.31	±0.58	±1.10
Pb	4.79	3.22	2.51	0.99
	±4.14	±1.58	±1.57	±0.72
Sb	0.19	0.17	0.09	0.03
	±0.11	±0.11	±0.06	±0.03
Se	0.13	0.08	0.12	0.05
	±0.08	±0.04	±0.06	±0.02
Zn	23.76	10.31	10.80	6.65
	±17.16	±5.55	±9.22	±7.44

surfaces and open funnels, since the calculation shows that 70 - 90 % of the dry deposition flux is made up of particle diameters > 1.5 µm. The poor statistical base when sampling the relatively few coarse particles may dominate the variation of the measured deposition if the sampling interval and surface is too small.

For the observed As, Cu, Ni, Pb, Sb, Se and Zn concentrations the dry deposition flux from the atmosphere to the sea surface was estimated (Table 2.4.7). The calculation is based on the average mean concentration in each transport sector and the average dry deposition velocities from size separated aerosol samples in each transport sector (Kriews 1992) using the model of Slinn and Slinn (1980).

These estimates and the percentage contribution to air pollution from each transport sector (Table 2.4.3) show the relative importance for the dry deposition fluxes in the southern North Sea area.

2.4.6.2 Total/Wet Deposition

The scavenging of particles and water-soluble gases by rain droplets has been postulated to be the most important process for cleansing the atmosphere. The process itself is not well understood, and most long-range modelling approaches therefore utilize average elemental scavenging ratios. In experimental studies, these were derived from rain concentrations and concurrent ambient air concentrations measured at ground level. With respect to precipitation in the marine environment, many measurements have been made at coastal sites or on islands and must therefore be regarded as disturbed due to convective forcing and additional rainfall at the coastline. Up to now, the determination of the precipitation height (h_i) and, consequently, the deposition above the oceans and marginal seas suffers from shortage of reliable data, especially when regional atmospheric input assessments are to be made.

The atmosphere serves as a reservoir of pollutants, and successive rain storms cleanse the same air mass consecutively. When the wet deposition flux F_{wet} is to be derived from the rain concentrations (c_i) of any pollutant, its accumulation from all successive rain events, i, has to be considered:

$$F_{wet} = \sum_i c_i h_i .$$

The chosen sampling procedure therefore has to ensure that both parameters are measured representatively all the time. While this is not state of the art in all aspects, it was decided to rely on different systems to measure the wet deposition.

Wet-only samplers (WOS) controlled by a rain sensor are said to collect rain samples undisturbed by dry deposition, thus giving the true rain concentration. Their performance, however, relies on the sensor and on an undisturbed flow around the sampler as well as in the vicinity of the site. Wet-only samplers, currently in use at many monitoring sites on land, first had to be technically improved before they were able to withstand the conditions at sea. Since such an instrument could not be mounted atop the mast of the research platform, it was installed on a stage 4 m above the main deck. In Westerhever it was possible to mount the sampler right behind the dike, freely exposed to the west.

As a second instrument, a simple and reliable total deposition sampler (TDS) was designed, consisting of a plastic funnel connected to an exchangeable bottle. Evaporation was found to be negligible and therefore rain concentrations could be determined as well. Alhough in the permanent open funnel dry fall-out is also collected, the data for the heavy metals of interest showed that on the yearly average the calculated dry deposition flux is about 20 % of the measured total deposition fluxes. In the following text the TDS "rain" concentrations of the trace metals were therefore assumed to be valid to allow for an easy intercomparison. The light construction of the sampler allowed it to be mounted atop the mast of the research platform almost 20 m above the main deck and 45 m above sea level. In Wester-hever, the same sampler was mounted beside the wet-only sampler.

Time series of weekly measured depositions and rain concentrations for Pb, Cd, Cu, Ni, Cr and As (Schulz 1992) as well as nitrate and ammonium (Schwikowski 1991) from both stations and both types of samplers at the same time are available for the period from 8.8.89 to 31.7.90. Evaluation is therefore based on this data set. As an example, the lead deposition is depicted in Fig. 2.4.18. These time series are summarized in Table 2.4.8.

The different collectors show sufficiently corresponding averages for a period of one year. Except for the wet-only sampler at the platform, all samplers collected 70-90 % of the rain amount sampled at the same time at the most adjacent island sites. The reason for this finding may be the diminished collection efficiency under high wind speeds, e.g. at the exposed mast-top of the research platform. Unfortunately, no correction function is available to assess this effect quantitatively. On the other hand, as indicated above, few such series exist for the North Sea and a comparable decrease in precipitation is often assumed.

The deposition data derived from the wet-only samplers differ to an unexpected degree from each other and from the TDS. The small difference between the component ratios at the FPN and in Westerhever indicates that contamination on the mast-top of the FPN is improbable. Trace metal deposition measurements in the FPN-WOS seem to be disturbed. To study the FPN wet-only sampler performance, a second TDS of the same design as the one on top of the mast was mounted beside the WOS for 10 months. Comparison showed that this sampler also collected only 56 % of the precipitation amount determined by the mast-top sampler. The structure of the platform is thus believed to distort the flow field and to catch part of the rain, and because of that, sampling on the stage position can only be repre-sentative at low wind speeds.

The differences in deposition between the TDS and the WOS in Wester-hever seem to suggest a large proportion of dry deposited material. How-ever, if the difference between TDS and WOS could be attributed totally to dry deposition, then the dry flux estimates given above are almost three times too small. We believe that the flow distortion around the bigger WOS structure itself changes the sampling efficiency for different rain events. Especially rain events with weak intensity, concentrated small droplets and

Table 2.4.8. Deposition measurements of lead and nitrogen with two types of collectors at the coastal site Westerhever/lighthouse and the research platform *Nordsee* (FPN) (8.8.89 - 31.7.90)

	Precipitation amount (mm)	Pb deposition $(mg/m^2/a)$	NH_4^+-N $(g/m2/a)$	NO_3^--N $(g/m^2/a)$
FPN				
Total mast-top	408	5.3	500	720
Wet-only	207	6.1	360	280
Westerhever				
Total	480	5.0	880	670
Wet-only	495	2.9	660	380
Precipitation recorded for the same period by the German Weather Services				
Helgoland	668			
List/Sylt	556			

high wind speeds might be missed when measuring the real deposition flux into the sea.

However, the deposition characteristics of each rain event together with the collector performance determine the deviations between the samplers in the weekly data to a high degree. Up to now, these two effects could not be distinguished from one another.

As can be seen from the time series in Fig. 2.4.19 wet-only deposition in Westerhever exceeds "total" deposition for some of the weeks. Weeklylead concentrations in rain and wet-only as well as total depositions as a function of precipitation amount are shown for all measurements of the period (Figs. 2.4.20a,b). The inverse relationship between rain concentration and precipitation amount might explain that total deposition of lead and nitrogen at Westerhever and at the platform are almost equal (Table 2.4.8). Light rains prevailing at sea may scavenge pollutants more efficiently from an air mass. As a consequence, the strong deposition gradients predicted by models with constant scavenging ratios cannot be found, even if precipitation amount decreases to the open sea.

The similar deposition pattern found for lead on the mast-top of the research platform and at the coastal station puts a question mark on the relevance of the process of resuspension of trace metals in the North Sea region. This process has gained some prominence in recent publications, and it has been suggested that the sea would be a source for some trace metals. This may be of importance in remote regions with substantially reduced atmospheric concentrations.

The few values of high deposition along with low precipitation suggest that contamination at the platform, especially at the top of the mast, was not the reason for the high concentrations found (Fig. 2.4.20b). The scatter in deposition versus precipitation for larger rain amounts reflects the variability of rain storms scavenging the lead either from polluted air masses or, e.g. from clean North Atlantic air.

The scatter of concentrations of trace metals in rain as well as their wet deposition lead us to the conclusion that simple arithmetic averages of rain concentrations should not be used to assess the wet deposition flux. Also, deposition or precipitation weighted concentration averages will continue to bear uncertainties until further investigations have shown how different

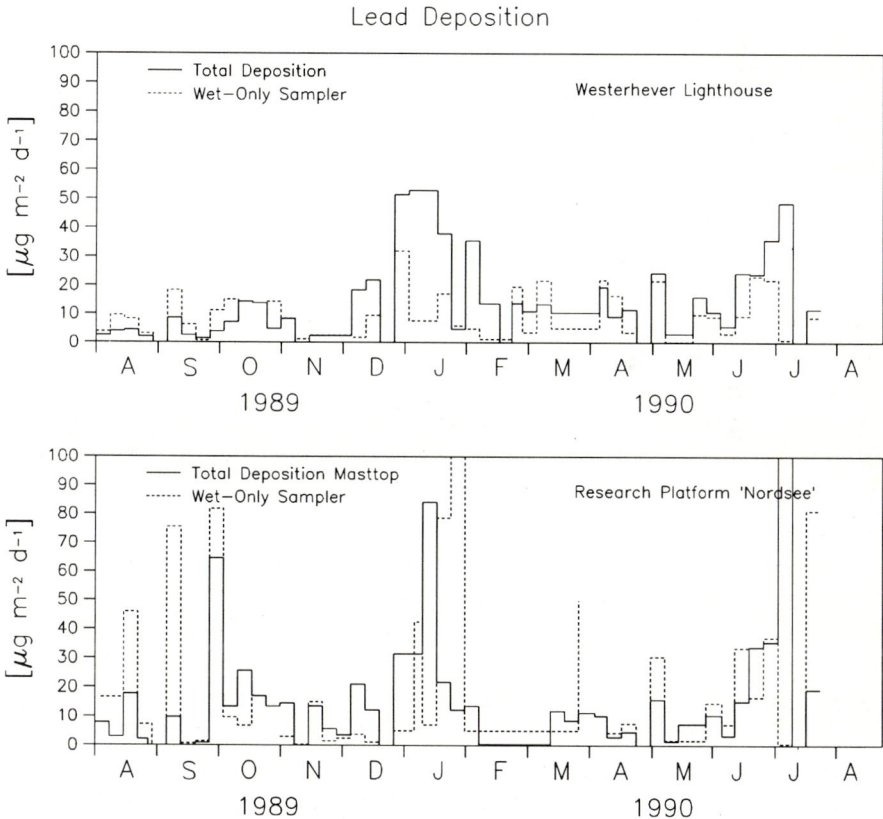

Fig. 2.4.19. Time series of lead deposition measured at two stations in the German Bight

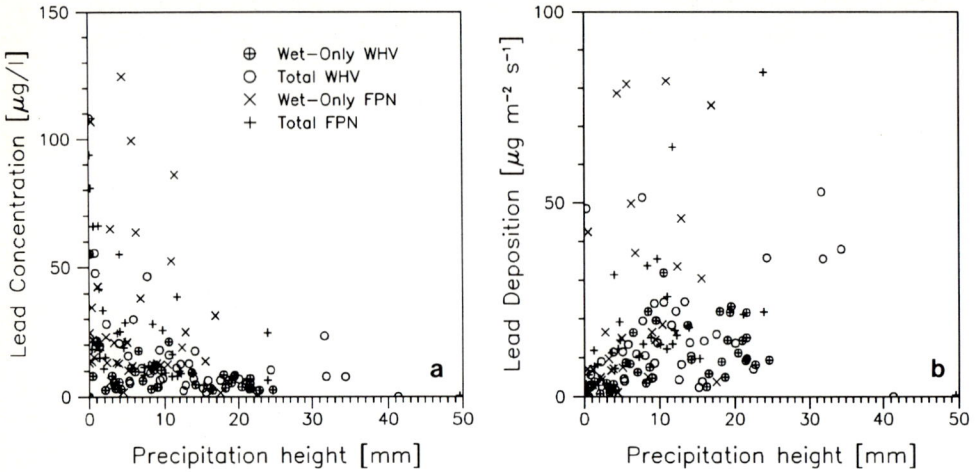

Fig. 2.4.20a,b. Pb rain concentrations and deposition versus precipitation during sampling period 8.8.89 - 31.7.90 in Westerhever (WHV) and at the research platform *Nordsee* (FPN)

collection efficiencies for different wind speeds affect the precipitation determination. For the calculation of the wet deposition flux the median of the heavy metal concentrations in rain, as determined from 161 samples at the FPN and in Westerhever, was taken to be representative. In Table 2.4.9 atmospheric input estimates from this analysis for an area of the size of the North Sea (525,000 km^2) are shown. This was done, although extrapolation to such a great area might be considered to be not fully justified. Deposition in the northern part of the North Sea is probably less, although the data presented here show small differences between a coastal and a sea-based station along with large uncertainties.

In comparison with dry deposition estimates given above, the importance of wet deposition is evident. These estimates are well in the range of the latest GESAMP report (Duce et al. 1990), the investigations made by the GKSS (Stössel 1987; Graßl et al. 1989) and Belgian projects (Baeyens et al. 1990; Injuk et al. 1990). However, our estimates have the advantage of utilizing a sea-based station in direct comparison with a coastal station, as discussed above. The results presented here weaken recent criticism on

Table 2.4.9. Rain concentrations of TDS and WOS samples from the German Bight (N = 161) and the annual atmospheric input inferred from these data, presuming 500 mm precipitation. For comparison, total deposition fluxes were calculated from the Helgoland ambient concentrations, dry deposition velocities (see Table 2.4.6) and scavenging ratios as given in the GESAMP-report (Duce 1990) combined with weekly rainfall at Helgoland. For Cd at Helgoland see aerosol part for discussion

	Median concentration WHV and FPN (μg/l)	Deposition flux into North Sea WHV/FPN data (t/a)	Deposition derived from Helgoland data (t/a)
Pb	12.2	3200	3500
Cd	0.32	84	
Ni	2.70	710	500
Cu	19.5	5100	670
As	0.60	160	160
Cr	2.0	530	400
Zn			4200
NH_4-N		263,000	-
NO_3-N		378,000	-

measurements at coastal sites, but disclose additional uncertainties in the determination of trace metals. With respect to the short-term impact atmospheric input may have, it becomes clear that a considerable range in input by single rain events has to be faced. An unfavourable event might very well affect the marine environment. For example, it is realistic that the atmospheric input within a few days may account for 500 μg/m^2 of lead and 100 mg/m^2 of nitrogen. In a water column of 30 m, this results in a seawater concentration of 15 ng/l Pb and 0.2 μmol/l N after mixing, which is in the order of magnitude encountered in coastal waters of the German Bight.

2.4.7 Aircraft Measurements

To understand contaminant fluxes in a large marine region like the North Sea, it would be of help to know to what extent different areas are affected by atmospheric pollution. Up to now, it has been difficult to obtain realistic estimates and seasonal averages for open sea regions, since measurements have not been performed there. Gross estimates can be made, however, if atmospheric transport conditions can be characterized and if the significance of the deposition rate is known. Part of our experimental studies have therefore focussed on the marine atmospheric boundary layer and on the change in continental air masses when transported over sea. Utilizing an aircraft enables the study of certain transport conditions in a quasisynoptic way.

The research aircraft (DO-128) of the TU Braunschweig was equipped with filtration units for trace constituents of the aerosol to be analyzed later in the laboratory and with sensors for the aerosol number concentration and size distribution (laser particle counter for particles with a diameter of 0.1-7.5 µm), condensation nuclei and nitrogen oxides. Meteorological sensors were utilized to measure temperature, humidity and wind at a rate of 25 Hz.

In May and June 1989, seven flight missions over the North Sea starting at Nordholz were performed under almost cloudless conditions and southerly to easterly winds. The weather conditions on the days of the flight missions were certainly not representative in a climatological sense. These weather conditions, however, have been identified as those bringing high pollutant loads to the German Bight (see above). The atmospheric input into the sea in spring and early summer might substantially alter and affect the marine ecosystem.

The most offshore position of the aircraft, at distances from the coast ranging from 250 to 360 km, depended on aircraft resources and flight regulations in the area. The flight pattern consisted of continuous ascents and descents to survey the whole mixing layer 'along trajectory'. It included also a few crossings to estimate lateral homogeneity and to measure turbulent exchange of water vapour, momentum and heat.

Meteorological data and part of the chemical data were averaged to 10-s intervals to correspond to the temporal resolution of the laser particle counter data aquiring system. The vertical profiles of particles and nitrogen oxides could be resolved in 10-50 m steps depending on the aircraft rates of climb. These highly time resolved measurements were subsequently interpreted by correlation with NH_4^+, NO_3^-, SO_4^{2-} and ambient trace metal concentrations derived from filter analysis. Turbulent fluxes were

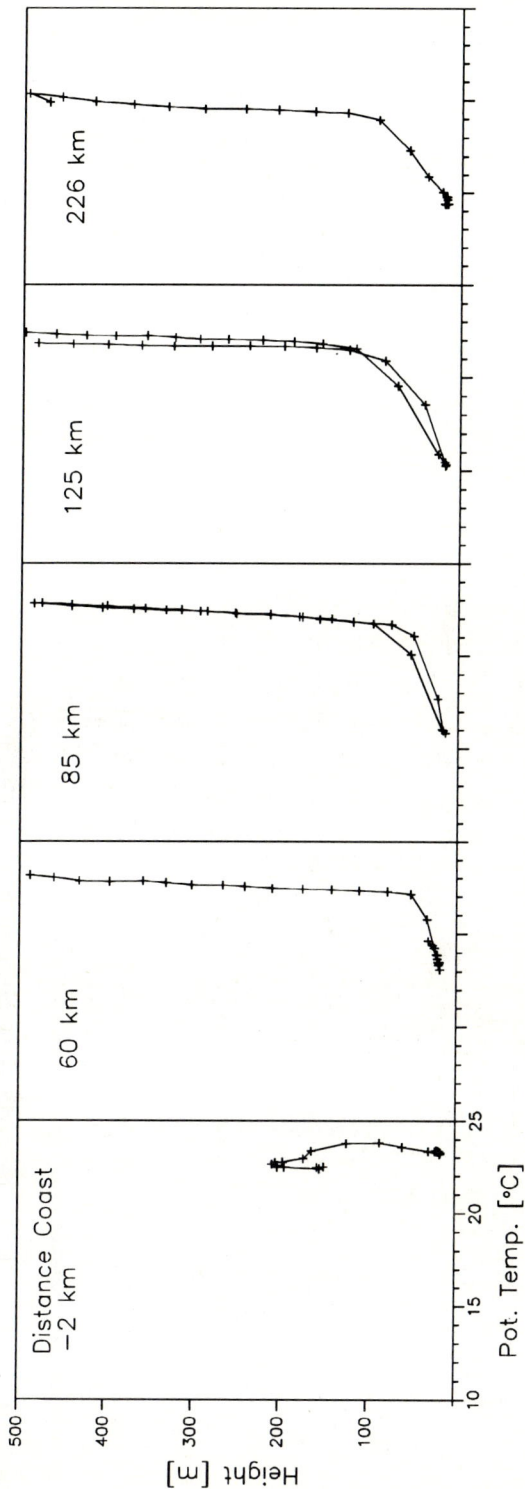

Fig. 2.4.21. Evolution of the marine boundary layer in the first 500 m along the outward flight track on 23rd of May in the afternoon

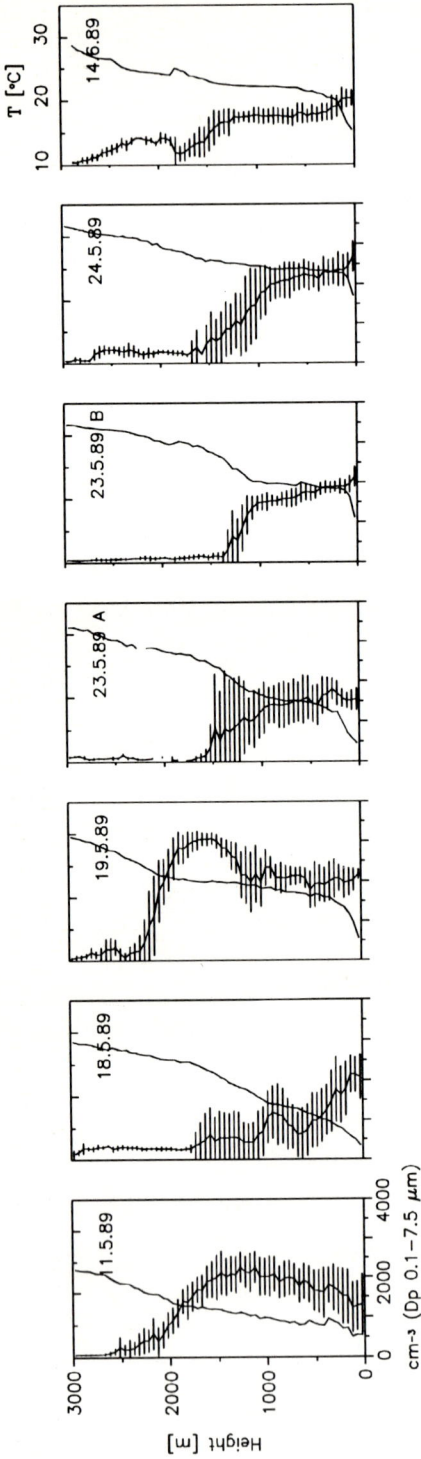

Fig. 2.4.22. Mean vertical profiles of particles (D_p 0.1- 7.5 µm) as measured with the laser particle counter aboard the aircraft. Standard deviation of particle concentration in all profiles along the flight track, calculated for 50 m height intervals. Potential temperature plotted as a clear line (Scale for temperature *above the last plot*)

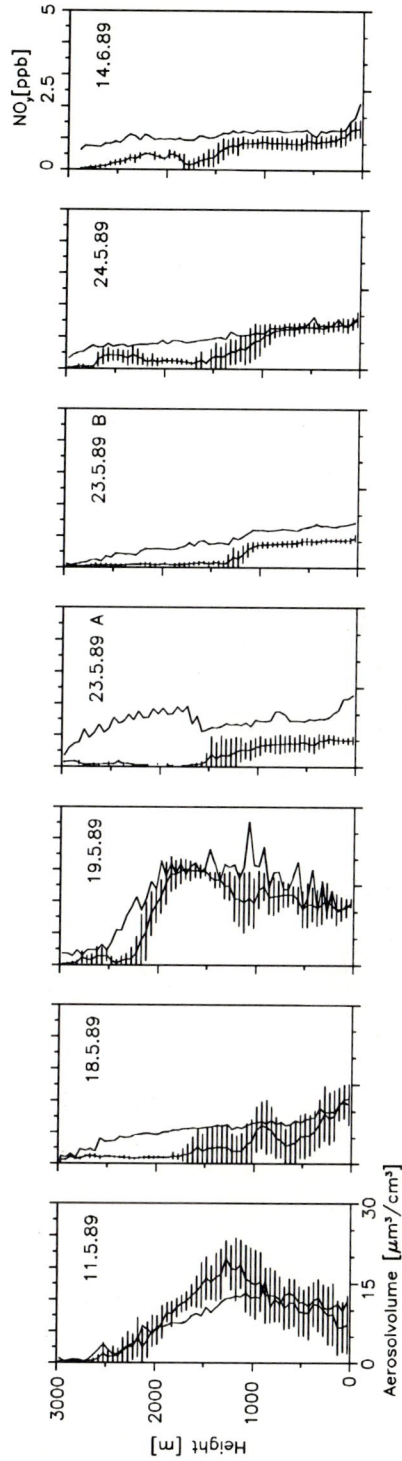

Fig. 2.4.23. Mean profiles as in Fig 2.4.22 but for the aerosol volume of the fine particle mode (D_p approx. 0.1 - 1μm) with its standard deviation and NO_y concentrations (*clear line*)(scale above last plot)

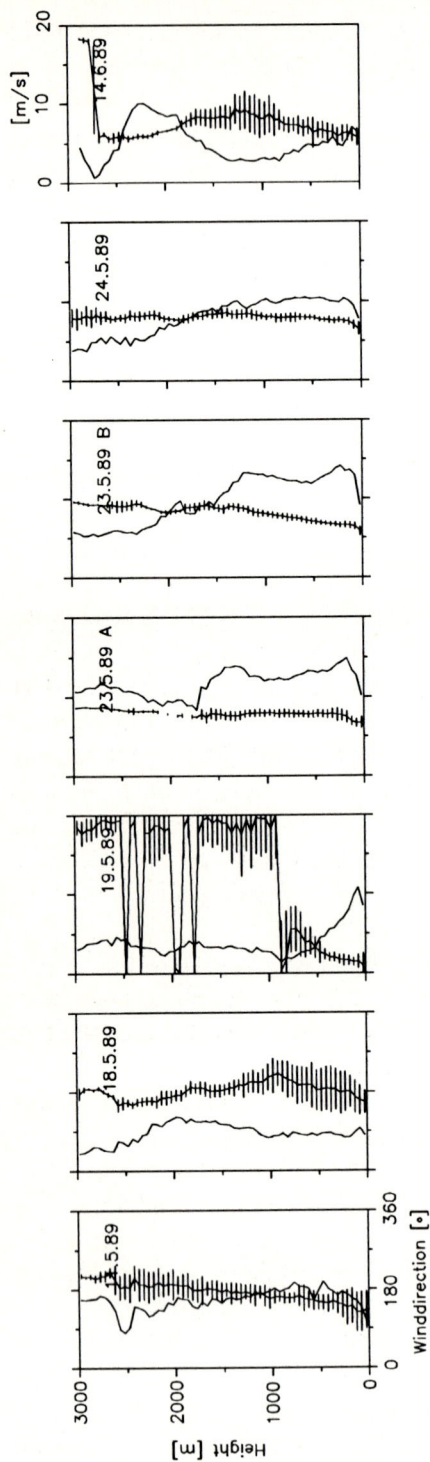

Fig. 2.4.24. Mean profiles as in Fig 2.4.22 but for wind direction with its standard deviation and the wind speed (*clear line*) (scale above last plot)

calculated from the 25 Hz data for selected periods. While evaluation of the huge data set is far from being complete, dominant features of the measurements can be summarized as follows.

Associated with high pressure systems built up over central or northeastern Europe, stable stratifications were observed in the boundary layer. Air-sea temperature differences were positive for all the days (ranging from $+0.9$ to $+7.0$ °C). In all cases, a very stable internal marine boundary layer (MBL) with 100 - 350 m vertical extent close to the sea surface could be observed in the boundary and mixing layer advected from land (its height ranging from 500 to 2100 m). In the morning, the stable surface MBL had its origin often in a nocturnal inversion over land which was advected to the sea. In the afternoon, when intensive mixing over land occurred, it evolved within a few kilometers of the coastline. The latter case is illustrated for the 23rd of May in Fig. 2.4.21. Cooling from beneath decouples the upper part of the boundary layer effectively from the sea surface. By then, vertical transport is strongly inhibited through the temperature inversion layer. This stabilization, with minor diurnal disrupture, enables effective long range transport over large sea areas when air pollutants are "trapped" in the upper boundary layer.

The chemical and meteorological data for all profiles of each flight were averaged for 50 m height intervals (see Figs. 2.4.22-2.4.24). The standard deviations calculated for these intervals certainly do not represent a random variable, since time and space dependent trends can be observed. It is obvious that the spatial averages obtained this way for the outward and return 200 km flight tracks are dominated by different processes. The mean profiles from day to day differ substantially from each other, indicating the dominant features of general transport conditions.

A dominant feature is the influence of stability on the vertical distribution of particles and nitrogen oxides. It seems that over hundreds of kilometers a layered structure of the boundary layer separated by temperature inversions is conserved. The internal MBL can be clearly identified also by physico-chemical parameters. It seems to be affected by sources and/or transformation processes of the chemical constituents different from the rest of the boundary layer. On two flights (May 23 and May 11), lower concentrations were observed in the MBL, whereas on the other days slightly higher concentrations were found. These could be attributed to the heavy ship traffic in the German Bight, being e.g. responsible for NO_y-emissions. The higher particulate load could either be due to direct particle emissions from ships or due to secondary gas-to-particle formation of particles.

The reduced potential of vertical exchange can also be seen from Table 2.4.10, where the measured momentum and water vapour fluxes are documented. Several of the flight legs from which turbulent fluxes are calculated were flown closely above the temperature inversion zone and calculated flux values are therefore very low and associated with large uncertainties. As expected, the results from closer to the sea surface show significantly higher fluxes of water vapour and momentum.

A close inspection of the concentrations of aerosol constituents found revealed that the different atmospheric pollutants are differently distributed with height (Schwikowski et al. 1990). We assumed that specific sources and characteristic transformation processes of each pollutant were responsible for this finding. While particulate nitrate in previous campaigns at the research platform *Nordsee* has been found to be present at the same level of concentration as sulphate, results obtained from the aircraft showed an excess of sulphate in the whole mixing layer due to its relative dominance in the upper layers. This was explained by the generally lower emission height of the main source of NO_y, traffic emissions, and the more rapid conversion to nitrate, SO_2, on the other hand, being mainly emitted from tall stacks having a greater potential to long-range transport.

Gas-to-particle conversion has a considerable influence on fluctuations of the fine particle aerosol concentration, since it is a steady process, depending on the history of the air mass and its contents of precursors. The change in particulate load due to gas-to particle conversion could be estimated to be 10%/h in the German Bight for the 23rd of May. Observations on this day could be evaluated, because flights in the morning and in the afternoon could be compared and because meteorological conditions stayed sufficiently stationary. It was possible to compare outward and return flights along-trajectory after correction for the distance air masses had travelled in the meantime. In Fig. 2.4.25, measurements in the lowest 100 m have been plotted along a transformed distance scale. Any position where a measurement at any time of the day had taken place was transformed to that position where the air mass had been at 6:45 GMT in the morning, assuming a transport along trajectory with the mean wind. Data points above each other should represent the same air mass. A steady increase in concentration of the fine particles with time can be seen.

Another dominant source of fluctuations in the aerosol is the diurnal cycle of emissions. Also in Fig. 2.4.25 a substantial change in concentration along-trajectory can be observed. This might be attributed to the change in source strength of the city of Hamburg, accounting for a certain delay due to transport to the German Bight. The air mass associated with a peak in concentration at about -100 km had been at approximately 06.00 h in the city of Hamburg, which tends to be the start of the rush hour.

Secondly, an atmospheric conversion rate of NO_y to nitrate could be calculated by evaluating the outward and return flight legs. Comparison of

Table 2.4.10. Turbulent fluxes of momentum, heat and water vapour as calculated from 25 Hz data from the flights with the Do-128 in May and June 1989 in the German Bight and mean meteorological parameters for the lateral crossings

	Height asl (m)	Wind dir. (°)	Wind speed (m/s)	Air temp. (°C)	Rel. humid. (%)	Mixing ratio (g/kg)	$<u'w'>$ (m^2/s)	$<T'w'>$ (K m/s)	$<m'w'>$ (g m/s)
11.5.									
15:08:40-15:20:19	440	134	8.8	5.6	52.5	3.1	-0.0014	-0.0001	0.0001
15:22:49-15:29:30	910	151	8.5	2.9	70.4	3.7	0.0002	-0.0002	-0.0002
15:51:30-16:03:59	140	109	9.5	10.4	62.2	5.0	0.0006	0.0001	-0.0005
16:29:50-16:43:09	240	168	12.9	12.5	54.4	5.1	-0.0020	0.0012	-0.0022
16:44:00-16:48:59	480	168	11.9	10.6	59.9	5.1	-0.0073	0.0026	-0.0022
18.5.									
11:29:30-11:33:19	160	207	6.2	12.4	39.5	3.5	0.0000	-0.0003	-0.0002
11:48:16-11:57:58	330	203	5.6	12.5	27.2	2.5	-0.0001	0.0001	0.0001
11:58:00-12:08:14	320	192	5.4	11.6	56.9	4.9	-0.0001	0.0000	0.0001
19.5.									
15:04:14-15:12:09	920	43	1.8	12.4	56.5	5.6	0.0005	0.0001	0.0004
16:04:50-16:15:39	25	21	7.7	13.7	82.5	8.0	-0.0113	0.0004	0.0005
16:18:09-16:23:10	450	54	5.1	15.6	48.3	5.5	0.0002	0.0001	0.0000
16:54:34-16:57:34	500	32	6.4	17.6	48.6	6.4	-0.0012	0.0002	-0.0001
23.5.A									
08:48:59-08:56:59	160	143	14.9	11.6	68.4	5.8	-0.0102	-0.0013	0.0012
10:05:00-10:15:50	100	130	13.7	13.3	64.9	6.1	-0.0261	0.0012	0.0020
10:17:29-10:24:40	300	133	15.9	15.9	28.4	3.2	0.0001	-0.0001	0.0000
23.5.B									
14:38:00-14:48:00	20	98	8.2	16.2	51.1	5.8	-0.0024	0.0000	0.0001
14:48:59-14:59:00	150	113	14.2	19.5	26.0	3.7	0.0005	0.0000	-0.0001
15:30:00-15:43:00	20	127	8.9	15.3	54.5	5.8	-0.0109	-0.0001	0.0006
15:44:29-15:52:00	170	136	14.6	18.3	28.1	3.7	-0.0001	0.0000	0.0000
16:54:00-17:00:59	20	111	8.6	18.7	46.8	6.2	-0.0010	0.0001	-0.0002
17:03:20-17:07:59	150	118	13.6	21.5	23.3	3.7	-0.0002	-0.0001	-0.0001
24.5.									
13:16:00-13:29:09	25	106	7.3	19.0	44.7	6.0	-0.0004	-0.0004	0.0002
13:31:49-13:42:59	150	128	10.2	22.1	24.8	4.1	0.0002	-0.0001	-0.0001
14:23:59-14:30:20	150	160	11.3	19.2	37.9	5.3	0.0005	-0.0001	0.0000
14:32:30-14:42:00	25	115	6.0	14.2	74.7	7.5	-0.0002	0.0001	-0.0001
15:15:59-15:25:20	25	111	7.5	17.0	60.0	7.1	0.0006	-0.0001	0.0000
15:26:29-15:37:00	150	128	10.9	22.1	26.9	4.5	0.0020	-0.0001	0.0000
16:05:40-16:09:34	150	140	10.5	24.5	18.8	3.6	-0.0046	-0.0003	0.0007
14.6.									
10:39:49-10:48:39	250	108	7.4	17.2	38.3	4.7	0.0000	0.0000	0.0000
10:53:09-10:58:00	25	114	6.4	15.2	85.9	9.1	-0.0130	-0.0003	0.0004

Flights German Bight 23rd May 1989

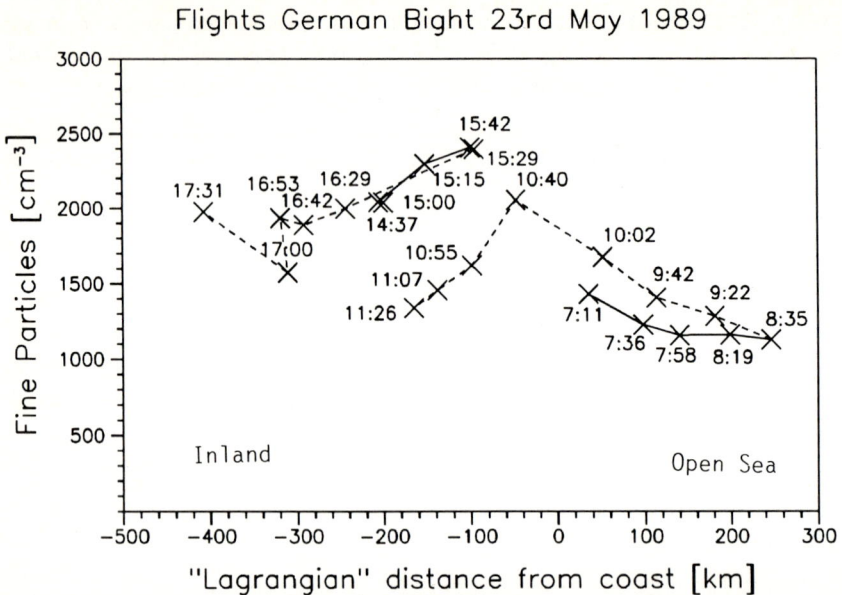

Fig. 2.4.25. Particle concentrations (with particle diameter $D_p = 0.1\text{-}7.5$ μm) in the atmospheric marine boundary layer along the "Lagrangian" distance scale (see text for explanation): distance corresponds to air mass position at 6:45 GMT near the coast at Nordholz. *Labels* indicate the actual time of measurement with the aircraft

the "same" air mass after some hours of transport gave estimates of conversion rates of approximately 15%/h. This corresponded reasonably well to the concurrent particle volume formation rate of 10%/h at the same time. Besides providing modelling approaches with simple parameterizations, this could be used to estimate how far out in the sea a reservoir of slowly depositing and water-insoluble NO_y in an air mass may contribute to nitrate particle formation. By assuming the above-mentioned conversion rate and a wind speed of 10 m/s, an air mass arriving at the central North Sea (400 km from the coastline) would still contain 40% of the NO_y present at the coast, and if little rain had affected the air mass, the particulate nitrate would also remain in the air mass.

The vertical and spatial distribution observed suggests different pathways for the air pollutants when transported over sea. Dry deposition diminishes only the small reservoir of pollutants within the MBL. This will mainly affect coastal areas. Sedimentation of coarse particles (> 1 μm) will not be inhibited and might add mainly to the pollution of coastal regions, as it is a first order exponential process.

Gases and fine particles (with diameter 0.1-1 µm), associated with ammonium sulphate and nitrate, trace metals and organic loads, may essentially stay in the upper layers. Gas-to-particle conversion for gases like NO_y steadily add to the fine particle load in the upper parts of the mixing layer. Alhough only present in half of the profiles, it was a dominant feature in the profiles measured away from the coast that higher particle concentrations could be observed in the upper part of the boundary layer.

Consequences for the distribution of atmospheric pollution are obvious. Polluted air masses can travel with small deposition losses to distant regions until, for instance, frontal passages decompose the vertical structure. Associated rainfall will then effectively cleanse the air mass. The upper layers of the mixing layer may thus form a reservoir and the rainfall and its regional distribution may determine the actual pollutant deposition.

References

Baeyens W, Dehairs F, Dedeurwaerder H (1990) Wet and dry deposition fluxes above the North Sea. Atmos Environ 24A:1693-1703

Cambray RS, Jefferies DF, Topping G (1975) An estimate of the input of atmospheric trace elements into the North Sea and the Clyde Sea (1972-3). AERE-Rep R 7733

Dedeurwaerder HL, Artaxo P (1987) Composition of ambient aerosol above the North Sea. In: Lindberg SE, Hutchinson TC (eds) Proc Int Conf Heavy metals in the enviroment, vol 2. New Orleans 1987. CEP Consultants, Edingburgh, pp 131-133

Duce RA, Liss PS, Merrill JT, Atlas EL, Buat-Menard P, Hicks BB, Miller JM, Prospero JM, Arimoto R, Church TM, Ellis W, Galloway JN, Hansen L, Jickells TD, Knap AH, Reinhardt KH, Schneider B, Soudine A, Tokos JJ, Tsunogai S, Wollast R, Zhoa M (1990) GESAMP-Reports and Studies No 38. WMO, Geneva

Graβl H, Eppel D, Petersen G, Schneider B, Weber H, Gandraβ J, Reinhardt KH, Wodarg D, Flieβ J (1989) Stoffeintrag in Nord- und Ostsee über die Atmosphäre. GKSS Rep No 89/E/8, Geesthacht

Hicks BB, Wesely ML, Durham JL (1980) Critique of methods to measure dry deposition. Workshop summary, Rep No EPA-600/9-80-050, Washington DC

Injuk J, Otten Ph, Rojas C, Wouters L, Van Grieken R (1990) Atmospheric deposition of heavy metals (Cd, Cu, Pb and Zn) into the North Sea. Final rep, Univ Antwerp

Kemp K (1984) Long Term Analysis of Marine and Nonmarine Transported Aerosols. Nucl Inst and Meth B 3:470

Krell U, Roeckner E (1988) Model simulation of the atmospheric input of lead and cadmium into the North Sea. Atmos Environ 22:375-381

Kretzschmar JG, Cosemans G (1979) A five year survey of some heavy metal levels in air at the Belgian North Sea coast. Atmos Environ 13:267

Kriews M (1992) Charakterisierung mariner Aerosole in der deutschen Bucht sowie Prozeßstudien zum Verhalten von Spurenmetallen beim Übergang Atmosphäre/Meer. In: Dannecker W (ed) Schriftenreihe Angewandte Analytik, Bd 15 Inst Anorganische und Angewandte Chemie, Univ Hamburg

Kriews M, Naumann K, Dannecker W (1988) Aerosol specification in the southern North Sea region by wind dependent sampling and multielement analysis. J Aerosol Sci 19: 1051-1054

Kriews M, Naumann K, Dannecker W (1989) Einsatz atomspektrometrischer Methoden zur Multielementbestimmung in marine Aerosolen. In: Welz B (ed) 5. Colloquium Atomspektrometrische Spurenanalytik, pp 633-646, Bodenseewerk Perkin Elmer GmbH, Überlingen

Luthardt H (1987) Analyse der wassernahen Druck- und Windfelder aus Routinebeobachtungen. Hamburger Geophys Einzelschriften A, H 83, Eigenverlag, Hamburg

Peters G (1991) Sodar, ein akustisches Meßverfahren für die untere Atmosphäre. Promet 21:55-62

Schulz M (1992) Die räumliche und zeitliche Verteilung des atmosphärischen Schadstoffeintrages in die Nordsee. Diss FB Chemie, Univ Hamburg

Schulz M, Steiger M, Schwikowski M, Kriews M, Naumann K, Dannecker W (1988) Variability of ambient trace element concen-trations at the North Sea with respect to air mass history. J Aerosol Sci 19:1171-1174

Schwikowski M (1991) Untersuchungen der Konzentrationen von Spurenstoffen - insbesondere Stickstoffverbindungen - in der Atmosphäre und im Niederschlag zur Abschätzung des Eintrages in die Nordsee. In: Dannecker W (ed) Schriftenreihe Angewandte Analytik, Bd 12, Inst Anorganische und Angewandte Chemie, Univ Hamburg

Schwikowski M, Schulz M, Steiger M, Naumann K, Dannecker W (1990) Transformation and transport of nitrogen compounds above the North Sea investigated by aircraft measurements. Aerosol Sci 19:1311

Slinn SA, Slinn WGN (1980) Predictions for particle deposition on natural water. Atmos Environ 14:1013-1016

Sorbjan Z, Counter RL, Wesley ML (1991) Similarity scaling applied to SODAR observations of the convective boundary layer above an irregular hill. Boundary-Layer Meteorol 56:33-50

Steiger M (1991) Die anthropogenen und natürlichen Quellen urbaner und mariner Aerosole charakterisiert und quantifiziert durch Multielementanalyse und chemische Receptormodelle. In: Dannecker W, (ed) Schriftenreihe Angewandte Analytik, Bd 11 Inst Anorganische und Angewandte Chemie, Univ Hamburg

Steiger M, Schulz M, Schwikowski M, Naumann K, Dannecker W (1989) Variability of aerosol size distribution above the North Sea and its implication to dry deposition estimates. J Aerosol Sci 20:1229-1232

Stössel RP (1987) Untersuchungen zur Naβ- und Trockendeposition von Schwermetallen auf der Insel Pellworm. Diss, GKSS 87/E/34 Report, Eigenverlag, Geesthacht

Thorpe A, Trevor G (1977) The nocturnal jet. Q J R Meteorol Soc 103:633-653

Van Aalst RM, Van Ardenne RAM, de Kruk FJ, Lems T (1983) Pollution of the North Sea from the atmosphere, TNO Rep No CL 82/152

Van Jaarsveld JA, Van Aalst RM, Onderdelinden C (1986) Deposition of metals from the atmosphere into the North Sea: model calculations. RIVM-Rep No 842015002

2.5 Local Studies in the German Bight During Winter/Spring 1988/89

K. Heyer, M. Engel, U.H. Brockmann, H.-J. Rick,
C.-D. Dürselen, H. Hühnerfuss, U. Kammann,
H. Steinhart, W. Kienz, M. Krause, L. Karbe,
A. Faubel and S. Regier

2.5.1 Introduction

For the final stage of the ZISCH project, a series of field investigations in the inner German Bight was planned with the object of obtaining more detailed information in space and time in a specific region of the North Sea ecosystem.

Hydrographic parameters, distributions of nutrients, seston, phyto- and zooplankton, benthos organisms and contaminants were investigated in various ecosystem compartments.

From December 1988 through May 1989, six monthly cruises were carried out with the aim of resolving one part of the seasonal cycle of those parameters (station net: see Fig. 2.5.1 and dates of the cruises: Table 2.5.1). This effort was conceived to give for the first time a consistent picture of the simultaneously measured relationship between hydrographic, biological and chemical parameters and their temporal and spatial variability within the German Bight ecosystem. It forms the basis for further process studies concerning contaminant fluxes.

This chapter gives only a brief summary of the measurements carried out during this period. The discussions of the specific details and interpretations are given in further chapters.

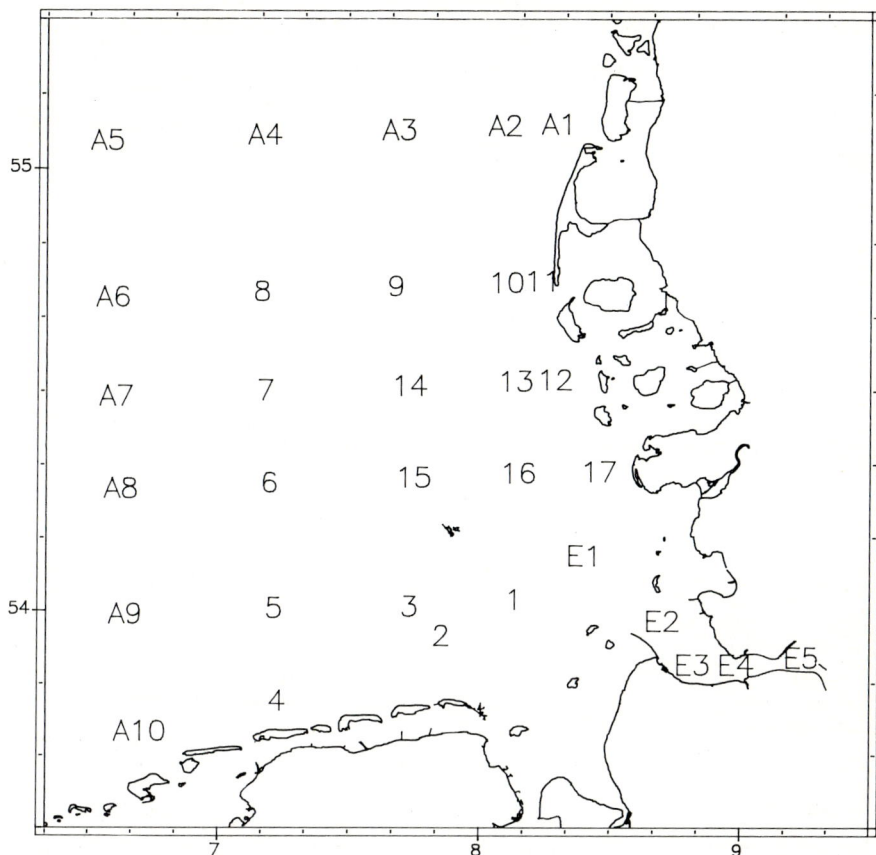

Fig. 2.5.1. Positions of the stations of the ZISCH surveys from November/December 1988 to May 1989

2.5.2 Ecological Situtation in the German Bight During the Experiment

2.5.2.1 The German Bight and its Hydrography

The German Bight is a rather shallow area of the North Sea. Water depths in the investigated region do not exceed 45 m. Its hydrography is subject to influences from the central North Sea, the Wadden Sea and the fresh-water discharges of the rivers Ems, Weser and Elbe, but also from the English Channel and the river Rhine.

The tides, mainly M_2 and S_2, and the fast and often predominant reaction to meteorological conditions determine its dynamical behaviour. Tidal currents vary between 0.2 and 2 knots in the outer parts and the Elbe river

Table 2.5.1. Dates of the ZISCH cruises and research vessels (RV): *Valdivia* (University of Hamburg); *Gauss* (Bundesanstalt für Seeschiffahrt und Hydrographie); *Victor Hensen* (Alfred-Wegener-Institut für Polar- und Meeresforschung, Bremerhaven); *Senckenberg* (Institut Senckenberg, Wilhelmshaven)

Cruise no.	Date	Research vessel	Stations (see Fig. 2.5.1)
1	28.11.-09.12.88	*Valdivia, Gauss*	1-17 and A1-A10
2	19.01.-24.01.89	*Valdivia*	1-17 and E1-E5
3	27.02.-04.03.89	*Gauss*	1-17 and E1-E5
4	21.03.-25.03.89	*Valdivia*	Half of the stations due to bad weather
5	10.04.-15.04.89	*Victor Hensen, Senckenberg*	1-17 and E1-E5
6	02.05.-05.05.89	*Valdivia, Gauss*	1-17 and E1-E5

mouth, respectively. However, these currents are often surpassed by wind driven motions which sometimes lead to dangerous storm surges. Residual currents, averaged over two tidal cycles for instance, can exceed 1 knot and vary in all directions. In addition to wind induced currents, strong horizontal density gradients and topographical effects yield further contributions to considerable vertical current shears. Due to all these conditions, it is practically impossible to talk about a mean situation of any parameter in the German Bight. In order to gain insight into the relationships between biological, chemical and physical processes, simultaneous measurements must therefore be carried out. This biologically very productive area of the North Sea is endangered by pollutants carried in from rivers, coastlines and the atmosphere.

During the period of our experiments, the rough overall comparison of temperature and salinity with long term means in the German Bight (Engel 1983) and in Helgoland reveals a significantly higher temperature, as can be seen in Fig. 2.5.2a,b (taken from Radach and Bohle-Carbonell (1990), with our findings at stations 15 and 16 inserted). Starting with almost mean relationships in November/December 1988, the temperature from January 1989 through May 1989 lies about 2-3 °C above the mean values, whereas

CENTIGRADE

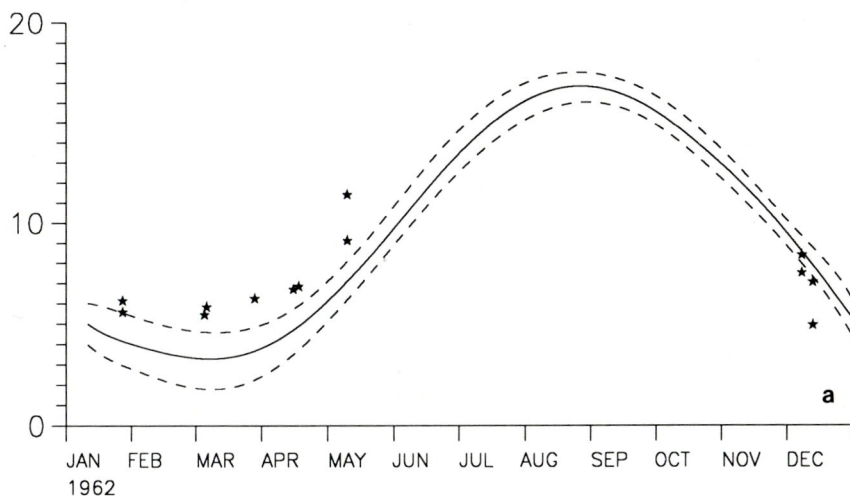

SURFACE TEMPERATURE
MEAN SEASONAL SIGNAL
CREWS OF M.B. 'ELLENBOGEN' AND 'AADE' / BAH

$^0/_{00}$

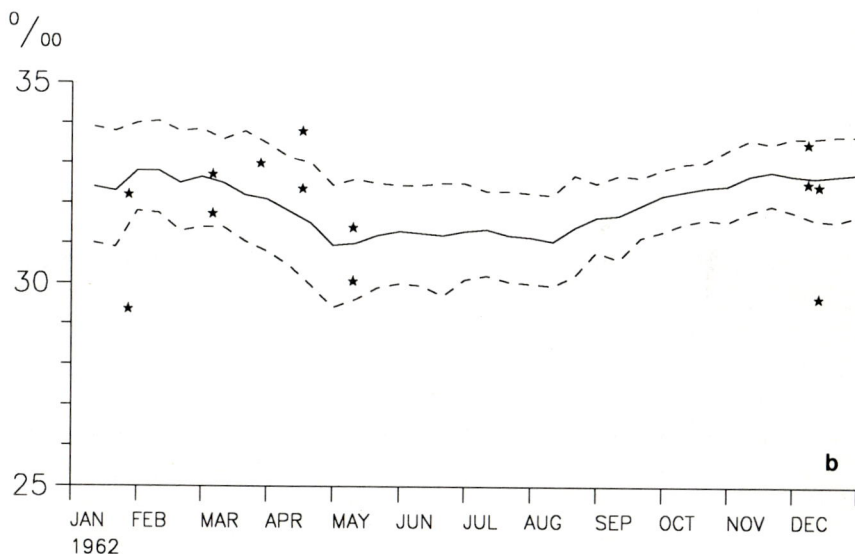

SALINITY
MEAN SEASONAL SIGNAL
B. METSCHNER, K. TREUTNER / BAH

Fig. 2.5.2. Temperature and salinity annual cycles, 23 years daily observation means and standard deviations with ZISCH measurements at station 15 and 16

the salinity distribution generally fits well into mean pictures. This temperature anomaly was due to unusual meteorological conditions. In November, the mean air temperature was about normal, but from December 1988 until May 1989 the air temperatures were 2-4 °C above the long term mean. From January through March 1989, the warmest 3-month mean ever observed was registered. In April the temperatures were slightly lower than usual, but in May, again, they exceeded the mean observations. May 1989 was also characterized by extremely low precipitation and very much sunshine.

Another factor important for the hydrography of the shallow inner German Bight is the Elbe freshwater discharge. Only in December 1988 and January 1989 did it exceed the long-term mean by about 250 m^3/s and 400 m^3/s, respectively, with a maximum discharge of about 1650 m^3/s at the beginning of January 1989. In November 1988 and February through May 1989 the monthly mean discharges were 200 m^3/s and in March up to 500 m^3/s below the long-term mean (ARGE Elbe 1988, 1989).

Besides CTD profile casts, continuous CTD sonde measurements at 5 m depth (ship bottom) yielded information about the horizontal near surface temperature distribution. A ship-mounted Acoustic Doppler Current Profiler (ADCP, with 600 kHz for shallow water) was run on RV *Valdivia* with 2 m vertical resolution of velocity profiles and temporal averages over 5 min. These measurements permit a broader insight into the vertical structure of the velocity fields in the German Bight.

2.5.2.2 Nutrients

The nutrient concentrations measured in the German Bight between December 1988 and May 1989 were mainly influenced by advection and discharges. Since generally the same station nets were investigated during the monthly cruises, a rough comparison of means and standard deviations is possible (Table 2.5.2).

During the last two cruises (April and May) a distinct decrease of silicate, phosphate and nitrate was observed, caused by the nutrient uptake during the phytoplankton spring bloom. High standard deviations are a consequence of strong horizontal gradients characterizing the nutrient regime in the German Bight.

The reasons for this are the dominating nutrient inputs from rivers and the Wadden Sea. The nutrients are diluted towards the open water. A detailed trend analysis from these nutrient data alone is not possible because the horizontal gradients are mainly controlled by highly variable advection. This and the biological nutrient conversion do not allow a detailed compari-

Table 2.5.2. Mean nutrient concentration (μmol/l)in the German Bight, excluding the four stations within the mouth of the Elbe

Date	NH_4^+	Si	PO_4^{3-}	NO_2^-	NO_3^-
11.88	4.14±1.29	13.05±7.84	1.05±0.46	1.29±1.29	18.48±15.11
1.89	10.29±12.72	32.56±43.21	1.55±0.74	1.00±1.35	66.44±97.35
2.89	6.04±9.79	31.40±43.20	1.49±0.66	0.91±1.13	62.64±74.95
3.89	10.13±16.96	39.08±59.28	1.64±1.26	1.17±1.14	109.01±154.03
4.89	3.03±1.53	22.68±32.59	1.25±0.76	0.64±0.35	70.55±85.05
5.89	1.15±0.48	6.07±20.25	0.56±1.00	0.27±0.18	37.95±64.36

son of the non-coherent data sets. Therefore, only some examples are discussed. The nitrate concentrations during January 1989 were mainly conservatively diluted, resulting in a significant negative correlation ($r=0.98$) with salinity (Fig. 2.5.3). Therefore - due to prevailing easterly winds - a river plume (Fig. 2.5.4) characterized by high nitrate concentrations was spreading into the German Bight (Fig. 2.5.5). This effect was intensified by the river discharge (Brockmann and Eberlein 1986), since during January the discharge of the Elbe already reached its annual maximum for 1989 with more than 1500 m^3/s (ARGE Elbe 1990).

During this cruise, the ammonium concentrations were also highest in the river plume, since at that time the nitrification in the Elbe river was limited by the low temperature, causing a significant nitrogen discharge of ammonium in winter (Fig. 2.5.6). Due to nitrification and preferred ammonium consumption by phytoplankton as source for nitrogen, the ammonium concentrations in the German Bight were diminished in May to values less than 1 μmol/l.

On the other hand, ammonium is the first remineralization product released by decomposers. Very active areas for decomposition in the German Bight are the Wadden Sea, where the ammonium concentrations reach their highest values, along with some central parts of the German Bight during November/December 1988 (Fig. 2.5.7). Due to the reduced primary production in winter (limitation by light, this effect being enhanced by the increased turbidity during winter storms) ammonium will not be utilized so fast, and sources of remineralization can be detected.

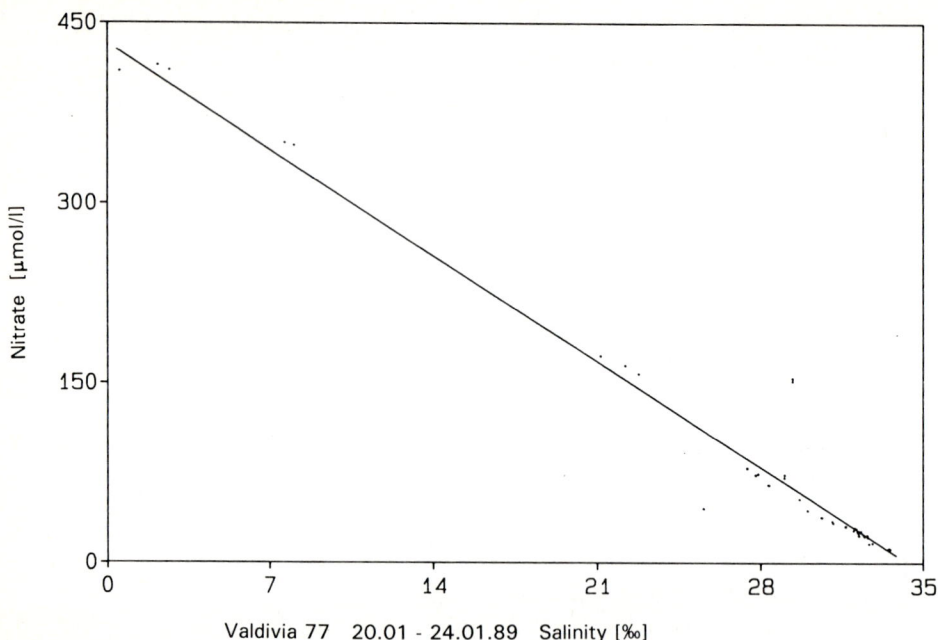

Fig. 2.5.3. Correlation nitrate with salinity in January 1989

2.5.2.3 Phytoplankton

Compared with 1987, the production and standing crop of the phytoplankton was high. The production rate was two to three times higher than measured in 1987, probably due to better light conditions (not measured) and extreme water temperatures (Table 2.5.3).

With the exception of two stations in March 1989 (station 4, the southernmost, and station 5, north of station 4) the biomass and the production rate of the primary producers showed the same trend (Figs. 2.5.8 and 2.5.9). At stations 4 and 5, in March high biomass and low production rates were measured due to the dominant diatom species *Coscinodiscus wailesii* reaching its stationary phase of development.

The allochthonous *Coscinodiscus wailesii* was the dominant species during the March investigation period. It has been observed since 1977 in the German Bight (Boalch and Harbour 1977). Investigations have shown that *Coscinodiscus wailesii* and another introduced species, *Thalassiosira punctigera*, are tolerant to heavy metals (Rick 1990), so they possibly have an advantage compared to the autochthonous species (Chap. 4.3). After the breakdown of the *C. wailesii* bloom a typical spring diatom bloom developed with *Thalassiosira polychorda*, *T. nordenskiöldii* , *T. punctigera*,

Table 2.5.3. Water temperature and production rate of the phytoplankton

Date	Temperature °C	Production rate mgC/m^2/h
29.5.-01.6.86	10.0±1.9	330
16.2.-21.2.87	2.0±1.2	27
19.1.-24.1.89	6.2±0.7	71
27.2.-04.3.89	5.6±0.4	48
21.3.-25.3.89	6.2±0.1	76
10.4.-14.4.89	6.6±0.1	133

T. rotula und *Lauderia gracilis*. At most of the stations the Prymnesiophyte *Phaeocystis globosa* was found. Until May, especially in the southwestern part of the German Bight, *P. globosa* increased to maximal values of 88 x 10^6 cells/l (Station 4). The northeastern part of the investigation area was dominated by diatoms of the summer plankton (e.g. *Odontella sinensis, O. regia, Guinardia faccida, Thalassionema nitzschiodes*).

2.5.3 Contaminants

2.5.3.1 Organic Contaminants

Cyclic organochlorines are environmentally persistent and widely distributed in marine ecosystems. Long-term observations by Ernst et al. (1988) and by Hühnerfuss und Weber (unpubl. data) allow some conclusions with regard to the distribution pattern of hexachlorocyclohexane (HCH) in the North Sea during nearly one decade: in the German Bight in comparison to investigations of 1979, an increase in γ-HCH levels (Lindane) and a decrease in α-HCH have been observed, while in the central and northern North Sea the changes were less significant. No other comparable long-term (at least one decade) data of other contaminants are available that represent the complete North Sea area.

Water. Short-term concentration variations of different organic contaminants in water were investigated in the German Bight during six cruises between

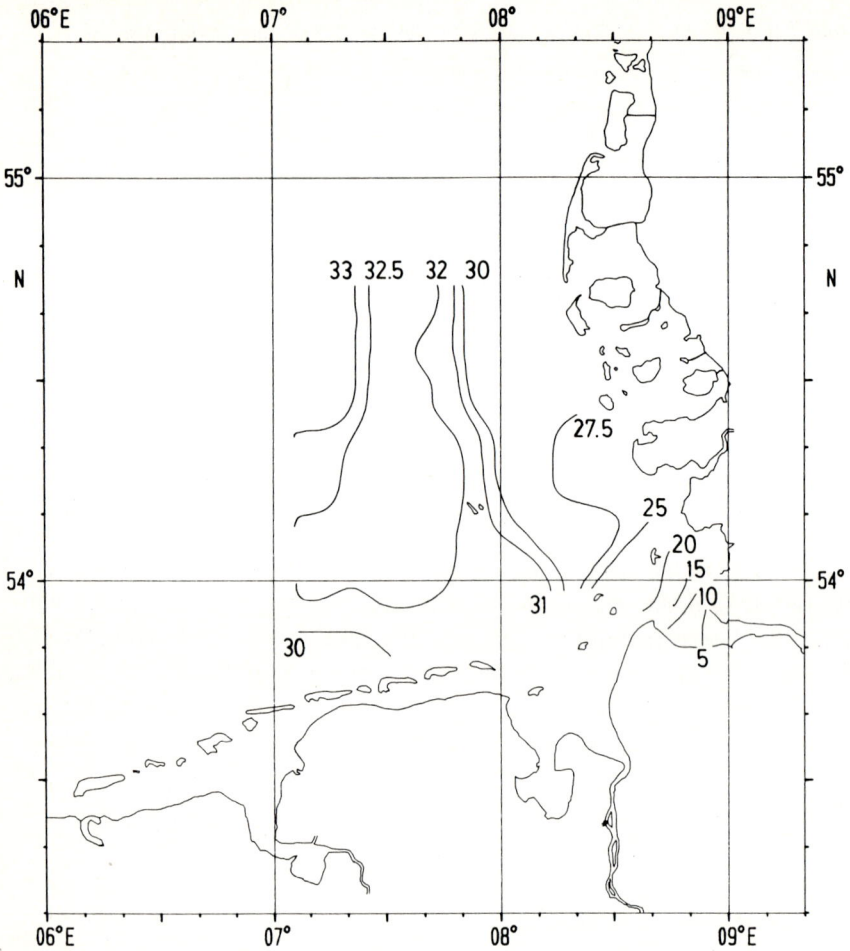

Fig. 2.5.4. Distribution of salinity in the German Bight in January 1989

November 1988 and May 1989. Data about short-term (hours and days) variability of organic contaminants are scarce in the literature.

The data set, which includes halogenated phenols, organophosphates, low-volatile halogenated hydrocarbons and high-volatile halogenated hydrocarbons, reveals that all investigated anthropogenic substances had a well-known and continuously active source: the river Elbe was clearly the dominating source for these substances (Hühnerfuss et al. 1990). Presumably, the ZISCH stations in the German Bight were too far away from the river Rhine to detect its influence. Pentachlorophenol (PCP) in a

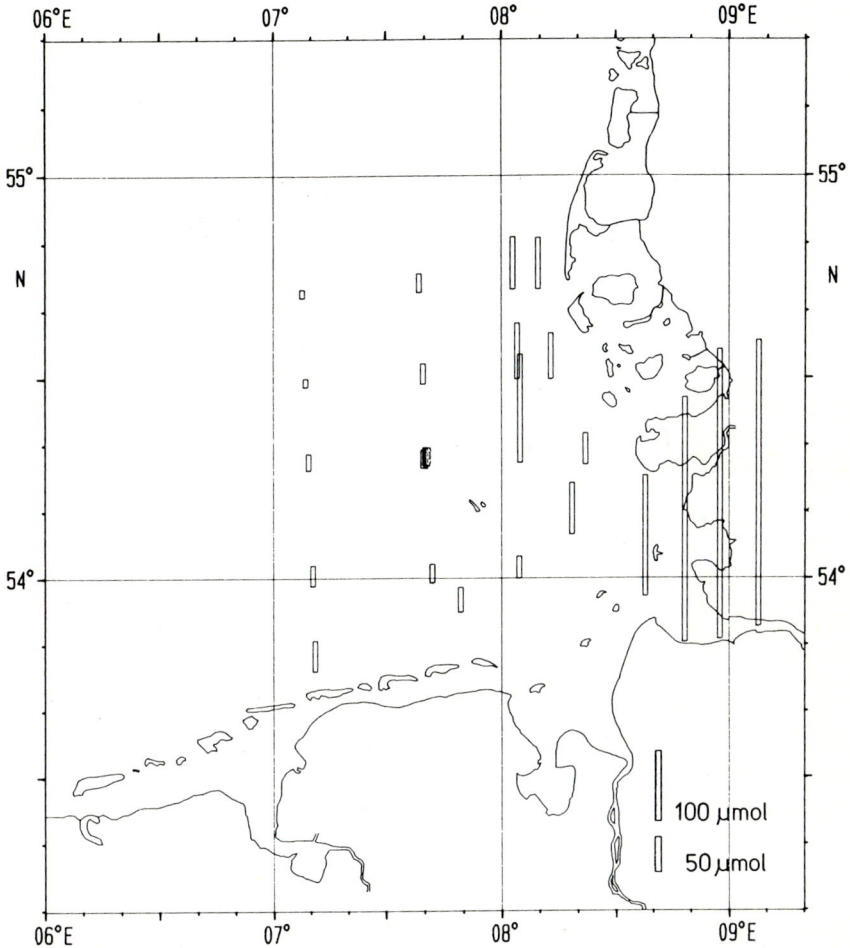

Fig. 2.5.5. Nitrate concentration (μmol/l) at the surface in January 1989

water depth of 10 m is shown as one example (Fig. 2.5.10). Decreasing concentrations were measured with increasing distance from the river Elbe. It is remarkable that no temporal trend is visible. Seasonal variability of HCH was observed by Hühnerfuss und Weber (unpubl. data). For γ-HCH they showed significantly higher concentrations in the coastal regions of the North Sea in early summer 1986 in comparison to the winter 1987, due to the seasonal agricultural application of Lindane. The concentrations of the other organic contaminants were relatively constant in time. The data set of the six cruises (Figs. 2.5.10 and 2.5.11) supplies experimental evidence that changing meteorological and oceanographic conditions may cause concentration variations of contaminants by nearly one order of magnitude. This may

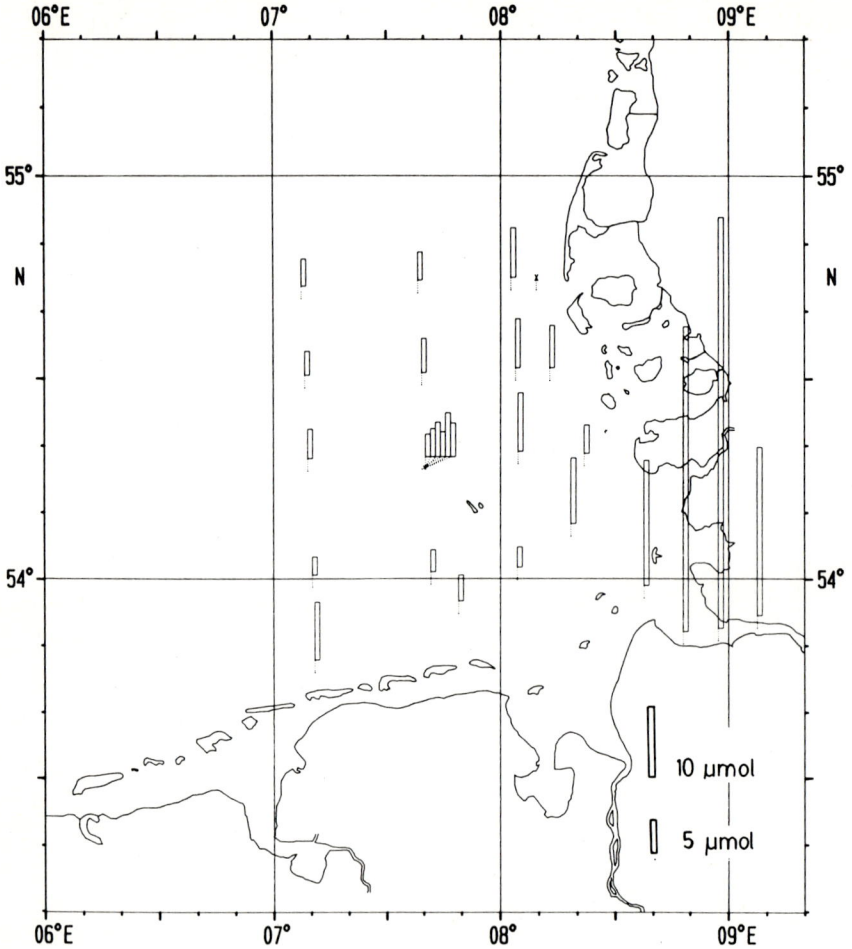

Fig. 2.5.6. Ammonium concentration (μmol/l) at the surface in January 1989

possibly cause additional stress to the marine ecosystem. An extensive discussion of these phenomena can be found in two recent publi-cations by Hühnerfuss et al. (1990) and Pfaffenberger et al. (1992). For example, in Fig. 2.5.11, at station 8 the ratio of γ/α-HCH was 2.4 during Cruise 1 (November/December 1988) and 2.9 during Cruise 2 (January 1989) and during the storm (Cruise 1a) the ratio dropped to 0.86. However, these variations are considered to be confined to North Sea areas that are suffi-

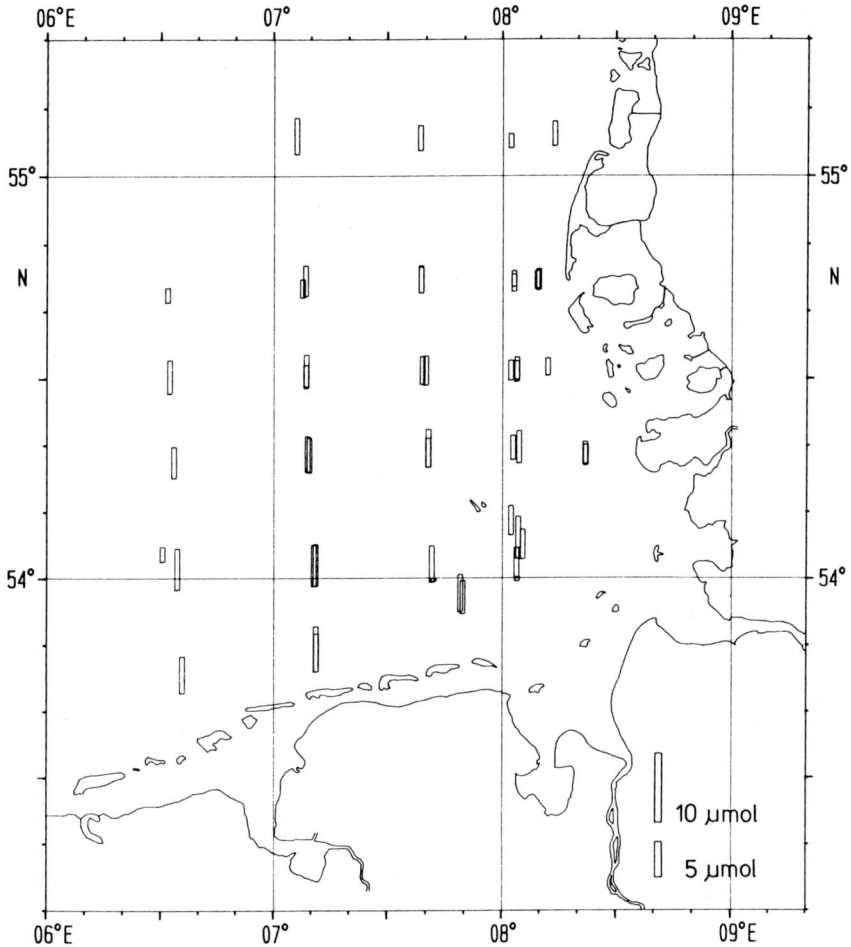

Fig. 2.5.7. Ammonium concentration (µmol/l) at the surface in November/December 1988

ciently close to the sources of the respective contaminants. During the first cruise, a drastic decrease in the concentration of anthropogenic substances was measurable during a storm. Therefore, the theoretical approach by Hainbucher et al. (1987), that changing meteorological conditions lead to a variation of concentrations of one order of magnitude, is only true near the sources of the contaminants.

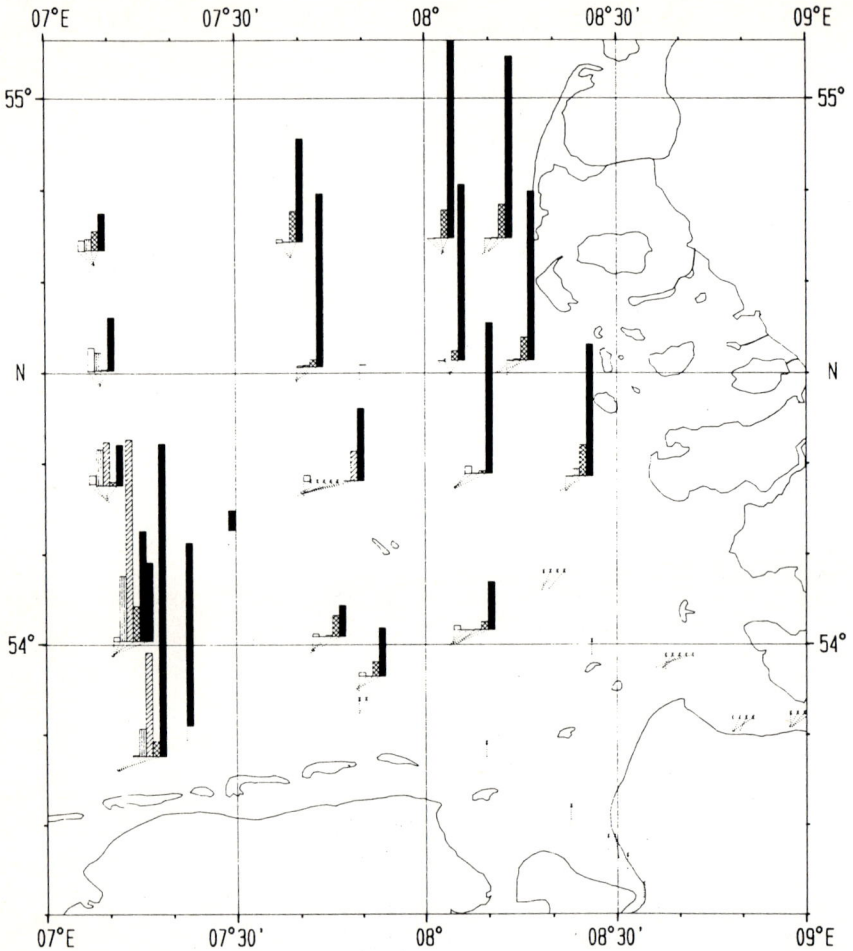

Fig. 2.5.8. Phytoplankton biomass (μg C/l) from January 1989 to May 1989 - values integrated over the water column. Depth interval: 0-20 m

Benthos. Seasonal events of the contamination in the dab *Limanda limanda*, the whelk *Buccinum undatum* and the sea star *Asterias rubens* with several organochlorine compounds were shown with samples collected between December 1988 and May 1989 in the German Bight (Kammann et al. 1990; Knickmeyer and Steinhart 1990a; Knickmeyer et al. 1993). The method used has been described in detail by Knickmeyer and Steinhart (1989). The body load of cyclic organochlorines changed with time, but this change was different for different compounds at the same sampling station and also for

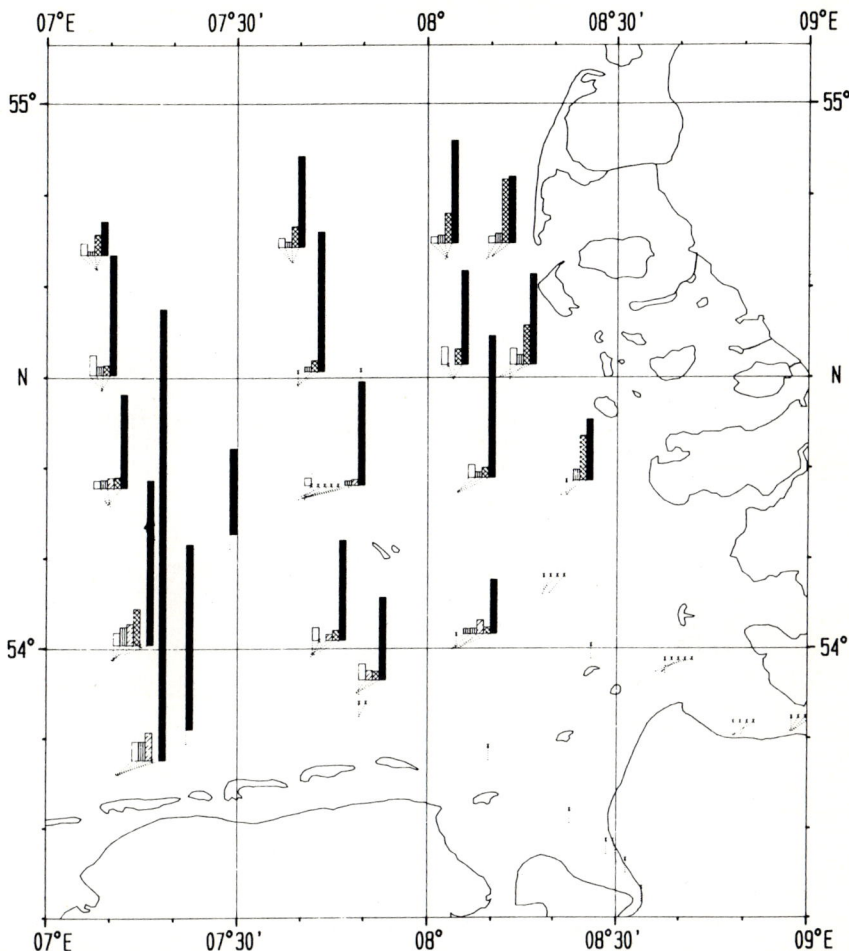

Fig. 2.5.9. Primary production (mg C/m^{2}h) from January 1989 to May 1989 (not fractionated)

the same compound at different sampling stations. Some typical results for the seasonal varability described above are shown in Fig. 2.5.12a-c.

The different patterns of aromatic organochlorines and Lindane are attributable to their sources and physico-chemical properties, such as uptake by food, loss by faeces and reproduction activities. Equilibrium partitioning of the compounds between ambient water and body lipids and the lipophilicity of the chemicals appear to be the most important processes with regard to bioaccumulation.

The water solubilities of α-HCH and Lindane are about two to four orders of magnitude higher than those of most aromatic compounds under

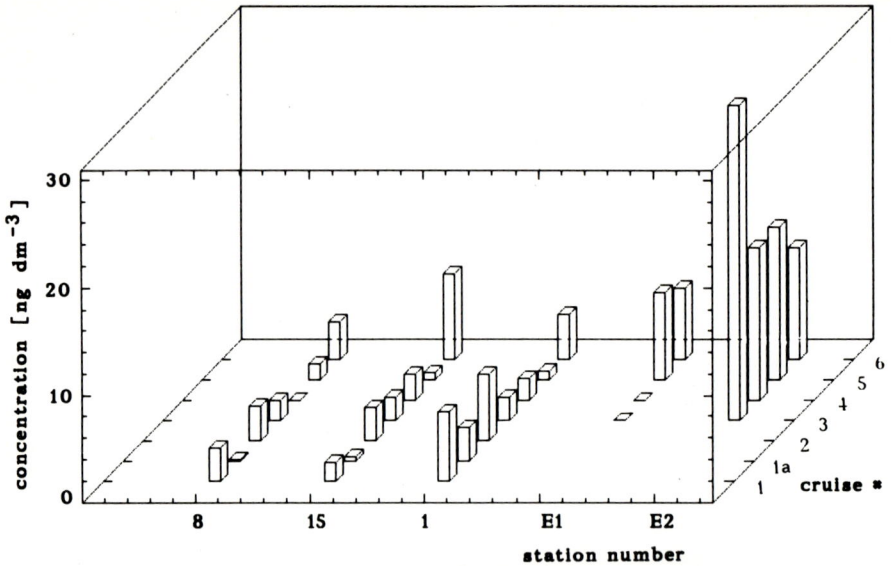

Fig. 2.5.10. Chronological development of the concentration of pentachlorophenol (PCP) in a water depth of 10 m for a transect through the stations E2/E1/1/15/8 (see Fig. 2.5.1)

investigation (Brodsky and Ballschmitter 1988) (cf. Chap. 4.5, Table 4.5.1). Laboratory experiments conducted by Ellgehausen et al. (1980) and Hansen (1980) demonstrate that less lipophilic chemicals such as HCHs are both taken up and lost by various organisms within a few hours. Clearance half-lifes of hexachlorobiphenyls in fish, e.g. guppies (*Poecilia reticulata*) vary from 35 to 50 days, the half-life in rainbow trout (*Salmo gairdneri*) has been calculated to be of the order of 60 to 70 days (Bruggeman et al. 1984; Norheim and Roald 1985). Therefore, concentrations of Σ-PCB, HCB and p,p'-DDE in the benthic organisms under investigation show time integration characteristics to a certain degree. Since HCHs were both taken up and lost from the lipid pool within a few hours, these compounds may be considered to reflect sudden changes in water masses.

Investigations on α-HCH and Lindane content in different samples from the North Sea indicated that changing γ/α-HCH ratio can be a useful indicator for different water bodies (Knickmeyer and Steinhart 1988, 1990b; König et al. 1989; Knickmeyer et al. 1992). The γ/α-HCH ratio in sea stars (Fig. 2.5.13a-c) from the German Bight reflects such short-term events as the changing influence of water from the rivers and from the central North Sea. Figure 2.5.13a depicts a decreasing γ/α-HCH ratio running from the

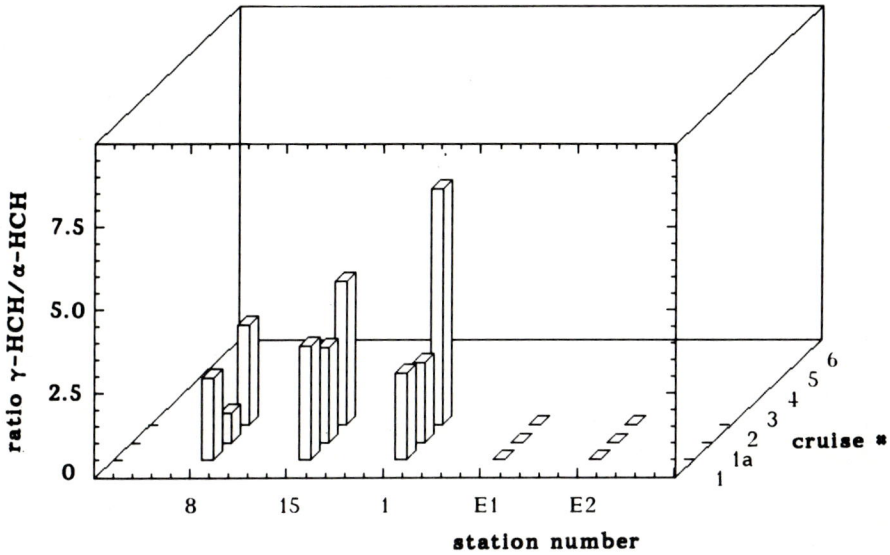

Fig. 2.5.11. Chronological development of the relation of the concentrations γ-HCH/α-HCH in a water depth of 10 m for a transect through the stations E2/E1/1/15/8 (see Fig. 2.5.1)

Ems-Dollart-region to the central North Sea. A plume of coastal water with a γ/α-ratio of 11 at station 12 (see Fig. 2.5.1), as well as of river water with a γ/α-ratio ≥ 5.0 coming from the southwest is obvious in Fig. 2.5.13b. In May, a more homogeneous situation was established (Fig. 2.5.13c).

Distinct seasonal variations in lipid content as well as in organochlorine contamination of male and female whelks, dabs and also sea stars sampled at the same position in the German Bight could be observed due to changes in the biology of the organisms. These changes are attributed to the reproductive cycle and feeding activities of the animals.

The sea stars are very adaptable to food and therefore are examples of feeders which may use any food source that becomes available during the season (Anger et al. 1977). The principle spawning time of the sea star in the German Bight is from February to April (Kowalski 1955).

Female whelks exihibited higher contamination with Σ-PCB, HCB and p,p'-DDE in April (Knickmeyer and Steinhart 1988, 1990a). These findings correspond with a maximum feeding rate of the whelks recorded in April (Hancock 1967).

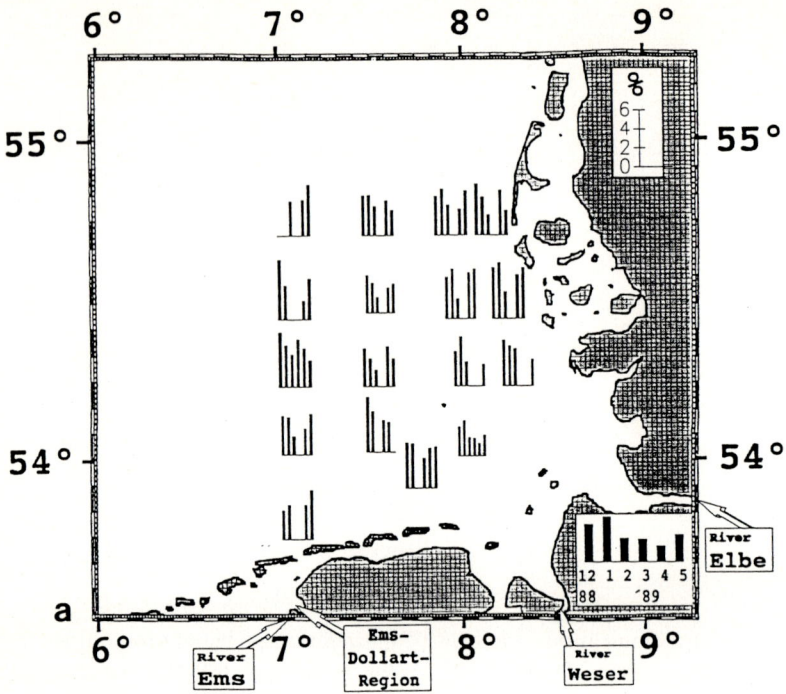

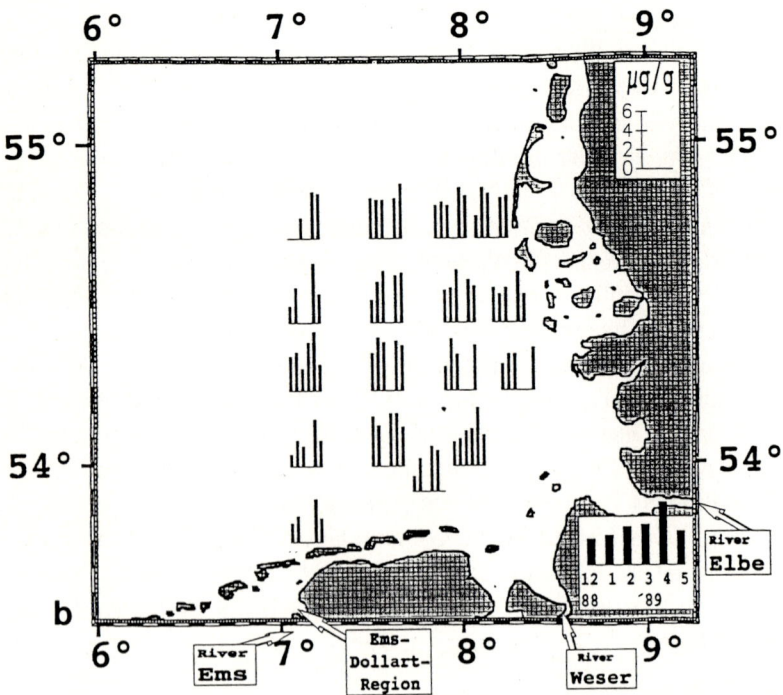

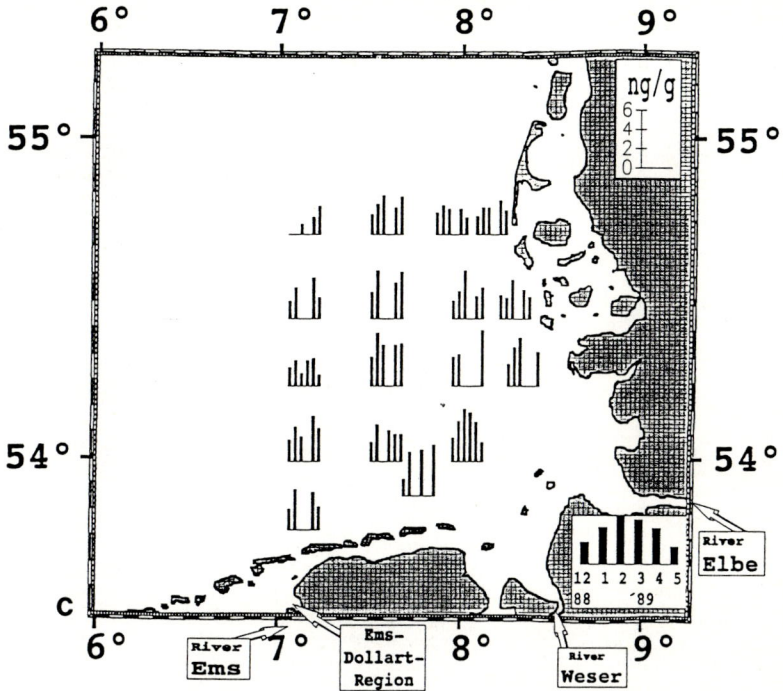

Fig. 2.5.12. *Asterias rubens*, December 1988 to May 1989. *a* Lipid related to dry matter (%). *b* Σ-PCB (sum of 35 individual congeners) in μg PCB/g n-hexan extractable lipid. *c* p,p'-DDE in ng p,p'-DDE/g n-hexan extractable lipid

As was the case in female dabs, the annual reproductive cycle causes a mobilization of body reserves, which are partly utilized for the anabolism of the eggs. An increase in the gonosomatic index during the spawning season of dabs caused a decrease in the hepatosomatic index, which reached its minimum in the postspawning period (Htun-Han 1978b).

Büther (1988) found a decrease of lipids in female dab livers of around 20% (fresh weight) in December to 6% in May. Figure 2.5.14 illustrates the seasonal cycle of lipids in female dab livers, which is in accordance with these findings. The neutral lipid fraction (storage lipids) decreases continuously parallel to the total lipid content, reaching its minimum in April, whereas the polar lipid fraction appears to be constant. At the beginning of the spawning period in February (Htun-Han 1978a), PCB and HCB contamination of livers exhibited their lowest levels (Fig. 2.5.15); however, Σ-PCB and HCB in ovaries increased. The maximal Σ-PCB and HCB load in ovaries coincides with a minimum of neutral lipids in female livers. These findings may be explained by a transfer of organochlorines from

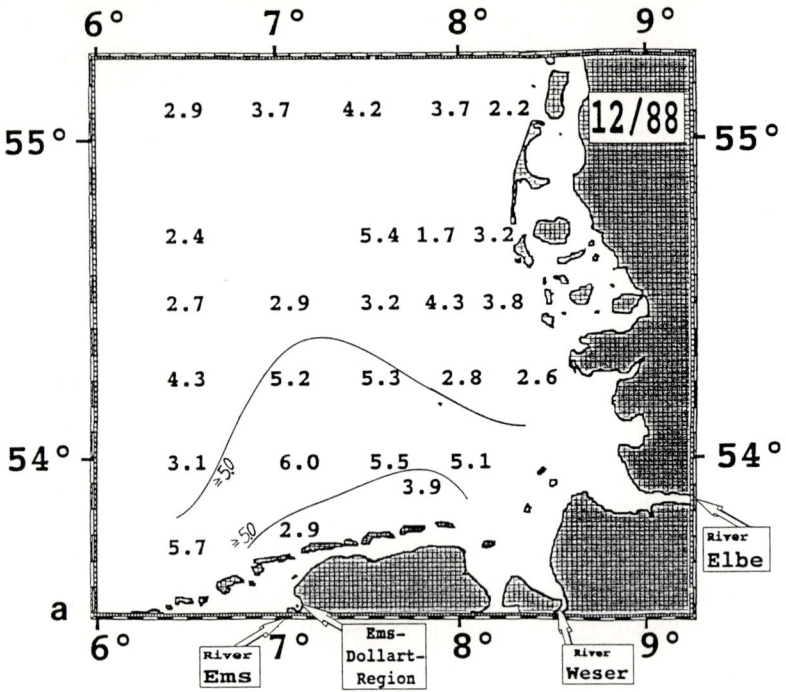

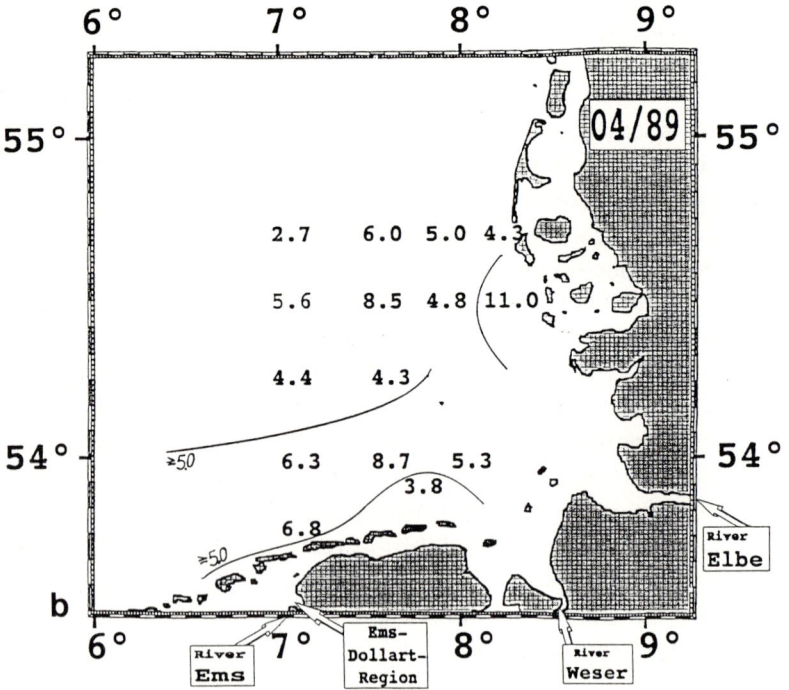

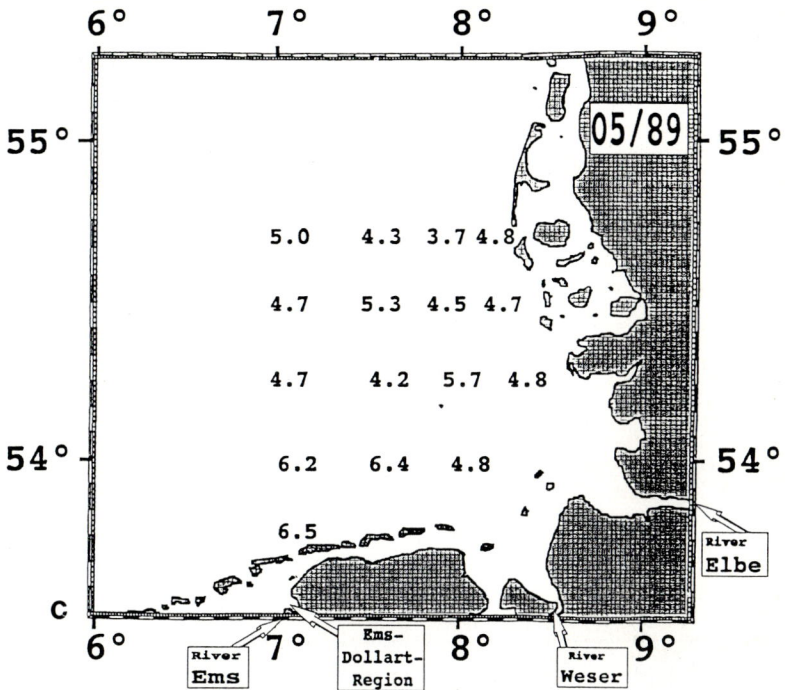

Fig. 2.5.13. *Asterias rubens*, ratio γ/α-HCH during December *(a)*, April *(b)* and May *(c)*

livers to ovaries during periods of unfavourable food conditions (Kammann et al. 1990).

2.5.3.2 Heavy Metals

Water. The water samples were taken from a rubber boat by a Go-flow-sampler. The sampling depth had been 7 to 8 m. On shipboard, the water samples were worked up, and they were analyzed in the home laboratory (Mart 1979a,b; Rick 1990). In the German Bight, the dissolved heavy metals differ in time and space. Copper, cadmium and mercury in filtered seawater showed the highest concentrations in coastal regions, especially near the North Frisian Islands (Figs. 2.5.16, 2.5.17 and 2.5.18). The rivers Elbe and Weser were the important sources of these heavy metals. For example, Karbe and Hablizel (1990) determined higher Cd concentrations in mussels (*Mytilus edulis*) coming from Weser and Elbe in contrast to the rivers Jade and Ems. High concentrations of zinc were determined in offshore regions as well (Fig. 2.5.19).

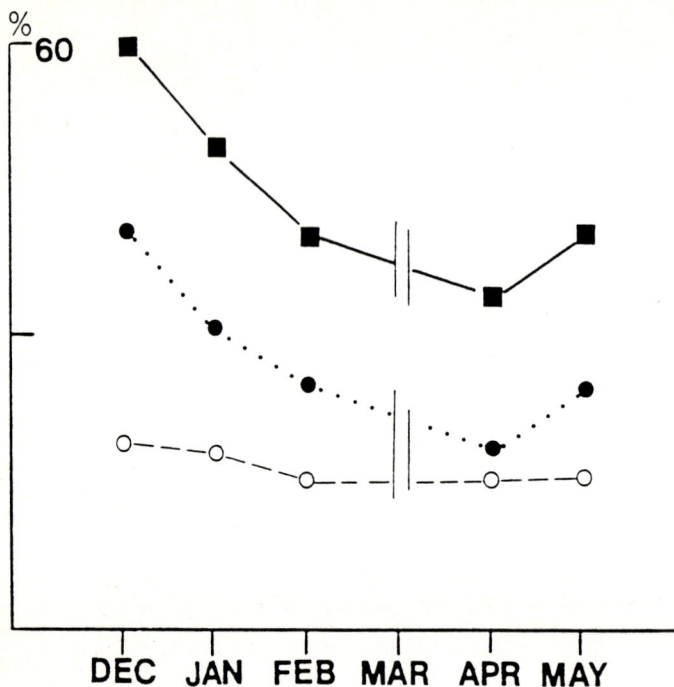

Fig. 2.5.14. Variation in lipids related to dry matter in female dab livers (station 12, December 1988 to May 1989; 10 individuals pooled). ■ total lipids (chloroform/methanol extractable); ○ polar lipids; ● neutral lipids

Since salinity is a good tracer for water bodies, the metal concentrations were correlated with this parameter.

With the exception of February, Hg in filtered as well as in unfiltered water samples correlates significantly negatively (Spearman rank correlation, $p < 0.05$) with salinity or the water temperature. Cd (Fig. 2.5.20) and Cu showed little correlation for low salinities but no correlations for higher salinity. Higher concentrations were measured in water with lower salinity. For Zn and Pb, no correlation was calculated over the total range of salinity. These results are comparable to Duinker et al. (1982a,b), who described distinct dependences between salinity and metal concentration for salinities lower than 10 to 15‰.

A direct comparison of results for one station from different cruises is difficult because of the variable hydrographic conditions in the German Bight. This is obvious in the great range of the measured concentrations - high metal concentrations were measured close to low concentrations.

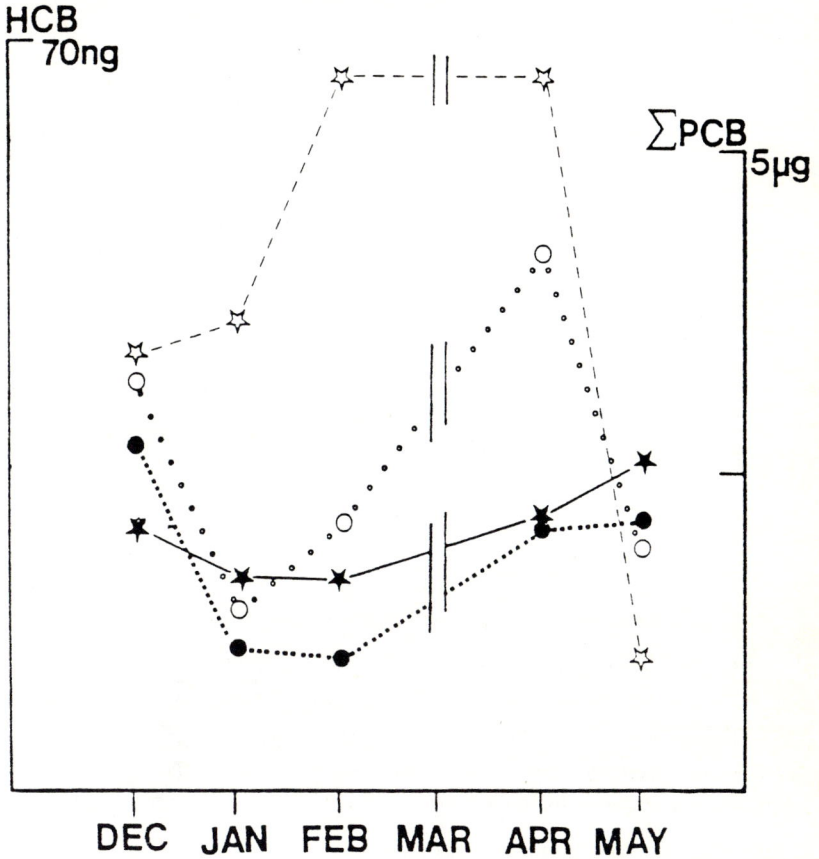

Fig. 2.5.15. Variation in organochlorines related to lipids in female dab livers and ovaries (station 12, December 1988 to May 1989, 10 individuals pooled). ● Σ-PCB in female livers; ○ = Σ-PCB in ovaries; ★ HCB in female livers; ☆ HCB in ovaries

Therefore, a comparision of the measured concentrations with those of other authors is difficult since other regions of the North Sea or the estuaries of the rivers Elbe or Weser were sampled. Nevertheless, published concentrations for the German Bight by other authors were in the same order of magnitude (Rick 1990).

The concentrations of most of the heavy metals decreased from January to May 1989 (cf. Chap. 4.2, Table 4.2.3) due to a lower discharge via the rivers from January to May (ARGE Elbe 1990) and an increasing metal sorption by the phytoplankton (Rick 1990). In the German Bight, e.g. the mean value for Hg in filtered water decreased from 0.19 ng/l in January to

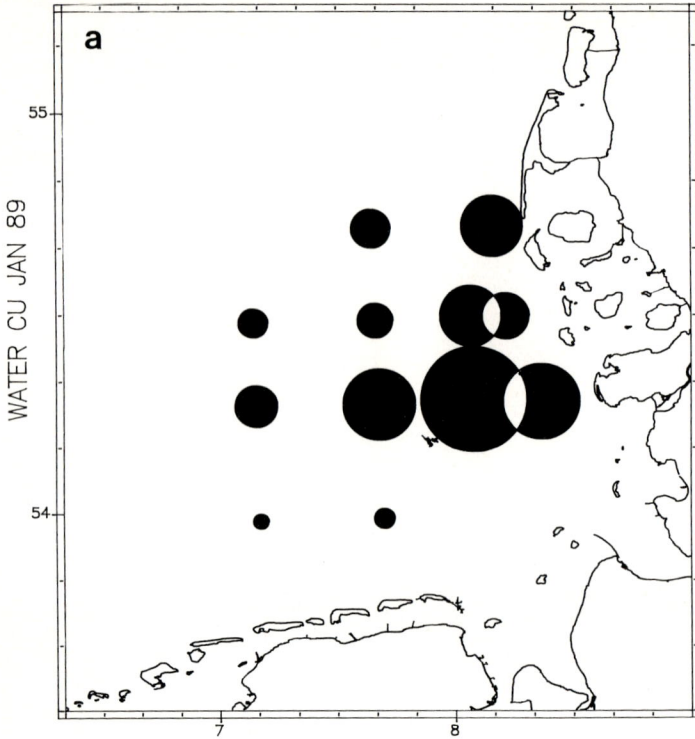

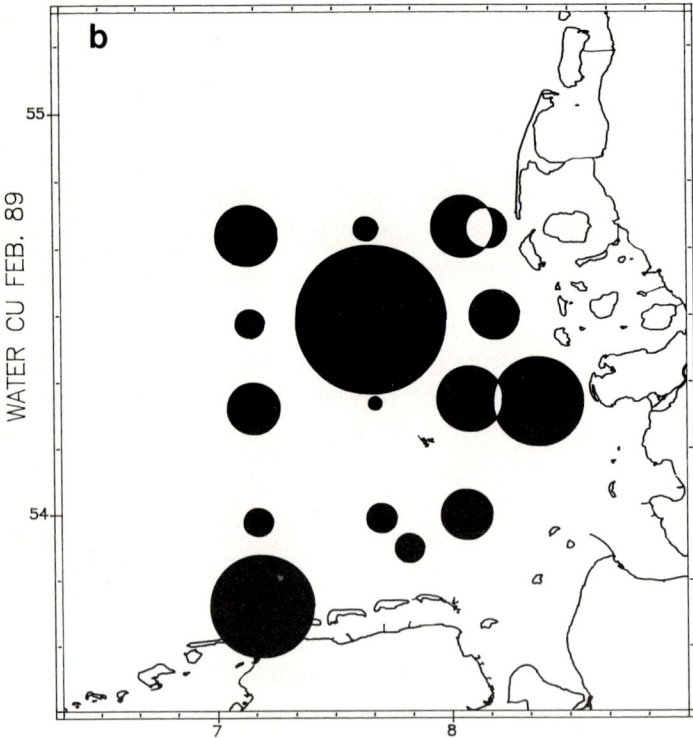

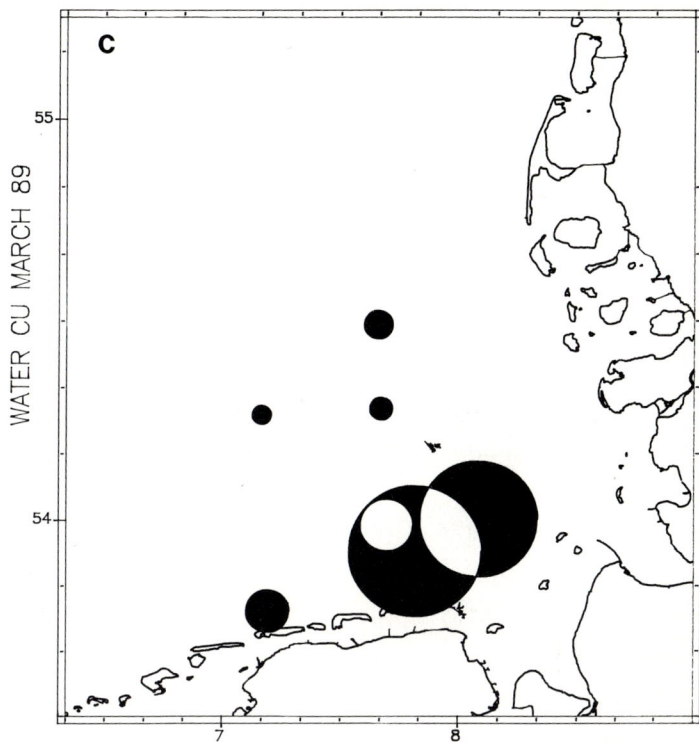

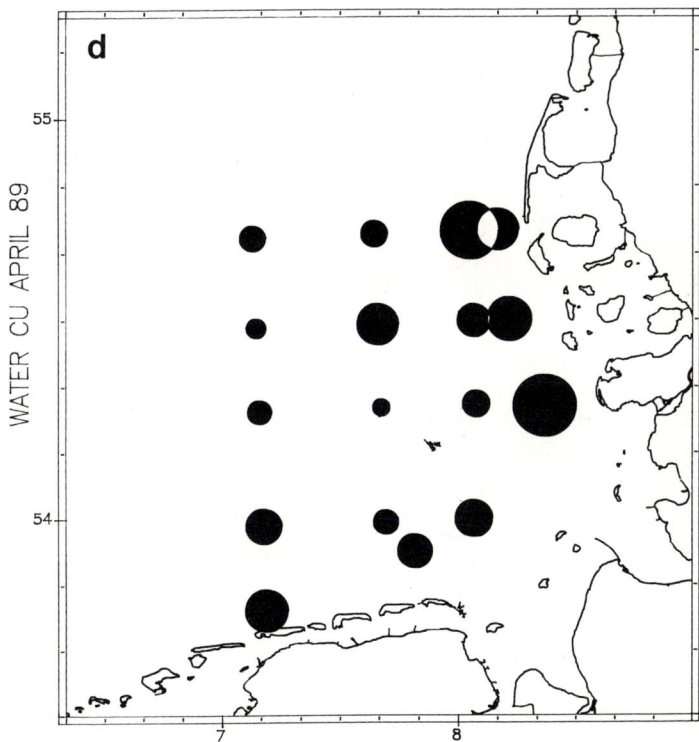

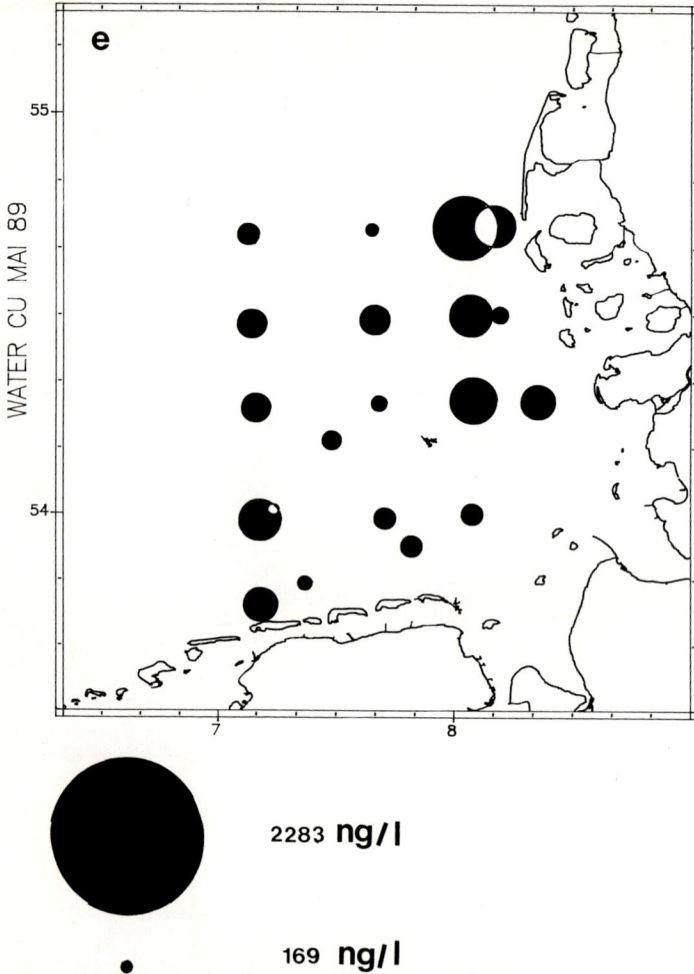

Fig. 2.5.16 *a–e*. Copper in filtered seawater (ng/l) from January 1989 to May 1989. Circles are concentrated around the stations

0.1 ng/l in May, and in unfiltered seawater from 1.5 ng/l in January to 0.21 ng/l in May.

Despite the high variability in the hydrographic conditions in time and space, these results show that the pollutant input by the rivers Elbe and Weser strongly affects the total German Bight.

Seston. Samplings of seston were carried out in December 1988, March 1989 and May 1989 on the ZISCH station grid. All samples were collected

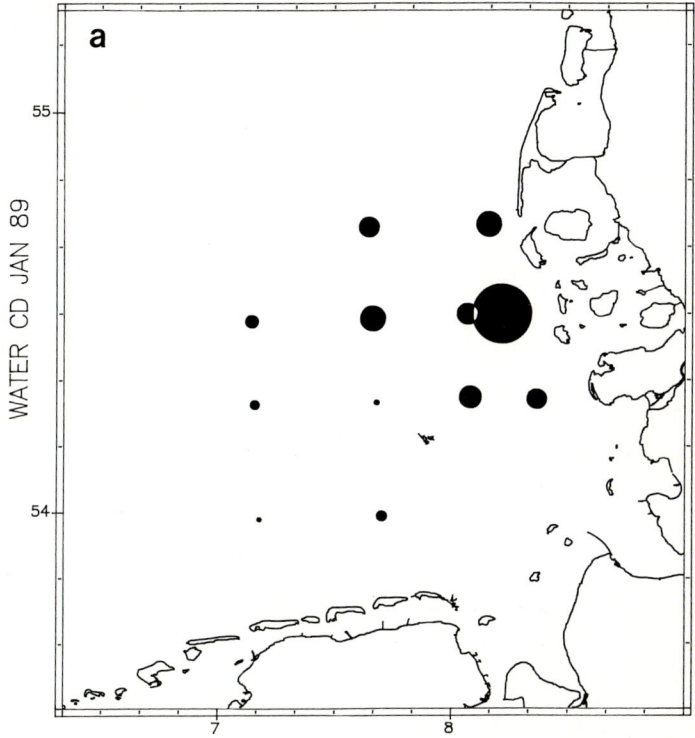

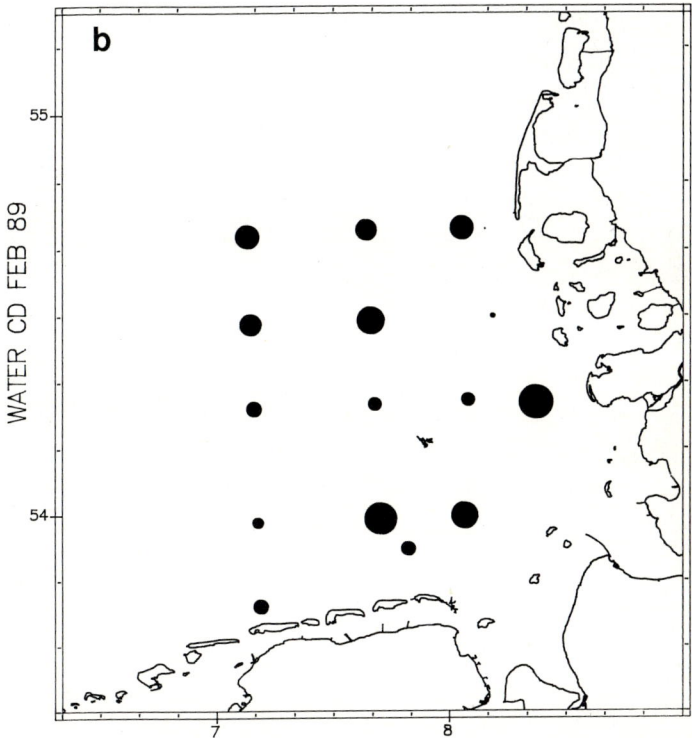

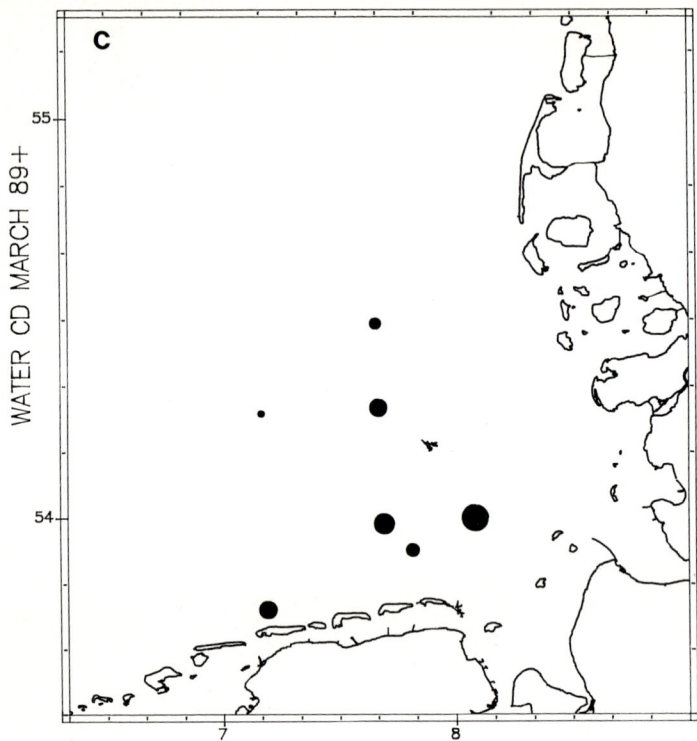

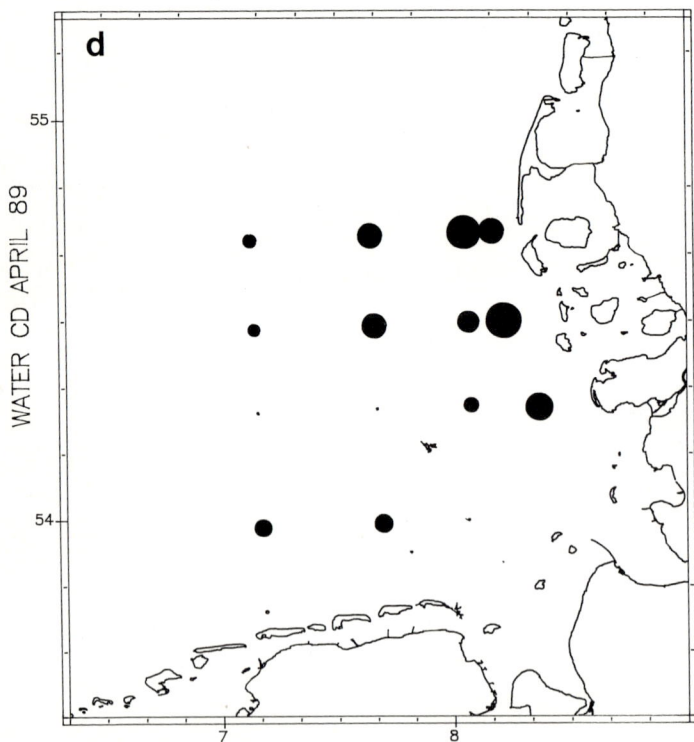

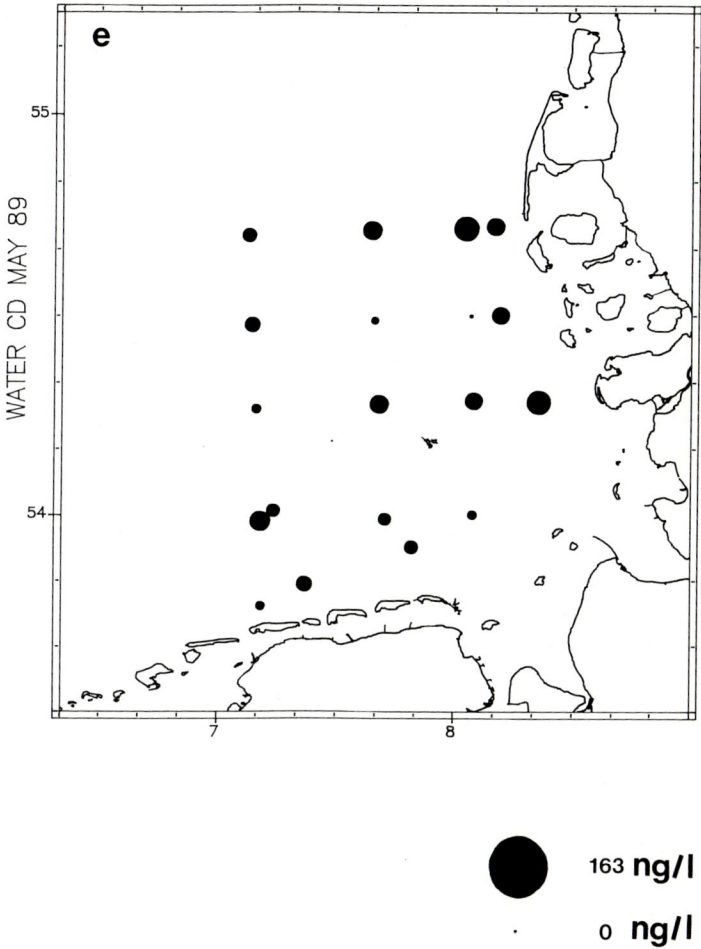

Fig. 2.5.17 a-e. Cadmium in filtered seawater (ng/l). For explanation see Fig. 2.5.16

by a noncontaminating pump from 10 m depth and subsequently centrifugated. The method is described in detail by Kersten et al. (1990). Al, Fe, Ca, Mn and the trace elements V, Zn, Cr, Ni, Cu, Hg were analyzed by AAS, AES-ICP and UV-VIS.

In Table 2.5.4, average concentrations and concentration ranges are given for the three cruises. Al, Fe, Ca, Mn, V, Cr, Ni and Cu showed significantly higher values in December and March than in May. Only for Zn were increasing concentrations observed in May.

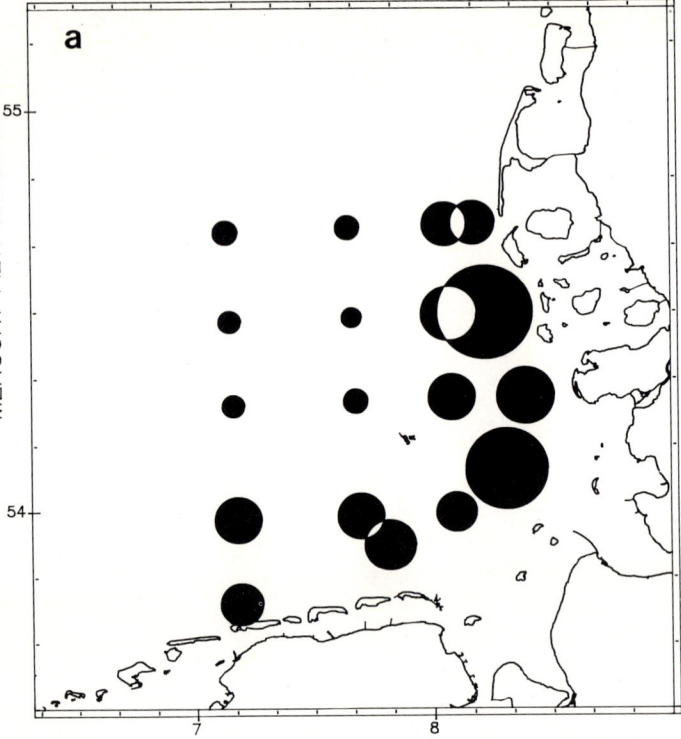

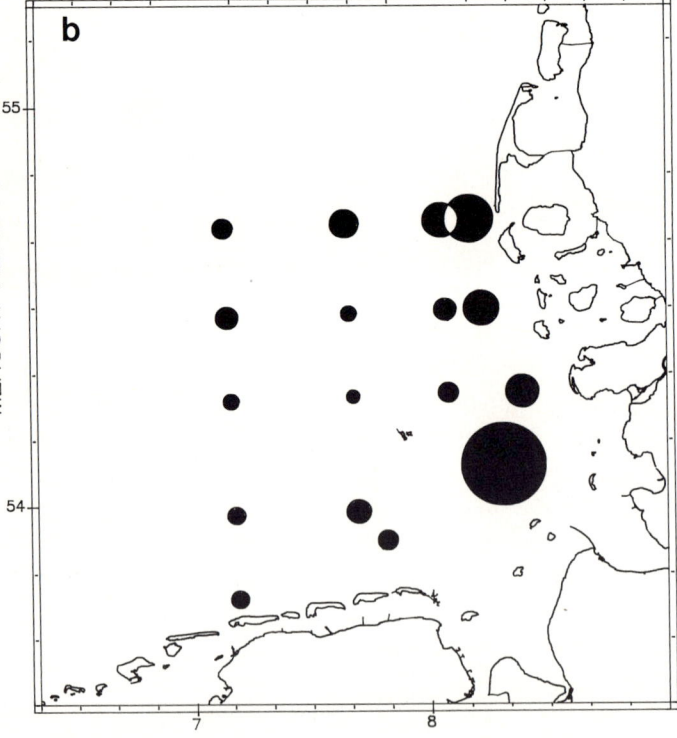

MERCURY FILT. WATER MARCH 89

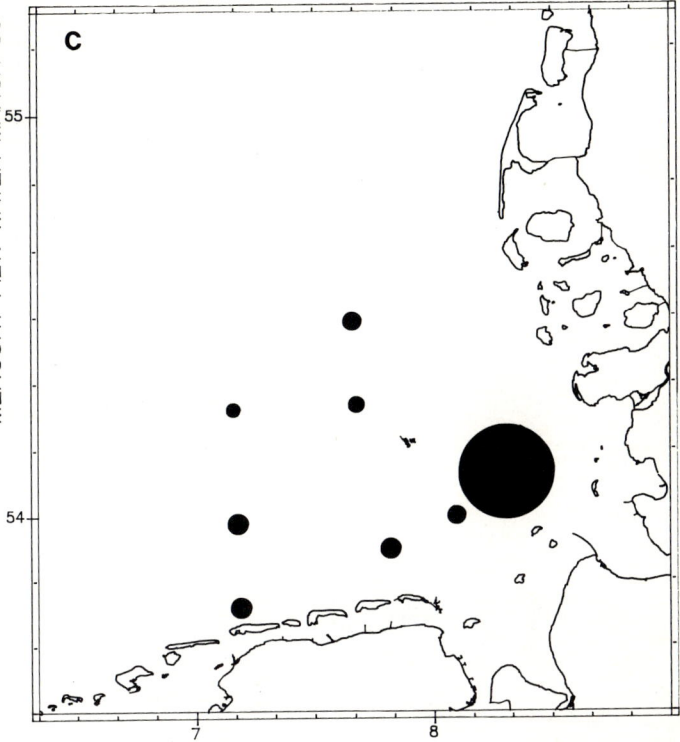

c

MERCURY FILT. WATER APR.89

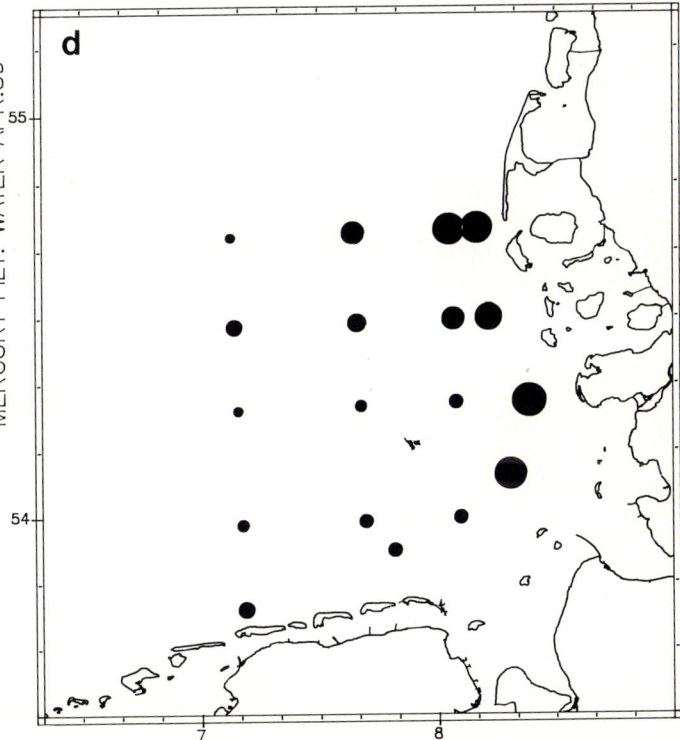

d

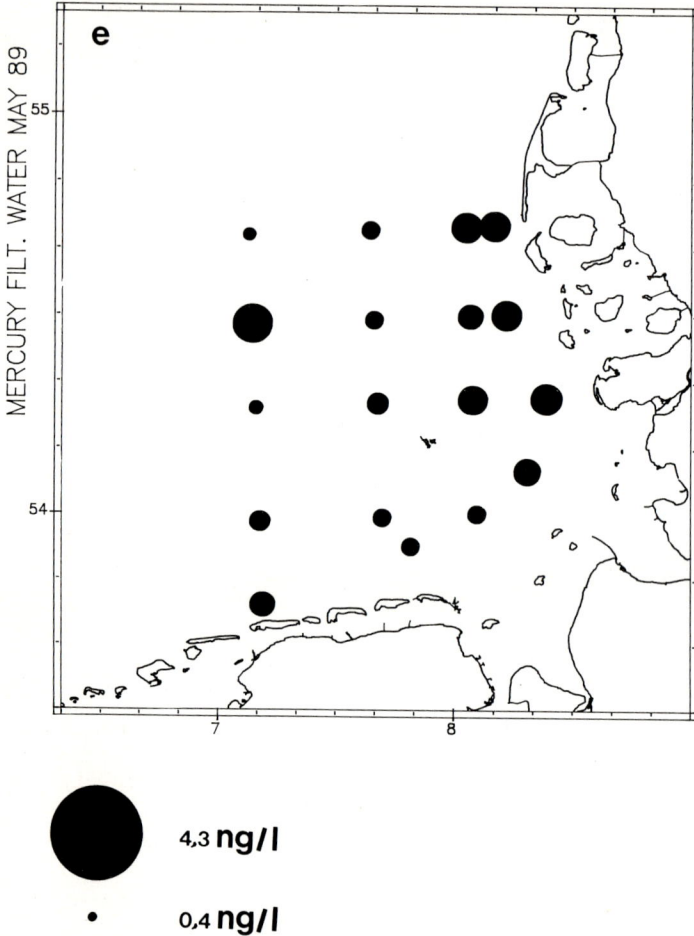

Fig. 2.5.18 *a-e*. Mercury in filtered seawater (ng/l). For explanation see Fig. 2.5.16

The regional distribution of Hg in the German Bight is shown in Fig. 2.5.21. Generally higher values were found in December and March in the eastern part of the German Bight between the coast and Helgoland (468-750 µg/kg) than in the western region (277-444 µg/kg). An inhomogeneous Hg distribution with local minima and maxima was found in May.

The elemental composition of suspended particulate matter (SPM) in coastal areas is strongly influenced by river water, resuspension of bottom material at low water depth as well as supply from coastal erosion (Nolting and Eisma 1988; Dehairs et al. 1989). Metal concentrations depend on the

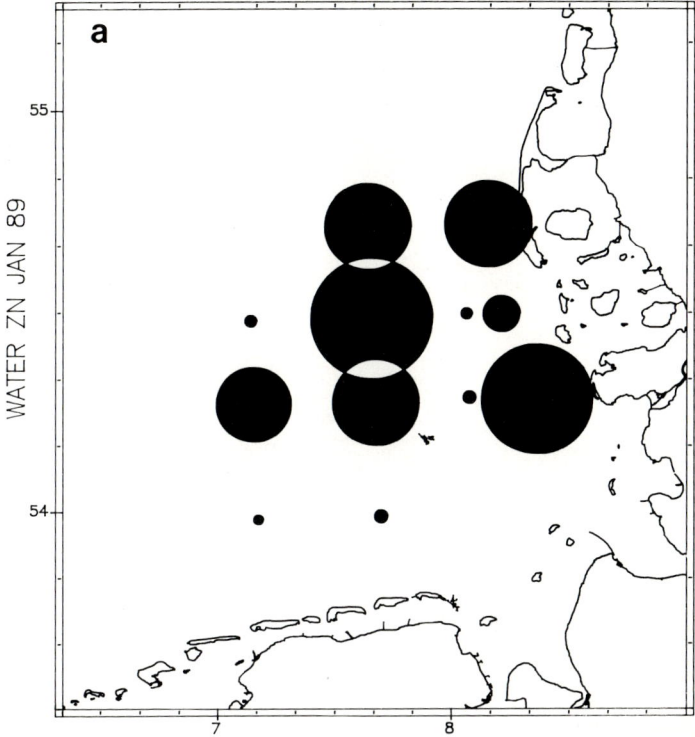

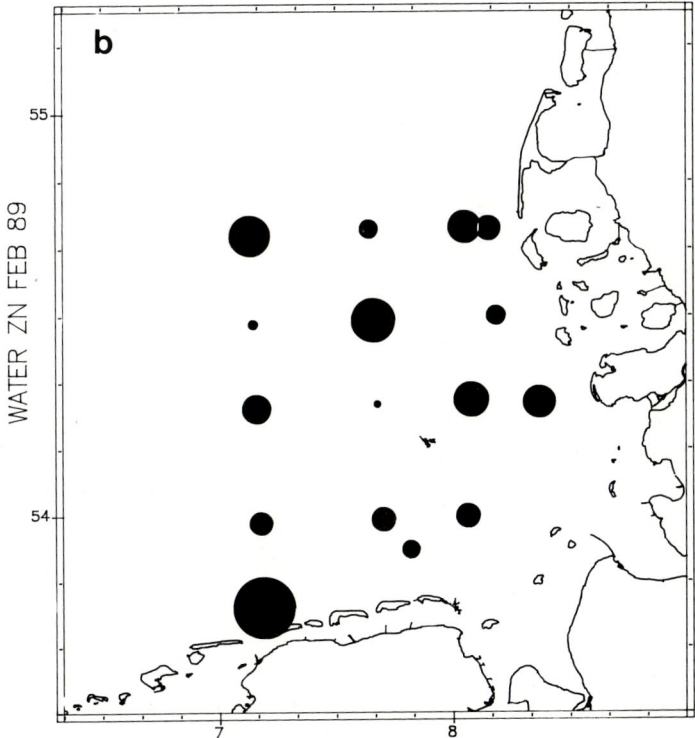

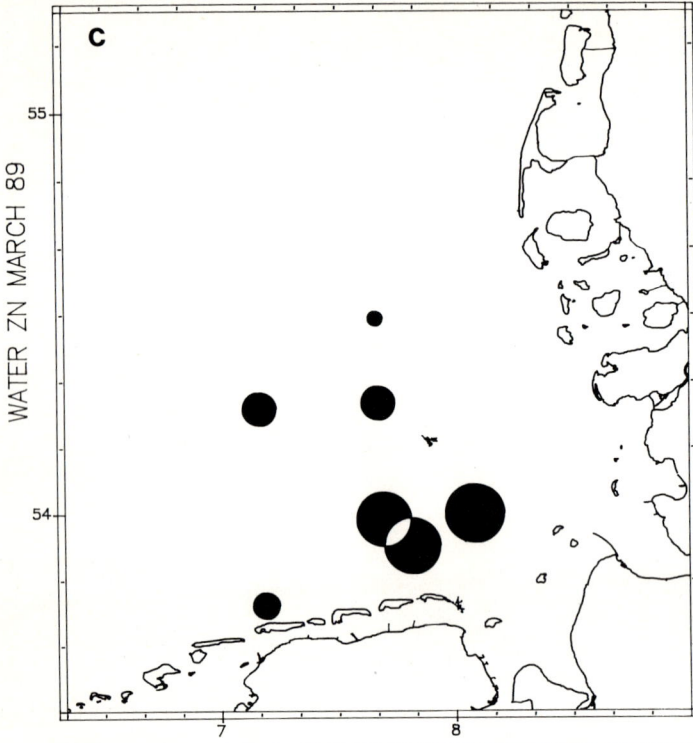

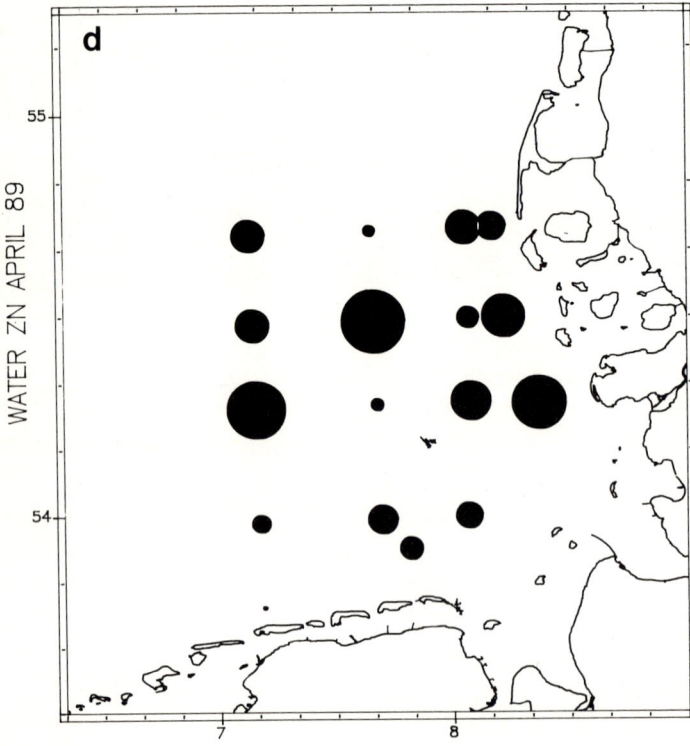

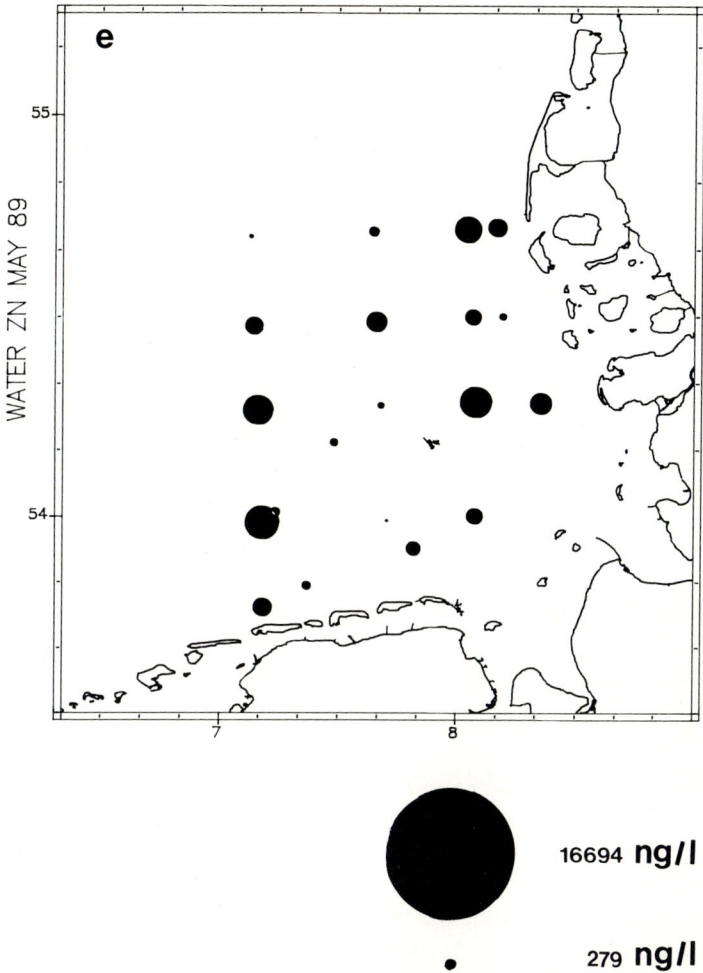

Fig. 2.5.19 *a-e*. Zinc in filtered seawater (ng/l). For explanation see Fig. 2.5.16

type, composition and surface structure of the suspended matter. Metals are strongly adsorbed by clay minerals, Fe-, Mn-oxides and organic matter. In Table 2.5.5 correlations of all measured elements with Al (usually used as independent variable) were listed separately for the three cruises. Hg, Fe, Mn, V, Cr had about the same positive relation with concentrations of Al. Ni showed good linear correlation in March and May, Ca only in May, whereas for Zn a relationship was found for December and March. Cu did not show any correlation. It can be assumed that elements which are strongly correlated with Al are probably associated with aluminosilicate material; and, conversely, elements which are weakly associated with Al

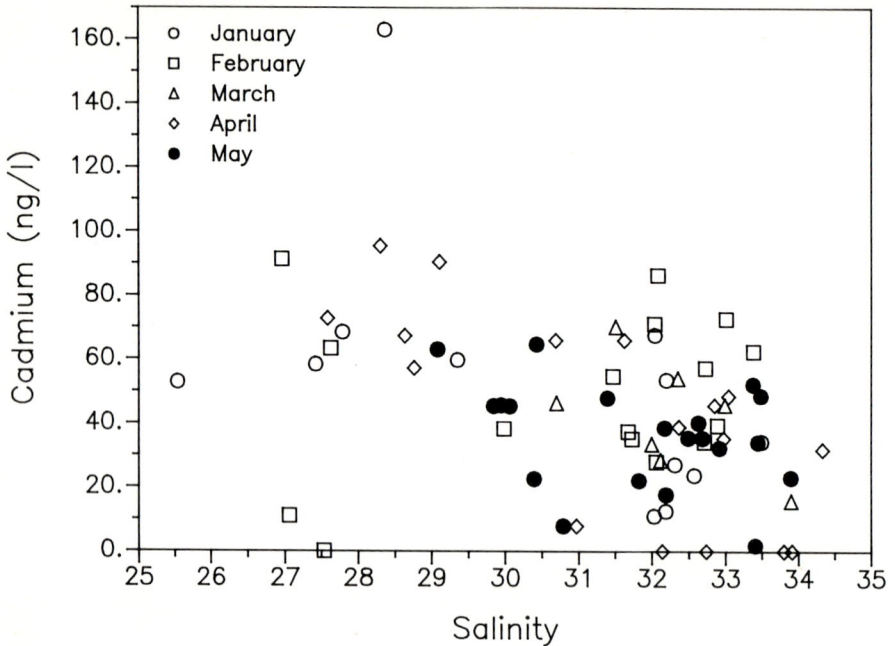

Fig. 2.5.20. Correlation of Cd with salinity from January to May 1989

are probably regulated by other particulate phases, such as biogenic matter or authigenic minerals (Feely et al. 1981).

The load of suspended particulate matter has been demonstrated to exert an important control on the particulate metal concentrations (Duinker 1983; Balls 1988; Kersten et al. 1988). A relation was found between the suspended matter concentration and the Al content in December and March. This reflects the predominance of mineral particles at high suspended matter concentrations. The sigificant negative correlation with salinity suggests that these Al-bearing mineral particles are provided by freshwater inputs, i.e. originate from the estuaries. Particulate Hg concentrations are thus higher in near-shore regions (Table 2.5.6). West of Helgoland lower amounts of suspended material were found and the Hg concentrations associated with detritus decreased. In this area, elemental composition of the suspended matter shifts from anorganic to more organic material.

Compared with the cruises in December and March, measurements made during May showed higher SPM concentrations in offshore regions. Organic material from primary production dominated the amount and elemental composition of seston. The presence of phytoplankton can complicate the interpretation of trace metal data, since such particles are unlikely to have the same properities as clay minerals (Balls 1989).

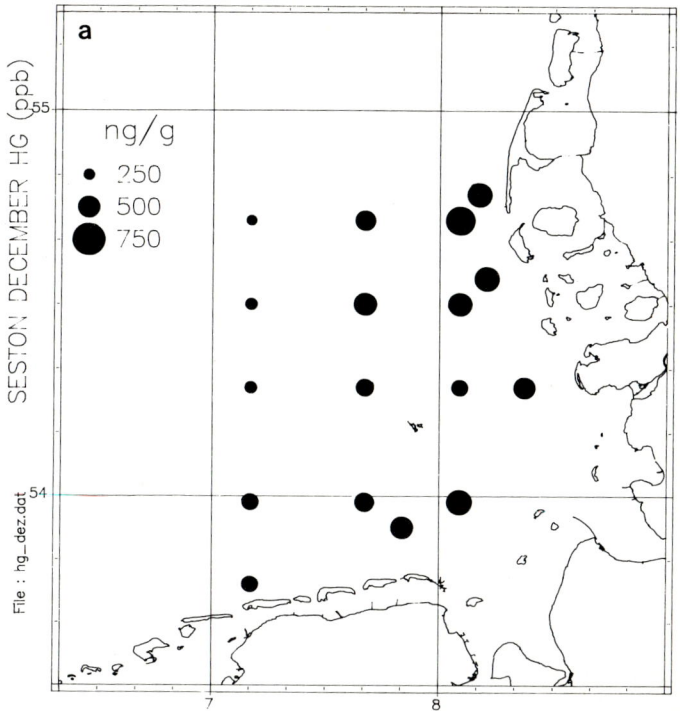

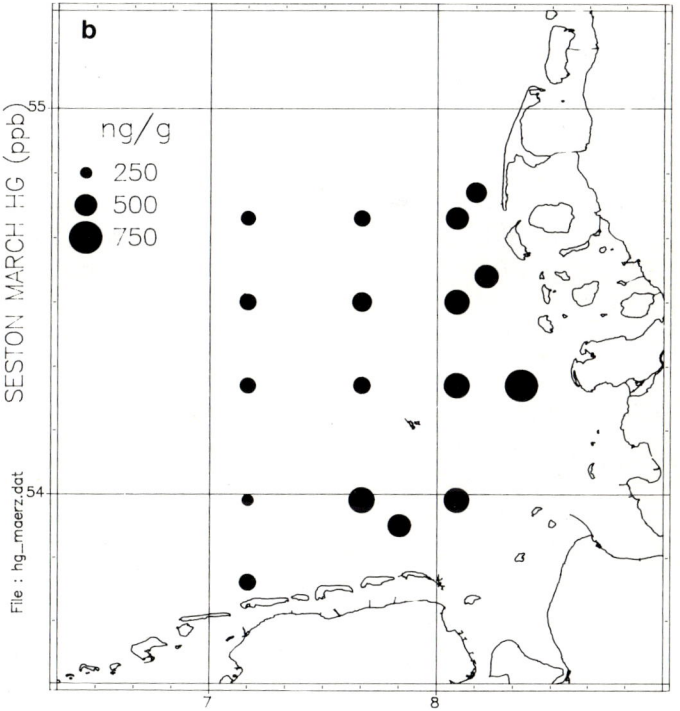

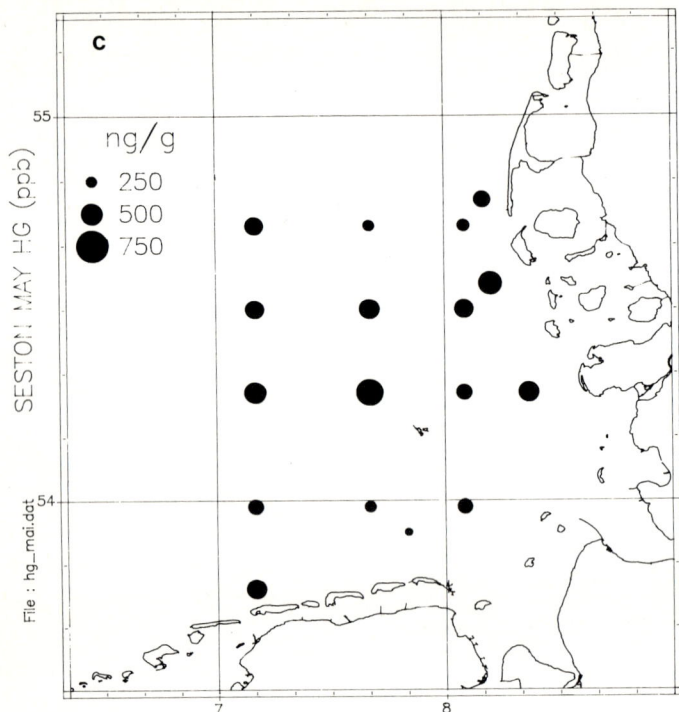

Fig. 2.5.21 *a-c*. Mercury in seston (ppb) in December 1988 (*upper left*), March 1989 (*upper right*) and May 1989 (*left*)

Differences in metal and SPM concentrations between offshore and near-shore areas became insignificant in May (Table 2.5.6). It can be supposed that control of Hg and Al concentrations by supended matter load is not valid in the presence of phytoplankton. Only a low correlation (r = 0.18) between SPM and Al was found in May (Table 2.5.5).

Phytoplankton. Dissolved heavy metals become adsorbed to the surface of the phytoplankton. The concentrations of Zn, Cd, Pb and Cu (Hg was not determined in phytoplankton) were negatively correlated with the phytoplankton biomass (cf. Chap. 4.3, Table 4.3.3). The adsorption capacity of the metals by the phytoplankton surface was calculated (Table 2.5.7). The bioaccumulation factor for Pb was not calculated because in many cases the total lead was bound to seston. The high values of bioaccumulation for Cu, Cd and Zn by the phytoplankton show the important role of the phytoplankton in the contaminant transfer. High station to station and month to month variability of the bioaccumulation of

Table 2.5.4. Metal contents of centrifuged suspended matter samples (n = 17)

	December		March		May	
	Mean	Range	Mean	Range	Mean	Range
Al (%)	2.22	(1.77-2.60)	2.10	(1.46-2.49)	1.58	(0.83-2.02)
Fe (%)	3.42	(2.68-3.98)	3.25	(2.35-4.00)	2.45	(1.21-3.18)
Ca (%)	9.92	(8.10-12.2)	10.3	(8.76-13.4)	8.30	(6.50-9.68)
Mn(%)	0.13	(0.07-0.17)	0.11	(0.06-0.17)	0.07	(0.02-0.13)
V (mg/kg)	84.7	(68.3-97.9)	77.6	(51.2-95.4)	60.2	(25.6-78.8)
Zn (mg/kg)	173	(124-204)	176	(132-241)	205	(153-288)
Cr(mg/kg)	85.3	(64.6-112)	80.5	(57.3-97.0)	59.6	(31.1-75.0)
Ni (mg/kg)	26.4	(22.8-32.0)	25.0	(19.0-32.1)	19.1	(13.4-26.5)
Cu (mg/kg)	28.6	(21.1-49.1)	29.6	(20.8-41.0)	22.2	(17.2-29.2)
Hg (µg/kg)	459	(277-667)	472	(258-750)	395	(174-618)

the phytoplankton was due to changing species composition, water temperature and interrelationships between nutrients and heavy metals. For further details and discussion see Chapters 4.3, 4.5 and 4.6.

Zooplankton. Data sets for heavy metal concentrations in zooplankton from field investigations found in recent literature mostly originate from mixed zooplankton samples. They are subject to considerable fluctuations. Moreover, zooplankton samples gained by net catches, especially in the southern and central North Sea, contain admixtures of slimy material (e.g. remainders of ctenophores and excreted substances from algae like *Phaeocystis*) as well as of phytoplankton, detritus and resuspended matter in variable qualities. These admixtures prevent exact determinations of heavy metals in zooplankton organisms.

For this reason, during our cruises in Helgoland Bight we began to produce samples for heavy metal measurements by separating relevant zooplankton species to enrich them in special reaction vessels after short washes in aqua quadro dest. This work was done on board the research vessel at a clean bench to prevent contamination with dust from the ship.

Table 2.5.5. Correlation coefficients between Al (%) and concentrations of other elements. Cruises December 1988, March 1989, May 1989. (* = significance level < 0.001)

Al	December	March	May
n	17	13	17
Fe	0.89*	0.95*	0.98*
Mn	0.71*	0.81*	0.85*
V	0.83*	0.82*	0.95*
Zn	0.84*	0.82*	-0.21
Cr	0.75*	0.90*	0.93*
Ni	0.19	0.70*	0.65*
Cu	-0.01	-0.15	-0.29
Hg	0.79*	0.48	0.69*
Ca	-0.18	-0.30	0.79*
Salinity	-0.82*	-0.80*	-0.62*
SPM	0.48*	0.53	0.18

Table 2.5.6. Comparison of SPM, Al, Hg and Cu concentrations between coastal and offshore regions on the ZISCH station grid (mean values, coastal: n = 11, offshore: n = 8)

Cruise	German Bight region	SPM (mg/l)	Al (%)	Hg (μg/kg)	Cu (mg/kg)
December 88	Coastal	4.78	2.35	514	29
	Offshore	2.14	2.03	381	28
March 89	Coastal	5.83	2.28	551	25
	Offshore	2.43	1.85	360	34
May 89	Coastal	6.74	1.48	369	21
	Offshore	6.37	1.71	431	23

Table 2.5.7. Heavy metal concentration in phytoplankton (10^{-10}ng/μm^3) and bioaccumulation (BAC) during the investigation period. For location of the stations, see Fig. 2.5.1

St. no.	Cadmium Feb Conc	Feb BAC	Mar Conc	Mar BAC	Apr Conc	Apr BAC	May Conc	May BAC	Zinc Feb Conc	Feb BAC	Mar Conc	Mar BAC	Apr Conc	Apr BAC	May Conc	May BAC
1							3.6	15894							896	43384
2							1.6	4386							463	25326
3					11.3	23214	12.0	37036					1539	38068	716	256344
4	0.8	1978	0.2	509			13.6	61451	185	2206	43	1198			507	21205
5	0.1	411	0.1				10.5	20077	67	2178	58				323	7272
6	0.2	625	0.5	3186			5.7	24708	106	2761	100	2185			497	12699
7	0.5	858			3.6	11182	7.9	20505	81	6663			1005	22191	557	24245
8	0.1	123			1.2	3535			71	1295			253	5658		
9					0.7	1129							462	29649		
10							2.4	3666							571	16294
11							0.8	1853							201	8503
12							1.5	3192							157	18793
13					1.7	2915	1.8	22354					535	18203	251	12172
14					1.3	1984	1.1	6197					676	7795	238	8971
15					2.1		2.6	5452					779	46157	522	71061
16					4.3	10972	0.8	1778					699	12843	237	5760
17					1.2	1698	2.0	3123					407	5518	370	13148
x̄	0.3	799	0.3	1848	3.0	7079	4.5	15446	102	3021	67	1692	706	20676	434	36345

St. no.	Lead Feb Conc	Mar Conc	Apr Conc	May Conc	Copper Feb Conc	Feb BAC	Mar Conc	Mar BAC	Apr Conc	Apr BAC	May Conc	May BAC
1				33.8							32.1	10067
2				43.2							19.9	6191
3			14.1	7.4					16.9	4464	32.4	10088
4	27.2	1.3		5.9	19.2	1230	1.2	190			10.5	2068
5	5.0	8.1		1.4	11.4	2628	5.3				8.2	1310
6	19.7	5.3		11.7	13.0	1648	2.3	832			19.8	4615
7	1.3		51.6	9.5	9.7	2252			14.9	5060	4.9	1114
8	2.9		15.9		5.9	631			9.5	2440		
9			44.4						20.0	4973		
10				143.2							52.6	5405
11				31.0							7.8	1223
12				20.0							3.5	1405
13			80.0	50.3					34.1	6763	10.0	1564
14			24.8	33.7					11.8	1866	8.0	1784
15			117.2	2.7					24.2	9284	17.3	7471
16			99.9	12.3					35.6	8581	6.5	928
17			70.9	47.9					29.1	3061	18.2	3538
x̄	11.2	4.9	59.2	30.7	11.8	1678	2.9	511	21.8	5166	16.8	3918

During six short cruises, altogether 140 samples were prepared in this way; a further 56 mixed zooplankton samples were taken for a comparison.

The samples, which had been frozen in the sampling vials (closable Eppendorf vials, 1 ml or 2 ml) were lyophilized and dry weight was determined. They were rendered soluble either in the same vials, according to Sperling (1984) and Zauke et al. (1985) with a H_2SO_4/HNO_3 mixture (1 + 4 v/v) for 4 h at 96 °C in a hot air cabinet, or transferred into test tubes with $HClO_4$ in a heating block (incineration program: 1 h 50 °C, 1 h 70 °C, 2 h 90 °C, 2 h 110 °C, 5 h 150 °C and 10 h 180 °C) and filled to a defined volume with fourfold distilled water. The Cd and Pb measurements were carried out with an Hitachi-AAS with Zeeman background compensation in tube cuvettes with the platform according to the standard addition method. For reference measurements a copepode homogenate (Ma-A-1 TM) of the IAEA was used.

The results indicate that heavy metal concentrations of single species were in the lower range of the concentration spectrum found in the literature, whereas mixed zooplankton samples (anew) showed extremely high and varying concentrations, especially for lead (0.08 to 1920.1 ppm). Based on present data material, it further might be stated that the investigated species and groups of zooplankton showed different ranges of metal concentrations, irrespective of regions and seasons (Fig. 2.5.22a,b). Thus, *Mysidacea* and fish larvae always had the lowest values of metal concentrations, whereas bottom invertebrate larvea (e.g. from the polychaete *Lanice conchilega*) and *Cumacea*, which stay in the water column only for some hours per day, contained highest concentrations of lead and cadmium. On the other hand, the copepods (e.g. *Calanus* spp.), forming the biggest proportion of zooplankton, could be found in the midfield.

Because of the fragmentary sampling grid and low data material at the moment, it is still impossible to detect seasonal or regional trends of heavy metal concentrations in zooplankton in the Helgoland Bight (see Tables 2.5.8 and 2.5.9). Further measurements are necessary to obtain reliable results about fluctuations in time and space. Nevertheless, it can be stated that single species or groups of zooplankton should be used for this type of investigation, since (apart from disturbing admixtures) mixed zooplankton samples generate variabilities of heavy metal concentrations due to changing species composition alone. Furthermore, it might be assumed that interspecific differences in heavy metal concentrations probably are of greater significance than a single species can disclose in time and space.

Benthos. As described in Chapter 2.3, the hermit crab *Pagurus bernhardus* was analyzed for heavy metals (Hg, Cd, Zn, Cu, Ag, Mn, Fe and Pb)

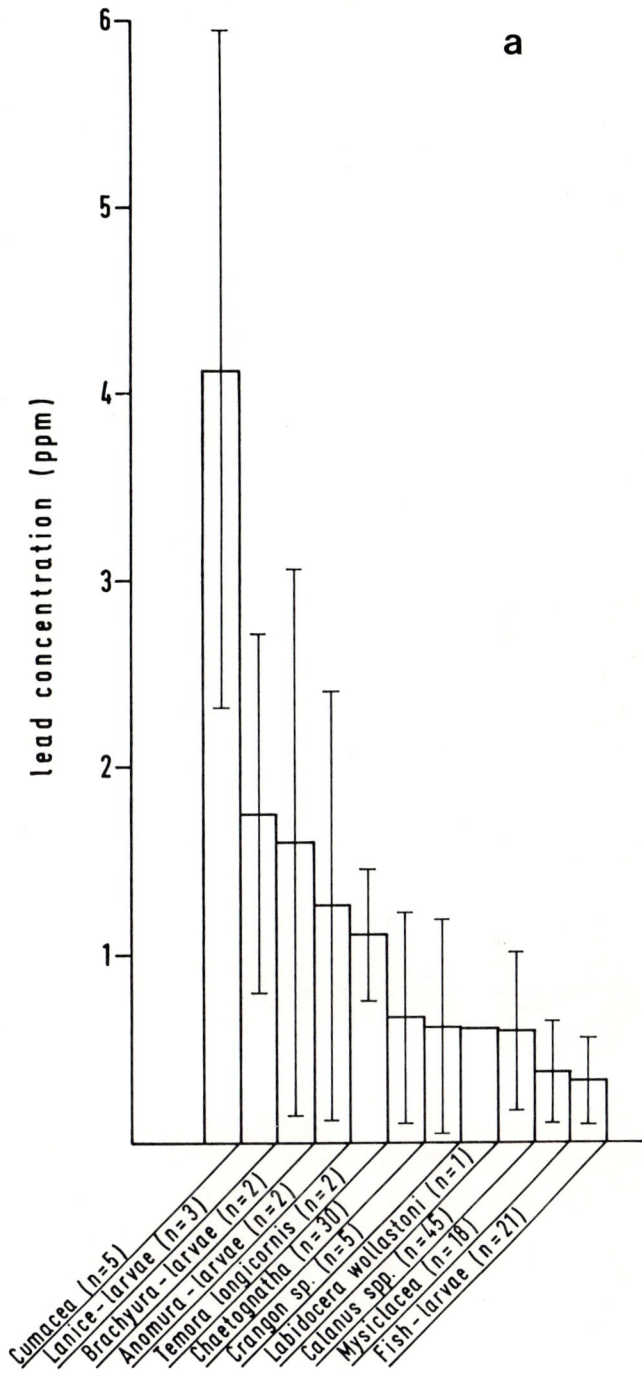

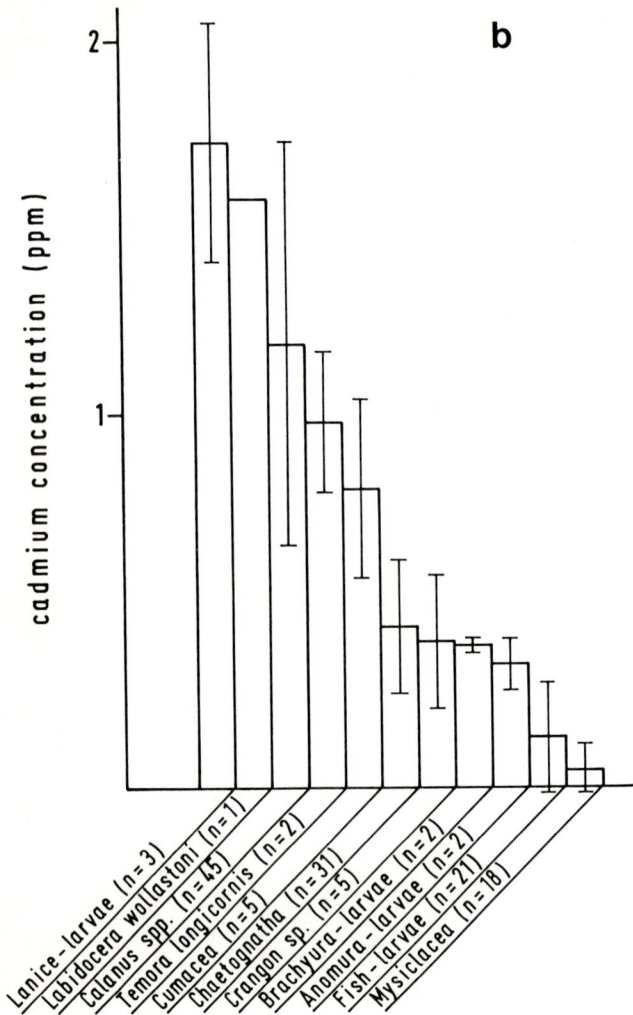

Fig. 2.5.22. Lead *(a)* and cadmium *(b)* concentration in zooplankton. Mean values of all surveys and sampling stations

during five cruises from December 1988 to May 1989. The Hg- and Cd concentrations of *Pagurus bernhardus* were chosen for a detailed description.

With the exception of April, the highest concentrations of mercury in *Pagurus bernhardus* were found near the coast, especially west of the North

Table 2.5.8. Mean values and standard deviation of Pb concentrations in zooplankton expressed in ppm (N = number of samples)

Sampling month	Calanus spp.	N	Sagitta spp.	N	Fish larvae	N	Mysidacea	N	Cumacea	N	Mixed samples	N
Dec.	1.12±0.59	8	1.15±0.44	7	0.34	1	-	0	-	0	-	0
Jan.	0.43±0.16	15	0.57±0.37	6	0.15	1	0.47±0.26	5	2.81	1	-	0
Feb.	0.55±0.20	8	1.57±0.63	2	0.30±0.28	7	0.43±23	6	5.29	1	11.45±20.01	10
Mar.	-	0	0.39±0.49	3	0.43±0.34	2	-	0	-	0	19.72±26.80	14
Apr.	0.41±0.32	10	0.48±0.48	7	0.24±0.18	7	0.24±0.16	4	6.84	1	114.29±383.45	24
May	0.65±0.33	4	0.16±0.09	5	0.17±0.07	3	0.15±0.03	3	2.83±1.16	2	2.59±2.61	7

Table 2.5.9. Mean values and standard deviation of Cd concentrations in zooplankton expressed in ppm (N = number of samples)

Sampling month	Calanus spp.	N	Sagitta spp.	N	Fish larvae	N	Mysidacea	N	Cumacea	N	Mixed samples	N
Dec.	1.16±0.35	8	0.65±0.13	7	0.05	1	-	0	-	0	-	0
Jan.	1.04±0.48	15	0.49±0.13	6	0.06	1	0.07±0.06	5	0.69	1	-	0
Feb.	1.79±0.65	8	0.51±0.13	2	0.18±0.24	7	0.02±0.01	6	1.28	1	0.94±0.75	10
Mar.	-	0	0.25±0.05	4	0.18±0.04	2	-	0	-	0	1.24±1.53	14
Apr.	1.06±0.37	10	0.26±0.05	7	0.11±0.05	7	0.01±0.01	4	0.63	1	0.51±0.32	25
May	0.91±0.18	4	0.39±0.06	5	0.06±0.03	3	0.01±0.00	3	0.70±0.03	2	0.19±0.08	7

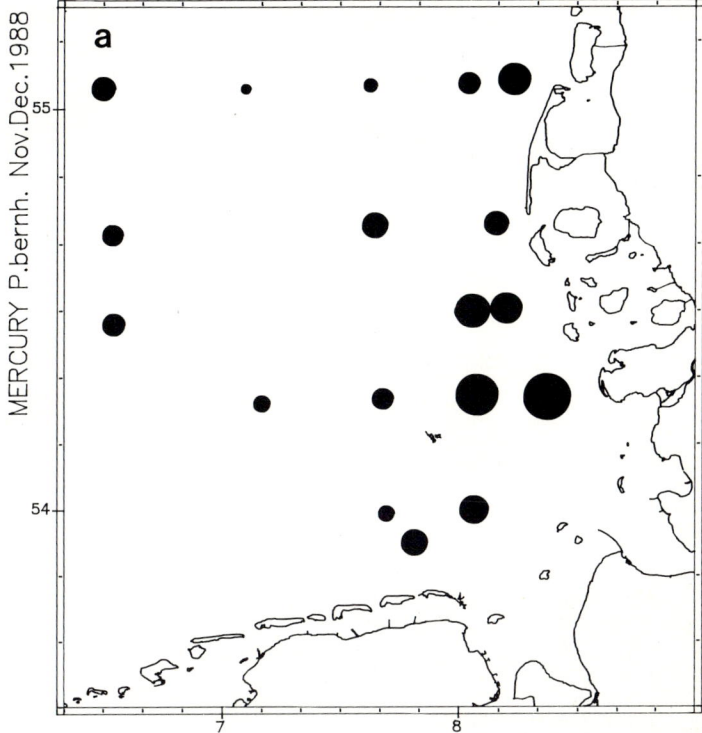

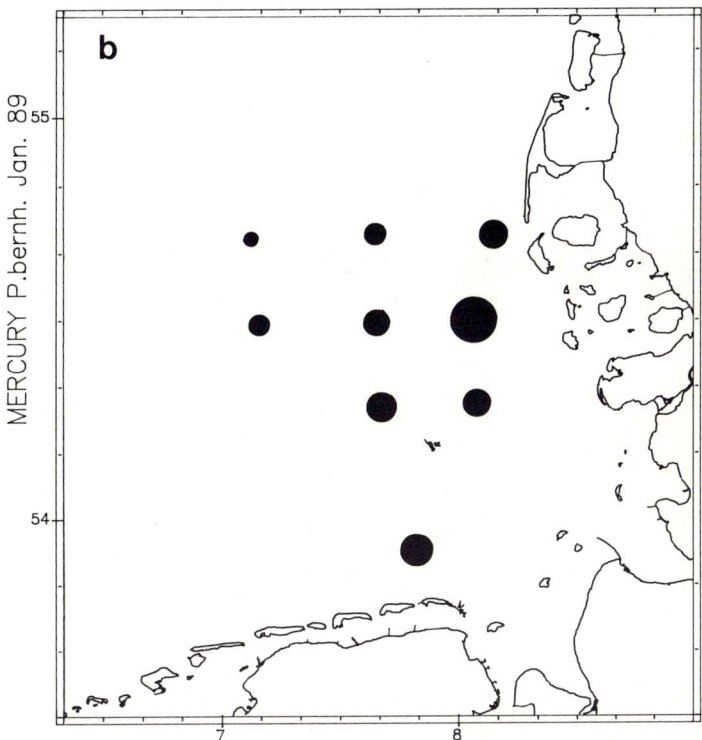

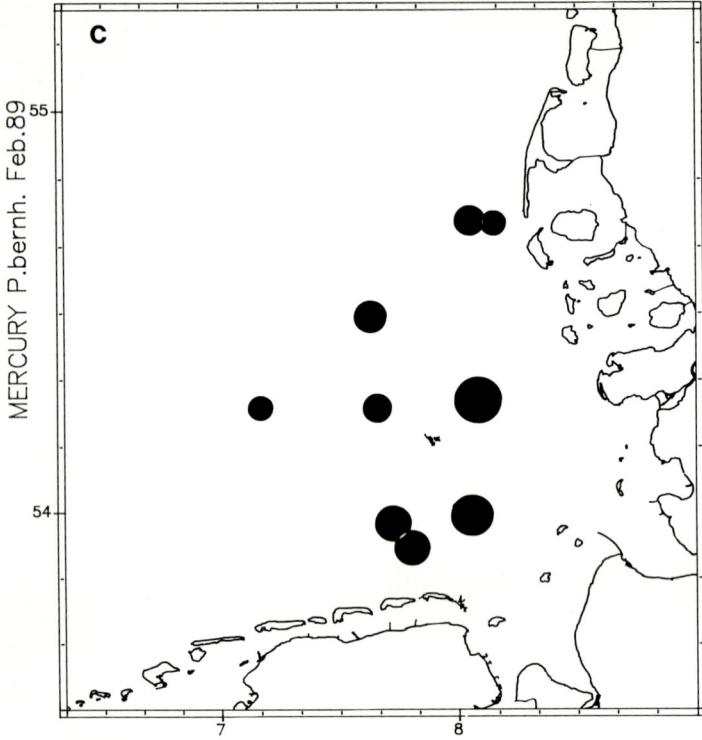

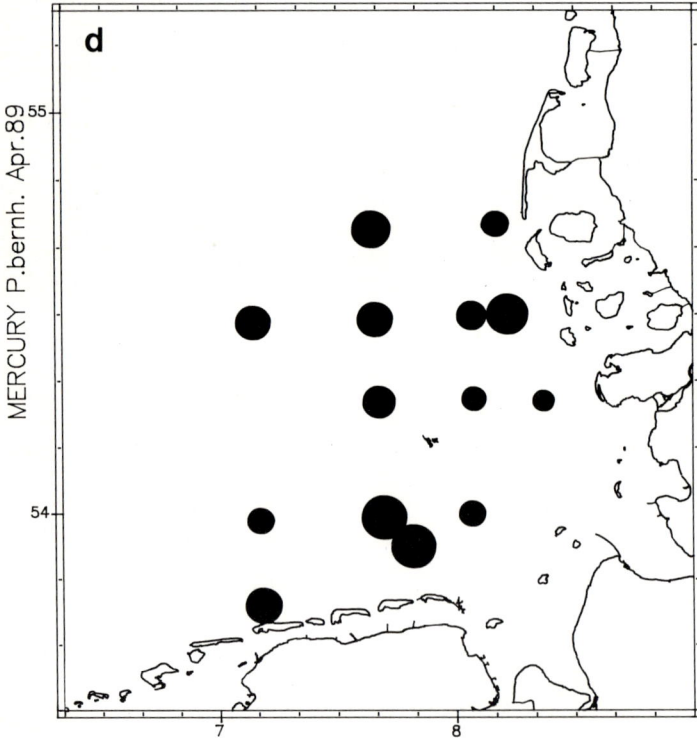

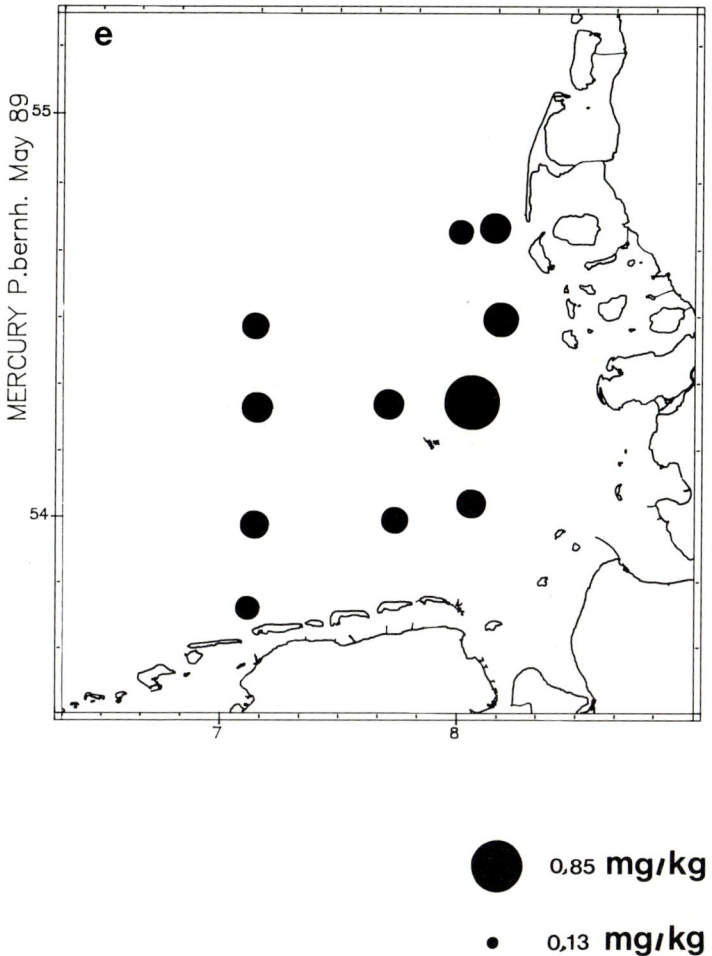

Fig. 2.5.23 a-e. Mercury in *Pagurus bernhardus* from December 1988 *(a)*, January *(b)*, February *(c)*, April *(d)* and May 1989 *(e)*

Frisian Islands, and the lowest Hg values were found in offshore regions of the German Bight (Fig. 2.5.23a-e). These results correspond with findings for the mussel *Mytilus edulis* of Borchardt et al. (1988), who described decreasing Hg concentrations with increasing distance from the Elbe estuary. In April, high mercury concentrations of *P. bernhardus* were measured west of the 8th meridian with a concentration peak north of the East Frisian Islands (Fig. 2.5.23d).

As described for the Hg concentrations in water, the distribution pattern of mercury in the hermit crab could be due to the discharge of the river

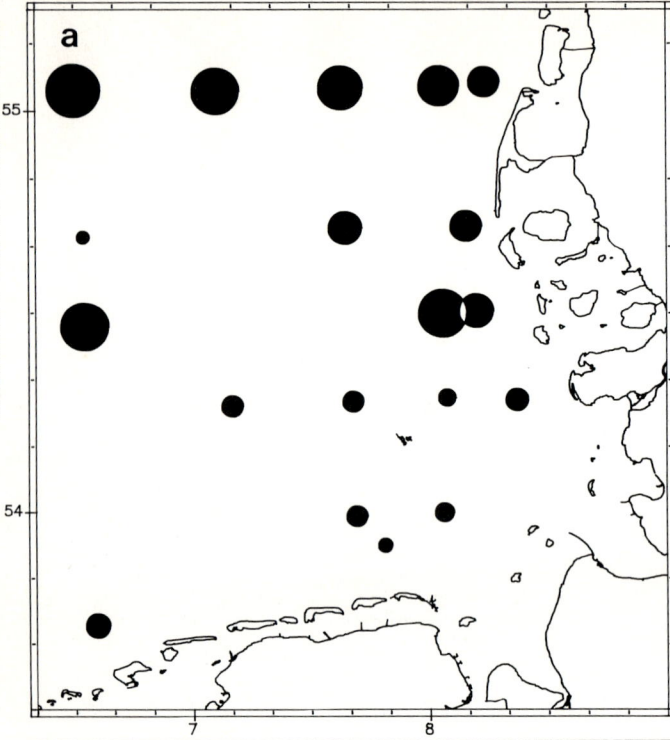

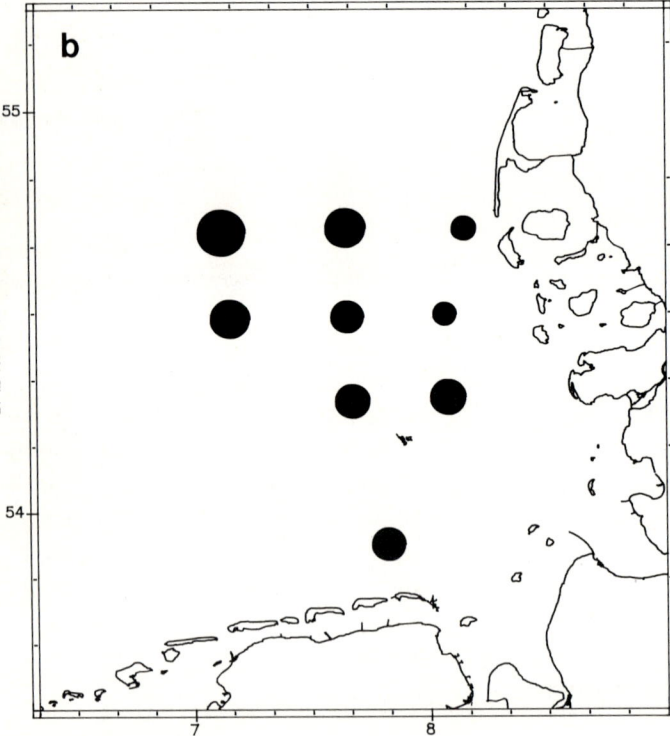

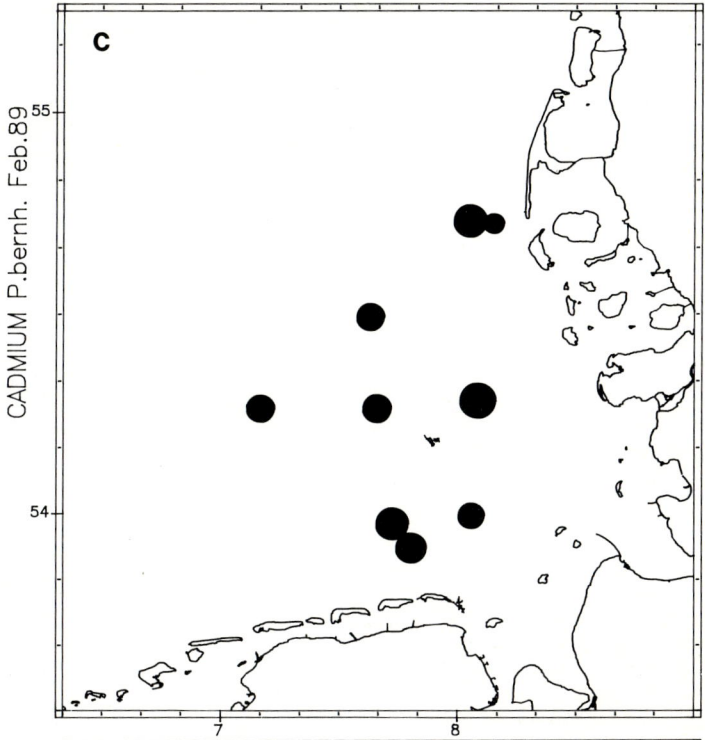

c   CADMIUM P.bernh. Feb.89

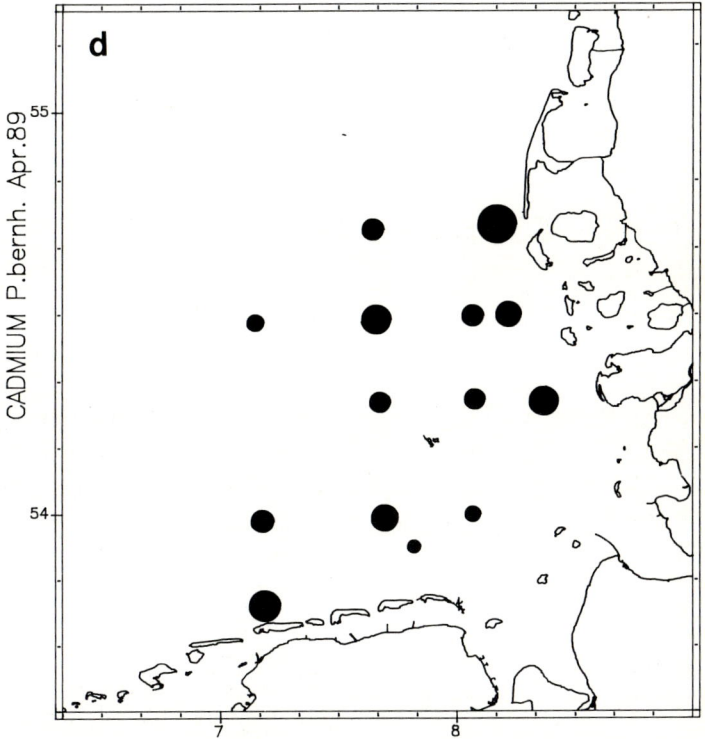

d   CADMIUM P.bernh. Apr.89

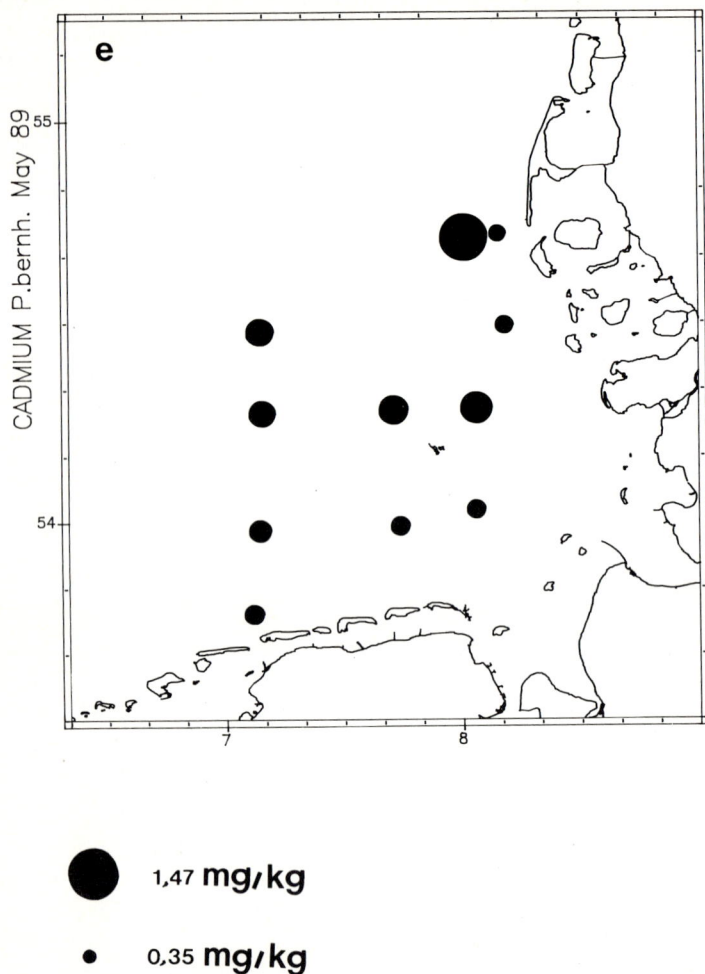

Fig. **2.5.24** *a-e*. Cadmium in *Pagurus bernhardus* from December 1988 *(a)*, January *(b)*, February *(c)*, April *(d)* and May 1989 *(e)*

Elbe. Westerly winds cause the Elbe plume to flow along the North Frisian Islands, whereas during easterly winds, the Elbe plume was found further to the west. In April, the Hg distribution in *Pagurus bernhardus* may be due to a previous east wind situation. Ten days before sampling, an easterly storm had taken place, and the following days were characterized by an easterly wind with lower speed. During the other months, the patterns were due to hydrographic conditions resulting from westerly wind situations. It seems that the response of *P. bernhardus* to changing metal concentration in the water is more rapid than expected, so that monthly sampling of *P.*

STATION Z6 : Aluminium

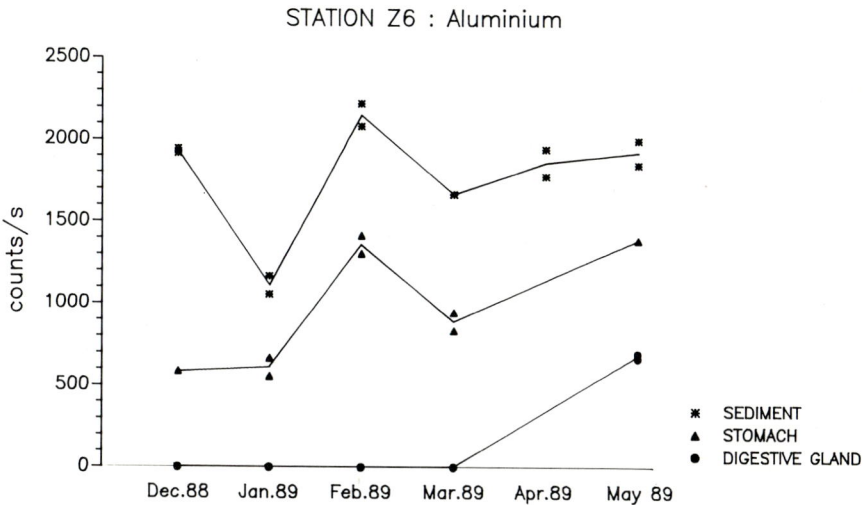

Fig. 2.5.25. Aluminium in sediment, stomach and digestive gland of *Pagurus bernhardus* during the investigation period from December 1988 to May 1989 at station 6

bernhardus detects these changes. Only in January were the Hg concentrations in *P. bernhardus* negatively correlated with the salinity measured during the samplings. The cases of no correlation between metal concentration and salinity show that it takes some time (a few days) until the change in metal concentration in *P. bernhardus* is measurable.

In contrast to Hg (see above) the Cd concentrations of *Pagurus bernhardus* and in water did not correspond (Fig. 2.5.24). The Cd concentrations in water and in the sediment (Irion and Müller 1987; Anonymous 1988) decrease, while in *Pagurus bernhardus* they increase with distance from the coast. Increasing Cd concentrations from the German Bight to the Dogger Bank were also shown by Kröncke (1988) for the polychaete worm *Nephtys hombergii* and Claussen (1988) for the dab *Limanda limanda*.

Only in April did the Cd concentration in water and in *Pagurus bernhardus* correlate significantly positively (Spearman rank correlation, p < 0.05). This relation is not a real coincidence: because of the east wind conditions at the beginning of April, the Cd pattern of *Pagurus bernhardus* might be explained by low Cd concentrations in the Elbe plume. The concentration pattern of Cd in water in April reflects the situation for west wind conditions and is not the explanation for the Cd concentration in *Pagurus bernhardus*.

STATION Z6 : Iron

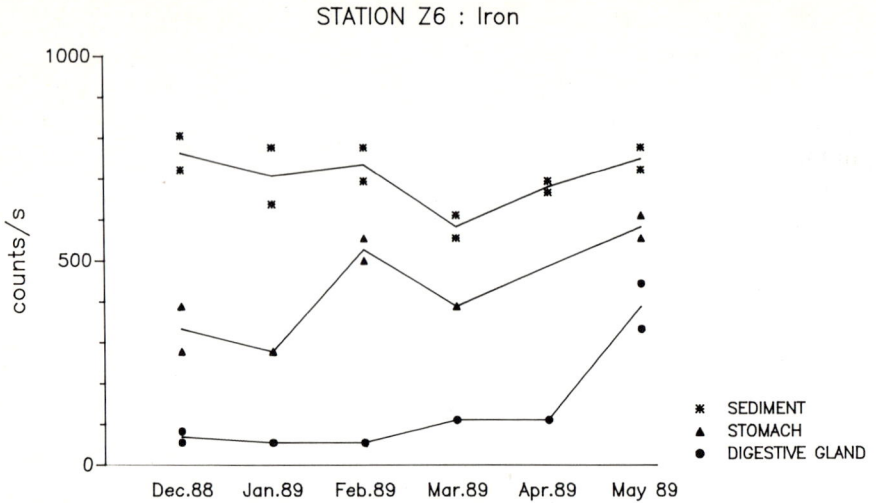

Fig. 2.5.26. Iron in sediment, stomach and digestive gland of *Pagurus bernhardus* during the investigation period from December 1988 to May 1989 at station 6

The results from *Pagurus bernhardus* could be explained by different bioavailiability of Cd in different regions of the German Bight. It seems that the uptake of Cd by *Pagurus bernhardus* is more effective if Cd is bound to organic material, for instance to sedimented phytoplankton. The Cd concentrations in the phytoplankton (this Chapter) in April and May were higher in the western and southwestern stations than at stations which were influenced by the Elbe outflow. In May, the Cd concentration in plankton was positively correlated with the salinity, and the Cd-bioaccumulation of the phytoplankton was negatively correlated (Spearman rank correlation, $p < 0.05$) with the Cd concentration in water.

With the results from the analysis of the elemental composition of sublittoral surface sediments and the ingested and partly assimilated food in the digestive tract of *P. bernhardus* by energy-dispersive X-ray analysis (EDAX), only in May could an uptake of metals be shown, although the amount of metals in the sediment did not change. For this investigation, the digestive tract of the hermit crab was divided into its functional compartments, the stomach, the digestive gland and the hindgut. The elements Cd, Pb and Hg could not be detected with the EDAX method because the concentrations present were too low (Miyoshi et al. 1982). The elements Na, Mg, S, Cl, K and Ca were detected as distinct peaks in the curves obtained for samples from different parts of the digestive tract of *P. bernhardus* and for the surface sediments at all stations and on each sampling date. These elements are known to contribute the main ions to the sea water (Smith 1974) and cellular fluids of crustacea (Nicol 1960).

Differences in occurrence were found for the elements P, Al, Si and Fe. Phosphorus was detected only in the samples of *P. bernhardus* and never in the sediments, which can be attributed to the fact that phosphorus is an important element in the metabolic cycles of organisms. Each month during the investigation period Al (Fig. 2.5.25), Si and, Fe (Fig. 2.5.26) were determined in the surface sediments, in the stomach, and in the hindgut samples. In the digestive glands, an uptake of Al (Fig. 2.5.25) and Si could be demonstrated only in May, when increasing assimilation of Fe was also detected (Fig. 2.5.26). Since only in May the amounts of Al, Si, and Fe in the digestive gland of *P. bernhardus* were above the detection limits of the EDAX method, it might be concluded that the absolute amount of each of these elements in the surface sediments is not the predominant factor affecting their uptake by the digestive gland, but perhaps the form in which these elements were bound: the spring plankton bloom increases the amount of organic particles in the water column, thus providing a larger surface for the adsorption of Al and Fe to diatom detritus and other organic particles. During the investigation period, Dürselen (1990) detected an enormous increase in the amount of phytoplankton in May, the phytoplankton consisting mainly of diatoms. Diatom species growing in April and May displayed the greatest sorption capacity for heavy metals (Rick 1990). Similar differences in the sorption capacity of diatom species for other metals seem likely. By sedimentation, the plankton and other organic particles as well as the adsorbed elements reach the sea floor and are integrated into the food web of the benthos. This might explain the uptake of Al, Si, and Fe in May by the digestive gland of *P. bernhardus*. The results of this study seem to hint that *P. bernhardus* can only assimilate Al and Fe in the digestive gland if they are bound to sedimented phytoplankton or other organic particles. The assimilation of Al, Si and Fe was detected only at two of the four selected stations. This might be explained by an uneven sedimentation to the sea floor. Frauenheim (1991) observed significant differences in the chlorophyll concentration at the sediment surface due to a great mesoscale (10 nm^2) variability, especially during sedimentation.

2.5.4 Conclusions

The German Bight is a region which is highly variable with regard to meteorological and hydrographic conditions. In the different compartments of the ecosystem, different temporal and spatial scales and variabilities of the measured compounds were observed. These variabilities are due to abiotic and/or biotic factors. An abiotic seasonal effect is, for example, the change in the outflow of the rivers. In the German Bight the regional distribution of nutrients, seston and pollutants in water show the great coastal influence; especially the rivers Elbe and Weser were a distinct source for the contaminants.

In the hermit crab, for instance, the highest mercury concentrations were measured in the Elbe plume. For nutrients, the Wadden Sea is an additional source. The maximal outflow of the Elbe is measured in January (ARGE Elbe 1990). This corresponds with the findings of higher nutrient and mercury concentrations in water and iron in seston in this month. The organic pollutants in water seemed not to be to influenced by the changing outflow of the Elbe. Only in spring were higher concentrations of Lindane measured, but these differences are due to the application of this substance in agriculture. This result could also be demonstrated in sea stars by their quick uptake and loss of these substances. For the organic contaminants in water, it is shown that short-term variabilities, e.g. storms, lead to significant changes of measured concentrations. The small fluctuation range of the measured heavy metal concentration in the benthos shows that the benthos is time integrating and that it takes some days to months until changes become measurable.

Seasonal changes in the biological activity from winter to spring lead to changes in the pollutant transfer. The phytoplankton biomass and production increased from January to May, as did the bioaccumulation of heavy metals by the phytoplankton. Therefore, the phytoplankton plays an important role in the uptake of dissolved contaminants into the marine foodweb. It binds the dissolved contaminant and makes it available for higher members of the ecosystem. Sedimented phytoplankton makes organically bound contaminants available to benthic organisms. Distinct seasonal differences in the contaminant concentration of benthic organisms could only be shown for organic pollutants due to changing reproduction cycles and feeding activities of the animals. Therefore, for the interpretation of the results it is necessary to take the biology of the species into account, e.g. the phase of reproduction.

Another important result is the high interspecies variability, shown for phyto- and zooplankton with regard to heavy metals and for the benthos with regard to organic contaminants.

An important question is the representativity of the large scale surveys (Chaps. 2.2 and 2.3) because of the large distance between the stations. Therefore, the results of these surveys must be compared with the six monthly small scale surveys to obtain information about the spatial and temporal variability.

Small scale structures could be detected for nearly all measured components due to the source of the river Elbe. These structures could not be seen in the results of the large-scale surveys. The observed differences of the concentration values in the small scale surveys were lower than differences between the southern and northern parts of the North Sea. The sampling strategies of the large as well as of the smaller sampling regions did not permit detection of short-term variability. The consequences of only one storm event were demonstrated by the organic contaminants in water (see above). These significant changes were expected for all compartments in the pelagic environment, especially near the sources. The benthic organisms turned out to be time-integrating. Despite the difficulties concerning the interpretation of measurements in such a highly variable area like the German Bight, this interdisciplinary work has given some encouraging insights into the behaviour of its ecosystem and the variety and interactions of its compartments.

References

Anger K, Rogal U, Schriever G, Valentin C (1977) In-situ investigations on the echinoderm *Asterias rubens* as a predator of soft-bottom communities in the western Baltic Sea. Helgol Wiss Meeresunters 29:439-459

Anonymous (1988) Überwachung des Meeres. Bericht für das Jahr 1986, Teil II - Daten. DHI Rep Hamb ISSN 0724-0449

ARGE (Arbeitsgemeinschaft für die Reinhaltung der Elbe) (1988, 1989, 1990) Wassergütedaten der Elbe von Schnackenburg bis zur See - Zahlentafeln 1989-1989

Balls PW (1988) The control of trace metal concentrations in coastal seawater through partition onto supended particulate matter. Neth J Sea Res 22:213-218

Balls PW (1989) The partition of trace metals between dissolved and particulate phases in european coastal waters: a compilation of field data and comparison with laboratory studies. Neth J Sea Res 23 (1):7-14

Boalch GT, Harbour DS (1977) Unsual diatom off the coast of southwest England and its effect on fishing. Nature 269:687-688

Borchardt T, Burchert S, Hablizel H, Karbe L, Zeitner R (1988) Trace metal concentration in mussels: comparision between estuarine, coastal and offshore regions in the southeastern North Sea from 1983 to 1986. Mar Ecol Proc Ser 42:17-31.

Brockmann UH, Eberlein K (1986) River input of nutrients into the German Bight. In: Skreslet S (ed) NATO ASI Ser vol G7, The role of freshwater outflow in coastal marine ecosystems. Springer, Berlin Heidelberg New York

Brodsky J, Ballschmitter K (1988) Reversed phase liquid chromatography of PCBs as a basis for the calculation of water solubility and log K_{OW} for polychlorobiphenyls. Fresenius Z Anal Chem 331:295-301

Bruggeman WA, Opperhuizen A, Wijbenga A, Hutzinger O (1984) Bioaccumulation of super-lipophilic chemicals in fish. Toxicol Environ Chem 7:173-189

Büther H (1988) Distribution of chlorinated organic compounds in livers of dab (*Limanda limanda*) of the southern and central North Sea. Mitt Geol Paläont Inst Univ Hamb 65:497-541

Claussen T (1988) Levels and spatial distibution of trace metals in dabs (*Limanda limanda*) of the southern North Sea. Mitt Geol Paläont Inst Univ Hamb 65:467-496

Dehairs F, Baeyens W, Van Gansbeke D (1989) Tight coupling between enrichment of iron and manganese in North Sea suspended matter and sedimentary redox processes: evidence for seasonal variability. Estuar Coast Shelf Sci 29:457-471

Duinker JC (1983) Effects of particle size and density on the transport of metals to the oceans. In: Wong CS, Boyle E, Bruland CW, Burton JD, Goldberg ED (eds) Trace metals in sea water. Plenum, New York, pp 209-226

Duinker JC, Hillebrand MTJ, Nolting RF, Wellershaus S (1982a) The River Elbe: Processes affecting the behaviour of metals and organochlorines during estuarine mixing. Neth J Sea Res 15:141-169

Duinker JC, Hillebrand MTJ, Nolting RF, Wellershaus S (1982b) The River Weser: Processes affecting the behaviour of metals and organochlorines during estuarine mixing. Neth J Sea Res 15:170-195

Dürselen (1990) Untersuchungen zur Schwermetallakkumulation von Phytoplanktongemeinschaften der Deutschen Bucht mit ergänzenden Laborversuchen zur Deutung der Ergebnisse. Dipl Arbeit Math Naturwiss Fak Aachen, 127 pp

Ellgehausen H, Guth JA, Esser HO (1980) Factors determining the bioaccumulation potential of pesticides in the individual compartments of aquatic food chains. Ecotoxicol Environ Saf 4:137-157

Engel M (1983) Evaluation of North Sea hydrocasts for modelling purposes. Summary of a poster presentation. In: Sündermann J and W Lenz (eds) North Sea dynamics, Springer, Berlin Heidelberg New York

Ernst W, Boon JP, Weber K (1988) Occurrence and fate of organic micropollutants in the North Sea. In: Salomons W, Bayne B, Duursma E, Foerstner U (eds) Pollution of the North Sea: an assessment, Springer, Berlin Heidelberg New York

Feely RA, Massoth GJ, Landing WM (1981) Major- and trace metal composition of suspended matter in the north-east Gulf of Alaska: relationships with major sources. Mar Chem 10:413-453

Frauenheim K (1991) Entwicklung einer Probennahmestrategie zur quantitativen Erfassung des Benthos. Berichte aus dem Zentrum für Meeres- und Klimaforschung, Univ Hamburg, Bd 22, ISSN 0936-949X, 171 pp

Hainbucher D, Pohlmann T, Backhaus J (1987) Transport of conservative passive tracers in the North Sea: first results of a circulation and transport model. Cont Shelf Res 7(10):1061-1079

Hancock D (1967) Whelks. Ministry of Agriculture, Fisheries and Food, Laboratory Leaflet (NS) No 15, Essex: Fisheries Laboratory, Burnham on Crouch, 14 pp

Hansen PD (1980) Uptake and transfer of the chlorinated hydrocarbon Lindane (γ-HCH) in a laboratory fresh water food chain. Environ Pollut A21:97-108

Hensen V (1887) Über die Bestimmung des Planktons oder des im Meer treibenden Materials an Pflanzen und Thieren. Ber Komm Wiss Unters Dtsch Meere 5(12-16:1882-1886)1-108 Anh I-XIX

Htun-Han M (1978a) The reproductive biology of the dab (*Limanda limanda* L.) in the North Sea: Seasonal changes in the ovary. J Fish Biol 13:351-359

Htun-Han M (1978b) The reproductive biology of the dab (*Limanda limanda* L.) in the North Sea: gonosomatic index, hepatosomatic index and condition factor. J Fish Biol 13:369-378

Hühnerfuss H, Dannhauer H, Faller J, Ludwig P (1990) Concentration variations of anthropogenic and biogenic organic substances in the German Bight due to changing meteorological conditions. Dtsch Hydrogr Z 43:253-272

Irion G, Müller G (1987) Heavy metals in surficial sediments of the North Sea. In: Lindberg SE, Hutchinson TC (eds) Proc Int Conf Heavy metals in the environment vol 2, New Orleans 1987. CEP Consultants, Edinburgh, pp 38-41

Kammann U, Knickmeyer R, Steinhart H (1990) Distribution of polychlorobiphenyls and hexachlorobenzene in different tissues of the dab (*Limanda limanda* L.) in relation to lipid polarity. Bull Environ Contam Toxicol 45:552-559

Karbe L, Hablizel H (1990) Biologisches Schadstoff-Monitoring in der Jade mittels exponierter Miesmuscheln. Mitt Niedersächsischen Landesamt für Wasser und Abfall, Hildesheim. Heft 1, 152 pp

Kersten M, Dicke M, Kriews M, Naumann K, Schmidt D, Schulz M, Schwidrowski M, Steiger M (1988) Distribution and fate of heavy metals in the North Sea. In: Salomons W et al. (eds) Pollution of the North Sea. Springer, Berlin Heidelberg New York, pp 300-347

Kersten M, Kienz W, Koelling S, Schröder M, Förstner U (1990) Heavy metal contamination of suspended particulate matter and sediments of the North Sea. Vom Wasser 75:245-272

Knickmeyer R, Steinhart H (1988) Cyclic organochlorines in the hermit crabs *Pagurus bernhardus* and *Pagurus pubescens* from the North Sea. A comparison between winter and early summer situation. Neth J Sea Res 22:237-251

Knickmeyer R, Steinhart H (1989) On the distribution of polychlorinated biphenyl congeners and hexachlorobenzene in different tissues of dab (*Limanda limanda*) from the North Sea. Chemosphere 19:1309-1320

Knickmeyer R, Steinhart H (1990a) Seasonal variations and sex related differences of organochlorine in whelks (*Buccinum undatum*) from the German Bight. Chemosphere 20:109-122

Knickmeyer R, Steinhart H (1990b) Concentrations of organochlorine compounds in the hermit crab *Pagurus bernhardus* from the German Bight, December 1988-May 1989. Neth J Sea Res 25(3):365-376

Knickmeyer R, Landgraf O, Steinhart H (1992) Cyclic organochlorines in the sea star *Asterias rubens* from the German Bight, December 1988-May 1989. Mar Environ Res 33:127-143

König W, Hühnerfuss H, Lange P, Lutz S (1989) Transport und Umsatz organischer Schadstoffe. In: Sündermann J, Beddig S (Hrsg) Zirkulation und Schadstoffumsatz in der Nordsee (ZISCH) - Zwischenber, Univ Hamburg, pp 97-106

Kowalski R (1955) Untersuchungen zur Biologie des Seesterns *Asterias rubens* im Brackwasser. Kiel Meeresforsch 11:201-213

Kröncke I (1988) Heavy metals in North Sea macrofauna. Mitt Geol Paläont Inst Univ Hamb 65:455-465

Mart L (1979a) Prevention of contamination and other accuracy risks in voltammetric trace metal analysis of natural waters. Part I. Preparatory steps, filtration and storage of water samples. Fresenius Z Anal Chem 296:350-357

Mart L (1979b) Prevention of contamination and other accuracy risks in voltammetric trace metal analysis of natural waters. Part II. Collection of surface water samples. Fresenius Z Anal Chem 299:97-102

Miyoshi S, Miyoshi E, Sato T (1982) Energy dispersive X-ray analysis of some hard tissues. In: The Congress Organizing Committee (ed) Electron microscopy 1982, vol 3 (Biology), 10th Int Congr Electron Microscopy, Hamburg, 1982, Deutsche Gesellschaft für Elektonenmikroskopie, Frankfurt Main pp 395-396

Nicol JAC (1960) The biology of marine animals. Sir Isaak Pitman, London, pp 707

Nolting RF, Eisma D (1988) Elementary composition of suspended particulate matter in the North Sea. Neth J Sea Res 22:219-236

Norheim G, Roald SO (1985) Distribution and elimination of hexachlorobenzene, octachlorostyrene and decachlorophenyl in rainbow trout, *Salmo gairdneri*. Aquat Toxicol 6:13-24

Pfaffenberger B, Hühnerfuss H, Kallenborn R, Köhler-Günther A, König WA, Krüner G (1992) Chromatographic separation of the enantiomers of marine pollutants. Part 6. Comparison of the enantioselective degradation of α-hexachlorocyclohexane in marine biota and water. Chemosphere 25:719-725

Radach G, Bohle-Carbonell M (1990) Strukturuntersuchungen der meteorologischen, hydrographischen, Nährstoff- und Phytoplankton-Langzeitreihen in der Deutschen Bucht bei Helgoland. Ber Biologischen Anstalt Helgoland, No 7

Rick HJ (1990) Ein Beitrag zur Abschätzung der Wechselbeziehung zwischen den planktischen Primärproduzenten des Nordseegebietes und den Schwermetallen Kupfer, Zink, Cadmium und Blei auf Grundlage von Untersuchungen an natürlichen Planktongemeinschaften und Laborexperimenten mit bestandsbildenden Arten. Diss Math Naturwiss Fak TH Aachen, 234 pp

Smith FGW (1974) Handbook of marine science, vol 1. CRC, Cleveland, Ohio, 627 pp

Sperling K-R (1984) Determination of Cd traces in environmental samples. In: Welz B (Hrsg) Fortschritte in der atmospektrometrischen Spurenanalytik, Bd 1. VCH, Weinheim

Zauke G-P, Jacobi H, Gieseke U, Sängerlaub G, Bäumer H-P, Butte W (1985): Sequentielle Multielementbestimmung von Schwermetallen in Brackwasserorganismen. In Welz B (Hrsg) Fortschritte in der atmspektrometrischen Spurenanalytik, Bd 2. VCH, Weinheim.

2.6 Measurements of Suspended Matter Dynamics in the German Bight

P. KÖNIG, A. FROHSE and H. KLEIN

2.6.1 Introduction

By definition, Suspended Particulate Matter (SPM) comprises all suspended particles with a diameter of > 0.4 µm. It is the amount of all material in a water sample which remains after filtration on a filter with a pore size of 0.4 µm.

Nowadays, it is widely known that SPM is an important carrier of pollutants. Contaminants such as heavy metals are taken up and transported by SPM (Vollbrecht 1980; Duinker 1985; Kersten et al. 1988). A critical review of empirical and conceptual adsorption models is given by Bourg (1987).

The distribution, organic content and particle size of SPM in the North Sea have been discussed elsewhere (Eisma and Kalf 1987a,b; Wirth and Seifert 1988; Eisma 1990). The results of our investigations, as discussed in this Chapter, are restricted to the German Bight area.

The transport, distribution and variability of SPM are controlled mainly by diffusive and advective processes. Resuspension and deposition of SPM depend not only on surface waves, tides and wind-driven currents but also on the characteristics (e.g. size and shape distribution, settling velocity, content of organic matter, flocculation) of the SPM present at any one time and local bottom sediments.

2.6.2   Methods

2.6.2.1   Field Measurements (Sampling Stations, Equipment and Techniques, Sampling Methods)

The bathymetry and the areas of study in the German Bight are shown in Fig. 2.6.1. Experiments in 1985 and 1986 took place on the Research Platform *Nordsee* (FPN). At this position the Bundesamt für Seeschiffahrt und Hydrographie (BSH) routinely takes wave measurements. In December 1987 the distribution of SPM in the German Bight was measured with a horizontal resolution of 5 nautical miles during the 106th cruise by the RV Gauss. A west-east transect near 55° N (northern boundary of the German Bight) was examined in 1989 (*Gauss* cruise 128). A year before, in March / April 1988, the western boundary of the German Bight near 6° 20' E was examined in a similiar experiment (*Gauss* cruise 111).

On both transects, Aanderaa current meters and Deployable Backscattering Meters (DBM) were deployed at a depth of 10 m and near the bottom at four locations. During the experiments water samples and CTD-profiles were taken along the transect and also at anchor stations near the moorings (see Table 2.6.1 for *Gauss* cruise 128).

SPM and Particulate Organic Carbon (POC) were collected at several depths (mainly around 10 m below the surface and, in most experiments, also near the bottom) using Niskin bottles and special near-bottom water samplers. Water samples were usually taken at each station at hourly intervals at least over 26.5 h in order to determine the varibility of a whole tidal cycle.

SPM dry weight was determined by suction-filtration through a pre-weighted 47 mm diameter 0.4 µm pore size polycarbonate membrane (Nuclepore). Sea salt was removed from the filters by rinsing them with distilled water. After sampling and returning to the laboratory, the filters were dried at 80 °C for 1 h before the SPM weight was measured.

Table 2.6.1. Deployment locations and sampling depths (*Gauss* cruise 128)

Station no.	Geographical position	Water depth	Sampling depths		
B	54° 59.8' N , 7° 54.2' E	17 m	10 m	5 m	80 cm ab[a]
C	55° 0.6' N , 7° 36.0' E	26 m	12 m	19 m	80 cm ab
D	55° 0.0' N , 7° 20.0' E	31 m	10 m	21 m	80 cm ab
F	55° 0.0' N , 6° 45.0' E	36 m	11 m	21 m	80 cm ab

[a]ab = above bottom.

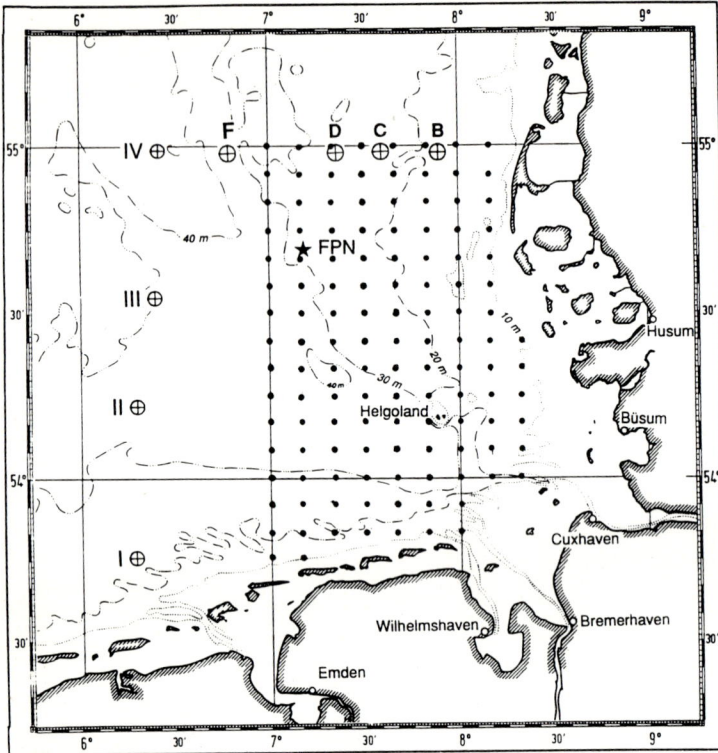

Fig. 2.6.1. Bathymetry of the German Bight and locations of the field experiments: ★ indicates Forschungsplattform "NORDSEE" (FPN); ⊕ indicates current meter, DBM and SPM sampling stations during March/April 1988 and March 1989; ● indicates SPM sampling stations, December 3-8, 1987. Depth contours are given in m

Particle size was measured using a Model TA II Coulter Counter until 1987 and after that a Coulter Counter Multisizer. This kind of measurement underestimates the size of particles. Especially in the presence of flocs, it "sees" only fragments of the real "in-situ" structures (Eisma 1987).

POC content was determined by suction-filtration through a 47 mm diameter glass fibre filter (Whatman GF/C with a pore size around 1.2 μm). These filters were stored frozen in Petri dishes. Combustion (950 °C, O_2 added) of the samples and analysis of the evolved carbon dioxide using a calibrated infrared gas analyzer (Astro) was done in the laboratory on land, after the inorganic carbonate had been removed from the samples (by use of HCl).

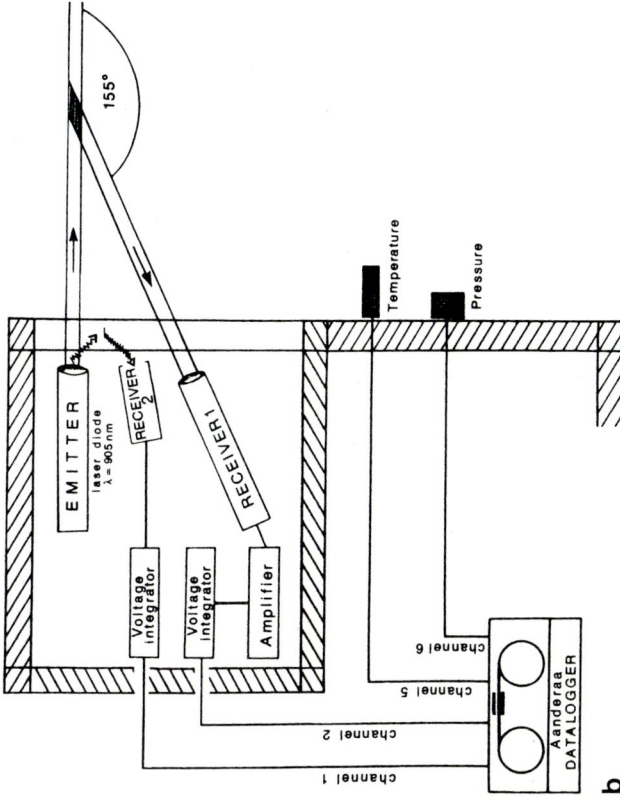

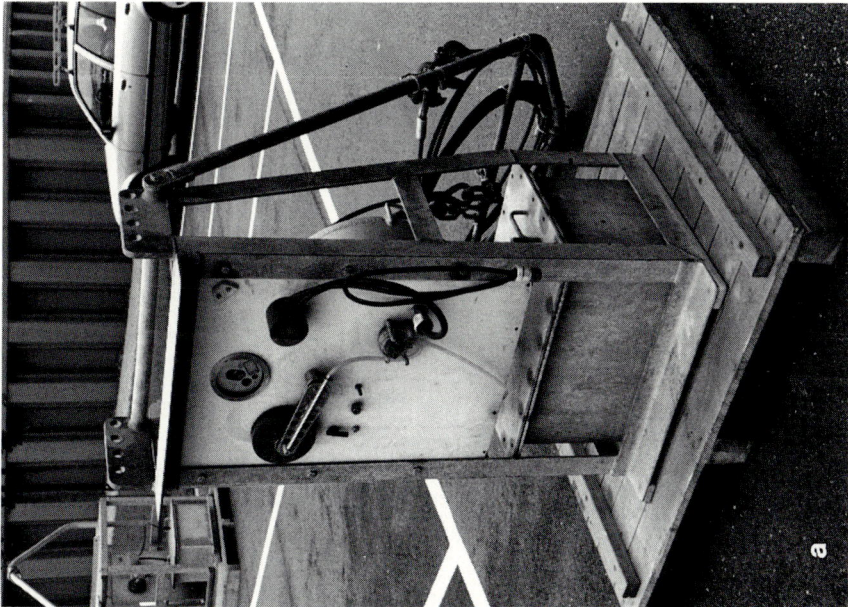

Fig. 2.6.2. *a* Deployable Backscattering Meter (DBM) with mooring frame. *b* Principle DBM instrument set-up

The DBM (see Fig. 2.6.2) consists of a modified EOS STM 1 turbidity meter coupled to an Aanderaa datalogger (Gienapp et al. 1981; Gienapp 1985). The backscattered light of a laser in continuous operation (λ = 905 nm in air, angle of backscattering: 155°) is measured and integrated over 5 min. This value is then stored by the datalogger together with those of the current pressure and temperature. The DBM can be deployed for up to 4 weeks. To avoid marine growth and film deposition, a rotating brush cleans the glass surface of the instrument every 6 h. Laboratory calibrations with Latex particles showed a high linear correlation between the DBM signal and mass concentration. At field calibrations in the German Bight, the slope of the linear regressions varied by a factor of 4, depending upon location and weather conditions. These variations were caused by changes in the optical properties (i.e. size distribution, scattering efficiency) of the particles present in the sea water under examination. During all field experiments with moored DBMs, calibrations in terms of SPM concentration were done by a simultaneous determination of SPM in water samples. Water samples were usually taken hourly over 26.5 h periods. During each cruise, two or three calls were made at each station.

Current measurements were taken using Aanderaa RCM-4 and RCM-5 current meters which recorded temperature, current speed and direction every 5 min. Some instruments also recorded conductivity and/or pressure. Current meters attached to the DBMs did not record current direction. The instruments were mainly deployed at a depth of about 10 m and 1.5 m above the sea bottom. Time-series data on turbidity and current speed were filtered with a Gaussian low-pass filter (Schönwiese 1985) to remove tidal and higher frequency signals. The cut-off frequency of the low-pass filter was 0.0403 cph.

Data from a WAVEC buoy near the FPN were provided by the wave group of the BSH (R.Berger, K.Richter, D.Schrader). Measurements over a 30 min period were usually taken every 3 h. The data set includes wave heights, directions and periods as well as other variables. For further information see Bundesamt für Seeschiffahrt und Hydrographie (1990).

2.6.2.2 Computer Simulations

To interpret the field measurements and to calculate the water mass transport across the boundaries of the German Bight, the results from an operational numerical model of the German Bight, which has included the Wadden Sea since October 1988, were used (Soetje and Brockmann 1983; Müller-Navarra and Mittelstaedt 1987; Dick and Soetje 1990). The model is operated by the model group of the BSH (S. Dick, K. Huber, S. Müller-

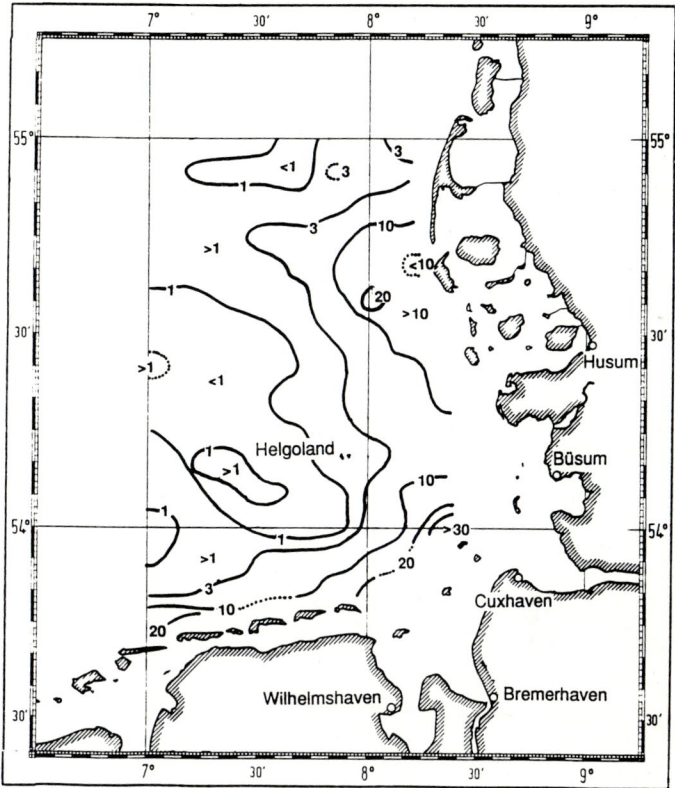

Fig. 2.6.3. Near-surface distribution of suspended particulate matter. Isoline intervals are not equal. German Bight (126 sampling locations, 3.-8.12.1987)

Navarra). It is based on the Navier-Stokes equations expressed in a finite difference form. Horizontal resolution is about 10.7 km (1.8 km in the Wadden Sea) and up to four levels are resolved in the vertical, depending on water depth. Tides and wind fields are the main data input for the model. Wind fields are calculated from predicted surface pressure fields (Deutscher Wetterdienst, Offenbach). Baroclinic effects are not considered. For each grid point, the model provides water level and three-dimensional current values.

Table 2.6.2. Mean SPM concentration (mg/l) per day (1.10.-10.10.1985) at three depths. Samples were taken each day between 08.00 and 20.00 h approximately hourly

Date	Position: Forschungsplattform *Nordsee* (54° 42.1' N ; 7° 10.2' E)			
	Suspended matter concentration (mg/l) Mean ± standard deviation			
	Depth: 8.5 m	Depth: 17.0 m	Depth: 24.7 m	All depths
1.10.85	0.73 ± 0.19	0.50 ± 0.18	0.62 ± 0.13	0.62 ± 0.19
2.10.85	0.48 ± 0.11	0.43 ± 0.13	0.55 ± 0.10	0.49 ± 0.12
3.10.85	0.50 ± 0.16	0.51 ± 0.18	0.55 ± 0.13	0.52 ± 0.15
4.10.85	0.57 ± 0.08	0.50 ± 0.12	0.60 ± 0.09	0.56 ± 0.10
5.10.85	0.72 ± 0.18	0.60 ± 0.10	0.85 ± 0.19	0.72 ± 0.19
6.10.85	0.51 ± 0.15	0.45 ± 0.07	0.58 ± 0.14	0.51 ± 0.13
7.10.85	0.50 ± 0.10	0.55 ± 0.23	0.55 ± 0.17	0.54 ± 0.17
8.10.85	0.51 ± 0.11	0.39 ± 0.09	0.39 ± 0.08	0.43 ± 0.11
9.10.85	0.45 ± 0.13	0.45 ± 0.15	0.43 ± 0.12	0.45 ± 0.13
10.10.85	0.55 ± 0.07	0.56 ± 0.11	0.76 ± 0.17	0.62 ± 0.15
1.10. to 10.10.85	0.55 ± 0.15	0.49 ± 0.15	0.58 ± 0.18	0.54 ± 0.17

2.6.3   Results

2.6.3.1   German Bight (December 1987, *Gauss* 106)

Figure 2.6.3   shows a "snapshot" of the near-surface distribution of suspended matter in the German Bight (see Fig. 2.6.1 for sampling locations). On five consecutive days in December, 1987, SPM samples were taken at 126 locations. Corrections to eliminate the aliasing effect of tidal oscillations (Brockmann and Dippner 1987) have not been applied to the data set. Weather conditions were stable over the whole period without any significant meteorological events. A light northeasterly wind predomi-

Table 2.6.3. Suspended matter concentration, standard deviation and range (mg/l) below the research platform *Nordsee* for three depths. 26.11.-10.12.1986; number of samples: 169; observation interval: 2 h)

Depth (m)	Suspended matter concentration ± standard deviation (mg/l)	Suspended matter concentration range (minimum, maximum) (mg/l)
6.0	2.0 ± 1.3	0.5, 6.3
15.0	2.1 ± 1.4	0.5, 6.7
23.5	2.4 ± 1.6	0.5, 6.7

nated. The SPM distribution is characterized by high values near the coast. Relatively cold water masses from the North Frisian Wadden Sea with high SPM content extended further to the west (G. Wegner, pers. comm.). A maximum SPM concentration of 31 mg/l was measured near the *Elbe 1* light vessel. This maximum was caused by the freshwater runoff from the river Elbe which contained a high level of SPM. The turbid Elbe plume can be recognized from the river mouth up to the southern end of the island of Sylt.

2.6.3.2 Forschungsplattform *Nordsee* (October 1985)

The daily mean SPM at the FPN over a 10 day period is listed in Table 2.6.2. Samples were taken each day between 08.00 and 20.00 h on an approximately hourly basis. Calm weather conditions prevailed before and during the experiment. The wind came predominantly from the SW with a speed range of 8 to 14 m/s (5 Bft, fresh breeze) with light seas with waves of 1 to 2 m. SPM content over the whole period was 0.5 (±0.2) mg/l at all three sampling depths. This value can be considered the minimum value for SPM content in the German Bight.

2.6.3.3 Forschungsplattform *Nordsee* (November/December 1986)

In November 1986 a second experiment took place at the FPN. Samples were taken every 2 h at three depths, with 40 min between sampling at each depth. One of the results is presented in Fig. 2.6.4, where the time-series of SPM content at 15 m depth is shown. In the first 5 days of calm weather (the wind from the southwest had dropped to 7 m/s) allowed the SPM content to decrease to 0.5 mg/l, which is about the same minimum value as observed a year before. The desired storms came with the onset of Decem-

Research platform NORDSEE

November/December 1986

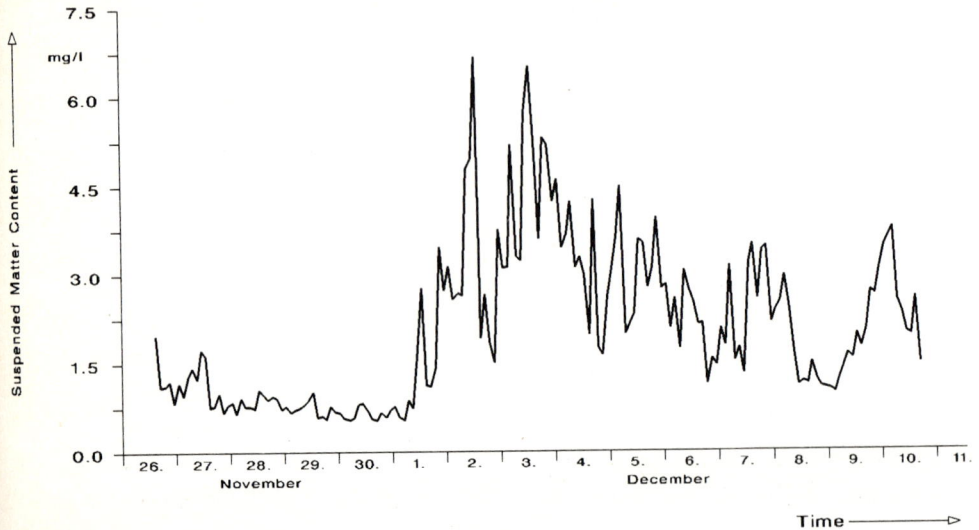

Fig. 2.6.4. Time-series of suspended matter concentration at 15 m depth. (26.11. to 10.12.1986, Research Platform *Nordsee*)

ber with wind speeds up to 21 m/s and significant wave heights above 5 m. The SPM content rose to 7 mg/l. An increase in particle volume per liter sea water from about 0.5 mm^3 to 2.5 mm^3 was observed using a Coulter Counter. Mean values and ranges of the SPM concentration for the 14 day period of Fig. 2.6.4 are listed in Table 2.6.3. A slight increase of SPM concentration and variability with depth can be seen.

Table 2.6.4. Number and total volume of particles for calm and stormy weather conditions. (Forschungsplattform *Nordsee*, 6 m depth, November/December 1986)

	Calm weather		Stormy weather	
	Size range 3.1-3.8 μm	Size range 9.6-12.2 μm	Size range 3.1-3.8 μm	Size range 9.6-12.2 μm
Number of particles per ml seawater	1796	79	3904	332
Total volume of particles per liter sea water (mm^3)	.038	.054	.082	.225

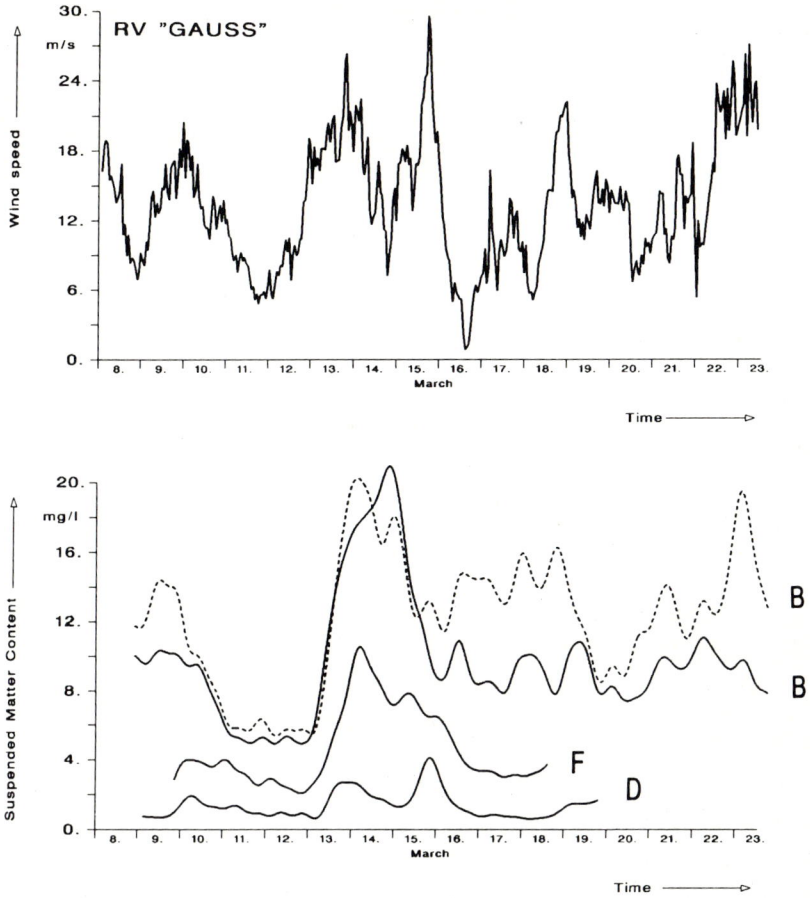

Fig. 2.6.5. Time series of *a* wind speed (aboard RV *Gauss*) and *b* low-pass filtered SPM concentrations at stations *B*, *D* and *F* near the northern boundary of the German Bight. *Solid line* 10 m depth; *dashed line* near the bottom

Table 2.6.4 shows how the differences in particle volume per liter sea water depend upon weather conditions. Water samples were taken at a depth of 6 m on November 30 (calm weather) and on December 3 (stormy weather). A significant change in the weather situation greatly influenced the particle size spectra: smaller particles doubled in number and larger ones even quadrupled.

The mean current speed for the time period 13.11. to 11.12.86 at 11 m depth was 0.26 ($\pm$0.07) m/s [range 0.07 - 0.48 m/s], while at 23.5 m depth 0.16 ($\pm$0.05) m/s [range 0.03 - 0.35 m/s] was observed. The dominant signal is the semi-diurnal tide. Cross-correlations (Schönwiese 1985) of current speed with turbidity (DBM data) at three water depths indicated no significant linear relationship between these parameters ($r^2 < 0.08$), while turbidity and significant wave heights were linear correlated ($r^2 = 0.35$). For the most part, the data show that the flood current from the central North Sea carries less turbid water to the FPN area than the ebb current from the coastal regions.

2.6.3.4 Northern Boundary of the German Bight (March 1989, Gauss 128)

During a cruise with RV *Gauss* in March 1989, SPM dynamics and transports at the northern boundary of the German Bight near 55°N were investigated (see also Sect. 2.6.2 and Table 2.6.1). Wind speed (Fig. 2.6.5a) and direction were registered aboard the ship every hour. General weather conditions were characterized by strong and stormy winds from southwesterly to northwesterly directions, caused by an eastwards moving low-pressure system. For short periods wind speeds up to 31.5 m/s (whole gale/violent storms) were measured.

As a result of stormy weather, SPM can increase sharply within a short time. The time-series of lpf SPM in Fig. 2.6.5b shows for March 12-14 at Station B at 10 m depth a nearly linear increase of SPM content about 16 mg/l over a period a 44.5 h, which is the equivalent of a change of 0.36 mg/l per hour. This was caused by a storm with wind speeds up to 26 m/s (Fig. 2.6.5a) with significant wave heights of about 5 m near the FPN (see also Fig. 2.6.6a). After the wind abated, the SPM content decreased linearly to 8.6 mg/l in 29.5 hours, corresponding to a change of -0.41 mg/l per hour. Figure 2.6.5b shows that the possible increase in SPM varied between 0.2 and 0.8 mg/l per hour, while the decrease ranged between 0.2 and 0.6 mg/l per hour.

Figure 2.6.6b-c displays for station D the near-bottom time series of SPM concentrations (from DBM data) and current speed 1.5 m above the sea bottom. Times of "low water Helgoland + 15 minutes" are marked by "x". Both curves show semi-diurnal tidal oscillations. Usually, two maxima per day can be seen in the SPM data, while at the same time, the current speed shows four maxima (two for the ebb and two for the flood currents). Although the ebb and flood currents reach nearly the same peak values, no SPM maxima are found at the corresponding time of the flood current. It may be concluded from this that the current speed did not exceed the threshold velocity required to start resuspension of SPM, otherwise the same number of maxima would be expected for SPM content and current speed. For calm weather periods, the semidiurnal peaks in the SPM data

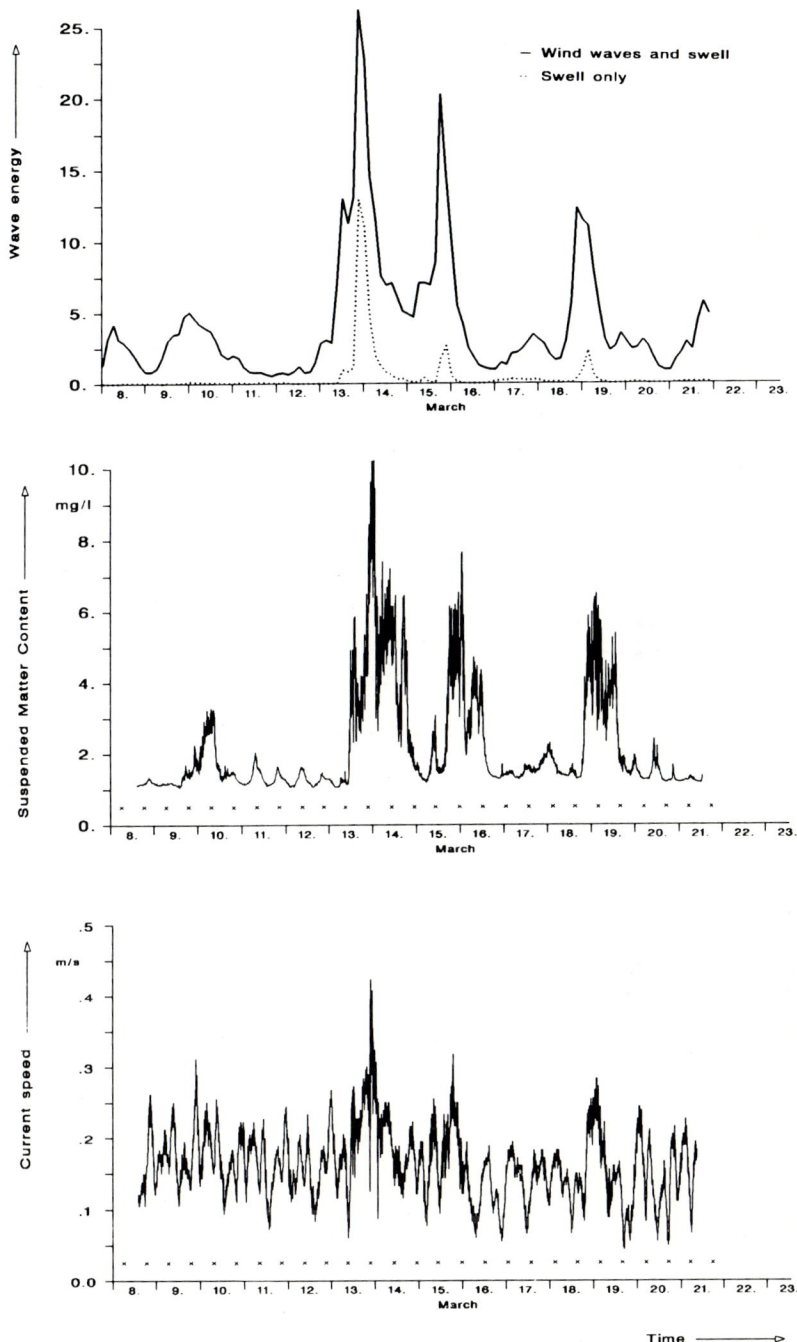

Fig. 2.6.6. Time series of *a* wave energy (near FPN), *b* SPM concentration and *c* current velocity near the bottom at station D. (*x* indicates times of low water Helgoland + 15 min). Data collected in March 1989

Table 2.6.5. Particulate organic carbon (POC) at four stations and two depths in the German Bight (measured on research cruise *Gauss 128* / March 1989)

Research cruise: *Gauss 128*			Particulate Organic Carbon (POC) ± standard deviation			
			Depth: 10-12 m		Depth: 80 cm ab[a]	
Station	Date	n	(mg/m³)	% of SPM	(mg/m³)	% of SPM
B	18./19.3.89	12	298 ± 109	4.4 ± 0.7	570 ± 239	3.8 ± 0.5
C	17.3.89	13	154 ± 45	6.6 ± 1.6	219 ± 44	6.1 ± 1.9
D	16.3.89	13	178 ± 20	9.5 ± 4.2	222 ± 38	8.2 ± 2.5
F	14.3.89	13	218 ± 38	2.6 ± 0.4	351 ± 41	2.4 ± 0.5

[a]ab = above bottom

may be explained only by the prevailing of advective processes. If the weather conditions are marked by storms, time series of SPM content and wave energy correspond (cf Fig. 2.6.6a,b). Cross-correlations give a maximum coefficient of determination of $r^2 = 0.77$ (time lag: 1 h).

At stations B, C and D times of high SPM concentration are in phase with times of low water densities and low water levels. Advection of turbid water from the coastal areas and clear water form the central North Sea generate periodic changes of SPM concentration. In contrast, at station F a time lag of about 6 h can be observed between the SPM maxima and the minima of water density and water level. This can be explained by an upwelling of water masses at the eastern boundary of the old Elbe river bed, which enriches the clear water with higher amounts of SPM; but further research on this is necessary.

During the *Gauss* cruise 128, POC was determined hourly at every station over a 12 h period. Table 2.6.5 lists mean values (± standard deviation) and the percentage of SPM content for every station. POC content mostly ranges between 100 and 400 mg/m³ with higher values near the bottom. On March 18, station B showed a short-term peak of 1250 mg/m³ POC during the passage of a density front moving eastwards at 20 UTC.

On March 12-13, 1989, measurements were conducted over a 26.5 h period at station B. Some results are shown in Figs. 2.6.7 and 2.6.8. Figure 2.6.7a-d represents the time series of particle volume per liter sea water, SPM concentrations, lpf density parameter gamma (S,T,p) of sea water, and lpf near-bottom pressure data. Measurement depths are given in Fig. 2.6.7a-c: dashed line - 5 m depth; solid line - 10 m depth and dotted line - 80 cm above bottom. The near-bottom pressure data are used as an indicator of the tidal phase. Periods with constant lpf pressure values in

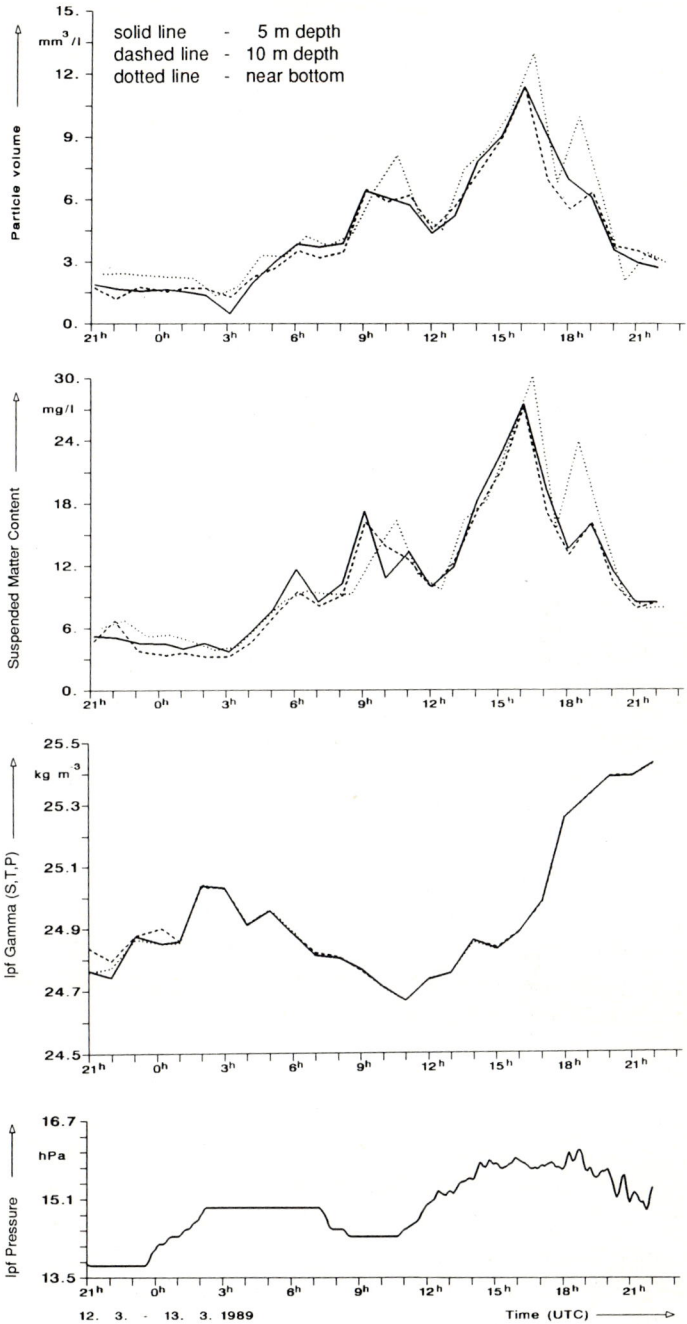

Fig. 2.6.7. Time series of *a* particle volume (per liter sea water), *b* SPM concentrations, *c* low-pass filtered density (gamma (S,T,p)) and *d* low-pass filtered near bottom pressure for station B. (For further details see text)

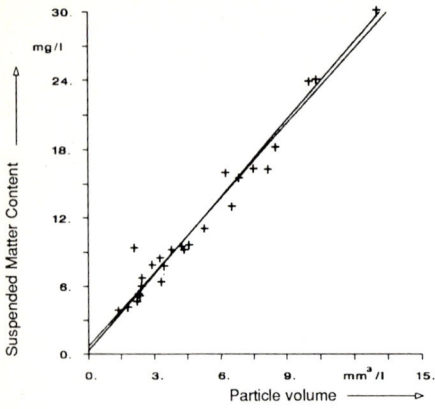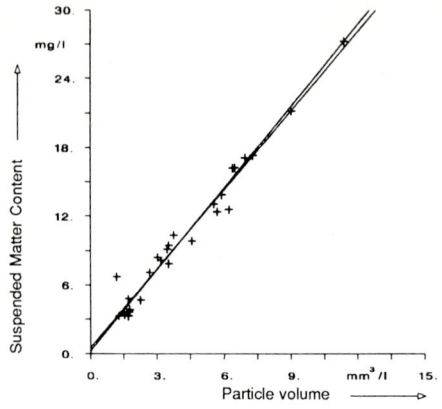

Depth: near bottom
Time interval: 1 hour
Time range: 26 hours

Depth: 5 m
Time interval: 1 hour
Time range: 26 hours

Fig. 2.6.8. Scatter plots of SPM concentrations versus particle volume at 5 m depth and near the bottom at station B

Fig. 2.6.7d are caused by the combined effects of a weak variability of the pressure values and the curve filtering procedure (low-pass filter, cut-off frequency: 2 cph). Strong variability in total particle volume and SPM content was observed. The range of total particle volume was 1.2 - 13.2 mm^3/l and that for SPM content was 3.2 - 30.1 mg/l (within 13.5 h). At station B from 12-13 March, both variables were strongly correlated with a coefficient of determination $r^2 > 0.94$ at all three depths (Fig. 2.6.8). "Apparent" density, which can be defined as "dry" SPM weight divided through "wet" total particle volume (Richardson 1987), was nearly constant at about 2.5 g/cm^3. High apparent density and low POC content (about 4 %, for details see Table 2.6.5) indicate that the mineral constituents of the suspended particles greatly predominated.

The weather conditions were determined by a low-pressure area moving from southern Scandinavia to the Baltic. Winds from the west increased to 26.3 m/s, while wave height at the station reached about 6 m. Maxima are reported for 21.00 UTC (13.3.91). At the FPN a maximum significant wave height of 5.1 m was registered by a WAVEC buoy at about the same time (see also Fig. 2.6.6a).

Surprisingly, the peak values for SPM and total particle volume occurred about 5 h earlier than those for wind speed and wave height. This can be explained by considering the time-series for salinity, temperature and near-bottom pressure. These data clearly show that SPM maxima coincide with minima of sea water density, salinity and near-bottom pressure. This effect was observed at stations B, C and D over the entire deployment period in March 1989. Advection of low-salt water masses from the shallow areas near the coast and from the Wadden Sea transported large masses of SPM to the stations, while water masses coming from the open sea were characterized by low SPM content and high salinity (density). At anchor station B the effect of advection on SPM content is evident. Local resuspension as a result of the storm produced a much weaker signal.

Profiles of velocity and SPM concentrations have to be combined to give the sediment flux throughout the water column. Because our data (research cruise: *Gauss 128*) are from two specified depths only, we had to use models to estimate the vertical distributions of SPM and velocity.

One possible approach when developing the vertical SPM profile is to use the theoretical Rouse-von Karman sediment-concentration-depth equation which can be written as follows:

$$C(z) = C_a \cdot \left(\frac{a}{h-a} \cdot \frac{h-z}{z} \right)^\zeta$$

(Rouse 1938; Komar 1976), where

z height above bottom
C(z) concentration at height z above the bottom
C_a reference concentration at z = a
h water depth

$$\zeta = \frac{w_s}{\kappa u_*}$$
Rouse number ζ
w_s particle settling velocity
u_* shear stress velocity
κ von Karman's constant.

To estimate the unknown Rouse number, the SPM content measured (from DBMs) 1.5 m above the bottom and at about 10 m depth were used. The near-bottom DBM data were taken as reference concentrations. Mean Rouse numbers for the stations of the *Gauss* 128 research cruise were calculated as described by McCave (1979) and resulted in values between 18×10^{-6} and 45×10^{-6}. Estimating the shape of the concentration profiles from these numbers resulted in nearly uniform distributions over the whole depth range. Therefore, we had to use another approach to model the vertical SPM profile for our SPM flux calculations. We simply regarded the data from 10 m depth as being representative of the upper 80% of the water column, while the near-bottom data were taken for the remaining 20%.

Highly resolved vertical turbidity profiles from the surface down to about 5 m above the bottom, which were taken hourly at each anchor station, did not usually show any significant vertical variations. The turbidity profiles were measured using an EOS backscattering meter, which was interfaced with a conductivity-temperature-depth profiler (CTD).

While we find a good exponential fit for the vertical velocity profile (analogous to a van Veen profile) for two stations at the western boundary of the German Bight (cruise *Gauss 111*), this is not the case for the stations at the northern boundary. This is why we assumed the current meter at 10 m depth to be good for the upper two thirds of the water column, and the near-bottom instrument for the lower third.

The northward transport of water masses and SPM were calculated based on these assumptions (see Table 2.6.6 and the dashed line in Fig. 2.6.9 for the results). In addition, the current meter data were replaced by the results of computer simulations (see Sect. 2.6.2). The amount of SPM transported was then estimated again (dotted line in Fig. 2.6.9). The computer model did not simulate the effects of the storms with its usual precision. Wind data from aboard RV *Gauss* (Fig. 2.6.5a) and wave data from the FPN (Fig. 2.6.6a) show that from March 11 to 16, 1989, the weather situation was marked by two storms with wind speeds about 26 m/s from W and 30 m/s from WNW, respectively. During this time, large amounts of SPM (up to more than 300,000 tons per 24.75 h) were transported northwards by the water masses. On March 16 a weak wind from the NE allowed SPM to be transported southwards for a short time.

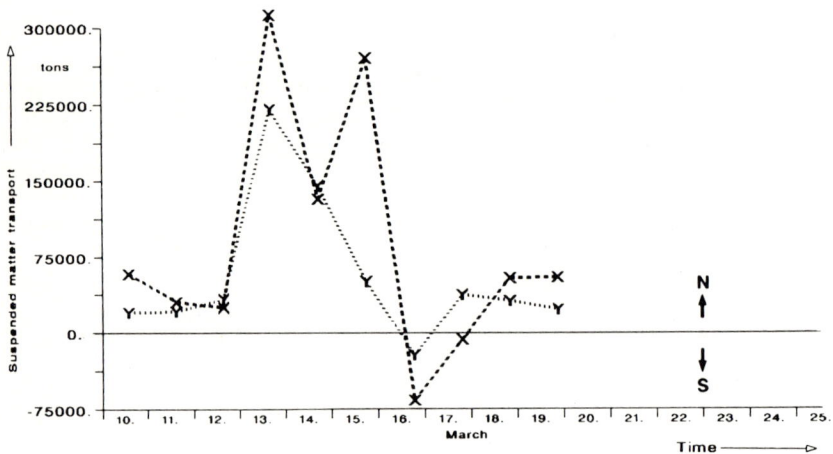

Fig. 2.6.9. Time series of SPM transport across the northern **boundary of** the German Bight. Velocity data: *dashed line* Aanderaa current meters, *dotted line* numerical model simulations

Table 2.6.6. Northward transport per tidal cycle of 24.75 h through the northern boundary of the German Bight (cross-section at 55°N ; 6° 19.6' E - 8° 18.2' E)

Start of a tidal cycle (day.month.yr; time UTC)	Water volume (Giga m³)			Suspended matter[d] (tons)	
	CM[a]	M1[b]	M2[c]	CM	M2
10.3.89; 2:45	16.0	20.5	20.9	58,481	19,947
11.3.89; 3:15	10.3	5.9	5.6	30,369	21,188
12.3.89; 3:45	13.1	3.0	3.0	25,090	33,118
13.3.89; 4:15	45.3	34.5	34.8	313,043	220,637
14.3.89; 4:45	15.2	29.0	28.6	132,176	145,479
15.3.89; 5:15	37.3	58.7	57.4	270,705	50,759
16.3.89; 5:45	-13.3	1.8	-1.1	-65,802	-21,663
17.3.89; 6:15	-3.3	-3.1	-4.3	-6,167	37,511
18.3.89; 6:45	8.2	3.3	2.6	54,057	31,432
19.3.89; 7:15	25.6	22.6	21.8	54,623	23,096
Sum over total time range: 10 days 7.5 h	154.3	176.2	169.4	866,576	561,505

[a]CM - velocity data from current meters.
[b]M1 - velocity data from model simulations (all grid points on cross-section).
[c]M2 - velocity data from model simulations [only four grid points (near current meter positions)].
[d]Suspended matter data were taken from DBMs; - sign indicates transport to the south.

Eisma (1990) estimated the amount of SPM supplied to the North Sea by rivers, coastal and seafloor erosion, primary production, atmosphere and from the English Channel to be at least approximately 35.6 x 10⁶ tons/yr, while the deposition in the German Bight is about 5.3 x 10⁶ tons/yr. The corresponding value (supply-deposition) for a period of 10 days 7.5 h would be 857,490 tons, which is of the same magnitude as our estimate of the northward transport across the northern boundary of the German Bight.

2.6.4 Data from Various Experiments

The size distribution of natural particles suspended in sea water is most frequently formulated using the power-law equation:

$$N = k \, d^{-b}$$

(Bader 1970), where N is the cumulative number of particles greater than size d (d := particle diameter in μm). b and k are constants. In plotting log N against log d we obtain a straight line with a slope of

$$m = -b = \frac{d \log N}{d \log d}$$

In our investigations the slope m = -b of the straight line was determined for each depth by linear regression analysis. Table 2.6.7 lists the estimated values of gradient b for various stations in the German Bight. Smaller values of b equal a higher number of particles with a larger diameter. Simpson (1982) gives a summary of data for the gradient b for different seas and oceans. No value is listed for the North Sea. For the N.W. Atlantic, he cited Sheldon (in McCave 1975) with a mean gradient b = 3.12 (range: 2.88-3.65) for depths between 5 and 100 m and a diameter range 4-20 μm. This value for b is very similiar to the one determined for the FPN (see Table 2.6.7). For the stations along the northern boundary of the German Bight we found lower values for b, caused by the higher amount of larger particles near the coast.

Table 2.6.7. Calculated values of the gradient, b, for various stations in the German Bight

Station(s)	Range of diameter (μm)	Depth (m)	b ± SD	No. of samples	Type of instrument	Date
B, C, D and F	2.86 - 13.96	10 nb[a]	2.73 ± 0.39 2.41 ± 0.34	309 297	M[b] M	March 1989
FPN	1.5 - 39.1	6.0 15.0 23.5	3.11 ± 0.23 3.11 ± 0.23 3.09 ± 0.25	166 166 168	TA II[c] TA II TA II	Nov./- Dec. 1986

[a]nb   =   80 cm above bottom.
[b]M   =   Coulter Counter Multisizer.
[c]TA II   =   Coulter Counter Model TA II.

2.6.5 Discussion

Our investigations describe the high variability of SPM concentrations in the German Bight. The normal range of SPM content outside the coastal region starts at about 0.3 mg/l and reaches values of more than 35 mg/l. SPM content responds quickly to stormy winds. We observed an increase of about 20 mg/l within 4.5 h over the entire water column at station B, for instance. When the storm abated, SPM content decreased even faster.

A semi-diurnal tidal signal is always superimposed over the weather-dependent SPM signal. The advection of water masses in the German Bight, caused by the tides, is the main reason for this. Variation of the SPM tidal signal is usually much weaker than that produced by stormy weather events. Tidal currents at our stations in the German Bight mostly failed to exceed the threshold velocity required to start resuspension. In areas with stronger tidal currents, such as the southern German Bight, this may be different.

During plankton blooms, organic material increases the amount of SPM. The highest POC portion of SPM measured during our research cruises was 51% ($\pm$18%) on March 30, 1988 at an anchor station near the *Borkumriff* light vessel. The maximum absolute value reached 0.936 mg/m^3 POC.

References

Bader H (1970) The hyperbolic distribution of particle sizes. J Geophys Res 75 (15):2822-2830

Bourg ACM (1987) Trace metal adsorption modelling and particle-water interactions in estuarine environments. Continent Shelf Res 7(11/12):1319-1332

Brockmann CW, Dippner JW (1987) Tidal corrections of hydrographic measurements. Dtsch Hydrogr Z 40:241-260

Bundesamt für Seeschiffahrt und Hydrographie (1990) Seegangsmessungen in der Deutschen Bucht im Jahre 1989. Meereskundliche Beobachtungen und Ergebnisse 68, Bundesamt für Seeschiffahrt und Hydrographie, Hamburg

Dick S, Soetje K (1990) Ein operationelles Ölausbreitungsmodell für die Deutsche Bucht. Dtsch Hydrogr Z Erg-H A:16

Duinker JC (1985) Chemical pollutants in the marine environment, with particular reference to the North Sea. In: Nürnberg HW (ed) Pollutants and their ecotoxicological significance. Wiley, New York

Eisma D (1987) Flocculation of suspended matter in coastal waters. SCOPE/UNEP Sonderbd, Mitt Geol Paläont Inst Univ Hamb, Heft 62

Eisma D (1990) Transport and deposition of suspended matter in the North Sea and the relation to coastal siltation, pollution, and bottom fauna distribution. Aquat Sci 3 (2/3):181-216

Eisma D, Kalf J (1987a) Distribution, organic content and particle size of suspended matter in the North Sea. Neth J Sea Res 21:265-285

Eisma D, Kalf J (1987b) Dispersal, concentration and deposition of suspended matter in the North Sea. J Geol Soc Lond 144:161-178

Gienapp H (1985) Mehrwöchige Registrierung mit auslegbaren Trübungsmessern in der Deutschen Bucht 1983 und 1984. Dtsch Hydrogr Z 38:233-241

Gienapp H, Carstens U, Schomaker K (1981) Eine Methode zur Reinigung und Reinheitskontrolle optischer Glasoberflächen in Seewasser und Anwendung des Verfahrens zum Bau eines auslegbaren Trübungsmessers. Dtsch Hydrogr Z 34:295-301

McCave IN (1975) Vertical flux of particles in the ocean. Deep Sea Res 22:491-502

McCave IN (1979) Suspended sediment. In: Dyer KR (ed) Estuarine hydrography and sedimentation. Cambridge University Press, Cambridge

Kersten M, Dicke M, Kriews M, Naumann K, Schmidt D, Schulz M, Schwikowski, Steiger M (1988) Distribution and fate of heavy metals in the North Sea. In: Salomons W, Bayne BL, Duursma EK, Förstner U (eds) Pollution of the North Sea - an assessment. Springer, Berlin Heidelberg New York

Komar PD (1976) Boundary layer flow under steady unidirectional currents. In: Stanley DJ, Swift DJP (eds) Marine sediment transport and environmental management. Wiley, New York, pp 91-106

Müller-Navarra SH, Mittelstaedt E (1987) Schadstoffausbreitung und Schadstoffbelastung in der Nordsee - eine Modellstudie. Dtsch Hydrogr Z Ergänzungsh B:18

Richardson MJ (1987) Particle size, light scattering and composition of suspended particulate matter in the North Atlantic. Deep Sea Res 34:1301-1329

Rouse H (1938) Experiments on the mechanics of sediment suspension. Proc 5th Int Congr Appl Mech, Cambridge, Mass

Schönwiese CD (1985) Praktische Statistik für Meteorologen und Geowissenschaftler. Borntraeger, Berlin

Simpson WR (1982) Particulate matter in the oceans - sampling methods, concentration, size distribution and particle dynamics. Oceanogr Mar Biol Annu Rev 20:119-172

Soetje K, Brockmann C (1983) An operational numerical model of the North Sea and the German Bight. In: Sündermann J, Lenz W (eds) North Sea dynamics. Springer, Berlin Heidelberg New York

Vollbrecht K (1980) Schwebstoff- und Schwermetallführung der Unterelbe. Dtsch Hydrogr Z 33:91-109

Wirth H, Seifert R (1988) Mineralogy, Geochemistry and SEM Observations of suspended particulate matter from the North Sea, Spring 1984. SCOPE/UNEP Sonderbd, Mitt Geol Paläontol Inst Univ Hamburg, Heft 65

2.7 Local Variability of Surface Currents Based on HF-Radar Measurements

F. Schirmer, H.-H. Essen, K.-W. Gurgel, T. Schlick and K. Hessner

2.7.1 Introduction

Surface currents which transport oil and surface pollutants, whether shorewards or out to sea, depend on many factors. These include winds, tides, waves, the coastline contour and geophysical forces such as gravity and the Coriolis force. Thus, near-surface offshore currents will be highly variable from area to area and at different times. While the ultimate objective is to predict these currents as a function of the driving forces, their accurate measurement is the necessary first step in the development and testing of such a prediction model. In fact, when near-surface currents are known, it is theoretically possible to numerically calculate the currents and circulation all the way down to the ocean floor.

Yet these near-surface currents are the most difficult to measure. Nearly all available techniques are Lagrangian in nature, meaning that they measure the trajectory of a parcel of water near the surface, thus obtaining one or more (current) streamlines vs. time. The method used in our case is a Eulerian one. It measures surface currents at points spaced approximately 3 km from each other in a rectangular grid.

2.7.2 The Method Used to Measure Surface Current Velocities by Electromagnetic (HF) Waves

The mechanism behind the HF-radar system - originally constructed by Barrick and Evans (1976) - permitting the measurement of near-surface currents, involves the physics of scatter of electromagnetic waves from ocean waves. Following Barrick and Evans, ocean waves (and a specific wavelength in particular) are the "target" for the radar system. Since this target is moving, it produces a Doppler shift on the received echo; this merely means that the signal returns at a slightly different frequency than that transmitted because of the radial rate of closure of the target. The radar receiver measures this Doppler shift, and this permits the determination of the radial component of the current velocity by the following process: the ocean wave motion - producing the measured Doppler shift - consists of two linear combinations of velocity. One is due to the "intrinsic" velocity of any deep-water gravity surface wave; from hydrodynamic theory, this phase velocity to first order is precisely

$$V_{phase} = \sqrt{\frac{gL}{2\pi}} \qquad (1)$$

where $g = 9.81$ ms^{-2} (the acceleration of gravity), and L is the wavelength of the water wavetrain. The second velocity component of the wave is due to current; it is as though the wave were being transported in a Cartesian coordinate system attached to the mean water transport. Thus, by measuring the total Doppler shift - and hence the total radial ocean wave velocity - and precisely knowing the first component (due to intrinsic wave velocity), one determines the second term, which is the desired radial component of surface current.

The logical question at this point is: how does one know precisely which wavetrain out of the entire spectrum constituting the ocean surface is doing the scattering? The physical mechanism behind HF scatter from the sea was discovered by Crombie (1955). By looking at ocean waves with a shore-based HF-radar and spectrally analyzing the received echoes, he observed two dominant "spikes" or peaks, symmetrically spaced about the transmitted carrier frequency (Fig. 2.7.1). These spikes indicated that of all the randomly time-varying ocean waves being observed by the radar, only two sets were significant, just as though two discrete targets were producing all of the scatter. By deducing the velocity of these "target" ocean waves from their symmetrical Doppler shifts, and thence the wavelength of the scatter-

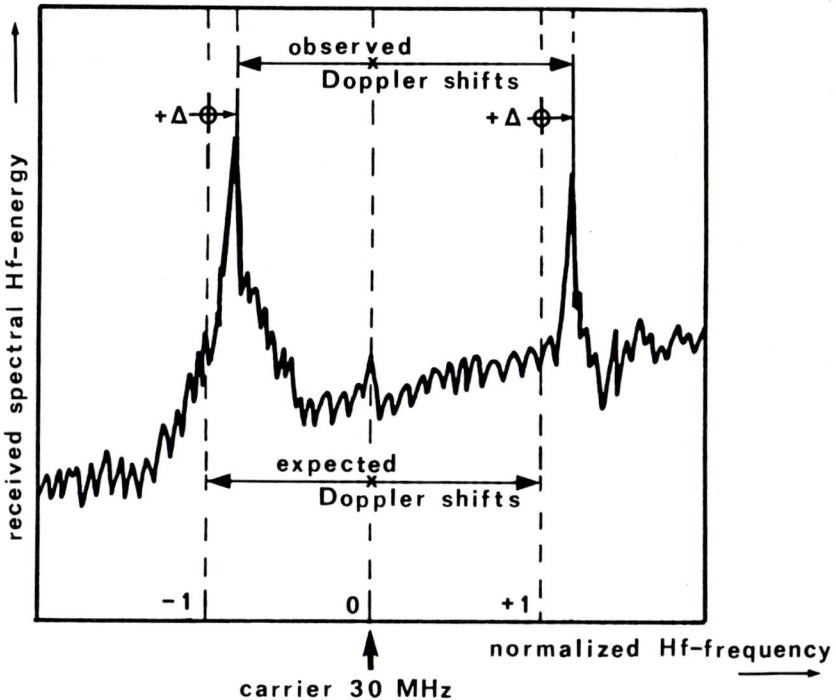

Fig. 2.7.1. Spectrum of received electromagnetic waves showing the Doppler shifts due to wave velocity and surface current (Δ)

ing wavetrains, he found that the wavelength was precisely one-half of the radar wavelength - for backscatter at grazing incidence along the sea. This is shown schematically in Fig. 2.7.2. Thus, Bragg scatter - or the diffraction-grating effect - is the responsible interaction mechanism. Therefore, out of all the wavetrains present on the surface, only the two wavetrains (or gratings) with a period precisely one-half the radar wavelength and moving radially towards and away from the radar produce the scatter. Crombie confirmed this explanation experimentally by noting that the unique "square-root" relationship between this Doppler shift and the radar carrier frequency followed the square-root water-wave dispersion relationship (between its phase velocity and its wavelength).

Crombie also first observed (Crombie et al. 1970) that these echo peaks are often offset by a discrete amount from their symmetrical positions about the carrier frequency. This is illustrated in Fig. 2.7.1. In the sample spectrum of Fig. 2.7.1 the centre tic on the abscissa represents the transmitter frequency - in our case 30 MHz - normalized to zero Hz. Delta represents the normalized Doppler offset contribution due to the radial component of the near-surface currents, where ± 1 are the expected (normalized) positions of the first-order peaks. Thus, the radial component of current is deduced from such a radar record using the equation $v_{cr} = \Delta x v_{ph}$, where

$$v_{ph} = \sqrt{\frac{g\lambda}{4\pi}} \qquad\qquad (2)$$

is the phase velocity of the first order Bragg-scattering waves, λ being the radar wavelength.

Crombie showed that the depths of the current flow that influence the Doppler offset lie within one twenty-fifth radar wavelength of the mean water surface. At 30 MHz, for example, currents to a depth of about 0.3 m would be detected by Bragg scatter.

We are using electromagnetic waves of 10-m wavelength - corresponding to 30 MHz - thus, water waves of 5-m length only are "being seen" by our radar. In the North Sea, such waves are present nearly throughout the entire year. We use them solely to measure the near-surface currents, not the orbital motion itself; also, the height of the waves, i.e. sea state, is not a subject of consideration.

Each radar station can only "see" that vectorial part (projection of the current) which is pointed radially towards the transmitter station. Therefore, at least two transmitting stations are required on land, operating simultaneously. The vectorial addition of both radar pictures results in the total surface current. The resolution of azimuth of this method is selectable by

Fig. 2.7.2. Electromagnetic waves from the land antenna are resonantly backscattered to land from a sea surface wavetrain

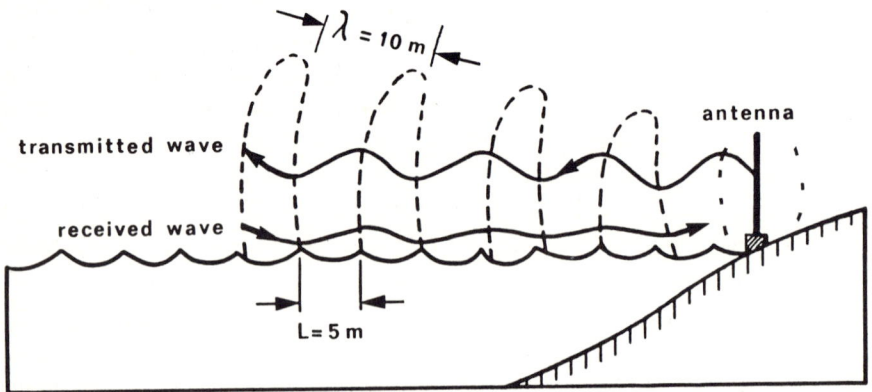

choice. In our case, it is 1°, i.e. the sea returns are grouped into beam fans of 1° width. The length of the range cells, or bins, is 1.2 km. Each time, one of the stations was operated for 18 min, the other one to follow immediately thereafter. From every pair of observations an overall picture is figured, thus representing a mean of a period over 40 min.

In order to show the current vectors in an orthogonal grid, a circle was drawn around each grid point with a radius of 3.0 km encompassing all radial vectors.

Among the first users of the technique described are Stewart and Joy (1974), and a working group at NOAA, USA with Barrick et al. (1977). Also, in France P. Broche and J.C. Maistre of the Laboratore de Sondages Electromagnetiques de L'Environment Terrestre, Toulon, have developed an HF instrument. During the international experiment MARSEN 1979, the instruments of the above groups and, in addition, one radar of the Institut Francais du Pétrole were compared with each other and with records of moored current meters (Janopaul et al. 1982). Among themselves, the HF current meters showed only slight differences, which may be explained from averaging by each instrument over sea areas of different size. There is no standard yet. Also, the "penetration" of the technique depends on the wave lengths, which were also different. Thus, the depth of measurements varied between 0.5 and 1 m. The results of such measurements always agree with records of moored current meters only to a limited extent, because current meters cannot be operated immediately beneath the sea surface due to variations of sea state and tidal range.

The accuracy of the HF backscattering method is about ± 5 cm/s. This, too, is the limit of resolution.

2.7.3   Current Variation in Time and Space: Observations

The area surveyed is located between Helgoland and the East-Frisian Islands (Fig. 2.7.3). One HF station had been installed on the island of Helgoland and the reciprocal station on the island of Wangerooge. Figures 2.7.4 - 2.7.7 show surface currents at intervals of 1 h. The cycle starts on 19 Dec. 1987, 15.00 UTC, indicating a well established ebb current. On this day, at the tide gauge of Helgoland low tide was 16.58 UTC (Fig. 2.7.8); the wind direction was from 40° (ENE) at 8 m/s. Only in the area of the mouths of Elbe and Weser rivers was there strong shearing due to run-off from the tidal flats and the rivers. This was observed during every ebb-phase independently of wind. One hour later, at 16.00 UTC (Fig. 2.7.5), the ebb current had significantly slackened, in the east, however, values were still about 0.5 m/s, because water continues to run off the ex-

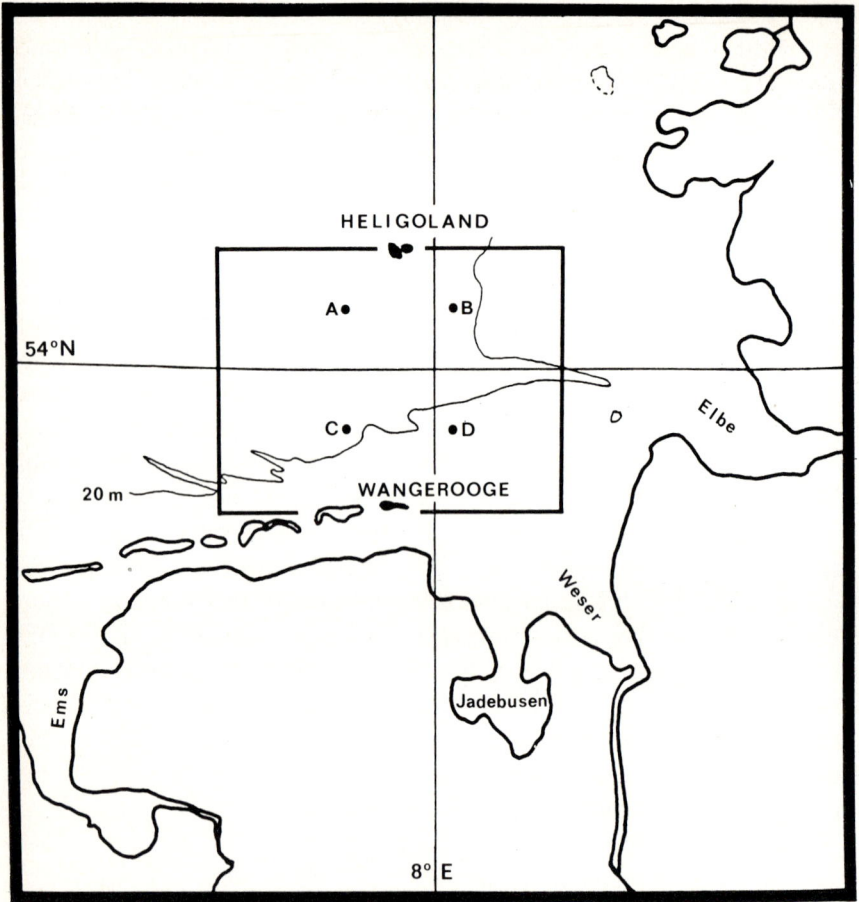

Fig. 2.7.3. The area under surface current observation

tensive tidal flats and shallow water areas. Even though we only see the motion at the surface, the current direction corresponds to the slope of the bottom; current vectors are mainly aligned perpendicular to the isobaths. About the time of the turn of the tide at low tide, 17.00 UTC (Fig. 2.7.6), in a limited area, about 8° E, 54° N, a state of rest occurred.

South of Helgoland, larger depth of water, up to 50m, (Helgoland Hole) evokes a reduction in current velocity and a slight shift of direction. One hour later, at 18.00 UTC (Fig. 2.7.7), a significant flood current has developed; it flows from the WNW, later only from the W. As long as there is no strong wind, patterns as shown in Figs. 2.7.4 through 2.7.7 are typical for the progression of the tides.

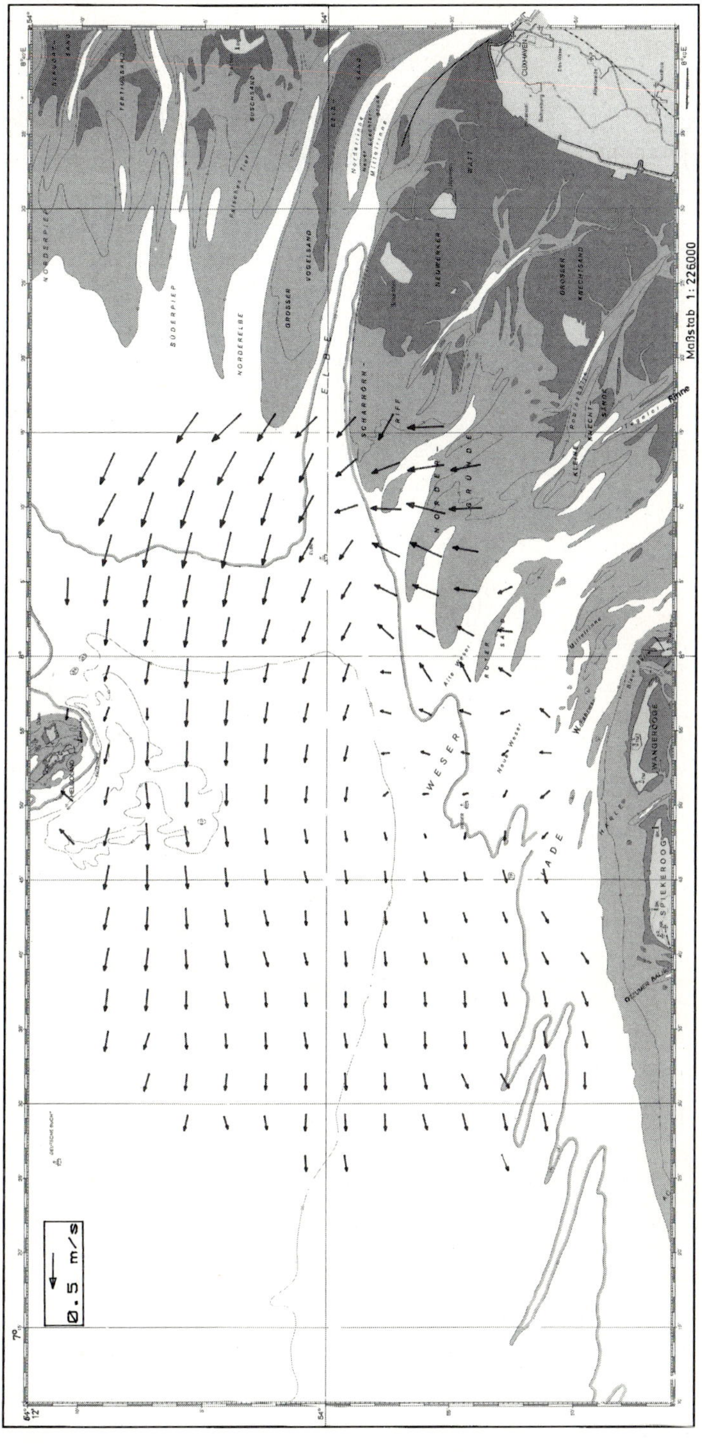

Fig. 2.7.4. Surface currents on 19 Dec. 1987, 15:00 UTC

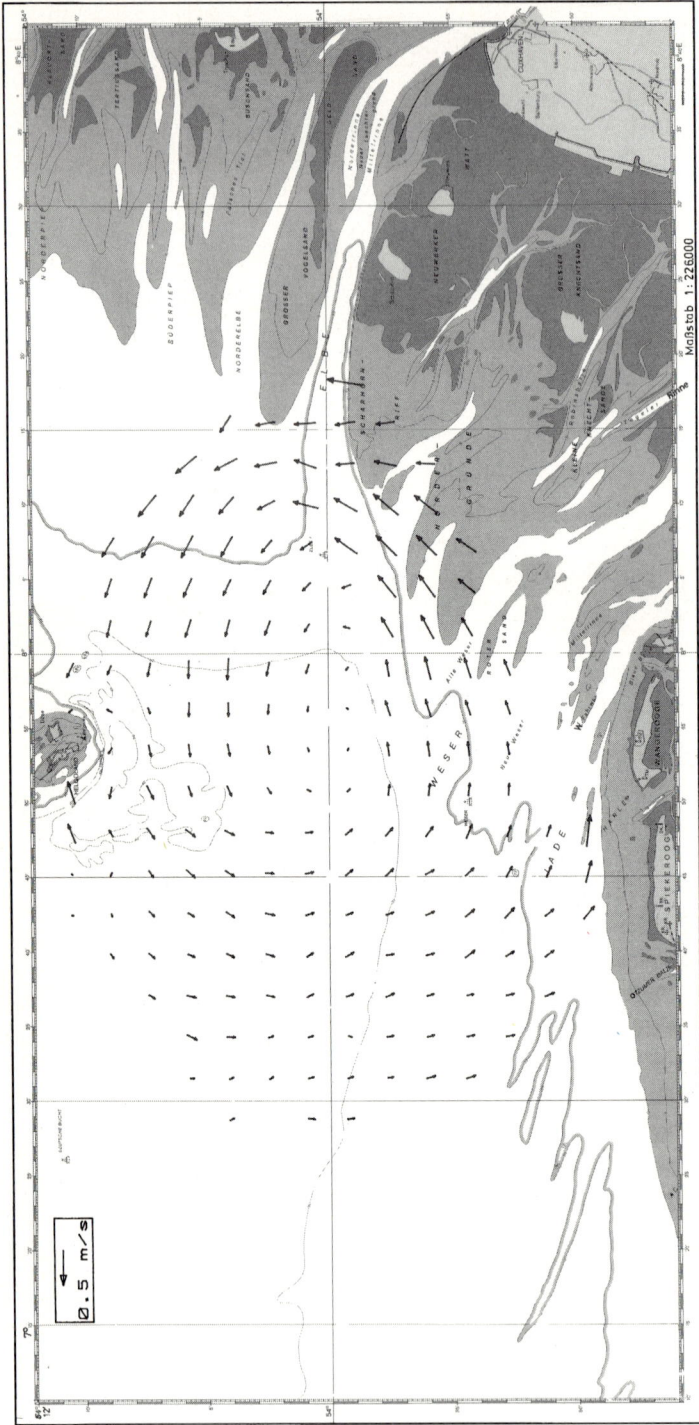

Fig. 2.7.5. Surface currents on 19 Dec. 1987, 16:00 UTC

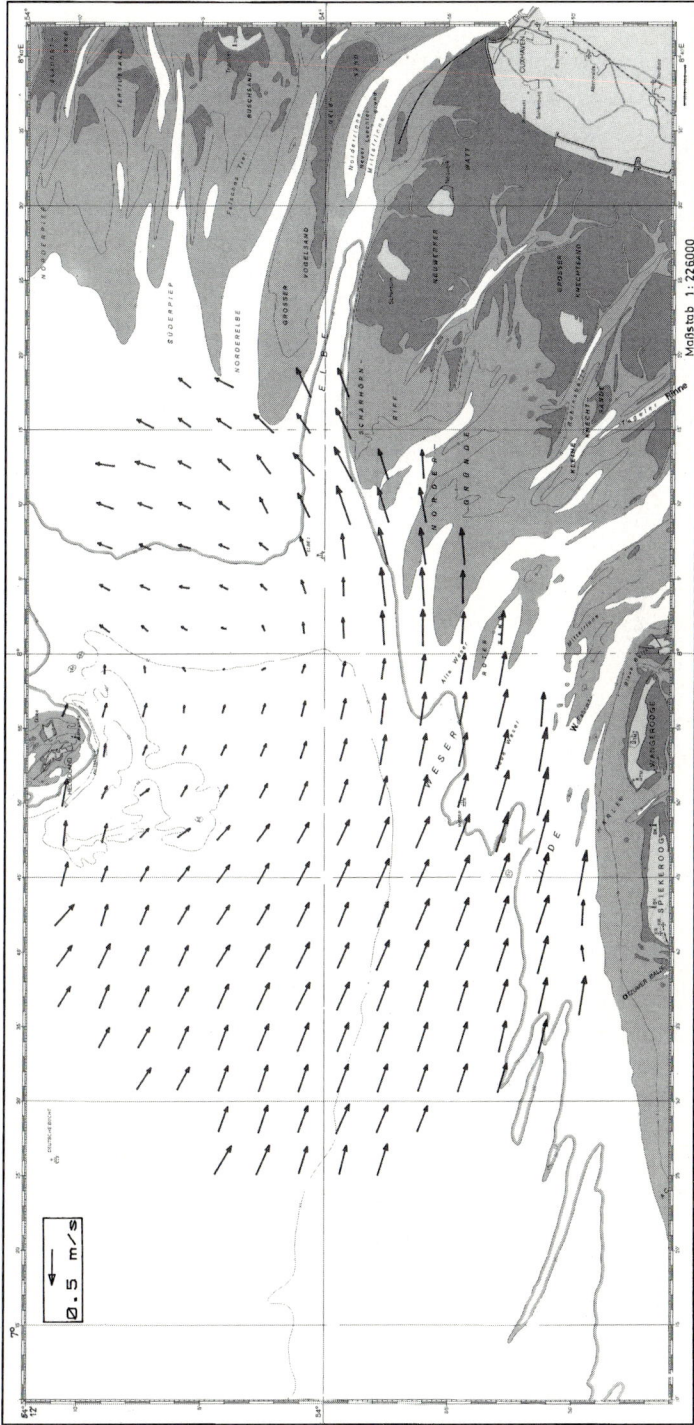

Fig. 2.7.6. Surface currents on 19 Dec. 1987, 17:00 UTC

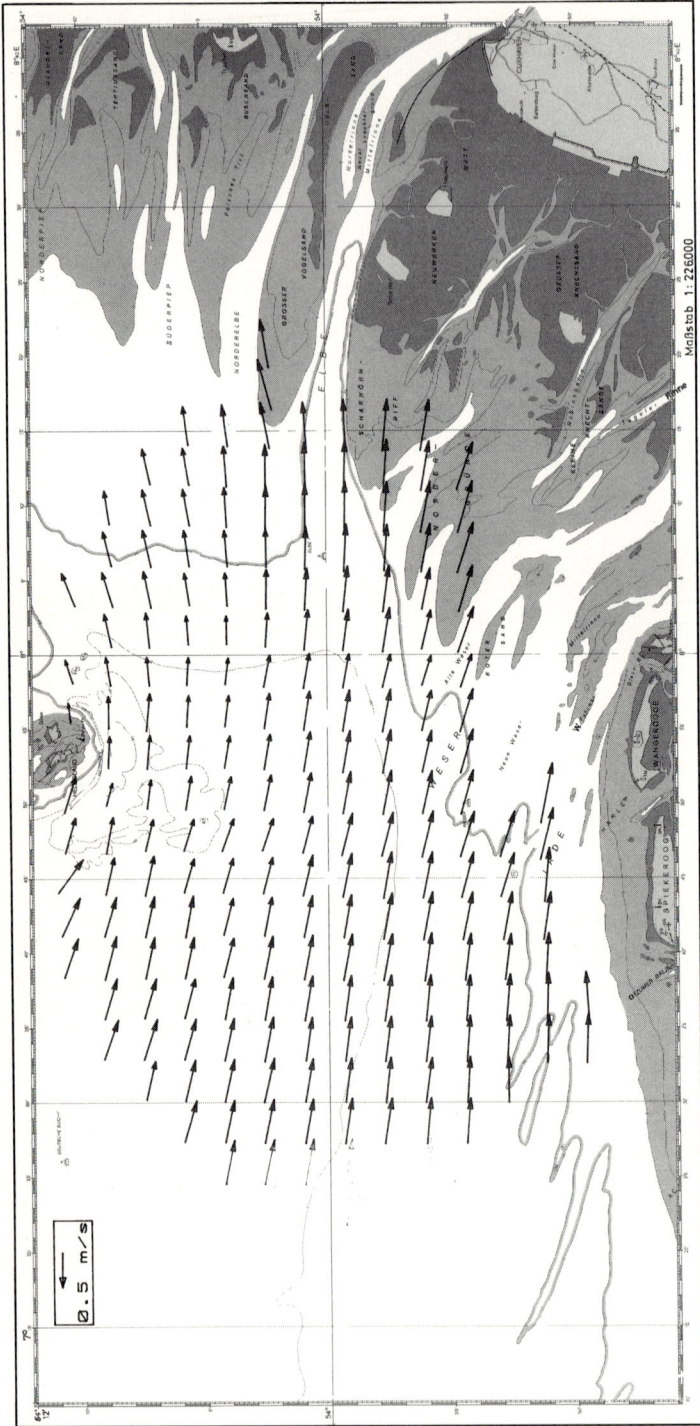

Fig. 2.7.7. Surface currents on 19 Dec. 1987, 18:00 UTC

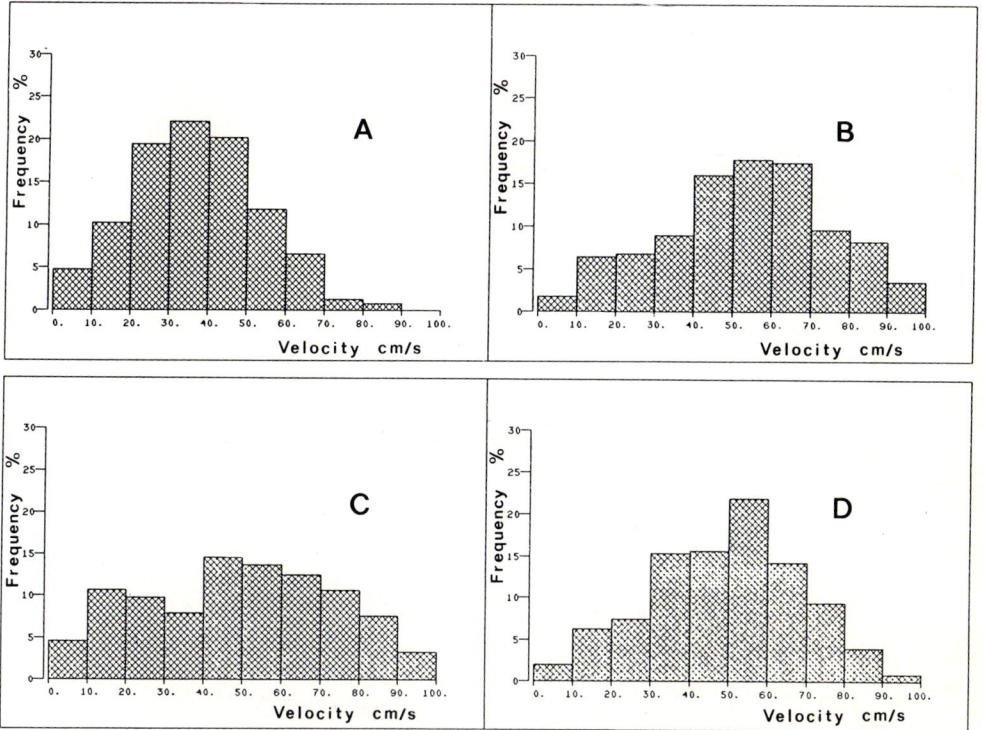

Fig. 2.7.8. Normalized frequency distribution of surface current velocity at positions *A* to *D* during Jan. 1988

2.7.3.1 Local Variability

The local variability of the magnitude of the current velocity was analyzed for the month of January, 1988, and is given for four locations as histograms. Figure 2.7.8 shows the respective relative frequency distributions for the locations A-D (Fig. 2.7.3). Besides nearly normally distributed frequencies around central values of 30 to 60 cm/s, in location C almost all velocity classes are found at equal weight. The four histograms are normalized to equal areas. The tidal ellipses in the next section give an idea about the absolute values of the current velocities. Prior to a discussion, though, the dependence of the current velocity on the local topography shall be demonstrated: computing the mean of the current velocity over January 1988 at each grid point (Figs. 2.7.4-2.7.7) the similarity between the isotachs and the isobaths is significant (Fig. 2.7.9). An exception is a small area off Wangerooge, where the water flowing into and out of the Jadebusen and the Weser flows significantly faster in elongated troughs.

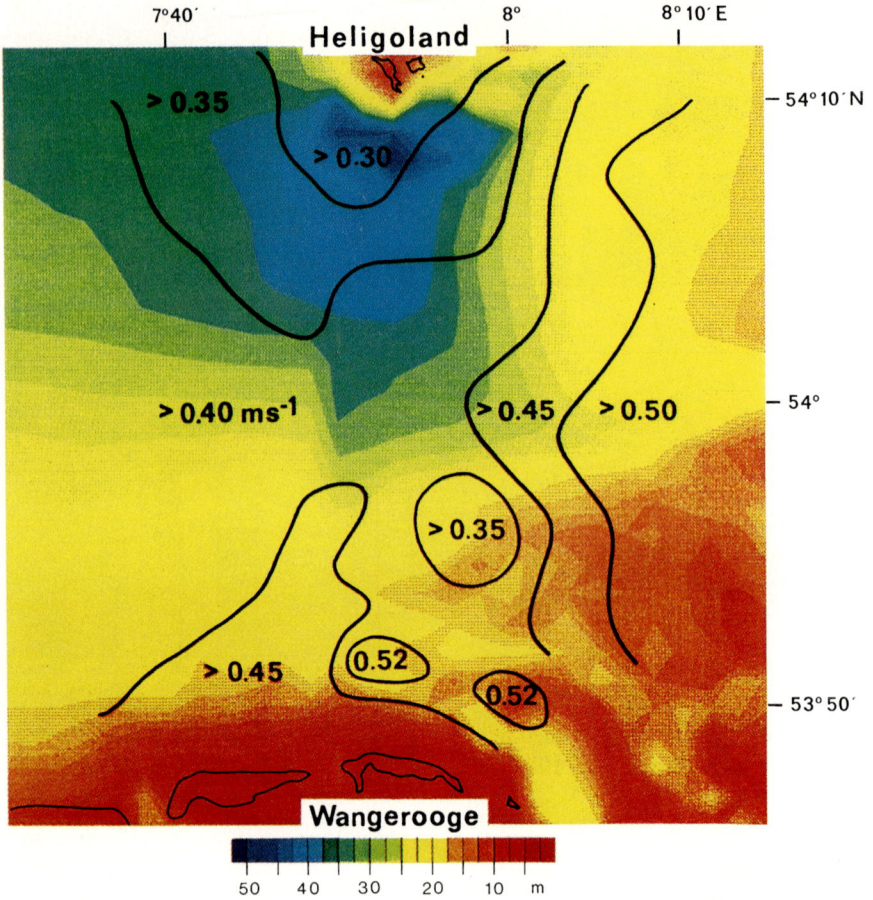

Fig. 2.7.9. Mean velocity of surface current given in cm/s in relation to depth topography

2.7.3.2 Tidal Signals

In our area of observation, the periodical part of the current includes the half-daily lunar tide, M_2. The tidal wave propagates from W to E. Therefore, at the western margin of the observed area we see another phase of the M_2-tide that is 15° less than in the east. The tidal curve of the Helgoland gauge, shown in Fig. 2.7.10, is used for comparison, since the longitude of the island runs exactly through the center of the observational area. The times of measurements used in Figs. 2.7.4 through 2.7.7 are marked. It is a day with medium wind, i.e. with a rather normal tide. The record of the Helgoland anemometer is shown in Fig. 2.7.11.

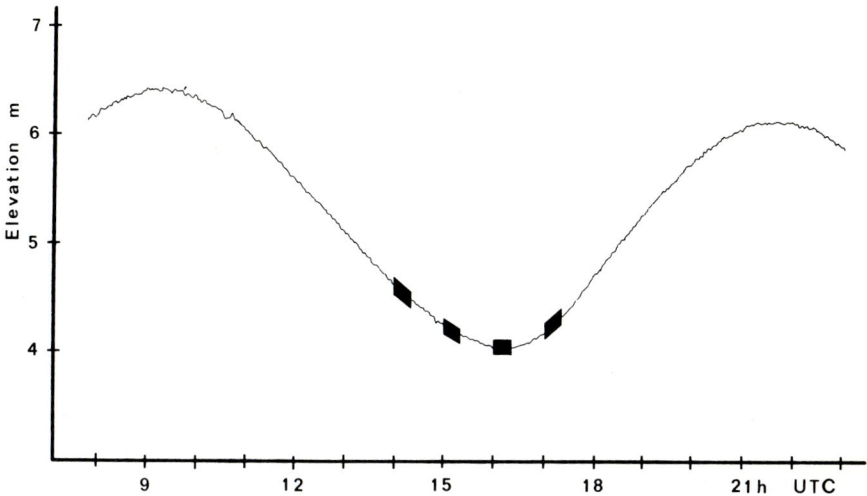

Fig. 2.7.10. Observed tide of 10 Oct. 1987 at Helgoland with marked HF observation periods of Figs. 2.7.4 to 2.7.7

Which partial tides contribute notably to the motion of the sea? To resolve this question, at all grid points the harmonic currents of the individual tides were fitted to the observed surface currents applying the least-squares method. Table 2.7.1, exemplary for one grid point, shows the (absolute) amount of the partial currents derived by this method. In E-W direction the M_2-tide generates about 52 % of the current, whereas the K_1-tide contributes only 3.2 % to the total current. The results are similar at all grid points in the area under observation.

Following general practice, we present (tidal-) current vectors at 1-h intervals. If one connects the endpoints of successive vectors, one obtains an ellipse, which, however, is not a hodograph. Yet, it permits one to readily read the current velocity and direction. The direction of rotation of the current vector is presented by arrows.

Figures 2.7.12 to 2.7.15 show the observed surface-current ellipses for the four strongest tidal components. Aside from the dominant E-W direction of their main axis, a band is observed along which the ellipses are distorted to lines. To the east and to the west of this band, the direction of rotation of the current ellipses differs. The equally distant HF grid permits an easy reading of the location of such directional reversals.

Mittelstaedt et al. (1983) showed from current meter moorings that in the area east of 8° and south of 54° the M_2-current vectors near the bottom rotate counter-clockwise or simply to and fro. According to their observations, there is an additional change in the sense of rotation of the current

Table 2.7.1. Tidal harmonics and their velocity amplitudes from least squares fit

Tidal harmonic	Period (h)	Velocity (m/s) east amplitude	Velocity (m/s) north amplitude
M_2	12.42	0.512	0.087
S_2	12.00	0.117	0.005
N_2	12.66	0.175	0.025
O_2	25.82	0.037	0.005
μ_2	12.88	0.075	0.035
M_4	6.21	0.041	0.032
K_1	23.94	0.032	0.002

vectors northeast of the M_2 ellipses in Fig. 2.7.12. Previously, Prandle and Matthews (1990) observed from OSCR measurements a reversal of direction of the partial-current vectors in areas with clearly simpler topography and straight coastlines. In our area, topography and a complicated coastline, forcing a Kelvin wave, contribute to the reversal of the sense of rotation of the ellipses.

2.7.3.3 Current Streamlines

In order to determine the traces of a water particle under the influence of tidal and residual currents, including wind, three particles were followed computationally on their trajectories through the field of observed Eulerian current vectors. Figure 2.7.16 shows their traces with prominent to-and-fro motions particularly reflecting the M_2-tide. The three particles start simultaneously on 18 Jan. 1988 at 00.00 UTC. Times are marked at 6-h intervals. The wind on Helgoland is given in Fig. 2.7.11. Even though the

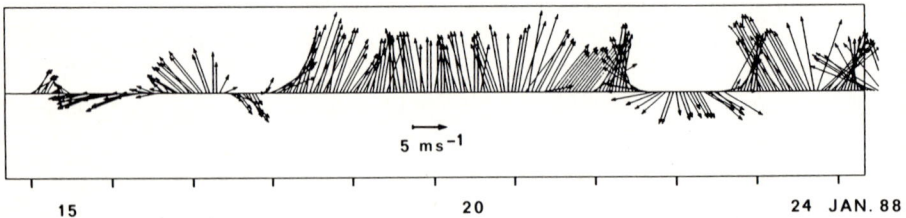

5 ms^{-1}

15   20   24 JAN. 88

Fig. 2.7.11. Observed wind at Helgoland

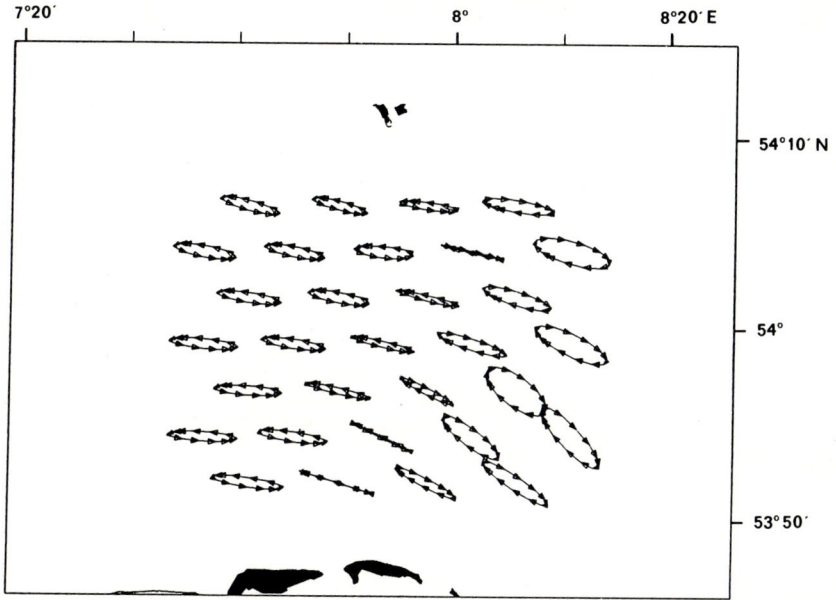

Fig. 2.7.12. Ellipses of the M_2 tide during Jan. 1988. Observe change of the direction of rotation

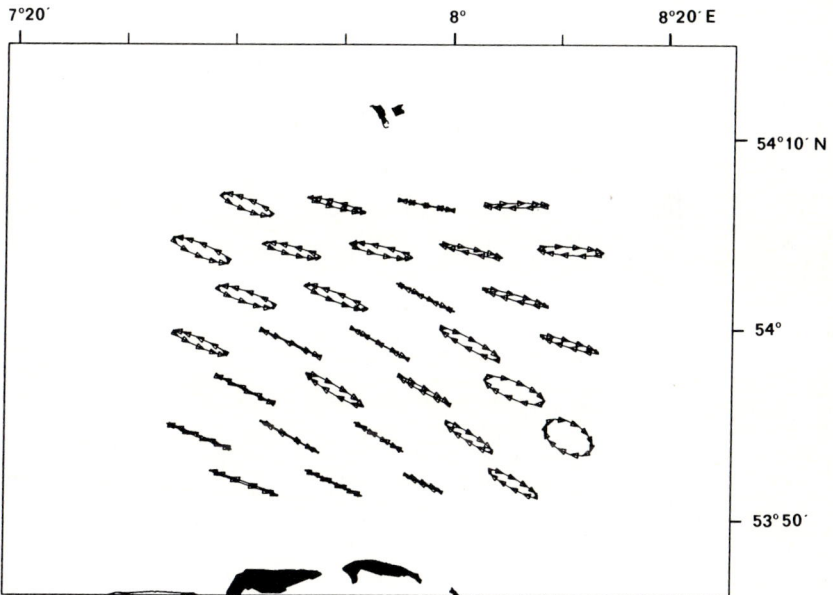

Fig. 2.7.13. Ellipses of the S_2 tide. Observe change of the direction of rotation. The scale is four times that of Fig. 2.7.12

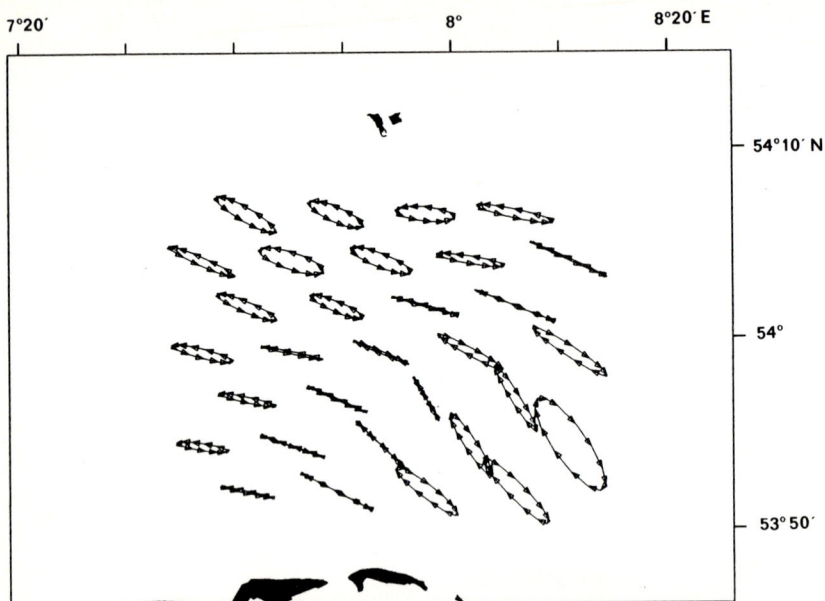

Fig. 2.7.14. Ellipses of the N_2 tide in Jan. 1988. The scale is three times that of Fig. 2.7.12

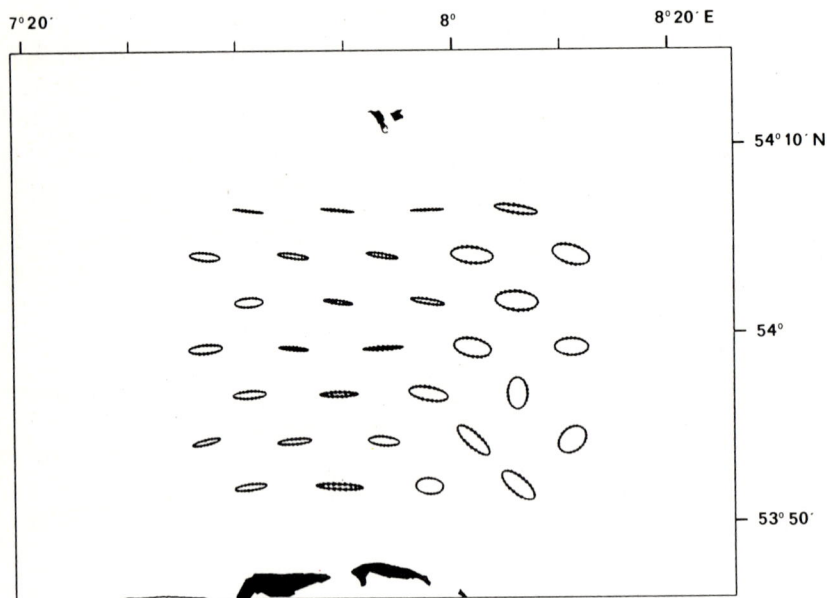

Fig. 2.7.15. Ellipses of the O_1 tide in Jan. 1988. The scale is ten times that of Fig. 2.7.12

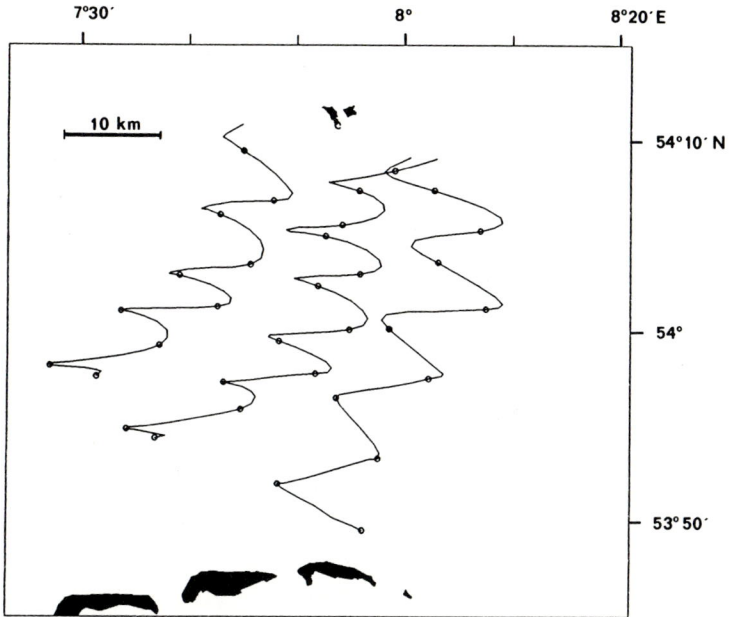

Fig. 2.7.16. Trajectories of three surface water particles due to tide and wind. Motion from south to north. *Time marks:* every 6 h from 17 Jan. to 22 Jan. 1988

traces of the particles are parallel, on the average, the speed of the particles is different. The traces appear to cross south of Helgoland; whereas the eastern particle has already left the area after 9.5 times 6 h, the particle on the centre trace requires 12.5 times 6 h to pass. Relative to time of observation of 2 months and to strong residual current the range of the HF-radar stations is much too short, for in 3 days a body of water has already left the area.

Looking at the motion of the water in the lower layers a similar picture develops, even though the effect of the wind is less. Similar to the investigated segment of the inner German Bight the water circulates with a to-and-fro motion fairly uniformly throughout the entire North Sea. A fair impression may be gained from models of Backhaus (1985) and Pohlmann (1991). Chapter 4.1 gives an example by Pohlmann for 19 January 1988. Hainbucher et al. (1986) have shown that the circulation not only varies seasonally but yearly due to large-scale variations of the weather. Considerable consequences follow regarding concentration and distribution of pollutants.

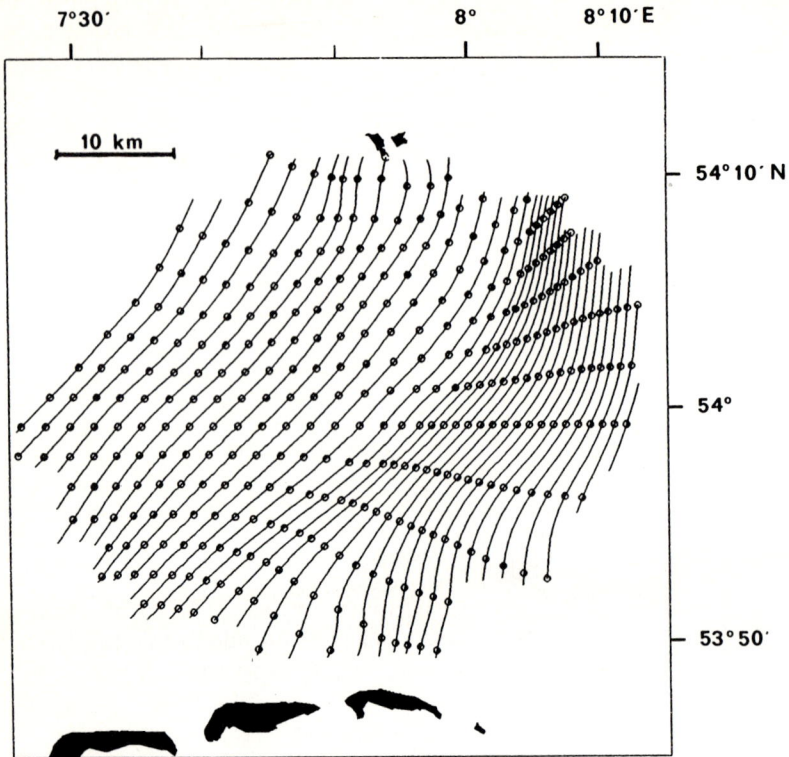

Fig. 2.7.17. Residual current stream lines from south to north computed from the current mean over Jan. 1988. *Time marks:* every 6 h

2.7.4 Residual Current

Subtracting the periodic part from the motion of the three particles in Fig. 2.7.16 results in a median trace, the residual current. In order to display these traces for many particles the first seven partial tides were determined for each grid point using least-squares calculations - much like computing tidal ellipses - and substracted from the currents. Then, the remaining residual current at each grid point was assumed to be constant with respect to time. The Lagrangian trajectories of Fig. 2.7.17 retrace these residual current vectors. In our example, the particles start in the south; one might just as well have let them start in the west. The time marks correspond to intervals of 3 h. The figure conveys the idea of the current pushing through the inner German Bight. The outflow of the Elbe and the Weser is contained in the current, much like the input of the wind during the period of observation from 1 Jan. to 31 Jan. 1988.

References

Backhaus JO (1985) A three-dimensional model for the simulation of shelf sea dynamics. Dtsch Hydrogr Z 38; 4:165-187

Barrick DE, Evans MW (1976) Implementation of a coastal current-mapping HF-radar system. Progress report No 1, NOAA Techn Rep ERL 272-WPL47, Boulder, 64 pp

Barrick DE, Evans MW, Weber BL (1977) Ocean surface mapped by radar. Science 198:138-144

Crombie DD (1955) Doppler spectrum of sea echo at 13.56 Mc/s. Nature 175:681-682

Crombie DD, Watts JM, Berry WM (1970) Spectral characteristics of HF ground-wave signals backscattered from the sea. Electromagnetics of the sea, AGARD Conf Proc No 77, Access No AD716305, NTIS, Springfield, VA

Hainbucher D, Backhaus JO, Pohlmann T (1986) Atlas of climatological and actual seasonal circulation patterns in the North Sea and adjacent shelf regions. Techn Rep 1-86, Inst Meereskunde, Univ Hamburg (unpublished)

Janopaul MM, Broche P, Maistre de JC, Essen HH, Blanchet C, Grau G, Mittelstaedt E (1982) Comparison of measurements of sea currents by HF radar and by conventional means. Int J Remote Sensing 3:409-422

Mittelstaedt E, Lange W, Brockmann C, Soetje KC (1983) Die Strömungen in der Deutschen Bucht. Bundesamt für Seeschiffahrt und Hydrographie (former DHI) Publ No 2347, Hamburg

Pohlmann T (1991) Untersuchung hydro- und thermodynamischer Prozesse in der Nordsee mit einem dreidimensionalen numerischen Modell. Diss Fachbereich Geowiss Univ Hamburg

Prandle D, Matthews J (1990) The dynamics of nearshore surface currents generated by tides, wind and horizontal density gradients. Continent Shelf Res 10:665-681

Stewart RH, Joy JW (1974) HF radio measurements of surface currents. Deep-Sea Res 21:1039-1049

2.8 Background Concentrations for Metals in the North Sea: Sediment, Water, Mussels and Atmosphere

M. KERSTEN, P.W. BALLS, R.J. VAN ENK, N. GREEN, K.J.M. KRAMER, M. KRIEWS, F. MONTENY and J.J.G. ZWOLSMAN

2.8.1 Introduction

Within the last decade, understanding of the biogeochemistry of trace metals in the marine environment has increased considerably. At the beginning of the 1980s, reliable sampling and analytical techniques became available to detect the relatively low metal concentrations in sea water. At the same time, concern over the possible adverse effects of environmental contamination by metals led to increased research effort in the field. The presence of detectable metal concentrations in the environment, however, is not necessarily indicative of contamination. With the exception of certain radionuclides the ubiquitous presence of metals in water, sediment and the atmosphere is entirely anticipated and arises from their natural occurrence in the earth's crust.

Metals, at their natural concentration, play an essential role in many biochemical processes in organisms, and this concentration is called the background concentration. Any concentration lower or higher than this background can be toxic. The backgound concentration is, however, not a fixed value. Changes in physical and chemical environment, e.g. changes in pH due to primary productivity or changes in salinity due to freshwater inputs, change bioavailability of metals. Organisms are able, at least partly, to adapt themselves to changing metal concentrations.

Due to human activities, metal concentrations have increased in the marine environment of the North Sea, which has often been regarded as the most heavily polluted shelf sea in the world. The result of a transect across the shelf edge presented by Kremling (1983) was one of the first pieces of evidence that the entire North Sea is affected by trace metal increases. Reviews of the present concentrations were recently published by Kersten et al. (1988) and Kramer et al. (1991b). Donze (1990) gives a compilation of the metal concentrations in European watersheds and Fowler (1990) gives a worldwide overview of present metal pollution in marine environments. The possible adverse effect of anthropogenic metal input increases has led to political action in many countries. In the North Sea Action Plan introduced at the Third North Sea Conference ministers of the individual adjoining countries agreed to a reduction of 50 % of the anthropogenic input load of trace metals (based on the 1985 data) before the year 1995. To accurately establish such a reduction, it is essential to know the background concentration of the individual metals. Moreover, in water quality models to evaluate potential reduction measurements and for an adequate application of policy regarding reduction measures, it is essential to know the backgound concentrations as well.

When determining reference data, we often rely on what unfortunately has become a current slogan in marine pollution research, "baseline studies". In principle, this type of approach is based on determining concentrations of certain constituents above a "natural background level" and attributing the excess to anthropogenic sources. However, in this chapter we argue that a true natural background (i.e. "geogenic" or "non-anthropogenic") concentration can be established only for sediments of appropriate "old" age, because other compartments of the ecosystem may be influenced to a greater or lesser extent by human activities. No "old" material for water or mussels is available. In the latter cases we have to rely on more loosely defined terms such as "natural reference levels" or "presumed high background levels at diffuse loading" (Knutsen and Skei 1992). For environmental pollution assessment and baseline studies, the present concentrations in these cases frequently are compared with concentrations in "pristine" areas, which then represent the natural reference level. In other circumstances, the reference value may simply represent the pre-existing concentration which indicates the pristine situation prior to the current disposal practice or pollution investigation. This may have a different value from the true natural background and can be defined as a benchmark value (JMG 1990) or target value (Jonkers and Everts 1992). It should be noted in general, however, that prior to the 1980s no reliable trace metal data are available except some few for sediments. During the last decade, a clearer and more coherent picture of the abundances and biogeochemical cycling of trace metals in the marine environment has emerged. During the same period there has been an increased application of modelling as a means to construct regional mass balances. Based on both the field and modelling data, a discussion was recently initiated whether the current monitoring programs are in fact capable of assessing progress towards some "benchmark" or "background concentration" in water or sediment etc. (e.g. Balls 1989; Larsen and Jensen 1989; Kersten et al., Chap. 4.6, this Vol.). This discussion is beyond our terms of reference, but we feel that a radical reconsideration is required if monitoring programs are going to be successful in this objective.

A further note should be added to the problem of terminology. Clearly, in science the correct terminology is a prerequisite to both precision and accuracy, but the importance of correct terminology is underscored by administrative initiatives for classification. In definitions of sediment and water quality criteria or environmental product standards, reference concentrations are used as an upper limit for the unpolluted class I target values. In this chapter we understand natural background concentrations as a level at "zero human activity". We will not attempt to discuss the classification problem, nor should this chapter be understood as an endorsement of environmental quality objectives in favour of the precautionary principle (i.e. scientific vs. political philosophy). Clearly, it is timely to come up with internationally accepted quality criteria for the various biotic and abiotic compartments of the North Sea, but we feel that these must be accompanied by internationally accepted uniform emission standards based on best available technology for an improvement of the pollution situation beyond that 50 %-benchmark. However, this discussion is also beyond our terms of reference (cf. Gray 1990; Lutter 1990). We hope that the present report based on trace metal group discussions during the International Workshop on Background Concentrations of Natural Compounds (Laane 1992) will provide the North Sea Task Force (NSTF) and other appropriate forums a scientifically sound base for future standard-setting activities and the development of a long-term emission reduction strategy as an effort towards a cleaner North Sea.

2.8.2   Background Concentrations in Sediments

For any study of sediments, a basic amount of information on their physical and chemical characteristics is required before an assessment can be made on the presence or absence of anomalous metal concentrations. A relevant compilation of these parameters is given by the *Guidelines for the Sampling and Analysis of Sediments under the Joint Monitoring Programme* (JMG 1990), which have been established after thorough discussions in several ICES working groups. The type of information that can be gained by the utilization of these guidelines is extremely useful for the interpretation of the complexity and diversity of situations encountered in the sedimentary environment. In our attempt to define background metal concentrations for sediments, we will refer only to those data which were gained according to these general JMG guidelines.

In terrigenous sediments, at least 25-50 % of the background trace metal content is likely to be bound to silicates and minor quantities of other resistate accessory minerals such as zircon, rutile, garnet and magnetite (Campbell et al. 1988). This association arises principally from substitution of the trace metals into the lattices of either primary minerals that have survived weathering or secondary silicates produced during weathering and diagenesis, but may also reflect physical inclusion or encapsulation of another phase, e.g. pyrite within silicate, or incorporation in biogenic silica derived from radiolaria or diatoms. Most procedures for total digestion and dissolution of silicate minerals are based on mixed-acid treatments including HF at high temperatures. Exclusion of HF from the acid digestion leads to incomplete extraction of some trace metals. The resistate portion may usually be neglected in case of Cd, Cu, Pb and Zn (Krumgalz and Fainshtein 1989), but is significant at least for Cr, Ni and Fe (Campbell et al. 1988). The guidelines stress that only total content of metals should be considered, i.e. the concentrations determined by total digestion by HF plus aqua regia of the sediment, or by a non-destructive instrumental analysis technique such as INAA or XRF.

Grain size is one of the most important parameters controlling the distribution of both natural and anthropogenic components in sediments. A general JMG recommendation is therefore that only sediments with more than 20 % of < 63 µm sediment fines should be considered for trend monitoring purposes, because they are sufficiently muddy to carry an appreciable contaminant load and are sufficiently homogeneous to minimize within-sample variability. Although our approach is predominantly aimed at baseline and monitoring studies, there is no necessity to generally adopt this criterion, because regional background concentrations can be determined from sandy sediment cores as well. Nevertheless, trace metals are associated with particle surfaces, and differences in metal concentrations among sites or even within a sediment core of the same site can be generated simply by differences in particle size distribution of the sediment samples. To compensate for this, sediment data can be normalized by dividing the raw concentration data by the fraction by weight of the fine-grained fraction. This is equivalent to assuming that no contaminant metals are associated with sand-sized particles and that the only effect of sand in a sample is to dilute its level of contamination. However, granulometric normalization alone is inadequate to explain all the natural trace metal variability in the sediments. In order to interpret better the compositional variability, it is also necessary to distinguish the sedimentary components with which the metals are associated throughout the grain-size spectrum. Since effective separation and analysis of individual sediment components is extremely difficult (Kersten and Förstner 1989), consideration of such associations must rely on indirect evidence by elemental co-relationships.

Various geochemical approaches used for the normalization of trace metal data have been extensively reviewed by Loring and Rantala (1992) and are considered by the JMG guidelines (JMG 1990). Since trace metals are mainly associated with the clay minerals, iron and manganese oxi-hydroxides, and organic matter (with decreasing abundance in that order), more information can be obtained by measuring the concentrations of elements representative of these components in the samples. An inert element such as aluminium, a major constituent of clay minerals, may be selected as an indicator of that fraction. It may be considered as a conservative element, that unlike iron and organic carbon (TOC) is not affected significantly by early diagenetic processes and strong redox effects frequently observed in estuarine and coastal sediments (e.g. Kersten and Förstner 1991; note, however, that in anoxic strata, the relationship between the concentration of trace metals and that of Al may be weak due to entrainment of trace metals in secondary sulphide phases: Huerta-Diaz and Morse 1992). Normalization to the TOC concentration is not appropriate for our approach, because TOC is also acting as a contaminant itself. TOC concentrations in sediments are usually lognormally distributed and skewed towards low concentrations, just as are contaminants. The Al normalization (e.g. Windom et al. 1989) is based on the fact that there are natural ratios between trace metals and Al that exist in the absence of any human influence (since Al is a major component of clays its concentration is assumed always to be a natural concentration). This method is similar to the normalization based on the fine-grained fraction because Al concentrations follow the grain-size distribution (note, however, that in sandy sediments, or in cases where glacier erosion is a major source of sediment, Li may more suitable for normalization: Loring and Rantala 1992). The correlations of trace metals with the fine-fraction are often higher, because metal to Al ratios vary regionally due to compositional variations. This normalization method is preferred for our approach, because once natural trace element to Al ratios are established for a specific site, any higher ratio is a measure of the extent of contamination. This normalization allows, therefore, the definition of an enrichment factor:

$$EF = (M/Al)_{\text{sediment}} / (M/Al)_{\text{background}},$$

where M/Al refers to the ratio of the concentration of a given metal M to that of Al in the sample and background, respectively. Departures from unit values indicate either contamination of the sediment or local mineralization anomalies.

Since about all present-day deposited sediments are probably to some extent contaminated by anthropogenic trace metals, natural background concentrations can only be established in sediments which were deposited in pre-industrial times (i.e. pre 1850). These layers are best preserved in rapidly accumulating laminated sediments with little or no physical or biological mixing and low intensity of diagenetic processes affecting the vertical distribution of the metal of interest. However, such sediments are not common in the North Sea. Most of the North Sea is non-accumulating with respect to sediments, is heavily trawled and is subject to frequent storms that resuspend the (mostly coarse-grained) bottom sediments. As a result, the JMG guidelines would argue against establishing background concentrations for temporal trend monitoring in most of the North Sea. Exceptions may be found in the net depositional areas off estuaries, the Wadden Sea and the Norwegian Trench. In such cases, however, detailed down-core sedimentological, geochemical, and radiochemical work is still neccessary to compensate for mixing effects (e.g. Kramer et al. 1991a). As discussed above, geochemical normalization permits compensation for the local and temporal variability of the sedimentation process and is therefore mandatory for the North Sea. Given these criteria, the relevant data base is surprisingly limited to only a few dated sediment cores which penetrated pre-industrial sediment layers in these areas. Three examples have been selected to represent the range of natural background element/Al ratios in the entire North Sea area (Table 2.8.1).

Levels of anthropogenic enrichment of intertidal sediments in the Humber estuary have been assessed relative to a natural background provided by sediments from intertidal exposures of Scrobicularia clays laid down in former channels of the Humber approximately 5000 years B.P. (Middleton and Grant 1990). From the elemental composition determined by X-ray fluorescence (XRF) analysis of total sediment subsamples, element/Al ratios were calculated and are listed in the first column of Table 2.8.1. Another core was collected in the western Dutch Wadden Sea (location Mokbaai) and dated by ^{210}Pb to construct pollution history profiles (Kramer et al. 1991a). No excess ^{210}Pb was found below 30 cm depth. The element/Al ratios listed in the second column of Table 2.8.1 were calculated as means of four individual sediment layer analyses sampled below 43 cm depths in this core. Norwegian NSTF/JMG data from six sediment cores, sampled along the Norwegian coastal area in an area of low deposition rates (i.e. ≤ 1 mm/a at water depths 200-400 m), at 15-20 cm depths were taken as representative for pre-industrial composition. The averaged values are given in column 3. These concentrations and metal/Al ratios can be compared to those calculated for average continental crustal rocks (the so-called Clarke Values: Martin and Whitfield 1983) in order to provide a more global perspective for these results.

A comparison of the individual metal data given in Table 2.8.1 indicates that the background metal composition of sediments around the North Sea deviates by less than a factor of 3 from the expected continental sources. This is not surprising when considering the different geological and hydrodynamic conditions. This result indicates also that natural background levels are site-specific and may not apply for other areas. Extrapolation of the data from one site to adjacent areas, however, is recognized to be a valuable approach when geological and oceanological data indicate a similar source of the sediments. The crust composition cannot be taken as a reference for sites where no sediment data exist (e.g. for non-depositional areas). Recognizing the severe lack of relevant sediment core data even for the sediment-accumulating North Sea areas, we strongly recommended to establish more such data sets within the framework of national or NSTF monitoring programs carried out in line with the JMG guidelines (JMG 1990; Loring and Rantala 1992).

2.8.3   Background Concentrations in Water

2.8.3.1   Dissolved Metals

Rivers contain higher concentrations of most dissolved trace metals than oceans, and are generally the dominant source of dissolved metals for sea water (Martin and Whitfield 1983; Bewers and Yeats 1989). The available data for the major rivers flowing into the North Sea indicate in fact a quite broad range of trace metal concentrations. Cd values in freshwater of the Rhine, Elbe and Weser rivers can range from 90 to 600 ng/l and for Cu from 250 to 7500 ng/l (Duinker et al. 1982). Comparing these river water concentrations with those observed in the North Sea (Kersten et al. 1988) suggests a mixing gradient occuring from fresh to sea water which decreases the dissolved metal concentrations. Furthermore, and unlike the case for nutrients, advection of Atlantic water across the shelf break cannot be considered as a source of trace metals to the North Sea. Hydes and Kremling (1993) found that all the waters on the shelf contain higher con-

Table 2.8.1. Range of background concentrations C_M and element/Al ratios of preindustrial sediments from three different locations around the North Sea coast as determined by total HF digestion except for Hg (C_M of trace metals in µg/g, Al in %; for reference see text)

	Humber		Wadden Sea		Norwegian Sea		Earth Crust	
	C_M	M/Al	C_M ($\pm 1\sigma$)	M/Al	C_M ($\pm 1\sigma$)	M/Al	C_M	M/Al
Al	7.1		4.8 ± 0.2	1	5.8 ± 1.0	1	6.93	1
As	22	3.1	13.5 ± 1.3	2.8	-	-	7.9	1.1
Cd	-	-	0.5 ± 0.01	0.11	0.08 ± 0.02	0.013	0.2	0.03
Cr	99	14.0	84 ± 0.5	17.5	-	-	71	10.2
Cu	17	2.4	22 ± 2.0	4.6	17 ± 5.9	3.1	32	4.6
Hg	-	-	0.067 ± 0.009	0.014	0.04 ± 0.03	0.007	-	-
Ni	38	5.4	37 ± 1.2	7.7	-	-	49	7.1
Pb	22	3.1	37 ± 2.9	7.7	26 ± 9,5	4.6	16	2.3
V	109	15.4	-	-	-	-	97	14.0
Zn	84	11.8	103 ± 5.2	21.4	110 ± 27	19.7	127	18.3

centrations of the metals than did their source waters beyond the shelf edge. The shelf acts as a source of dissolved metals to the Atlantic Ocean. It is therefore likely to find still nearly pristine conditions in the Atlantic ocean which can be used as a reference for the open North Sea. However, it is improbable that metal concentrations in the Atlantic ocean represent true natural background values unaffected by human activities, especially for Pb due to its significant atmospheric pollution on a global scale (Kersten et al. 1992). Table 2.8.2 provides a revised assessment of sea water end-member concentrations of dissolved trace metals based on a survey of data reported for the northeast Atlantic Ocean and the Norwegian Sea. Our discriminatory evaluation of recent data results in major revisions to previous assessments of the average trace metal composition of shelf sea water (e.g. ICES 1991), especially in the case of Hg. For this and some other trace elements major improvements in sampling and analytical technology have been gained within the last few years.

Table 2.8.2. Natural reference values for dissolved trace metals in end-member salinity environments, i.e. off-shelf sea water (As, Sb, V: Middelburg et al. 1988; Cd, Co, Cu and Ni: Hydes and Kremling 1993; Cr: Yeandel and Minster 1987; Hg: Fileman and Harper 1989; Pb: Brügmann et al. 1985; Tl: Flegal and Patterson 1985; Zn: Danielsson et al. 1983) and natural river water (As, Co, Cr, Mo, Sb, Tl and V: Zuurdeeg et al. 1992; Cd, Cu, Ni, Pb and Zn: Bewers and Yeats; Hg: Cossa et al. 1988), in µg/l. Note that a seasonal variation has been reported for at least Cd and Zn

Element	Atlantic sea water mean ($\pm$ 1σ)	Natural river water		
		(Zuurdeeg et al. 1992)		(Bewers and Yeats 1989)
		Mean	Range	Mean
As	1.4 $\pm$ 0.1	1.24	0.28 - 5.42	-
Cd	0.004 - 0.009	0.12	0.04 - 0.35	0.01
Co	0.0035	0.031	0.010 - 0.098	0.08
Cr	0.16 $\pm$ 0.03	0.097	0.024 - 0.39	-
Cu	0.070 $\pm$ 0.013	2.0	0.8 - 5.3	2.0
Hg	0.0005 $\pm$ 0.0001	0.060	0.028 - 0.13	0.002
Ni	0.140 $\pm$ 0.015	3.6	1.0 - 13.3	0.5
Pb	(0.033 $\pm$ 0.015)	3.1	1.1 - 8.4	0.05
Sb	0.14 $\pm$ 0.007	0.42	0.16 - 1.09	-
Tl	0.013 $\pm$ 0.002	0.016	<0.01 - 0.035	-
V	1.66 $\pm$ 0.06	0.30	0.14 - 0.63	-
Zn	0.13 - 0.40	18.5	8.0 - 42.7	0.5

Reference data for river water in Table 2.8.2 are obtained from a compilation of natural background concentrations of trace and major elements in surface water in the Netherlands (Zuurdeeg et al. 1992). This study was carried out by a commission of the Dutch Ministry of Planning and the Environment and is based on the following assumptions: (1) the natural quality of freshwater in the Netherlands can be approached by natural water from geologically and morphologically similar areas in the northern European lowland (France to Bielo-Russia) and the bordering foothill zones; and (2) the concentration of a certain trace element in fresh water varies with the major element composition (i.e. the water type). After a thorough survey of (mostly) grey literature, a total of 11740 data sets were selected which (1) fulfilled the criteria for unpolluted to slightly polluted water based on oxygen balance (> 8 mg/l), nitrogen content (trace NH_4-N), and organic matter load ($BOD_5 < 1$ mg/l), i.e. originate from oligotrophic waters for both the sampling point and the entire drainage basin upstream of the sampling site, with no industrial or major agricultural activity at a distance of less than 50 km from the drainage area; and (2) was declared free of pollution (i.e. pristine) by the rapporteur or monitoring ageny itself. Major drawbacks for certain elements are seen in the fact that (1) atmospheric deposition was not considered as an important non-point pollution source in the drainage basin; (2) the data set may be biased towards northern Germany due to the inclusion of a large number of data from the BGR Hannover; (3) the data may be also biased to small rivers of low suspended particulate matter content, which may be not well comparable with the major rivers such as the Rhine; and that (4) no other evaluation of the analytical accuracy could be made than to exclude the data published before 1978. In addition, data sets with > 40 % of samples below the detection limit were rejected, and for the remaining populations these concentrations were replaced by an average value estimated from the lognormal data distribution curve. However, detection limits have been lowered considerably in routine monitoring, because major improvement in sampling and analytical technology for measurement of trace metals in marine waters has been paralleled by similar developments with respect to chemical measurements in freshwater. Considerable care must, therefore, be exercised in using data on the incidence of trace metals in rivers in much the same way as had to be done with respect to sea water data gained during the 1970s and early 1980s. Bewers and Yeats (1989) published such a discriminatory evaluation of recent data based on the known laboratory reputation which resulted in a major revision to previous assessments of the global average trace element composition of rivers, including one of the most cited (Martin and Whitfield 1983). The global average river concentrations of Cd, Ni, Pb, Zn and Hg (the latter taken from Cossa et al. 1988), are in fact by one order of magnitude lower than those for N-European lowlands given by Zuurdeeg et al. (1992). Although the selection criteria are not mentioned explicitly by Bewers and Yeats (1989), their data seem to be biased to major rivers of high particulate matter load (e.g. Mississippi or Amazon) or rivers which drain a base rock area (e.g. St. Lawrence). In fact, their data match quite well the Rhine river background values and rivers in southern Sweden (Göta and Nordre: Danielsson et al.

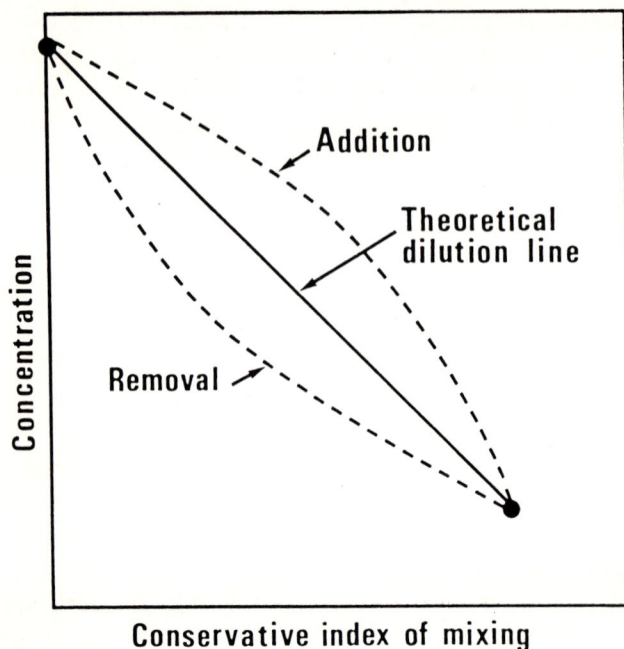

Fig. 2.8.1. Theoretical relationship between the composition of estuarine water and salinity; end-members constant over the flushing time of the estuary

1983). It can be recommended, therefore, to take the data of Bewers and Yeats (1989) as a freshwater background reference for the particle-reactive elements Cd, Cu, Ni, Pb, and Zn, the data of Cossa et al. (1988) for Hg, and the data of Zuurdeeg et al. (1992) for the other elements listed in Table 2.8.2.

When it comes to establishing reference values for the range of salinities in between both end-members, i.e. for the estuaries and coastal areas, we run into a problem. It is clear that no single trace metal reference value can be established for the entire salinity range. In identifying pristine or contaminated areas it is not sufficient merely to compare concentrations, the salinity must also be taken into account. Although the behaviour of any given metal can be interpreted differently, depending on the data set used, an overall pattern emerges from the combined data (ICES 1991). Variability in trace metal concentrations in the estuarine/near-shore region of the North Sea (i.e. at salinities less than 30) appears to be related to the relative influence of specific river inputs. Mixing curves which relate trace metal concentrations to salinity provide the principal practical method for describing metal behavior in these mixing zones. A linear mixing curve in-

ZINC

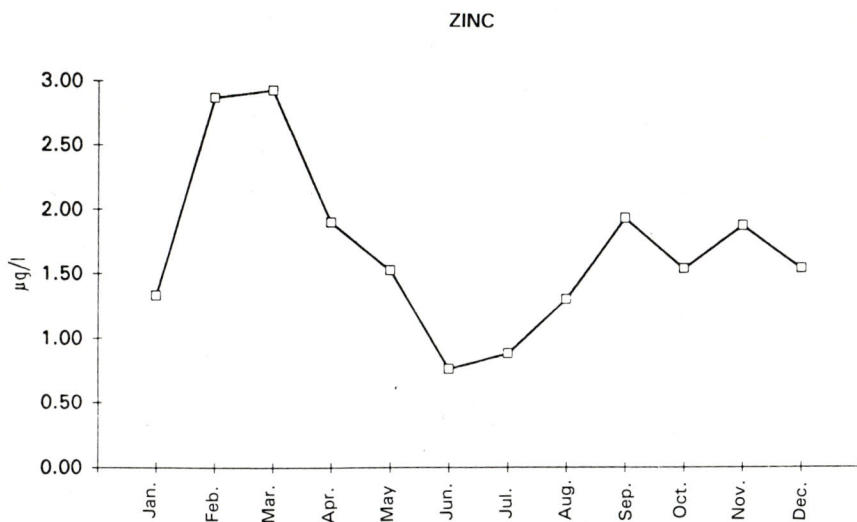

Fig. 2.8.2. Average concentrations of Zn (in μg/l), calculated for the individual months over the period 1980-1988 for the Dutch coastal zone. (Kramer et al. 1991b)

dicates conservative behavior of the metal upon mixing of fresh with seawater. Any other relationship (concave, convex, multiple inflections; cf. Fig. 2.8.1) indicates addition and/or removal of the metal from the dissolved phase. The results from a variety of coastal surveys suggest, in fact, that an inverse linear relationship between trace metals (Cd, Cu, Pb, Zn, Ni, but not Hg) and salinity occurs (at least in the salinity range of 30 to 36), but with slopes of the regressions varying from area to area (ICES 1991). It should be noted, however, that metal concentrations below that of the Atlantic end-member have also been observed in coastal areas due to enhanced internal fluxes of resuspended sediment particles or phytoplankton blooms (e.g. Balls 1987), which may lead to disequilibrium with respect to sorptive exchange of metals. Seasonal variability in the particulate matter composition has therefore also to be accounted for, especially in eutrophic shelf and coastal waters which are not buffered by high amounts of detrital particulate matter (i.e. < 5 mg/l; Balls 1989). Figure 2.8.2 shows such a strong seasonal variation for Zn in Dutch coastal waters (Kramer et al. 1991b).

In the salinity range below 30, non-conservative behaviour was found for most trace metals. When the metal-salinity relationships for offshore waters are extrapolated back to zero salinity, the resulting effective freshwater end-member concentrations exceed those found in rivers (GESAMP 1987). Non-conservative behaviour is caused by the large number of physical and chemical processes which affect the partitioning of trace metals between particulate and dissolved phases. Marked gradients in the master variables such as Cl content, particulate matter composition and redox conditions are the major controlling factors in the estuarine mixing zone, which render conservative trace metal behaviour a highly unlikely situation. Exchange reactions between solution and particulate matter may change with increasing salinity, causing removal or addition to the dissolved metal phase (Fig. 2.8.1), because both adsorption and especially desorption occur at rates which are significant with respect to characteristic estuarine and coastal water replacement times (Morris 1990). Whether or not non-conservative behaviour of any particular metal is observed will depend on the interaction between desorption kinetics (which are relatively slow), hydrodynamic residence time, and the chemical affinity for adsorption to the different types of particulate matter in the mixing zone. Non-linear geochemical models have been suggested to construct metal-salinity plots for the low salinity range (e.g. Windom and Smith 1985), but due to the considerable intra- and inter-estuarine variability and seasonal effects, the models cannot be used to compare systematically the trace metal concentrations in intermediate salinity mixing zones of widely differing characteristics. One general conclusion is therefore that for determining reference values at salinities above 30 linear mixing curves may be used, but that no such linear reference plots can be used for the mixing zone at a salinity range below 30 and, consequently, that no reference concentrations can be given for the low salinity range.

2.8.3.2 Particulate Metals

The load of suspended particulate matter (SPM) has been demonstrated to exert an important control both on the partitioning of a metal between dissolved and particulate phases (e.g. Balls 1988) and on particulate trace metal concentrations (Duinker 1983; Kersten et al. 1988, 1991a; Nolting and Eisma 1988). When plotting the particulate metal concentrations C_{PM} (on a mass basis) versus the SPM concentrations C_{SPM}, distinct curves are generally observed which fit a hyperbolic relationship (Fig. 2.8.3):

$$C_{PM} = A + \frac{B}{C_{SPM}}$$

where A and B are hyperbolic regression parameters. The hyperbolic relationship reflects the fact that the higher the SPM concentration in natural water is, the more the particulate metal concentration will tend to that of resuspended bottom sediment ($C_{PM} = A$). The y-axis intercept A may thus be interpreted as a measure of the bottom sediment composition. This is supported by the element/Al ratios in the SPM which are very similar to that of bottom sediment at high SPM concentrations (typically in the order of several mg/l). The resuspended fraction has higher contents of Al, As, Ca, Co, Cr, Fe, Mn, Pb, and V. The hyperbolic relationship for these elements is thus inverse with a negative regression coefficient B, which reflects the more or less significant negative correlation of these elements with total organic carbon (TOC). On the other hand, at low resuspended sediment values, the SPM composition is predominated by a "permanently suspended fraction" (Duinker 1983). The latter fraction has higher concentrations of TOC, Cd, Cu, Ni, Hg and Zn and tends to be predominated by organic matter in composition. The strong positive correlation between TOC and these elements might suggest their association with the organic carbon component of SPM. The regression coefficient B may thus be understood as a measure of the scavenging efficiency of organic matter, which is most important at low dilution by resuspended mineral particles. The slope of the metal vs. the inverse SPM concentration plots vary more or less seasonally due to the highly variable coefficient B which determines the slope in the low SPM concentration range (i.e. < 3 mg/l). It is recommended, however, not to use this approach for evaluation of phytoplankton blooms, for which this model might not be valid. Reference values for oceanic phytoplankton such as those reported by Collier and Edmond (1984) may be used to cover this case.

This non-linear regression model is supported by the results of several extended cruises covering the entire North Sea, including a range of low and high SPM concentrations at winter seasons (Kersten et al. 1988, 1991a; Nolting and Eisma 1988). For interpretation of particulate metal concentrations such metal vs. the inverse SPM concentration plots can be used to define a regional regression constant A and to compare it with background values from marine or fluviatile sediments in order to assess the degree of anthropogenic contamination. Clearly, this approach is only feasible if samples from an appropriate range of SPM concentrations have been derived. In case of high SPM concentrations in estuaries and rivers (i.e. > 3 mg/l), metal contents can be compared directly to the fluviatile sediment background concentrations. An example for fluviatile SPM background composition made by Zuurdeeg et al. (1992, see discussion of this paper in the previous section) is listed in Table 2.8.3.

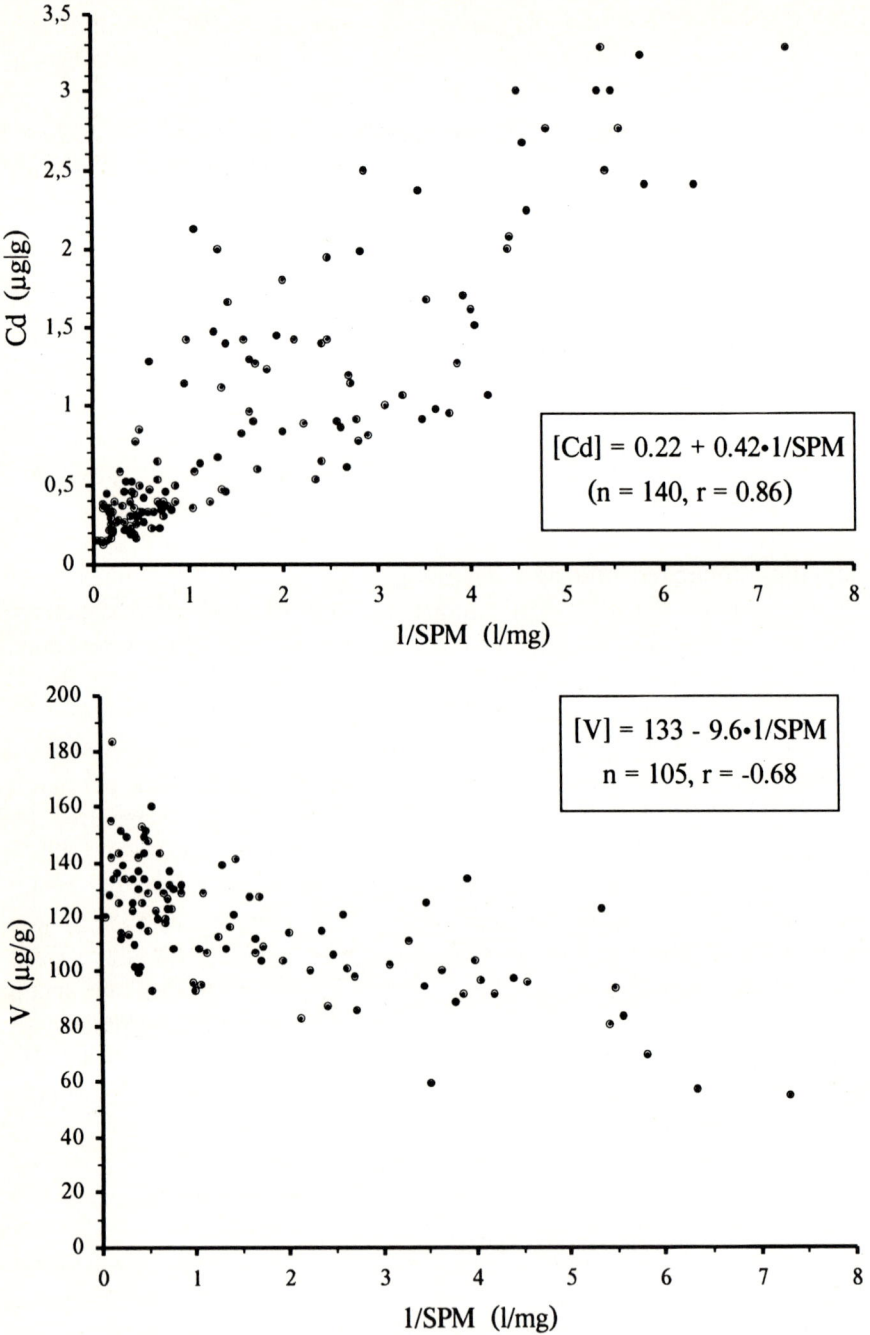

Fig. 2.8.3. Relationship of Cd and V (mass/mass basis) vs. the inverse suspended particulate matter (SPM) concentration (mass/volume basis) in surficial waters of the North Sea during winter season 1987. (Data from Kersten et al. 1991a)

Table 2.8.3. Assessment of the natural background composition of Rhine particulate matter with 20 wt. % organic matter (LOI), based on mediaeval river clay (sampled near Utrecht, The Netherlands)

Element	Si	Al	Fe	Ca	K	Mg	Na	S
Concentration (wt. %)	20.4	6.9	3.82	2.28	1.75	0.76	0.33	0.07
Element	Mn	Zn	Cr	Ni	Co	Cu	Pb	As
Concentration (μg/g)	1370	106	55	41	13	35	21	9
Element	Mo	Cd	Hg					
Concentration (ng/g)	1830	230	150					

2.8.4 Background Concentrations in the Atmosphere

2.8.4.1 Dry deposition

For any attempt to establish background concentrations for the atmosphere one has to recognize that it is an important pathway for trace metal pollution of the North Sea. The principal sources for most particulate trace metals in the marine atmosphere are low temperature crustal weathering (crustal source), sea-salt spreading (oceanic source) and a variety of mainly high temperature combustion processes (anthropogenic sources; natural sources such as volcanic activity and release from forest fires are of minor importance for the North Sea) (Chester and Murphy 1989). The most commonly used approach relating an element in atmospheric particulate matter (APM) to its source is by employing a source indicator element to calculate enrichment factors. For example, for the crustal APM Al is used as the indicator element and the average composition of crustal rocks is employed to calculate an enrichment factor EF analogous to that discussed in the sediment section. Trace metals in anthropogenic marine APM may have EF_{crust} values in the range > 10 to > 1000 (Fig. 2.8.4). This approach, however, cannot be readily used to calculate background values for air masses which originate from continental sources, mainly because APM from continental origin contains also varying amounts of non-lithogenic components such as ammonium sulphate and other salts. Other problems are (1) the varying particulate matter concentration in the marine atmosphere, (2) the fact that the average composition of crustal rocks does not necessarily reflect the composition of local source rocks, and (3) a sig-

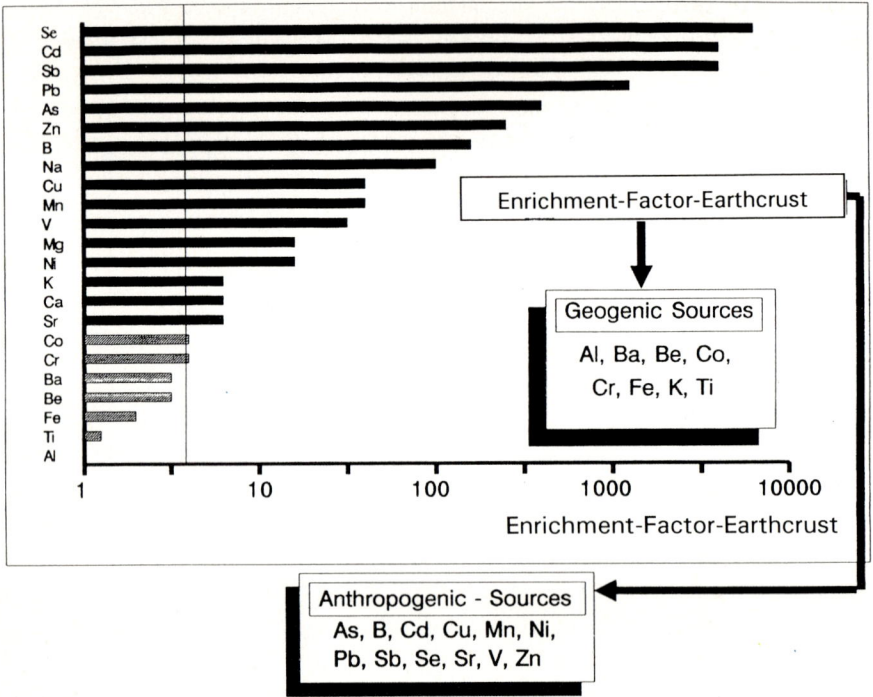

Fig. 2.8.4. Enrichment factors for the elements in a marine mixed aerosol sample collected at Helgoland. (Kersten et al. 1991b)

nificant anthropogenic input of Al by fly ash emission from combustion processes. A similar approach for air masses with APM of oceanic source could be considered using sodium as indicator element and the average composition of the Atlantic ocean, but this calculation is also not feasible at an acceptable accuracy, because trace metal concentrations at the air-water interface deviate significantly from those of the bulk sea water.

An alternative approach to determine the continental end-member APM concentration of trace metals is to use data from remote, undomesticated and otherwise pristine alpine areas, albeit there is no remote ecosystem in the northern hemisphere which is totally free of anthropogenic atmosphere pollution effects (Shirahata et al. 1980). Trace metal data from an APM sampling station on the Jungfraujoch mountain (Switzerland) are included in Table 2.8.4 as an example for this approach. Average dust concentrations at this station are quite low (3.1 µg/m³) and appropriate data for the North Sea have to be extrapolated for an approximately ten times higher average terrigenous dust concentration. For coastal sampling stations this approach is not feasible, because there is no land around the North Sea which is not to a certain degree industrialized (Pacyna et al. 1991).

Oceanic end-member concentration of the North Sea atmosphere can be similarly assessed by analyzing air masses which originate from remote and pristine marine areas. Because of its geographical location and temperate climate, the North Sea is affected by a variable wind regime. Despite this, it is possible to identify a number of end-member air masses on the basis of their wind direction and sources (Kriews et al. 1988; Chester and Bradshaw 1991; Kersten et al. 1992). The percentage frequency of the various surface wind directions is exemplified for the Helgoland sampling station in Table 2.8.5. In addition, the percentage contribution of the wind sectors to the total elemental composition of APM particles has been measured (Kriews et al. 1988). High concentrations of trace metals in airborne particulate matter were found in southern to eastern wind directions characterized by continental and anthropogenic air masses. As expected from the geographical distribution of the land masses around the North Sea, the lowest contribution to the total atmospheric trace metal concentrations were observed for the northwest wind sector, when the sampled air mass originates from off-shore regions. Trace metal concentrations in APM of air masses which end up in this wind direction were determined at Helgoland during the period of April 1987 to September 1988, thereby covering an entire seasonal cycle. The mean values of this sampling campaign are listed in the second column of Table 2.8.4. This example, however, should not imply that all air masses from the same wind direction have similar sources. It is the origin of the air masses themselves, and not the prevalent wind-direction at locations where they are sampled, which determines atmospheric element concentrations as discussed in detail by Chester and Bradshaw (1991). Sixty-hour back trajectories for near-bottom wind of representative samples indicate, however, that the air masses sampled from NW wind directions have indeed passed almost exclusively over the open North Sea during their transport to the German Bight. Gasoline Pb contributions to these samples could be excluded by $^{206}Pb/^{207}Pb$ isotope ratio analysis (Kersten et al. 1992). Chester and Bradshaw (1991) have found by a similar approach the oceanic end-member atmospheric concentrations listed in the third column of Table 2.8.4. Both data sets deviate by no more than a factor of 2, which encourages us to recommend these data as representative reference values for the oceanic end-member air masses.

Table 2.8.4. Reference values (mass/volume) of trace metals for the marine atmosphere of the North Sea

Element	1 (Jungfraujoch) ng/m^3	2 (NW-winds) ng/m^3	3 (Atlantic air masses) ng/m^3	4 (rainwater) μg/l
Al	51	110	90	0.79
As	0.23	0.26	-	-
Cd	0.5	0.1	-	5.2
Co	0.05	0.12	0.05	-
Cr	0.36	1.9	0.9	-
Cu	0.88	1.5	1.2	50
Fe	36	205	150	37
Mn	1.5	7.0	3.9	0.25
Na	-	6550	-	22,800
Ni	-	2.2	-	-
Pb	4.4	4.8	5.6	12
Sb	0.2	0.34	-	-
Se	-	0.33	-	-
Ti	2.4	17.6	-	-
V	0.29	1.6	-	-
Zn	9.9	16	8.1	366
SPM (μg/l)	3.1		-	-

Data for the four columns were derived as follows: (1) mean APM trace metal concentrations in air masses from continental sources represented by data from a remote undomesticated alpine area (Jungfraujoch, Switzerland; data from Dams and DeJonge 1976) at mean dust concentration of 3.1 μg/m^3; (2) mean trace metal concentrations in APM of marine source sampled from NW-winds (300° - 330° sampling sector) at Helgoland between April 1987 and September 1988 (Kriews 1992); (3) APM trace metal concentrations at four different offshore locations in the North Sea associated with air masses of Norwegian Sea origin (Chester and Bradshaw 1991); (4) dissolved metal concentrations in ambient North Sea rainwater at infinite precipitation volume (parameter A from Baeyens et al. 1990)

Table 2.8.5. Wind direction frequency distribution (1971-1980 and 1987-1989) and percentage contribution of source regions in different wind sectors related to long term average mean concentrations from integral aerosol samples (1986/1990). (Kriews et al. 1991)

	Percentage contribution of wind sectors to integral concentrations				Predicted	Measured mean 1986/90
Sector	1	2	3	4		
	30°-180° [%]	180°-240° [%]	240°-300° [%]	300°-330° [%]	(ng/m^3)	(ng/m^3)
Wind frequency	33.7	17.0	22.2	11.1		
As	79.4	11.5	7.6	1.4	1.7	1.7
Ca	57.0	14.6	16.9	11.5	483	400
Co	61.0	18.4	14.4	6.2	0.18	0.18
Fe	66.4	14.8	12.9	6.0	322	236
K	45.8	18.7	21.3	14.2	274	299
Mn	61.3	16.0	15.0	7.7	8.5	7.1
Na	21.0	27.2	32.0	19.7	3099	4108
Ni	54.4	17.9	19.6	8.1	2.5	2.6
Pb	62.3	22.7	12.5	2.5	18.0	20.2
Sb	56.3	19.9	20.1	3.7	0.86	0.96
Se	66.2	19.1	10.9	3.7	0.82	0.98
Sr	42.2	17.2	21.8	18.8	4.6	4.3
Ti	66.1	11.3	14.5	8.2	20.2	15.7
V	49.2	26.4	21.4	3.0	5.0	5.8
Zn	70.8	13.9	11.7	3.6	42.3	38.2
TSP $(\mu g/m^3)$	47.2	22.0	19.9	10.9	26.8	31.8

2.8.4.2 Wet Deposition

Wet deposition is also an important source of trace metals to the North Sea ecosystem. Baeyens et al. (1990) found by plotting the dissolved metal concentrations in rain versus the precipitation volumes a dilution profile which fits a hyperbolic relationship similar to that described for SPM composition:

$$C_{wet} = A + \frac{B}{P}$$

where C_{wet} is the dissolved metal concentration in rainwater, P the precipitation volume, and A,B hyperbolic regression parameters. This relationship reflects the fact that the higher the precipitation volume is, the more the dissolved metal concentration in rain water will tend to be initial or cloud vapour concentration ($C_{wet} = A$). A is a measure of the initial concentration of the dissolved metals in raindrops. The regression coefficient B may be interpreted as a measure of the scavenging efficiency, which means the additional amount of the metal which will be incorporated and solubilized in the rain drops during their fall. The maximum scavenging amount is reached immediately after rain has started and is important only at low precipitation volume. The dissolved metal concentration will then rapidly decrease with increasing precipitation volume. Average values for the regression constants A given by Baeyens et al. (1990) are used as an approximation of natural reference values for rainwater above the North Sea in Table 2.8.4. It should be stressed that this is the only data set we rely on, and that a better understanding of the atmospheric processes controlling the initial cloud vapour concentrations as well as the scavenging processes will allow us to deduce more refined, appropriate dilution parameters. Unless more reliable and direct data are available on natural trace metal concentrations in cloud vapour and rainwater, we recommend using this dataset, but with caution.

2.8.5 Background Concentrations in Mussels

There is some evidence that the general trend of decreasing trace metal concentrations from estuarine to coastal environments does not hold true for all benthic organisms. Borchardt et al. (1988) were able to show, after a normalization to the water depth, that there is a regional trend of increasing concentrations in indicator species such as the blue (common) mussel

(*Mytilus edulis*) from south to north and from east to west for Cd in the North Sea. This frequently observed anomalous pattern suggests that conditions governing contaminant uptake or release in mussels like food availability, hydrographical conditions, genetic differences and regional differences in bioavailablility of particulate metals may overshadow changes due to temporal trend or spatial trace metal gradients. This is corroborated by an extensive available data base and suggests that no natural background or reference values can be established at sufficient accuracy for organisms in the North Sea. It is recommended to put some effort in the near future to establishing local natural reference values. This concept was recently elaborated in Norway in an attempt to classify trace metal concentrations according to occurrence (excess concentrations in relation to "normal" concentrations; Knutsen and Skei 1992). Obviously, this concept cannot be very precise, but may yield the best compromise for lack of true natural background data (Table 2.8.6).

Table 2.8.6. Proposal for "presumed high background levels at diffuse loading", i.e. commonly found concentrations of metals in the blue mussel (*Mytilus edulis*) which cannot be traced to any known sources. Data for Norwegian coastal waters (Knutsen and Skei 1992)

Elements	Concentration (μg/g dry weight)
As	< 10
Cd	< 2
Cr	< 3
Cu	< 10
Pb	< 5
Hg	< 0.2
Ni	< 5
Ag	< 0.3
Zn	< 200

Another potential concept for obtaining reference data for mussels is the equilibrium partitioning method as a means to relate concentrations of trace pollutants in water and exposed organisms (Kooij et al. 1991). Although an equilibrium partitioning of trace metals is expected between water (C_W) and mussel tissue (C_M), expressed by the bioconcentration factor:

$$BCF = \frac{C_M}{C_W}$$

this does not imply that background tissue concentrations can readily be calculated from the proposed background C_W data. Seasonality of environmental factors such as food quality and availability render this approach too weak for an estimation of background reference values.

2.8.6 Summary and Recommendations

True natural background concentrations of trace metals, i.e. those prevailing in an ecosystem at zero-human activity, can nowadays only be found in sediment layers deposited in ancient times. Trace metal concentrations determined from these sediments in line with the JMG guidelines (JMG 1990; Loring and Rantala 1992) may be used as regional natural background values (Table 2.8.1). All other compartments of the North Sea environment (water, biota, atmosphere) are more or less polluted. Less polluted, near pristine conditions can be found in the Atlantic Ocean. Sea water and air masses which originate from the open ocean may be thus considered as a reference for the North Sea (Tabs. 2.8.2 and 2.8.4). No reference values can be given for the mixed zones with salinities below 30. For background concentrations of particulate trace metals in North Sea water, the sediment data may be used at SPM concentrations > 3 mg/l (Table 2.8.3). A non-linear geochemical model may be applied to tackle both the case of low SPM concentrations in sea water and for wet atmospheric deposition. Even less well defined are reference values for biota. The best compromise for the lack of natural background is to use normal concentrations at still pristine coastal stations, where the trace metal concentrations cannot be traced to any known point sources (Table 2.8.6).

Acknowledgements. Dr. R.W.P.M. Laane is thanked for his invitation and hospitality during the 1st International Workshop on Background Concentrations of Natural Compounds in the North Sea, held at Rijkswaterstaat, The Hague, 6-10 April 1992. Mrs. I Akkerman and Drs. J.H.J. Ebbing, A. Morris and B.W. Zuurdeeg provided helpful suggestions during the trace metal group discussions at that workshop.

References

Baeyens W, Dehairs F, Dedeurwaerder H (1990) Wet and dry deposition fluxes above the North Sea. Atmos Environ 24A:1693-1703

Baeyens W, Panutrakul S, Elskens M, Leermakers M, Navez J, Monteny F (1991) Geochemical processes in muddy and sandy tidal sediments. Geo Mar Lett 11:188-193

Balls PW (1987) Dispersion of dissolved trace metals from the Irish Sea into Scottish coastal waters. Continent Shelf Res 7:685-698

Balls PW (1988) The control of trace metal concentrations in coastal seawater through partition onto suspended particulate matter. Neth J Sea Res 22:213-218

Balls PW (1989) Trend monitoring of dissolved trace metals in coastal sea water - a waste of effort? Mar Pollut Bull 20:546-548

Bewers JM, Yeats PA (1989) Transport of river-derived trace metals trough the coastal zone. Neth J Sea Res 23:359-368

Borchardt T, Burchert S, Hablizel H, Karbe L, Zeitner R (1988) Trace metal concentrations in mussels: comparison between estuarine, coastal and offshore regions in the southeastern North Sea from 1983 to 1986. Mar Ecol Progr Ser 42:17-31

Boyle E, Collier R, Dengler AT, Edmond JM, Ng AC, Stallard RF (1974) On the chemical mass-balance in estuaries. Geochim Cosmochim Acta 38:1719-1728

Brügmann L, Danielsson LG, Magnusson, B Westerlund S (1985) Lead in the North Sea and the north east Atlantic Ocean. Mar Chem 16:47-60

Campbell PGC, Lewis AG, Chapman PM, Crowder AA, Fletcher WK, Imber B, Luoma SN, Stokes PM, Winfrey M (1988) Biologically available metals in sediments. NRCC Publ No 27694, Ottawa, Canada, 298 pp

Chester R, Murphy KJT (1990) Metals in the marine atmosphere. In: Furness R, Rainbow P (eds) Heavy metals in the marine environment. CRC, Boca Raton, pp 27-49

Chester R, Bradshaw GF (1991) Source control on the distribution of particulate trace metals in the North Sea atmosphere. Mar Pollut Bull 22:30-36

Collier R, Edmond J (1984) The trace element geochemistry of marine biogenic particulate matter. Prog Oceanogr 13:113-199

Cossa D, Gobeil C, Courau P (1988) Dissolved mercury behaviour in the Saint Lawrence estuary. Estuarine Coastal Shelf Sci 26:227-230

Dams R, DeJonge J (1976) Chemical composition of Swiss aerosols from the Jungfraujoch. Atmos Environ 10:1079-1084

Danielsson L, Magnusson B, Westerlund S, Zhang K (1983) Trace metals in the Gota river estuary. Estuarine Coastal Shelf Sci 17:73-85

Donze M (1990) Aquatic pollution and dredging in European community. Delwel, The Hague, 184 pp

Duinker JC (1983) Effects of particle size and density on the transport of metals to the oceans. In: Wong CS, Boyle E, Bruland KW, Burton JD, Goldberg ED (eds) Trace metals in sea water. Plenum, New York, pp 209-226

Duinker JC, Hillebrand MIJ, Nolting RF, Wellershaus S (1982) The river Elbe and Weser: processes affecting the behaviour of metals and organochlorines during estuarine mixing. Neth J Sea Res 15:141-195

Fileman C, Harper D (1989) Depth profiles of Cd, Cu, Pb and Hg in seawater at the ICES reference station, 60° 30' N, 5° 00' W. ICES CC 1989/E: 14, Copenhagen, 16 pp

Flegal AR, Patterson CC (1985) Thallium concentrations in seawater. Mar Chem 15:327-331

Fowler SW (1990) Critical review of selected heavy metals and chlorinated hydrocarbon concentrations in the marine environment. Mar Environ Res 29:1-64

GESAMP (1987) Land-sea boundary flux of pollutants. United Nations Joint Group of Experts on the Scientific Aspects of Marine Pollution (GESAMP). Rep Stud Ser 32, UNESCO, Paris, 172 pp

Gray JS (1990) Statistics and the precautionary principle. Mar Poll Bull 21:174-176

Honeyman BD, Santschi PH (1988) Metals in aquatic systems. Environ Sci Technol 22:862-871

Huerta-Diaz MA, Morse JW (1992) Pyritization of trace metals in anoxic marine sediments. Geochim Cosmochim Acta 56:2681-2702

Hydes DJ, Kremling K (1993) Patchiness in dissolved metals in North Sea surface waters: seasonal differences and influence of suspended sediment. Continent Shelf Res (in press)

ICES (1991) A review of measurements of trace metals in coastal and shelf sea water samples collected by ICES and JMG laboratories during 1985-1987. ICES, Copenhagen

JMG (1990) Guidelines for the sampling and analysis of sediments under the Joint Monitoring Programme. In: Oslo and Paris Commissions/Principles and Methodology of the Joint Monitoring Programme. Monitoring Manual A13.1/90-E, 8/5/2

Jonkers DA, Everts JW (1992) SEAWORTHY - derivation of micropollutant risk levels for the North Sea and Wadden Sea. Rep to the Netherlands Ministry of Housing, Physical Planning and Environment (VROM-DGM), The Hague, 74 pp

Kersten M, Förstner U (1989) Speciation of trace elements in sediments. In: Batley G (ed) Trace elements speciation: analytical methods and problems. CRC, Boca Raton, pp 245-317

Kersten M, Förstner U (1991) Geochemical characterization of pollutant mobility in cohesive sediments. Geo Mar Lett 11:184-187

Kersten M, Dicke M, Kriews M, Naumann K, Schmidt D, Schulz M, Schwikowski M, Steiger M (1988) Distribution and fate of heavy metals in the North Sea. In: Salomons W, Bayne BL, Duursma EK, Förstner U (eds) (1988) Pollution of the North Sea - an Assessment. Springer, Berlin Heidelberg New York, pp 300-347

Kersten M, Irion G, Förstner U (1991a) Particulate trace metals in surface waters of the North Sea. In: Vernet JP (ed) Heavy metals in the environment, vol 1. Elsevier, Amsterdam, pp 137-159

Kersten M, Kriews M, Förstner U (1991b) Partitioning of trace metals released from polluted marine aerosols in coastal seawater. Mar Chem 36:165-182

Kersten M, Förstner U, Krause P, Kriews M, Dannecker W, Garbe-Schönberg CD, Höck M, Terzenbach U, Grassl H (1992) Pollution source reconnaissance using stable lead isotopes (^{206}Pb/^{207}Pb). In: Vernet J-P (ed) Impact of heavy metals in the environment, Elsevier, Amsterdam, pp 311-325

Knutsen J, Skei J (1992) Preliminary proposals for classification of marine environmental quality respecting micropullutants in water, sediments and selected organisms. Norwegian Institute for Water Research, Rep O-862602/O-89266, Oslo, 22 pp

Kramer KJM, Misdorp R, Berger G, Duijts R (1991a) Maximum pollutant conentrations at the wrong depth: a misleading pollution history in a sediment core. Mar Chem 36:183-198

Kramer KJM, Scholten MCTh, van der Wal JT, van der Vlies EM (1991b) De Verspreiding van Spoormetalen in de Noordzee II. Statistisch onderzoek naar temporele en geografische trends 1980-1990. TNO-Rapp Nr R 91/358, Delft, 105 pp

Kremling K (1983) Trace metal fronts in European shelf waters. Nature 303:225-227

Kriews M (1992) Charakterisierung mariner Aerosole in der Deutschen Bucht sowie Prozeßstudien zum Verhalten von Spurenmetallen beim Übergang Atmosphäre-Meerwasser. Ph D Thesis, Univ Hamburg

Kriews M, Naumann K, Dannecker W (1988) Aerosol specification in the southern North Sea region by wind dependent sampling and multielement analysis. J Aerosol Sci 19:1051-1054

Kriews M, Bergmann J, Naumann K, Dannecker W (1991) Charakterisierung des Helgoländer Aerosols durch gezielte Probennahme und atomspektrometrische Multielementanalytik. In: Welz B (ed) 6. Colloquium Atomspektrische Spurenanalytik. Perkin-Elmer, Überlingen, pp 725-740

Krumgalz BS, Fainshtein G (1989) Trace metal contents in certified reference sediments determined by nitric acid digestion and atomic absorption spectrometry. Anal Chim Acta 218:335-340

Laane RWPM (1992) Background concentrations of natural compounds in rivers, sea water, atmosphere and mussels. Rep to the Netherlands Ministry of Transport, Public Works and Water Management (DGW-92.033), The Hague, 84 pp

Larsen B, Jensen A (1989) Evaluation of the sensitivity of sediment stations in pollution monitoring. Mar Pollut Bull 20:556-560

Loring DH, Rantala RTT (1992) Manual for the geochemical analysis of marine sediments and suspended particulate matter. Earth-Sci Rev 32:235-283

Lutter S (1990) Comment on: statistics and precautionary principle. Mar Pollut Bull 21:547-548

Martin JM, Whitfield M (1983) The significance of the river input of chemical elements to the ocean. In: Wong CS, Boyle E, Bruland KW, Burton JD, Goldberg ED (eds) Trace metals in sea water. Plenum, New York, pp 265-296

Middelburg JJ, Hoede D, van der Sloot HA, van der Weijden CH, Wijkstra J (1988) Arsenic, antimony and vanadium in the North Atlantic Ocean. Geochim Cosmochim Acta 52:2871-2878

Middleton D, Grant A (1990) Heavy metals in the Humber estuary: *Scrobicularia* clay as a pre-industrial datum. Proc Yorkshire Geol Soc 48:75-80.

Morris AW (1990) Kinetic and equilibrium approaches to estuarine chemistry. Sci Total Environ 97/98:253-266

Nolting RF, Eisma D (1988) Elementary composition of suspended particulate matter in the North Sea. Neth J Sea Res 22:219-236

Pacyna JM, Münch J, Axenfeld F (1991) European inventory of trace metal emissions to the atmosphere. In: Vernet JP (ed) Heavy metals in the environment, vol 1. Elsevier, Amsterdam, pp 1-20

Shirahata H, Elias RW, Patterson CC (1980) Chronological variations in concentrations and isotopic compositions of anthropogenic atmospheric lead in sediments of a remote subalpine pond. Geochim Cosmochim Acta 44:149-162

van der Kooij LA, van de Meent D, van Leeuwen CJ, Bruggeman WA (1991) Deriving quality criteria for water and sediment from the results of aquatic toxicity tests and product standards: application of the equilibrium partitioning method. Wat Res 25:697-705

Windom HL, Smith RG Jr (1985) Factors influencing the concentration and distribution of trace metals in the South Atlantic Bight. In: Atkinson LP, Menzel DW, Bush KA (eds) Oceanography of the southeastern U.S. Continental Shelf. AGU, Washington, DC, pp 141-152

Windom HL, Schropp SJ, Calder FD, Ryan JD, Smith RG Jr, Burney LC, Lewis FG, Rawlinson CH (1989) Natural trace metal concentrations in estuarine and coastal marine sediments of the southeastern United States. Environ Sci Technol 23:314-320

Yeandel G, Minster JF (1987) Cromium behavior in the ocean: global versus regional processes. Global Biogeochem Cycles 2:131-154

Zuurdeeg BW, van Enk RJ, Vriend SP (1992) Natuurlijke achtergrondgehalten van zware metalen en enkele andere sporenelementen in Nederlands oppervlaktewater. GEOCHEM Research Rep by commission of VROM, The Hague (unpublished)

3   Model Experiments

3.1   Mean and Local Transport in Air

K.H. Schlünzen and U. Krell

Atmospheric input can make up more than half of the integral input of contaminants into the ocean (see Chap. 1). The input data are an essential parameter for modelling of tracer transports in the ocean (Chap. 3.2). The objective of this section is to resolve the atmospheric input of trace substances into the North Sea in space and time and to discuss local features.

Up to now, most of the input figures refer to the mean yearly input over the whole North Sea area (see, e.g. Table 1.1 of Chap. 1). Temporal or regional variations are not specified. This aim is achieved by use of two different atmospheric pollution transport models. A large-scale model is used for the calculation of monthly and yearly input values for selected airborne trace substances. A mesoscale model is used for the calculation of hourly and daily input values and for case studies to determine extreme input situations.

3.1.1   Models for Calculation

For calculation of the atmospheric input into the North Sea, the main factors influencing the transport of trace substances in the atmosphere and tracer deposition at the ground have to be taken into account in the models used. The factors can be roughly separated into two groups: (1) meteorological factors, and (2) tracer factors. Figure 3.1.1 gives a schematic view of all factors which are important for the calculation of the atmospheric input.

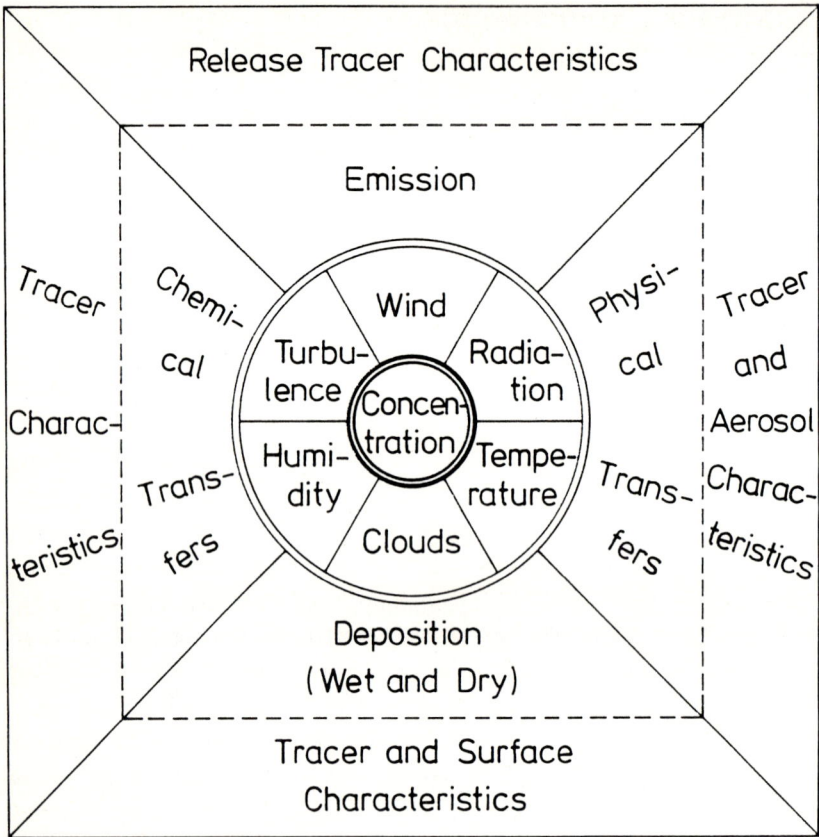

Fig. 3.1.1. Schematic view of important factors influencing the transport of trace substances and their deposition

For determination of the atmospheric input the values for dry and wet deposition are summed up. Naturally, the deposition strongly depends on the concentration of the considered trace substance. The concentration itself is influenced by meteorological parameters, e.g. the wind, temperature, humidity and turbulence fields, the radiation budget, cloud formation and precipitation. The concentration is also changed by dry and wet deposition, and by chemical and physical transformations. The mean concentration level is mainly influenced by the source strength. Further emittant characteristics, e.g. the emission temperature and the physico-chemical state of the contaminant at the source, characterize the concentration pattern in the first few tens of kilometres near the source (e.g. Bigalke 1991, 1992).

The above-mentioned processes, namely emission, chemical and physical transformations, dry and wet deposition, depend on the contaminant characteristics (arranged in the outer rectangle in Fig. 3.1.1) and, again, on meteorological parameters (arranged in the outer circle in Fig. 3.1.1). Dry deposition, for instance, is controlled by tracer characteristics (e.g. particle size and shape, solubility in water) and surface characteristics (e.g. stomata resistance, wetness of the ground, absorbability of the surface for the specified tracer) (Hosker and Lindberg 1982). The wet deposition mainly depends on the precipitation rate but also on the type of cloud and the solubility of the tracer in water. For tracers which are bound to particles, e.g. Cd and Pb, both deposition processes strongly depend on the particle size.

The mentioned processes are interdependent and influence the total deposition of tracers into the North Sea. The models used in this chapter simulate most of the processes influencing the dry and wet deposition directly or in a parameterized way. Chemical reactions have been neglected up to now since they are of secondary importance for the considered trace elements. Most of the anthropogenic trace metals have a residence time in air of more than a week (Wiman et al. 1990), because they are bound to aerosols of the accumulation mode corresponding to a particle size of 0.1 to 1.0 µm. Therefore, the calculation of mean monthly input values requires a model area of a few thousand square kilometres.

The large-scale model used for this purpose can determine the yearly and monthly mean input values only with a large horizontal resolution. It cannot describe in detail local transport and deposition processes. For that a mesoscale model can be used which has a horizontal extension of a few hundred kilometres. With this kind of model, daily input values can be calculated and case studies performed.

3.1.1.1 Large-Scale Model (HHLRT)

For calculating the monthly-mean atmospheric deposition of lead into the North Sea, a three-dimensional stochastic trajectory model based on the Monte-Carlo technique is used (Krell and Roeckner 1988). A reasonably good agreement between calculated and measured monthly mean lead concentrations in surface air and wet deposition has been shown by the authors. The model HHLRT (Hamburg Long Range Transport) has also been successfully applied to the simulation of the long-range transport of atmospheric sulphur in an episodic and monthly time scale (Lehmhaus and Roeckner 1986).

The dispersion of a trace substance is simulated directly by releasing a large number of particles from a specified emission source. The three-dimensional trajectories of these particles are independently calculated from mean wind fields in five layers (Pettersen 1956). Transport due to horizontal turbulent diffusion is neglected compared to transport due to the mean wind. Atmospheric stability is considered in the parameterization of the vertical turbulent diffusion using a stability-dependent constant exchange coefficient within the planetary boundary layer, as well as in the parameterization of the dry deposition at the top of the constant-flux layer. A constant empirical dry deposition velocity of 0.2 cm/s is used at the surface. The wet deposition is parameterized simply in terms of the precipitation rate, the height of the planetary boundary layer, and a constant scavenging factor of 500,000. Using those two constant values for deposition modelling, we take into account only the major size range of lead loaded aerosols between 0.2 and 1.0 µm (Milford and Davidson 1985).

At the top of the planetary boundary layer, the height of which may change with time, total reflection of the particles is assumed. The change of the planetary boundary layer height is the only mechanism acting to transfer particles between the planetary boundary layer and the free atmosphere, where particles are transported with the mean wind only and no turbulent diffusion takes place. Chemical reactions, sedimentation and transformations of particle size have been neglected for simplicity.

The spatial resolution of the model depends on the resolution of the input data - here 1.5° in longitudinal and meridional directions. The meteorological data (e.g. wind and temperature data, surface heat flux, friction velocity, precipitation rate) is obtained from the circulation model of the European Centre for Medium Range Weather Forecast (ECMWF, Reading). The precipitation rate is one of the most important and critical parameters in this model simulation. The values used were compared with an estimate based on multi-year measurements (RSU 1980). For the 4 months under consideration there is good agreement between the estimated (57 mm/month) and the calculated (64 mm/month) monthly mean of the precipitation intensity. The height of the planetary boundary layer is the result of an objective analysis of radiosonde data by the Norwegian Weather Institute (Oslo).

The yearly mean emission of lead is given by Pacyna (1985) (see Fig. 3.1.2). The values are estimates for 1980, and in this first approach we assumed they are valid for 1984, to which the simulations refer, neglecting any temporal variation in emission rates. All model input data had to be interpolated to the model timestep of 2 h.

Pb [t/a]

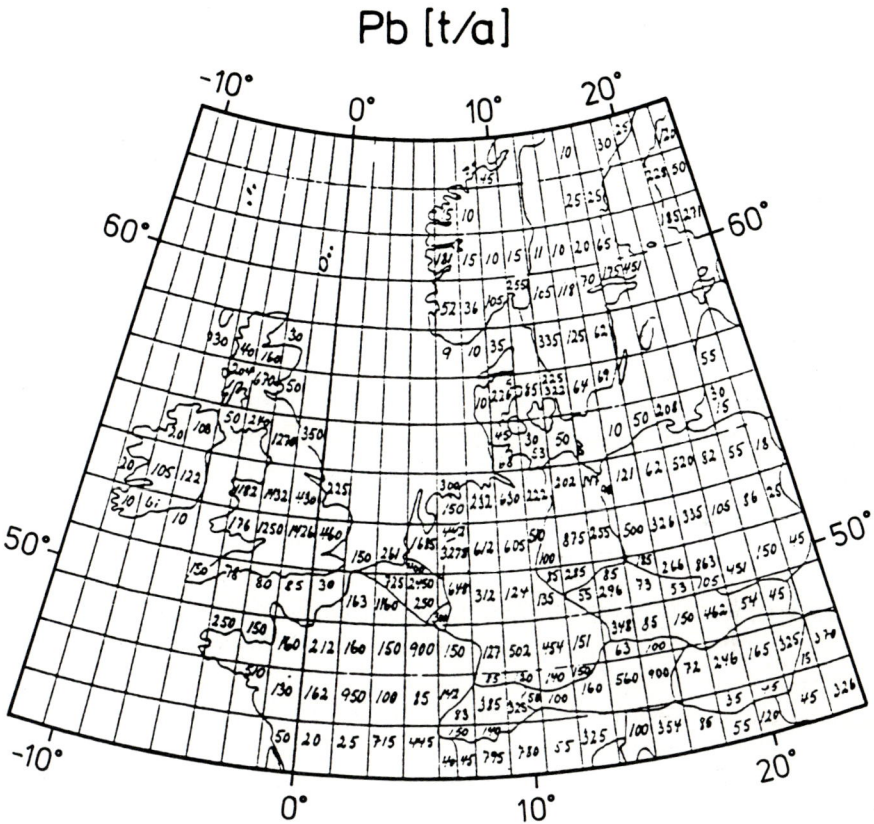

Fig. 3.1.2. Spatial distribution of lead emission (t/a) in Europe for the year 1980 in the 1.5° HHLRT grid net. (Pacyna 1985)

3.1.1.2 Mesoscale Model (METRAS)

For determination of extreme input situations and to study the local effects of the coastline on the atmospheric input into the North Sea, the three-dimensional, non-hydrostatic prognostic mesoscale model METRAS (Mesoscale Transport and Stream) is used (Schlünzen 1988). It was applied to simulate coastal effects (Schlünzen 1990, Schlünzen and Pahl 1992) as well as to study the dependence of the dry deposition of SO_2 on surface characteristics (Pahl and Schlünzen 1990).

The wind, temperature, water vapour, and cloud/rain water fields used in the mesoscale model as well as the concentrations of trace substances are calculated from prognostic equations, whereas pressure and density are determined by diagnostic equations. Only the anelastic approximation is applied in the model, the hydrostatic assumption is not used. With these characteristics, the model is applicable to a wide range of mesoscale phenomena with typical horizontal extensions from a few to more than a hundred kilometres (see, e.g. Wippermann 1981). The influences of clouds, urban heat islands, gravity waves, topographically induced effects, land-sea breezes and other mesoscale phenomena (e.g. Orlanski 1975) can be directly simulated in the model calculations. Their influences do not have to be parameterized.

To reduce the computational effort but retain high resolution in interesting areas (e.g. at a coastline) a non-uniform grid is used in horizontal and vertical directions. Terrain-following coordinates simplify the formulation of surface boundary conditions. All subgrid-scale turbulent processes are parameterized utilizing a first order closure hypothesis with the exchange coefficient determined according to the method of Dunst (1982), with some modifications. The validity of surface layer similarity theory is assumed for the calculation of surface fluxes below a height of 10 m. Cloud microphysical processes are included by use of a modified Kessler-Scheme (Kessler 1969). The dry deposition is modelled dependent on the concentration of the trace substance and the deposition velocity which itself is calculated following the resistance model concept (for details on the calculation, see Pahl and Schlünzen 1990; Schlünzen and Pahl 1992). Both the concentration and the deposition velocity of the trace substance are controlled by the atmospheric stability and the wind field, which are strongly affected by terrain-forced local circulation systems.

The forecast time of the mesoscale model is restricted to a few days at maximum for input scenarios using emission data comparable to Fig. 3.1.2 and an outlined model area of a few hundred square kilometres. Tracer transport through the lateral boundaries of the model has to be taken into account for longer integration times. In the present model calculations, mainly the influence of different large-scale and mesoscale situations on the local trace element transport at the coastline were studied and idealized sources close to the coast were assumed.

For initialization of the mesoscale model, the large-scale meteorological situation was prescribed stationary and horizontally homogeneous. For different large-scale pressure gradients and correspondingly different geostrophic winds within a variable atmospheric stratification, several case studies were performed. They all focus on the effects of topography, roughness changes, and surface temperature differences on tracer transport.

The main characteristics and differences of the large- and the mesoscale models are summarized in Table 3.1.1. The large-scale model is used for calculation of mean transport in air, whereas local transport is calculated by use of the mesoscale model.

Table 3.1.1. Characteristics of a typical large-scale transport model and a mesoscale transport and stream model

Model characteristics	Large-scale model (e.g. HHLRT)	Mesoscale model (e.g. METRAS)
Horizontal resolution	$1.5° \times 1.5°$	1×1 km^2 → 10×10 km^2
Vertical resolution of the planetary boundary layer	Mean wind in two layers, particle transport continuous	14 to 20 levels, boundary layer vertically resolved
Temporal resolution	Monthly values (→ yearly mean input values)	Hourly values (→ daily mean input values)
Meteorology	Diagnostically calculated from synoptic data	Prognostically calculated
Dry deposition	Simple approach with deposition depending on atmospheric stability and height	Dry deposition model (calculated deposition dependent on atmospheric stability, tracer and surface characteristics)
Wet deposition	Scavenging factor, precipitation rate from synoptic data	Scavenging factor, precipitation rate calculated in the model (parameterized cloud micro-physics)
Deposition close to source	Estimated as constant part of emission	Explicit calculation with plume rise model

3.1.2 Model Results for the Mean Transport in Air

The simulations with the large-scale model HHLRT had to be restricted to 4 months, since the Monte-Carlo technique used consumes a large amount of computer time. Considering possible seasonal changes in the atmospheric deposition over the North Sea, the months January, April, July and October 1984 were selected.

The simulated spatial distribution of the lead deposition over the North Sea does not show characteristic seasonal changes (see Fig. 3.1.3). A gen-

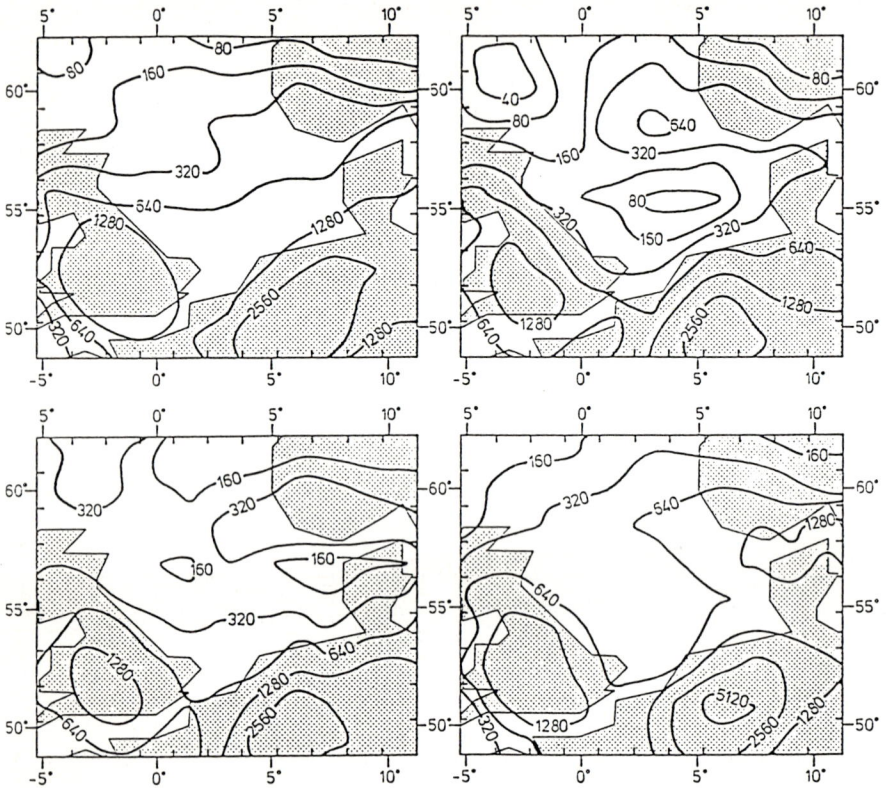

Fig. 3.1.3. Total deposition of lead (μg Pb/m^2) for January (*top, left*), April (*top, right*), July (*bottom, left*), and October (*bottom, right*) calculated with the model HHLRT, using meteorological data for 1984 and emission data for 1980. Land areas overlayed with *dots*

eral pattern for the lead deposition can be recognized though the input differs from month to month. Every input pattern shows two areas with absolute maxima of Pb deposition, one in the south and southwest of the North Sea which corresponds to the main emission areas (see Fig. 3.1.2), and the other in the northeast which corresponds to high precipitation areas (see below). The deposition decreases from south to north and in particular over the southern North Sea. The input values over the northern region reach only one half or one quarter of the values calculated over the southern part.

The highest lead depositions are calculated for the southern North Sea close to the main emission areas (see Fig. 3.1.3). In most of the large-scale models, e.g. in model HHLRT as well as in model EMEP (e.g. Graßl et al. 1989; Petersen et al. 1989), the deposition close to the sources is parameterized in a very simple way (see Table 3.1.1). In the EMEP simulations, the deposition close to sources and restricted to coastal regions makes up to 20% of the integral atmospheric input into the North Sea (Krell 1991). To obtain more detailed information on the temporal and spatial patterns of coastal deposition, the values should be localized by use of mesoscale models (see Sect. 3.1.3).

The general pattern of lead deposition described (Fig. 3.1.3) is mainly caused by the distribution of the Pb emission (Fig. 3.1.2). Motor traffic is the major source for atmospheric lead in Europe, but in the two main emission areas smelters for pyrometallurgical non-ferrous metal production are the main sources (Pacyna 1983). No lead emission is given over the North Sea itself and resuspension from the water surface is neglectable for Pb.

Other trace elements have emission distributions different from lead, and therefore their input values might not fit the general deposition pattern described. For example, the North Sea itself is a considerable source for anthropogenic mercury, since resuspension takes place (Lindquist and Schroeder 1989). Trace elements with emission distributions similar to Pb will have calculated deposition patterns similar to the ones for lead. The cadmium emission data, for instance, resemble lead emission data but the emission rates are reduced by about a factor of a hundred or even more. Consequently, Cd deposition is roughly a factor of 100 lower than the Pb deposition but the distributions are qualitatively the same (Krell and Roeckner 1988).

Variations in the monthly deposition of lead (Fig. 3.1.3) are mainly caused by differences in the precipitation rate. In this regard, the North Sea can be divided into three regions: south, central and northeast. Over the southern North Sea, the air concentration of Pb is always high because the main emission regions are nearby. Thus, enough lead is available and large precipitation amounts always induce a large wet deposition whereas small precipitation rates might reduce the deposition. In any case, the concentration of lead in air is always high enough over the southern North Sea to result in considerable input values.

In northern regions of the North Sea the lead concentration is much lower than in the south. Therefore, even large precipitation rates do not always result in large wet deposition values. When highly polluted air masses are transported to regions with large precipitation, regional deposition maxima can occur (see Fig. 3.1.3, top right). Due to orographic effects, large precipitation amounts often occur over the northeastern North Sea, southwest of the Norwegian coast (e.g. Grønås 1989).

The precipitation over the central North Sea seems to be lower than in other regions most of the time (Barrett et al. 1991). Consequently, the wet deposition in this area is low too. In some months regional minima of total deposition can even be distinguished, e.g. in April and October 1984 (Fig. 3.1.3).

The integral total deposition over the North Sea area gives the calculated monthly atmospheric lead input (see Table 3.1.2). The values depend on the distribution of the monthly precipitation. The high input values in January and October 1984 are mainly a response to the effectiveness of high precipitation amounts. In January, the high deposition values over the southern North Sea lead to the high input (Fig. 3.1.3), whereas in October the high wet deposition over the northeastern North Sea causes the high values for the integrated deposition.

Table 3.1.2. HHLRT results for the monthly atmospheric input of lead for the North Sea area

	January 1984	April 1984	July 1984	October 1984	Mean 1984
Atmospheric input of Pb (t/month)	288	133	158	309	222
Precipitation (mm/month)	108	32	23	94	64
Proportion of wet to total deposition (%)	91	81	79	87	85

The importance of the precipitation and, consequently, of the wet deposition for the atmospheric input is documented by the large proportion of wet to total monthly deposition in Table 3.1.2. Wet deposition is by far the most effective process for deposition of lead over the North Sea. The proportion of wet to total deposition increases with increasing distance from the main emission areas. In the northern North Sea the wet deposition is a factor 4 to 8 higher than the dry deposition. All model simulations (van Jaarsveld et al. 1986; Krell und Roeckner 1988; Petersen et al. 1989) agree that wet deposition is the most important factor for the atmospheric input of Pb into the North Sea (for details see Krell 1991). The proportion of both deposition processes may be of the same order for other ocean areas, as shown by Asman and Runge (1991) for nitrogen input to the Baltic Sea. On land close to the sources or at the coast, dry deposition might exceed wet deposition, as measurements carried out by Michaelis (1988) and Stößel (1987) on the island Pellworm pointed out.

The calculated atmospheric input of lead varies strongly between the 4 months examined in 1984. The input is low during spring and summer and high during autumn and winter. Such seasonal changes are well known for the SO_2 air concentration and deposition (e.g. Lehmhaus and Roeckner 1986). Seasonal changes in the SO_2 deposition are caused by extreme meteorological situations, e.g. vertically limited air exchange in winter with smog. Contrary to this, the apparent seasonal changes in the calculated atmospheric input of lead into the North Sea are due to the different precipitation distributions. Multi-year simulations with the large-scale model EMEP (Petersen et al. 1989) found that the seasonal changes are typical for the year 1984 but are not always observed for other years (1980, 1985) since the precipitation does not always follow a seasonal trend. At present, no seasonal trend can be postulated for lead.

For the monthly atmospheric input of lead the large-scale model HHLRT gives a range of 100 to 300 tons per month and the model EMEP based on 3 years' simulations a range of 100 to 500 tons. The annual atmospheric input varies less (Petersen et al. 1989). The extrapolation of the mean monthly input of lead leads to an annual atmospheric input into the North Sea of 2664 tons Pb based on the model HHLRT (Table 3.1.3). This estimate is just a little lower than the one based on EMEP (Petersen et al. 1989) and a little higher than the estimate from van Jaarsveld et al. (1986), calculated with the TREND model. The input values given by Krell and Roeckner (1988) were based on the HHLRT model results for 2 months only. Due to the distinct influence of precipitation amount on deposition, the data base from 2 months has been too small.

The estimates based on the different model results are compiled in Table 3.1.3. The values are pretty close together and give a range for the atmospheric input into the North Sea of 2500 to 3000 tons of lead per year. The agreement of the different model estimates is remarkably good, because large differences exist concerning the simulation periods and input data used by the models as well as the theoretical approaches of the different models. For example, the estimates with TREND (van Jaarsveld et al. 1986) used meteorological data over 20 years from one meteorological station in the Netherlands. In contrast, the input values in this paper are based on 4 months only but take into account regional changes.

The model estimates only differ a little, particularly in comparison to the estimates based on measurements (Table 3.1.3). These estimate the atmospheric input from concentration in air and precipitation, measured particularly at the western and southern coasts, and from the annual precipitation rate assumed over the North Sea; see Chap. 2.4 on the uncertainties in the calculation of the integrated input values from local measurements.

The estimated input based on measurements is normally higher than the input calculated by models. This can be explained with the aid of the model results (see Fig. 3.1.3). Concentrations measured near the main emission areas are much higher than over the North Sea. Therefore, their use must lead to an overestimation of the atmospheric input. Petersen et al. (1989) showed this using model calculated concentrations at the coast for the estimation of atmospheric input, comparing this value with the whole model calculated input.

Table 3.1.3. Estimates of the yearly atmospheric input of lead for the North Sea area

Atmospheric input (t Pb/a)	Author	Method
2664	This study	Model
2878	Petersen et al. 1989	Model
1531[a]	Krell and Roeckner 1988	Model
2600	van Jaarsveld et al. 1986	Model
3200-3500	Chapter 2.4	Measurements
2310-4830	INC 1990	
2400	Otten et al. 1989	Measurements/aircraft
1800-6400	Stößel 1987	Measurements
3600-13000	van Aalst et al. 1983	Measurements
8100	Cambray et al. 1979	Measurements
5800	Cambray et al. 1975	Measurements

[a]Calculation based on 2 months only.

For the input estimates based on measurements, it is a problem to get concentrations which are representative for the whole North Sea area (see Chap. 2.4 on this topic); but even if this were possible the question on the annual precipitation over the North Sea would not be answered. For the input estimates based on large-scale model results, the main problem is to obtain reliable emission and precipitation data.

3.1.3   Model Results for the Local Transport in Air

The problem in obtaining reliable emission data exists not only for large-scale but also for mesoscale models. In these models, the demands on spatial and temporal resolution of the data are even higher. However, to study the modification of the large-scale input situation due to local and regional scale meteorological phenomena in general, an idealized source distribution is at first sufficient.

The main parameters influencing tracer transport at the coastline are changes in roughness length, horizontal temperature gradients, developing clouds, the prevailing large-scale situation and topographic influences. The latter can be very important for the precipitation close to the Scottish and especially the Norwegian coasts (e.g. RSU 1980) but are of less importance in the southern part of the North Sea, where the highest inputs are calculated (see Fig. 3.1.3). The other parameters mentioned are important at the whole North Sea coast. The investigations performed with the mesoscale model focus on them and aim at the determination of the main input situations for coastal waters. Although a definite conclusion cannot be drawn from the present case studies, a first attempt can be made.

In the simulated area, mudflats lie between land and water (Wadden Sea). This is characteristic for the German as well as for the Dutch and Danish coastlines. At 54.5° N the Wadden Sea has a typical width of 20 km at low tide and the coastline is orientated approximately north-south. Thus, in idealized model calculations as presented here, a laterally homogeneous coastline is assumed for simplicity. In realistic case studies a three-dimensional topography has to be used (e.g. Schlünzen et al. 1992). In Fig. 3.1.4 the simulated domain, the surface characteristics and the non-uniform grid structure are schematically outlined.

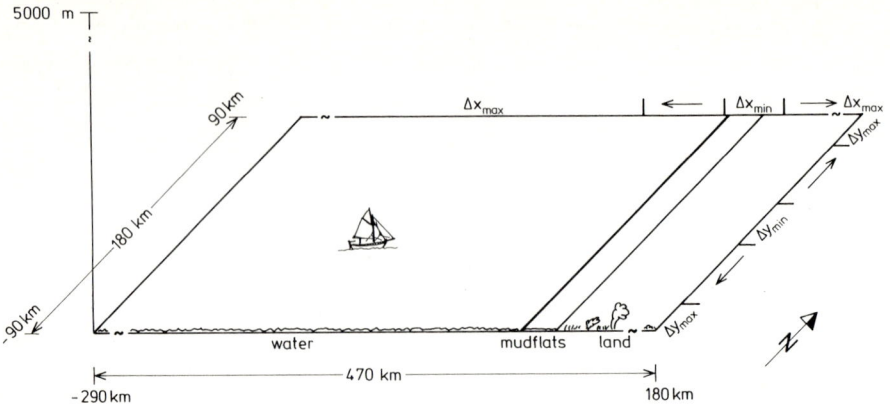

Fig. 3.1.4. Simulated domain, surface characteristics and horizontal grid size. The grid size changes horizontally from $\Delta x_{min} = \Delta y_{min} = 1$ km, over the mudflats and the centre of the model area, in longitudinal and lateral direction to $\Delta x_{max} = \Delta y_{max} = 10$ km over land and water; vertically close to the surface from $\Delta z_{min} = 20$ m (for $z \leq 40$ m) to $\Delta z_{max} = 500$ m (for $z \geq 2500$ m) (Schlünzen and Pahl 1992)

3.1.3.1 Mesoscale Circulation Systems for Large-Scale Offshore Wind

With the aim of determining the main input situations for coastal waters, only large-scale offshore wind situations are focused on. The differences between results from the large-scale model and the higher resolving mesoscale model are pointed out first.

No Surface Temperature Gradients. A situation corresponding to the large-scale model is given, if horizontal temperature gradients are neglected, and only the roughness step at the coastline is included in the model simulation. The influence of the roughness step on the developing local wind field is quite small as can be seen from Fig. 3.1.5. Naturally, the wind velocity close to the ground is lower over land compared to values at the same height over water. The roughness change at the coastline does not cause local circulation systems and thus the tracer transport is relatively undisturbed at the coastline and mainly governed by the large-scale situation. However, the deposition velocity and, accordingly, the dry deposition may be much higher over water than over land, as will be discussed later (see Fig. 3.1.10).

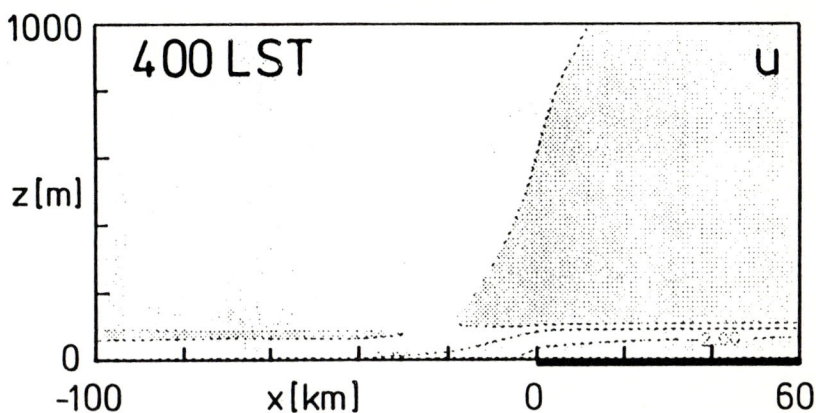

Fig. 3.1.5. Vertical cross-section of the simulated domain for the east-west wind component (u) at 04.00 LST (Local Standard Time). Stably stratified atmospheric boundary layer with a prevailing offshore geostrophic wind of 3 m/s, roughness step at the coastline considered

Surface Temperature Gradients - Sea Breeze. Generally, sea and land surface temperatures are different. The resulting mesoscale wind and temperature fields are totally different and no longer comparable with the large-scale results. Local circulation systems can be generated by the surface temperature gradient; with an offshore geostrophic wind a sea breeze might occur (for a review see Atkinson 1981). In the southern North Sea area, a sea breeze generally develops in May and June, in case of a high pressure system with low pressure gradients which may last for a few days over northern Europe. Measurement analyses have shown that sea breezes can be found both on sunny and on cloudy days (Anto 1977). While it may be the predominant local circulation system in May and June, it is less frequent in other months. Nevertheless, the situation is important for the ecosystem. Since the growth rate of the phytoplankton is at its maximum in spring and a nutrient limitation might occur thereafter (see Chap. 2.2 for details), the atmospheric input becomes important in the late spring and early summer. At this time sea breezes can be found quite often (e.g. Anto 1977; Meesters et al. 1989).

In Fig. 3.1.6 the development of a sea and land breeze circulation for a zero geostrophic wind is presented schematically. The increase in surface temperature over land due to solar heating causes vertical mixing and a horizontal pressure gradient (0900 LST=Local Standard Time) which results in a convergence zone (cold front) at the coast at noon. The front separates relatively cool air advected from sea from the warmer air over

P₃ ————————————
P₂ ————————————
P₁ ————————————
P₀ ～～～～～～///////////// 6

Atmosphere at rest

P₃ ————————————
P₂ ————————————
P₁ ————————————
P₀ ～～～～～～///////////// 18

Radiational cooling becomes dominant
over solar heating; sea breeze winds
remove pressure gradient

P₃ ————————————
P₂ ————————————
P₁ ————————————
P₀ ～～～～～～///////////// 9

Mass mixed upwards over land

P₃ ————————————
P₂ ————————————
P₁ ————————————
P₀ ～～～～～～///////////// 21

Sinking as air cools by radiative
flux divergence; downward mass flux

P₃ ————————————
P₂ ————————————
P₁ ————————————
P₀ ～～～～～～///////////// Noon

Convergence zone develops over land

P₃ ————————————
P₂ ————————————
P₁ ————————————
P₀ ～～～～～～///////////// Midnight

Divergence zone develops over land;
upwind over water

P₃ ————————————
P₂ ————————————
P₁ ————————————
P₀ ～～～～～～///////////// 15

Inland penetration of the
sea–breeze front; downward
transport over water

P₃ ————————————
P₂ ————————————
P₁ ————————————
P₀ ～～～～～～///////////// 3

Shallow land–breeze – more stable
at night

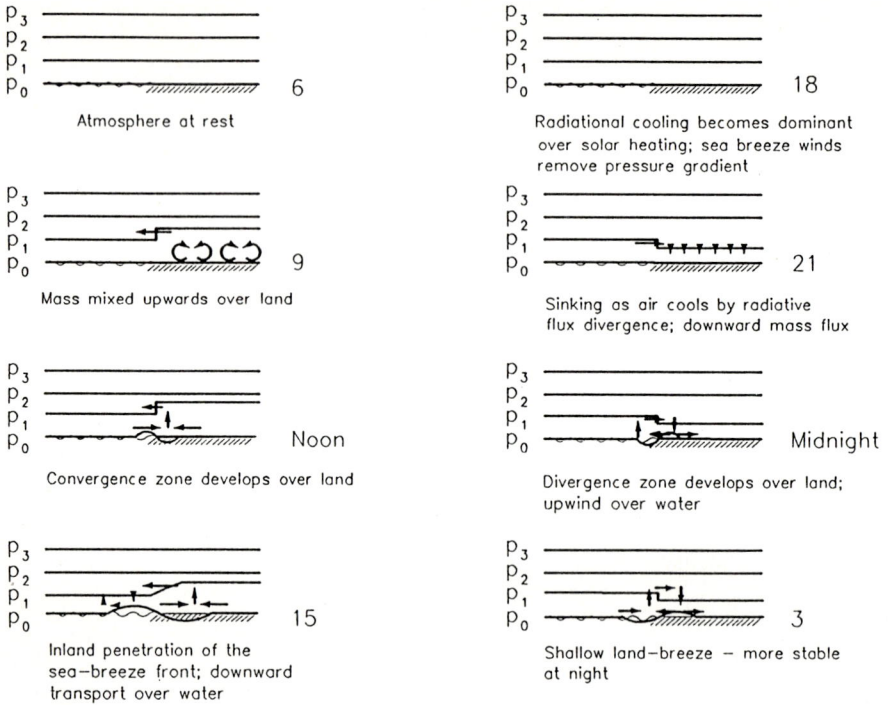

Fig. 3.1.6. Schematic view of the diurnal evolution of the sea and land breeze with a zero geostrophic wind. (After Pielke 1984)

land. The sea breeze front penetrates inland during the day (e.g. 1500 LST). Over water, the circulation extends up to 200 km offshore (Atkinson 1981) bounded by a weak downward vertical wind zone. In the evening, the cooling of the land surface removes the surface pressure gradient and during night an opposite but weaker circulation system is generated. A land breeze develops close to the ground at midnight with an upwind zone over water and a subsidence over land. Thus, during nighttime, tracers are transported offshore. Observational studies on pollution transport verify that a developing sea and land breeze can influence the transport of tracers and that the tracers can even be locked into the circulation (e.g. Lyons and Olsson 1973; Shair et al. 1982).

In Fig. 3.1.7 the diurnal evolution of a sea breeze at the North Sea is displayed. It is calculated for a large-scale situation of zero geostrophic wind with the model METRAS for the 21st of June. The results agree with the expected development of the system (see Fig. 3.1.6) as well as with measurements (e.g. Simpson et al. 1977) and other model studies on this topic (e.g. Physick 1980; Kessler et al. 1985; Savijärvi and Alestalo 1988;

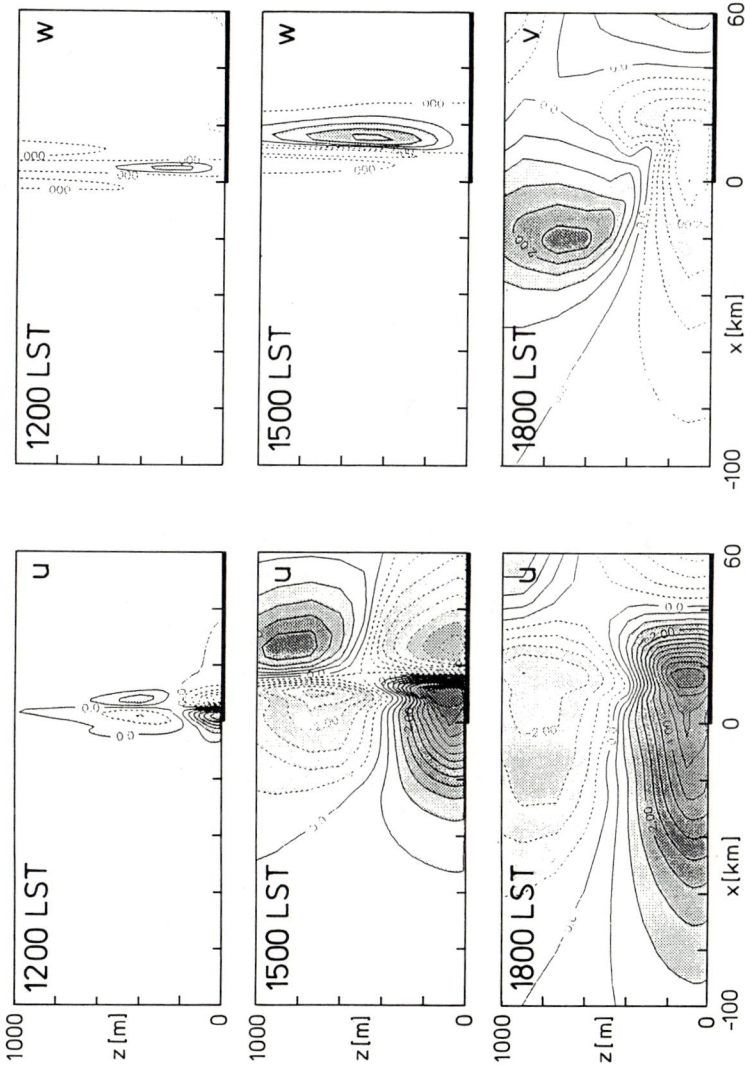

Fig. 3.1.7. Vertical cross-section of the simulated domain for the wind components in a sea-breeze calculated with a prevailing zero geostrophic wind and considering temporal changes in surface temperature. *Top* east-west (*u*) and vertical wind (*w*) at 12.00 LST; *middle* east-west (*u*) and vertical wind (*w*) at 15.00 LST; *bottom* east-west (*u*) and north-south wind (*v*) at 18.00 LST. *Dashed contours* indicate negative values (e.g. downward, easterly, northerly wind); *contour interval for u, v 0.5 m/s, for w 0.1 m/s*

Avissar and Pielke 1989). The rotation of the circulation system due to the Coriolis effect causes a northerly wind component v which is still present in the evening (Fig. 3.1.7 18.00 LST) and during the night.

The evolution of the sea breeze circulation is important for atmospheric input simulations in several connections:

1.   At daytime the surface heating enhances the vertical turbulent mixing and thus reduces the concentration over land in lower levels.
2.   At daytime, in the case of a low large-scale pressure gradient, the onshore wind zone (sea breeze) limits the offshore tracer transport to upper levels above the sea breeze and to areas inland of the sea breeze front.
3.   At nighttime the surface cooling reduces the vertical turbulent mixing and enhances the concentration over land in lower levels.
4.   At nighttime, in the case of a low large-scale pressure gradient, the offshore wind intensifies the offshore tracer transport.

From these qualitative statements it becomes clear that at daytime an area of reduced atmospheric input may exist over land, mudflats and water in the area of the sea breeze. The horizontal extension of this area depends on the inland penetration of the sea breeze front and the offshore extent of the sea breeze. The dependency of both on the large-scale situation (geostrophic wind) and local features (Wadden Sea and tide) will be discussed in the following.

Inland Penetration of Sea Breeze Front. In Fig. 3.1.8 the inland penetration of the main sea breeze front is presented as a function of time for the different case studies. In general, the inland penetration of the front becomes faster during the day, with its actual position depending on the large-scale situation. In case of a zero geostrophic wind, the sea breeeze area reaches up to 32 km inland at 1800 LST (thick dashed line). With a prevailing onshore geostrophic wind of 3 m/s, the front reaches this area about 5 h earlier, resulting in a wide area over land which is influenced by the sea breeeze during the day (light dashed line). With an offshore geostrophic wind of the same strength, the front remains closer to the coast; it is only 16 km inland at 1800 LST (dashed line). As confirmed by these model results, the prevailing geostrophic wind can accelerate or decelerate the inland penetration of the front (see also Savijärvi and Alestalo 1988).

The displacement of the sea breeeze front depends not only on the large-scale situation but also on local features. Condensation and cloud for-

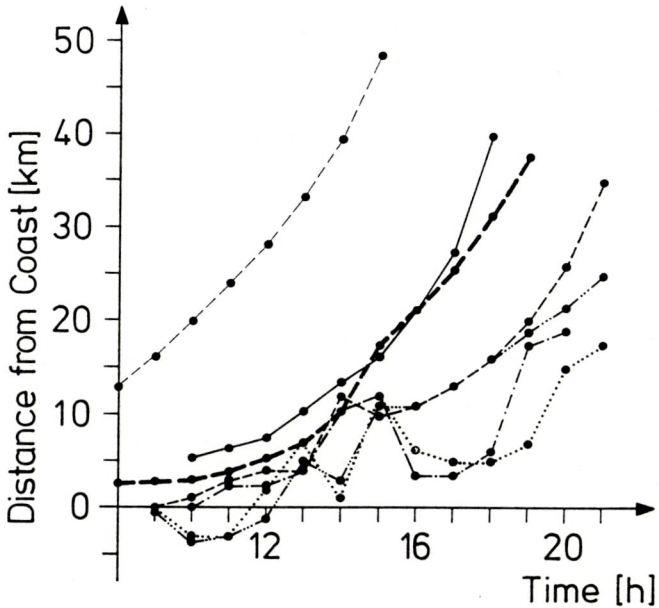

Fig. 3.1.8. Inland penetration x (km) of the main sea breeze front for different meteorological situations dependent on time. *full straight line* including clouds; *dashed line* no mudflats; *dotted line* open mudflats; *dashed-dotted line* low tide at 15.00 LST; *dashed-triple-dotted line* high tide at 15.00 LST; *thick line* zero geostrophic wind; *thin line* onshore geostrophic wind of 3 m/s; *other line thickness* offshore geostrophic wind of 3 m/s

tion over land, for instance, add an additional horizontal temperature difference to the difference caused by radiational heating of the ground, therefore accelerating the inland penetration of the front (solid line in Fig. 3.1.8).

Further changes in the displacement of the sea breeeze front arise from the tidally flooded mudflats, which cause a temporally periodical change in surface temperature. A generally unflooded mudflats area (dotted line in

Fig. 3.1.8) splits the large horizontal temperature gradient at the coastline into two lower temperature steps. With tidal influences, the mudflat area periodically changes, which additionally modifies the inland penetration of the front. For details on the effect of the tides and mudflats on the sea breeze circulation see Meesters et al. (1989) and Meesters and Vugts (1991) and on the inland penetration of the sea breeeze front see Schlünzen (1990).

The inland penetration of the sea breeeze front is mainly important for air pollution studies over land. Within the sea breeeze circulation only comparatively unpolluted air from the sea is advected, apart from tracers emitted from sources which are situated between coastline and sea breeeze front. Thus, the concentration of tracers in air is reduced and an area of comparatively low deposition can be expected in coastal waters and over land. Even sources between the coastline and the sea breeeze front and therefore within the sea breeeze circulation only secondarily influence the deposition in the area of the sea breeeze circulation. These sources mainly increase the concentration in the area of the return current and thus affect the atmospheric input at the offshore extent of the sea breeeze. Nonetheless, sources within the sea breeeze result in higher concentrations within the circulation system than sources outside. Thus, the lowest atmospheric input into coastal waters (up to 100 km offshore) can be expected when the sea breeeze front remains close to the coast (dotted and both dashed-dotted lines in Fig. 3.1.8) and most of the tracer sources are situated outside.

Offshore Extent of the Sea Breeze Circulation. The other important parameter for the calculation of the atmospheric input into the coastal North Sea is the offshore extent of the sea breeeze circulation. From the case studies performed, those were selected which show the lowest inland penetration of the sea breeeze front and thus havee the lowest emissions within the circulation. The offshore extent for these case studies is given in Fig. 3.1.9 as a function of time. The sea breeeze area has an offshore extent between only a few kilometres in the morning and up to hundred kilometres in the late afternoon. These values lie within the range given by measurements (see Atkinson 1981).

The calculated offshore extent is similar for the four case studies, all of which were performed with an offshore geostrophic wind of 3 m/s and different local features (Wadden Sea, tide). Only in the afternoon is the offshore extent different. A few hours before the system collapses (18.00 LST) the offshore extent is largest for the two case studies performed without mudflats (dashed line) and with open mudflats (dotted line). The

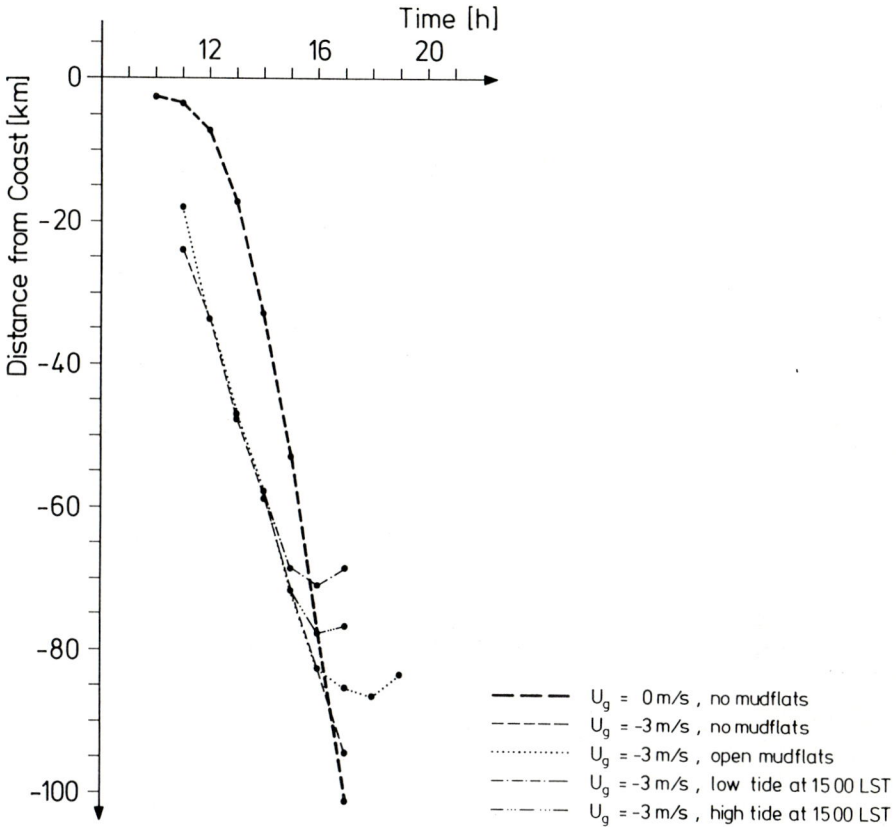

Fig. 3.1.9. As in Fig. 3.1.8 but offshore extent x (km) of the sea-breeze circulation

extent is lower for the case study with low tide in the afternoon (dashed-dotted line) or for the opposite tidal cycle (dashed-triple-dotted line). The tidal cycle seems to slow down the offshore movement of the sea breeeze circulation.

In general, the offshore extent is lower for the case study performed with a zero geostrophic wind (thick dashed line) (see also, e.g. Arritt 1989). The sea breeeze circulation is expanding further offshore in the evening, reaching a distance of 200 km from the coast at 23.00 LST. Later, the circulation collapses, whereas with the offshore geostrophic wind its intensity is reduced 3 h earlier.

In the case study performed with a zero geostrophic wind, the generally lower offshore extent is connected with a larger inland penetration of the sea breeeze front (see Fig. 3.1.8). The whole circulation system lies further

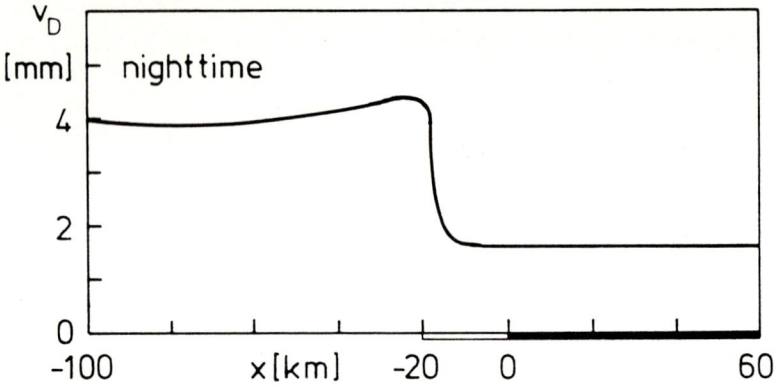

Fig. 3.1.10. Horizontal variance of the deposition velocity for SO_2 (mm/s) in a stably stratified atmospheric boundary layer

inland. Thus, the number of sources within the sea breeeze circulation with a zero geostrophic wind is higher compared to the case study performed with a geostrophic offshore wind, and, consequently, a higher tracer concentration has to be expected close to the surface.

In conclusion, the concentration of tracers in the atmosphere at the surface up to 100 km offshore might be lowest during daytime for a sea breeeze situation with an onshore or a weak offshore geostrophic wind. This case can be characterized as a minimized input situation. The maximum input might develop during nighttime or in the morning with a land breeze generated (see Fig. 3.1.6), which advects polluted air from land to the ocean.

3.1.3.2 Atmospheric Input Characterized by Dry Deposition

The dry deposition is not only affected by the meteorological situation and the tracer concentration but also dependent on the deposition velocity v_D (see Sect. 3.1.1). The most investigated atmospheric deposition process is that of SO_2 (for an overview on measurements see for instance: McMahon and Denison 1979; Sehmel 1980). For this reason, most models select SO_2 for deposition modelling (e.g. Aritt et al. 1988; Kitada and Kitagawa 1990; Schlünzen and Pahl 1992). The tracer is relevant in connection with acid rain, but it is itself harmless for the ocean. The emitted SO_2 becomes important for the aquatic system after forming sulphate particles in the at-

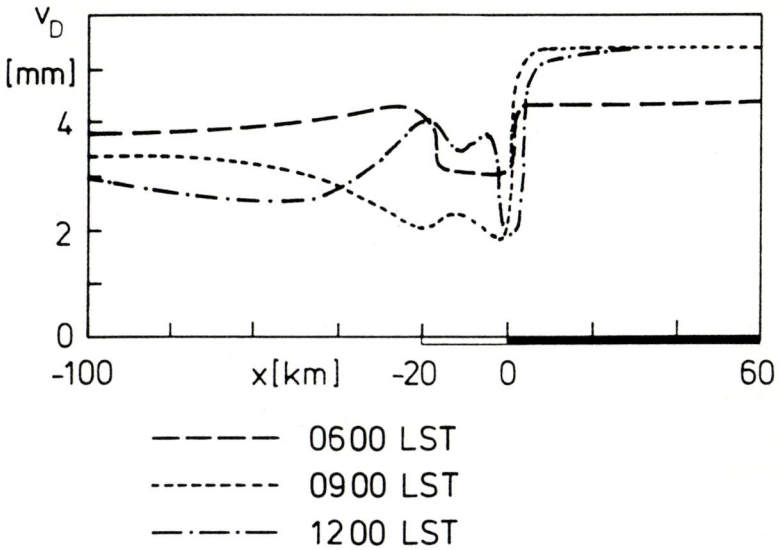

Fig. 3.1.11. Horizontal variance of the deposition velocity for SO_2 (mm/s) at 06.00, 09.00, 12.00 LST

mosphere. Together with nitrate particles they make up about 80% of the atmospheric particles (see Chap. 2.4).

The deposition velocity of SO_2 is about doubled over water compared to the values over land, caused by the high solubility of SO_2 in water (see Fig. 3.1.10). In case of a roughly constant concentration field at the coast, which might be true for tracers with distant sources, the calculated deposition velocity corresponds to about doubled input values over water in comparison to those over land.

In Fig. 3.1.11 the horizontal variance of the deposition velocity is presented for three times. The values over land are about tripled at daytime compared to the values during night (see Fig. 3.1.10). When plants' stomata are closed, the tracer concentration reaching the coastal North Sea can be higher during the night than during daytime. The main dry deposition to coastal waters is to be expected during nighttime and in the early morning.

The presented results of the mesoscale model METRAS were derived for particular meteorological situations causing a sea breeeze. Before these results can be generalized for other gases than SO_2 or for aerosols, the tracer characteristics have to be taken into account, too (see Fig. 3.1.1). Diurnal variations in emission rates, which might be lower during nighttime compared to daytime for certain trace elements (e.g. car exhaust gases, power plant emissions), are still not well specified. Even for constant emission rates the concentration might not remain constant during the day. Chemical transformations and the deposition of tracers depend on insolation and enhanced vertical mixing, factors which might reduce the surface concentration.

A concluding statement concerning the main atmospheric input situation for different tracers cannot yet be given. A tentative conclusion can be derived for SO_2, assuming constant emission rates. The maximum dry deposition in coastal regions might occur during nighttime and in the morning in the case of a developing land breeze or for offshore winds. Three-dimensional case studies performed by Schlünzen et al. (1992) confirm this statement. For the specified conditions not only maxima in atmospheric input of SO_2 but also values higher than calculated with a large-scale model are to be expected.

3.1.4 Mean Features of the Atmospheric Pollution Transport

For the atmospheric input of pollutants into the North Sea, it can be stated that, besides the amounts emitted, the input is mainly dependent on the meteorological situation and the precipitation rate. The yearly mean input values for lead derived from several large-scale models are quite similar although using different meteorological and emission data. The values derived from measurements show a comparatively wide range (see Table 3.1.3). For lead, a mean yearly input of 2664 tons is calculated by the model HHLRT, based on input data for 4 months in 1984.

The calculated input values show temporal and spatial differences which lead to a division of the North Sea into three areas. In the south, high input values develop due to the comparatively high concentrations which are always high enough to result in considerable input values. Further north, the lead concentration is lower. Thus, the atmospheric input decreases with increasing distance from the source areas. Due to the high precipitation rate at the southwest Norwegian coast, wet deposition is very effective in the northeast of the North Sea. In this area the atmospheric input of lead can be higher compared to the central North Sea but is usually lower than in the south of the North Sea.

This large-scale input pattern is modified by mesoscale processes which can change the deposition values, especially close to the coastline. The development of local circulation systems is dependent on the large-scale meteorological situation (e.g. the geostrophic wind) and on local features (e.g. width of mudflats, tidal cycle, atmospheric stability). With regard to minimum and maximum input situations, it can be concluded from the present sea breeeze case studies that the maximum dry deposition for SO_2 into coastal waters might occur at night or in the morning. The lower daytime input to coastal waters might be reduced further due to rain and wash-out effects in coastal fronts over land (e.g. van de Berg and Oerlemans 1985). These conclusions assume that emission and transformation rates are constant. A more general conclusion cannot be derived at present. Even if the main atmospheric input up to 100 km offshore into the North Sea would occur during nighttime, this does not mean that this is the most severe input situation for the ecosystem. This decision can only be made with regard to the circumstances that influence the further tracer transport in the water and the tracer uptake by suspended matter and plankton.

References

Anto AF (1977) Observational studies on land-sea breeze phenomena around the island of Sylt. Meteorol Rundsch 30:118-122

Arritt RW (1989) Numerical modeling of the offshore extent of sea breezes. Q J R Meteorol Soc 115:547-570

Arritt RW, Pielke RA, Segal M (1988) Variations of sulfur dioxide deposition velocity resulting from terrain-forced mesoscale circulations. Atmos Environ 22:715-723

Asman WAM, Runge EA (1991) Atmospheric deposition of nitrogen components to Denmark and surrounding sea areas - a preliminary estimate. Rep NFAC R-91-1, Netherland Foundation for Atmospheric Chemistry

Atkinson BW (1981) Meso-scale atmospheric circulations. Academic Press, London

Avissar R, Pielke RA (1989) A parameterization of heterogenuous land surfaces for atmospheric numerical models and its impact on regional meteorology. Mon Wea Rev 117:2113-2136

Barrett EC, Beaumont MJ, Corlyon AM (1991) A satellite-derived rainfall atlas of the North Sea 1978-87. Remote Sensing Unit, Dept of Geography, Univ Bristol, UK

Bigalke K (1991) Interaktive Modellkopplung zur Berücksichtigung heißer Punktquellen in Gitterpunktsmodellen und Einfluß der Kopplungsstufe auf die Immission. Thesis, Univ Hamburg, Berichte aus dem ZMK 12, 141 pp

Bigalke K (1992) A new method for incorporating point sources in Eulerian dispersion models. In: van Dop H, Kalles G (eds) Air pollution modelling and its application, vol IX. Plenum, New York (in press)

Cambray RS, Jefferies DF, Topping G (1975) An estimate of the input of atmospheric trace elements into the North Sea and the Clyde Sea (1972-3). AERE-Rep R 77, AERE, Horwell

Cambray RS, Jefferies DF, Topping G (1979) The atmospheric input of trace elements to the North Sea. Mar Sci Commun 5:175-195

Dunst M (1982) On the vertical structure of the eddy diffusion coefficient in the PBL. Atmos Environ 16:2071-2074

Graßl H, Eppel D, Petersen G, Schneider B, Weber H, Gandraß J, Reinhardt KH, Wodarg D, Fließ J (1989) Stoffeintrag in Nord- und Ostsee über die Atmosphäre. GKSS-Ber 89/E/8, GKSS, Geesthacht

Grønås S (1989) Flow over and around southern Norway as described by the Norwegian high resolution model. IAMAP-Conf, Reading, Pap MF26

Hosker RP, Lindberg SE (1982) Review: atmospheric deposition and plant assimilation of gases and particles. Atmos Environ 16:889-910

INC (1990) Zwischenbericht 1990 über den Qualitätszustand der Nordsee. Ministry of Transport and Public Works, The Hague, The Netherlands

Kessler E (1969) On the distribution and continuity of water substance in atmospheric circulations. Meteor Monogr No 32. American Meteorological Society, Boston

Kessler RC, Eppel D, Pielke RA, Mc Queen J (1985) A numerical study of the effects of a large sandbar upon sea breeze development. Arch Met Geoph Biocl Ser A 34:3-26

Kitada T, Kitagawa E (1990) Numerical analysis of the role of sea breeze fronts on air quality in coastal and inland polluted areas. Atmos Environ 24A:1545-1559

Krell U (1991) Vergleich von Modellen für den Eintrag von Spurenstoffen aus der Atmosphäre in die Nordsee am Beispiel Blei. Thesis Univ Hamburg

Krell U, Roeckner E (1988) Model simulation of the atmospheric input of lead and cadmium into the North Sea. Atmos Environ 22:375-381

Lehmhaus J, Roeckner E (1986) Entwicklung von Ausbreitungsmodellen zur Analyse und Prognose von Emissionsepisoden beim mittel- und großräumigen Transport von Luftverunreinigungen. F+E-Vorhaben 10402618, Umweltbundesamt, Berlin

Lindquist O, Schröder WH (1989) Cycling of mercury in the environment with emphasis on the importance of the element in acid rain studies. In: Pacyna J, Ottar B (eds) Control and fate of atmospheric trace metals, NATO ASI Series. Reidel, Dordrecht, C 268 303-310

Lyons A, Olsson LE (1973) Detailed mesometeorological studies of air pollution dispersion in the Chicago lake breeze. Mon Wea Rev 101:387-403

Mc Mahon TA, Denison PJ (1979) Empirical atmospheric deposition parameters - a survey. Review Paper. Atmos Environ 13:571-585

Meesters AGCA, Vugts HF (1991) Sea breeze in the Dutch Wadden Sea area: observations and modelling. Beitr Physik Atmos 64:313-328

Meesters AGCA, Vugts HF, van Delden AJ, Cannemeijer F (1989) Diurnal variation of the surface wind in a tidal area. Beitr Physik Atmos 62:258-264

Michaelis W (1988) Experimental studies on dry deposition of heavy metals and gases. In: van Dop H (ed) Air pollution modelling and its application VI, vol 11. Plenum, New York, pp 61-74

Milford JB, Davidson CI (1985) The sizes of particulate trace elements in the atmosphere - a review. J Air Pollut Control Assoc 35:1249-1260

Orlanski I (1975) A rational subdivision of scales for atmospheric processes. Bull Am Meteoro Soc 56:527-530

Otten P, Rojas C, Wonters L, van Gieken R (1989) Atmospheric deposition of heavy metals (Cd, Cu, Pb and Zn) into the North Sea. Rep 2, UTAC, Dept of Chemistry, Univ Antwerpen

Pacyna JM (1983) Trace element emission from anthropogenic sources in Europe, Norwegian Institute for Air Research, Rep No 10/82, Lillestrøm

Pacyna JM (1985) Spatial distribution of the As, Cd, Cu, Pb, V and Zn emissions in Europa within a 1.5° grid net. Norwegian Institute for Air Research, Rep No 60/85, Lillestrøm

Pahl S, Schlünzen H (1990) Parameterisierung der trockenen Deposition in einem mesoskaligen Transport- und Strömungsmodell. Bayer Landwirtsch Jahrb 67, Sonderheft 1:65-76

Petersen G, Weber H, Graßl H (1989) Modeling the atmospheric transport of trace metals from Europe to the North Sea and Baltic Sea. In: Pacyna J, Ottar B (eds) Control and fate of atmospheric trace metals, NATO ASI Series. Reidel, Dordrecht, C 268 15-32

Pettersen S (1956) Weather analysis and forecasting, vol 1. McGraw-Hill, New York, 27 pp

Physick WL (1980) Numerical experiments on the inland penetration of the sea breeze. Q J R Meteorol Soc 106:735-746

Pielke RA (1984) Mesoscale meteorological modeling. Academic, New York, 612 pp

RSU (Rat der Sachverständigen für Umweltfragen) (1980) Umweltprobleme der Nordsee. Gutachten. Kohlhammer, Stuttgart

Savijärvi H, Alestalo M (1988) The sea breeze over a lake or gulf as the function of the prevailing flow. Beitr Phys Atmos 61:98-104

Schlünzen KH (1988) Das mesoskalige Transport- und Strömungsmodell METRAS - Grundlagen, Validierung, Anwendung. Hamb Geophys Einzelschriften A 88

Schlünzen KH (1990) Numerical studies on the inland penetration of sea breeze fronts at a coastline with tidally flooded mudflats. Beitr Phys Atmos 63:243-256

Schlünzen KH, Pahl S (1992) Modification of dry deposition in a developing sea-breeze circulation - a numerical study. Atmos Environ 25A:51-61

Schlünzen KH, Bigalke K, Niemeier U (1992) Local atmospheric input patterns in the German Bight. ICES CM 1992/E23, Copenhagen

Sehmel GA (1980) Particle and gas dry deposition: a review. Atmos Environ 14:983-1011

Shair FH, Sasaki EJ, Carlan DE, Cass GR, Goodin WR, Edinger JG, Schacher GE (1982) Transport and dispersion of airborne pollutants associated with the land breeze-sea breeze system. Atmos Environ 16:2043-2053

Simpson JE, Mansfield DA, Milford JR (1977) Inland penetration of sea-breeze fronts. Q J R Meteorol Soc 103:47-7

Stößel RP (1987) Untersuchungen zur Naß- und Trockendeposition von Schwermetallen auf der Insel Pellworm. GKSS-Bericht 87/E/34, GKSS, Geesthacht

van Aalst RW, van Ardenne RAM, de Kruk JF, Lems T (1983) Pollution of the North Sea from the atmosphere. TNO Rep C182/152, Delft

van de Berg LCJ, Oerlemans J (1985) Simulation of the sea-breeze front with a model of moist convection. Tellus 37A:30-40

van Jaarsveld JA, van Aalst RM, Onderdelinden D (1986) Deposition of metals from the atmosphere into the North Sea: model calculations. Rep No 842015002. National Institute of Public Health and Environmental Hygiene, Bilthoven

Wiman BLB, Unsworth MH, Lindberg SE, Bergkvist B, Jaenicke R, Hansson H-C (1990) Perspectives on aerosol deposition to natural surfaces: interactions between aerosol residence times, removal processes, the biosphere and global environmental change. J Aerosol Sci 21:331-338

Wippermann F (1981) The applicability of several approximations in mesoscale modelling - a linear approach. Beitr Phys Atmosph 54:298-308

3.2 Currents and Transport in Water

T. POHLMANN and W. PULS

The distribution of substances in the sea depends to a great extent on the characteristics of advective and diffusive transport. This chapter deals with the numerical simulation of the water movement in the North Sea. Both mean and extreme conditions are presented and discussed. The water movement is the basis for simulating the transport of (1) conservative and passive substance and (2) suspended particulate matter. Concerning the latter, settling, deposition and erosion are also considered.

3.2.1 Simulation of the North Sea Circulation

The motions in the sea can be described by using a system of mathematical equations. Since this system is generally too complex to be solved analytically, the equations are transformed in such a way that the system can be solved numerically with the help of a computer. Errors occur not only when setting up the equations but also when transforming them. These errors always have to be considered when discussing the results.

3.2.1.1 Driving Forces

The forces that drive the model consist of three components, astronomical tides (commonly the dominating M_2-tide is taken), the density gradient and the wind and air pressure distributions over the North Sea.

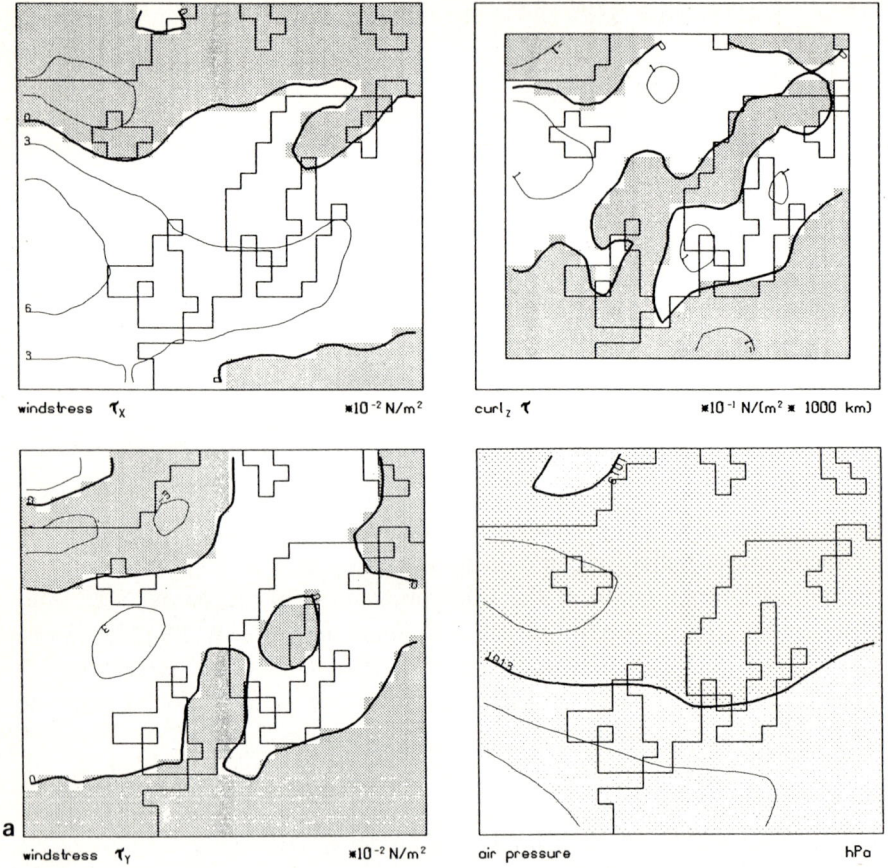

Fig. 3.2.1a. Climatological means of weather patterns over Northern Europe in summer

The North Sea tide is an external tide that is defined at the open boundaries of the model. The density gradients in the North Sea are induced by the temperature and salinity distribution. They are both subject to strong seasonal variations. In addition, the variabilities of the temperature and salt distributions are not in phase, so the resulting density structure is exposed to more complex variations.

The temporal and spatial distributions of salt and temperature were already discussed in Chap. 2.1.

Once again, the important role of the summer thermocline has to be mentioned. It occurs in the northern and central areas of the North Sea. It nearly cuts off the whole heat and momentum fluxes between the warm surface water and the cold bottom water. This fact has quite important con-

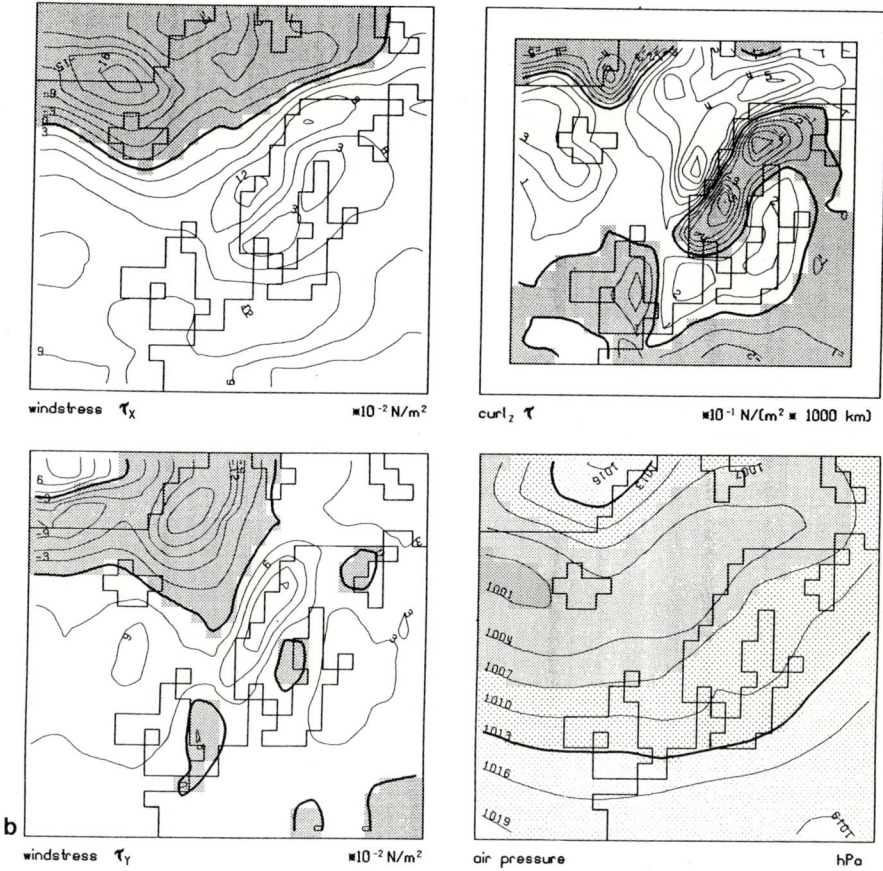

windstress τ_x $\times 10^{-2}$ N/m^2 curl$_z$ τ $\times 10^{-1}$ N/(m^2 $\times$ 1000 km)

windstress τ_Y $\times 10^{-2}$ N/m^2 air pressure   hPa

Fig. 3.2.1b. Climatological means of weather patterns over Northern Europe in winter

sequences for the biological and chemical processes in the North Sea. Tide- and storm-induced turbulence is responsible for the vertically homogeneous water column in the coastal and shallow parts of the North Sea. In the transition zones between the stratified and the unstratified water, so-called thermal fronts exist.

In the simulation, the density driven processes, especially those depending on the thermocline, are simplified. Here, only two different temperature and salt distributions are used. They are based on climatological means for summer and winter, respectively. For this reason, the water and temperature budget of the North Sea is reproduced in a very simple way.

The data for the atmospheric forcing of the model experiment consists of 6-hourly wind stress and air pressure distributions (Backhaus et al. 1985). These were provided by the Norwegian Meteorological Institute.

Figure 3.2.1a,b shows calculated climatological means of these atmospheric fields (air pressure, wind stress, curl), comparing summer and winter.

In agreement with the general expectation one can see that in the winter mean (Fig. 3.2.1b) higher air pressure gradients exist with stronger wind stress than during summer (Fig.3.2.1a).

As mentioned above, in the model simulation only the changing wind and air pressure distribution can cause a temporal variability of the flow. This is superimposed by a semi-annually changing density driven circulation. In this investigation we are only interested in low frequency variabilities with periods of more than 1 day. For this reason, the influence of the tide can be considered as constant. Due to these restrictions, the simulated variability of the North Sea is smaller than can be expected in reality.

3.2.1.2 Description of the Circulation Model

The purpose of the circulation model (Backhaus 1985), which is based on the laws of hydrodynamics, is to calculate the reaction of the North Sea system to the driving forces described earlier. The case in question is too complex to be solved for the complete space-time continuum. By using numerical calculations, it is possible to reproduce the natural processes in space and time on a four-dimensional grid. The configuration of this grid is dependent on the scales of the processes on the one hand and limited by the computer capacity on the other.

The circulation model covers the whole northwestern European Shelf as well as a part of the northeastern Atlantic (Fig. 3.2.2; Backhaus and Hainbucher 1987). The average grid distance is about 20 km. In accordance with the seasonally varying vertical stratification, the resolution in the vertical is 12 layers in summer and 7 layers in winter. The thickness of the layers increases from the surface to the bottom of the sea, due to the smaller vertical scale near the surface. The simulation was carried out for a period of 14 years with a time step of 40 min.

3.2.1.3 The Mean Annual Cycle of the Circulation

Figure 3.2.3 demonstrates the simulated seasonal variability of the circulation for the climatological average (Hainbucher et al. 1986). The cir-

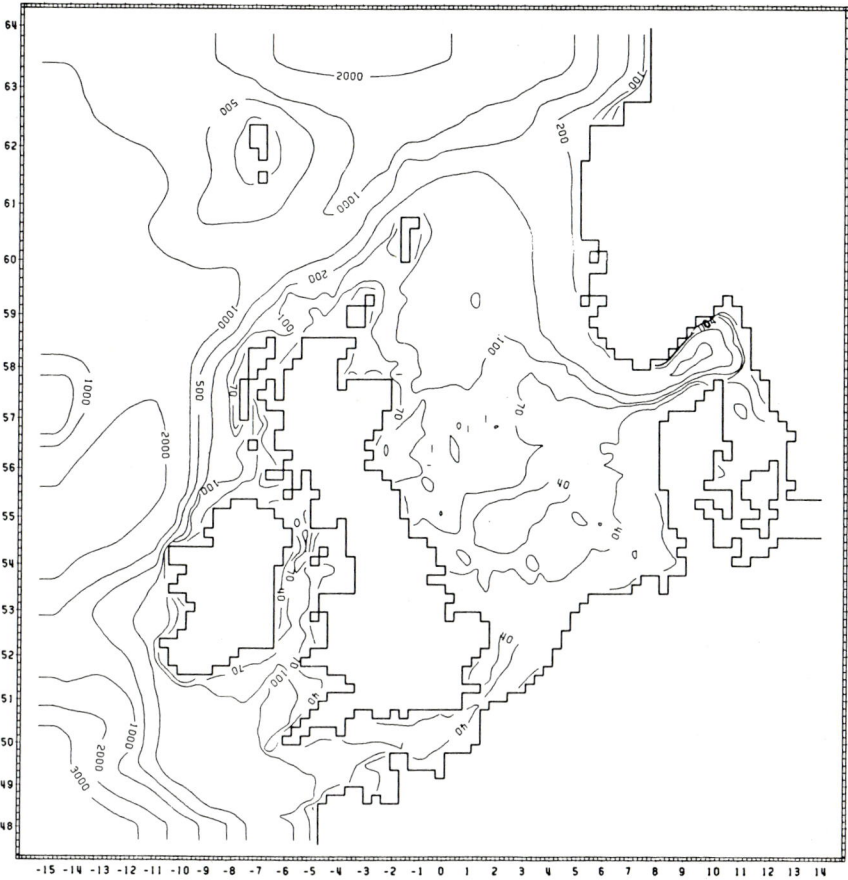

Fig. 3.2.2. Topography of the North Sea and adjacent regions defined on discrete model grid (depth in m)

culation is displayed by means of the streamfunction in Sverdrup units (1 Sv = 10^6m^3/s). The most intensive flow is found along the continental slope. The location of the 1 Sv line is almost identical with the shelf edge. Due to larger water depths, the water mass transport rates reach values up to 6 Sv towards the open ocean, whereas on the shelf the transport rates are less then 1 Sv. A considerable amount of North Atlantic water enters the North Sea between the Orkney and the Shetland Islands. About 75% turns to the east already in the northern North Sea and flows back into the Atlantic via the Norwegian Coastal Current. It is obvious that the seasonal variations in this northern recirculation cell reach up to 0.5 Sv, whereas the variability in the eastern parts of the cell are extremely high. The variations in the southern North Sea reach values only up to 0.2 Sv.

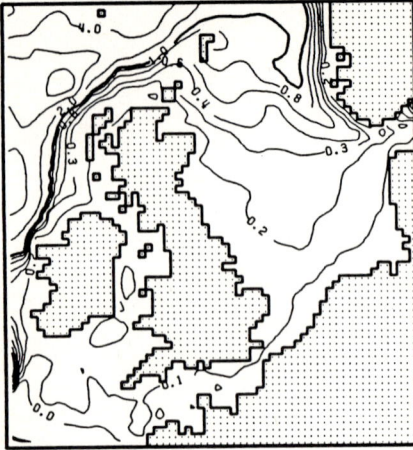

Spring

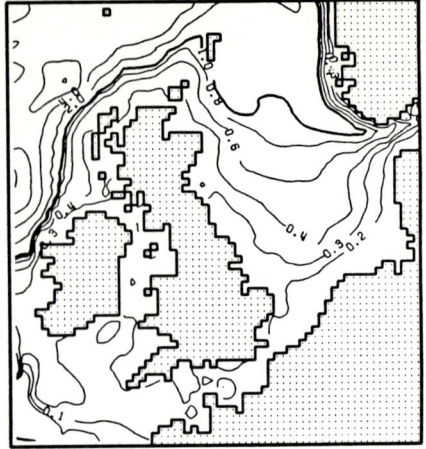

Autumn

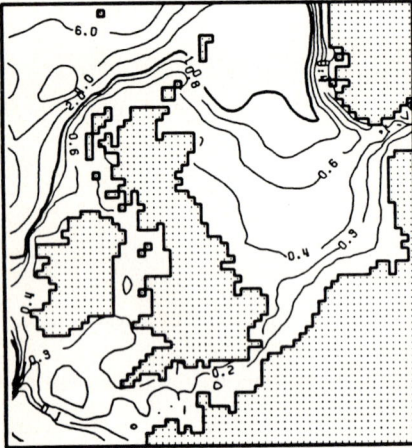

Winter

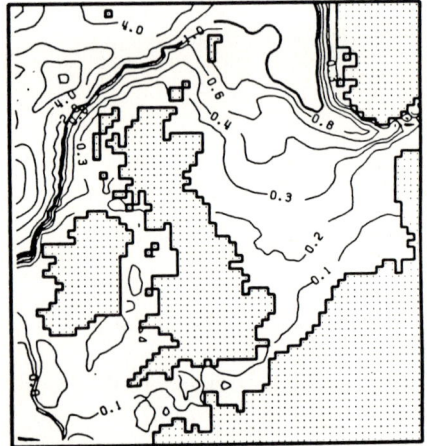

Summer

Fig. 3.2.3. Climatological seasonal means of water mass transport in the North Sea (in Sv). The stream function of the flow pattern is shown. (The contours indicate the routes of the water masses, their densities are a measure for the speed of the flow)

A remarkable feature is the occurrence of the highest flow rates in the northern North Sea during summer and autumn. In the southern North Sea, on the contrary, the circulation reaches its maximum intensity in winter. These differences cannot be explained by spatial changes of the atmospheric forcing, since they are not very pronounced. This fact leads to the conclu-

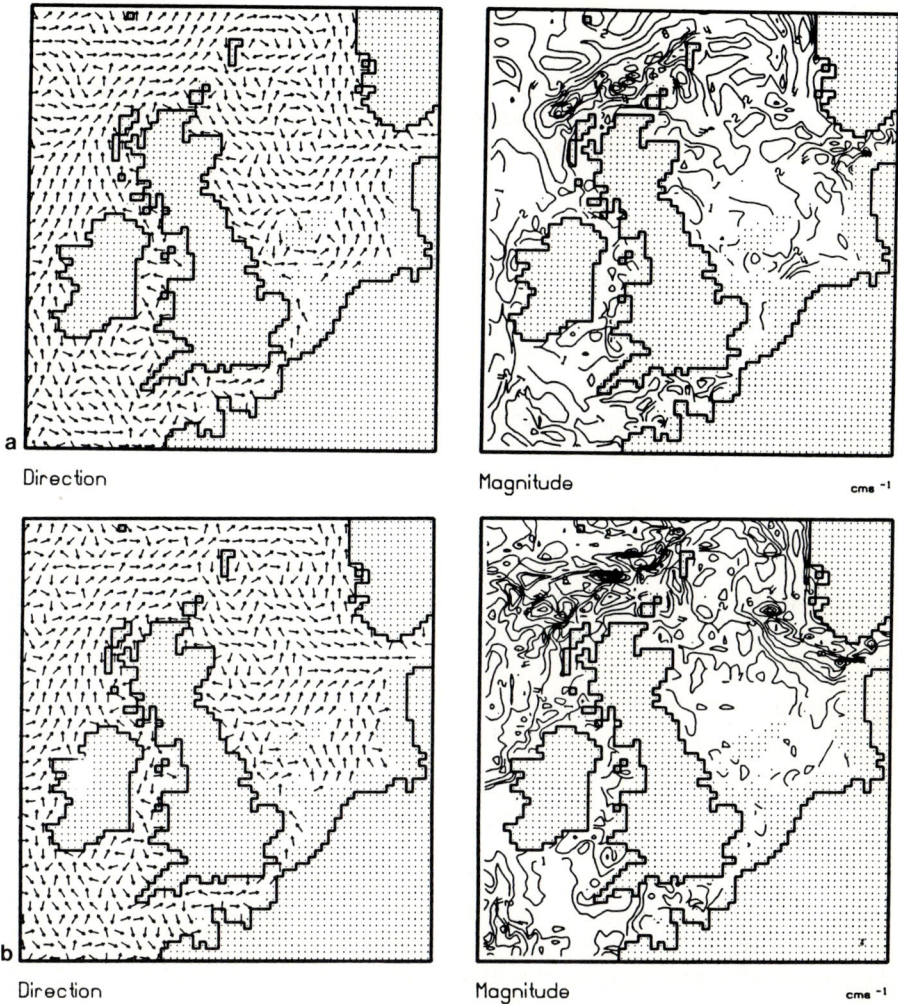

Direction   Magnitude   cms⁻¹

Direction   Magnitude   cms⁻¹

Fig. 3.2.4. Climatological mean of bottom circulation *a (above)* in winter, *b (below)* in summer (depth of bottom layer: 30 m)

sion that, in the northern North Sea, density variations dominate the annual cycle, whereas the more shallow southern parts of the North Sea reflect the annual cycle of the wind forcing.

The influence of the open boundaries is most obvious when looking at transport rates through selected sections. Figure 2.1.15 of Chap. 2.1 shows the climatological mean of the net mass transports through the boundaries of the ICES boxes (km³/day) (Lenhart 1990). Through the northwestern entrances about 100 m³/day reach the North Sea from the North Atlantic. Already 66 km³/day of this amount recirculates north of 7.5° North, and most of the rest, 34 km³/day, recirculates in the central North Sea. Five km³/day, i.e. only 5% of the incoming amount of water, reaches the southern North Sea. There, these water masses encounter water that flows through the English Channel. The climatological mean transport rate of this incoming Atlantic water is 8 m³/day. The water that reaches the southern North Sea leaves it through the Skagerrak and it finally leaves the North Sea via the Norwegian Coastal Current.

To give a more detailed estimate of the circulation and the resulting dispersion of pollutants, in the following the currents of the bottom layer are presented. The bottom layer is defined as a layer of 30 m thickness that follows the topography. If the water depth is less than 30 m, it will not be taken into consideration. Figure 3.2.4 shows the climatological mean of the bottom circulation - direction and magnitude - during the winter and during the summer season, respectively.

In winter as well as in summer the flow north of the Dogger Bank is being deflected towards the east. In addition, one can recognize that in the winter an extensive cyclonic circulation exists around the Dogger Bank which disappears in summer. In the southern North Sea, the magnitude of the velocities in summer reaches less than 1 cm/s, while in the northern North Sea velocities of more than 6 cm/s are present. In contrast, in the winter season the north-south velocity gradient has disappeared. All over the North Sea maximum velocities of up to 4 cm/s are reached.

3.2.1.4 Extreme Situations in Comparision

To demonstrate the range of the variability of the circulation in the North Sea, the calculated circulation patterns of spring 1974 and winter 1974/75 will be compared (Fig. 3.2.5a,b). During winter 1974/75 there existed a basinwide cyclonic circulation with recirculation branches in the northern, central and southern part of the North Sea (Fig. 3.2.5b). In spring 1974, on the contrary, the overall circulation has broken down into smaller, mostly anticyclonic eddies (Fig. 3.2.5a). These immense variations of the circulation can have an important influence not only on the location but also on the concentration of released pollutants. If a large scale cyclonic circulation, as in winter 1974/75, is present, pollutants can leave the North Sea on the shortest way. This is inhibited during spring 1974 by the breaking down of the cyclonic circulation system. On the contrary, under these conditions a continuous inflow of pollutants can lead to their accumulation.

a   Direction

air pressure   hPa

b   Direction

air pressure   hPa

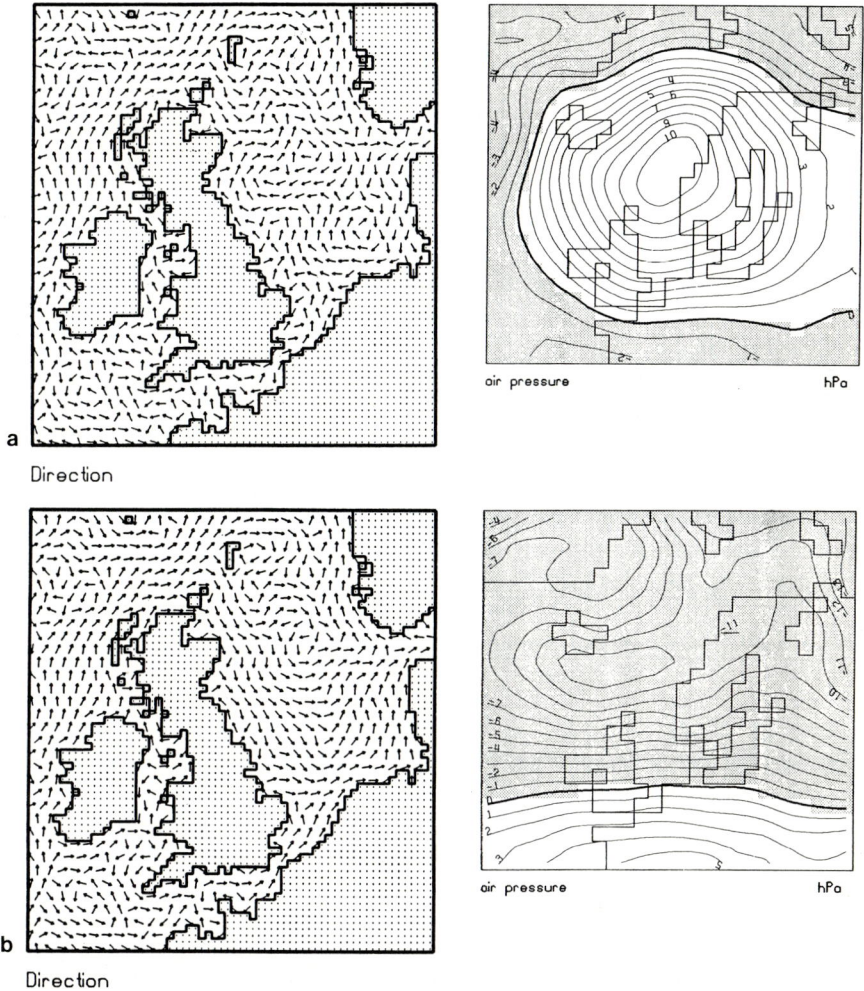

Fig. 3.2.5. Circulation and weather anomalies *a* in spring 1974, *b* in winter 1974/75

The calculated variability as described in Section 3.2.1.1 is caused by the wind- and air pressure distribution. The centre of a low pressure southeast of Iceland causes westerly winds over the whole North Sea during winter 1974/75, whereas in spring 1974, a high pressure system over the northern North Sea is the reason for easterly winds in the North Sea area. This example explains that westerly winds, which dominate the climatological mean situation, guarantee an optimum water mass flushing in the North

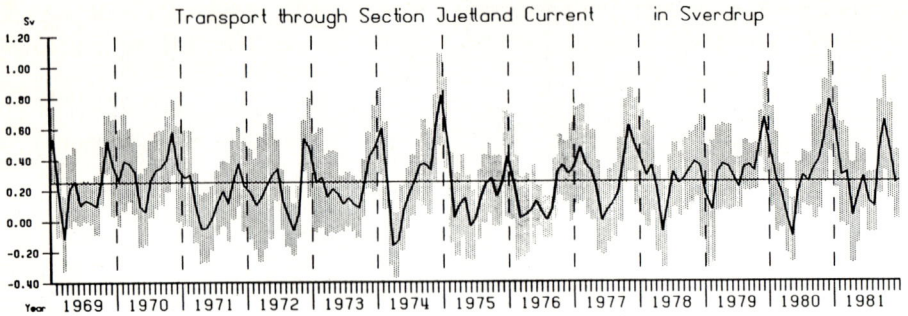

Fig. 3.2.6. Water mass transport through control section off Jutland (in Sv). *Shading* indicates high frequency fluctuations of the flow

Sea. Thanks to this fortunate circumstance, most probably the system North Sea has not collapsed completely up to now, in spite of the enormous damage caused by human impact.

3.2.1.5 Water Mass Transport Through Selected Sections

The observation of water mass transports through selected sections also shows strong temporal fluctuations. For example, the consequences of the extreme situations, as described in Section 3.2.1.4, are clearly visible in the transport rates through a control section of the Jutland Current (Fig. 3.2.6; Backhaus 1989). The curve represents fluctuations with periods above 1 month, whereas shorter term fluctuations appear as shaded areas. The negative values display a southward directed flow. In the year 1974 one can recognize two distinct extrema of the flow. The first is a negative anomaly, which lasts for about two months. It appears in conjunction with the occurrence of small scale anticyclonic eddies described in Section 3.2.1.4. The second event, in the winter 1974/75, is the overall extreme of the record. This coincides with the previously described strong cyclonic circulation driven by extensive west wind activities.

The comparison of these two extreme situations provides an impression regarding the possible range of the variability of the marine weather. It is likely that by taking thermal and haline processes fully into account, the variability of the circulation will increase considerably.

3.2.2   Conservative Transport of Substances

As a first step, the dispersion of conservative substances will be examined. This means that the substances react in the same way as the surrounding water. This assumption seems to be correct for all water soluble materials. In addition, these experiments also give a first approximation of the behaviour of all other substances in the water, because the circulation and the turbulent processes always have a considerable influence on the dispersion of matter.

To be able to drive a transport model for real substances, more information is needed about the properties of these substances, such as biological and chemical reactions, sinking or rising, grazing by zooplankton as well as accumulation on suspended matter. The last two processes make it necessary to gain more knowledge about zooplankton and the dynamics of suspended matter. The latter problem will be discussed in detail in Section 3.2.3.

3.2.2.1   Description of the Transport Model

The model for simulating the dispersion of matter (Hainbucher et al. 1987) has to be provided with information about the physical conditions influencing the dispersion of matter, i.e. the existing circulation and the turbulence. Together, these parameters define the advection and diffussion processes. Presently, these two parameters cannot be measured in a sufficient spatial and temporal resolution, so the input data are provided by the circulation model described in Section 3.2.1.2.

To make the transfer of the parameters as easy as possible, the same grid structure as in the circulation model is used. Due to the length of the simulation period, it is impossible to store the results of every time step. Only daily mean values of the circulation and their daily temporal variance can be transmitted to the transport model.

The advective part of the dispersion depends only on the velocities of the flow field. All processes with a time scale less then 1 day will be parametrized with the help of an eddy diffusion coefficient. This coefficient is a function of the variance of the circulation.

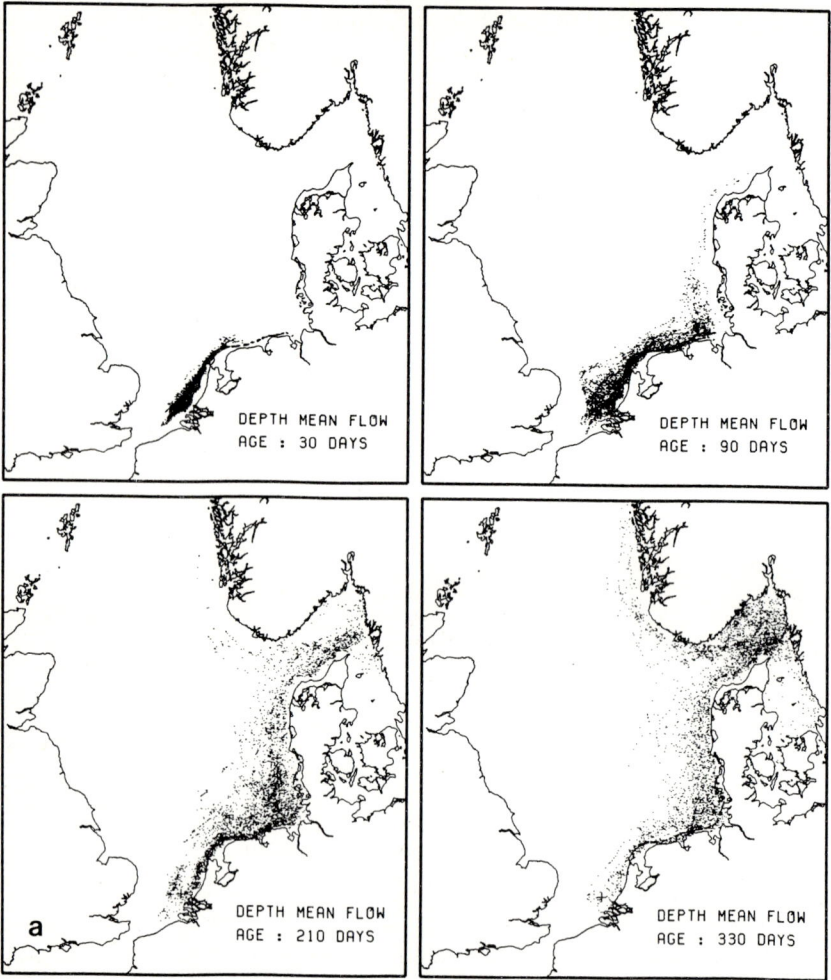

Fig. 3.2.7a. Probability of stay of water particles during 1969-1982. Source: Rhine

3.2.2.2 Single River Releases

The dispersion of matter during the simulation period (1969-1982) was based upon the following concept. Every day of the simulation period, water particles were released at selected rivers. These particles were traced over a 2 years' period. Figure 3.2.7a,b shows the calculated location of every particle, corresponding to the number of days it has already been in the water. This yields the desired distributions for the probability of stay of a particle ensembled at a certain age level.

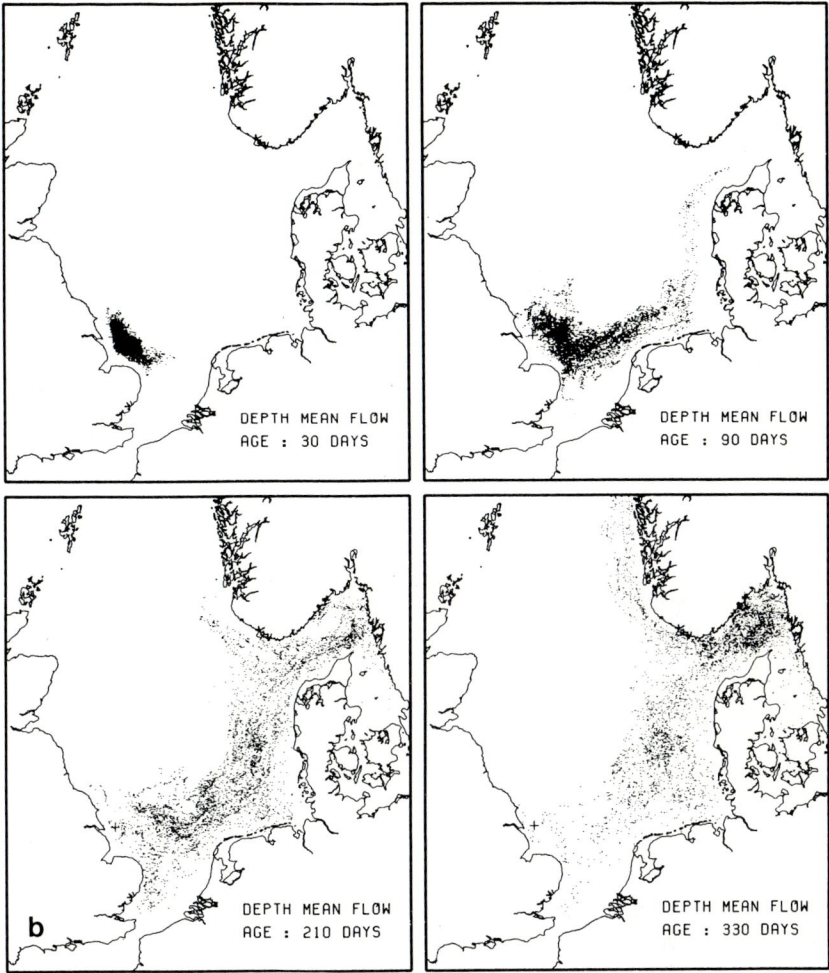

Fig. 3.2.7b. Probability of stay of water particles during 1969-1982. Source: Humber

The difference in the position of particles of the same age is a measure of the variance of the existing circulation during the time of simulation, because in this simulation the circulation is the only quantity that is subject to a temporal variability.

To reduce the effort of the calculations, diffusion processes were neglected and the dispersion of the particles was calculated with the help of vertically integrated circulation patterns.

As an example, the probabilities of stay of particles with ages of 30, 90, 210 and 330 days are shown. They were released at the mouths of Rhine and Humber. Figure 3.2.7a shows for the Rhine release that 90 days after their input, the particles can have very different positions. The largest number is still in the area of the release point, however a small proportion have already reached the German Bight. After 210 days, the possible locations of the particles extend from the Rhine to the Skagerrak, with a maximum concentration in the German Bight. It is obvious that the particles stay in a relatively small band along the continental coast.

Ninety days after the input of the particles at the Humber (Fig. 3.2.7b), a large number of particles still stay near the British coast. Only a small number has travelled through the southern North Sea into the German Bight. After 210 days, the cloud which shows the possibility of stay covers the complete southern and eastern North Sea, but a considerable number of particles are still present off the British coast. Especially high particle concentrations are found in the area of the Dogger Bank which is characterized by a high biological activity. Obviously none of the particles released at the Humber reach the inner parts of the German Bight. This fact proves also true for all the other British rivers. Therefore, it is likely that the German Bight is mainly contaminated by the rivers Rhine and Elbe.

3.2.2.3 The Overall Contamination by All Rivers

The simulated probability of stay for all rivers can be superimposed to give an overall mean of the contamination by all rivers. In accordance with the proportion of the different input magnitudes, the particles of the sources enter the concentration calculations with different weihgts. According to the report of the *Quality Status of the North Sea* (DHI 1984), the following proportions of input rates were chosen in relative units: Rhine 1.00, Elbe/Weser 0.36, Tyne: 0.12, Firth of Forth 0.12, Thames 0.08 and Humber 0.06. Figure 3.2.8 shows the mean contamination for a period from 1969 to 1982 and the corresponding standard deviation. The chosen shading intervals are logarithmic.

The maximum concentrations are found in an area along the continental coast. They are caused by the high amount of more than 50% that the river Rhine contributes to the overall contamination. As already explained in Section 3.2.2.2., the particles released at the mouth of the Rhine form a narrow band along the continental coast.

The British rivers have relatively low input concentrations and therefore only a minor contribution to the maximum contamination; but they alone are responsible for the contamination of the central North Sea, and especially of the Dogger Bank area.

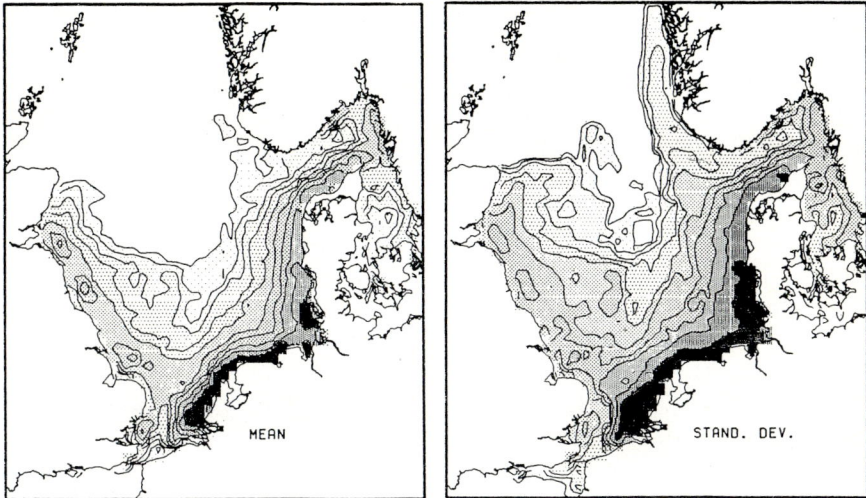

Fig. 3.2.8. Relative contamination of the North Sea by river input. *Left* mean; *right* standard deviation (logarithmic shading intervals)

The standard deviation shown in Fig. 3.2.8 is not a measure of a statistical error, but it demonstrates the variability of the contamination in the North Sea. The latter is caused by the variability of the circulation which, again, is driven by the variations of the atmospheric forcing. The distribution of the standard deviation has a pattern similar to the mean values. Again, the values along the continental coast are extremely high. They are even higher than the maximum mean concentration.

This means that along the continental coast we can find not only the highest mean but also the highest peak contaminations. This is a very important fact for the ecosystem, which is influenced far more by extremes than by the mean situation.

3.2.2.4 Dispersion of Atmospherically Deposited Lead

The input of contaminants through the atmosphere represents an important contribution to the pollution of the North Sea. Figure 3.2.9a shows the atmospheric input of lead in January, 1984 (Krell and Roeckner 1988). It is the result of the meteorological transport model that is described in Section 3.1. In addition to the meteorological conditions, industrial and traffic statistics as well as emission data were used to drive the model. The main feature of Fig. 3.2.9a is the increasing deposition rate near the continental coast.

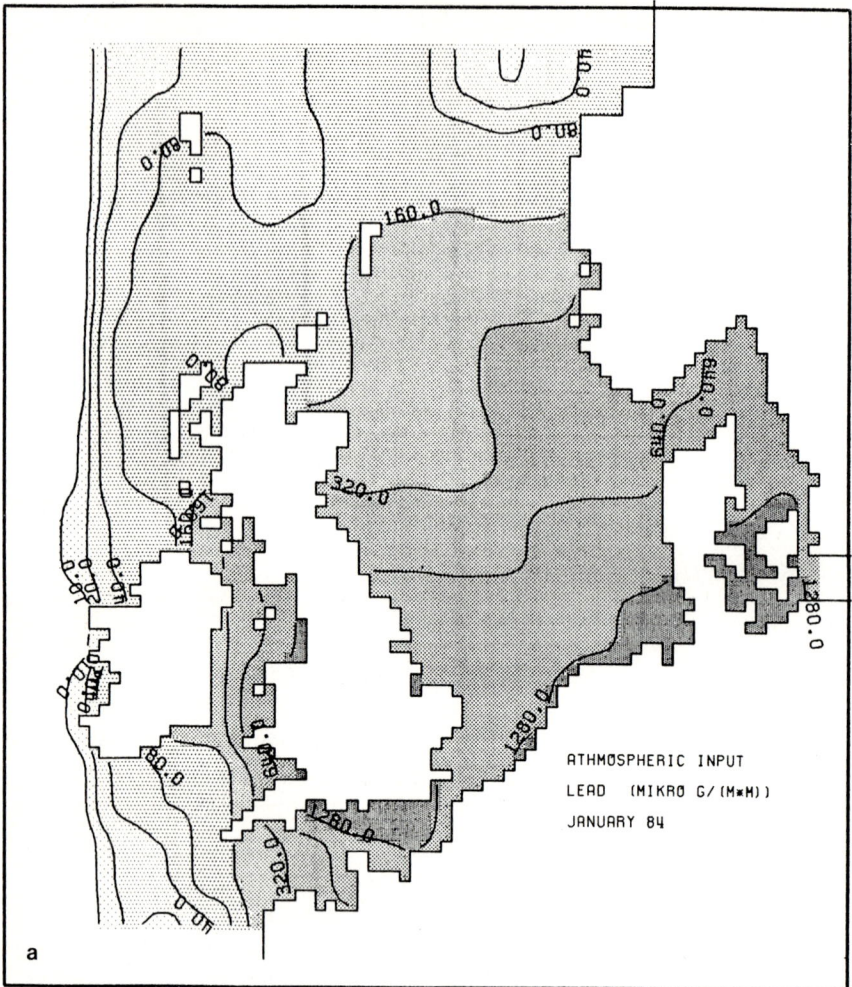

Fig. 3.2.9a. Atmospheric input of lead in the surface layer in January 1984 $(\mu g/m^2)$

The linkage between the atmospheric and the oceanographic model was carried out in the following way. During January 1984 the calculated daily deposition rates over the North Sea were used as input data for the transport model. At the end of this month the input of lead stops. In the following months only the dispersion of lead is observed. This is a rather unrealistic case study but it helps to understand the existing processes.

Figure 3.2.9b-d shows the simulated lead concentration in the surface layer at the end of the months January, March and June 1984. At the end of January (Fig. 3.2.9b) the structures of the atmospheric input are nearly

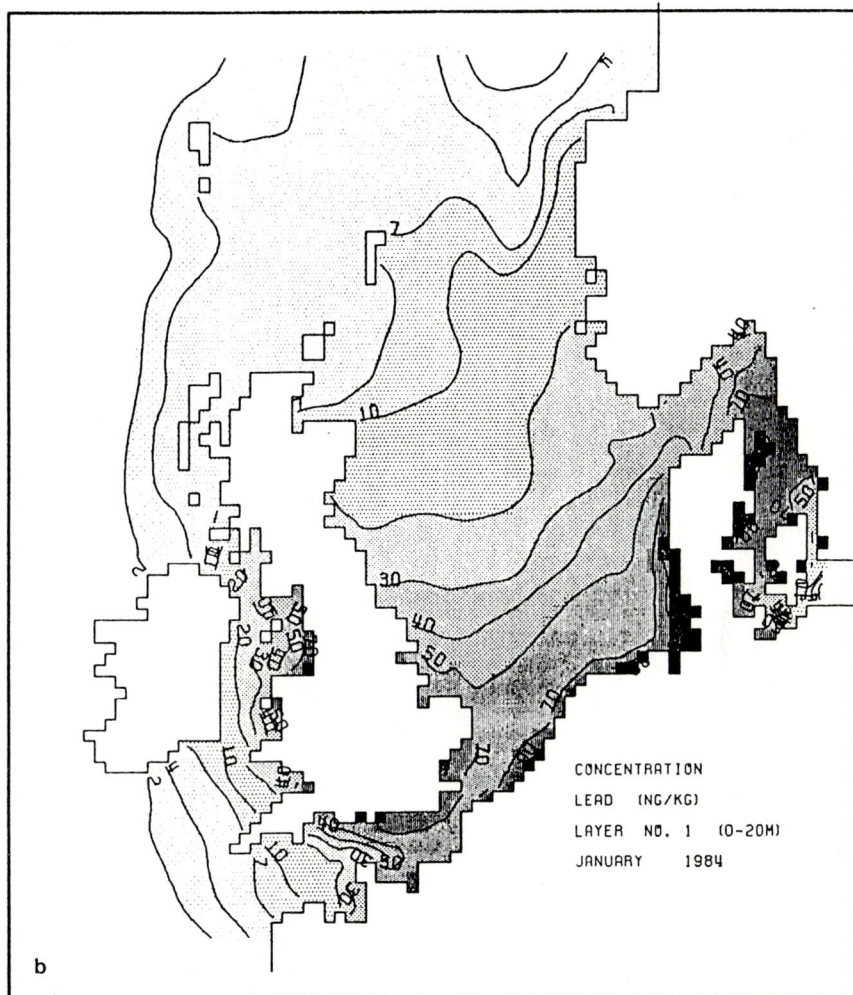

Fig. 3.2.9b. Resulting concentration of lead in the surface layer at the end of January 1984 (ng/kg)

unchanged. Noteworthy is the inflow of highly contaminated water through the English Channel. This is caused by the high atmospheric input in the English Channel area. At the end of March (Fig. 3.2.9c), the maximum concentration in the German Bight begins to decrease. Through the English Channel less contaminated water now enters the North Sea. At the end of June (Fig. 3.2.9d) the lead concentration in the North Sea increases from west to east. This distribution is a result of the dominating cyclonic circulation. Because less contaminated water enters the North Sea mainly in the northwest, the concentrations first begin to decrease in the western part of

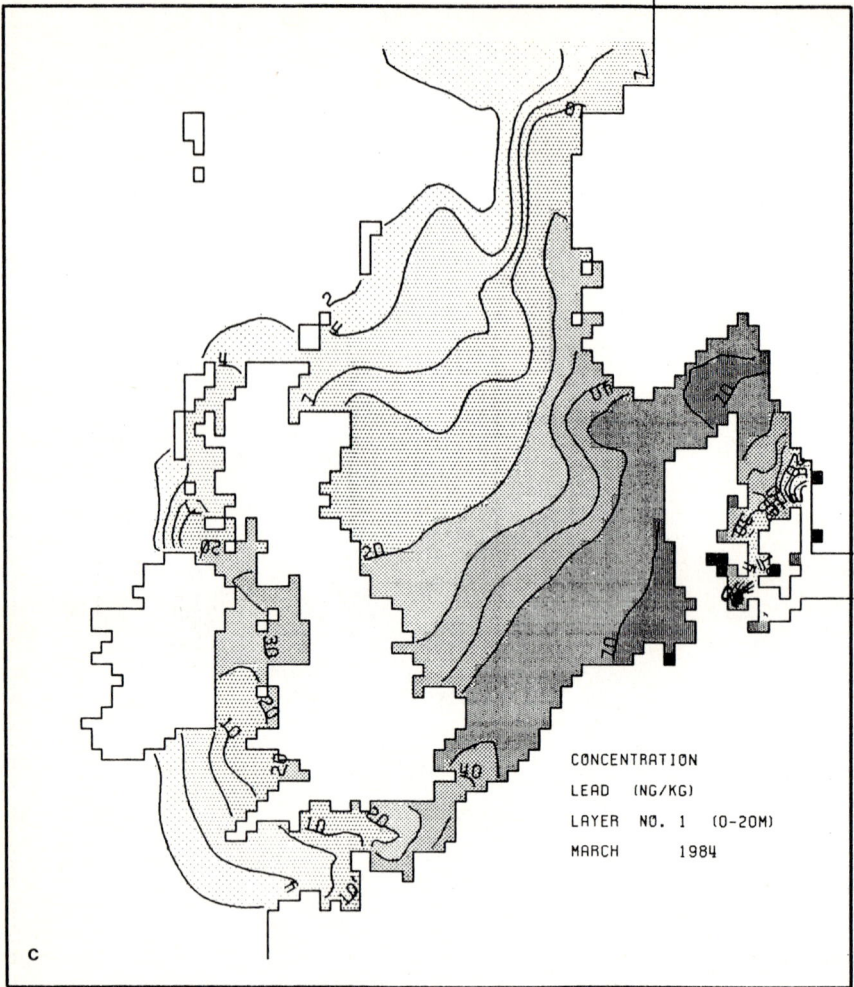

Fig. 3.2.9c. Resulting concentration of lead in the surface layer at the end of March 1984 (ng/kg)

the North Sea. Although the magnitude has decreased, the maximum concentrations are still found in the southeastern North Sea.

The investigations of this section have demonstrated that the area along the continental coast, which receives the highest pollution from river discharges, also is affected by the highest contamination due to the atmospheric deposition.

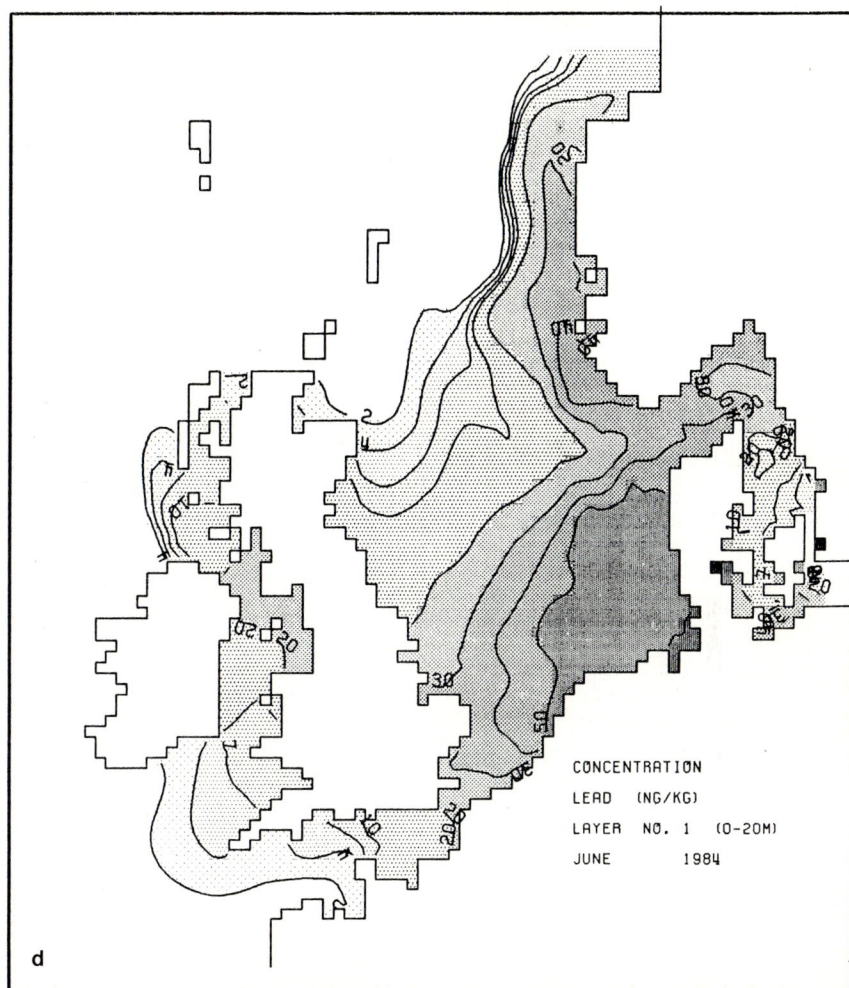

Fig. 3.2.9d. Resulting concentration of lead in the surface layer at the end of June 1984 (ng/kg)

3.2.3 Transport of Suspended Matter

The following sections deal with (1) describing the basics of a numerical model for suspended particulate matter (SPM) transport and (2) comparing model results with measured data. The geographic locations mentioned in the text are shown in Fig. 3.2.10. A comprehensive survey of a former model version is given by Puls and Sündermann (1990).

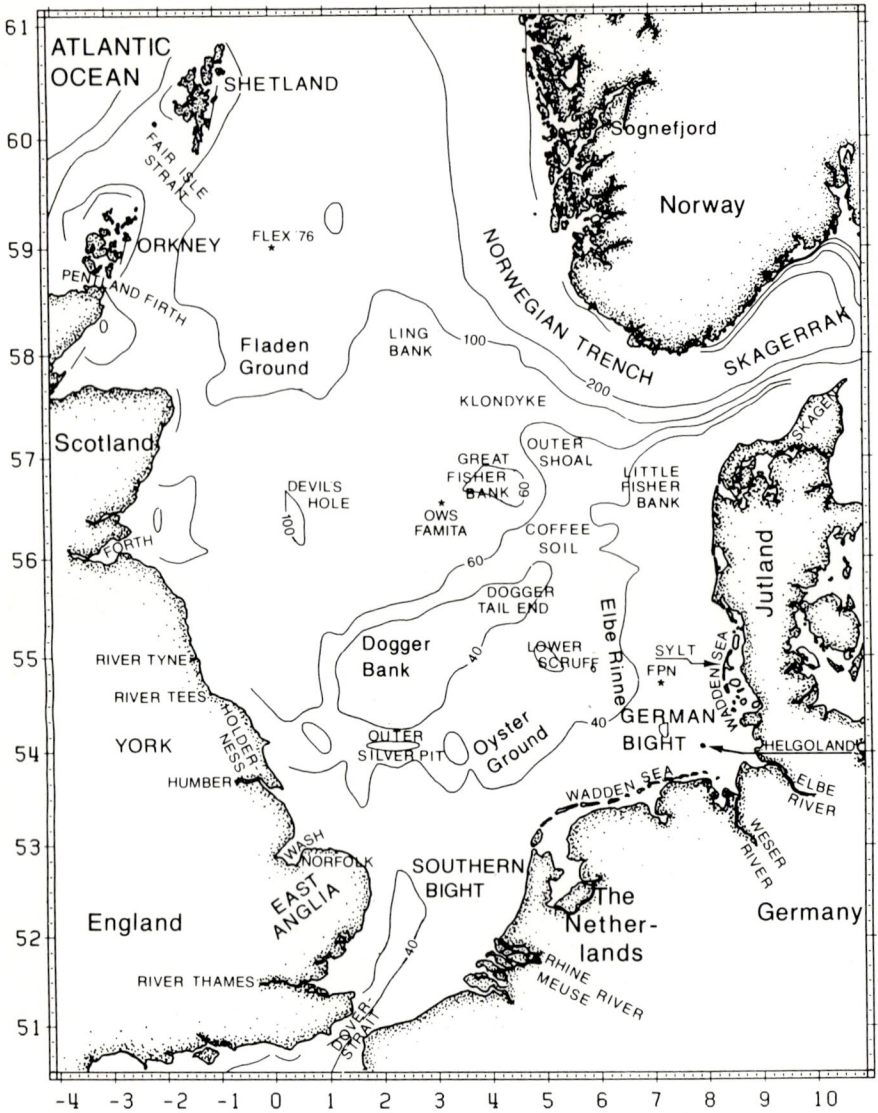

Fig. 3.2.10. Features and locations in the North Sea relevant for investigations of SPM dynamics

The model does not include the simulation of phytoplankton, which may dominate the appearance of SPM in spring, summer and autumn (e.g. Reid et al. 1990). This is not a severe shortcoming when regarding the long-term behaviour of SPM, because most of the phytoplankton, having been produced in spring and summer, is re-mineralized in autumn and winter. In

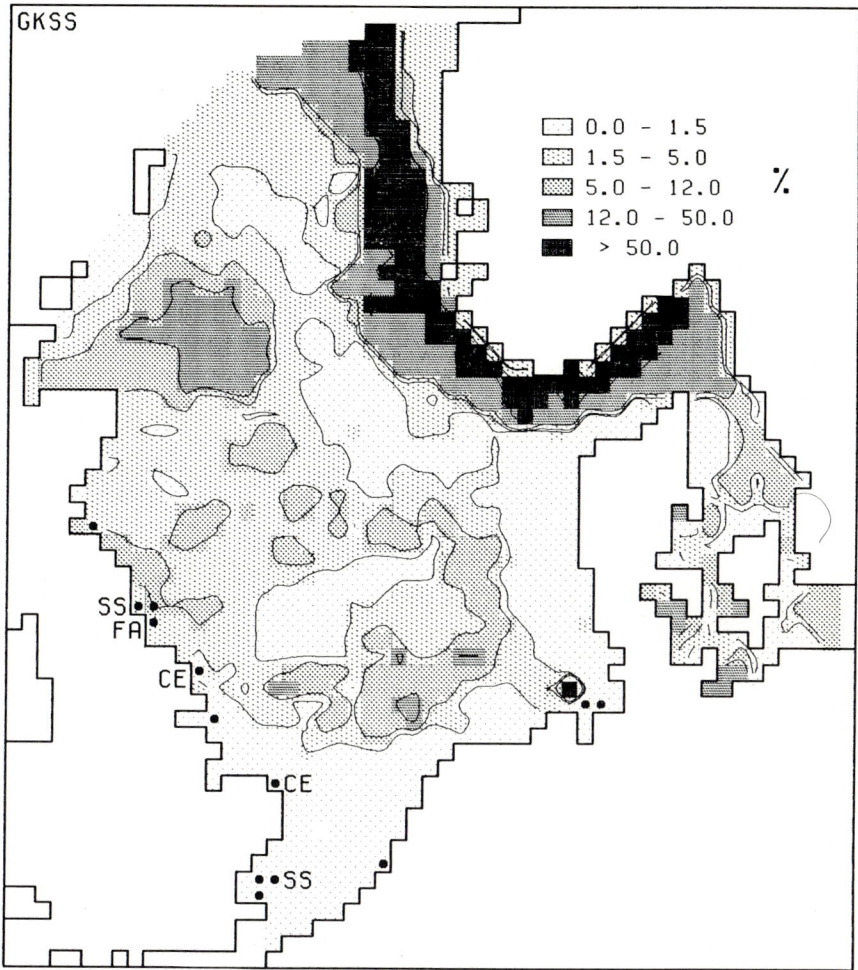

Fig. 3.2.11. Model area with the distribution of fine sediment content; *black spots* input of SPM and contaminants by cliff erosion (*CE*), dumping of sewage sludge (*SS*), dumping of fly ash (*FA*), river input (no notation); *two input points* for the Thames river

any case: a comparison of the simulated SPM concentrations in the water column with measured data (Section 3.2.3.8) is only "allowed" in winter, before the beginning of the first spring phytoplankton bloom.

SPM in the North Sea consists of microflocs of mineral particles and organic matter. In winter, the mass-proportion of organic matter (mainly detritus) in SPM is about 20 % (Schröder 1988). According to Eisma and Kalf (1987b) about 85 % of the mineral mass of grains was smaller than 20 µm in January 1980. The peak of the mineral grain size distribution was between 2 and 5 µm.

The transport model for SPM uses the North Sea part of the shelf model shown in Fig. 3.2.2. The North Sea model area is given in Fig. 3.2.11 which shows the content $c_{20µm}$ (in % of dry mass) of sediment with grain sizes smaller than 20 µm in the upper 2 cm of the sediment bed.

Also marked in Fig. 3.2.11 are the input-points of land-borne SPM. The annual input-masses are given in Section 3.2.3.7.

3.2.3.1 Transport Processes in the Water Body

Many investigators (e.g. Drake and Cacchione 1989; Lyne et al. 1990) use a relationship between the SPM concentration, on the one hand, and the bed characteristics and the present bed shear velocity v_*, on the other. The concentration of sediments in the water column is assumed to be in equilibrium with the local, instantaneous bed shear velocity. Therefore, the modelling of SPM concentration on the US east coast continental shelf by Lyne et al. (1990) is "limited to sediments 24 µm in size or larger (fall velocities > 0.04 cm/s)". For finer sediment, Lyne et al. (1990) argue that "advection from distant sources may be large".

In the model presented here, the transport of SPM 20 µm in size or smaller is simulated, because this is the fraction which carries the lion's share of the particulate contaminants. Advection is one of the dominant processes determining SPM transport. There is no equilibrium between suspended load and the bed shear velocity.

The movement of SPM is simulated by a tracer method. SPM in the North Sea is represented by 40,000 to 60,000 tracer particles. Each particle is "loaded" with a certain mass of suspended matter. New particles enter the model area from land (rivers, cliff erosion, dumping), from the adjacent seas and by bottom erosion. Particles can lose their "load" partly or totally when deposition of SPM occurs. Finally, particles can leave the North Sea by being transported over a seaward boundary. The time step for the transport computations is 40 minutes.

The SPM concentration in a grid box is simply determined by summing up the mass of all particles within the box and dividing by the box volume.

Within the model area, particles migrate through the three-dimensional grid by the following movements:

-     By the three-dimensional residual currents (wind- and density-driven) obtained from the shelf model described in Section 3.2.1. For every

model grid box, the daily residual currents in the two horizontal and in the vertical direction are provided.
- By the three-dimensional currents of the M_2- and the S_2-tide, provided every 40 min.
- By horizontal diffusion. With $A_H = 0.1$ m²/s (constant in space and time) the horizontal diffusion coefficient is rather small. It is only applied for the sake of "completeness". Diffusion processes are simulated in the model by a random walk of particles (e.g. Maier-Reimer and Sündermann 1982).
- By settling, with different settling velocities (see Table 3.2.1) for different sorts of suspended matter.
The settling velocities were partly deduced from measurements (König et al., Chap. 2.6) and partly from comparing the model results with measured data (tuning). Generally, knowledge about SPM settling velocities in the North Sea is very poor. The values determined by Rodhe (1973) in the Skagerrak-Kattegat area are higher than the values given in Table 3.2.1.
- By vertical diffusion with a diffusion coefficient A_z which is vertically constant and which is calculated from

$$A_z[\frac{m^2}{s}] = \begin{cases} 0.16 \cdot v_* & \text{if } v_* \geq 0.0125 \frac{m}{s} \\ 0.002 & \text{otherwise} \end{cases} .$$

The determination of the bed shear velocity v_*[m/s] is outlined in the following section. The water depth is not taken into account for calculating A_z, and the decrease of A_z by vertical density gradients (e.g. Hainbucher 1984) is also not considered.

By relating A_z to v_*, A_z depends both on the current velocities and on the surface water waves. Since v_* quantifies the impact of the water on the sea bottom, the relation $A_z \sim v_*$ is "most valid" near the bed.

The vertical diffusion coefficient A_z as described above is rather crude. Considering, however, that the vertical profile of SPM concentration depends on A_z and on w_s, and considering further the insufficient knowledge of w_s, a more sophisticated approach for A_z would not improve the model results. Moreover, specifying A_z as a function of the vertical coordinate z leads to problems when the SPM concentration is computed by a random walk procedure with a time step of 40 min (Mayer 1991).

Table 3.2.1. Settling velocities of SPM of different origin. The upper line indicates the SPM mass portion having the settling velocity given in the same column

Mass portion:	1/3	1/3	1/3	1/1
SPM origin	Settling velocity (10^{-6} m/s)			
Rivers, cliff erosion, dumping	2	12	80	
North Atlantic, Baltic Sea				2
Dover Strait	2	10	40	
Eroded from the sediment				300
Eroded from the mud blanket				80

3.2.3.2 Bed Shear Velocity

Deposition of SPM and sediment erosion depend on the bed shear velocity v_* which is calculated from:

- Wind and density driven currents plus M_2 and S_2 tidal currents. All currents were computed by the shelf model described in Section 3.2.1.2; a time series is shown in Fig. 3.2.20. The shear velocity is calculated from the currents in the bed-nearest model layer, from the water viscosity (which is 1.5×10^{-6} m^2/s) and from the bed roughness, which is 2 mm for gravel, 200 μm for sand and between 200 and 16 μm for $c_{20\mu m}$, increasing from 2 % to 50 %.

    Figure 3.2.12 shows the distribution of maximum shear velocities during February 1987, determined from the near-bed current velocities. In most parts of the North Sea and especially in the coastal zones of the UK, the Netherlands and Germany, the tidal current is dominant. Off Jutland, wind- and density--driven currents are of similar or higher importance.

    In the central and the northern North Sea, v_* due to currents is computed to be below 0.01 m/s, i.e. maximum current velocities 3 m above bottom are below 0.3 m/s. Tryggestad et al. (1983), however, report maximum current velocities 3 m above bottom of 0.5 m/s at Ekofisk and 0.45-0.8 m/s in the Norwegian Trench. This discrepancy is severe concerning sediment erosion. One reason for the difference between computed and measured near-bottom current velocities might be the coarse vertical discretization of the water column: for a water

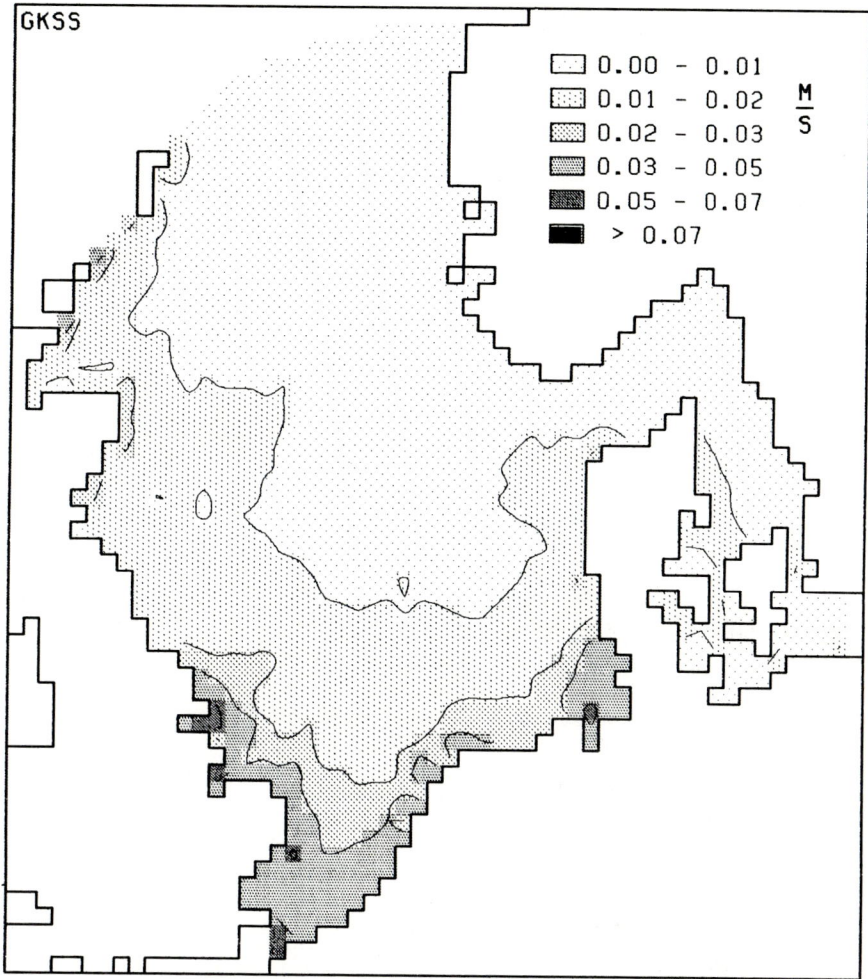

Fig. 3.2.12. Computed maximum bed shear velocity in February 1987, produced by currents (not a synoptic picture)

depth of 300 m, for instance, the bed-nearest model layer is 200 m thick for the model's winter version (50 m for the summer version).

- Sea and swell waves; the wave data were calculated and made available by the Amt für Wehrgeophysik of the German Armed Forces. The wave spectrum is provided every 2 h at 60 points in the North Sea, except for the Kattegat and the Western Baltic Sea, where the significant wave height is calculated from the wind velocity, the fetch length and the water depth (Stephan 1978). The wave height that

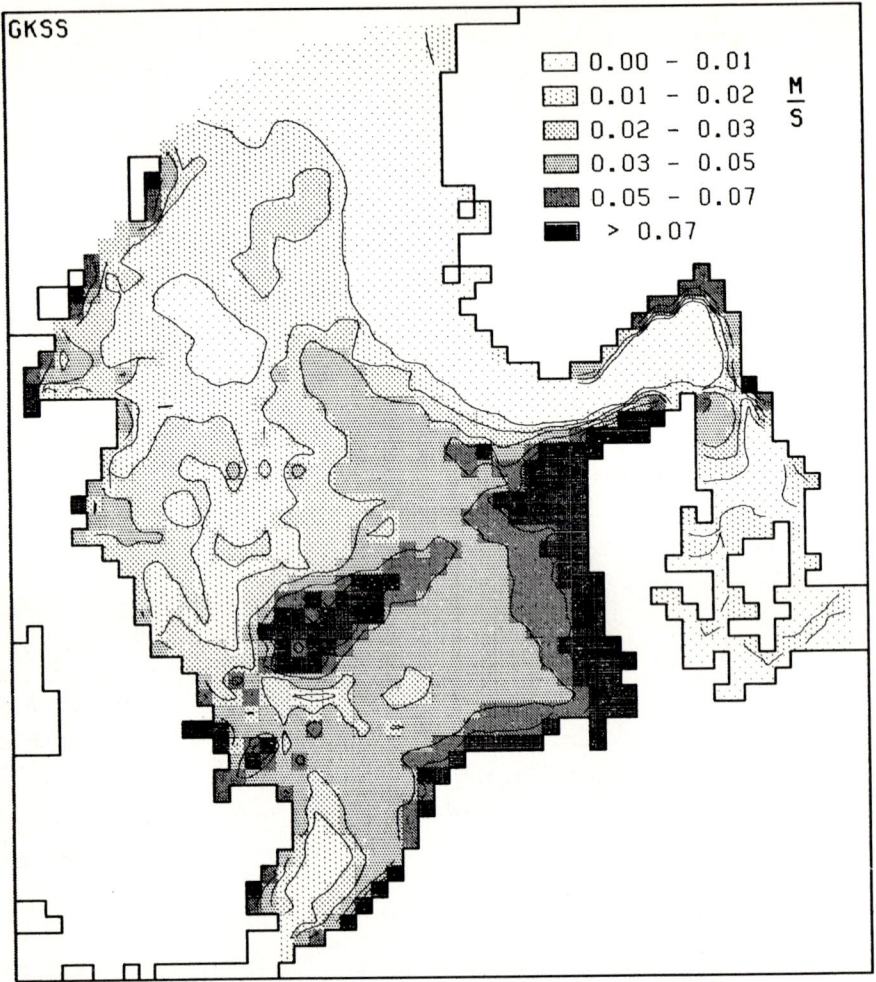

Fig. 3.2.13. Computed maximum bed shear velocity in February 1987, produced by waves (not a synoptic picture)

enters the model computations is the "most probable maximum wave height" (Kokkinowrachos 1980) in a 40-min interval (= model time step).

For computing the bed shear velocity generated by waves, the bed roughnesses given above are multiplied with the factor 2.5 (Dyer 1986).

Figure 3.2.13 shows the distribution of maximum shear velocities during February 1987, determined from sea and swell waves. In shallow water (coastal zones, Dogger Bank) the waves are mostly

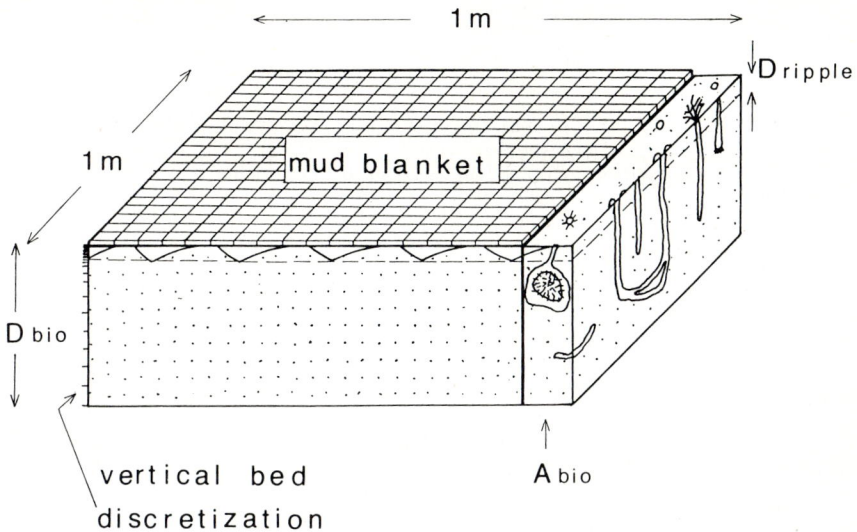

Fig. 3.2.14. Schematized partition of the sediment bed

effective. The predominance of westerly winds finds expression in higher v_*-values in the eastern part of the North Sea. Areas with small wave-generated v_*-values are the Norwegian Trench, the Fladen Ground, the Outer Silver Pit and the Southern Bight.

The bed shear velocity v_* that enters into the formulations for erosion and deposition is calculated from the combined action of waves and currents (Grant and Madsen 1979).

Figure 3.2.20 shows a time series of the computed bed shear velocity at the German North Sea research platform FPN. The v_*-time series is governed by (1) the semi-diurnal (M_2) tidal cycle, (2) the fortnightly spring-neap cycle and (3) several storm events.

3.2.3.3 Bed Characteristics

At the bottom of each water column of the model, the sediment bed is represented by a quadratic bed section of one square meter, see Fig. 3.2.14. The sediment bed thickness which is reworked by bottom fauna is taken as $D_{bio} = 0.2$ m. Nearly all the benthic biomass is found in that layer (e.g. de Wilde et al. 1984). The maximum bioturbation depth in the North Sea, however, extends to more than 0.5 m depth (e.g. due to the crustacean *Callianassa*).

The bed is vertically divided into 43 layers. The layer thicknesses increase from 10^{-4} m near the bed surface to 0.024 m in the lower half of D_{bio}.

The bed in the North Sea is partitioned into three classes:

-    a non-mud bed, having a fine sediment content of $c_{20\mu m} \leq 25$ %
-    a mud bed, having a fine sediment content of $c_{20\mu m} > 25$ %
-    a mud blanket on the surface of the sea bottom.

The mud bed has a uniform dry density ρ_{dry} of 0.55 t/m^3. This value is based on own density measurements of "undisturbed" Skagerrak-mud. In the mud patch in the German Bight (see Fig. 3.2.11) the dry density was found to be 0.4 t/m^3.

The model's non-mud bed has a dry density ranging from 1.85 t/m^3 for pure sand to 0.8 t/m^3 for sediment with $c_{20\mu m} = 25$ %. Similar values result from water contents of North Sea sediment which were measured by Wirth and Wiesner (1988).

The mud blanket is specified by the mass of fine sediment per square meter of the sea bottom. It consists of deposited SPM that is not (yet) mixed into the sediment bed. A mud blanket of 100 g/m^2 is about 0.5 mm thick (assumed dry density: 0.2 t/m^3, Floderus and Hakanson 1989).

Another bed characteristic that influences the fate of suspended matter are current ripples. Current ripples are "defined" here as small-scale transverse bed features with a maximum height of 6 cm (Reineck 1984) that are produced by currents.

A bed square metre is horizontally partitioned (see Fig. 3.2.14) into a portion A_{bio} which is colonized by benthic macrofauna and a portion $(1-A_{bio})$ with "sterile" sediment. In reality the benthic macrofauna is of course more or less uniformly distributed over the square metre. A schematic division as given in Fig. 3.2.14, however, facilitates the mathematical simulation of processes.

The quantification of A_{bio} depends on the biomass of macrobenthos in the sediment. Based on biomass numbers given by Rachor (1982) and Heip et al. (1990), the biomass in the North Sea model's sediment increases with decreasing water depth and increasing content $c_{20\mu m}$ of fine sediment in the bed. The presence of ripples decreases the biomass.

The macrobenthos biomass (and thus A_{bio}) is not constant during the year: it is maximum in the middle of October and minimum (= half of the maximum value) at the end of April (Rachor 1991, pers. comm.). The seasonal variation between 0.5 and 1 times the local maximum value, however, is only valid for a water depth H < 50 m. For H > 80 m, there is no seasonal variation and for H between 50 and 80 m there is a transition.

The maximum value of A_{bio} is $A_{bio} = 0.25$. This value is valid in October in the southern North Sea (H < 50 m) in a muddy bed ($c_{20\mu m}$ > 5 %). The minimum value is $A_{bio} = 0.023$ in the northern North Sea (H > 80 m) in a sandy bed ($c_{20\mu m}$ < 1 %).

3.2.3.4 Deposition

There are two modes of SPM deposition: by settling and by macrobenthos filtration.

Deposition by Settling. The mass $c_b \cdot w_s$ reaches the bed per second and square meter, where c_b is the near bed SPM concentration, determined from the settling velocity and the vertically constant eddy diffusivity A_z. The results do not differ substantially from c_b-values resulting from an expression developed by Teeter (1986) assuming (1) a parabolic vertical distribution of diffusivity and (2) a dependence on the deposition probability p (see below).

Whether deposition occurs or not depends on the bed shear velocity v_* and the critical bed shear velocity for deposition $v_*^{cr,d}$: if v_* is greater than $v_*^{cr,d}$, no deposition occurs; if v_* is smaller than $v_*^{cr,d}$, the portion p = 1 - $v_*^2 \cdot (v_*^{cr,d})^{-2}$ of $c_b \cdot w_s$ is deposited (Krone 1962).

Deposition tests of Creutzberg and Postma (1979) with Oyster Ground mud yielded $v_*^{cr,d} = 0.008$ m/s. This is in agreement with flume tests of Hydraulics Research (1979) which yielded $v_*^{cr,d} = 0.007$ m/s. Rodger and Odd (1985) used $v_*^{cr,d} = 0.01$ m/s for modelling SPM transport in Liverpool Bay.

The values used in this model are

$$v_*^{cr,d} = \begin{cases} 0.008 \dfrac{m}{s} & \text{if } w_s \le 5 \cdot 10^{-5} \dfrac{m}{s} \\[2mm] 0.023 \dfrac{m}{s} & \text{if } w_s \ge 3 \cdot 10^{-4} \dfrac{m}{s} \\[2mm] 0.094 + 0.02 \cdot \log_{10}(w_s) \dfrac{m}{s} & \text{otherwise .} \end{cases}$$

The value $v_*^{cr,d} = 0.023$ m/s is valid for SPM that was eroded from the sediment. The selection of such a high value was necessary for simulating the measured time series of SPM concentration shown in Fig. 3.2.20. In any case, caution is advised concerning the high value of $v_*^{cr,d} = 0.023$ m/s. High $v_*^{cr,d}$-values, however, were also found by Mehta and

Partheniades (1975): in tests using kaolinite; they measured $v_*^{cr,d}$-values ranging from 0.013 to 0.033 m/s (cited in Mehta et al. 1989).

SPM that settles in the A_{bio}-portion of the bed is assumed to be immediately consumed by deposit feeders and stored in the sub-bottom layer 0 - 5 cm in the A_{bio}-portion of the bed (see Fig. 3.2.14). The layer 0 - 1 cm receives 60 % of the deposited material, the layer 2 - 5 cm receives 10 %.

SPM that deposits on the bed area "1 - A_{bio}" forms a mud blanket on top of the sediment (Fig. 3.2.14). If this mud blanket is not re-suspended, it is gradually consumed by macrobenthos, i.e. it is transported into the bed portion A_{bio}. The intensity of this gradual consumption depends on the present macrobenthos biomass and on the proportion of deposit feeders. The maximum value is a substantial decrease of the mud blanket within 2 h (muddy bed, southern North Sea). On a sandy bottom in the southern North Sea the decrease takes place within 1 - 2 days, and in the northern North Sea within about 10 days. The storage in the bed is the same as for the material that is directly deposited.

The comparatively long residence time of mud blankets on sandy bottoms is compatible with observations of Floderus and Hakanson (1989) and Eisma and Kalf (1987a). Bottom photography in 2000 m depth in the Porcupine Seabight, North Atlantic, (Rice et al. 1986) showed the disappearance of a detrital layer within at least 12 days. This disappearance, however, might mainly be due to the degradation of organic matter by bacteria.

Deposition by Filtration. This is proportional

1. to the near bed SPM concentration c_b,
2. to the instantaneous local macrobenthos-biomass, expressed by "A_{bio}",
3. to the proportion SF of suspension feeders in the macrobenthos. This proportion is high (90 %) in a sandy bottom and moderate (30 %) in a muddy bottom. The proportion of deposit feeders in the macrobenthos is "1 - SF"; the biomass of predators is not taken into account.

The maximum filtration rate takes place in the sandy part of the southern North Sea, its seasonal average is 100 liter/m²/day. The minimum (mud, northern North Sea) is about 10 liter/m²/day.

A filtration rate has the same dimension as a settling velocity: 100 liter/m²/day is equivalent to about 10^{-6} m/s. This is smaller than the settling velocities given in Table 3.2.1. It must be taken into account, however, that deposition by filtration takes place continuously, while deposition by settling only happens when the bed shear velocity is below the threshold value

$v_*^{cr,d}$. A filtration rate of 100 liter/m^2/day means the filtration of a 10 m high water column within 100 days.

The mass of suspended matter that is deposited by suspension feeders per square metre and day is

$$\text{"filtration rate"} \cdot c_b.$$

The filtered SPM is stored in the sub-bottom layer 0 - 5 cm, distributed in the same way as the material deposited by settling.

3.2.3.5 Bioturbation

Vertical Bioturbation. Within the sediment bed there is vertical mixing by bioturbation only in the bed portion A_{bio}. This mixing is simulated by a diffusion process, the mixing coefficient is A_z^{bio}.

Pheiffer Madsen and Larsen (1986) report that in the Kattegat the vertical mixing in 50 % of the examined sediment cores could be described by a coefficient A_z^{bio} of more than 10 cm^2/year. Albrecht (1991, pers. comm.) estimated values between 0.5 and 12 cm^2/year in the North Sea from Pb-210-profiles. Boudreau (1986) and Officer and Lynch (1989) use values of the same order for their numerical investigations.

The above given values are "effective" diffusion coefficients; they stand for bioturbation of the whole sediment, i.e. for the full square metre shown in Fig. 3.2.14. Those effective values are multiplied with A_{bio}^{-1} in order to obtain the diffusion coefficients for the simulation of vertical mixing in the bed portion A_{bio}. With the average value $A_{bio} = 0.1$, the model's diffusion coefficient A_z^{bio} is 100 cm^2/year. This value is spatially and temporarily constant, representing the intensity of vertical mixing in the bed portion A_{bio}. Local and seasonal variations of the mixing effectivity are due to biomass variations, i.e. to variations in the size A_{bio} of the vertically mixed bed portion.

Because about 90 % of the macrofauna biomass is situated in the sub-bottom layer 0 - 5 cm (Kröncke 1991, pers. comm.), the mixing coefficient for vertical bioturbation must decrease with increasing sub-bottom depth. There is not enough data to make any definite statement on the vertical distribution of the bioturbational mixing function (Boudreau 1986). As a first approximation, the mixing coefficient $A_z^{bio} = 100$ cm^2/year is taken as the bed surface value, and the coefficient at the sub-bottom depth D_{bio} is chosen to be one order of magnitude smaller than the surface value. The decrease of A_z^{bio} from 100 cm^2/year at the bed surface to 10 cm^2/year at the sub-bottom depth D_{bio} is assumed to be linear.

Horizontal Bioturbation. The sediment portion $(1 - A_{bio})$ is not affected by vertical bioturbational mixing. It is, however, interacting with the sediment portion A_{bio} by intersectioning (horizontal mixing) of the two bed portions. Physically, this is due to horizontal burrowing (e.g. by *Echinocardium cordatum*) or to a change of position of benthic fauna.

The diffusion coefficient for horizontal mixing A_{xy}^{bio} varies with sub-bottom depth: it is assumed to be one order of magnitude smaller than the <u>effective</u> vertical mixing coefficient at the same sub-bottom depth.

The maximum value of A_{xy}^{bio} at the sediment surface applies to a muddy bed $(c_{20\mu m} > 5 \%)$ and $H < 50$ m in October (seasonal maximum): $A_{xy}^{bio} = 2$ cm^2/year. For the same conditions, but at the sub-bottom depth D_{bio}, A_{xy}^{bio} is 0.2 cm^2/year. In a sandy bottom in the northern North Sea those values are one order of magnitude smaller.

For the sake of simplicity, the two-dimensional horizontal mixing is simulated in the model by a one-dimensional horizontal shifting velocity v_{hs}. Per time step the sediment part A_{bio} (virtually) moves the distance $v_{hs} \cdot \Delta t$ into the part "$1-A_{bio}$", the part "$1-A_{bio}$" moves the same distance in the opposite direction. This results in an intersectioning (mixing) of the two sediment parts. The compatibility of the two processes (two-dimensional diffusion on the one hand and one-dimensional shifting on the other) was shown by numerical tests; details cannot be given here. A result of the tests: a horizontal mixing coefficient A_{xy}^{bio} of 1 cm^2/year is equivalent to $v_{hs} \approx 1$ m/year.

3.2.3.6 Erosion

The model conception for the erosion of a non-mud bed is: a surface sand layer of the thickness D_e is stirred up when v_* exceeds $v_*^{cr,e}$, which is the threshold shear velocity for erosion at the sediment surface. Only the present content of fine sediment leaves the bed and can be transported away by the currents. The fate of the fraction > 20 μm is not considered because it is not a "pollution carrier". Only the fine sediment is of interest for the model computations.

The value used for $v_*^{cr,e}$ on a sandy bottom is 0.028 m/s; this is the minimum value for simulating the measured time series of SPM concentration shown in Fig. 3.2.20.

In a sediment mixture of sand and mud the model's value of $v_*^{cr,e}$ had to be increased, resulting from the observation that fine sediment is accumulated in the southern part of the Oyster Ground in a water depth between 30 and 40 m (Zuo et al. 1989). During storms, v_* in that area increases above 0.028 m/s; the mud would gradually be eroded away. To avoid this erosion, $v_*^{cr,e}$ had to be increased to a maximum value of 0.040 m/s for a bed with a high admixture of fine sediment:

$$v_*^{cr,e} = \begin{cases} 0.028 \dfrac{m}{s} & \text{if } c_{20\mu m} \leq 2 \% \\[2mm] 0.040 \dfrac{m}{s} & \text{if } c_{20\mu m} \geq 10 \% \\[2mm] 0.0015 \cdot c_{20\mu m} + 0.025 \dfrac{m}{s} & \text{otherwise} . \end{cases}$$

Recent investigations in 1991 with the EROMES system near Helgoland (sediments with $c_{20\mu m}$ between 5 and 50 %) show $v_*^{cr,e}$-values between 0.02 and 0.03 m/s at the sediment surface. Sediment with $c_{20\mu m}$ of about 15 % shows the highest $v_*^{cr,e}$-values. A description of EROMES is given by Schünemann and Kühl (1991).

Remark: the Shields diagram (Shields 1936) that is based on grain sizes, gives too small $v_*^{cr,e}$-values, because it does not include bed armouring (e.g. Führböter 1983).

The erosion formula for non-mud is:

$$D_e = 0.0014 \frac{v_*^2 - (v_*^{cr,e})^2}{(v_*^{cr,e})^2}, \qquad v_* > v_*^{cr,e}$$

where D_e [m] is the erosion depth and v_* is in m/s. Erosion of non-mud is an instantaneous process. The coefficient of 0.0014 in the D_e-formula was deduced from observed SPM concentration increases above a sandy bottom during storm events (see Fig. 3.2.20).

Assuming a maximum bed shear velocity of about 0.1 m/s, maximum D_e-values are about 16 mm. Drake and Cacchione (1989) assume an erosion depth during the largest storms of the order 3 - 5 mm. Lyne et al. (1990) use the erosion depth "as an adjustable parameter, typically 2 - 10 mm of the seabed".

A particular erosion mode: when ripples (height $D_{ripple} \geq 0.5$ cm) exist on the bed, and if $D_e < D_{ripple}$, the eroded mass of fine sediment is (as usual) computed to be the mass in the layer D_e. But instead of removing this sediment mass from the eroded layer D_e, it is removed evenly from the sub-bottom depth 0 - D_{ripple}.

Erosion of a mud bed is a time dependent process, i.e. the bed is eroded successively layer by layer. Mud erosion is calculated by the formula

$$\epsilon = M \cdot \left[v_*^2 - (v_*^{cr,e})^2 \right], \qquad v_* > v_*^{cr,e}$$

where ϵ is the erosion rate (t/m^2/s). The formula is based on flume experiments carried out by Hydraulics Research (UK) with estuarine muds (Puls 1984; Rodger and Odd 1985). M is the erosion constant and is equal to M = 10^{-4} t/s/m^4.

Based on erosion tests, the same value for $v_*^{cr,e}$ was taken for a mud and a non-mud bed in a former model version (Puls and Sündermann 1990). Recent EROMES tests in the German Bight have shown that the muddiest sediment ($c_{20\mu m} > 25$ %) is easier to erode than sediments with $c_{20\mu m} <$ 25 %. So we keep to the "concept" of the former model version and use $v_*^{cr,e} = 0.028$ m/s for a mud bed, which is the same value as for a sandy bed ($c_{20\mu m} \leq 2$ %).

A mud blanket is eroded en bloc if v_* exceeds $v_*^{cr,e}$. The threshold shear velocity for erosion is 0.028 m/s.

Sediment Erosion by Trawling. Erosion by trawling is switched off now in the model. Especially in areas with a low bottom stress (e.g. the northern North Sea) erosion by beam trawling can outweigh erosion by currents and waves. The model uses a trawling penetration depth of 5 mm in sand ("hard ground") and 30 mm in mud ("soft ground").

According to Rauck (1985), in certain areas of the southern North Sea "each square meter on average is fished three to five times a year". A survey of trawling effects in the North Sea is given by ICES (1988) and Anonymous (1990). Dronkers et al. (1990) emphasize the importance of human activities (trawling, dredging, dumping) on the sediment budget in the southern North Sea. Churchill (1989) reports on the effect of trawling over the continental shelf off Long Island Sound. Trawling in the Baltic Sea is reported on by Krost et al. (1990).

Bio-Erosion. An erosion mode that is not simulated by the model is bio-erosion. It is, for instance, produced by the ophiuroid *Amphiura filiformis* in the muddy parts of the southern North Sea. *Amphiura* whirls up mud from the bottom and filters the highly concentrated near-bottom suspension. Sediment is also whirled up by fishes looking for food (e.g. the haddock) or by hiding in the sediment (flatfish). The crustacean *Callianassa* excavates the sediment in sub-bottom depths of more than 35 cm and "throws" this sediment to the bed surface, where it is exposed to the attack of currents and waves. The same is true for sediment that is egested by infaunal species (e.g. the worm *Echiurus echiurus*) to the sediment surface.

3.2.3.7 Input of Suspended Matter

The model's locations of suspended matter input are drawn in Fig. 3.2.11. The input masses (in 10^3 t of fine sediment with grain size < 20 µm/year) are listed below:

Elbe: 500

Thames: 80

Tyne/Tees: 30

Norfolk cliffs: 400

Weser: 268

Humber: 50

Forth: 20

Holderness cliffs: 500

Rhine/Meuse/Scheldt: 750

London sewage sludge: 160

North England sewage sludge: 17

North England fly ash: 340

The river data was obtained from SRU (1980), from McCave (1987), from Landesamt für Wasser und Abfall Nordrhein-Westfalen (1989) and from Hupkes (1990). The cliff erosion data was given by McCave (1987), the dumping information by Wood and Franklin (1983) and by the Bundesminister für Umwelt, Naturschutz und Reaktorsicherheit (1987).

The input of colliery waste (North England) was not taken into account because it mainly consists of rocks. Dredged sediment from harbours and estuaries is also not considered, because it is mainly of marine origin and therefore not an "original" input.

The SPM input is evenly distributed over the year for the UK rivers and for dumping. For the Rhine river the annual variation of suspended load has been measured directly (Landesamt für Wasser und Abfall Nordrhein-Westfalen 1989); the input is highest in January, March and April.

For the Elbe and the Weser rivers, the greater probability of high fluvial water discharges (Kracht und Dietze 1990) leads to higher input numbers in winter and spring (seaward migration of the estuarine turbidity maximum). Cliff erosion is mainly due to wave impact; it is assumed to occur only between October and April.

At the seaward model boundaries the concentration of suspended matter is prescribed. Whether SPM enters the North Sea or not depends on the direction of the currents and on the efficiency of horizontal diffusion. The following concentrations are used:

Dover Strait: 3.2 mg/l in January - March
                     1.5 mg/l in April - June
                     1.3 mg/l in July - September
                     2.8 mg/l in October - December
Pentland Firth: 0.6 mg/l
Fair Isle (between Orkney and Shetland): 0.4 mg/l
Norwegian Sea (between Shetland and Norway): 0.3 mg/l
Baltic Sea: 1.0 mg/l.

The numbers for the Baltic Sea and for the northern boundaries were taken from Eisma and Kalf (1987a); additional information about the Baltic Sea is given by Brügmann et al. (1985) and Ingri et al. (1991). The Dover Strait numbers come from (1) CNEXO-measurements, cited in van Alphen (1990), (2) from an annual mud supply of 17×10^6 t/year (van Alphen 1990) through the Dover Strait and (3) from assuming that 70 % of SPM in the Dover Strait is smaller than 20 µm (Eisma and Kalf 1979).

There is a discussion about the correct net supply through the Dover Strait. Eisma (1981) gives 10×10^6 t/year. Postma (1990) argues that the net supply "would in any case be significantly smaller than 10×10^6 t/year" - he estimates 2.5×10^6 t/year.

The supply numbers include organic matter; the material entering the North Sea from the Norwegian Sea, for instance, has an organic mass content of about 50 % in winter (Eisma and Kalf 1987b).

3.2.3.8 Comparison of Measured and Simulated SPM Concentrations

The model results are compared with five measurements of SPM concentration. The measurements took place:

-   in May and June 1986, three satellite pictures of CZCS; this comparison is done in Chapter 4.1,
-   in January - February 1980, Eisma cruise, reported by Eisma (1981),
-   in October 1985 at the German North Sea research platform (FPN) in the German Bight,
-   in November - December 1986, time series at FPN,
-   in January - February - March 1987, ZISCH star cruise.

For comparing the model results with the measured data, the following two periods were simulated: January 1978 - February 1980 and June 1984 - March 1987.

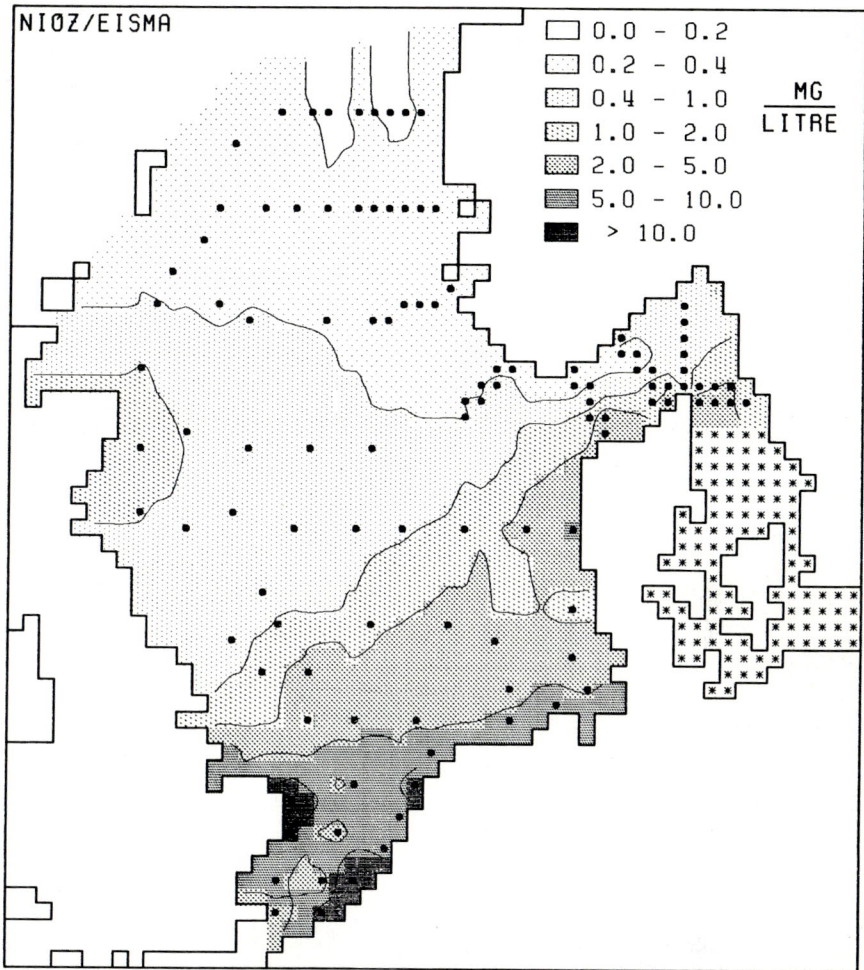

Fig. 3.2.15. SPM concentration (vertically averaged), measured between January 3 and February 11 1980 by Eisma (1981)

For each of the two simulated periods the computation started with a SPM distribution, a mud blanket and a fine sediment distribution that resulted from a "warming up" computation of at least 2 years.

Eisma Cruise 1980. A cruise in the North Sea was carried out between January 3 and February 11 1980 by Eisma (1981). Figure 3.2.15 shows the measured (vertically averaged) SPM distribution. The spots indicate the measurement's positions - the spot's density is highest in the Norwegian Trench. Between the spots the SPM concentration was inter- or extrapol-

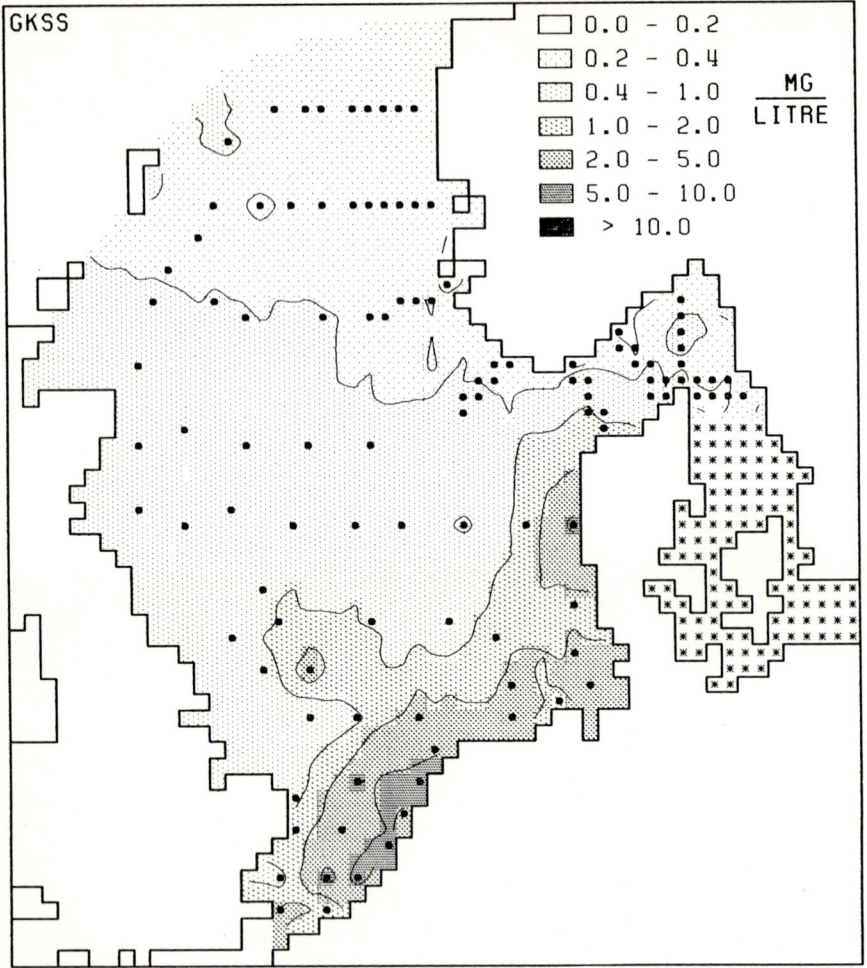

Fig. 3.2.16. Computed SPM concentration (vertically averaged), Eisma cruise 1980

ated, respectively. In areas with a "complicated" pattern of SPM concentration, like the coastal zones off Humber and Wash, this extrapolation is not reliable, of course. There are no measurements in the Kattegat and in the Baltic Sea; this area is covered by stars. It is self-evident that the distribution shown in Fig. 3.2.15 is not synoptic.

In order to compare the model results with the measured data, a method was applied which is best described by an example: one of the Eisma cruise measurements was carried out SE of Helgoland [position in the model: grid point (43,38)] on January 25, 1980, 8.00 GMT. When the simulation run passed that particular time, the computed, depth averaged

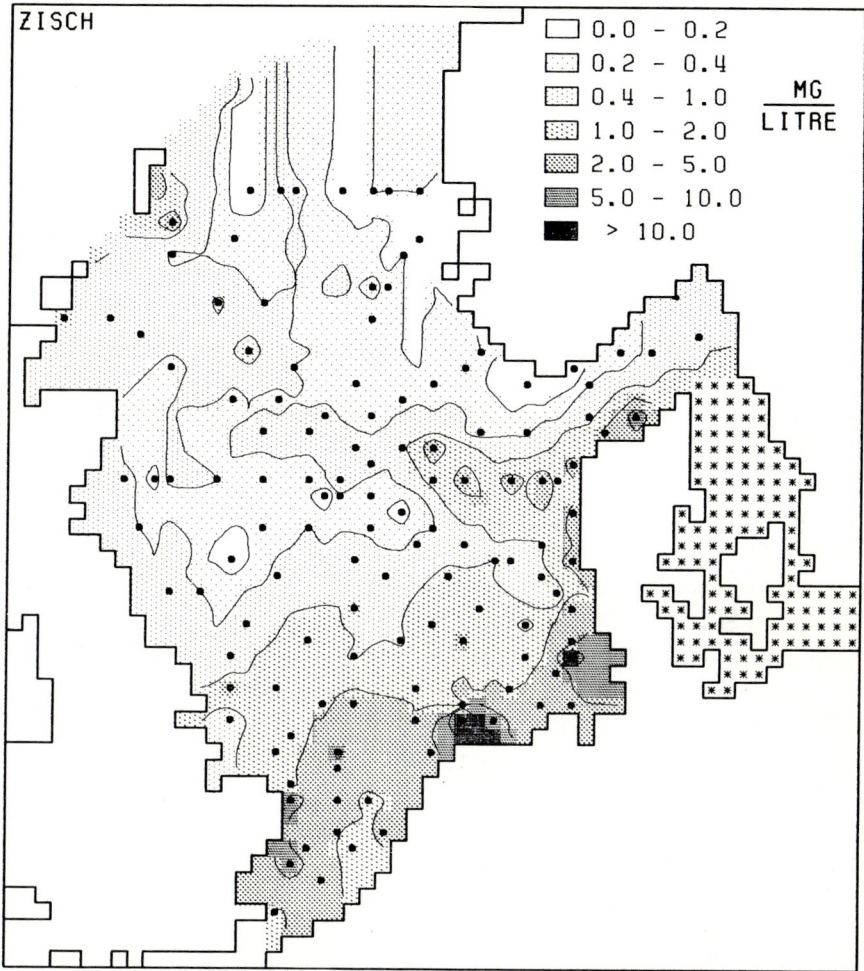

Fig. 3.2.17. SPM concentration (water depth: 10 m), measured between January 28 and March 3 1987 by Schröder (1988)

suspended matter concentration at grid point (43,38) was recorded and stored in a file. After having simulated the whole period of the Eisma cruise, the file contained for each measured SPM concentration a corresponding computed value.

Figure 3.2.16 shows the computed (vertically averaged) SPM distribution. The comparison of measured and computed SPM concentrations shows:

-     a good agreement of the 0.4 mg/l-line in the northern North Sea

-     too small computed values east of Skagen
-     in the southern North Sea the computed concentrations are generally
      smaller than the measured ones. This is most marked off East Anglia.

The last item reveals one of the model's weak points: the non-simulation of
the SPM accumulation in coastal zones. SPM accumulations off East Anglia
are supposed to be produced by onshore directed bottom currents (e.g.
Eisma and Kalf 1987a). The shelf model is too coarse for simulating such
effects.

The question arises whether the SPM shown in Fig. 3.2.16 is mainly
eroded fine sediment or whether it is SPM from external sources. In this
chapter, allochthonous SPM is defined as SPM having entered the North
Sea from external sources (rivers, cliff erosion, dumping, adjacent seas) and
having existed in the North Sea only in the water body or in the mud
blanket. Autochthonous SPM is eroded fine sediment. Particularly original
allochthonous SPM that was incorporated in the sediment and then
re-eroded is thereafter counted as "autochthonous" SPM.

In the central and the northern North Sea, the computed SPM (Fig.
3.2.16) consists of allochthonous material. Autochthonous matter prevails
only along the Jutland west coast and at the isolated increased concentration
on the Dogger Bank. In the coastal zone between the Rhine and Sylt Island,
allochthonous and autochthonous SPM is of equal importance.

ZISCH Star Cruise. The next comparison of measured and computed SPM
concentrations is done for the ZISCH star cruise in January - March 1987.
Fig. 3.2.17 shows the concentrations measured in a water depth of 10 m.

The computed values (Fig. 3.2.18) are averaged over the water column
from the water surface down to the 20-m horizon (or down to the sea floor
if the total water depth is less than 20 m). The comparison shows too small
computed SPM concentrations

-     in the Skagerrak
-     off north-west Jutland
-     off the Dutch and the German Wadden Sea
-     off East Anglia

There is one measured SPM concentration that again hints at a weak
point of the model: the concentration of more than 10 (exact value: 32)
mg/l off the Dutch Wadden Sea. This high concentration reflects the
influence of the Wadden Sea. The model, however, simulates no interaction
between the Wadden Sea and the open sea.

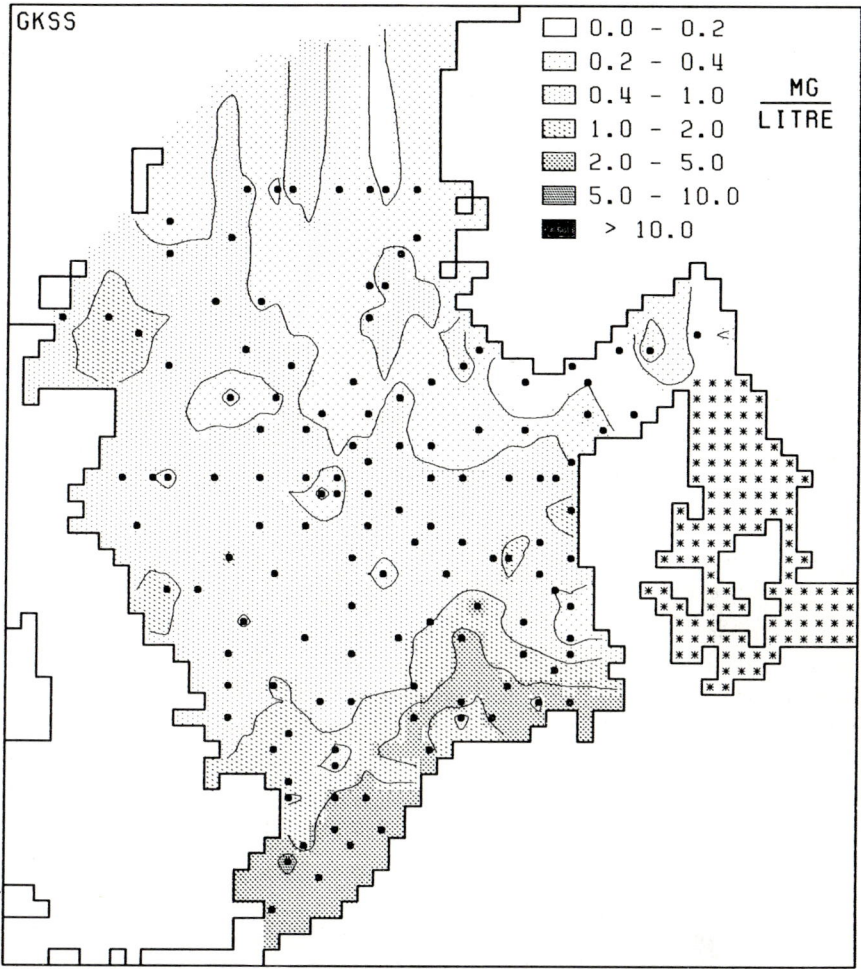

Fig. 3.2.18. Computed SPM concentration, vertically averaged over the upper 20 m of the water column, ZISCH star cruise 1987

As for the Eisma cruise, the SPM in Fig. 3.2.18 consists mainly of allochthonous matter. Autochthonous SPM dominates in the coastal zones of the southern North Sea.

FPN Measurements in October 1985. From the 1st to the 10th of October 1985, SPM concentrations were measured at the German North Sea Research Platform (FPN) in the German Bight. The time series of the measured values is given in Table 2.6.2 of Chapter 2.6. In Fig. 3.2.19 the

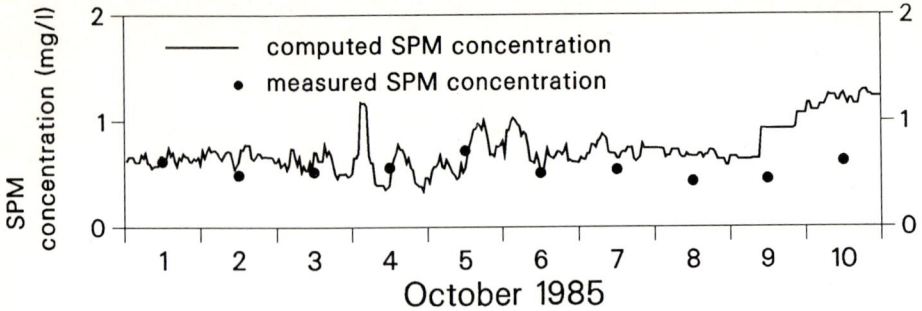

Fig. 3.2.19. Time series (October 1-10 1985) of measured and computed SPM concentration; measured SPM concentration from König et al. (Chap. 2.6, Table 2.6.2)

measured data points for "all depths" are plotted together with the time series of the computed depth averaged SPM concentrations.

Both the measured and the computed SPM concentrations are fairly constant with time - the level of the computed concentrations is higher. The model simulates a slight erosion event on October 5th, resulting in a slight (bifurcated) increase of the computed SPM concentration. The measured data also show an increase on that day. On October 9th the computed concentration increases due to a greater SPM inflow from the south. The measured time series does not increase significantly on that day. However, not the details of the time series are important, but the general levels of the measured and the computed concentrations. These levels are not exactly identical, but the result is satisfying, above all when considering the differing SPM concentration levels during the non-storm conditions shown in Fig. 3.2.20.

The vertical distribution of the measured SPM concentrations is nearly uniform. This is also true for the model's results.

FPN Measurements in November-December 1986. Figure 2.6.4 of Chapter 2.6 shows SPM concentrations measured at FPN between November 26 and December 10 1986 by water sampling; the samples were taken at a depth of 15 m (total water depth: 28 m). That data is shown once more in Fig. 3.2.20, complemented by SPM concentrations at the same water depth, based on turbidity measurements between November 13 and November 25. The whole period is characterized by four erosion events, beginning on the 18th and the 23th of November and on the 1st and the 9th of December, respectively.

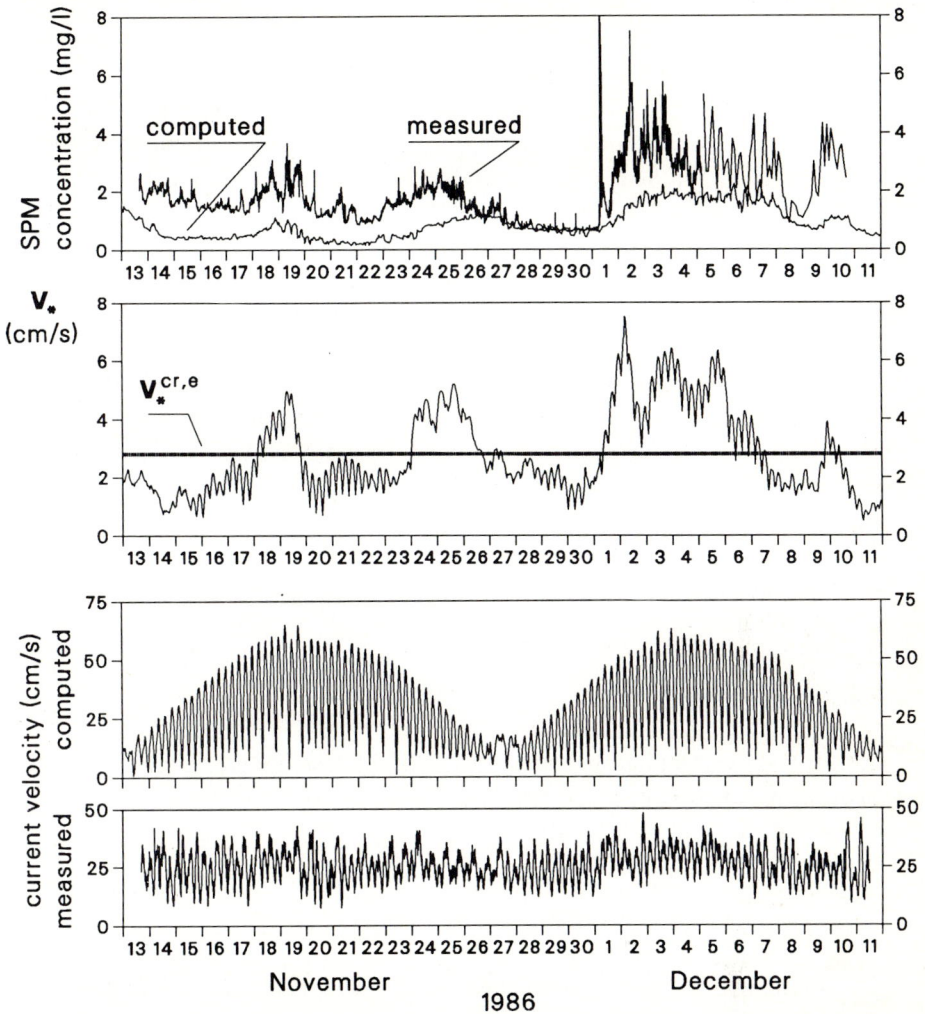

Fig. 3.2.20. Time series (November 14 - December 10 1986) of SPM concentrations (measured and computed), current velocities (measured and computed) and bed shear velocity (computed only); measured data from König et al. (Chap. 2.6)

The time series of the depth averaged SPM concentrations, computed at FPN for the period November 13 to December 10 1986, is plotted together with the measured data. Figure 3.2.20 also shows (1) the computed bed shear velocity v_*, (2) the computed current velocity (for the surface layer 0-20 m) and (3) the measured current velocity at a water depth of 11 m. The threshold bed shear velocity for erosion $v_*^{cr,e} = 0.028$ m/s was deduced from the computed v_*-time series and the observed erosion events; $v_*^{cr,e}$ is plotted in the v_*-diagram as a thick line.

Comparing the measured and the computed SPM concentrations in Fig. 3.2.20, three observations can be made: (1) Three of the computed erosion events start at the same time as observed, the second event (November 23th) starts with a delay of 1 day. (2) The basic level of SPM concentrations (i.e. during calm weather) was measured to be between 1 and 2 mg/l; the basic level of computed SPM concentration was only between 0.2 and 1 mg/l. (3) The computed increases of the SPM concentration during storms are generally smaller than the observed increases.

The comparison of computed and measured current velocities is disappointing. The observed current velocities are between 10 and 50 m/s with no distinct time variability, while the computed values are between 0 and 70 m/s with a pronounced spring-neap cycle. The comparison reveals a main shortcoming of the model's present stage: the M_2- and the S_2-tide are computed independently and are thereafter simply superposed.

3.2.3.9 Accumulation and Erosion Areas

Concerning accumulation and removal of fine sediment, there are three types of sediments in the North Sea:

a) Areas with relict sediment (mostly till) which (1) resists erosion and where (2) no SPM is accumulated. Often, the fine particles have been washed away, leaving the coarse sediment behind.

b) Areas with a fine sediment content $c_{20\mu m}$ that is the result of a dynamic equilibrium of erosion, bioturbation and deposition; this equilibrium developed during the last millenia. The equilibrium can schematically be described by

$$D_e \cdot 0.01 \cdot c_{20\mu m} \cdot \rho_{dry} = \text{"settling rate"} \cdot T_e$$

where "settling rate [g/m^2]" includes both normal settling and filtration by benthic fauna, T_e is the typical time period between two successive erosion events (storms or peak tidal currents) and D_e is the typical erosion depth of such an event.

The fine sediment fraction $c_{20\mu m}$ at a certain site is determined from the above "equilibrium equation". In most cases, this sediment type is

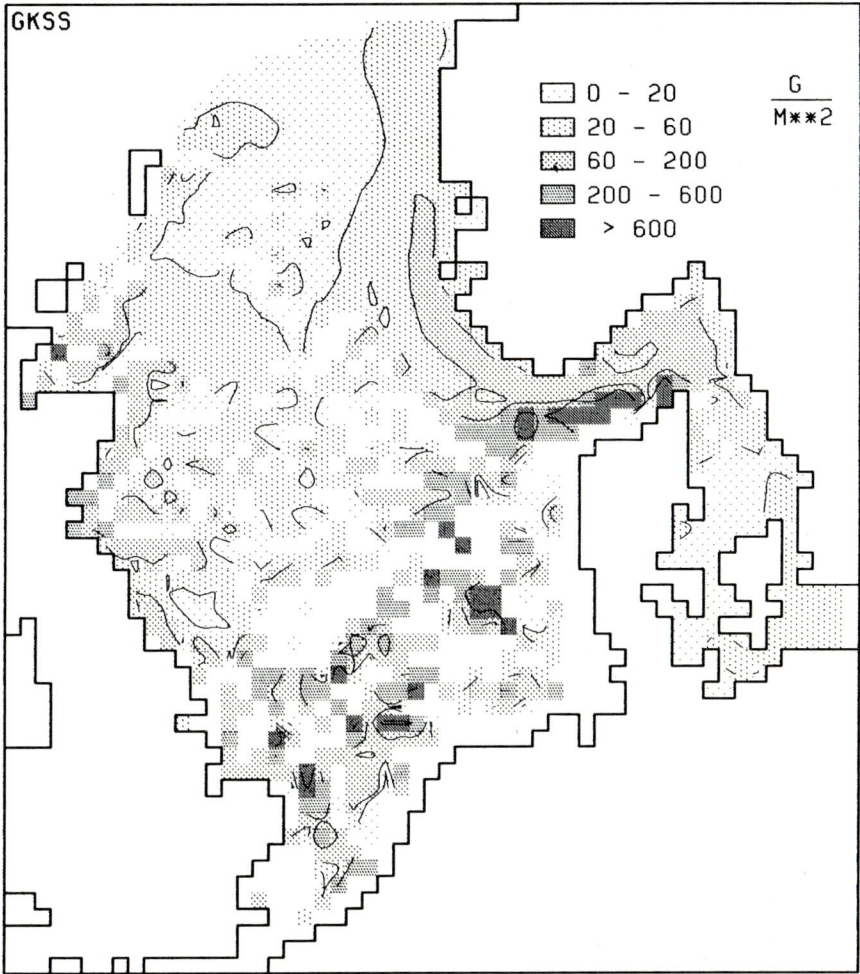

Fig. 3.2.21. Computed annual deposition of fine sediment; average distribution of the 3 years 1979, 1985, 1986

sand with $c_{20\mu m}$ < 2 % and a small sorting coefficient. The equilibrium level of $c_{20\mu m}$, however, may also be larger.

c)   Muddy areas where SPM is accumulating.

Figures 3.2.21, 3.2.22 and 3.2.23 show the computed distributions of the annually deposited and eroded fine sediment in the North Sea. The results of the years 1979, 1985 and 1986 have been averaged. Figures 3.2.21 and 3.2.22 show the complete sediment "reworking". Figure 3.2.23

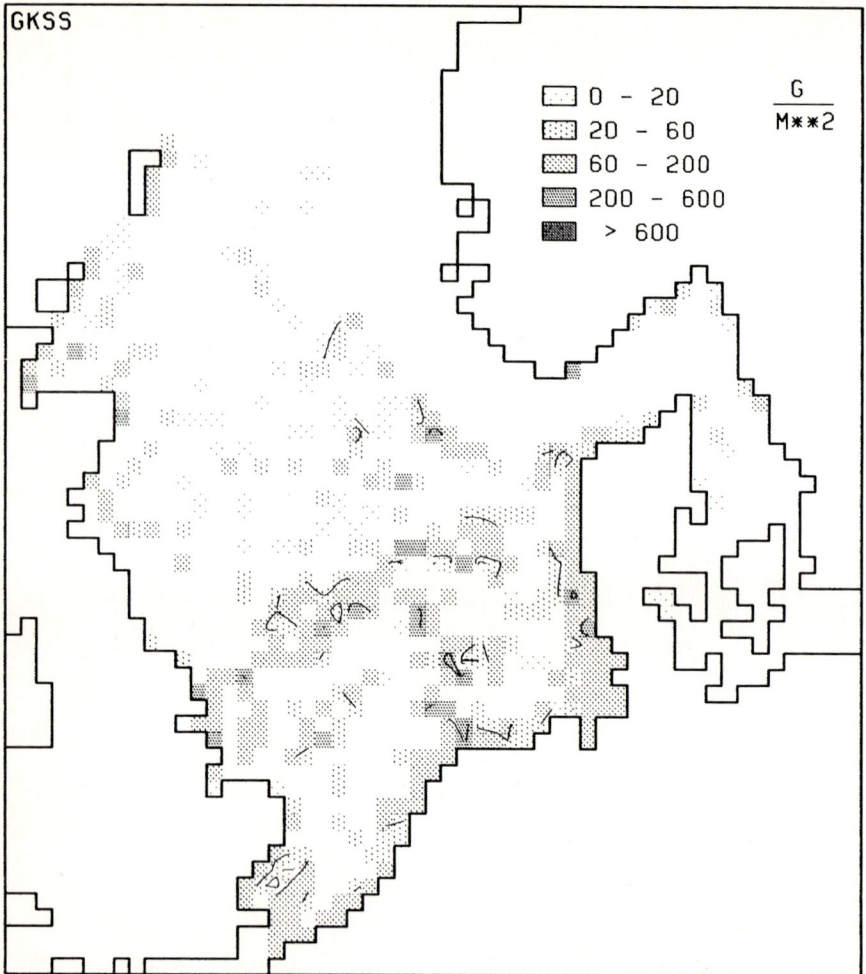

Fig. 3.2.22. Computed annual erosion of fine sediment; average distribution of the 3 years 1979, 1985, 1986

shows only the accumulation of "allochthonous SPM" which is the SPM that enters the model from outside (adjacent seas, rivers, dumping, cliff erosion). At the start of a 1-year simulation run, the initial state of the bed is "relict fine sediment only". Compared to that initial state there can only be accumulation and no erosion of allochthonous SPM.

Three remarks concerning the distributions in Figs. 3.2.21 and 3.2.22:

1.   The patchy appearance of erosion and deposition areas is due to the coarse bathymetry distribution of the model. A local depression is a

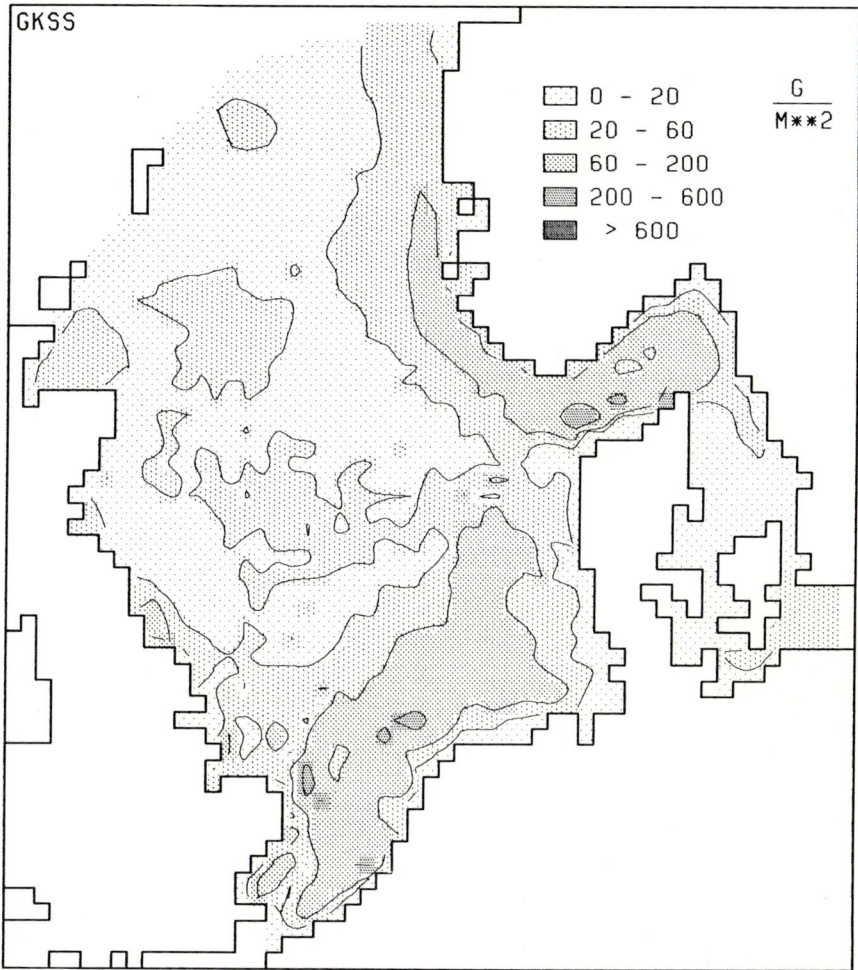

Fig. 3.2.23. Computed annual accumulation of allochthonous fine sediment, average distribution of the 3 years 1979, 1985 and 1986

sink for SPM that is eroded in the neighbourhood (remember the high settling velocity of eroded SPM, Table 3.2.1). The pattern of deposited allochthonous SPM is much smoother (see Fig. 3.2.23).

2. The mud area (about 500 km^2) in the inner German Bight SE of Helgoland (see Fig. 3.2.11) is located at a water depth of about 20 m. It is an accumulation area (e.g. Irion et al. 1987; Eisma and Irion 1988). In the model runs, this mud area would continuously be eroded, resulting in an unrealistic increase of SPM concentrations in the German Bight. To avoid this, the mud area is treated as a sandy area.

The continuous erosion of a mud accumulation area is one of the defects which point out the model's limits.

3. According to Eisma and Irion (1988), the Wadden Sea and the estuaries are accumulation areas. The model cannot simulate the small scale and complicated processes in the near coastal zone.

The model results given in Figs. 3.2.21 and 3.2.22 can be divided into three categories:

- Erosion areas
  Strong and continuous erosion takes place in the near-coast zones of the southern North Sea, on the Dogger Bank and near the Dover Strait. This cannot be realistic, because then the question arises why there is still fine sediment in the bed at all in those zones.
- Accumulation areas
  These areas are characterized by extensive fields in which accumulation takes place only (Fladen Ground, Skagerrak, Norwegian Trench).
- Ambiguous areas
  These areas are characterized by a pattern of adjoining erosion and deposition zones; they encompass most parts of the North Sea. Within the ambiguous areas, there can be tendencies towards accumulation or erosion. There is a tendency towards accumulation (1) generally in the central North Sea between Fladen Ground and Dogger Bank and especially off Tyne/Tees, (2) in the Coffee Soil and the Elbe Rinne, (3) on the south "slope" of the Dogger Bank (Duineveld et al. 1987) observed high densities of the mud-loving *Amphiura filiformis* in that area), (4) in the southern Oyster Ground (5) in the Southern Bight and off East Anglia. The last item is unrealistic: there is sand only in the Southern Bight and off East Anglia.

  Tendencies for erosion are computed (1) for the Lower Scruff and southeast of it, (2) for the area east of Coffee Soil and (3) for the Ling Bank and Klondyke/Outer Shoal SW of the Norwegian Trench.

The accumulation rates computed by the model are compared with measured data:

Skagerrak and Norwegian Trench. Van Weering et al. (1987) determined annual sedimentation rates in the Skagerrak between 0.1 and 0.4 cm/year. Highest values were found in the northeastern Skagerrak. Other authors measured sedimentation rates of the same order. Rodhe (1973) measured high settling rates in the Kattegat east of Skagen.

No measured sedimentation rates are available for the Norwegian Trench. The thickness of fine grained deposits decreases from maximum values of 127 m in the Skagerrak to 15 m off Sognefjorden (Eisma and Kalf 1987a).

Salge and Wong (1988) claim that recent sedimentation in the Skagerrak is restricted to deep water area (> 200 m), but according to van Weering et al. (1987), there is deposition in areas with a water depth < 100 m.

Liebezeit (1988) concludes from sediment core data, that "sedimentation on the southern slope of the Skagerrak does not occur in a continuous mode but is rather controlled by episodic events, e.g. heavy storms".

The computed sedimentation rates in the Skagerrak are highest on the southern slope (see Fig. 3.2.21) at water depths between 40 and 200 m. The highest annual accumulation rate is 3500 g/m^2. The SPM that deposits on the southern slope consists to about 90 % of autochthonous (eroded) SPM with "high" settling velocities, having been deposited after storm events.

In the deeper parts of the Skagerrak, the computed accumulation rates are much lower than on the southern slope: the annual values range between 50 and 100 g/m^2. The same values are also computed in the deeper parts of the Norwegian Trench. East of Skagen, the model computes only slight accumulation. The SPM that deposits in the deeper parts of the Skagerrak and the Norwegian Trench consists of allochthonous SPM only.

With a sediment dry density of 0.55 g/cm^3, the maximum computed sedimentation rate on the southern slope of the Skagerrak is 0.64 cm/year. The values for the deeper parts are between 0.01 and 0.02 cm/year. Compared to the measured values, the computed values in the deeper parts are more than one order of magnitude lower. This discrepancy is enormous; speculations about model deficits concern (among others) (1) too high settling velocities for eroded SPM or (2) an insufficient simulation of the tracer particle movements.

Fladen Ground. Johnson and Elkins (1979) measured a sedimentation rate in the Fladen Ground of 0.005 - 0.006 cm/year during the last 5000-7000 years. Investigating a 5.7-m-long vibrocore, Long et al. (1986) observed a 20-40 cm thick facies of "sandy silt" which "represents the onset of the Holocene with conditions similar to those occurring at present". Assuming an accumulation time of 8000 years, the annual deposition rate is

0.0025-0.005 cm. From sedimentation studies during a phytoplankton bloom, Cadee (1986) estimated a sedimentation rate of 0.004 cm/year.

The annual accumulation rate computed by the model in the Fladen Ground is about 25 g/m^2. A bulk density of 1.5 g/cm^3 (own measurements) means a dry density of 0.8 g/cm^3 - that dry density results in a sedimentation rate of 0.003 cm/year.

The SPM that deposits in the Fladen Ground has entered the North Sea from the Atlantic and is assumed to have a grain size < 20 µm. On the other hand, there is only a $c_{20\mu m}$-content of 25 % in the Fladen Ground.

Coffee Soil. From a sediment core taken in a water depth of 50 m, Heinrich (1991, pers. comm.) determined a settling rate of 200 cm within 9000 years. The weight content < 63 µm is 25 %, which means, according to Wiesner et al. (1990), a content of 12 % < 20 µm. With a sediment dry density of 1.15 g/cm^3, the annual deposition of SPM with grain sizes below 20 µm is about 30 g/m^2.

The model computes a heterogeneous pattern of erosion and accumulation in the Coffee Soil, with a tendency towards accumulation. The annual accumulation rate of allochthonous SPM is about 60 g/m^2.

Southern Oyster Ground. Zuo et al. (1989) investigated a narrow zone of muddy deposits in the southern Oyster Ground in a water depth between 30 and 40 m. The average mud accumulation rate was determined as 0.39 cm/year. With $c_{20\mu m}$ = 14 % and a sediment dry density of 1.1 g/cm^3, this means an annual accumulation of 600 g/m^2.

The model computes a value of about 650 g/m^2-year. The one half of this material is allochthonous SPM, the other half is eroded and re-deposited fine North Sea sediment.

In the main part of the Oyster Ground, Cadee (1984) found little sedimentation. According to Eisma and Irion (1988) there is an exchange of relict and recent material, but the net effect is approximately zero. In the model results, however, deposition is dominating, not all over the area, but in most parts.

3.2.3.10 SPM Budget

Table 3.2.2 gives a survey of the computed SPM-masses that enter ("input") and leave ("output") the North Sea water body per year (average of the years 1979, 1985, 1986). The major sinks for SPM are (1) the North Atlantic between Shetland and Norway, due to the northward directed

Norwegian Coastal Current (see Sect. 3.2.1.3), and (2) the sea floor, mainly the Skagerrak and the Norwegian Trench.

Table 3.2.2. Annual input (supply or erosion) and output (outflow or deposition) of SPM masses (in 10^6 t) for the North Sea water body, computed average of the years 1979, 1985 and 1986. The "net amount" is the sum of "input" and "output"

Location	Input	Output	Net amount
Pentland Firth and Fair Isle Strait	34.2	21.8	12.4
North Atlantic between Shetland and Norway	42.4	47.6	-5.2
Dover Strait	29.8	16.2	13.6
Baltic Sea	2.4	1.5	0.9
Rivers, cliff erosion, dumping	3.1	-	3.1
Deposited allochthonous SPM	-	23.7	-23.7
Erosion (input) and deposition (output) of fine sediment	133.5	134.0	-0.5

Comments and supplements to Table 3.2.2:

1. The sum of the "net amount" is not exactly zero which is due to temporal changes of the SPM mass in the North Sea water body.

2. The total amount of allochthonous SPM in the North Sea water body is in the order of 30×10^6 t. The content of eroded fine bottom sediment varies: it was computed to be about 1×10^6 t during during calm weather and 50×10^6 t during a severe storm.

3. The turnover (input or output) of SPM from adjacent seas is in the order of 100×10^6 t/year. This is in the same order as the turnover (erosion and re-deposition) of fine sediment from the sea floor.

4. The sum of the net amounts of allochthonous SPM, entering from adjacent seas and from land sources, is 24.8×10^6 t. This number is higher than the allochthonous SPM (23.7×10^6 t) that is deposited on the sea floor, which simply means that the allochthonous SPM content of the North Sea water

body has increased on the average by 1.1×10^6 t between January 1 and December 31 in the 3 years considered.

5.   There is a net output (deposition) of fine bottom sediment of 0.5×10^6 t which is a coincidence, resulting from different wave and current conditions at the start- and the end-dates of the budget. The outflow of eroded, autochthonous SPM from the North Sea to adjacent seas is negligible ($\approx 0.1 \times 10^6$ t/year), which is the result of the high settling velocity of eroded SPM, see Table 3.2.1.

6.   At the North Atlantic boundary between Shetland and Norway the annual net outflow is 5.2×10^6 t; 2.5×10^6 t of this is SPM from Dover Strait, Baltic Sea, rivers, cliff erosion and dumping, leaving the North Sea with the Norwegian coastal current. The sum of net input of those sources is 17.6×10^6 t. Subtracting the outflowing mass of 2.5×10^6 t, the annually deposited SPM from those sources is in the order of 15×10^6 t.

7.   The separate budget for SPM from the North Atlantic is as follows: There is a net input through Pentland Firth and Fair Isle Strait of 12.4×10^6 t and a net output between Shetland and Norway of 2.7×10^6 t. This means an annual deposition of about 10×10^6 t.

3.2.4 Summary and Conclusion

The first part of this Chapter outlines the basic principles of the three-dimensional water circulation model. The currents are driven by (1) the M_2-tide, (2) density gradients and (3) daily changing wind and air pressure fields. A mean water circulation is presented separately for the summer and the winter. Two selected circulation patterns are shown that are produced by extreme meteorological conditions. Finally, a time series covering the years 1969-1981 of the intensity of the Jutland current is discussed.

The second part deals with the transport of dissolved conservative matter. The probability density of substances released by the rivers Rhine and Humber are calculated, as well as the relative "contamination" of the North Sea by all rivers. A further illustration of the transport characteristics is the simulation of lead concentrations after a 1-month deposition from the atmosphere.

The third part of the Chapter deals with the transport of suspended particulate matter (SPM). After describing the model (including the simulation of the transport processes erosion, settling, deposition and bioturbation), results of the SPM simulations are (1) compared with the measured concentrations of two North Sea cruises and two time series in the German Bight and (2) compared with observed mud accumulation rates.

The numerical computation of the water movement in the North Sea is a well developed and accepted procedure. The data of the long-term current simulations (daily mean values are stored) are an excellent basis for transport computations, especially (combined with tidal current data and wave data) for the highly intermittent transport of SPM.

The SPM model is too coarse for simulating the near-coast accumulation of SPM in the water body. Generally, the simulation of SPM transport is still far away from being "perfect"; it must be improved with respect to several processes, mainly concerning erosion and deposition.

Acknowledgements. We thank the Bundesamt für Wehrtechnik und Beschaffung (Koblenz) and the Amt für Wehrgeophysik (Traben-Trarbach) for supplying the wave data.

References

Anonymous (1990) Effects of beam trawl fishery on the bottom fauna of the North Sea. BEON-Rapport 8, Netherlands Institute for Sea Research, Texel, 57 pp

Backhaus JO (1985) A three-dimensional model for the simulation of shelf sea dynamics. Dtsch Hydrogr Z 38:165-187

Backhaus JO (1989) The North Sea and the climate. Dana 8:69-82

Backhaus JO, Hainbucher D (1987) A finite difference general circulation model for shelf seas and its application to low frequency variability on the North European Shelf. In: Nihoul JCJ, Jamart BM (eds) Three-dimensional models of marine and estuarine dynamics. Elsevier, Amsterdam (Elsevier Oceanography Ser 45) pp 221-244

Backhaus JO, Bartsch J, Quadfasel D, Guddal J (1985) Atlas of monthly surface fields of air pressure, wind stress and wind stress curl over the North Eastern Atlantic Ocean: 1955-1982. Inst Meeresk Univ Hamburg, Techn Rep 3-85:251 pp

Boudreau BP (1986) Mathematics of tracer mixing in sediments: I. Spatially dependent, diffusive mixing. Am J Sci 286:161-198

Brügmann L, Danielsson L-G, Magnusson B, Westerlund S (1985) Lead in the North Sea and the north east Atlantic Ocean. Mar Chem 16:47-60

Bundesminister für Umwelt, Naturschutz und Reaktorsicherheit (1987) Bericht der Bundesregierung an den Deutschen Bundestag zur Vorbereitung der 2. Internationalen Nordseeschutz-Konferenz (2.INK). Drucksache 11/878, Deutscher Bundestag, 11. Wahlperiode

Cadee GC (1984) Macrobenthos and macrobenthic remains on the Oyster Ground, North Sea. Neth J Sea Res 18(1/2):160-178

Cadee GC (1986) Organic carbon in the water column and its sedimentation, Fladen Ground (North Sea), May 1983. Neth J Sea Res 20(4):347-358

Churchill JH (1989) The effect of commercial trawling on sediment resuspension and transport over the Middle Atlantic Bight continental shelf. Continent Shelf Res 9(9):841-864

Creutzberg F, Postma H (1979) An experimental approach to the distribution of mud in the Southern North Sea. Neth J Sea Res 13(1):99-116

de Wilde PAWJ, Berghuis EM, Kok A (1984) Structure and energy demand of the benthic community of the Oyster Ground, central North Sea. Neth J Sea Res 18 (1/2):143-159

DHI (1984) Gütezustand der Nordsee. Deutsches Hydrographisches Institut, Hamburg, Meereskundliche Beobachtungen und Ergebnisse Nr 55, 135 pp

Drake DE, Cacchione DA (1989) Estimates of the suspended sediment reference concentration (C_a) and resuspension coefficient (γ_0) from near-bottom observations on the California shelf. Continent Shelf Res 9(1):51-64

Dronkers J, van Alphen JSLJ, Borst JC (1990) Suspended sediment transport processes in the southern North Sea. In: Cheng RT (ed) Residual currents and long-term transport. Coastal and Estuarine Studies 38. Springer, Berlin Heidelberg New York, pp 302-320

Duineveld GCA, Künitzer A, Heyman RP (1987) *Amphiura filiformis* (Ophiuroidea: Echinodermata) in the North Sea. Distribution, present and former abundance and size composition. Neth J Sea Res 21 (4):317-329

Dyer KR (1986) Coastal and estuarine sediment dynamics. Wiley, Chichester, 342 pp

Eisma D (1981) Supply and deposition of suspended matter in the North Sea. Spec Publ Int Assoc Sediment 5:415-428

Eisma D, Irion G (1988) Suspended matter and sediment transport. In: Salomons W, Bayne BL, Duursma EK, Förstner U (eds) Pollution of the North Sea. Springer, Berlin Heidelberg New York, pp 20-35

Eisma D, Kalf J (1979) Distribution and particle size of suspended matter in the Southern Bight of the North Sea and the eastern Channel. Neth J Sea Res 13 (2):298-324

Eisma D, Kalf J (1987a) Dispersal, concentration and deposition of suspended matter in the North Sea. J Geol Soc Lond 144:161-178

Eisma D, Kalf J (1987b) Distribution, organic content and particle size of suspended matter in the North Sea. Neth J Sea Res 21 (4):265-285

Floderus S, Hakanson L (1989) Resuspension, ephemeral mud blankets and nytrogen cycling in Laholmsbukden, south east Kattegat. Hydrobiologia 176/177. In: Sly PG, Hart BT (eds) Sediment/water interaction. Kluwer, Dordrecht, pp 61-75

Führböter A (1983) Über mikrobiologische Einflüsse auf den Erosionsbeginn bei Sandwatten. Wasser Boden 3:106-116

Grant WD, Madsen OS (1979) Combined wave and current interaction with a rough bottom. J Geophys Res C4:1797-1808

Hainbucher D (1984) Parametrisierung des vertikalen Impulstransfers in einem dreidimensionalen Zirkulationsmodell der Nordsee. Diplomarbeit im Fach Ozeanographie, Univ Hamburg, 173 pp

Hainbucher D, Backhaus JO, Pohlmann T (1986) Atlas of climatological and actual seasonal circulation patterns in the North Sea and adjacent shelf regions: 1969-1981. Inst Meeresk Univ Hamburg, Techn Rep 1-86:201 pp

Hainbucher D, Pohlmann T, Backhaus JO (1987) Transport of passive tracers in the North Sea: first results of a circulation and transport model. Continent Shelf Res 2:1161-1171

Heip C, Herman PMJ, Craeymeersch J, Soetaert K (1990) Statistical analysis and trends in biomass and diversity of North Sea macrofauna. ICES C.M.1990/MINI:10, 21 pp

Hupkes R (1990) Pollution of the North Sea imposed by West European rivers (1984-1987). ICWS-rep 90.03, 102 pp

Hydraulics Research (1979) Properties of Belawan mud. Hydraulics Research Rep EX 880, Wallingford, UK

ICES (1988) Report of the study group on the effects of bottom trawling. ICES C.M. 1988/B:56, 30 pp

Ingri J, Löfvendahl R, Boström K (1991) Chemisty of suspended particles in the southern Baltic Sea. Mar Chem 32:73-87

Irion G, Wunderlich F, Schwedhelm E (1987) Transport of clay minerals and anthropogenic compounds into the German Bight and the provenance of fine-grained sediments SE of Helgoland. J Geol Soc Lond 144:153-160

Johnson TC, Elkins SR (1979) Holocene deposits from the northern North Sea: evidence for dynamic control of their mineral and chemical composition. Geol Mijnb 58:353-366

Kokkinowrachos K (1980) Hydromechanik der Seebauwerke. In: Wendel K (ed) Handbuch der Werften Band XV. Schiffahrts-Verlag Hansa, Hamburg, pp 15-167

Kracht U, Dietze W (1990) Hoch- und Niedrigwasserwahrscheinlichkeiten der Oberwasserzuflüsse in den Tidebereichen von Ems, Weser und Elbe. Dtsch Gewässerkundl Mitt 34 (5/6):185-195

Krell U, Roeckner E (1988) Model simulation of the atmospheric input of lead and cadmium into the North Sea. Atoms Environ 22(2):375-381

Krone RB (1962) Flume studies of the transport of sediment in estuarial shoaling processes. Final Report, Hydraulic Engineering Laboratory and Sanitary Engineering Research Laboratory, Univ California, Berkeley, 110 pp

Krost P, Bernhard M, Werner F, Hukriede W (1990) Otter trawl tracks in Kiel Bay (Western Baltic) mapped by side-scan sonar. Meeresforschung 32:344-353

Landesamt für Wasser und Abfall Nordrhein-Westfalen (1989) Deutsches Gewässerkundliches Jahrbuch, Rheingebiet Teil III, Abflußjahr 1987. Landwirtschaftsverlag Hiltrup, D-4400 Münster-Hiltrup

Lenhart HJ (1990) Phosphatbilanz der Nordsee - eine Abschaetzung auf der Grundlage der ICES-Boxen. Dipl Thesis, Inst Meeresk Hamburg, 111 pp

Liebezeit G (1988) Early diagenesis of carbohydrates in the marine environment - II. Composition and origin of carbohydrates in Skagerrak sediments. Org Geochem 13(1-3):387-391

Long D, Bent A, Harland R, Gregory DM, Graham DK, Morton AC (1986) Late quaternary palaeontology, sedimentology and geochemisty of a vibrocore from the Witch Ground basin, central North Sea. Mar Geol 73:109-123

Lyne VD, Butman B, Grant WD (1990) Sediment movement along the U.S. east coast continental shelf - II. Modeling suspended sediment concentration and transport rate during storms. Continent Shelf Res 10(5):429-460

Maier-Reimer E, Sündermann J (1982) On tracer methods in computational hydrodynamics. In: Abbott MB, Cunge JA (eds) Engineering applications of computational hydraulics vol 1. Pitman, Boston, pp 198-217

Mayer B (1991) Schwebstoffe in der Nordsee: Untersuchungen transportrelevanter Prozesse in der Wassersäule und im Sediment mit einem numerischen Vertikalmodell. Diplomarbeit im Fach Ozeanographie, Univ Hamburg, 136 pp

McCave IN (1987) Fine sediment sources and sinks around the East Anglian Coast (UK). Journal of the Geol Soc Lond 144:149-152

Mehta AJ, Partheniades E (1975) An investigation of the depositional properties of flocculated fine sediments. J Hydraul Res 13(4):361-381

Mehta AJ, Hayter EJ, Parker WR, Krone RB, Teeter AM (1989) Cohesive sediment transport I: Process description. J Hydraul Engin 115(8):1076-1093

Officer CB, Lynch DR (1989) Bioturbation, sedimentation and sediment-water exchanges. Estuarine Coastal Shelf Sci 28:1-12

Pheiffer Madsen P, Larsen B (1986) Accumulation of mud sediments and trace metals in the Kattegat and the Belt Sea. Rep no 10, Marine Pollution Laboratory, Charlottenlund, Denmark, 54 pp

Postma H (1990) Transport of water and sediment in the Strait of Dover. In: Ittekkot V, Kempe S, Michaelis W, Spitzy A (eds) Facets of modern biogeochemistry. Festschrift for ET Degens. Springer, Berlin Heidelberg New York, pp 147-154

Puls W (1984) Erosion characteristics of estuarine muds. Hydraulics Research Rep IT 265, Wallingford, UK

Puls W, Sündermann J (1990) Simulation of suspended sediment dispersion in the North Sea. In: Cheng RT (ed) Residual currents and long-term transport. Coastal and Estuarine Studies 38. Springer, Berlin Heidelberg New York, pp 356-372

Rachor E (1982) Biomass distribution and production estimates of macro-endofauna in the North Sea. ICES C.M.1982/L:2, 10 pp

Rauck G (1985) Wie schädlich ist die Seezungenbaumkurre für Bodentiere. Infn Fischw 32(4):165-168

Reid PC, Lancelot C, Gieskes WWC, Hagmeier E, Weichart G (1990) Phytoplankton of the North Sea and its dynamics: a review. Neth J Sea Res 26(2-4):295-331

Reineck H-E (1984) Aktuogeologie klastischer Sedimente. Senckenberg-Buch 61. Kramer, Frankfurt

Rice AL, Billett DSM, Fry J, John AWG, Lampitt RS, Mantoura RFC, Morris RJ (1986) Seasonal deposition of phytodetritus to the deep-sea floor. Proc R Soc Edinb 88B:265-279

Rodger JG, Odd NVM (1985) Sludge disposal in coastal waters. Hydraulics Research Rep SR 70, Wallingford, UK

Rodhe J (1973) Sediment transport and accumulation at the Skagerrak-Kattegat border. Report no 8, Institute of Oceanograpy, Univ Gothenburg, 32 pp

Salge U, Wong HK (1988) The Skagerrak: a depo-environment for recent sediments in the North Sea. In: Kempe S, Liebezeit G, Dethlefsen V, Harms U (eds) Biogeochemistry and distribution of suspended matter in the North Sea and implications to fisheries biology. Mitt Geol Paläontol Inst Univ Hamb 65:367-380

Schröder M (1988) Untersuchungen zur Chemie von Schwebstoffen in der Nordsee von Januar bis März 1987. Diplomarbeit (Teil 1) des Geologisch-Paläontologischen Instituts, Univ Hamburg, 77 pp

Schünemann M, Kühl H (1991) A device for erosion measurements on naturally formed, muddy sediments: the EROMES-System. GKSS rep 91/E/18, 28 pp

Shields A (1936) Anwendung der Ähnlichkeitsmechanik und der Turbulenz-forschung auf die Geschiebebewegung. Mitteilungen der Preußischen Versuchsanstalt für Wasserbau und Schiffbau, Berlin, Heft 26, 26 pp

SRU (1980) Umweltprobleme der Nordsee. Sondergutachten des Rats von Sachverständigen für Umweltfragen. Kohlhammer, Stuttgart, 503 pp

Stephan H-J (1978) Verfahren zur Voraussage der kennzeichnenden Größen des Seegangs. Mitt. des Leichtweiß-Instituts für Wasserbau der TU Braunschweig 63:219-270

Teeter AM (1986) Vertical transport in fine-grained suspension and newly deposited sediment. In: Mehta AJ (ed) Estuarine cohesive sediment dynamics. Lecture notes on coastal and estuarine studies 14. Springer, Berlin Heidelberg New York, pp 170-191

Tryggestad S, Selanger KA, Mathisen JP, Johansen O (1983) Extreme bottom currents in the North Sea. In: Sündermann J, Lenz W (eds) North Sea dynamics. Springer, Berlin Heidelberg New York, pp 148-158

van Alphen JSLJ (1990) A mud balance for Belgian-Dutch coastal waters between 1969 and 1986. Neth J Sea Res 25 (1/2):19-30

van Weering TCE, Berger GW, Kalf J (1987) Recent sediment accumulation in the Skagerrak, northeastern North Sea. Neth J Sea Res 21(3):177-189

Wiesner MG, Haake B, Wirth H (1990) Organic facies of surface sediments in the North Sea. Org Geochem 15(4):419-432

Wirth H, Wiesner MG (1988) Sedimentary facies in the North Sea. In: Kempe S, Liebezeit G, Dethlefsen V, Harms U (eds) Biogeochemistry and distribution of suspended matter in the North Sea and implications to fisheries biology. Mitt Geol Paläontol Inst Univ Hamb 65:269-287

Wood PC, Franklin A (1983) The bioavailability of metals in fly ash and colliery waste dumped at sea. ICES CM 1983/E:15, 3 pp

Zuo Z, Eisma D, Berger GW (1989) Recent sediment deposition rates in the Oyster Ground, North Sea. Neth J Sea Res 23(3):263-269

3.3 Phytoplankton Modelling in the Central North Sea During ZISCH 1986

A. MOLL and G. RADACH

3.3.1 Introduction

In this Chapter we will show how a "simple model" of the phytoplankton dynamics is able to give valuable insight which may serve for interpretation of data and for planning of further experiments. We will take the ZISCH North Sea data to give an example of interpretation.

The setting up of our phytoplankton dynamics model is guided by the intention to simulate the most important features of the annual plankton dynamics (as described in the introduction of Chapter 2.2), but at the same time to make the model as simple as possible. Thus, we intentionally avoid much well-known physical, chemical and biological detail in order to elucidate the dynamics by introducing only a few of the various processes that act on the phytoplankton standing stock. We do not deal with different phytoplankton species or groups of species and we use only one nutrient as a triggering state variable, in spite of the fact that nature is much more complicated. We hope to describe with our model the characteristic features of the annual cycle of phytoplankton biomass formation and decay (Fig. 3.3.1), with a spring bloom at the right time and with the maximum of the spring bloom in the right order of magnitude, with a quasi-steady state in summer due to nutrient depletion and only occasional nutrient injections from below the thermocline, and possibly a fall bloom after remineralization has provided again sufficient nutrients, reaching the upper layers by intensified vertical mixing.

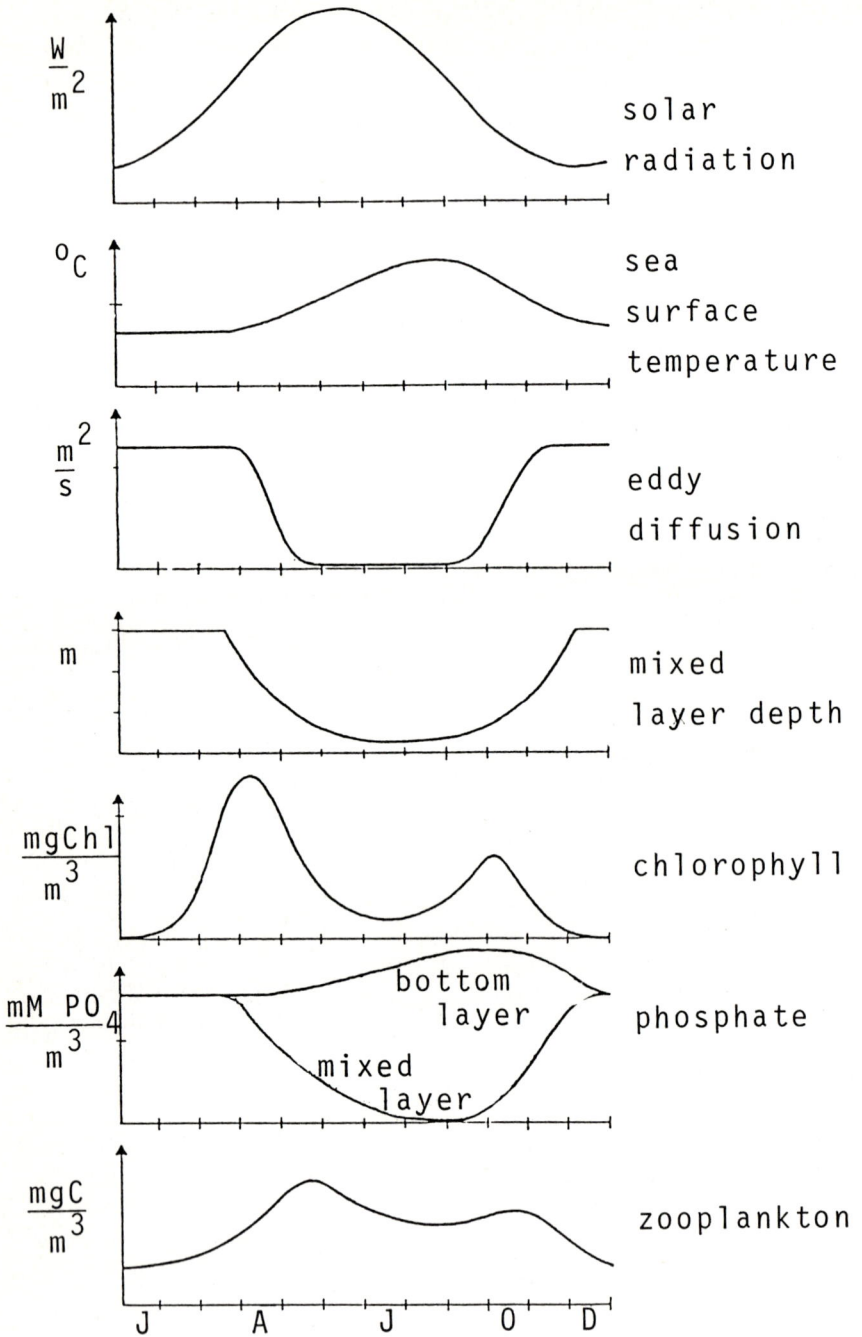

Fig. 3.3.1. Schematic representation of the annual cycle of phytoplankton and the main influencing factors, for mean latitudes. (Radach 1990)

All this will be based on the knowledge of the physical environment. Here we restrict ourselves to the dynamics of the water column, i.e. the development of the annual temperature and turbulence structures in the vertical. The development and decay of the thermocline plays a dominant role in this dynamical frame. In Fig. 3.3.1 we describe the main influencing factors for the annual phytoplankton cycle for mean latitudes. Radiation is warming the upper layers of the sea until (usually) in April the sea surface temperature starts to increase noticeably and the mixing rate is decreasing, which gives rise to a shallowing upper mixed layer. Due to phytoplankton growth, nutrient concentrations decrease in the upper layers. The lower layer nutrient content remains high and becomes even higher in late summer and in fall due to the remineralization of organic material. The storms in fall stir up the water column down to regions where the nutrients have accumulated and mix the water column during events of overturning, thus homogenizing the water column not only with respect to heat but also to nutrients.

The annual cycle of phytoplankton dynamics is strongly coupled to the annual cycles of nutrient availability and of zooplankton. Until now it is not clear to which degree the spring blooms in the North Sea, for example, are generally terminated by nutrient depletion and/or by zooplankton grazing. In any case, the simulations of the phytoplankton cycle may not neglect the possible role of zooplankton. The pattern of the (mean) annual cycles of phytoplankton within the North Sea is coupled to the annual cycles of zooplankton (Fig. 3.3.2). The northern North Sea is characterized by a strong phytoplankton spring peak (in the mean) and decaying biomass until fall. In the central and eastern North Sea two pronounced peaks emerge for the mean situation, one in spring and one in fall. In the southern North Sea the magnitude of spring and fall maxima shrinks relative to the summer state: although still visible in the mean, the maxima together with the summer biomass form one large cycle culminating in summer. The zooplankton cycle is everywhere nearly the same (Fig. 3.3.2), possibly indicating a relatively loose coupling to phytoplankton.

We have good reasons to deal with phosphorus only and ignore the nitrogen cycle. Several investigations have shown that phosphorus can be the limiting nutrient in spring, but also in summer. This has been observed in many different parts of the North Sea and for many time periods (Radach et al. 1984; Riegman et al. 1990). During the spring bloom in 1976 (Fladen Ground Experiment FLEX'76), for example, the time shift between nitrate and phosphate limitation was less than a few days and the ratio P:N was nearly constant from March to June (Eberlein et al. 1980). The difference between the annual cycles of phosphate and nitrate may be ignored for the sake of simplicity to get the 0-order effects. Thus we use phosphate as the

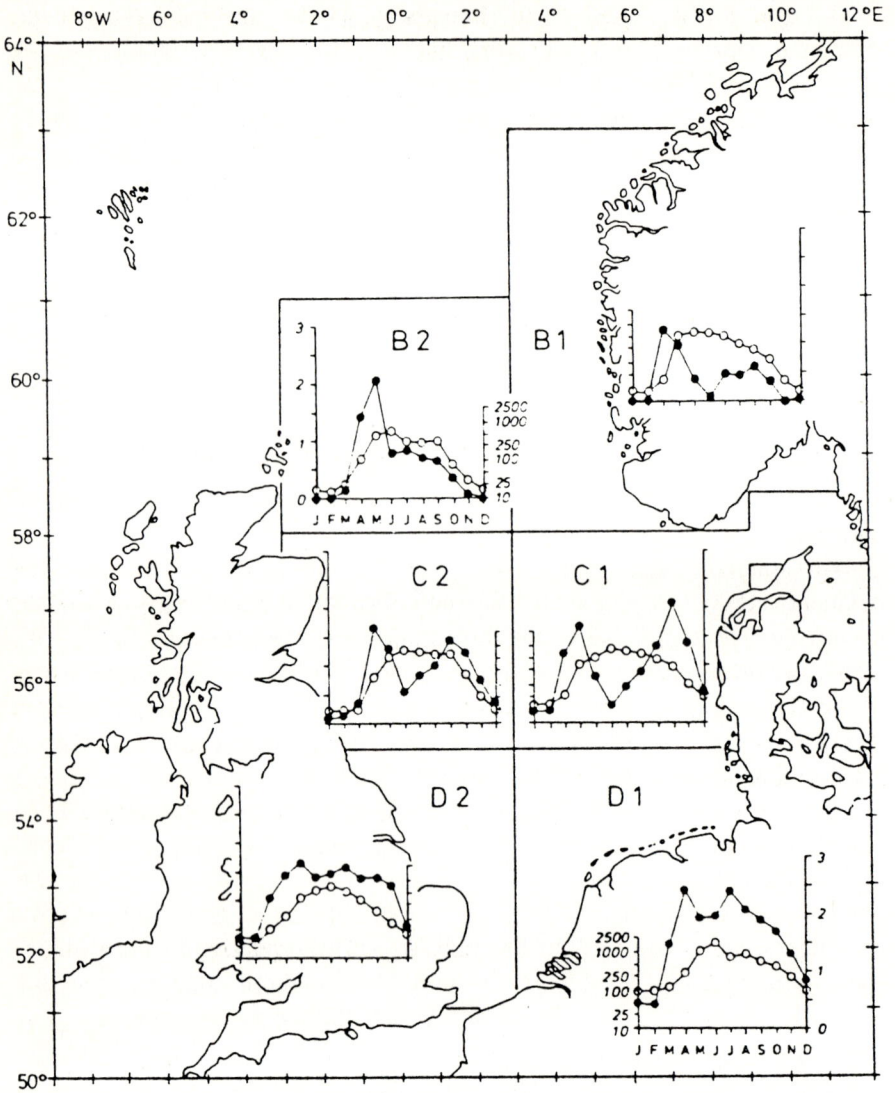

Fig. 3.3.2. Annual cycles of phytoplankton (*filled circles*) and zooplankton (*open circles*) in six areas of the North Sea, obtained from the Continuous Plankton Recorder Survey data. (Colebrook 1979)

only limiting nutrient here. The complex phosphorus cycle for the shelf sea is represented in the model in the "rudimentary" version shown in Fig. 3.3.3.

In our modelling project we started with a 1-D model version. This type of model can contribute to a series of important questions, because the 1-D model is suited to investigating principal questions relating to the phytoplankton dynamics such as the following (Harris 1980; Denman and Powell 1984; Legendre and Demers 1984; Fransz et al. 1991):

- Which are the main determining local agents for the evolution of the phytoplankton biomass, given a certain biological dynamical frame?
- How does the same biological dynamical system react to different levels of nutrient availability?
- To which degree can local meteorological variability explain biological variability in the sea?

We will show how far we get in answering questions of this type with 1-D modelling. After explaining the strategy of the setup of the model, we report about the validation efforts performed so far. We then apply the model to the central North Sea under meteorological forcing of the year 1986 and compare the simulations to the data of the ZISCH summer cruise. Next we deal with the problem of eutrophication by investigating how changed winter conditions for nutrients influence the annual cycle of phytoplankton biomass development. In the last section we show that a large part of the variability of the phytoplankton-phosphate system can be reproduced by imposing nothing but meteorological variability.

3.3.2 Design of the 1-D "Phytoplankton-Phosphate-Detritus-Model" for the Annual Cycle

The state-of-the-art in algal bloom modelling has been presented by Radach (1990), Radach and Moll (1990a), Radach et al. (1993) and Fransz et al. (1991). There, the advantages of the water column models are discussed in comparison to various one-layer and two-layer box models (e.g. Fransz and Verhagen 1985). For our purpose of resolving instantaneous mixing in its vertical extent, it is necessary to resolve the full water column, both in the physical and in the biological model.

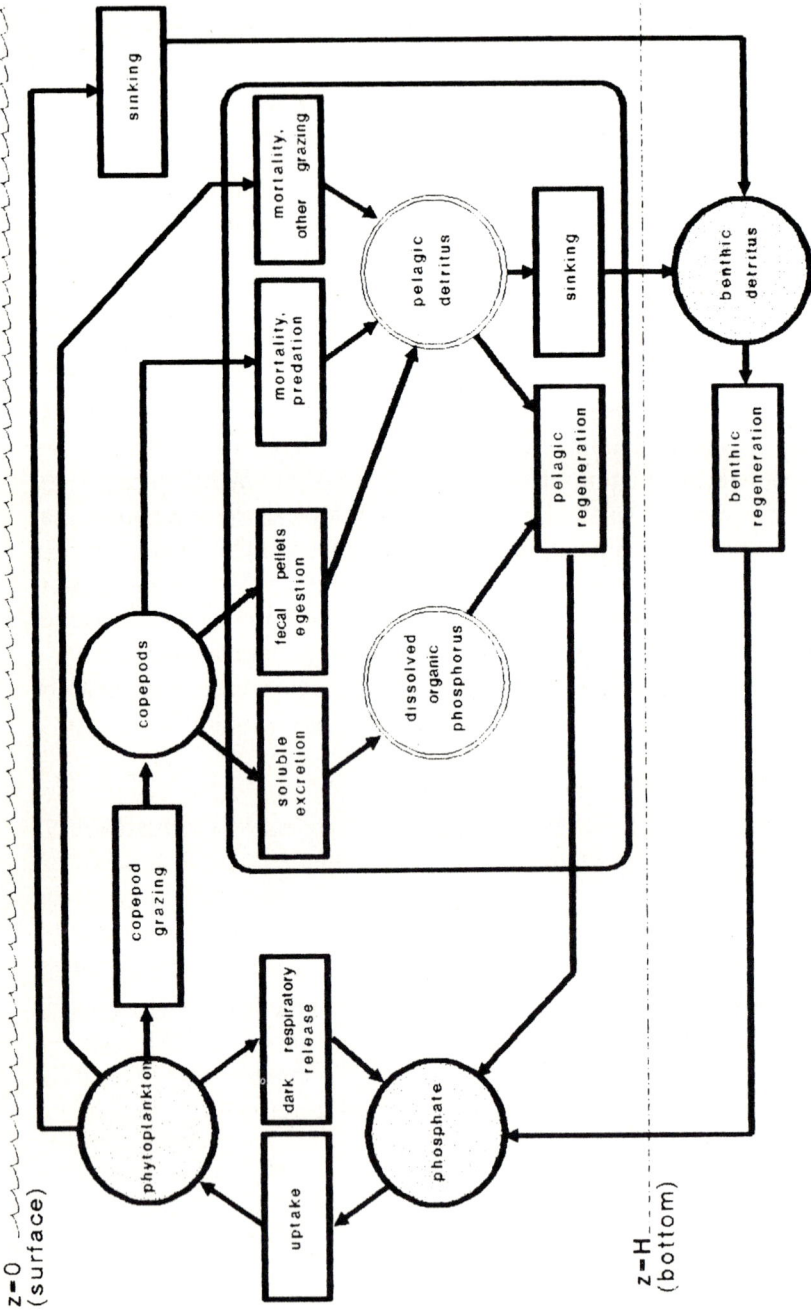

Fig. 3.3.3. Interaction diagram for a simplified phosphorus cycle of the biological upper layer model. The diagram separates prognostic variables (*shaded circles*), diagnostic variables (*white circles*), indirectly modelled variables (*double circles*) and processes (*boxes*) (Radach and Moll 1993)

3.3.2.1 Model Concept

The concept of the model (Moll 1989; Radach and Moll 1990b; Radach and Moll 1993) is based on the assumption that physical, chemical and biological dynamics should be represented in a comparable complexity, or more adequately: simplicity. We intend a balance of the different parts of the model in a way such that relations of the simulation output variables to available data sets can be established.

In many models of phytoplankton dynamics the links to the higher trophic levels are realized, but the representation of the physical determinants is poor (Fransz et al. 1991). The importance of stratification for the development of phytoplankton blooms is regarded to be enormous (Kiefer and Kremer 1981; Radach 1983, 1990; Rothschild 1988; Stigebrandt and Wulff 1987; Radach and Moll 1990b), and thus the simulation of the thermal structure of the ocean in reaction to the actual temporal small-scale meteorological forcing is introduced as an essential ingredient into our simulation of phytoplankton dynamics. We simulate the physical upper layer dynamics with a differential upper layer model (Kochergin et al. 1976; Friedrich et al. 1981). The decay of plankton blooms, on the other hand, cannot be explained without admitting nutrient limitation of algal growth, which is realized by the Michaelis-Menten kinetics of nutrient uptake.

Dealing with annual cycles of phytoplankton, the replenishment of the depleted upper layer of the sea has to be taken into account. The effect of the small food web is parameterized by the fact that a certain percentage of dead organic matter being generated in the water column is immediately recycled into the nutrient pool. The bulk part of the detritus, however, is immediately falling down to the bottom into the detritus pool, which is gathering dying phytoplankton cells, faecal pellets, and carcasses of zooplankton, as was described in Chapter 2.2.7. The detritus pool at the bottom serves to regenerate phosphate by bacterial activity, thus causing a delay in making the nutrient available again, firstly by the slow remineraliation and secondly by the necessary transport time for the replenishment of the upper layers. In addition, the simplification chosen does not necessitate solving a third advection-diffusion-reaction equation for detritus but restricts the problem to solving an additional ordinary differential equation. The regenerated phosphate is fed from the bottom into the water column by adequate boundary flux conditions for the state variable phosphate.

The evolution of phytoplankton biomass is brought about by the combined action of the processes of turbulent diffusion, photo-synthesis-related growth, respiration, mortality, grazing and sinking. The evolution of the nutrient phosphate is caused by turbulent diffusion, uptake of nutrient by algae, remineralization of part of the detrital material in the water column (as the effect of the small food web) and excretion of phosphate. The evolution of detritus at the bottom is resulting from the flux of the larger part of live and dead phytoplankton as well as detritus (zooplankton feces and carcasses) from the water column to the bottom, and its remineralization at the bottom. The flux out of the water column is realized by the bottom boundary condition for phytoplankton. The flux of phosphate from the bottom into the water column is achieved by the bottom boundary condition for phosphate. For phytoplankton biomass and phosphate, sequences of profiles result from the simulations. For detritus at the bottom, we get a time series of carbon per unit area. The vertical resolution of the water column is 2.5 m. The differential equations are solved by finite difference methods. A detailed description of the model is given in Radach and Moll (1993).

3.3.2.2 Annual Cycles of Forcing Functions

It has been shown (Radach 1983; Klein and Coste 1984) that temporal small-scale variability of meteorological forcing functions has a great effect on the development of phytoplankton and nutrient dynamics. We thus decided to use a forcing derived from standard meteorological observations for our studies.

Appropriate data sets of annual meteorological cycles at suitable stations are rare. For the year 1986 no measurements of the meteorological forcing data in the form of annual cycles exist for the central North Sea. Therefore, we have to use forcing data from other stations for our simulation. The best-suited station we could find is the LV ELBE 1, although data are missing during October and November due to a collision with the light vessel (Moll and Radach 1990a). Fig. 3.3.4a-c gives the annual cycles from 1986 for the parameters wind velocity, direct and diffusive solar radiation and the daily heat balance at the sea surface. It may be pointed out here that one can clearly notice the storm in the beginning of June that the ZISCH cruise experienced.

At LV ELBE 1 in the German Bight data are also available from 1962 to 1986. We used this data set to derive the needed forcing functions, i.e. latent and sensible heat flux, back radiation and global radiation (Moll and Radach 1990a). The latter quantity was simulated according to the model by Dobson and Smith (1988). The model was applied to the meteorological data of LV ELBE 1 and tested against daily sums of global radiation at

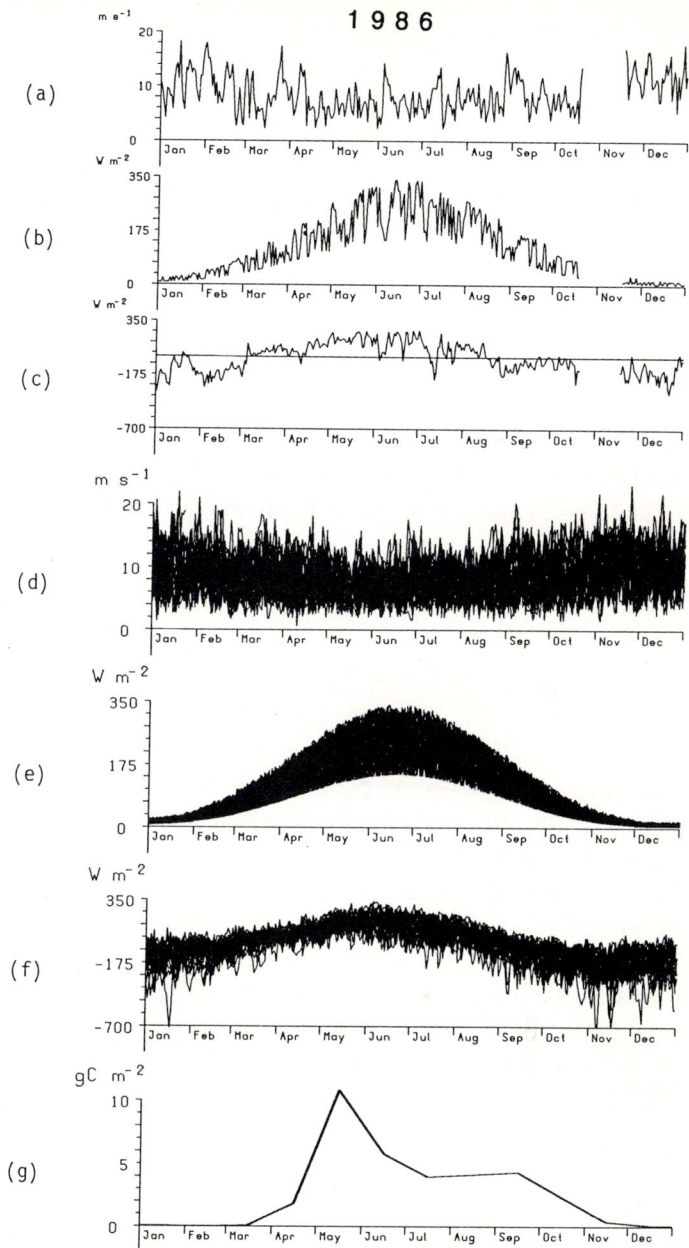

Fig. 3.3.4. The annual cycles of the meteorological forcing parameters at Light Vessel ELBE 1 for the year 1986 (*a-c*), with a storm at the beginning of June, and no data from mid-October until mid-November. 25 annual cycles of the meteorological forcing parameters at Light Vessel ELBE 1 projected into one year (*d-f*), for the parameters *a,d* wind velocity, *b,e* direct and diffusive solar radiation, *c,f* daily heat balance at the sea surface from Moll and Radach (1990a). *g* Mean annual cycle of copepod biomass derived from Continuous Plankton Recorder Data. (Moll 1989)

Helgoland to yield global radiation values at any instant (Moll and Radach 1991).

To relate this single year 1986 to the variability of the meteorological forcing during 25 years (1962-1986), we present all the data projected into 1 year in Fig. 3.3.4d-f. There is a considerable variability in the annual cycles for all parameters. Wind velocity exhibits the least variability in summer, during May to August. The lower values of radiation are model-specific. The daily heat balance values range from -385 to 35 W/m^2 in winter and from 0 to 350 W/m^2 in summer. The daily heat balances show that the establishment of the thermocline, caused by the change in sign of the heat balance at the surface, may occur within about two months, from the beginning of March to the end of April.

The zooplankton biomass was prescribed for all simulations discussed here by the mean annual cycle in the central North Sea (Fig. 3.3.4e) derived from the Continuous Plankton Recorder survey data (Colebrook 1979). The derivation is given in Moll (1989).

3.3.2.3 Validation of the Model

The aim of the validation procedure is to show that the model is able to reproduce the characteristics of the annual cycle of phytoplankton dynamics. These characteristics are the starting point of the spring bloom, the magnitude of the spring bloom, a quasi-steady state in summer, and occasional blooms in late summer and fall. We should obtain mean annual cycles for the central North Sea, a total primary production in the order of measured values and realistic P/B ratios as given in the literature.

A prerequisite for the simulation of the phytoplankton characteristics is the correct simulation of the temperature development, the depth of the mixed layer and the depth of the euphotic layer.

The validation of a model can be achieved only by testing the model against several independent data sets which originate from the same dynamics, but not necessarily from the same area. This fortunate situation is but rarely given. Our attempts to validate the model refer to one experiment so far, the international, interdisciplinary Fladen Ground Experiment (FLEX'76) which took place in 1976 in the northern North Sea to investigate the spring phytoplankton bloom. The achievements from the test of the model against the FLEX'76 data can be judged from Radach (1983, Fig. 8), where the comparison is described in detail. For other times of the year no such complex data set exists which is suited to serve for validating the model.

To further check the validity of our model we used even more characteristics. Moll (1989) extrapolated the time span of model application from the
spring bloom to the full annual cycle for 1976. The degree of coincidence
of simulation and FLEX'76 measurements decreases, however, because the
initial and boundary conditions are prescribed from the weather ship
FAMITA data (see Sect. 3.3.3). Thus, we reach only a certain similarity
of the annual simulation results for 1976 during the FLEX'76 period with
the FLEX'76 observations. One has to keep in mind that uncertainties in the
forcing (due to different locations) which we use to force the simulations
will cause uncertainties in the simulation results.

Usually, model simulations are compared to measurements of the state
variables. Other criteria are the agreement between simulated and measured
flows of matter through the system and the capability of reproducing the
variability observed in the state variables and process contributions. We
validated the model-derived variability using local as well as integral
properties by simulating 25 annual cycles under realistic forcing conditions.
This validation step is reported in Radach and Moll (1993), where the
comparison of observed and simulated variability is described in detail.

3.3.3   Case Study: Phytoplankton Dynamics in 1986 for the Central North Sea

Simulations for the meteorological year 1986 were performed to attempt a
comparison of observed features of the plankton dynamics during the
ZISCH survey with the corresponding features of the simulations. The
radiative forcing data for 1986 were continually available only until mid
October (day 291), thereafter the simulation is continued by interpolated
values.

Of course, the two surveys of ZISCH in summer 1986 (2 May-13 June)
and winter 1987 (21 January-9 March) provided data which are resulting
from the full system's dynamics, including the advective and diffusive
lateral transports. The advantage of simulations like ours consists in their
potential to give patterns of explanations, even if they are not able to
provide quantitative simulated mirror pictures of the observations during the
ZISCH cruises. A data report for the two cruises mentioned above was
presented by Moll und Radach (1990b,c).

For our simulation of phytoplankton dynamics in the central North Sea
we use the configuration of the model for weather ship FAMITA (Table
3.3.1) applying the meteorological annual cycles of 1986 from LV ELBE
1 for calculation of the forcing (Moll and Radach 1990a). Weather ship

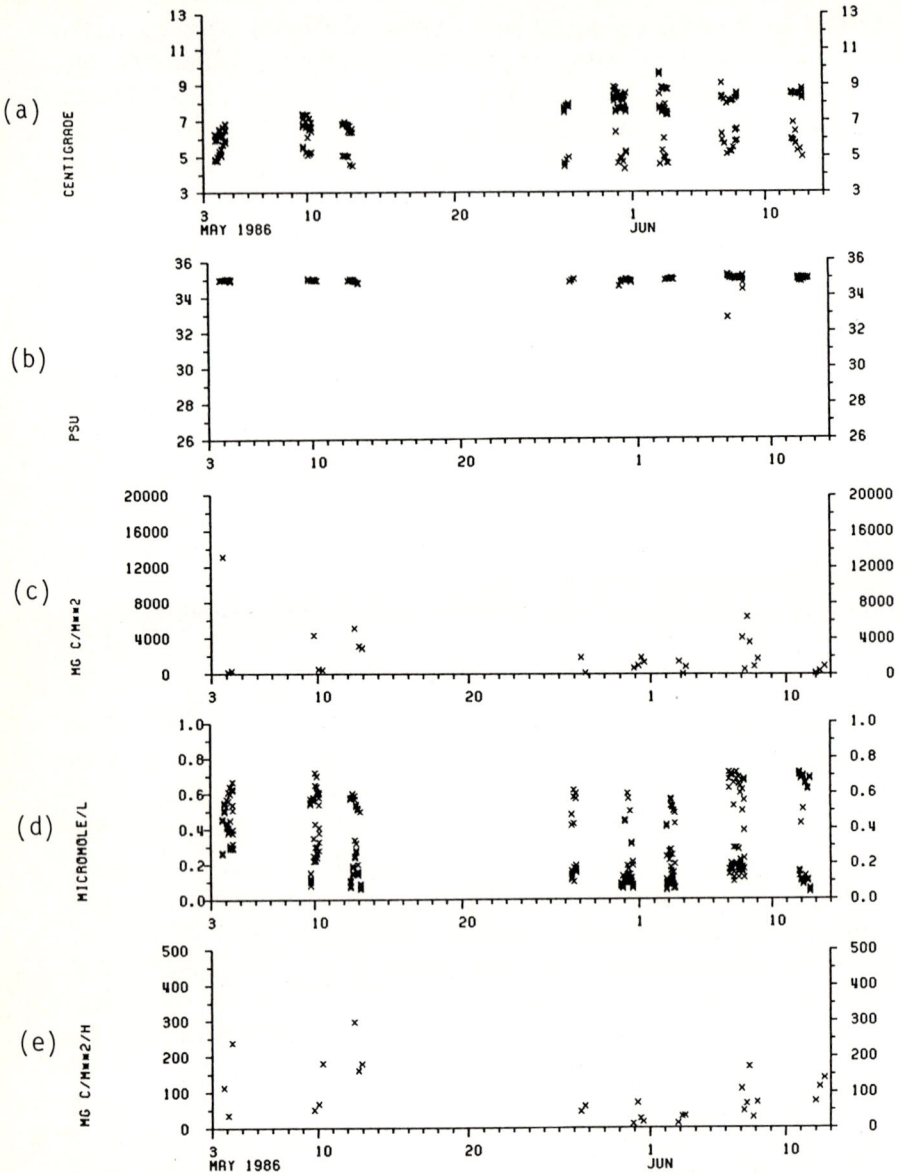

Fig. 3.3.5a-e. Data for phytoplankton and phosphate from the summer ZISCH cruise (3 May-13 June 1986) as functions of time during the path through the North Sea. Regional patterns are mapped as time changes. Data originates from different investigators (ZISCH PARAMETER REPORT by Moll and Radach 1990c, originators are cited therein). *a* Temperature profiles, *b* salinity profiles, *c* phytoplankton depth-integrated (0-20 m), *d* phosphate profiles, *e* primary production depth integrated (0-20 m). Nutrient depletion of the upper layer can be recognized from the lowest values in *d*

FAMITA is situated at 57°N, 3°E in the central North Sea, close to the centre of the star-shaped surveys of ZISCH (see Fig. 3.2.11 of the previous chapter). For comparison we use the ZISCH data obtained in the rectangle from 1° to 5° E and from 55°30' to 58°30'N. In Fig. 3.3.5 we give the data in this rectangle, projected onto the time axis during the summer survey in 1986. The figure gives an impression of the ranges of observed temperature, salinity, phytoplankton, phosphate and primary production during the summer cruise.

It can be seen that the temperature within the central region is about 6-8°C, and the salinity is about 35 for the full time span of the survey, indicating that the area had constant physical conditions. Thus, we may look at the dynamics in this area as if they were locally induced. The observa-

Table 3.3.1. Variables, dynamical constants, conversion factors for the phytoplankton-phosphate-detritus model for the central North Sea at weather ship *Famita* (water depth 60 m, vertical resolution 2.5 m).

	Quantity	Value	Unit
D	Detritus concentration	Variable	gC/m^2
D_o	Initial detritus concentration	0.0	gC/m^2
e	C/Chl ratio	50.0	gC/gChl-a
g_P	P/C ratio	0.6944	mmol/gC
I_o	Optimum light intensity (*10^{18})	123	quanta/m^2s
k_s	Half saturation constant for phosphate	0.06	$mmol/m^3$
N	Phosphate concentration	Variable	$mmol/m^3$
N_o	Initial phosphate concentration	0.6	$mmol/m^3$
n_E	Percentage soluble excretion	0.33	-
n_F	Percentage reminer. fecal material	0.33	-
n_Z	Percentage dead zooplankton	0.33	-
P	Phytoplankton concentration	Variable	gC/m^3
P_o	Initial phytoplankton conc.	0.001	gC/m^3
P_1	Threshold for zooplankton grazing	0.01	gC/m^3
p_F	Percentage reminer. fecal material	0.20	-
p_M	Percentage reminer. dead org. matter	0.20	-
P_s	Half-saturation constant for grazing	0.1	gC/m^3
p_Z	Percentage of reminer.	0.2	-
r_D	Remineralization rate of bottom detritus	0.0167	/d
r_M	Mortality rate	0.05	/d
r_P	Growth rate	1.5	/d
r_P	Percentage photorespiration	0.05	-
r_R	Percentage basic respiration	0.1	-
r_R	Temperature dependent respiration	0.069	/°C
r_Z	Grazing rate	0.5	/d
w_s	Sinking velocity	0.5	m/d

tions for the state variables phytoplankton, primary production and phosphate (Fig. 3.3.5) show changes in time. The surface concentrations of phosphate decrease from about 0.38 mmol/m^3 on 3 May to about 0.08 mmol/m^3 on 10 May, remaining at this level until the end of the cruise. Primary production decreases from about 100 to 225 mgC/m^2/h during the first half of May to levels of 50 to 100 mgC/m^2/h in late May and early June.

Let us now turn to the results of the simulations. The physical situation is characterized by cooling in February and the start of thermocline development in March (Fig. 3.3.6a). Until mid April a series of severe storms cause a deepening of the mixed layer regime (Fig. 3.3.6b). Not before day 124 does the 7°C isopleth deepen from the surface. There are storm events causing a cooling of the upper layers during summer, namely in early June, mid July and in early September, penetrating down to about 30 and then 40 m. The decay of the thermocline occurs late and gradually in October. The mixed layer depth decreases until mid April to about 10 to 15 m and increases again from August onwards (Fig. 3.3.7).

Phytoplankton standing stock and primary production (Figs. 3.3.6c-d) correspond closely to the development of turbulent mixing intensity (Fig. 3.3.6b). Parallel to the warming of the upper layers, a spring bloom is initiated which results in a first maximum around day 95. Then a storm disturbs this development by mixing the water column and distributing the algae over the water column. The bloom starts again and culminates without further disturbance in April. A summer storm (on days 152 to 160) deepens the mixed layer again, thus initiating a summer bloom due to nutrient entrainment into the upper layer (Fig. 3.3.6d,e). From the phosphate development, one can conclude that the storms are replenishing the upper layers with nutrients, and less production occurs below the nutricline (Fig. 3.3.6e). Strong events of this kind occur during 1986 in June (on day 156), in July (on day 191) and in August (on day 240). Not before the end of October is the phosphate accumulated by regeneration in the lowest layers distributed again over the full water column in this simulation.

The governing equations of the model (Moll 1989; Radach and Moll 1993) may serve for deriving a mass balance for the three state variables by integrating them over depth and a variable time T. The single cumulated (= time integrated) process terms in the equations are shown in Fig. 3.3.8a-c. The balance terms for phosphate (Fig. 3.3.8b) make it clear that the initial content within the water column is taken up 1.4 times by the algae during the full year. Until the peak of the bloom nutrient uptake is about half the annual value. Nearly the same amount is regenerated at the bottom and diffused upwards via the boundary condition into the water column. Gross annual production amounts to about 100 gC/m^2, while net

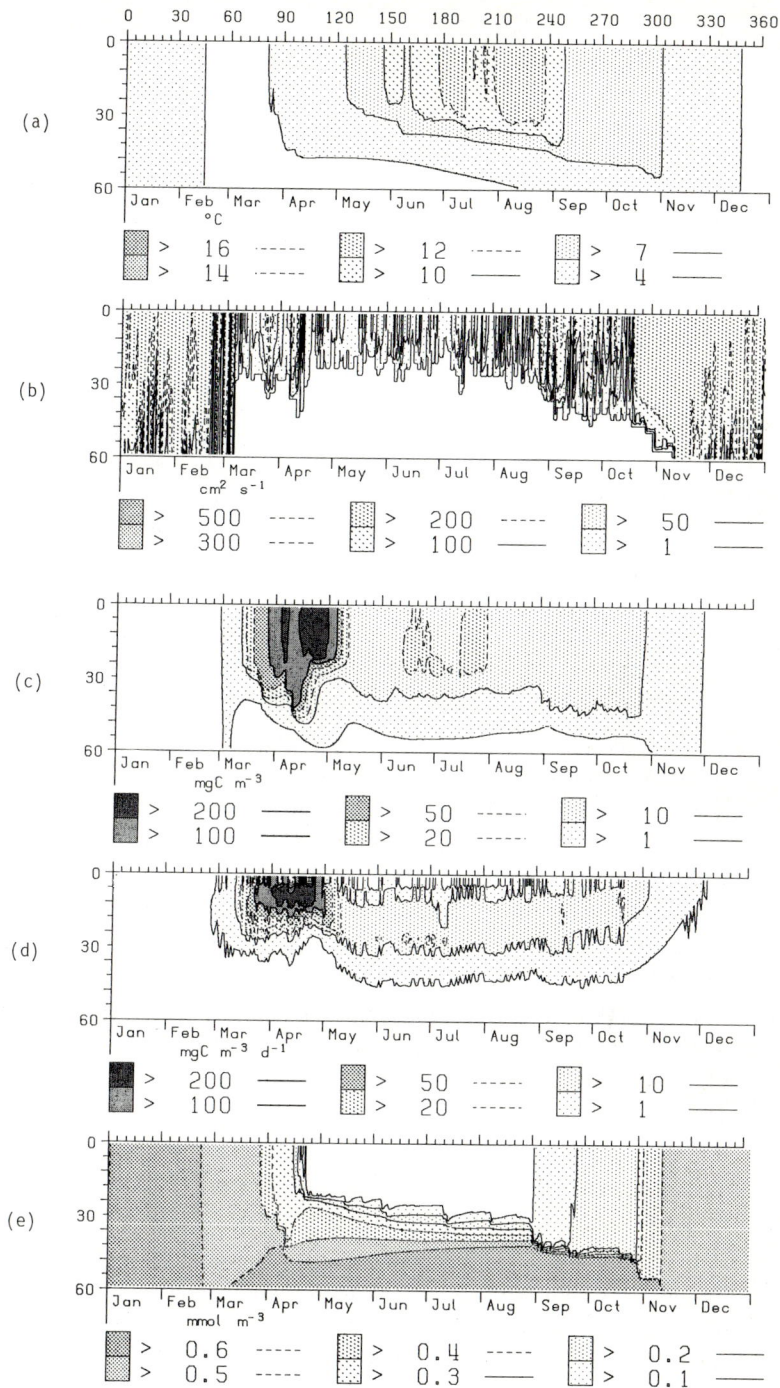

Fig. 3.3.6a-e. Simulated profiles (0-60 m) for *a* temperature, *b* turbulent diffusion coefficients, *c* phytoplankton standing stock, *d* primary production, *e* phosphate concentration for 1986

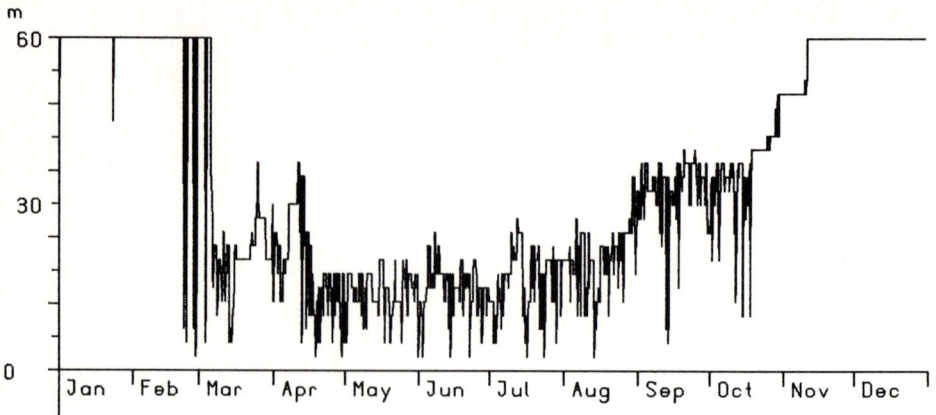

Fig. 3.3.7. Simulated mixed layer depth for the year 1986

production is half this value in this simulation. The standing stock is very small compared to the transfers leading to it (Fig. 3.3.8a).

The piling up of detritus at the bottom (Fig. 3.3.8c) happens to occur within about 1 month. Thereafter, the decay is faster than the filling up of the pool. The time shift between the two curves results according to the assumed remineralization rate of about 2 months.

Now we will look at the ZISCH measurements (Fig. 3.3.5) in the framework given by this simulation (Fig. 3.3.6). It can be seen that the observed phosphate concentrations are higher only in early May but originate in both cases from a situation of nutrient depletion. The values for phytoplankton biomass and primary production are in the same range at the times of observations.

We thus may state that the summer cruise of ZISCH in 1986 fell in the time of the decaying phase of the spring phytoplankton bloom in the central North Sea, right after its culmination, in accordance with the results of Chapter 2.2. After a short phase of recovery, primary production intensified again: this was due to the partial replenishment of the upper layer with phosphate caused by the stormy weather around 10 June.

From the results of the model study, we can obtain quantities of interest which could not be measured with a fine temporal resolution, i.e. contributions of certain processes to the plankton and phosphate dynamics. From the model results we estimated that during the time interval from 1 May to 15 June a total gross and net primary production of about 14 and 8 gC/m^2, and a phosphate uptake of about 7 $mmol/m^2$ was accomplished during the cruise. The sedimentation was about 8 gC/m^2 during this time. Compared to this, the simulated spring bloom before the cruise had resulted in 57 gross and 30 gC/m^2 net production, previous to 1 May.

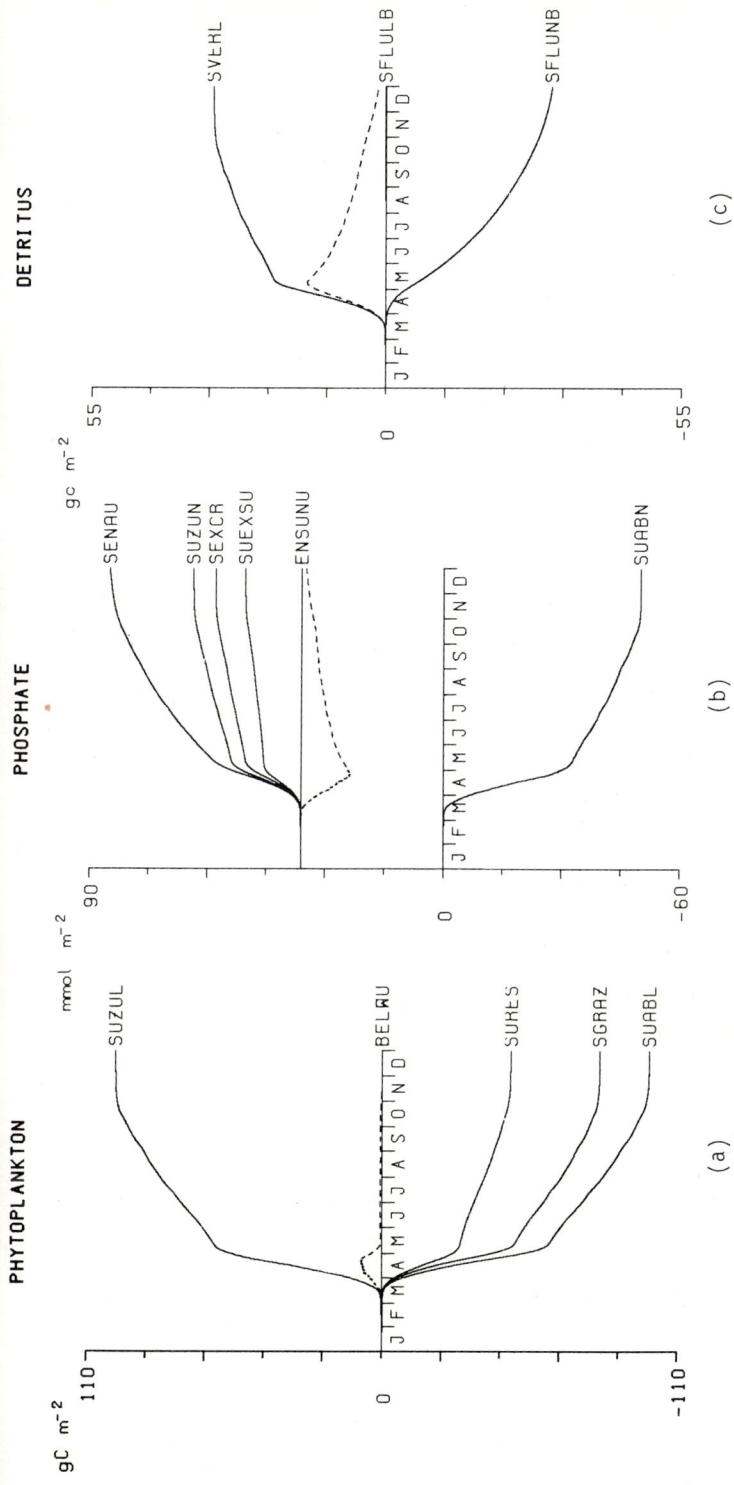

Fig. 3.3.8. Mass balances for the three state variables, integrated over depth and time. *a* Phytoplankton biomass (*dashed line*), and the cumulative process contributions primary production (*SUZUL*), respiration (*SURES*), mortality (*SUZUL*), grazing (*SGRAZ*) and sinking (*ELWA*), for 0-60 m. *b* Phosphate content (*dashed line*), and the cumulative process contributions uptake (*SUABN*), excretion (*SEXCR*), remineralization (*SUZUN*), exudation (*SUEXSU*) and diffusion (*SENAU*), for 0-60 m. *ENSUNU* is the initial value. *c* Detritus (*dashed line*), and the cumulative process contributions diffusion (*SFLULB*), sinking out of the water column (*SVERL*) and remineralization (*SFLUNB*). The cumulative positive and negative process contributions to the balance are added in the graph, respectively

Summarizing, we may state that the simulation for 1986 gives a useful reference frame for the ZISCH summer cruise and that this cruise has provided a second data set suitable for the successful (second) validation of our model.

3.3.4 Phytoplankton Variability Induced by Nutrient Variability (Eutrophication)

In this section we report on an investigation of scenarios of eutrophication to learn about the effects of varying phosphate values at the end of winter on the phytoplankton dynamics of the following year (Radach and Moll 1990a). We present simulation results of three scenarios which characterize different nutrient loading conditions, namely the phosphate loading in estuaries like the Elbe river mouth (2.4 mmol/m^3, case 1), in the continental coastal strip (1.2 mmol/m^3, case 2) and in the central North sea (0.6 mmol/m^3, case 3), to compare the annual cycles of phytoplankton dynamics under these conditions, using identical physical forcing.

The physical upper layer model was driven by 3-hourly wind, radiation and heat flux values derived from meteorological standard observations at weather ship FAMITA during 1976 (Moll 1989). Again, zooplankton is prescribed as the mean annual cycle of copepod biomass (Moll 1989) to formulate the grazing pressure.

The whole setting may be considered somewhat artificial, but it may serve to investigate the tendency of the consequences which result from increased nutrient loads under the assumption that "everything else" remains constant, i.e. the physical situation and the dynamic mechanisms.

Figure 3.3.9a shows the annual cycles of phytoplankton profiles for the initial phosphate values of case 1. For increasing phosphate loads (case 2 and 3), the spring bloom increases also (Fig. 3.3.9b- c), but mostly the summer blooms increase in duration and in magnitude. The beginning and the end of the vegetation period is not altered.

Figure 3.3.10 shows the corresponding phosphate development. The phase of phosphate depletion (< 0.2 mmol/m^3) shrinks from May-October (case 1) to July-September (case 2) and finally to one month, namely August (case 3). At the bottom, phosphate accumulates due to the regeneration process, especially from April to October, until storms mix the full water column again.

To detect the nonlinearity of the dynamics, the comparison of phosphate and phytoplankton distributions for the three cases can be accomplished by looking at the vertically integrated balance terms of the system. One should

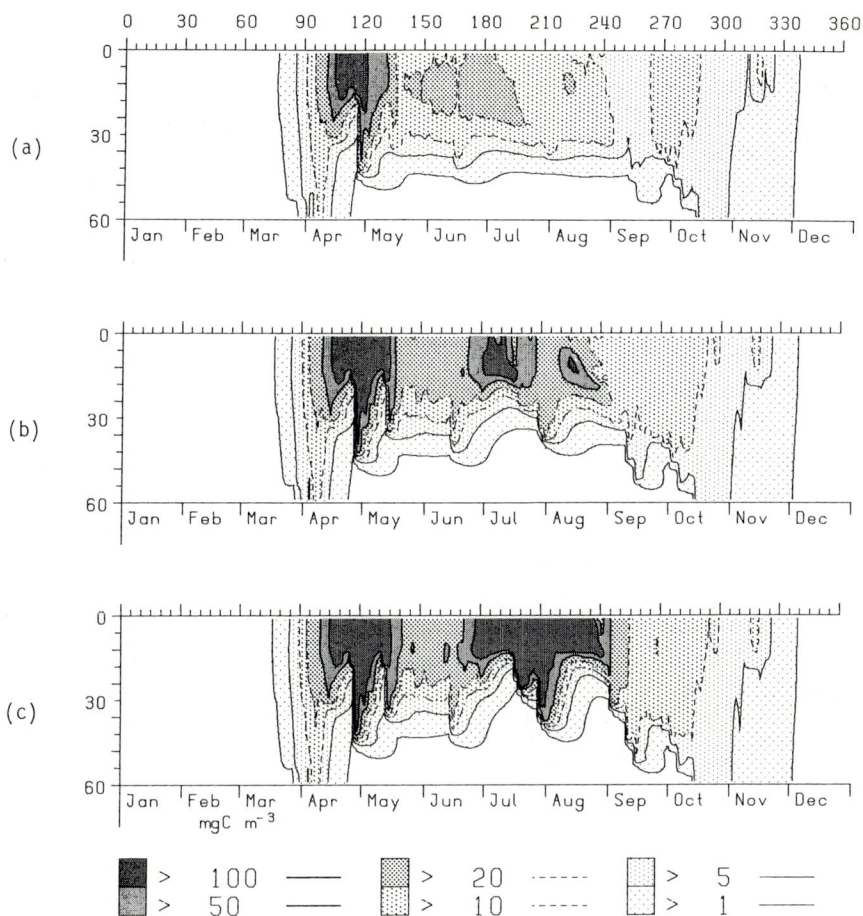

Fig. 3.3.9a–c. Simulated profiles of the annual cycles of phytoplankton, starting with different initial phosphate concentrations. *a* 0.6 mmol/m^3, case 3, *b* 1.2 mmol/m^3, case 2, *c* 2.4 mmol/m^3, case 1. (Radach and Moll 1990a)

notice that the increase of the winter phosphate load by a factor of 2 from case 1 to 3 does not result in a corresponding increase of phytoplankton. The peak spring concentrations rise from nearly 6 to 8, and further to about 10 gC/m^2 (Fig. 3.3.11). In addition, summer blooms develop from a level of 1 to 3 gC/m^2 and, for the highest nutrient load, up to 8.5 gC/m^2! The latter is nearly the same as during spring.

The difference in the annual cycles of phytoplankton development may be seen even more clearly in Fig. 3.3.12, where the ratios of the phytoplankton curves from Fig. 3.3.11 are plotted. While a linear relation-

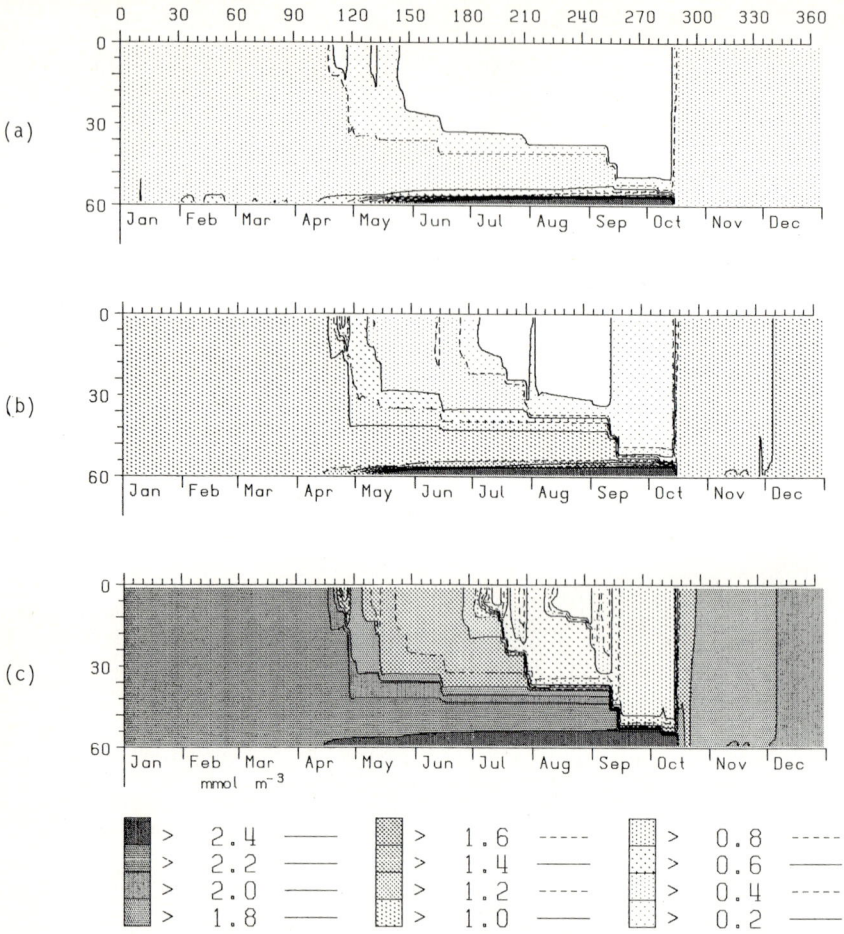

Fig. 3.3.10a–c. Simulated profiles of the annual cycles of phosphate, starting with different initial phosphate concentrations. *a* 0.6 mmol/m^3, case 3, *b* 1.2 mmol/m^3, case 2, *c* 2.4 mmol/m^3, case 1. (Radach and Moll 1990a)

ship between phosphate load and phytoplankton would cause a ratio of two, Fig. 3.3.12 shows that the relationship is strongly non-linear!

The strong increase of summer blooms is caused by the increase in the winter nutrient load, which cannot totally be consumed by the algae in spring due to light limitation by self-shading. The plentiful nutrient supply for a longer period in the upper layer and, of course, continually in the lower layer enables continued production. Later in the year, during summer, production is fed by an upward turbulent transport of phosphate through the thermocline. The balances taken from the surface down to the mean thermocline depth in summer, which is the upper 25 m, demonstrate

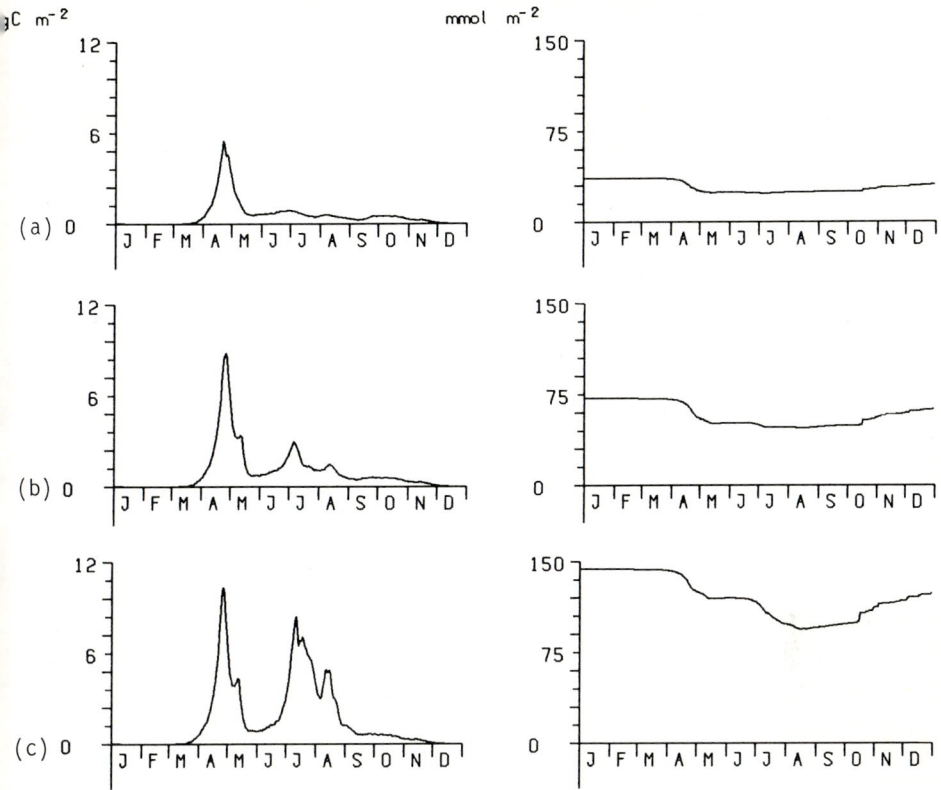

Fig. 3.3.11a–c. Simulated annual cycles of vertically integrated phytoplankton and phosphate for the three cases *a* 0.6 mmol/m^3, case 3, *b* 1.2 mmol/m^3, case 2, *c* 2.4 mmol/m^3, case 1. (Radach and Moll 1990a)

the importance of diffused nutrients into the euphotic layer: three times, five times, and four times the initial content of phosphate in the upper layer is diffused into the upper layer during the year for the cases 1 to 3, respectively.

The comparisons of ratios of standing stock and phosphate contents in the water column for the different nutrient loads teach that the phytoplankton-phosphate dynamics of this simple system are reacting in a strongly non-linear pattern to the increase of nutrient loads. The increase of a basic nutrient (here phosphate) load in winter enhances not only the spring bloom, but even more strongly the summer blooms. This is an effect which happens in spite of the crude nutrient dynamics included. The effect is brought about by the physical mixing process in combination with a chemical entity (here phosphate), which acts as a limiting nutrient for primary production.

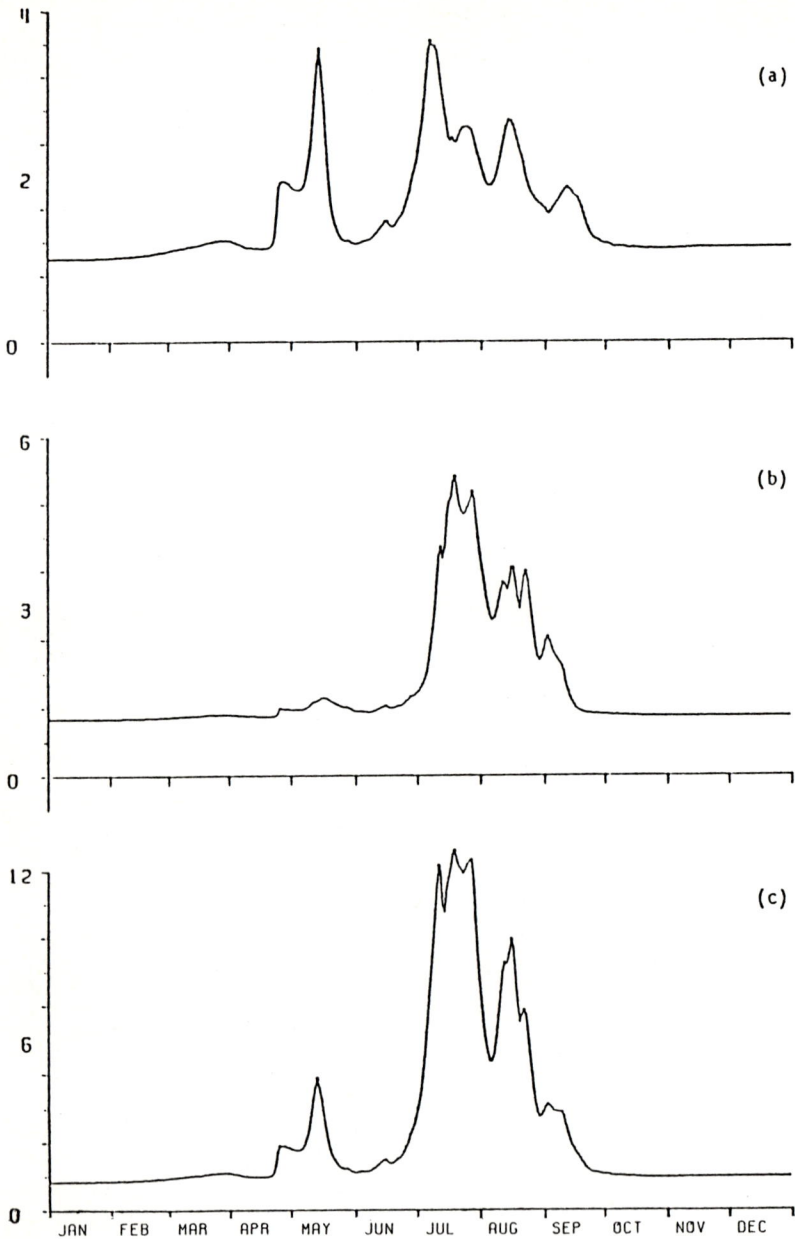

Fig. 3.3.12a-c. Ratios of simulated annual cycles of vertically integrated phytoplankton of Fig. 3.3.11 for three cases. *a* Simulation results decribed by Fig. 3.3.11b divided by simulation results described by Fig. 3.3.11a. *b* Simulation results decribed by Fig. 3.3.11c divided by simulation results described by Fig 3.3.11b. *c* Simulation results decribed by Fig. 3.3.11c divided by simulation results described by Fig 3.3.11a. (Radach and Moll 1990a)

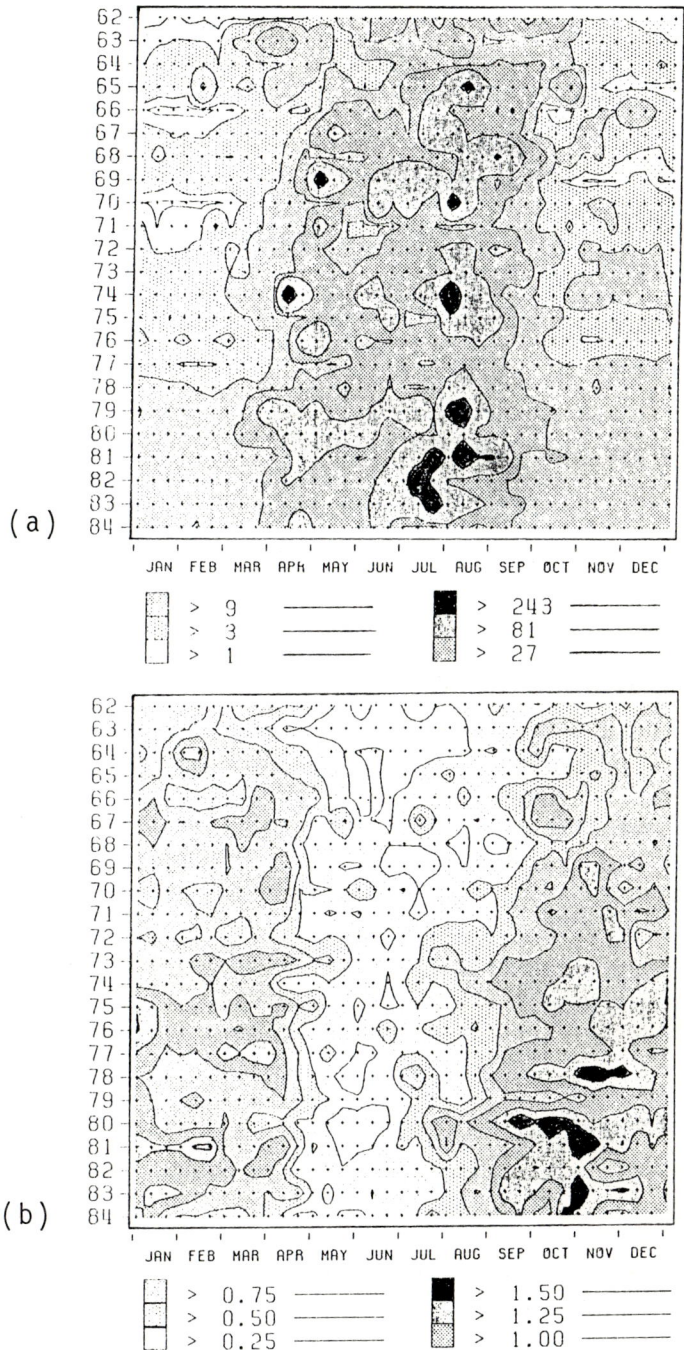

Fig. 3.3.13a,b. Long-term changes of the annual cycles of: *a* phytoplankton biomass (mg C m^{-1}), *b* phosphate concentration (mmol m^{-3}) at Helgoland Roads from 1962 to 1984. (Radach et al. 1990)

The appearance of the phytoplankton dynamics calls for a comparison of these results with the long term measurements at Helgoland (Radach et al. 1990), where the increase of nutrients over 23 years also has the effect of increasing algal blooms in summer (Fig. 3.3.13a). Although in our model we have only one phytoplankton compartment, namely biomass, and although the biomass increase at Helgoland is brought about by flagellates in summer, the similarity of the dynamics under increasing phosphate loads (Fig. 3.3.13b) is striking. The shortening of the phase of depletion of phosphate, however, develops from both ends, spring and fall, in the model results, whereas in reality the starting point of depletion remains in April.

3.3.5   Phytoplankton Variability Induced by Weather Variability

In a further study we investigated the effects of the variability of the weather on phytoplankton dynamics (Radach and Moll 1993). The assumption is that a great deal of biological variability is created solely by the physical variability impressed on the biological system. We thus used the 25 annual cycles (1962-1986) of 3-hourly standard meteorological observations at LV ELBE 1 to drive the physical and the biological model, without altering any of the dynamical constants or the structure in the biological model. The results are described in detail in Radach and Moll (1993); here we give a brief report of the main results only.

The physical upper layer simulations were performed with the same model as mentioned above. For the 25 years they result in a total variability of sea surface temperature of about 4.6 °C in winter and up to 9°C in summer (Fig. 3.3.14a). The mixed layer depth (Fig. 3.3.14b) is characterized by a broad band of possible positions of the thermocline (=turbocline), which is about 30 m deep in summer and can be situated very close to the surface. This is not quite in accordance with observations, and possibly the process of mixing the heat down into the water column is not strong enough in the model.

The euphotic depth (Fig. 3.3.14c) is situated at about 50 m in winter. It decreases during March and April and becomes shallow (about 10 to 20 m) in many years during the spring bloom, due to self-shading. In May it deepens, and it remains in about 35 to 40 m during summer.

Here, we cannot show the variability of the profiles. Instead, we consider the variability of two of the three state variables (phosphate and phytoplankton biomass) from the simulations and of two balance terms (gross production and diffused nutrients across the thermocline) in the verti-

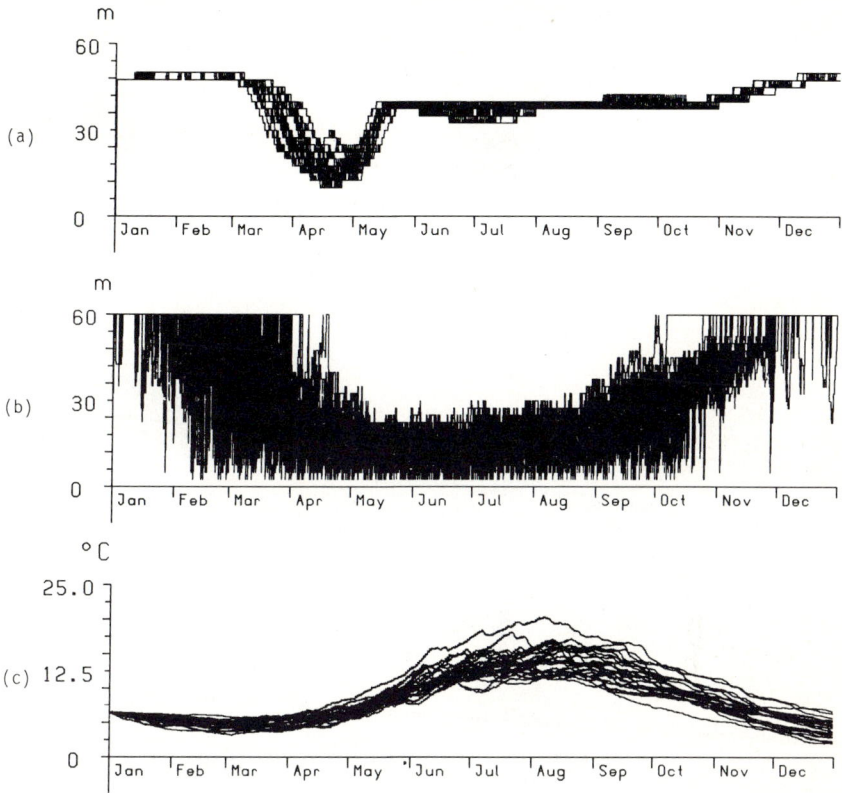

Fig. 3.3.14a-c. Simulated *a* euphotic depth, *b* mixed layer depth and *c* sea surface temperature for the 25 years, projected into one annual cycle. (Radach and Moll 1993)

cally integrated form for all of the 25 years of meteorological forcing (Figs. 3.3.15a-d) within the upper 27.5 m.

The phytoplankton concentrations during the spring blooms (Fig. 3.3.15c) do not exceed 9 gC/m^2, and small contents of not more than 1 gC/m^2 limit the range of variability to the lower end. Summer blooms are much smaller than spring blooms. They do not exceed 1 gC/m^2. The model seems not yet to be capable of reproducing blooms in fall realistically. Fall blooms seem to occur regularly in nature (compare Fig. 3.3.15c and Fig. 3.3.2, boxes B2, C2). The spring blooms generally do not start before March. The start of the phase of exponential increase may vary by 1 month.

Corresponding to the large variation in the biomass during spring blooming, there is a large variability of the phosphate content in the upper layer, varying extremely from April to June (Fig. 3.3.15a). The instant

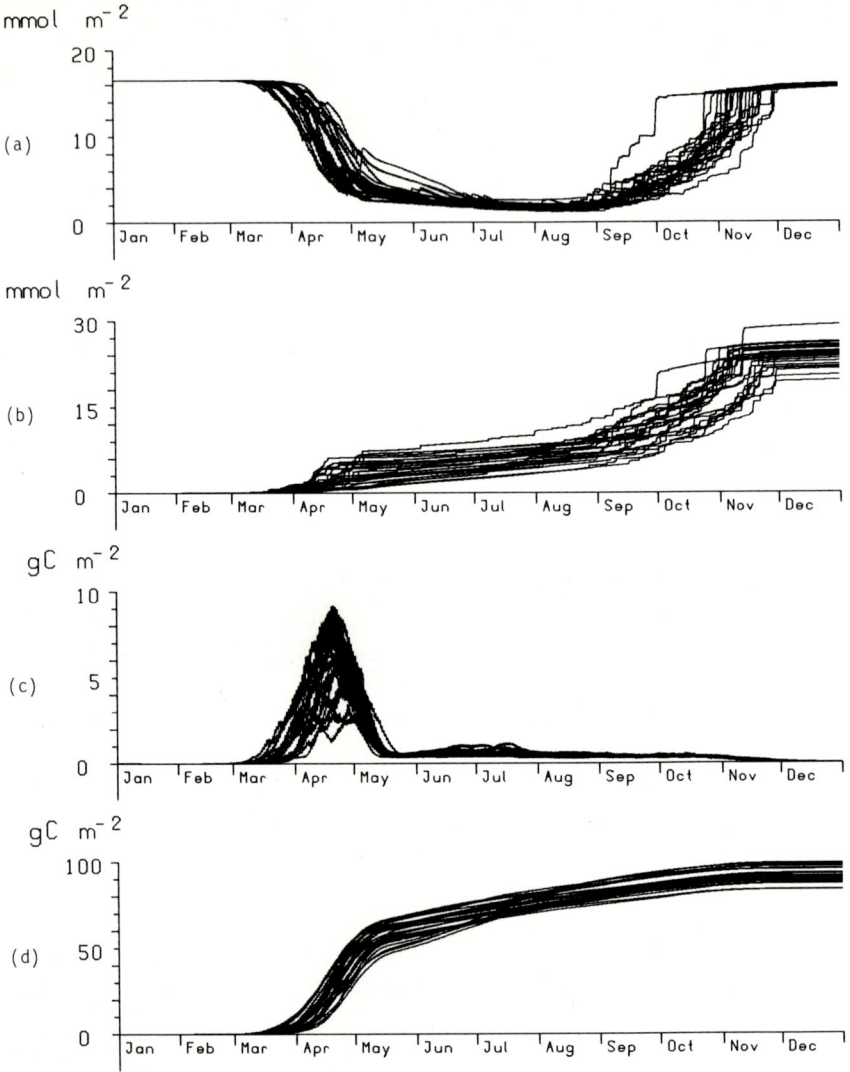

Fig. 3.3.15a-d. Simulated *a* phosphate content (0-27.5 m), *b* cumulated turbulent diffusion across 27.5 m, *c* phytoplankton content (0-27.5 m) and *d* cumulated gross production for 25 years projected into one annual cycle. (Radach and Moll 1993).

when a depletion of, say, 5 mmol/m² in the upper 27.5 m is reached may occur from mid-April until the end of June. The time for replenishment of the water column with phosphate varies by about 50 days (early summer until late October).

Production mainly takes place within the upper 30 m. There is nearly no difference between the cumulative production curves for the depth range of

0-27.5 m compared to the depth range of 0-60 m. Cumulative gross production (Fig. 3.3.15d) shows its highest variability during April and May. At the end of May cumulative gross production ranges from 51 to 72 gC/m^2. At the end of the year total production sums up to about 91 gC/m^2, within a bandwidth of only 16 gC/m^2. Thus, although production events are unevenly distributed over spring and summer according to the meteorologically favourable conditions, total production is much the same for all 25 years.

For all processes involved in the plankton dynamics of our model the ranges of variability can be given, for example, for the layer 0-27.5 m. The cumulative total water column respiration reaches annual values of around 38 gC/m^2, with a variability bandwidth of 15 gC/m^2. Cumulative grazing is of the same order of magnitude as respiration, reaching a mean annual value of 33 gC/m^2. The variability range is 25 to 40 gC/m^2.

Nutrient dynamics are a direct picture of phytoplankton dynamics. Cumulative nutrient uptake corresponds directly to phytoplankton gross production. The phosphate content (Fig. 3.3.15a) of the upper layer starts to decrease generally not earlier than March. The process of depleting the upper layer may last 2 months. Thus, the variability of the time of reaching a certain content, say half the initial content, has a variability of 3 weeks. Annual mean cumulative uptake amounts to 47 $mmol/m^2$, with a range of about 10 $mmol/m^2$.

An interesting quantity is the amount of phosphate diffused from the lower layer into the upper layer, thus fuelling phytoplankton production. This quantity may be added to the initial phosphate content in the upper mixed layer to serve as an estimate of phosphate being incorporated into new production of phytoplankton (for the discussion of new production in the ocean see e.g. Jenkins and Goldman 1985). The variability of diffused phosphate across the 27.5 m level is high. In Fig. 3.3.15b the amount of cumulative diffused phosphate across the depth of 27.5 m is plotted for 25 annual cycles. The annual amounts range from 20 to 29 $mmol/m^2$, which is 56 to 81 % of the initial content of the total water column of 60 m depth. The content of the upper layer of 27.5 m, namely 16.5 $mmol/m^2$, is replaced by diffusive transport 1.2 to 1.8 times during 1 year.

This study shows that the simulations of weather-induced variability of spring plankton blooms can explain most of the variability observed. This is not true for summer and fall bloom events. A possible reason could be the mixing mechanism at the bottom, which transports phosphate regenerated at the bottom up into the water column again. In the physical model used there is no tidal stirring effect included so far, as, e.g. in the models by James (1977) and van Aken (1984). We believe that the inclusion of the tidally driven bottom boundary layer would yield the observed fall blooms also in the model simulations, because the mixing within the tidal bottom boundary layer would have a larger vertical scale than the nutrient-rich water has in our model (Fig. 3.3.6e). Then, the nutrient would be distributed further up in the water column, and a storm in fall could entrain the nutrient into the depleted upper layer earlier. The next improvement concerns the introduction of a prognostic zooplankton equation.

3.3.6 Conclusions

The model simulations of phytoplankton and phosphate dynamics prove to be a very helpful means for the understanding and interpretation of phytoplankton and phosphate dynamics in the field.

The simulation for the year 1986 provides a good framework for the interpretation of the dynamic state during the ZISCH summer cruise. The cruise took place right after the culmination of the spring phytoplankton bloom in the central North Sea and covered a phase of decay and recovery of phytoplankton standing stock until a storm caused nutrient intrusions into the mixed layer, thus initiating enhanced primary production again.

The model studies using different phosphate loading in winter, as occurs in the North Sea in different regions, suggest that increasingly high loading causes increasing summer blooms. The resulting patterns of plankton dynamics resemble those which were observed in the German Bight at Helgoland where the annual cycles of plankton and phosphate changed drastically during the last 25 years due to increasing eutrophication.

The magnitude of variability due to different states of eutrophication is comparable in spring to the magnitude of the natural variability due to different weather. As the model study using 25 meteorological years at LV ELBE 1 suggests that the observed variability of plankton and nutrients may be large enough to explain most of the observed variability, it will be a problem to discriminate weather effects from eutrophication effects.

Concerning the extension of the water column model to a regional 2-D or 3-D North Sea model, we learn that the model design must be such that small-scale temporal forcing can be introduced and realized. This is an important demand for a corresponding spatial resolution, both vertically and horizontally. Secondly, for simulating annual cycles the vertical mixing in the bottom boundary layer will have to receive special attention.

Both the eutrophication scenarios and the weather scenarios were evaluated to obtain guidance for the planning of measurement campaigns in PRISMA, the follow-up of ZISCH, to investigate important processes within the ecosystem of the German Bight. For any such project in a region influenced both by eutrophication and by highly variable weather conditions, it would be a prerequisite to obtain knowledge about the variability occurring in this region to define the demands for the temporal and spatial distributions of the measurements in order to overcome errors due to variability. The simulations reported build part of the basis for further field research.

Acknowledements. We would like to thank Ms. O. Kleinow for her great help in programming and Mr. T. Soltau for technical assistance.

References

Colebrook JM (1979) Continuous plankton records: seasonal cycles of phytoplankton and copepods in the North Atlantic Ocean and the North Sea. Mar Biol 51:23-32

Denman KL, Powell TM (1984) Effects of physical processes on planktonic ecosystems in the coastal ocean. Oceanogr Mar Biol Annu Rev 22:125-168

Dobson FW, Smith SD (1988) Bulk models of solar radiation at sea. Q J R Meteorol Soc 114:165-182

Eberlein K, Kattner G, Brockmann U, Hammer KD (1980) Nitrogen and phosphorus in different water layers at the central station during FLEX'76. Meteor Forschunsergeb 22 (Serie A):87-99

Fransz HG, Verhagen JHG (1985) Modelling research on the production cycle of phytoplankton in the Southern Bight of the North Sea in relation to riverborne nutrient loads. Neth J Sea Res 19:241-250

Fransz HG, Mommaerts JP, Radach G (1991) Ecological modelling of the North Sea. Neth J Sea Res 28:67-140

Friedrich H, Kochergin VP, Klimok VI, Protasov AV, Sukhorukov VA (1981) Numerical experiments for the model of the upper oceanic layer. Meteorol Gidrol 7:77-85

Harris GP (1980) Temporal and spatial scales in phytoplankton ecology. Mechanisms, methods, models, and management. Can J Fish Aquat Sci 37:877-900

James ID (1977) A model of the annual cycle of temparature in a frontal region of the Celtic Sea. Estuarine Coastal Mar Sci 5:339-353

Jenkins WJ, Goldman JC (1985) Seasonal oxygen cycling and primary production in the Sargasso Sea. J Mar Res 43:465-491

Kiefer DA, Kremer JN (1981) Origins of vertical patterns of phytoplankton and nutrients in the temperate, open ocean: a stratigraphic hypothesis. Deep-Sea Res 28:1087-1105

Klein P, Coste B (1984) Effects of wind-stress variability on nu trient transport into the mixed layer. Deep-Sea Res 31:21- 37

Kochergin VP, Klimok VI, Sukhorukov VA (1976) A turbulent model of the ocean Ekman layer. Sb Chisl Metody Mekhan Sploshnoi Sredy 7:72-84

Legendre L, Demers S (1984) Toward dynamic biological oceanography and limnology. Can J Fish Aquat Sci 41:2-19

Moll A (1989) Simulation der Phytoplanktondynamik für die zentrale Nordsee im Jahresverlauf. Ber Zentrum Meeres- Klimaforsch Univ Hamb 2:1-139

Moll A, Radach G (1990a) Wärme- und Strahlungsflüsse an der Grenzfläche Wasser-Luft berechnet bei Feuerschiff FS ELBE 1 in der Deutschen Bucht: 1962-1986. Inst Meereskunde Univ Hamburg, Techn Rep 2-90

Moll A, Radach G (1990b) ZISCH Parameter Report. Compilation of measurements from two interdisciplinary STAR-shaped surveys in the North Sea. (Vol I: Graphic Reports). Inst Meereskunde Univ Hamburg, Techn Rep 6-90

Moll A, Radach G (1990c) ZISCH Parameter Report. Compilation of measurements from two interdisciplinary STAR-shaped surveys in the North Sea. (Vol. II: Data Lists). Institut für Meereskunde der Universität Hamburg, Techn Rep 7-90

Moll A, Radach G (1991) Application of Dobson and Smith's solar radiation model to German Bight data. Q J R Meteorol Soc 117:845-851

Radach G (1982) Dynamic interactions between the lower trophic levels of the marine food web in relation to the physical environment during the Fladen Ground Experiment. Neth J Sea Res 16:231-246

Radach G (1983) Simulation of phytoplankton dynamics and their interactions with other system components during FLEX'76. In: Sündermann J, Lenz W (eds) North Sea Dynamics. Springer, Berlin Heidelberg New York, pp 584-610

Radach G (1990) Coupling physical and biological upper layer dynamics (Lecture Notes. Intensive Course: Modelling of Marine Ecosystems, 11-29 Aug 1990, La Baume-les-Aix, Inst Meereskunde Univ Hamburg, 147 pp

Radach G, Moll A (1990a) State of the art in algal bloom modelling. In: Lancelot C, Billen G, Barth H (eds) Eutrophication and algal blooms in North Sea coastal zones, the Baltic and adjacent areas: prediction and assessment of preventive actions. Brussels, Commission of the European Communities. Wat Pollut Res Rep 12:115-149

Radach G, Moll A (1990b) The importance of stratification for the development of phytoplankton blooms - a simulation study. In: Michaelis W (ed) Estuarine water quality management - monitoring, modelling and research, coastal and estuarine studies vol 36. Springer, Berlin Heidelberg New York, pp 389-394

Radach G, Moll A (1993) Estimation of the variability of production by simulating annual cycles of phytoplankton in the central North Sea. Prog Oceanogr (accepted)

Radach G, Berg J, Heinemann B, Krause M (1984) On the relation of primary production to grazing during the Fladen Ground Experiment 1976 (FLEX'76).

In: Fasham MJR (ed) Flows of energy and material in marine ecosystems. Plenum, New York, pp 597-625

Radach G, Berg J, Hagmeier E (1990) Long-term changes of the annual cycles of meteorological, hydrographic, nutrient and phytoplankton time series at Helgoland and at LV ELBE 1 in the German Bight. Continent Shelf Res 10:305-328

Radach G, Regener M, Carlotti F, Kuhn W, Moll A (1993) Modelling water column processes in the North Sea, Philos Trans R Soc Lond (in press)

Riegman R, Colijn F, Malschaert JFP, Kloosterhuis HT, Cadee GC (1990) Assessment of growth rate limiting nutrients in the North Sea by use of nutrient-uptake kinetics. Neth J Sea Res 26:53-60

Rothschild BJ (ed) (1988) Toward a theory in biological-physical interactions in the world ocean (NATO ASI Series C, Mathematical and physical sciences, vol 239). Kluwer, Dordrecht

Stigebrandt A, Wulff F (1987) A model for the dynamics of nutrients and oxygen in the Baltic proper. J Mar Res 45:729- 759

van Aken HM (1984) A one-dimensional mixed-layer model for stratified shelf seas with tide- and wind-induced mixing. Dtsch Hydrogr Z 37:3-27

4 Interdisciplinary Evaluation of Field and Model Data

4.1 Evaluation of the North Sea, Joining in Situ and Remotely Sensed Data with Model Results

R. DOERFFER, W. PULS, D. PAN, H.-H. ESSEN, K.-W. GURGEL,
K. HESSNER, T. POHLMANN, F. SCHIRMER and T. SCHLICK

4.1.1 Introduction

The complicated distribution and the rapid changes of water masses in the North Sea make it difficult to interpret data of ship survey cruises as quasi synoptic observations. Isoline maps of any concentrations derived from cruises of several weeks' length do not show an actual situation but a mixture of the spatial distribution as well as the temporal development, which cannot be separated from each other. The plotted distribution includes effects of phytoplankton growth and the advective transport of water bodies with different constituents. Furthermore, the variability of other sources and sinks such as erosion of sediment and resuspension of particles or the input from rivers and through the atmosphere modify the distribution during the ship cruise, particularly in cases of strong meteorological events. In Chapter 4.7 Bohle-Carbonell analyzes the ZISCH-STAR cruise data set with respect to these shortcomings of ship observations. One possibility to improve the signification of observations is the additional use of remote sensing data. In the past 20 years, remote sensing techniques have been developed to observe and measure sea surface parameters such as temperature, sea state, currents and sea water constituents from aircraft, satellites and ship or ground based stations. Another possibility for assessing observed data is numerical modelling. This concerns especially the physical explanation for processes which have caused the observed distribution.

With respect to the overall goal of the ZISCH project, the remote determination of water constituents, particularly of suspended matter and of the currents which determine the distribution of water masses, was of major interest. Thus, during the ZISCH experiments, case studies were carried out employing the following two remote sensing techniques in order to assess their potential for scientific investigations and operational monitoring programs:

- satellite remote sensing of water constituents based on measurements of the backscattered solar radiation within different bands of the visible spectral range,
- ground based remote sensing of surface currents based on the backscattering of high frequency (HF) radio waves, which are transmitted and received from coastal stations.

For the determination of water constituents, data of the Coastal Zone Color Scanner (CZCS, satellite NIMBUS 7) have been analyzed. The surface currents in the German Bight were measured with a CODAR system.

Within this Chapter, results of both remote sensing techniques will be presented and compared with ship observations and model calculations. For the satellite remote sensing technique, a short description of the data evaluation method is included, since this technique has partly been developed during the evaluation of the data presented here. The CODAR technique and the results of the experiments with this system in the German Bight are described in Chapter 2.7, this Volume. In this Chapter the comparison with a three-dimensional hydrodynamic current model will be presented and discussed.

For the overall goal of ZISCH - investigation of the circulation and transport of contaminants in the North Sea - both these remote sensing techniques are of particular importance with respect to the distribution and transport of suspended matter, which is a main carrier of inorganic and organic pollutants. For some of the contaminants, the fraction which is carried by suspended matter exceeds 70 % (Eisma and Irion 1988).

4.1.2   Remote Measurements of Sea Water Constituents and Phytoplankton Chlorophyll

Data of the CZCS have been successfully utilized in the past to map the phytoplankton chlorophyll of the world oceans (Feldman et al. 1989) These satellite-derived concentration maps have been used to calculate the primary production, e.g. of the North Atlantic (Platt et al. 1991). The method for determining the chlorophyll concentration is based on the water colour, which shifts from blue to green with increasing concentration due to the absorption and scattering of light by phytoplankton. This green shift is determined as the ratio of the water leaving radiances in two spectral bands, in the case of the CZCS the ratio of band 1 (443 nm) to band 2 (520 nm) or band 3 (550 nm) is taken. A prerequisite for this algorithm is the correction of the atmospheric impact on the radiation field, since the atmospheric path radiance contributes by more than 85% to the radiance which is measured by the satellite.

The green-blue algorithm has been successfully applied to almost all open ocean areas, where only phytoplankton determines the optical properties of the sea water. Since suspended matter and gelbstoff changes the green-blue ratio as well, this algorithm does not apply to areas with high concentrations of these substances. For the major part of the shallow southern North Sea and particularly for the coastal areas, a retrieval method has to be applied which is capable of separating all the substances which contribute significantly to the upwelling radiance. For this purpose an inverse modelling procedure was developed for the CZCS data evaluation which allows us to determine the aerosol path radiance and the concentrations of suspended matter, phytoplankton chlorophyll and gelbstoff separately.

4.1.2.1 Evaluation of CZCS Data with the Inverse Modelling Procedure

The method of deriving the concentrations of different sea water constituents from radiance data of the CZCS has recently been developed by Fischer and Doerffer (1987) and Doerffer (1990). The basic idea of this technique is to simulate the remote sensing process, i.e. the radiative transfer of sunlight into the water and back to the sensor by a model, and to compare the results of the model calculations, i.e. the radiance at the sensor, with measured radiances in the first four spectral channels of the CZCS. Within an optimization loop, the concentrations as well as the aerosol path radiance are modified in order to find the minimum in the deviation between the measured and calculated radiances (Fig. 4.1.1). This procedure has to be applied pixel by pixel, while for each pixel the actual azimuth and zenith angles of the sun and the viewing direction has to be computed. To keep the computational effort small, a radiative transfer model based on a single scattering model for the atmosphere and a two-flow model for the underwater radiative transfer was developed (Doerffer 1992). The atmospheric part of the model assumes separate layers of aerosol, air molecules ("Rayleigh atmosphere") and absorbing gases; it includes the

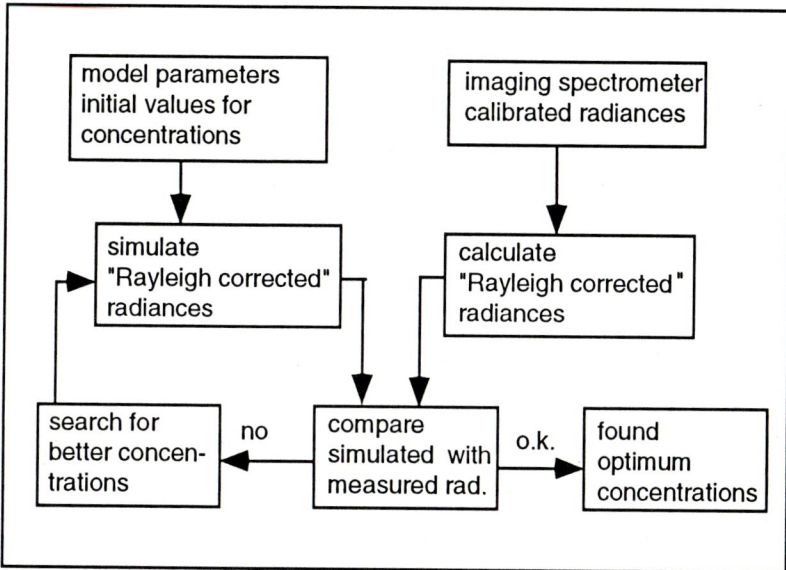

Fig. 4.1.1. Scheme of inverse modelling procedure

specular reflectance of skylight, a wind-dependent rough sea surface with direct sunlight reflectance (sun glitter) and diffuse reflectance by sea-foam. The atmospheric part is based on André and Morel (1989), Sturm (1980, 1981) and Iqbal (1983), the two-flow model on Joseph (1950) and Doerffer (1979). The specific optical properties of the water constituents and of pure water, as used in the model, are described by the absorption and the diffuse attenuation coefficient per concentration unit for each of the four CZCS channels. They are taken from Prieur and Sathyendranath (1981), Morel (1974) and Doerffer (1979). These coefficients are assumed to be constant for the whole North Sea; furthermore, a constant homogeneous vertical distribution of the sea water constituents is presumed for the signal depth z90, which is defined here as the depth above which 90% of the radiance leaving the water at a wavelength of 550 nm originates.

A particular problem is the determination of the spectral properties, i.e. the wavelength dependence of the aerosol path radiance. This can be described by either the Angstrom coefficient or by spectral factors, which have to be determined for each of the CZCS channels and then normalized to CZCS channel 4 (l=670 nm). Since no optical aerosol measurements are available for the period of the ZISCH-STAR experiment, this information had to be retrieved from the satellite data itself. For this task, the relative path radiances of all four channels were derived from radiance differences.

These differences were calculated from the highest and lowest radiances of a short horizontal profile across an aerosol front, which was selected by a visual inspection of the image. It is assumed that the water reflectance is constant and that only the amount of aerosol is changing along this profile. By normalizing these radiance differences with the spectrum of the sun and by the radiance difference at CZCS channel 4 , the required relative spectral factors could be computed. These spectral aerosol path radiance factors were assumed to be constant for the whole satellite scene. Atmospheric ground pressure and wind speed were taken from the *Berliner Wetterkarte*, the ozone concentration from climatological data (cited in Iqbal 1983).

Three different scenes of the ZISCH-STAR period (May 1-June 16, 1986) have been selected: May 1, May 22, June 16. The satellite data were provided by NASA Goddard Space Flight Centre and the receiving station at Dundee. The in situ data were taken from the ZISCH data base (Moll and Radach 1990) and from Hölemann (1988), the bathymetric data were provided by Hainbucher et al. (1986).

4.1.2.2 Results

All three images have been evaluated with the inverse modelling procedure as described above so that concentration maps of phytoplankton chlorophyll, suspended particulate matter (SPM), gelbstoff and the aerosol path radiance could be derived. Furthermore, the signal depth z90, i.e. the depth from which 90% of the water leaving radiance comes, was computed and mapped.

Since the distribution of suspended matter as a main carrier of contaminants is of particular interest for the goal of ZISCH, we will concentrate the further considerations on this important constituent. The derived suspended matter concentration maps are shown in Fig. 4.1.9(3) - 4.1.9(5). Details of the distribution in the southern North Sea on May 1 and the corresponding signal depth can be seen in Fig. 4.1.9(1) and 4.1.9(2). Areas which could not be evaluated due to cloud coverage, saturation of the sensor due to high aerosol path radiance or technical problems have been masked out and are coded as white areas. Furthermore, one has to consider that the radiances of water pixels adjacent to clouds are often not correct because of the saturated detector. Thus, the derived concentrations are not valid in the direct surroundings of clouds, as, e.g. in Fig. 4.1.9(3) along the clouds in the northeastern part of the image.

The map of the suspended matter distribution derived from data of the ZISCH-STAR cruise is presented in Fig. 4.1.9(7); samples are from 1-2 m depth. The simulated suspended matter distributions of May 1 and 22 and June 16 (same dates as satellite data) are shown in Fig. 4.1.9(8) - 4.1.9(10). All maps are plotted with the same projection and the same concentration colour scale. The bathymetric map of the North Sea [Fig. 4.1.9(6)] is presented in addition for discussing the influence of water depth on the SPM distribution.

4.1.2.3 The Problem of Validation of the Satellite Data

One general, important problem with remote sensing data of the ocean which is extremely difficult to solve is the validation of the results. In this study, the concentrations have been derived from the CZCS radiances only by model calculations without using any concentrations as ground truth to fit the results. Only the specific optical properties of suspended matter, phytoplankton chlorophyll and gelbstoff have been measured in the German Bight (but not in the same year) and have been used to validate ship-borne measurements of the radiance leaving the water (Doerffer 1979, Fischer 1984, Kronfeld 1988). The main problem for satellite data is that a comparable data set from ship sampling never exists: the water volume represented by one picture element is several orders of magnitude larger than the volume taken by the ship, the temporal difference between the water sample and the satellite overpass may run up to weeks as in the case of the ZISCH cruise of 1986. Thus, criteria other than a direct comparison have been used to validate the derived concentrations:

- The SPM concentrations derived from the CZCS data form structured patterns which are in agreement with SPM distributions derived from ship data. Since each of the pixels is evaluated independently with the same starting conditions in the optimization loop (Fig. 4.1.1), this test proves also that the optimization procedure is not running into side minima.
- Structures in the aerosol path radiance distribution are independent from structures in the distribution of water constituents in most regions. They cross coastlines, while structures in the distribution of water constituents follow coastlines and - in the shallow areas of the southern North Sea - also the sea bottom topography.
- Since aerosol, thin clouds (cirrus or jet trails) and suspended matter have very similiar optical properties, their separation is critical. The correctness of this separation was tested by evaluating the suspended matter concentration and the aerosol-path-radiance parameter along a horizontal profile through an image part, which is partly covered by a jet

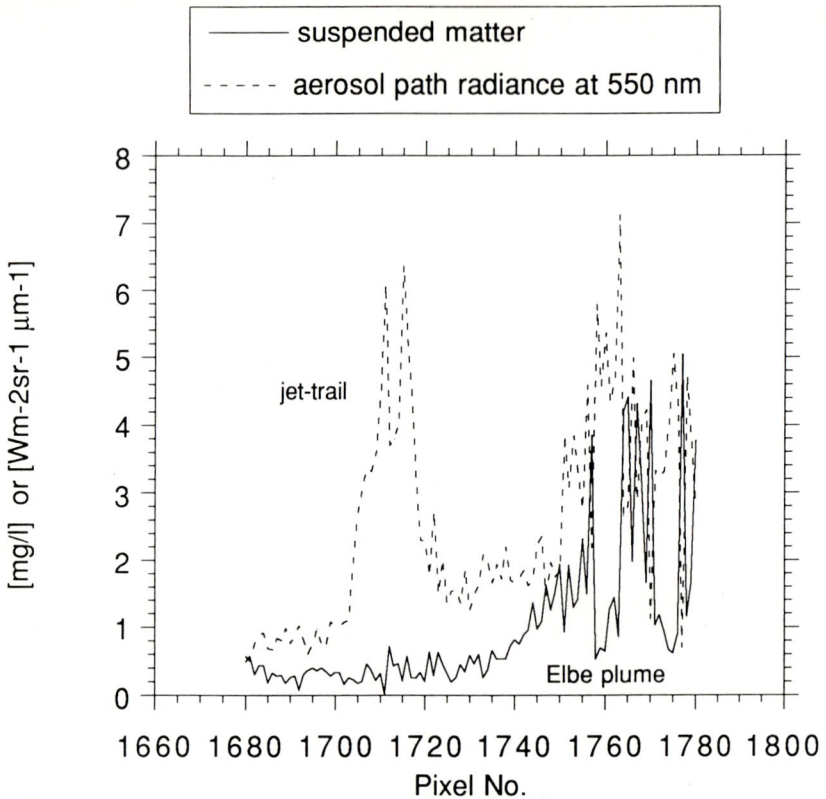

Fig. 4.1.2. Section through CZCS image of May 22 through the Elbe river plume showing the derived suspended matter concentration and the aerosol path radiance

trail. In Fig. 4.1.2 it can be clearly seen that the SPM concentration and the aerosol path radiance as derived by the inverse modelling procedure are totally independent.

Before comparing satellite, ship and model data, the history of weather and, particularly, the wind conditions in April, May and June 1986 in the North Sea have to be considered. From April 10 to May 6, the weather was calm. From May 7 to May 28, there were seven wind events, more or less regularly distributed over that period. The strongest wind event happened on May 22. After a week of calm weather, a storm started on June 4. That storm had its maximum on June 5, followed by a slighter storm event on June 8. From June 9 on, the weather was calm until the next wind event on June 19.

	Ship cruise	CZCS May 17, 86
mean	1.58 mg	1.68 mg
std. dev.	2.24 mg	1.81 mg
mode	0,8 mg	0.98 mg

Histogram of suspended matter ZISCH cruise

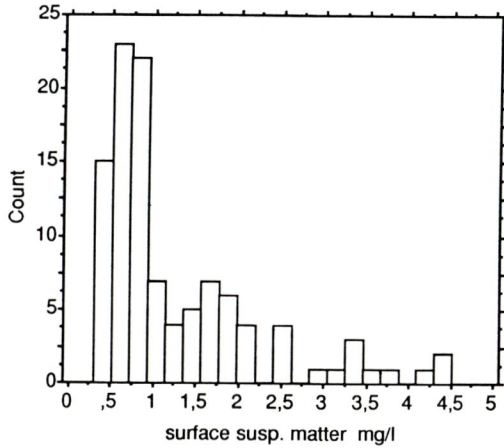

Histogram of suspended matter CZCS image

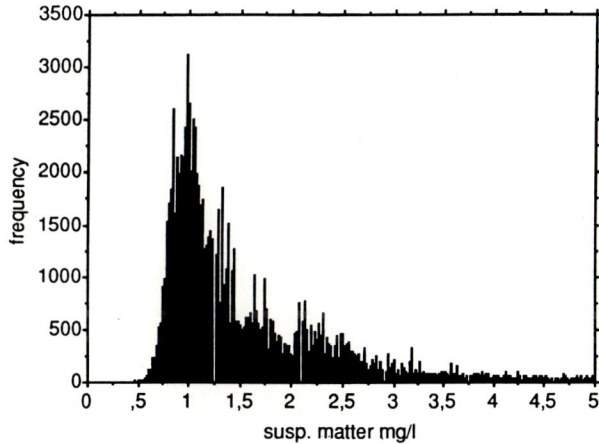

Fig. 4.1.3. Comparison between suspended matter concentrations of the North Sea derived from ship samples (ZISCH-STAR) and from the CZCS scene of May 22

4.1.2.4 Comparison with Ship Data

The main question, of course, concerns the comparison between the concentration maps derived from ship samples and satellite data. Since, as stated before, a pixel by sample comparison is not appropiate due to the extreme temporal differences, we compare the mean values and the histogram of the suspended matter as found with both methods for the whole North Sea. The result is given in Fig. 4.1.3. There is surprisingly good agreement in both the mean concentration and the shape of the histogram, although no procedures have been used to fit the satellite to the ship data. A similiarly good agreement between concentrations derived from CZCS data and ship samples was found for a data set from the German Bight of 1979 (Doerffer 1990). However, the statistical comparison of data sets which span different periods is not in any case appropriate for checking a method, particularly if events like strong plankton growth or storms have occured during the period of ship sampling. In this study, the CZCS scene of May 22nd was used in order to obtain adapt a sample in the middle of the ship cruise.

The horizontal distribution shows similiar general patterns in both data sets [Figs. 4.1.9(1)-4.1.9(5) and 4.1.9(7)]: there are large patches in the shallow southern North Sea with concentrations of more than 10 mg/l and low concentrations (< 1 mg/l) in the deep northern North Sea. Because of its high spatial resolution, the satellite images show many more details; also ,significant temporal differences [between Fig. 4.1.9(3)-4.1.9(5)] are obvious in some parts of the North Sea.

Of particular interest is a comparison of the fine structure in the actual suspended matter distribution of the satellite data compared to the ship survey. Very obvious are the complicated patterns in the southern North Sea. The long bands stretching from the southeast coast of England in the northeast direction are bounded to the north by a line which separates that part of the North Sea which is characterized by stratified water during the summer from the coastal and channel areas which are normally well mixed. The fine structure can be used to analyze the mechanism of the suspended matter transport, its sources and sinks, particularly in combination with model simulations. The main sources for the high concentrations in the southern North Sea are coastal or sea-floor erosion or the erosion of cliffs in the shallow water off the East Anglia coast. The material is then transported by the current according to the general circulation into the eastern and the northeastern direction. High concentrations on May 1 occur also along the coastline of Belgium and Holland and the English coast of the Channel, furthermore in the Wadden Sea of Holland, Germany and Denmark. These features are not so obvious in the ZISCH-STAR image.

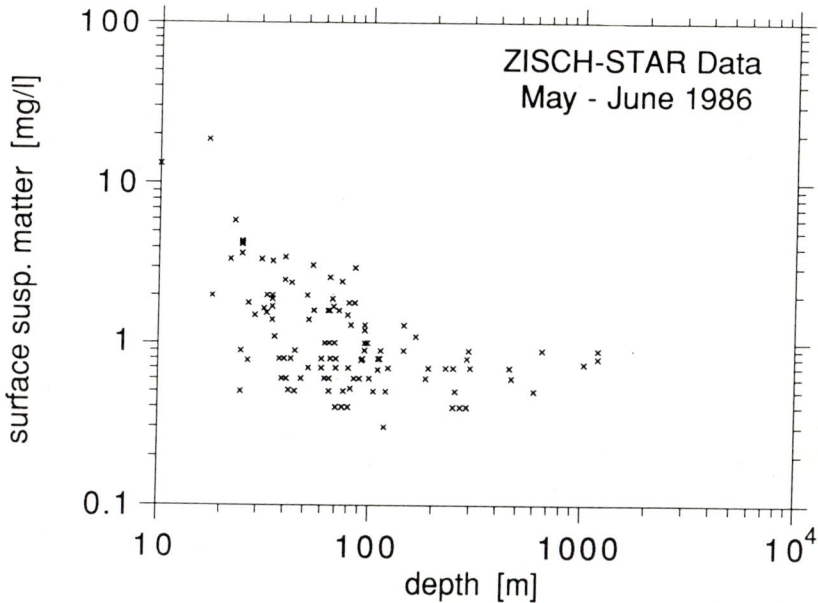

Fig. 4.1.4. Scatter plot of the relation between water depth and suspended matter concentration, data from ship samples of ZISCH-STAR cruise

The high concentration along the Dutch coast may be overrepresented by the interpolation of samples from the Wadden See.

The CZCS image of May 22 [Fig. 4.1.9(4)] shows very clearly the influence of the bathymetry [compare with Fig. 4.1.9(6), particularly the areas Dogger Bank and the Jutland coast]. The storm occurring on the same day has probably caused strong vertical mixing of suspended matter and a resuspension of sediment. The influence of the water depth is also obvious in the scatter plot (Fig. 4.1.4), which was derived from the ZISCH-STAR data. One can see that the suspended matter concentration over shallow water is very variable (range 0.3- 20 mg/l) while over the deep water of the northern North Sea and the Norwegian Trench the concentration varies only between 0.5 and 1 mg/l.

The CZCS image of June 16 [Fig. 4.1.9(5)] shows a completely different distribution in the central part of the North Sea. The long and partly meandering bands of high concentrations - one of which is also visible in the Moray Firth, Scotland, could also be caused by phytoplankton species with a high scattering coefficient, such as coccolithophorids with their calcium shells.

Comparing all three CZCS-derived suspended matter maps with that derived from the ZISCH-STAR cruise data, one can see that the general distribution obviously coincides. However, details of the structures, particularly in the central part of the North Sea, are different in each of the satellite derived maps. They do not compare with the ship data map, which represents a mixture of the spatial distribution and the temporal development of 6 weeks.

4.1.2.5 Comparison with Model Results

Another important information source which will give insight into the transport and distribution of suspended matter, are model simulations. A detailed description of the suspended matter transport model used is given in Chapter 3.2.

For the comparison with satellite data, Fig. 4.1.9(8)-4.1.9(10) show three model-simulated distributions of suspended particulate matter (SPM) in the North Sea. The SPM concentrations shown are vertically averaged over the upper 20 m of the water column. A description of the model and some results are shown in Chapter 3.2 for SPM and in Chapter 4.4 for lead. The computed SPM concentrations shown in Fig. 4.1.9(8)-4.1.9(10) are valid for the same clock time as the corresponding satellite data (maximum time deviation: 20 min).

The mean concentration of all three simulated distributions is about 0.8 mg /l, which is about half the value for the ZISCH-STAR cruise data and the data of the CZCS image of May 22. Irrespective of the lower mean concentration, the histogram of simulated SPM concentrations (Fig. 4.1.5) is similiar to the ship and satellite data. One reason for the lower calculated concentrations is that the model data do not include phytoplankton and other autochthonous organic material.

Both the satellite and the model data show SPM concentrations < 1 mg/l in the northern North Sea and concentrations > 1 mg/l in the southern North Sea. For the model data this separation is quasi time invariant. The separation line between high and low concentrations is between Humber and north Jutland (for the locations see Fig. 3.2.11 of Chap. 3.2).

Concerning the satellite data, the most striking feature in the northern North Sea appears on May 22, 1986: the high SPM concentrations (> 4 mg/l) along Jutland's west coast extend to the Skagerrak. The high SPM concentrations in the shallow zone west of Jutland can be attributed to the wind event on May 22. It is hard to believe, however, that the high SPM concentrations in the deep Skagerrak on that day result from a single wind event. In Fig. 4.1.9(9), showing the model simulations, the wind event on

model   May 22, 86

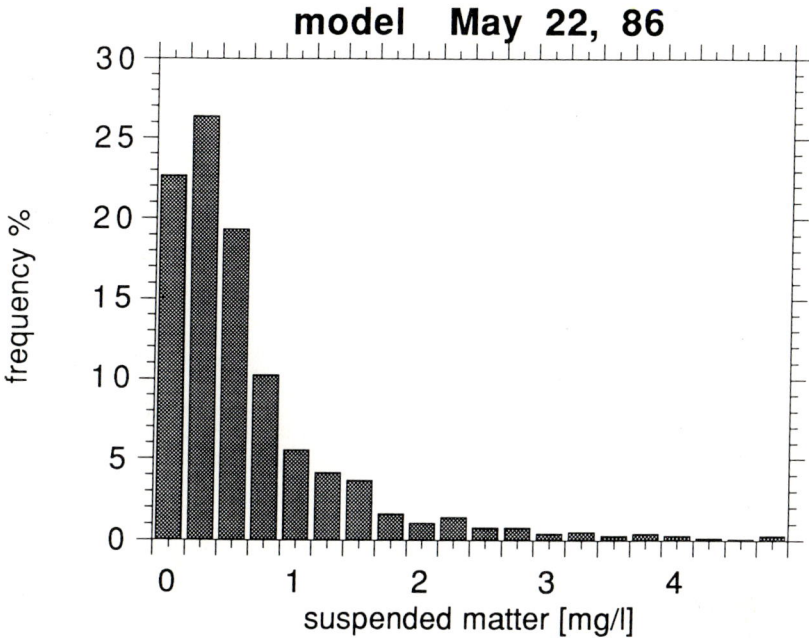

Fig. 4.1.5. Histogram of suspended matter from model calculation of May 22, 1986

May 22, 1986 increased the SPM concentrations only slightly; the increase is less than 1 mg/l.

Let us now regard the SPM concentrations in the southern North Sea. On May 1, weather conditions were calm, including the preceding 3 weeks. Thus, the SPM distribution in the southern North Sea on May 1 should be dominated by SPM from adjacent marine areas (e.g. the English Channel), from rivers, from waste dumping or from coastal zones (e.g. the Wadden Sea). The satellite picture of May 1 [Figs. 4.1.9(1) and 4.1.9(3)] shows the "typical pattern" of SPM concentrations that has been reported by several authors, e.g. Eisma (1981). The SPM concentration in the Southern Bight is highest at the Belgian and the Dutch coast on the one hand, and the coast of East Anglia (including the Thames estuary and the Wash), on the other. In the central part of the German Bight, the SPM concentration is comparatively low.

The most striking SPM feature in the southern North Sea is the plume of high turbidity that extends northeastward from East Anglia. This plume coincides approximately with the axis of minimal salinity (Lee 1980), indicating that the plume contains a high admixture of coastal water. The

Marine and Atmospheric Sciences Directorate (1992) reports an annual flux within this plume of 6.6 x 10^6 tons of suspended matter eastwards across the southern North Sea.

Compared to Fig. 4.1.9(3), the simulated SPM concentration on May 1 in the Southern Bight [Fig. 4.1.9(9)] shows a different pattern. The SPM in the Southern Bight originates mainly from the Dover Strait. It is transported north-eastward into the *central* part of the Southern Bight. Especially the simulated SPM concentrations around East Anglia are low, and the plume extending northeastward from East Anglia does not appear in the simulation. This shortcoming of the numerical model simulations was already discussed in Chapter 3.2. The main explanation is that the spatial resolution of the model is too coarse to simulate near-coast transport processes correctly.

The computed SPM concentrations on May 22 and on June 16, 1986, do not differ much from the concentrations on May 1, 1986.

As for the satellite data, large parts of the Southern Bight are covered by clouds on May 22. Generally, the SPM concentration on that day is increased in the southern North Sea compared with May 1. This is due to the wind event on May 22. The simulated SPM concentrations in the southern North Sea on May 22 are hardly increased. This indicates that sediment erosion is not adequately simulated in the model. It must be taken into consideration that a correct simulation of erosion also depends on a correct simulation of vertical sediment transport processes within the sediment bed, mainly caused by bioturbation.

On June 16, the satellite picture shows a zone of high SPM concentrations between northern East Anglia and the German Bight. Especially the SPM in the German Bight should be eroded bottom sediment, possibly as the result of the storm event between June 4-9. The simulated SPM concentrations do not show that zone of high SPM concentrations. The SPM that was eroded during the storm event of June 4-9 settled to the sea bottom within 3 days after the end of the storm in the model simulation.

4.1.3 Comparison of Surface Currents Derived from HF-Radar Measurements and a Hydrodynamic Model

The other remote sensing method which was used during the ZISCH project is the measurement of the surface current using the Doppler shift of backscattered electromagnetic waves in the HF band. The method is described in Chapter 2.7, this Volume; here, we will concentrate on the comparison of the measurements with the results of a three-dimensional hydrodynamic model, which is described in Chapter 3.2.

The simulated currents of January 19, 1988, calculated with the North Sea model are shown in Fig. 4.1.6. The currents of this day have been averaged over exactly two periods of the M_2-tide, i.e. 24 h, 50 min. Since this is the dominant tide in the North Sea, averaging lets the residual current stand out. It shows the slow circulation of the water in response to wind, differences of barometric pressure, and input by the Atlantic M_2-tide. Usually, the velocity stays below 0.5 m/s. This current transports heat, plankton, as well as contaminants through the North Sea. Figure 4.1.6 shows the 5-m thick surface layer only. Another 18 layers underneath display different, though partly similar current patterns. The results have been computed by Pohlmann (1991) using a model of Backhaus (1985). This is a modification of the three-dimensional baroclinic model described in Chapter 3.2 that is driven at its open northern margin by the M_2-tide, which is computed with a model of the NE Atlantic. The surface current goes largely with the wind, which is measured in 3-h intervals. For this comparison the wind data, together with barometric pressure of the 19th of January 1988, were entered into the model. Furthermore, the local salinity and surface temperature have to be taken into consideration. The salinity used is the climatological mean of January from 1968 to 1985, i.e. an average value. Temperatures, however, were taken from ships' observations of 19 January 1988. The dependence of this "climate" of the seasons is discussed more fully in Chapter 2.1.

For a comparison of the residual current in Fig. 4.1.6 with the residual current observed by remote sensing using electromagnetic waves, current vectors are required at identical positions. Thus, the remote sensing observation points were positioned on the grid points of the model. A radius of 10 km was selected for areal averaging. Furthermore, each point in the model and in the observed surface current, as described in Chapter 2.7, was averaged over the same two periods of the M_2-tide. Thus, both model and field data have identical temporal and spatial means.

The residual current of the HF observation area between Helgoland and the East Frisian Islands on the 19th of January 1988 is shown in Fig. 4.1.7 together with the model results: bold arrows show the model value, thin lines show the observed values. The terminal point of the measured current vector varies within the rectangle that is drawn instead of an arrowhead. This symbolizes quantitatively the established variability of the residual current, including instrumental uncertainties.

For a comparison of the model results and the HF-radar measurements, one has to consider the following differences in the determined quantities:

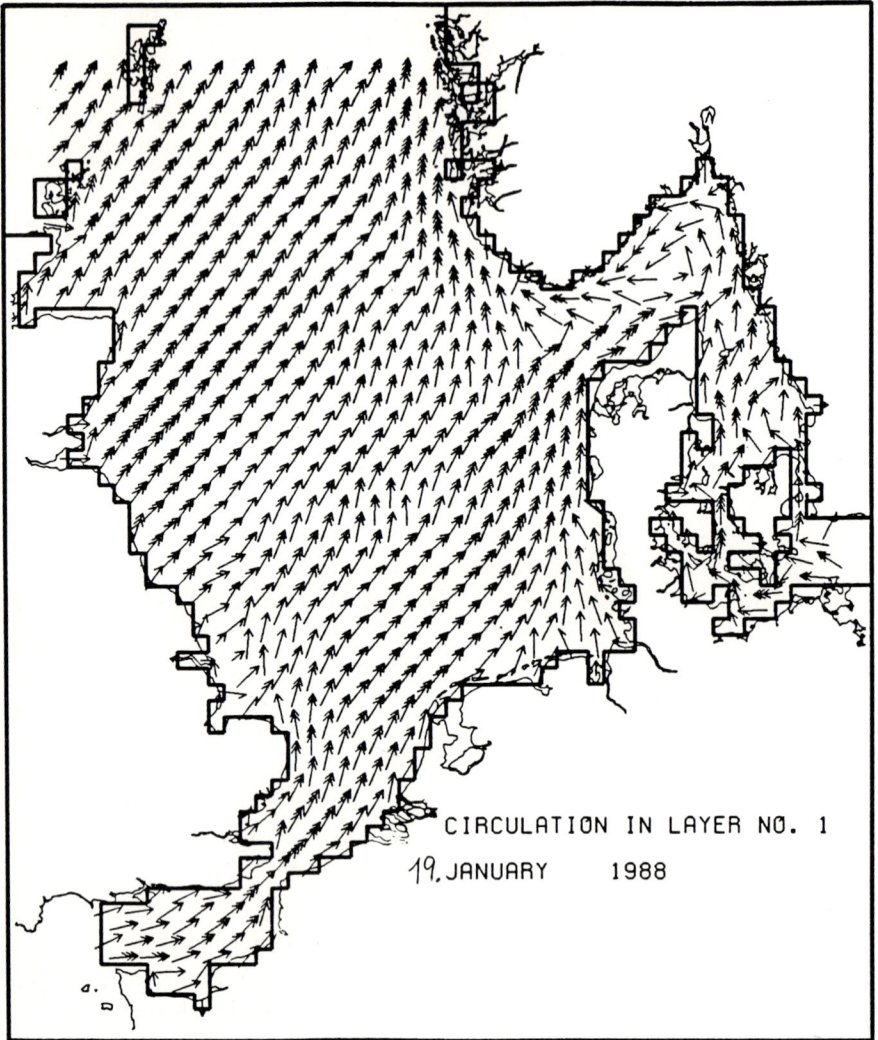

Fig. 4.1.6. The mean residual current of January 19th 1988. Modelled with forcing of true wind, surface temperature and salinity

- The HF-radar measures the velocity of the very surface of the water (first few decimeters), while in the model the velocity of the first 5 m of the water is regarded here. Depending on wind speed, direction and local topography in the shallow German Bight, considerable differences can be expected between both quantities.

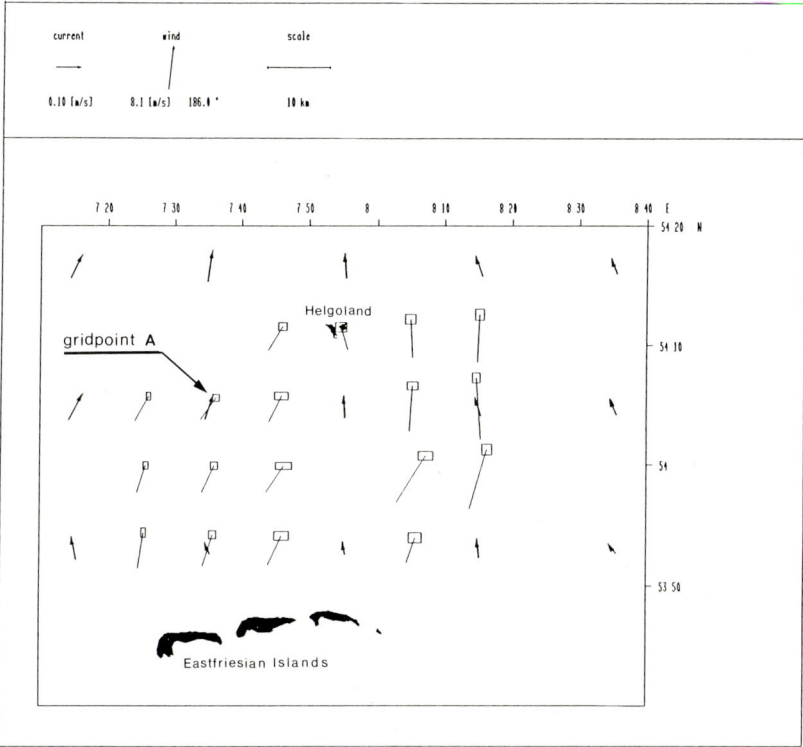

Fig. 4.1.7. The mean residual current of January 19th 1988 between the islands of Helgoland and Wangerooge. *Bold arrows* show the model value, *thin lines* measured current vector, with *rectangles* giving observed variability

- The model is driven by the large scale wind field, while the topical surface current of the area (determined with the HF-radar) is driven by the actual local wind.
- The spatial resolution of the model is coarse compared to the complex topography which is investigated here. Particularly the assumption of a constant depth of water during the tidal cycle is not satisfied in this shallow area.

However, in the vicinity of gridpoint A, the depth is about constant and the margin does not yet confound the model. Thus, this point was selected to compare the results on the temporal scale. For the period from 5 January to 27 January 1988 the vectors of the residual current at gridpoint A in Fig. 4.1.8 (centre trace) are plotted against the observed current (lower trace). The wind data (upper trace) are 24-h means of the station Helgoland. They

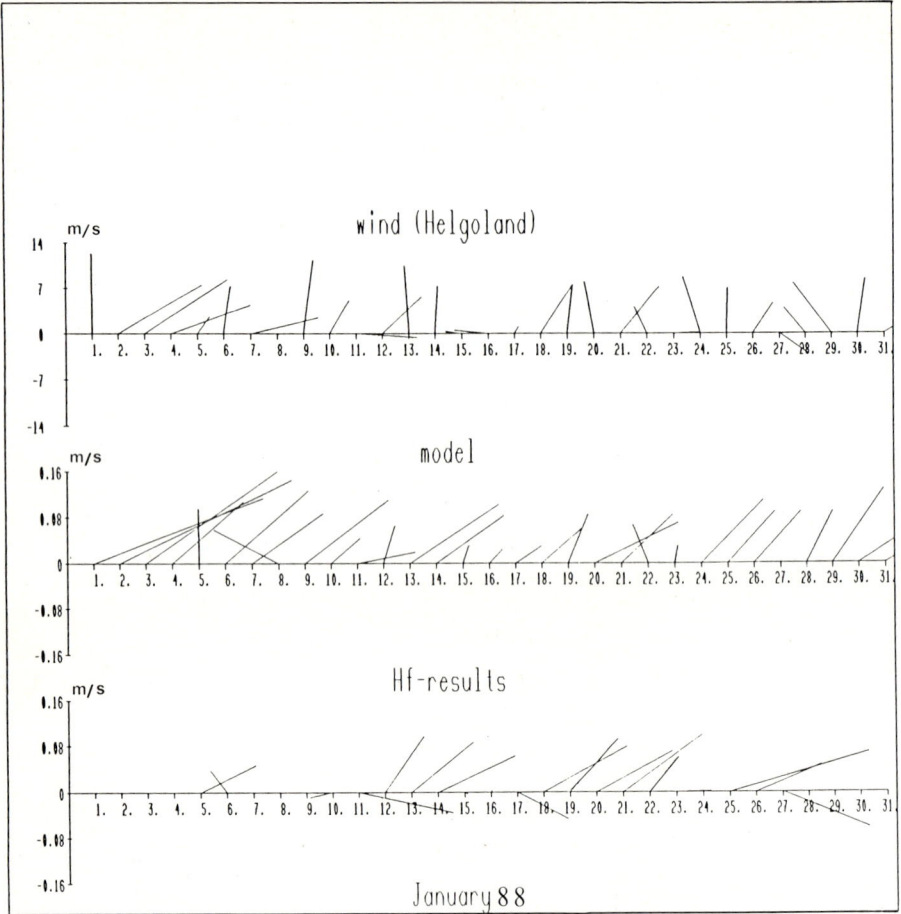

Fig. 4.1.8. Time dependence of residual current at gridpoint *A* of Fig. 4.1.7. *Lower trace* observed current; *center trace* modelled current; *upper trace* observed wind at Helgoland

state the local wind more accurately than the large-scale wind field driving the model. The agreement between modelled and measured current with regard to magnitude and direction as shown in Fig. 4.1.8 justifies confidence in the model for the residual current in the entire North Sea, at least in the 5-m surface layer, although the measured current vectors reflect the actual wind vectors much more than the current of the first 5 m, as calculated with the model.

This case study has shown that HF-radar measurements can be an important tool for validation and calibration of future high resolution models. They are able to provide current data in addition to models on a

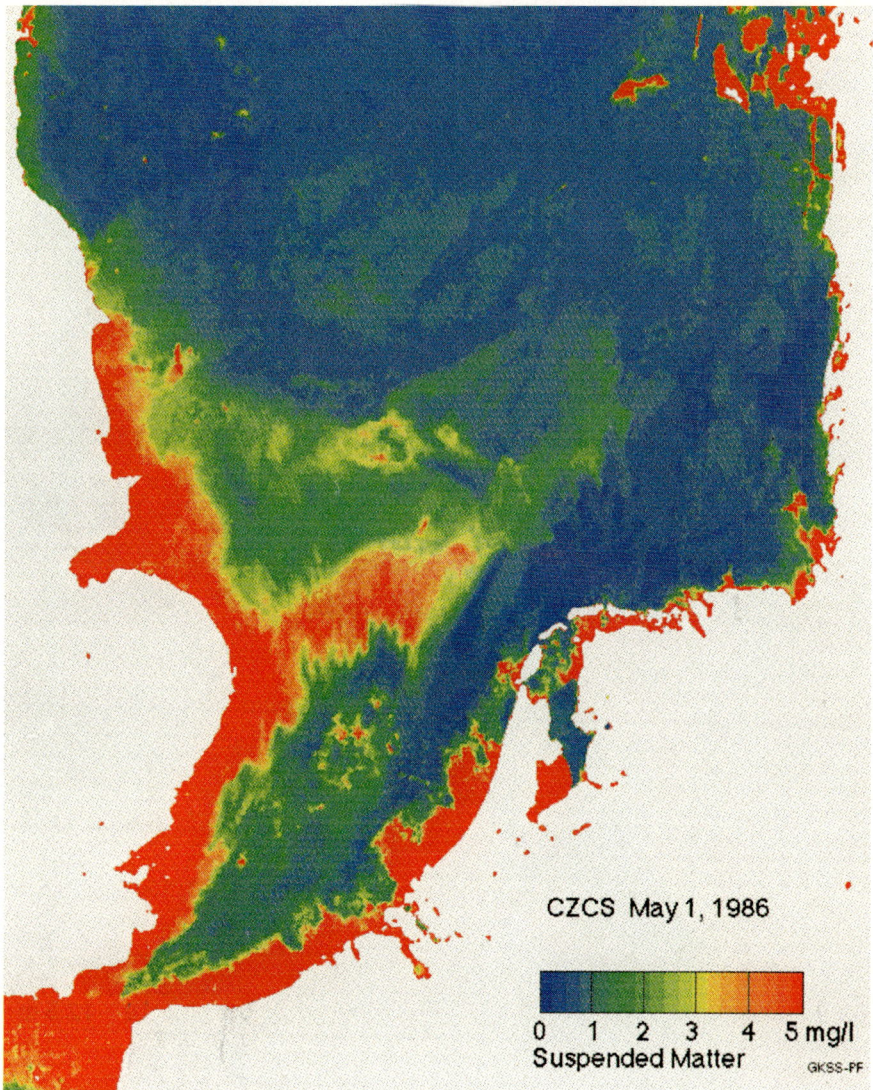

CZCS May 1, 1986

0   1   2   3   4   5 mg/l
Suspended Matter

GKSS-PF

Fig. 4.1.9. (1) Suspended matter distribution in the southern North Sea of May 22, 1986, derived from CZCS data

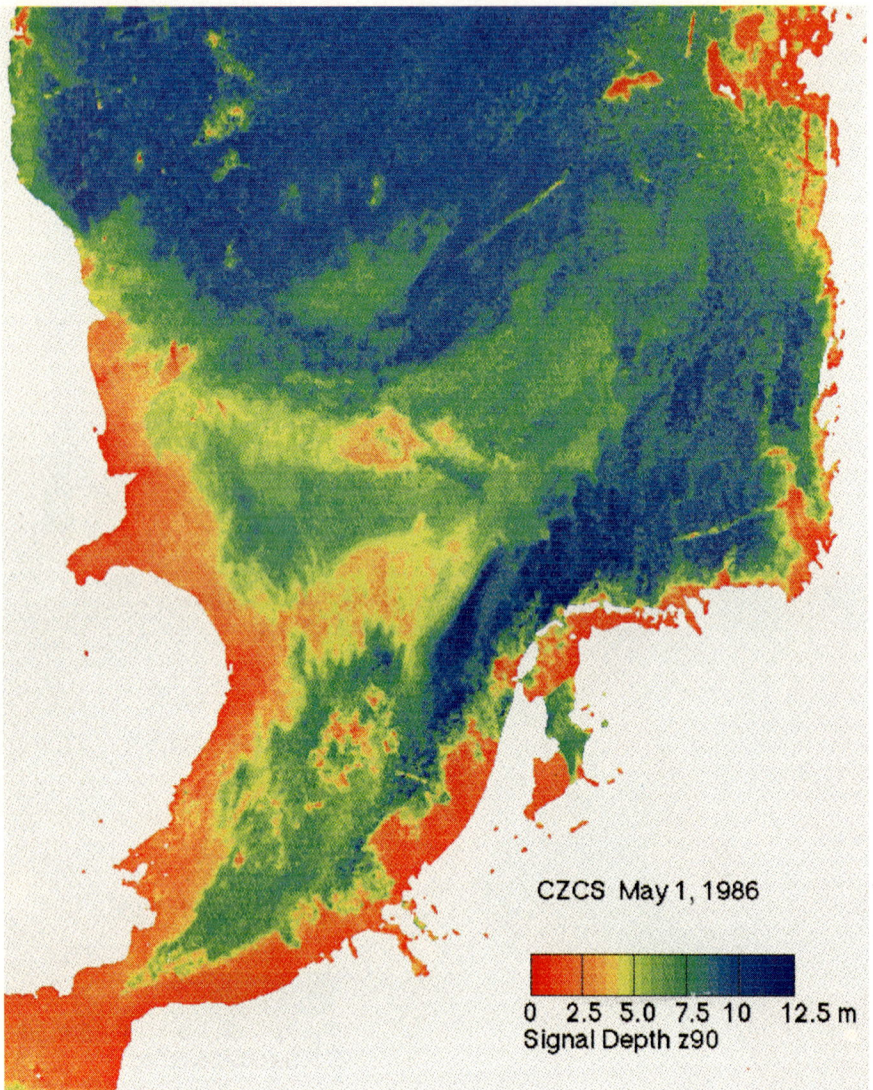

CZCS May 1, 1986

Signal Depth z90

0 2.5 5.0 7.5 10 12.5 m

Fig. 4.1.9. (2) Signal depth z90 showing the depth from which 90% of the water leaving radiance at CZCS spectral channel 3 (550 nm) stems

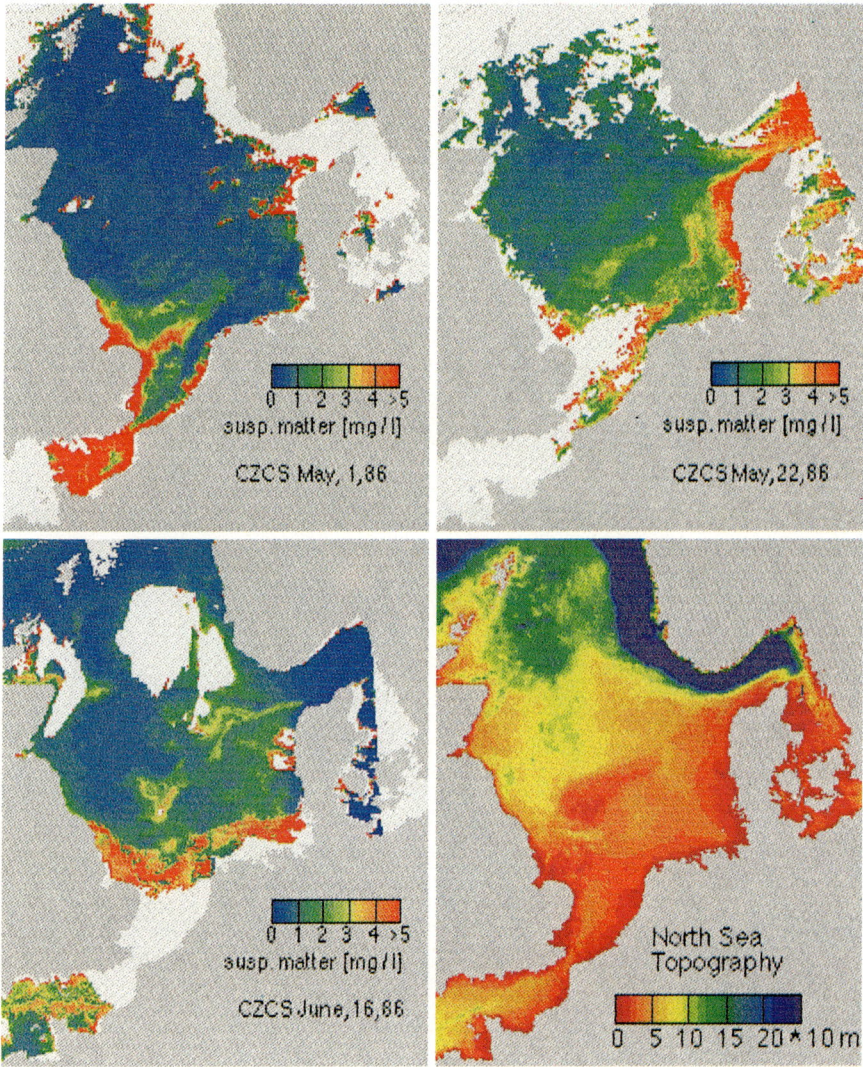

Fig. 4.1.9. (3)-(5) Suspended matter distribution in the North Sea derived from CZCS data, *white areas* indicate cloud coverage or areas where data are not valid. (6) North Sea topography

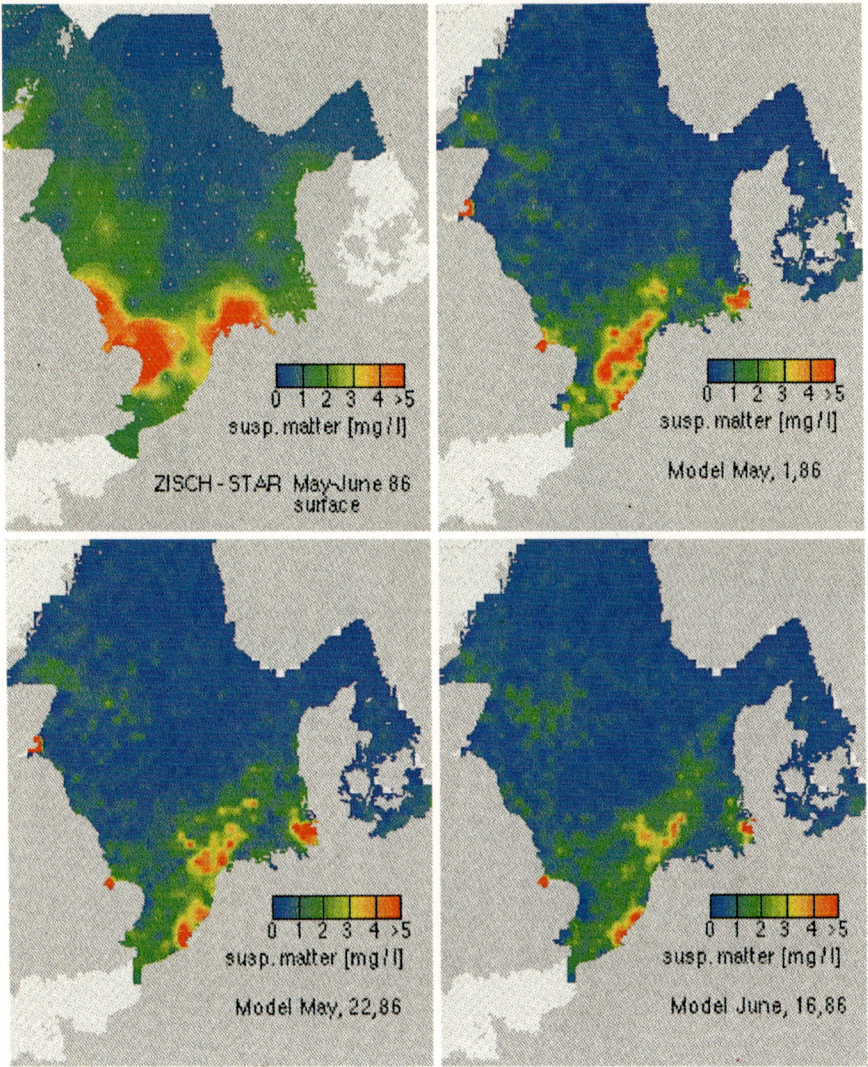

Fig. 4.1.9. (7) Surface suspended matter distribution in the North Sea derived from data of the ZISCH-STAR cruise May 2-June 13, white dots indicate sampling positions. **(8)-(10)** Suspended matter distribution of the upper 20 m, simulations with the suspended matter transport model

nearly permanent basis. In conjunction with remote measurements of sea water constituents, as described in the first part of this Chapter, these methods allow us to analyze not only the distribution but also the transport of dissolved and suspended material and of plankton. Since the HF-radar technique is independent of cloud cover, it may help to interpolate satellite data.

4.1.4 Conclusions

The results of both case studies confirm that remote sensing techniques are very efficient tools for the investigation of marine areas, particularly for studying the distribution, transport and the spatial and temporal development of sea water constituents, including phytoplankton. It is the only method which provides synoptic observations of at least the surface layer, which is a prerequisite for analyzing the temporal development of larger areas on different time scales.

Each remote sensing technique has its principal limitation; for example, satellite observations in the visible and infrared part of the spectrum depend on cloud-free weather. All satellite or airborne observation systems view only the water surface. Thus, they cannot replace in situ observations and samples taken from ships or moored platforms, but they complement the conventional techniques in an ideal way. In the future, a new generation of instruments such as imaging spectrometers and operational radar techniques will improve the existing potential of remote sensing. A big step forward can be expected from the combination of all available remote sensing techniques and from the joint evaluation with in situ observations and models. This integrated approach will lead us to a much better understanding of observed data and will allow us to monitor the environment much more effectively than is possible today. The use of remote sensing during the ZISCH project and the results presented here can be regarded only as a first step in this direction.

Acknowledgements. The CZCS data were kindly provided by the NASA Goddard Space Flight Centre data archive and by the satellite receiving station in Dundee, Scotland.

References

André JM, Morel A (1989) Simulated effects of barometric pressure and ozone content upon the estimate of marine phytoplankton from space. J Geophys Res 94(C1):1029-1037

Backhaus JO (1985) A three-dimensional model for the simulation of shelf sea dynamics. Dtsch Hydrogr Z 38(H4):165-187

Doerffer R (1979) Untersuchungen über die Verteilung oberflächennaher Substanzen im Elbe-Ästuar mit Hilfe von Fernmessverfahren. Arch Hydrobiol 43:119

Doerffer R (1990) How to derive concentrations of chlorophyll, suspended matter and gelbstoff from multispectral radiances of case II waters. ICES Statutory Meet, International Council for the Exploration of the Sea, Copenhagen, Theme Session P, Pap C M 1990 E:19

Doerffer R (1992) Imaging spectroscopy for detection of chlorophyll and suspended matter. In: Toselli F, Bodechtel J (eds) Imaging spectroscopy: fundamentals and prospective applications. Kluwer, Dordrecht, pp 215-257

Eisma D (1981) Supply and deposition of suspended matter in the North Sea. Spec Publ Int Assoc Sediment 5:415-428

Eisma D, Irion G (1988) Suspended matter and sediment transport. In: Salomons W, Bayne B L, Duursma E K, Foerstner U (eds) Pollution of the North Sea - an assessment. Springer, Berlin Heidelberg New York, 687 pp

Feldman G, Kuring N, Ng C, Esaias W, McClain C, Elrod J, Maynard N, Endres D, Evans R, Brown J, Walsh S, Carle M, Podesta G (1989) Ocean color: availability of the global dataset. Eos Transactions AGU 70 (23):634-635, 640-641

Fischer J (1984) Remote sensing of suspended matter, phytoplankton and yellow substances over coastal waters - Part 1 Aircraft measurements. Mitt Geol Paläontol Inst Univ Hamb SCOPE/UNEP Sonderbd 55:85

Fischer J, Doerffer R (1987) An inverse technique for remote detection of suspended matter, phytoplankton and yellow substance from CZCS measurements. Adv Space Res 7(2):21-26

Gordon HR, Morel A (1983) Remote assessment of ocean color for interpretation of satellite visible imagery: a review. In: Bowman M (ed) Lecture notes on coastal and estuarine studies, vol 4. Springer, Berlin Heidelberg New York, pp 1-114

Hainbucher D, Backhaus JO, Pohlmann T (1986) Atlas of climatological and actual seasonal circulation patterns in the North Sea and adjacent shelf regions: 1969-1981. Inst Meereskunde, Univ Hamburg, Techn Rep, pp 1-86

Hölemann J (1988) Suspension in der Nordsee: Konzentration, Hauptelementzusammensetzung und REM-Beobachtungen (Mai-Juni 1986). Diplomarbeit Fachbereich Geowissenschaften, Univ Hamburg

Iqbal M (1983) An introduction to solar radiation. Academic, Toronto

Joseph J (1950) Untersuchungen über Ober- und Unterlichtmessungen im Meere und über ihren Zusammenhang mit Durchsichtigkeitsmessungen. Dtsch Hydrogr Z 3:324-335

Kronfeld U (1988) Die optischen Eigenschaften der ozeanischen Schwebstoffe und ihre Bedeutung für die Fernerkundung von Phytoplankton, Rep GKSS 88/E/40, GKSS Forschungszentrum, Geesthacht, Germany, pp 153

Lee AJ (1980) North Sea: physical oceanography. In: Banner FT, Collins MB, Massie KS (eds) The north-west European shelf seas: the sea bed and the sea in motion II. Physical and chemical oceanography, and physical resources. Elsevier Oceanography Ser 24B. Elsevier, Amsterdam, pp 467-493

Marine and Atmospheric Sciences Directorate (1992) Newsletter No. 11, NERC North Sea Project. Proudman Oceanographic Laboratory, UK, 11 pp

Moll A, Radach G (eds) (1990) Compilation of measurements from two interdisciplinary STAR-shaped surveys in the North Sea. vols I and II. Techn Rep 6-90, Inst Meereskunde, Univ Hamburg

Morel A (1974) Optical properties of pure water and pure sea water. In: Jerlov NG, Steemann-Nielsen E (eds) Optical aspects of oceanography. Academic, London, pp 1-24

Platt T, Caverhill C, Sathyendranath S (1991) Basin-scale estimates of oceanic primary production by remote sensing: the North Atlantic. J Geophys Res 96, C8:15147 - 15159

Pohlmann T (1991) Untersuchung hydro- und thermodynamischer Prozesse in der Nordsee mit einem dreidimensionalen numerischen Modell. Dissertation, Fachbereich Geowissenschaften Univ Hamburg

Prieur L, Sathyendranath S (1981) An optical classification of coastal and oceanic waters based on the specific absorption curves of phytoplankton pigments, dissolved organic matter, and other particulate materials. Limnol Oceanogr 26:671

Sturm B (1980) The atmospheric corrections of remotely sensed data and the qualitative determination of suspended matter in marine water surface layers. In: Cracknell AP (ed) Remote sensing in meteorology, oceanography and hydrology. Horwood, Chichester, pp 163-197

Sturm B (1981) Ocean colour remote sensing and quantitative retrieval of surface chlorophyll in coastal waters using CZCS data. In: Gower JFR (ed) Oceanography from space, Marine Science Series, vol 13. Plenum, New York, pp 267-279

4.2 The Influence of Weather and Climate

H. GRAßL

4.2.1 Introduction

The short-term variability of temperatures, residual currents, sea level, nutrients and salinity on time scales of days, weeks and months is mainly driven by weather events, while the basic features and the long-term variability of the North Sea as, for instance, mean salinity distribution, mean currents, yearly nutrient input by river discharge, distribution of fish species and their year-to-year variation, are mainly due to climate, i.e. the statistics of weather. Any change in climate which will mean systematically changed weather thus may have a strong impact on the North Sea as a whole. Since climate will change even without external forcing due to the interaction of climate system components on all time scales, the long-term basic features of the North Sea also have to change continuously. This Volume, therefore, is mainly a report of the contemporary North Sea characteristics, which are a mixture of natural and man-made contributions, both to mean values and variability. The separation of man's contribution is still not possible in many cases. This statement also holds for some of the trends observed during the last decades or the last century.

This Chapter firstly (Sect. 4.2.2) gives examples of the impact of weather events on both the energy fluxes at the North Sea surface as well as on phytoplankton characteristics. Section 4.2.3 then discusses observed and anticipated long-term trends. The final Section (4.2.4) recommends main avenues of future research.

4.2.2   The Impact of Weather Events on the North Sea Energy Budget and on Phytoplankton Blooms

The large-scale atmospheric flow as evidenced in our latitudes by moving low and high pressure systems at the surface and planetary waves in the upper troposphere dominates the short-term variability of the energy and momentum budget at the surface. Thus horizontal as well as vertical fluxes of heat, momentum, water, water vapour, dissolved and suspended substances as well as the life cycle of organisms react to large-scale weather patterns. The surface temperature and thus part of the density stratification, the thickness of the mixed layer, the oxygen content of North Sea bottom water, the number density of a distinct species all depend on weather and thus change rapidly with time in a shallow sea.

4.2.2.1   The Surface Energy Budget of the North Sea and at *Elbe 1*

The data base for an evaluation of the energy budget at the surface of the North Sea has not existed long enough and to such a spatial resolution that a detailed picture can emerge. However, using basic meteorological data like near-surface wind speed, sea-surface temperature, air temperature, cloud amount and near surface humidity, parameterizations exist which give short-term variability, seasonal variation and a rough indication of year-to-year fluctuations of energy fluxes.

The energy budget of a sea surface element is composed of

- Net solar radiation flux density S_N which is always an energy gain;
- Net thermal radiation flux density F_N, which is mainly a function of surface temperature, atmospheric water vapour column content and cloud base temperature, and which constitutes an energy loss except for rare cases with cold water and very warm and humid air aloft;
- Latent heat flux E, which depends on saturation water vapour pressure at surface temperature, absolute humidity in near surface air and wind speed. In most cases (when water evaporates) E is an energy loss; however, for instance in spring in the German Bight, it is sometimes a small energy gain (when water condenses at the surface);
- Sensible heat flux H, which changes direction with thermal stratification in near-surface air and which is the dominant heat flux under strong cold air advection, when hundreds of watts per square metre may be lost to the atmosphere.

Adding all these fluxes and neglecting the small energy flux caused by precipitation with a temperature different from water temperature, we obtain the net energy flux Q to or from the sea

$$Q = S_N + F_N + E + H$$

Using the hitherto most comprehensive compilation of surface data and also deriving energy fluxes for the entire North Atlantic, Isemer and Hasse (1987) show that for the North Sea (see Fig. 4.2.1) $Q \approx 0$ on the average, i.e. no significant net heat input into the North Sea from the Atlantic exists. A similar result, namely a small energy input from the Atlantic, has already been given by Becker (1981) in his comprehensive study on the hydrography and the heat budget of the North Sea. Using Becker's values of 8 W/m^2 heat input into the North Sea by advection of warmer Atlantic water, derived from temperature differences between the main inputs and outputs we still must state that only a very small portion (a few percent) of the northward heat transport by the Atlantic at latitudes from 50 to 60°N is directed into the North Sea. As also shown already by Becker (1981), the German Bight is an area of nearly no heat loss or a small heat gain. Moll and Radach (1990) also came to a similar conclusion for the inner German Bight. As demonstrated in Fig. 4.2.2 for the observations at light vessel

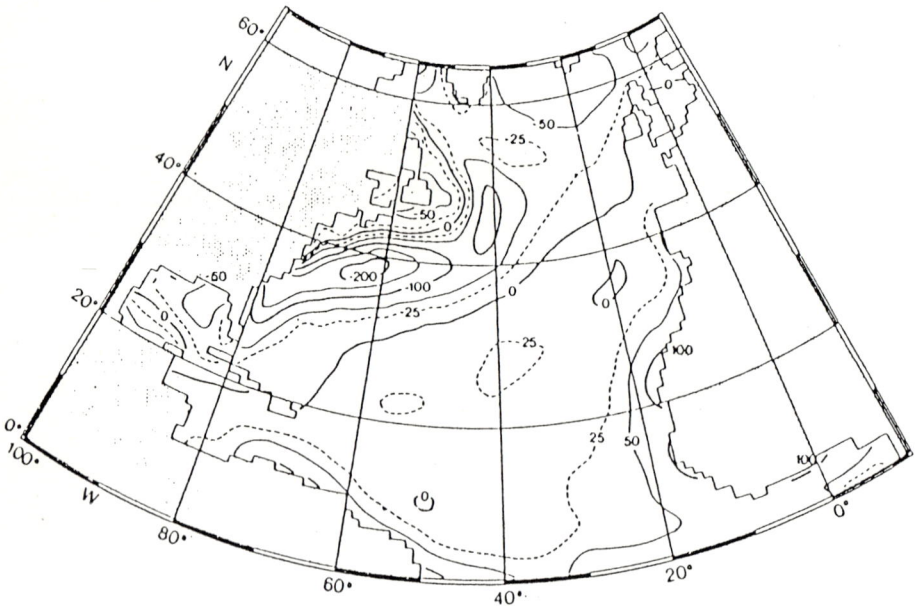

Fig. 4.2.1. Net energy flux Q at the ocean surface for the North Atlantic. (Isemer and Hasse 1987)

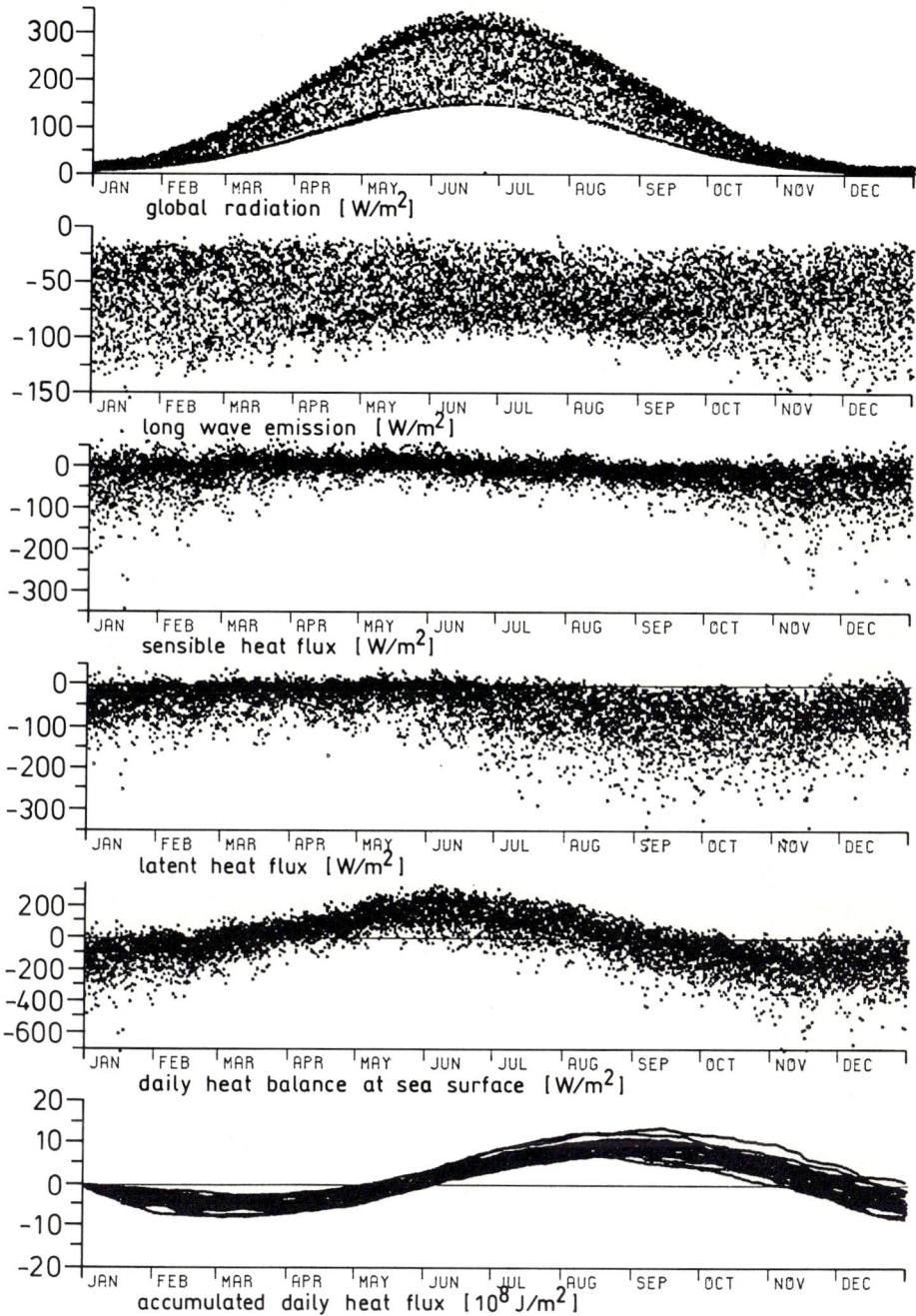

Fig. 4.2.2. The yearly integrated value of daily net surface energy flux at light vessel Elbe 1 in the inner German Bight during 25 years

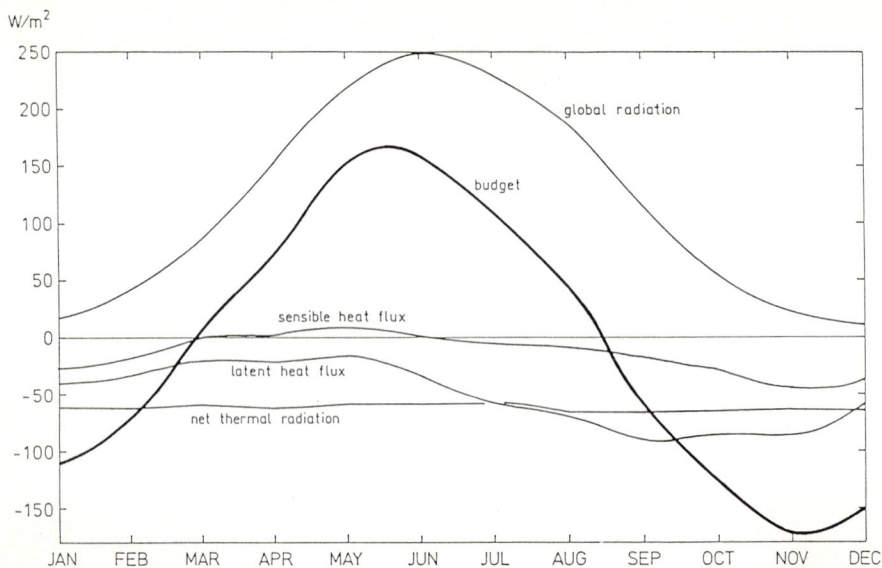

W/m²

Fig. 4.2.3. Smoothed average monthly means of surface energy budget components at light vessel *Elbe 1* in the inner Geman Bight for 1962-1986. (Moll and Radach 1990)

Elbe 1, the yearly integrated value of daily heat budgets is slightly negative except for the year 1971. In view of the uncertainties of the parameterizations used, only the statement that the North Sea water in the inner German Bight neither gets nor loses energy from or to the atmosphere in the yearly mean is justified.

If the monthly net heat flux values averaged over the years 1962 to 1986 are plotted, as in Fig. 4.2.3, a rather clear picture emerges. Net solar radiation flux density S_N reaches 250 W/m² in June and is as low as 11 W/m² in December. The relative standard deviation of daily averages is just above 40% in winter and reaches only 25% in summer; the absolute standard deviation of daily averages, however, climbs from 5 to 60 W/m² with the climbing sun. In contrast to net solar radiation flux density, the net thermal radiation flux density is nearly constant throughout a year if monthly averages are plotted. This unexpected result has a simple explanation: we live in just that climate zone where a surface emission increase is compensated by an atmospheric downward emission increase if surface temperature T_s rises. In winter a lower surface emission $F\uparrow = \varepsilon \, \sigma \, T_s^4$ (with ε = surface emissivity and σ = Stefan Boltzmann constant) is not hindered as much as in summer to reach space, because the main greenhouse gas, water vapour, is less concentrated than in summer, and the parameterization of downward thermal radiation flux density

$F\downarrow = \varepsilon\ \sigma\ T_a^3\ (T_a\ f + 4\ (T_s - T_a))$ shows a strong dependence on air temperature T_a, which normally is within a few degrees of surface temperature T_s. Water vapour pressure e and cloudiness N influence $F\downarrow$ also, here via the factor $f = 0.180 + 0.250\ \exp\ (+ 0.0945\ e)\ (1 - 0.736$ $N^2)$, where water vapour pressure e has to be given in hPa. The day to day variation of F_N is rather high, mainly driven by changing cloudiness N. It lies between 20 and 30 W/m². On the average, half the yearly mean solar input of ~ 120 W/m² is re-radiated by thermal radiation, i.e. only 60 W/m² can be left for latent (E) plus sensible (H) heat flux, if the net energy flux $Q \approx 0$.

The yearly course of latent heat flux E with a minimum in spring (only 16 W/m² in May) and a maximum in autumn (above 80 W/m² from September to November) is mainly driven by three physical parameters, wind speed, thermal stratification of air and the strongly non linear water vapour saturation pressure temperature dependence. Comparably low wind speed, stable stratification of air and still low temperatures cause the spring minimum. Above average wind speed, a convective planetary boundary layer of the atmosphere and comparably high temperatures lead to the autumn maximum. In October and November, latent heat flux E is the biggest single energy balance component. The relative day-to-day deviations are at least 55% of the mean in autumn and reach more than 100% in March to June. In April and May even the monthly mean may be zero in some years, pointing to many situations where cold water and warm humid air aloft cause sea fog or at least condensation at the surface.

The direct transport of heat to or from the North Sea water, expressed as sensible heat flux H, is rather small on the average and changes sign from day to day or even hour to hour, depending on the vertical temperature gradient in near surface air, whereby this gradient is changing sign with air mass changes. The relatively warm water in autumn causes a maximum of energy loss in November and the comparably cold spring water a smaller maximum energy gain in May. The standard deviation of daily averages is far larger than the monthly mean.

Summing all energy fluxes to obtain the energy balance leads to a strong seasonal cycle (Table 4.2.1) with a minimum (main loss) in November and a broad maximum (main gain) in May and June, differing by 330 W/m². Variability from day to day is high throughout the year and may reach more than 170 W/m² (December 1978). The special situation of a bight favours an especially strong change from August to September when the energy budget is lowered by more than 100 W/m² due to the combined action of less solar radiation, increased evaporation and direct heat loss.

Within the study on eutrophication funded by the Federal Environment Agency (Umweltbundesamt) Stengel and Grassl (1985 in Gerlach 1989)

Table 4.2.1. Mean monthly values of the net energy flux Q at *Elbe 1* and standard deviations S of daily means (taken from Moll and Radach, 1990); all values in W/m^2.

	Jan	Feb	Mar	Apr	May	June	Juli	Aug	Sept	Oct	Nov	Dec	mean
Q	-112	-73	+6	71	152	159	107	44	-59	-126	-171	-149	12≈
S	84	73	56	52	60	78	78	80	78	83	99	90	

evaluated the light vessel *Elbe 1* time series of meteorological parameters from 1946 to 1983 also for surface energy flux variations but mainly in the light of situations favouring the summertime stratification and, thus, oxygen depletion in bottom waters in the outer German Bight. At this site in the inner German Bight, still strongly influenced by the Elbe river plume, they found:

- The yearly mean surface energy flux is only slightly positive but varies from + 20 W/m^2 to -13 W/m^2 within the 1946 to 1983 time period, and there is no significant trend.
- The years 1981 and 1983 with oxygen depletion were not characterized by exceptionally high numbers of days with low wind speed.
- Also the duration of rather calm periods was not strongly above average for 1981 and 1983.

As a main conclusion, the following hypothesis is put forward: oxygen depletion following excessive summer algal blooms is not tied to the calmest summers; rather, it is tied to summers with calm periods and intermittent strong wind episodes not intense and long enough to mix the entire water column but bringing enough phosphate (and other nutrients) to the surface waters, which already get increasing amounts of nitrate from the atmosphere (30-200 mg N/m^2/a for the North Sea; Iversen et al. 1990).

If this hypothesis holds, excessive blooms as well as oxygen depletion will not remain a rare event, given the observed increase in nutrients in German Bight waters and the occurrence of typical calmer than average summers.

Since most of the evaluations presented up to now and in later sections put emphasis on vertical exchange, its relation to horizontal advection must be discussed. A thorough evaluation would need tested three-dimensional baroclinic (and biophysical) models of the North Sea. Since these do not exist, only the following weak justification can be given, especially if considering phytoplankton blooms. They grow and decay within weeks; thus, mostly vertical exchange processes dominate their magnitude.

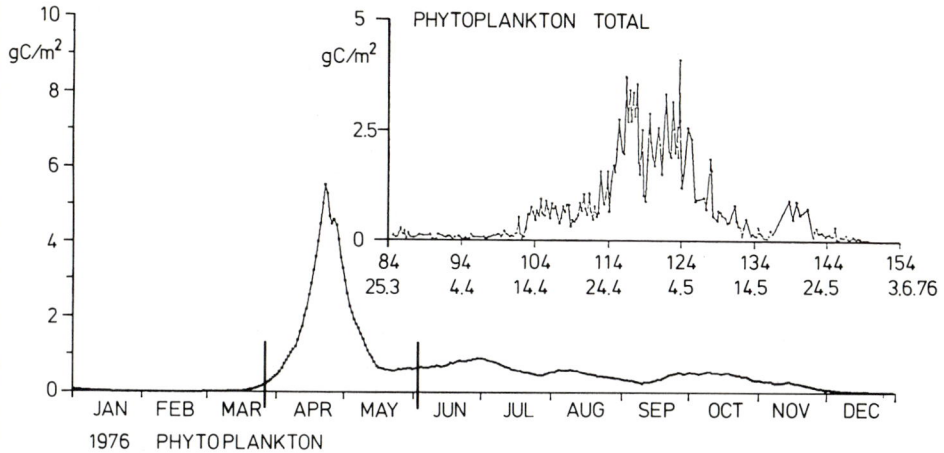

Fig. 4.2.4. Total phytoplankton mass in the water column in gC/m^2 during the spring bloom in 1976 in the northern North Sea; the model result is also shown. (After Radach 1983)

However, the long-term build-up of favourable conditions is also influenced by horizontal advection of nutrient-rich water. The plankton bloom, therefore, is not only an event driven by local meteorological conditions but also by slow advection. However, as shown by Radach and Moll (1990) for the central North Sea, the calculated and the measured annual course of phytoplankton mass agree quite well, despite the neglect of advection in the calculations.

4.2.2.2 When do Algal Blooms Occur?

The main causes of strengthened environmental research in the North Sea have been first indications that the burden by pollution from rivers, air, ships and waste disposal has caused detrimental changes, like fish poisoning by excessive algal blooms or oxygen depletion in bottom waters. Since the bloom of algae in spring constitutes the basis of the food web for another year, plankton blooms themselves are not at all of major concern; they are the basis of life in the ocean. Only excessive blooms cause damage to ecosystems as well as to economies. Although we know that only two main parameters, namely sunlight at wavelengths below 0.685 µm and some nutrients, determine plankton growth, we are far from understanding why excessive blooms of distinct plankton species occur in some years and not in others. In order to understand - not to forecast - a distinct bloom, we need to know nutrient load in winter river discharge, start of thermal stratification of surface waters in spring, intensity of earlier blooms, number of storm events prior to the bloom, remineralization of nutrients within the uppermost sediment layers, length of a calm period, cloud cover during the bloom development, short-cut recycling of nutrients in surface waters, number of cysts of a distinct species wintering in the sediment, zooplankton mass and species distribution and other parameters.

Since all these parameters are never known, the exact reconstruction of a bloom is impossible. Nevertheless, the basic features have clearly emerged: the more nutrients the more intense the bloom if meteorological conditions are favorable. An example is given in Fig. 4.2.4, where - at a given nutrient level - the carbon fixed in phytoplankton as measured and as calculated clearly depends on some local physical parameters, determined by weather.

4.2.3   Is There a Climate and Nutrient Trend in the North Sea?

Climate changes on all time scales, therefore also nutrient levels in a marginal sea, have to change continuously. If nutrient levels change phytoplankton mass, composition and productivity, as well as the annual cycle of these parameters, will also be shifted. This again has an impact on zooplankton mass and dynamics and, ultimately, determines the success of fisheries.The length of time series necessary to detect a significant trend is strongly dependent on natural variability. This variability is very high for meteorological parameters. Fortunately, meteorological time series are rather long, thus allowing a trend analysis. If time series of chemical and biological parameters are much shorter (the normal case), a trend analysis is only justified for those parameters with a meaningful time integration, i.e. seasonal means for one parameter or a calm period mean for another. For example, it would be unwise to only search for a trend in yearly average surface water phosphate concentration in the German Bight knowing that phytoplankton blooms deplete the surface layer of phosphate and changing river discharge adds or subtracts to it mainly in winter. Therefore it would be - if the simple trend analysis failed - meaningful to concentrate on winter water above a certain salinity threshold, because total mixture of North Sea water to the ground by winter storms and nearly complete lack of or small phytoplankton growth as well as avoidance of river plumes would then point to a possible trend, even in open North Sea water.

This section will report on available trend analyses for climate parameters, nutrients and phytoplankton characteristics. It will start, however, with a description of the problems in separating natural from anthropogenic trends.

4.2.3.1 The Basic Dilemma: Anthropogenic and Natural Within the Same Feedback Loops

The global carbon cycle strongly determines - via carbon dioxide concentration in the atmosphere and concomitant water vapour reactions - the greenhouse effect of the atmosphere. This effect, in turn, is responsible for basic features of atmospheric and oceanic circulation as well as temperature, precipitation and river runoff; and these basic climate parameters again fix vegetation belts and oceanic upwelling areas with strong phytoplankton blooms. Thus, carbon cycle components have the potential to change the carbon dioxide concentration in the atmosphere, completing the complex feedback loop. If man interferes somewhere in this feedback loop by changing some relevant parameter (methane concentration, for example), he might have a severe global or regional impact. However, a separation of his influence from natural variability is very difficult, and if such a separation were possible, the long time constants of some climate system components like oceans, forests and ice sheets will always show only parts of the changes to which we are already committed.

For a marginal sea like the North Sea, a possible solution would be offered by the following case: no climate parameter trends, no change in river discharge, but a significant trend in nutrient levels in river water and North Sea water. We will see whether this holds.

4.2.3.2 Sea Level Rise

An integral parameter of the water cycle on Earth is mean sea level. It depends on ocean bottom topography, volume of ice sheets, mean ocean temperature, glaciers, permafrost ice in soils, lakes, rivers, winter snow and atmospheric water vapour. It varies by about 100 m in a glacial/interglacial cycle, it has risen by approximately 120 m from the maximum glaciation 18,000 years B.P. to the disappearance of the Laurentian and the Scandinavian ice sheets 8000 years B.P. For a shallow sea like the North Sea, this means disappearance and rebirth within a glaciation cycle of typically around 100,000 years. Much smaller sea level changes, however, also have profound consequences on a shallow sea.

Before we discuss observations in the German Bight during the last 100 years, the global mean sea level trend analysis is reported. As reported in the WMO/UNEP (1990) scientific assessment of climate change, which constitutes a global scientific assessment, many researchers have found a rate of mean sea level rise between 10 and 25 cm per century from tide gauge time series of the last 100 to 150 years. The differences are mainly due to different tide gauge station selection criteria and different corrections applied to remove vertical tectonic displacements. As pointed out by the above-mentioned report, it is highly probable that global mean sea level has risen during the last 100 years by at least a decimetre.

Sea level rise in the German Bight is different from global mean sea level rise, because of tectonic movements, an unexplained strong increase of tidal amplitude during the last 40 years and the regionalization of sea level rise by ocean water warming. Figure 4.2.5, taken from Führböter and Jensen (1985), delineates mean high water level fluctuations in two different time averages. While the yearly mean clearly shows the strong interannual variations of up to 20 cm, the 5-year running mean points to a strong sea level rise (here represented by mean high water) of above 20 cm during the last 100 years. The lower part of Fig. 4.2.5 additionally emphasizes two points: the strong increase of tidal amplitude during the last 30 to 40 years and the monotonous increase of the 19-year running mean used.

The exceptionally low sea level in 1947 was caused by a $+10$ hPa anomaly of surface atmospheric pressure, equivalent to a 10 cm water column, and predominant easterly winds. Whether the tidal amplitude increase, seen at all German Bight tide gauge stations from Borkum Riff to List on the island of Sylt, is a resonance effect of a slightly deeper very shallow sea is not clear at present.

The sea level increase in the German Bight of more than 30 cm since 1850, if represented by the increase in high water level, which is the basic parameter for diking, has caused a steady shrinking of some North Sea islands. It has forced ever-increasing coastal protection work and has also threatened the Wadden Sea, the latter both directly by higher water levels and indirectly by increased tidal currents because of the protection of former tidal flats and marshlands by dikes.

Therefore, the further sea level rise forecast by climatologists as a consequence of global warming due to an enhanced greenhouse effect (WMO/UNEP 1990) is of vital importance for the North Sea coasts, which often need protection. Depending on greenhouse gas emission rates, i.e. mainly on mankind's behaviour in using fossil fuels, in developing agricultural practices and in halting forest destruction, the most probable sea level rise will range between 65 cm and 25 cm in the year 2100, with a strong further rise thereafter for a long time. The uncertainty in global

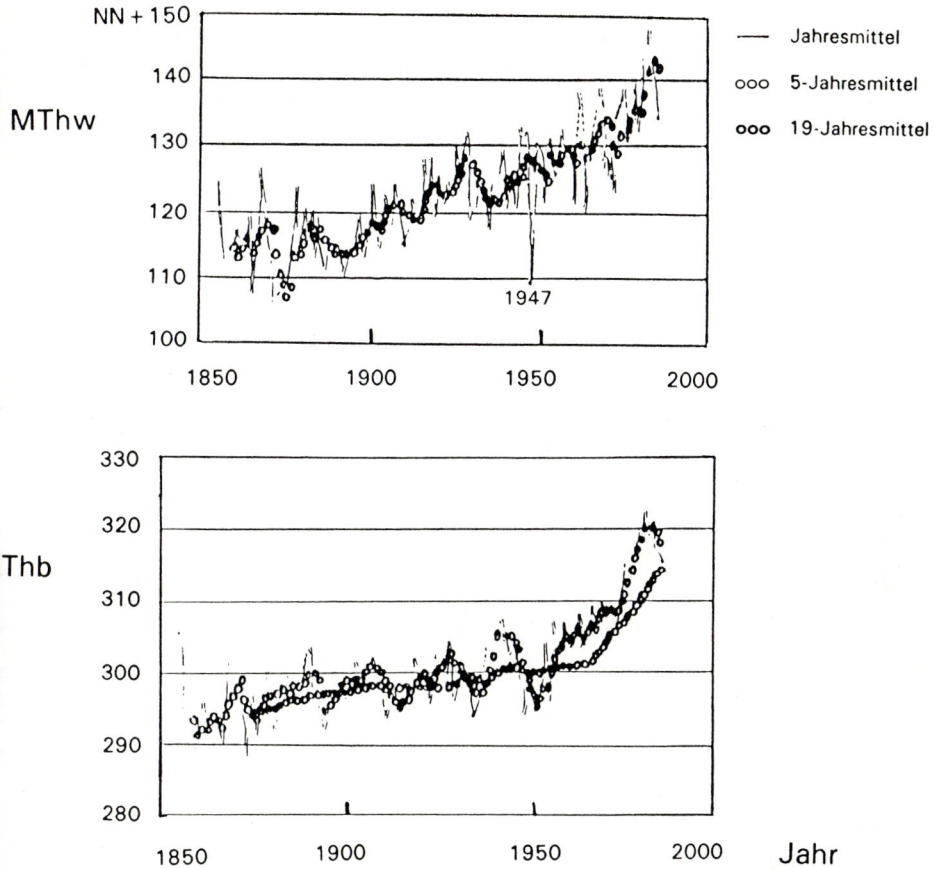

Fig. 4.2.5. Mean sea level rise in the German Bight. *Upper panel* yearly mean and 5-year running mean high water level for a 10-station average; *lower panel* increase in tidal amplitude for the same stations for the yearly mean and 5- as well as 19-year running means

water cycle estimates causes a high uncertainty range of approximately ±50%, as shown in Table 4.2.2. The numbers given account for four separate causes of sea level rise, namely - if ordered according to importance - warming of sea water (~60 cm/°C), melting of mountain glaciers, change of the Greenland ice sheet (0.3 ± 0.2 mm/year per degree warming), build-up of the Antarctic ice sheet (-0.3 ± 0.3 mm/year per degree warming). Since the upper as well as the lower estimates have the same probability, decisions on higher dikes and other coastal protection work by adopting the upper estimates should be the normal immediate reaction.

Table 4.2.2. Estimated sea level rise (WMO/UNEP 1990) for scenario A = Business as Usual, and scenario D = fast and global reduction of fossil fuel use with nearly complete renewable energy use after 2050; and separation into different causes for scenario A.

Scenario A: sea level rise until 2030 in cm					
	Thermal Expansion	Mountain Glaciers	Greenland	Antarctic	Total
High Estimate	14.9	10.3	3.7	0.0	28.9
Best Estimate	10.1	7.0	1.8	-0.6	18.3
Low Estimate	6.8	2.3	0.5	-0.8	8.7

Sea level rise until 2100		
	Scenario A	Scenario D
High Estimate	110	60
Best Estimate	66	33
Low Estimate	31	16

4.2.3.3 Is There a Temperature or Wind Trend?

Given rather precise measurements of meteorological parameters for a longer time period at a few stations, the most reliable trend analysis will be that for parameters with high spatial coherence like air temperature and geostrophic wind. Trend analyses for surface wind, water temperature and cloud amount are all subject to either small spatial coherence (and thus cannot be representative for an entire area) and/or suffer from subjective observation practices. Therefore, if trend analyses for air temperature, geostrophic wind (derived from surface pressure measurements) and sea surface temperature are reported here, only the first two would be valid for a larger area.

Air Temperature. The evaluation of some long term observations by the German Meteorological Service (Müller-Westermeier 1990) also contains with Bremen a station near to the North Sea. As Fig. 4.2.6 reveals, an

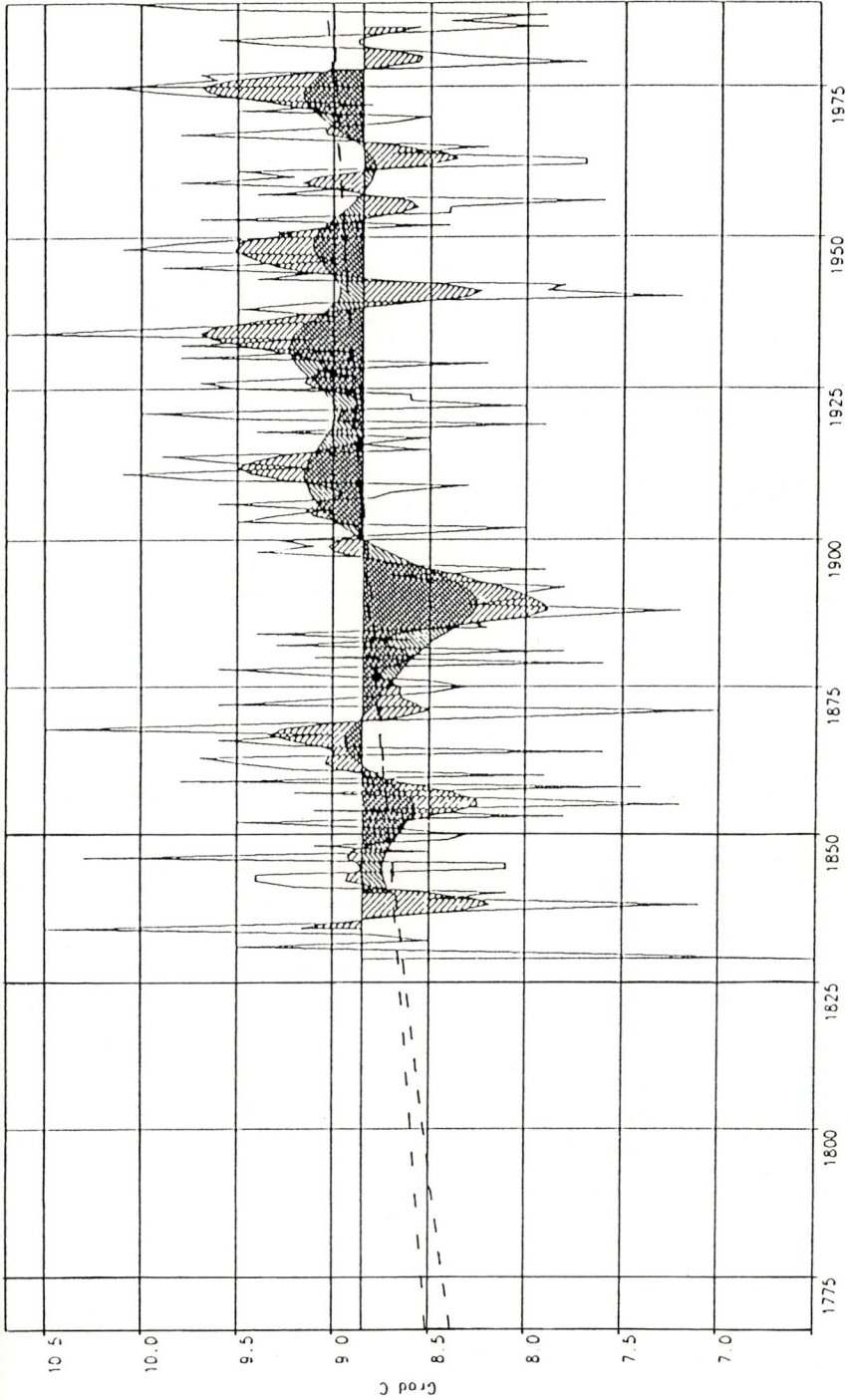

Fig. 4.2.6. Surface air temperature trend in Bremen: yearly averages, 10- and 30- year low pass filtered data; *dashed curves show* the linear as well as the quadratic upward trend. (Müller-Westermeier 1990)

upward trend of 0.35 °C is visible for yearly mean temperature during the last 160 years, the time series is correlated with a correlation coefficient of 0.83 with the Berlin time series, although there is no trend in Berlin. Since the station has been moved several times since 1829 and the city of Bremen has grown, causing a heat island effect, the trend found might contain a respective bias. More important is the strong interannual variability. Not only does yearly mean temperature vary from 7.0 to 10.7 °C, but also by more than 1 degree from one year to the next.

How does this analysis compare to a gridded global trend of 2 m air temperature, as reported (+ 0.5 °C/100 years for the global mean) in WMO/UNEP (1990)? The answer is: the agreement in magnitude and timing of the temperature rise is quite good. This allows a somewhat more detailed discussion for the North Sea. After a rather cold episode during the last 25 years of the 19th century, a strong warming of about 0.5 °C occurred until the 1940's, followed by a slight cooling into the early 1970's and again a warming in the 1980's.

Geostrophic Wind. The geostrophic wind follows from an idealiza-tion, the equilibrium between the pressure gradient force and the Coriolis force; it assumes stationarity of a pressure field, no influence by the curvature of isobars and, if applied to surface pressure observations, defines a surface geostrophic wind which in reality cannot exist, due to the importance of frictional forces in near surface layers of the atmosphere. Nevertheless, it is a good measure of mean wind speed in the lower atmosphere, far better suited for a trend analysis than values from direct wind observations, which suffer strongly from changes in observation practice.

Taking pressure observations at Borkum, Fanö and in Hamburg, which form a triangle with 200 km length having its centre in the inner German Bight, near to the former position of the light vessel *Elbe 1*, Schmidt (1990) calculated the horizontal pressure gradient and transformed it into geostrophic wind speed. Applying a method from Luthardt and Hasse (1983) he could show that:

- Yearly mean geostrophic wind speed was varying around 11.5 m/s from 1876 to 1976 with a maximum of 12.8 and a minimum of 10.0 m/s;
- No significant long-term trend exists;
- The strong wind periods with u_g > 15 m/s and wind directions of special concern for the German Bight and the estuaries (250 to 340°) were strongly changing (Fig. 4.2.7) with time. While the occurrence of stormy periods with wind directions causing surges was as low as 7.5% in a 5-year low pass filtered average around 1925, 17.5% were reached

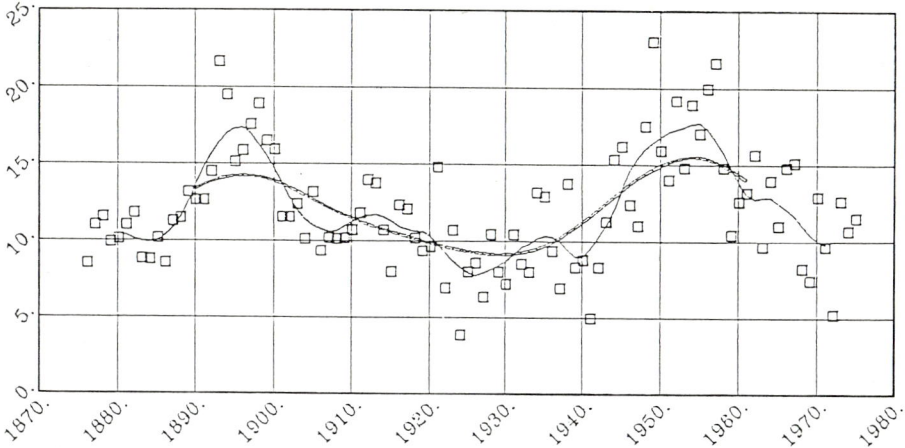

Fig. 4.2.7. Probability of strong winds (> 15 m/s causing surges in the German Bight (directions from 250 to 340 degrees) for 1876 to 1976. (Schmidt 1990)

around 1895 and 1955; i.e. more than a doubling or halving of situations with storm surges was observed in just 30 years.

Water Temperature in the German Bight. While air temperature measurements have reached a certain standardization, water temperatures have been and still are measured in at least two different ways: with unshielded or shielded thermometers in a bucket or continuously at thecooling water intake. However, the discrepancy found by Folland in air and water temperature data before 1950 has not been found by Baudner (1990) from 222,673 measurements reported by ships in the German Bight (53°-55°N, 5°-9°E). For the entire time series from 1902 to 1987, no steady water temperature trend has been detected. For the first 50 years, however, a significant upward trend of 0.5 $\pm$ 0.03 °C clearly emerged in a 30-year low pass filtered average and 89% of the variance could be explained. After 1952, a negative trend occurred, not as significant as the earlier upward one.

The strong temperature increase of 1 °C in 15 years after 1962 reported by Radach et al. (1990a) for the Helgoland Roads data is also partly visible in the German Bight analysis. The longer-term analyses by Baudner (1990) clearly show that climate parameter time series have to be far longer than 20 years for a correct description of variability.

4.2.3.4 Trends in Nutrient Levels

The unique data set of physical, chemical and biological parameters, as measured by the Biologische Anstalt Helgoland at Helgoland Roads in the German Bight since 1962 and as published in annual reports, has become the backbone of a trend analysis for nutrients as well as phytoplankton density in German Bight waters. After some preliminary evaluations of parts of the time series by different scientists, the eutrophication study started in 1983 after the widespread 1981 and 1983 oxygen depletion in the Baltic and the northern German Bight coordinated by Gerlach and stimulated by the Federal German Environmental Agency (Umweltbundesamt in Berlin) has led to a first intense evaluation of all measured parameters. The trends found for the nutrients phosphate, nitrate, silicate, ammonium and nitrite were significant or highly significant, even on the basis of yearly means, whereby the nutrient depletion (nearly complete for phosphorus) after intense plankton blooms has been included in the trend analysis (Gerlach 1989). The winter water (December to March) trend analysis shows similar results; however, a more thorough analysis would be needed, for a more reliable trend in open North Sea winter waters, since the high phosphorus levels seem to have shifted the onset of phytoplankton spring blooms into March and winter blooms, except in January, have also been recorded. Figure 4.2.8, with five year running means relative to the starting five years for phosphate, silicate, nitrate, nitrite, ammonium and total dissolved nitrogen, shows quite different trends.

The winter water values (Dec.-Feb.) developed as follows:

- Phosphate levels increased strongly from 0.77 µmol/l in 1962 to 1.20 µmol/l in 1984, with a levelling off since 1975.
- Silicate fell strongly to 5.22 µmol/l in 1984, starting with 13.1 µmol/l in 1966.
- Nitrate increased extremely strongly from 5.7 µmol/l in 1962 to 21.7 µmol/l in 1984, showing the most dramatic increase in the 1970s and early 1980s.
- Nitrite increased more slowly but earlier than nitrate from 1.0 µmol/l to 1.5 µmol/l.
- Ammonium fell to 4.4 µmol/l in 1984 from 11.1 µmol/l in 1962.
- Total dissolved nitrogen increased from 17.7 µmol/l to 27.6 µmol/l from 1962 to 1984.

Despite the indicated imperfections of the trend analysis, there is no doubt that German Bight waters show an increase in the main nutrients phosphate and nitrate, both of which are due to anthropogenic eutrophication, and a

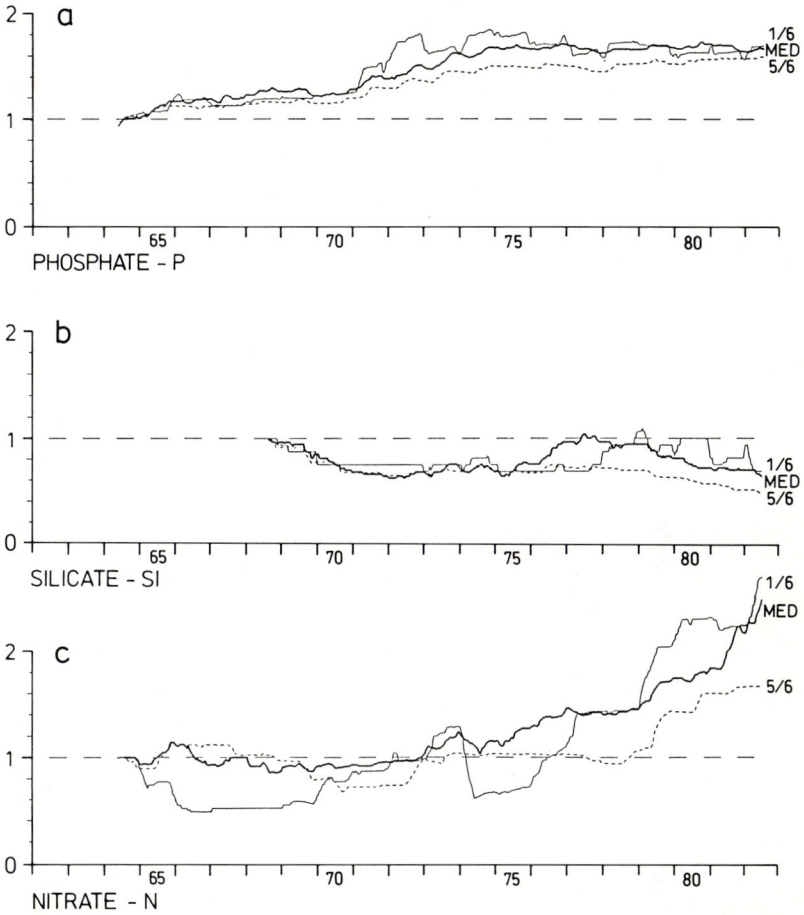

Fig. 4.2.8a-c. Five year running means (*central curve*) and 1/5 as well as 5/6 quartiles for phosphate (*a*), silicate (*b*), nitrate (*c*). (Radach and Bohle-Carbonell 1988). Normalized to the first 5-year mean as measured at Helgoland Roads up to 1983

decrease in silicate and ammonium during the last 20 years. Similar trend analyses in open North Sea waters and adjacent coastal waters do not exist, since time series are too short.

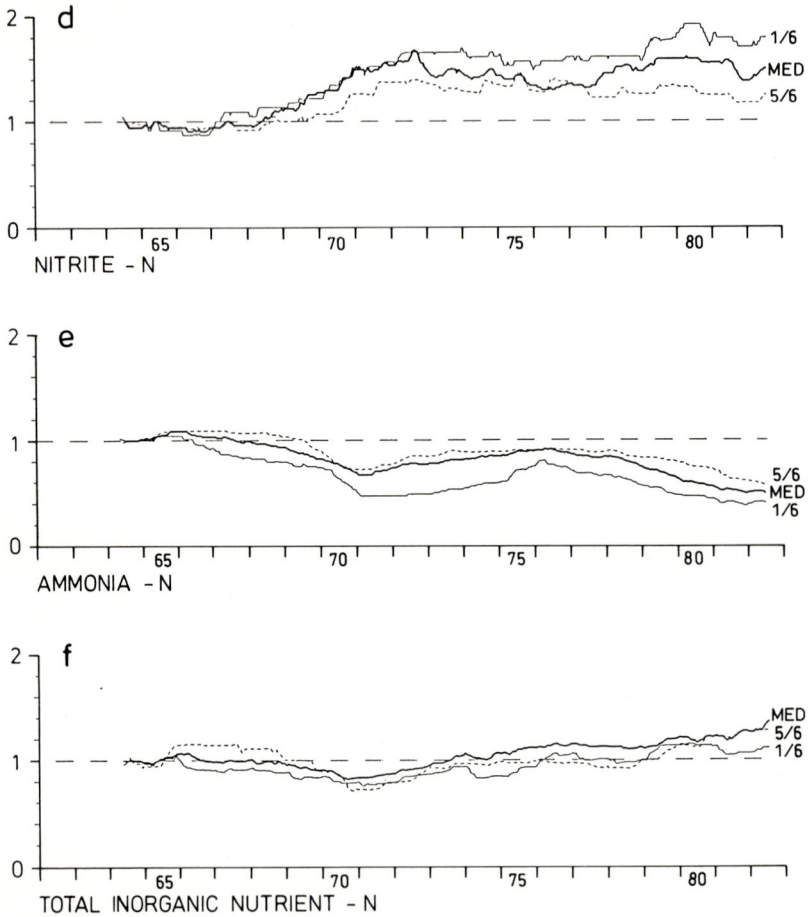

Fig. 4.2.8d-f. Five year running means (*central curve*) and 1/5 as well as 5/6 quartiles for nitrite (*d*), ammonium (*e*) and dissolved anorganic nitrogen (*f*). (Radach and Bohle-Carbonell 1988). Normalized to the first 5-year mean as measured at Helgoland Roads up to 1983

4.2.3.5 Changes or Trends in Phytoplankton

If there is no large solar radiation or wind speed change, as shown in Section 4.2.3.3, and if phosphate is often a limiting factor in phytoplankton growth, then the observed increase in wintertime phosphate amount has also to be transformed into an average phytoplankton standing stock increase.

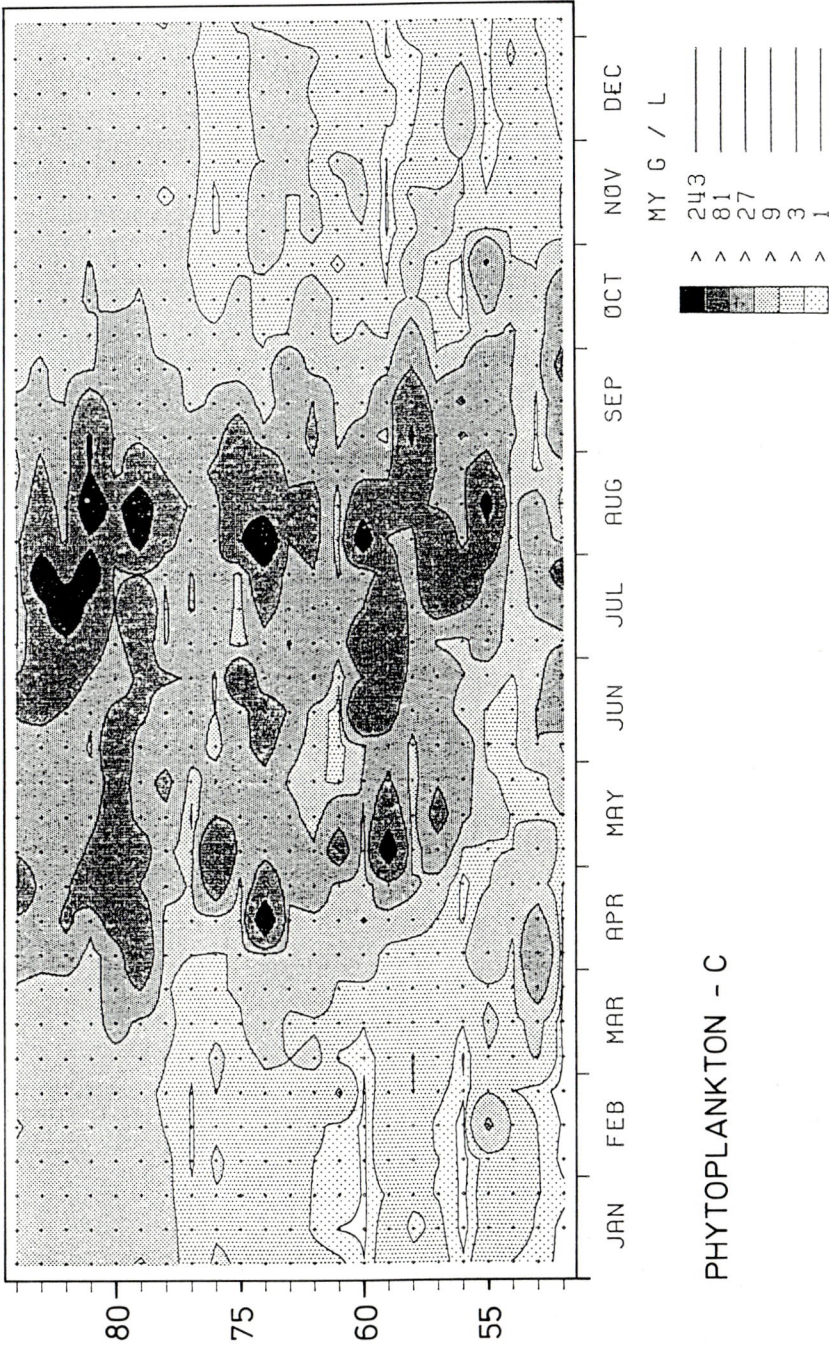

Fig. 4.2.9. Seasonal cycle of phytoplankton mass per volume (given in mg carbon per m^3) at Helgoland Roads from 1962 to 1984, taken from Gerlach (1989), evaluated by Radach et al.; please note the *logarithmic steps* in isolines

Figure 4.2.9 does not only show the strong increase throughout all months at Helgoland Roads, but also a shift to maximum values in the summer blooms and it points to an earlier onset of spring blooms. The relative increase is highest in winter or - in other words - typical summertime values of the early 1960s are found in winter in the early 1980s! Since there was a concomitant silicate decrease, a shift from diatoms to flagellates occurred especially in summer, the former increasing by a factor of 4.7, the latter falling to two thirds (Radach et al. 1990a).

Knowing the strong increase in nitrates in the 1980s, partly due to the unbroken upward trend in nitrous oxide emission ($NO_x = NO + NO_2$) from motor vehicles in Europe, and knowing the measures taken to reduce phosphate, it is easy to forecast a growing importance of phosphate as a limiting factor.

How far out into the open North Sea does this phytoplankton increase reach? The answer cannot be given as firmly as for the German Bight due to a lack of long-term direct observations. However, the wintertime surveys of nutrients have shown that only about 180 km from the inner German Bight to the northwest nutrient enriched zones exist. The probability of a restriction of a phytoplankton increase to the coastal zones is thus high. This is also underlined by a time series evaluation reported by Reid (1990), where a decline rather than an increase in phytoplankton is reported for open North Sea waters.

4.2.3.6 Modelling of Phytoplankton Blooms

The biomass production in the ocean, driven by the activity of 3 gigatons of carbon in global phytoplankton standing stock and fixing 30-60 gigatons of carbon per year, is a component of the global carbon cycle nearly as important in carbon fixation as that by all land plants. The global variability of this carbon cycle section is caused by atmospheric CO_2 level, world ocean circulation, runoff from continents and its nutrient load as well as deep water formation in polar latitudes. If only the phytoplankton dynamics of a certain region are considered for present-day conditions, then they are dominated by the immediate physical boundary conditions. Net solar radiation and wind speed determine via thermal stratification and depth of the well-mixed layer whether there is enough light and nutrients for a strong phytoplankton bloom. Therefore, prescribing these physical boundary conditions every 3 h from meteorological observations, the biomass production rate for parts of the North Sea, where advection is not dominating local conditions, should be accessible for model calculations. Exactly this was tried by Radach and Moll (1990), based on an earlier at-

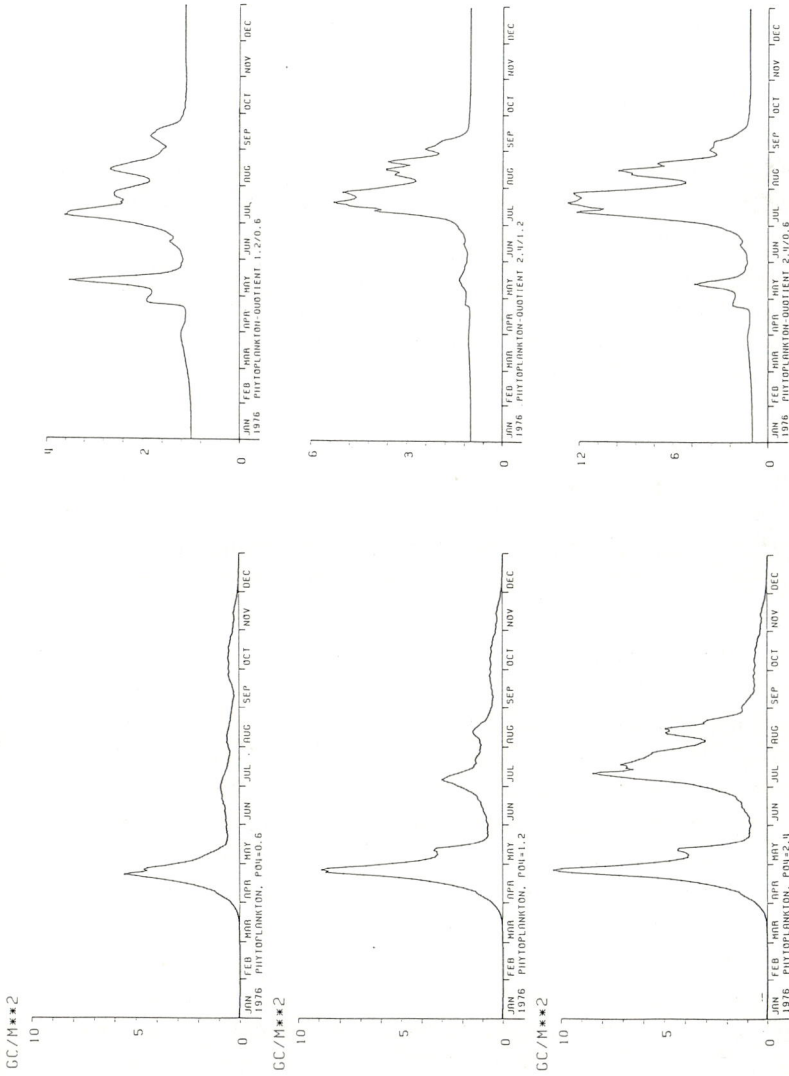

Fig. 4.2.10. Caclulated annual cycle of phytoplankton in gC/m^2 in the central North Sea for three different phosphate levels in winter water (0.6, 1.2, 2.4 mmol/m^3) using 1976 meteorological input data (*left*). Ratios of simulated annual cycles of vertically integrated phytoplankton for phosphate loads of 0.6, 1.2, 2.4 mmol/m^3 (*right*). (Radach and Moll 1990)

tempt by Radach (1983). Restricting the model to only one limiting nutrient, phosphate, which was found to limit phytoplankton growth during FLEX '76, they were able to simulate the annual course of phytoplankton activities. A big step forward has been the inclusion of detritus falling to the sea floor and its remineralization there, which allows a summer algal bloom after a short summer storm through a new nutrient input into the mixed layer. The neglect of advection, however, still is a major drawback for a more realistic simulation of phytoplankton characteristics (see Chap. 3.3).

In Fig. 4.2.10, a central result is presented. After complete mixing of the water column by winter storms and shortly after the first thermal stratification the spring bloom starts. In 1976 - meteorological data of this year were available and applied to central North Sea conditions - this bloom starts end of March/beginning of April and lasts until May. Both summer blooms and the autumn bloom fall into quiet time intervals after storm events, however, they do not reach the spring bloom standing stock, which had surmounted briefly 5 gC/m^2, the reason being the incomplete mixing of the water column. Assuming more than 0.6 mmol phosphate per m^3 (the open North Sea mean value), for instance 1.2 mmol phosphate per m^3 (German Bight conditions), the annual cycle is strongly changed and a complex structure emerges in summer, always a consequence of certainly not extreme meteorological events. The strong non-linearity is underlined by a factor between 1 and 3.6 in phytoplankton standing stock at a nutrient doubling. The autumn bloom stops despite higher phosphate level at nearly the same time in October under winter conditions because sunlight and strong mixing become the limiting factors. The comparably weak summer blooms are probably caused by the neglection of tidal mixing in near bottom layers, which has the capacity to lift remineralized detritus so that a short summer storm event is able to mix it up to the surface.

A phytoplankton bloom forecast would need a wealth of input data which at present are not available and will not be for a long time to come. In addition to a weather forecast, possible for a week now, nutrient input from rivers and remineralization levels would be needed. Nevertheless, observations and simulations point to intermittent storms in summer, high nutrient load in winter river discharge and strong tidal mixing of remineralized nutrients close to the ocean floor as the preconditions for exceptional plankton blooms and oxygen depletion in deeper layers. The upward trend for nitrate, nitrite, phosphate and the downward trend for silicate (described in Sect. 4.2.4 for the German Bight) increases the risk of excessive phytoplankton blooms in summer with all their negative effects like poisoned fish, oxygen depletion in deeper layers, and even total destruction of benthos communities during anaerobic episodes. It also points to a strong shift in the composition of plankton communities.

4.2.4   Conclusions and Recommendations for Research

Although natural variability of and anthropogenic influence on ecosystems is strong and the separation of different causes of observed changes difficult, the combined study of climate, nutrient and phytoplankton trends in the German Bight has brought a big step forward in understanding some portions of this complex marine ecosystem. Fortunately the analysis of the 20 year time series from Helgoland Roads and of meteorological observations of at least a century have allowed an at least partial answer to the question: is there an anthropogenic trend?

The different evaluations and trend analyses have revealed:

- The surface energy balance of the North Sea as a whole as well as the German Bight in particular is near to zero for the yearly average, i.e. no significant net heat import from the Atlantic exists.
- There is no significant trend of the climate variables except sea level investigated in the German Bight; neither water temperature nor geostrophic wind have systematically changed during the last 100 years. Geostrophic wind shows two marked maxima around 1890 and 1955. Water as well as air temperature have shown short period variations of up to 1 °C in 20 years at Helgoland Roads.
- There is a highly significant upward trend in nutrient concentration in coastal waters of the North Sea during the 1960s and 1970s for nitrate and phosphate. This upward trend continues at Helgoland Roads for nitrate into the 1980s, is even accelerated and has reached more than double (2.5), however, levels off for phosphate at a factor of 1.7. Silicate concentration shows a reduction at Helgoland Roads as well as in river discharge. That at least the bulk of these trends is anthropogenic, is clear.
- In open North Sea water a sound nutrient trend analysis is not possible. The sparse data suggest that there is no trend in nutrient concentration.
- At Helgoland Roads silicate concentration has fallen drastically until 1984 and, therefore, should have had an influence on phytoplankton composition.
- The annual cycle of all nutrients has also changed significantly at Helgoland Roads since 1962, the yearly amplitude of a 5-year running mean comprising two thirds of all values has risen from 0.6 mmol/m^3 for phosphate to almost 1.0 mmol/m^3 and from 16 mmol/m^3 to 24 mmol/m^3 for nitrate.
- Phytoplankton mass has increased at Helgoland Roads by a factor of 4.7 for flagellates as a consequence of increased nutrient levels, i.e. has

shown a strongly non linear reaction; at the same time, diatoms have decreased to two thirds, probably due to a far stronger silicate decrease.

As a central conclusion, regarding all these evaluations, therefore the following statement is put forward: the anthropogenic eutrophication of German Bight waters is a fact; however, the long term consequences of this fact are not known.

Since so many facets of biogeochemical cycles are still unknown for the North Sea and also the German Bight, it is clear that more research is needed. However, a good research strategy could not only accelerate our gain of knowledge but also reduce costs.

Components of such a strategy are:

- Further evaluation of existing time series.
  Example: phosphate trend analysis for quasi open North Sea water by excluding all measurements below a certain salinity and using only December and January data, thereby excluding very late or very early plankton blooms, as well as restriction to strong wind episodes,
- Combination of remote sensing and in-situ measurements.
  Example: ship measurements for process studies (and not surveys) for instance for nutrient mixing to the surface or river plume stratification, are supported by aerial surveys of sea surface temperature, ocean turbidity, wind speed or velocity and global radiation from satellites.
- Strengthening of phytoplankton modelling.
  Example: combination of a 3-D flow model with existing phytoplankton bloom subroutines accounting for remineralized nutrients mixed upwards by tidal currents and validated by tailored in-situ measurements as well as remote sensing surveys.
- Build-up of nutrient cycle models.
  Example: one-dimensional water column models with balance equations for phytoplankton, zooplankton, detritus, nutrients, oxygen driven by observed meteorological conditions.
- Continuous physical, chemical (and biological) monitoring of the entire water column at one station (in the outer German Bight for Germany).

The components of such research are interdisciplinary in nature and need the resources of some well equipped institutions cooperating under a general research plan. The planned Center for Marine and Atmospheric Sciences in Hamburg could devote a portion of its facilities to this topic and would constitute a rather inexpensive means to speed up research on the consequences of a eutrophication of a marginal sea and thus could give advice how to effectively reduce the burden to the North Sea.

References

Baudner H (1990) Jahresmittel der Wassertemperatur in der Deutschen Bucht von 1902 bis 1987. DWD intern Nr. 38, Offenbach/Main

Becker GA (1981) Beiträge zur Hydrographie und Wärmebilanz der Nordsee. Dtsch Hydrogr Z 34:167-262

Deutscher Wetterdienst (1990) Ist der anthropogene Treibhauseffekt in langen mitteleuropäischen Meßreihen nachweisbar? DWD intern Nr. 38, Offenbach/Main

Führböter A, Jensen J (1985) Säkularänderungen der mittleren Tidewasserstände in der Deutschen Bucht. Küste 42:78-100

Gerlach S (1989) Abschlußbericht des Teilvorhabens 9 des Projektes Eutrophierung der Nord- und Ostsee. Forschungsvorhaben 102 04 215, Inst Meereskunde, Kiel, Teil I (Text), Teil II (Abb)

Isemer H-J, Hasse L (1987) The Bunker climate atlas of the North Atlantic Ocean, vol II. Air-sea interactions. Springer, Berlin Heidelberg New York, 252 pp

Iversen T, Halvorsen NE, Saltbones J, Saudnes H (1990) Calculated budgets for airborne sulphur and nitrogen in Europe. EMEP/MSC-W Rep 2/90; Norwegian Meteorological Institute, Oslo (ISSN 0332-9879)

Luthardt H, Hasse L (1983) The relationship between pressure field and surface wind in the German Bight at high wind speeds. In: Sündermann J, Lenz W (eds) North Sea Dynamics. Springer, Berlin Heidelberg New York

Moll A (1989) Simulation der Phytoplanktondynamik für die zentrale Nordsee im Jahresverlauf. 2, Zentrum für Meeres- und Klimaforschung, Hamburg, ISSN 0936-949X, 139 pp

Moll A, Radach G (1990) Wärme und Strahlungsflüsse an der Grenzfläche Wasser-Luft berechnet bei Feuerschiff Elbe 1 in der Deutschen Bucht: 1962-1986. Inst Meereskunde Univ Hamburg, Techn Rep 2-90

Müller-Westermeier G (1990) Untersuchung langer mitteleuropäischer Temperaturreihen. DWD intern Nr 38, Offenbach/Main

Radach G (1983) Simulations of phytoplankton dynamics and their interactions with other components during FLEX '76. In: Sündermann J, Lenz W (eds) North Sea dynamics. Springer, Berlin Heidelberg New York, pp 584-610

Radach G, Bohle-Carbonell M (1988) Struktureigenschaften der Langzeitreihen von Helgoland Reede. Abschlußbericht zum UBA-Projekt 102 04 215, Teilvorhaben 20, Umweltbundesamt, Berlin

Radach G, Moll A (1990) State of the art in algal bloom modelling. In: Eutrophication and algal blooms in North Sea coastal zone, the baltic and adjacent areas. Water Pollution Research Report 12, In: Lancelot C, Billen G, Barth H (eds) European Community Directorate General for Science, Research and Development, Brussels

Radach G, Berg J, Hagmeier E (1990a) Long-term changes of the annual cycles of meteorological, hydrographic, nutrient and phytoplankton time series at Helgoland and at LV Elbe 1 in the German Bight. Continent Shelf Res 10:305-328

Radach G, Schönfeld W, Lenhart H (1990b) Nährstoffe in der Nordsee - Eutrophierung, Hypertrophierung und deren Auswirkungen. In: Lozan JL, Lenz W, Rachor E, Watermann B, von Westernhagen H (eds) Warnsignale aus der Nordsee. Parey, Hamburg, pp 48-65

Reid PC (1990) The dynamics of algal blooms in the North Sea. In: Lancelot C, Billen G, Barth H (eds) Water pollution research report 12. Commission of the European Communities, Brussels

Schmidt H (1990) 100 Jahre geostrophischer Wind in der Deutschen Bucht. DWD intern Nr 38, Offenbach/Main

WMO/UNEP (1990) Scientific assessment of climate change, policymaker's summary of the report of working group I to the Intergovernmental Panel on Climate Change, Geneva, July 1990

4.3 Seasonal Correlation Between Nutrients and Contaminants

U.H. Brockmann, M. Haarich, H.-J. Rick, H. Hühnerfuss,
D. Schmidt, M. Kersten, H. Steinhart, O. Landgraff,
L. Aletsee, C.-D. Dürselen and V. Becker

4.3.1 Introduction

Knowledge of the phase distribution of contaminants is essential for their transfer in the food-web as well as for their transport behaviour. The affinity of the numerous contaminants to the different phases - dissolved, coloids, particles and organisms of different quality, sediment and aerosols - is variable due to environmental conditions (pH, oxygen) and will change when the contaminants are modified by chemical reactions (speciation, conversion).

Trace elements, for instance, can be grouped with respect to their enrichment in different suspended particulate matter (SPM) components (Fig. 4.3.1; Duinker 1983). Group I elements such as Cu, Cd, Zn and Ni are associated with the mostly permanently suspended fraction of fine organic material. These are the so-called nutrient-type metals, which have often been found to be correlated with nutrients like phosphate and silicate (Saager et al. 1992). In regions of low SPM concentrations such as the northern North Sea, the small organic fraction predominates the SPM composition with increased particulate Cu, Cd and Ni concentrations (Kersten et al. 1988). Coarser and denser particles with increased concentrations of group II elements (conservative-type elements: Al, Fe, Ti, Cr, V, and As) represent the mineral fraction of the SPM. Increased SPM concentrations due to resuspension of sediments result in an increase of

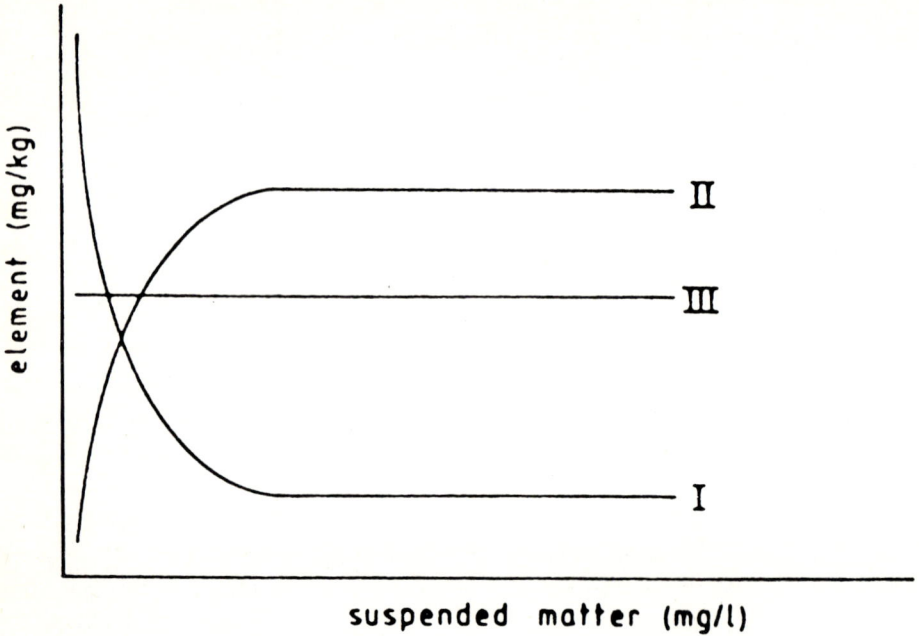

Fig. 4.3.1. Schematic relations of heavy metals and suspended matter concentrations of different group elements: I C, Cu, Cd; II Mn, Zn, Pb, Cr 77(after Duinker 1983); III Al, Fe

these elements. Metals of the group III (scavenged-type elements: Mn, Co and Pb) show no preference to any of the SPM fractions. These metals are redox-sensitive and are mostly remobilized from anoxic sediments by early diagenetic processes (Santschi 1988; Dehairs et al. 1989). Moreover, the relatively high affinity to all SPM fractions is probably caused by the oxidative uptake of these elements on the particle surfaces ($Mn^{2+} \rightarrow Mn^{4+}$, $Co^{2+} \rightarrow Co^{3+}$, $Pb^{2+} \rightarrow Pb^{4+}$; Whitfield and Turner 1987). This relationship between the elements and their dominant SPM host component was found to be helpful in describing variations of particulate metal distributions in the North Sea (Kersten et al. 1991).

Likewise, the broad variety of organic pollutants can be grouped corresponding to their amphipatic qualities, i.e. by their affinity to water and lipids or by their partition coefficients in octanol/water. Hydrophobic compounds will be adsorbed stronger at phase boundaries and organic particulate material. The water solubility of some selected compounds (Table 4.3.4) gives an indication of the great variability of adsorption behaviour.

The variety of organic contaminants is enormous; in a single sample 100-150 different organo-chlorines can be identified. Therefore, it is nearly impossible to draw any specific conclusions with respect to observed biological effects if the individual compounds are not dominating. On the other hand, it is possible to localize sources of compounds not only by gradients, but also by ratios of specific isomeres. The latter is only possible when the compounds are converted randomly in the sea to more stable isomeres.

Phytoplankton cells and organic detritus are known to act as reactive matrices, especially the matrices of fresh biological material which has been proven to be very effective in competition with other material in adsorption experiments (Calmano et al. 1990). The availability of these matrices underlies permanent formation and decomposition processes. Contrary to the winter situation with its minimum biological activity, during spring and early summer a maximum of biomass is distributed as single cells or colonies or other small planktonic organisms (e.g. copepods) with a large cumulative surface-potential for adsorption and biochemical reactions. Therefore, a scavenging effect was seen, especially during the late phytoplankton spring bloom and during specific events like the succeeding coastal *Phaeocystis* bloom (see Chap. 2.2), which significantly affects the distribution of contaminants in comparison to winter (see Chap. 4.6). In the nutrient-poor mixed layer, where the particulate organic material also had disappeared, scavenging had already occurred following the biomass formation and its sedimentation. On the other hand, in bottom layers where remineralization of sedimented material already proceeded during spring, a remobilization of contaminants should be detectable.

In summer, reactive matrices such as fresh phytoplankton cells change frequently regionally. For this reason, the biological effects with respect to the phase distribution of contaminants in the open sea will be better detectable by the state of nutrient conversion as an integrated process.

The data sets presented for the most part in Chapters 2.2 and 2.3 are used here for general and specific regional correlations between contaminants in the different phases versus the distribution of nutrients and nutrient elements as an integral of biological development.

4.3.2 Methods

Basic analytical methods are described in Chapters 2.2 and 2.3. Here, some additional information will be given which has to be considered, e.g. for interpretation of the calculated dissolved trace metal data.

For trace metal analysis of unfiltered seawater samples different analytical methods including separation and preconcentration steps have been used. All samples were acidified by addition of extremely purified concentrated hydrochloric or nitric acid and stored at -20 °C.

Reactive mercury (Chap. 2.3) was determined by Cold Vapor Atomic Absorption Spectrometry (CVAAS; Schmidt and Wendlandt 1987). Dissolved and labile complexed cadmium and lead (Chap. 2.3 and this Chap.) were analyzed by Difference Pulse Anodic Stripping Voltammetry (DPASV; Dicke et al. 1987). Graphite Furnace Atomic Absorption Spectrometry (GFAAS; Schmidt et al. 1986) was used for the determination of total cadmium in samples from the cruises in 1986 and 1987, and for manganese, iron, nickel and copper for the samples in 1986 only. In 1987 samples of total lead, manganese, iron, copper, nickel and zinc were analyzed by Total-Reflection X-Ray Fluorescence Spectrometry (TXRF; Freimann and Schmidt 1989).

Due to principal differences between the analytical procedures, the results achieved by DPASV measurements are lower than those by GFAAS or TXRF, since interference by dissolved organic material is more effective, especially when these organic compounds are able to form complexes with metal ions. In some instances, GFAAS and TXRF analysis of unfiltered samples from the summer cruise may be altered, in addition, by high concentrations of organic suspended matter.

The results presented here are sometimes inconsistent, since trace metals were analyzed first from unfiltered, small samples (1 1; Schmidt and Dicke 1990)) or from suspended material, collected from large volumes (200 1) by centrifugation (Kersten et al. 1988) and calculation of dissolved metal contents, or reference of particulate content to extracted volume (centrifugation) has been done by combining up to six different data sets.

4.3.3 Fixation of Nutrients and Contaminants Winter/Spring

One of the classical relations between trace metals and nutrients is that of cadmium and phosphate in the open sea (Boyle et al. 1981; Bruland and Franks 1983; Danielsson and Westerlund 1983). These elements will often be particularized and remobilized simultaneously. In this Chapter, simultaneous particularization and remobilization of phosphorus and heavy metals as well as conversion of organic contaminants will be discussed for the North Sea as a semi-enclosed shelf sea.

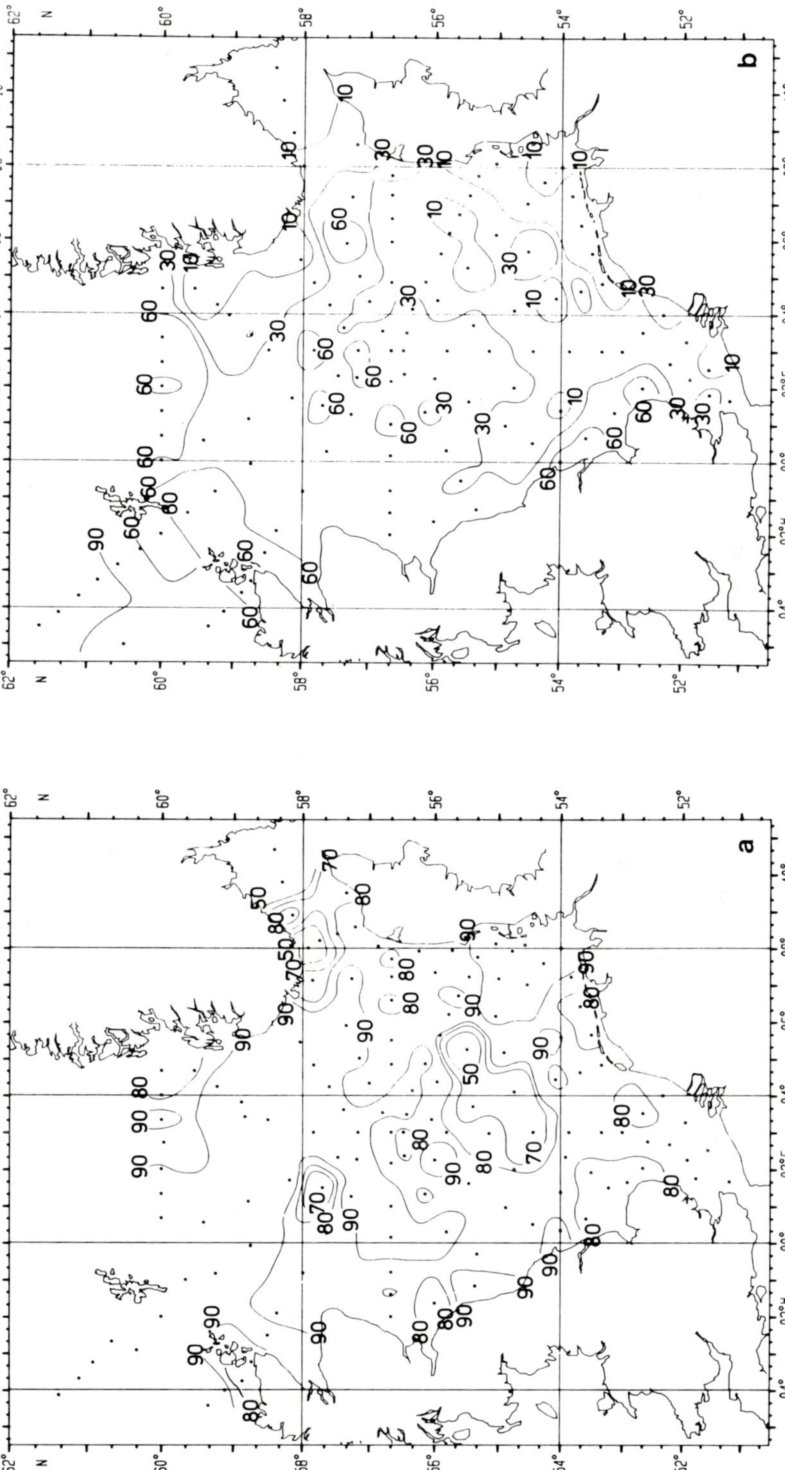

Fig. 4.3.2a,b. Phosphate as percentage of total P at the surface during winter 1987 (*a*) and spring 1986 (*b*). (Brockmann and Kattner 1993)

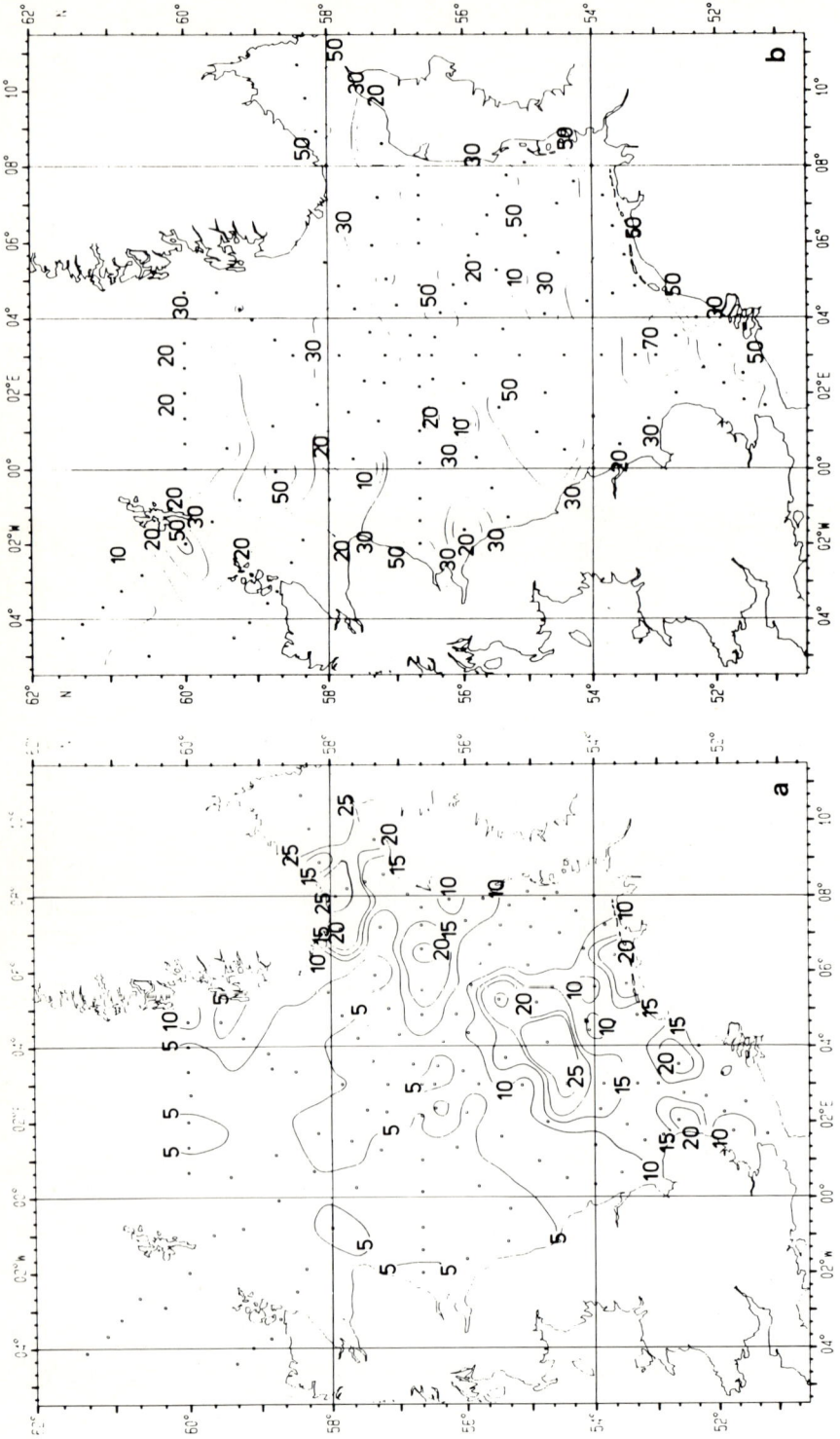

Fig. 4.3.3a,b. Particulate phosphorus as percentage of total P at the surface during winter (*a*) and spring (*b*). (Brockmann and Kattner 1993)

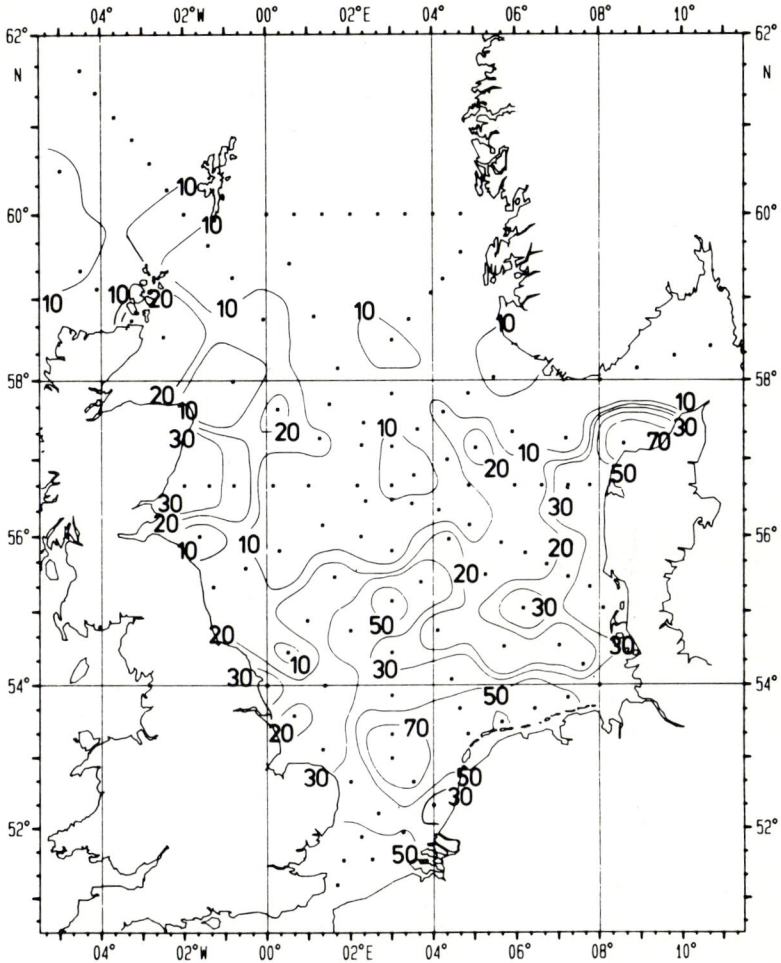

Fig. 4.3.4. Particulate phosphorus as percentage of total P near the bottom during spring

4.3.3.1 Phosphorus

The conversion of phosphate to particulate phosphoric compounds was clearly demonstrated by the distribution of phosphate as percentage of total phosphorus (comparison of winter and spring/summer situation; Fig. 4.3.2). During spring, only 10-30% of the detected phosphorus compounds (inorganic phosphate, dissolved organic and particulate phosphorus) were present as phosphate in surface waters. At some coastal sites and especially in the Atlantic inflow, phosphate contribution reached 60-70%. Concentrations of particulate phosphorus near the surface often ranged between 30 and 70% at individual locations (Fig. 4.3.3b). The inhomogeneous distribution of particulate phosphorus is a consequence of phytoplankton patchiness (Chap. 2.2) and the scavenging process after growth-stagnation. A more homogeneous distribution was found during spring near the bottom. Here, the percentage of particulate phosphorus was highest in the shallow, more turbulent southern North Sea, especially in a belt of coastal water reaching from Britain to Denmark. Here, particulate phosphorus accounted for more than 20%, peaking in some areas to more than 50% (Fig. 4.3.4). In the central and northern part of the North Sea, particulate phosphorus contributed less than 10%.

This means either that especially organic resuspended matter was high in the southern North Sea with a significant load of phosphorus or that in the shallow areas production was still ongoing due to short-cycled remineralization, e.g. documented by the bloom of *Phaeocystis globosa* (Chap. 2.2).

During winter, however, 80 and 90% of phosphorus was represented by phosphate (Fig. 4.3.2a-b; see also Chap. 2.2). Accordingly the particulate fraction reached concentrations just above 20% (Fig. 4.3.3a) only in some productive areas, e.g. the shallow Dogger Bank, the stratified Norwegian coastal current, and some coastal sites.

4.3.3.2 Cadmium

Cadmium concentrations were mostly measured near the surface at 10 m depth. A comparison of total cadmium concentrations between winter and spring (Chap. 2.3) reveals that at 10 m depth concentrations were generally below 10 ng/l in most parts of the North Sea in spring and in winter above 10 ng/l, respectively. In the continental coastal water the cadmium concentrations decreased in spring, corresponding to the decrease of nutrient concentrations. Cadmium concentrations greater than 20 ng/l at coastal sites indicated discharges or remobilization in shallow areas. For example, the spring data show that the Cd discharge near the Humber is comparable to the nutrient discharge.

In order to calculate the particulate cadmium load, suspended matter concentrations (SPM)/dry weight were converted via SPM/liter and referred to ASV unfiltered sample measurements. At 50% of the stations sampled in spring (Fig. 4.3.5b), cadmium concentrations in SPM were mostly above 10% of total Cd in 10 m depth. In winter (Fig. 4.3.5a), however,

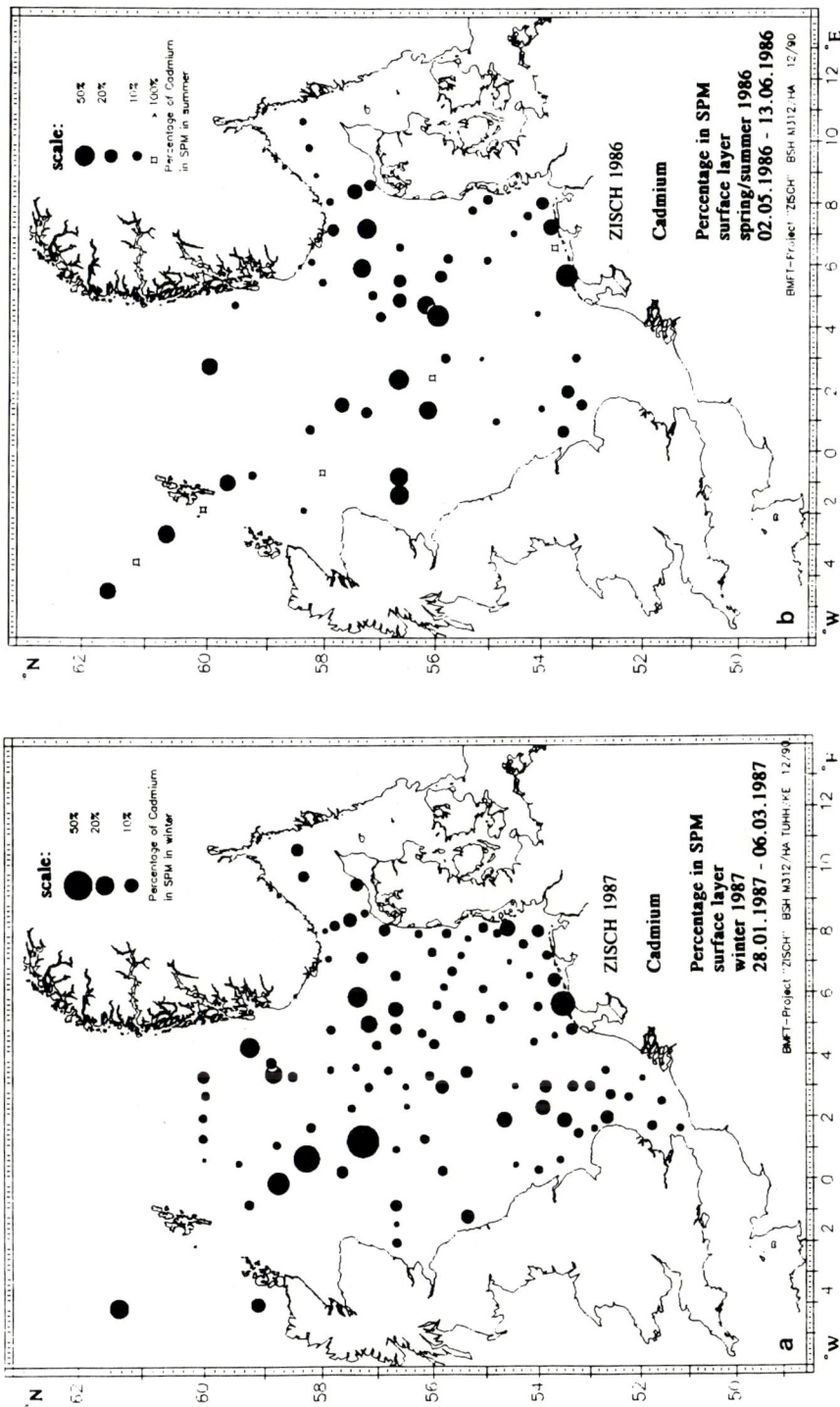

Fig. 4.3.5a,b. Cd concentration as percentage of suspended material in winter (a) and spring (b) (10 m layer)

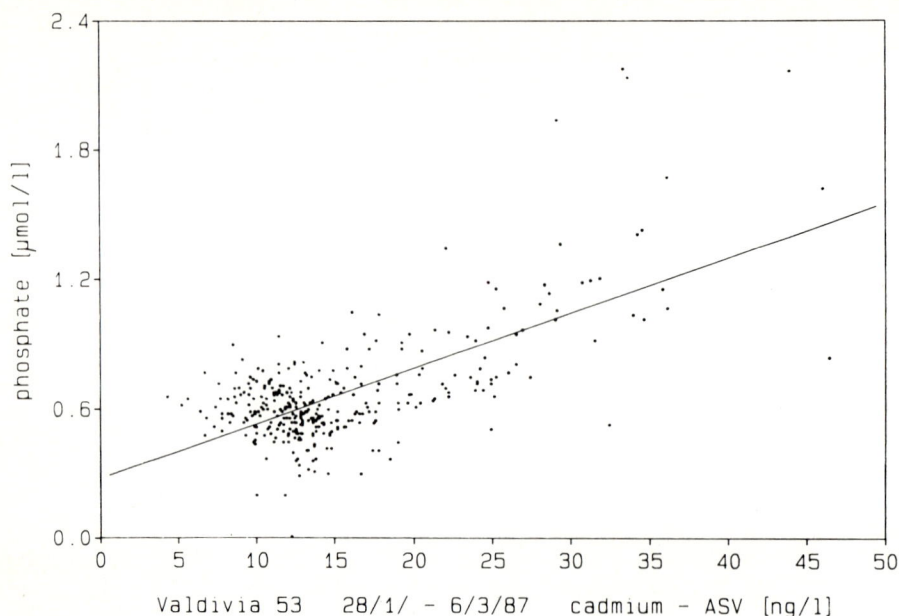

Fig. 4.3.6. Correlation between phosphate and total Cd during winter

analytical data based on the particulate cadmium were below 5% of total Cd in the surface layer at a majority of stations. This supports the hypothesis of a seasonal effect of biogenic particularization of cadmium in the North Sea and shows a parallel trend with phosphorus turnover (Fig. 4.3.3).

However, no significant correlation was found between particulate phosphorus distribution and total Cd, due to the patchy plankton distribution in spring and the analytical interferences by biogenic material. Only during the winter cruise was a linear, slighty significant correlation between phosphate and total cadmium calculated (Fig. 4.3.6), which explains the dilution gradients in the coastal water. During spring, there were no significant correlations between phosphate and total Cd, since concentrations of the latter in unfiltered samples were more inhomogeneously distributed. For these elements, the standard deviations in % of means were lower in winter (about 40%) than during spring (70-190%; Table 4.3.1). This was also an effect of the patchy plankton growth. Generally, however, the means decreased: phosphate from 0.7 to 0.3 µM and Cd from 15 to 11 ng/l.

The scavenging was more evident when phosphate and Cd concentrations were compared between bottom and surface layers during spring (Figs. 4.3.7 and 4.3.8). Phosphate was nearly depleted in the mixed layer during

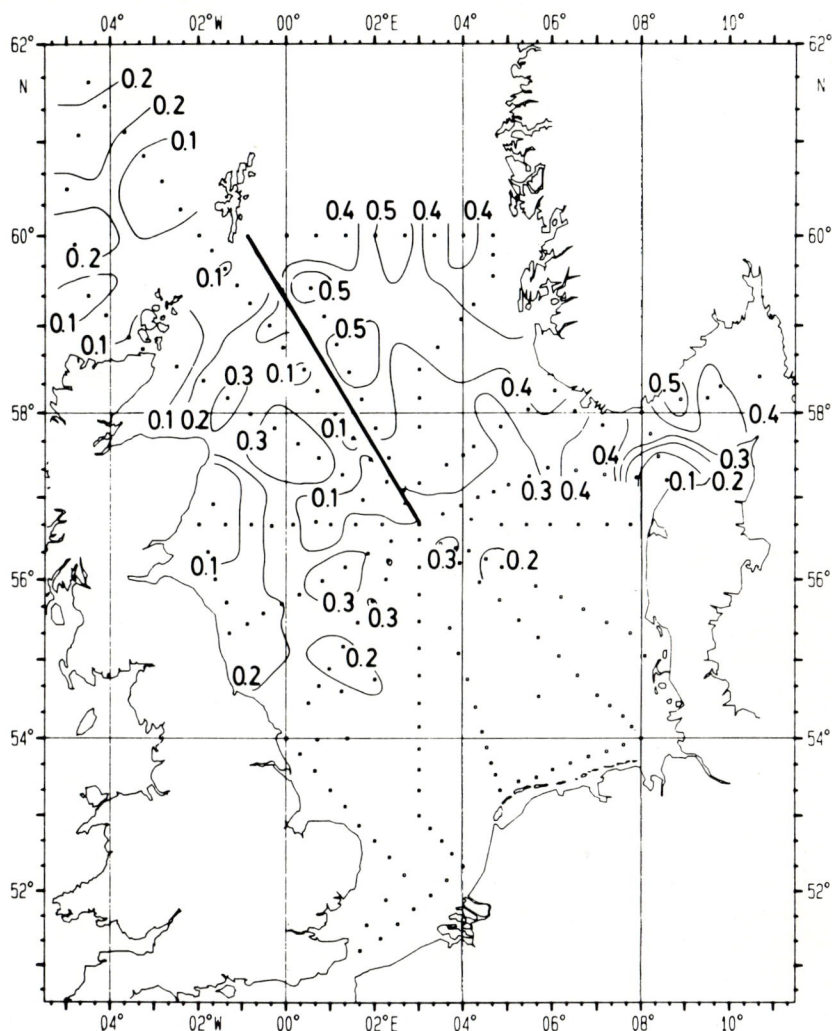

Fig. 4.3.7. Vertical differences of phosphate (μM) during spring

spring. Near the bottom, however, phosphate was still available and probably increased by remineralization of sedimented material (see Chap. 2.2); the difference between deep and surface water ranged between 0.3 and 0.6 μM. These differences were mainly found in the deeper part of the North Sea investigated.

Approximately in the same area, vertical gradients of total Cd were found with higher values in the bottom layer. The mean difference between

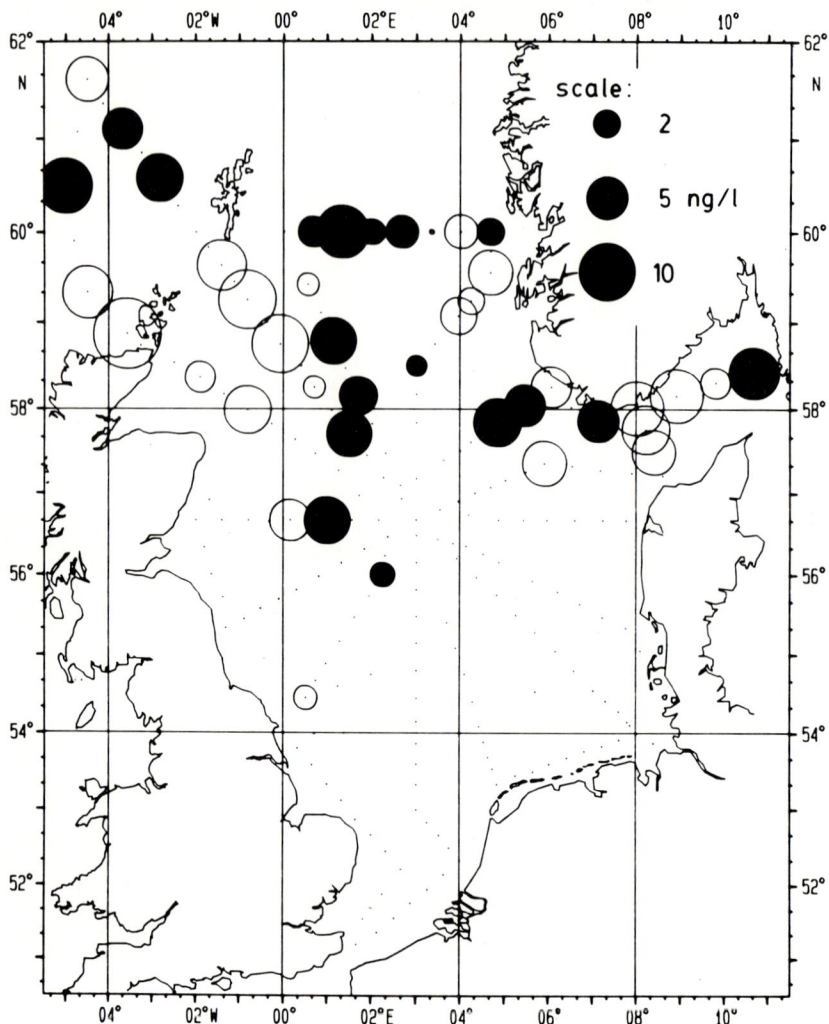

Fig. 4.3.8. Vertical differences of total Cd (ng/l) (ASV) during spring

surface and bottom water was 5 ng/l. At the adjacent stations, surface layer values were higher, due to actual discharges or unfinished adsorption and sedimentation. This is another indication of the scavenging effect, transporting Cd towards the bottom in the deeper parts of the North Sea. A similar situation was found in different areas.

Kremling et al. (1987) also found higher concentrations of dissolved Cd in lower layers of the North Sea in July 1984, but not in the area described

Table 4.3.1. Means and standard deviations of phosphorus and some trace metals

	Mean	SD (%)	No. of measurements considered
Winter			
PO_4 (μmol/l)	0.67	38.8	341
part. P (μmol/l)	0.60	115	314
Cd (ASV) (ng/l)	15.40	44.1	341
Pb (ASV) (ng/l)	87.00	111	314
Spring			
PO_4 (μmol/l)	0.31	90.3	347
part. P (μmol/l)	0.125	97.6	329
Cd (ASV) (ng/l)	11.1	65.8	347
Pb (ASV) (ng/l)	17.9	146	329

above. In his study, only vertical phosphate gradients were similar to ours. Kremling (1985) found significant Cd/P relations only in the adjacent Atlantic water during spring 1982. In shelf regions Cd/PO_4^{3-} relationships may be masked by discharges from rivers or atmospheric input. Thus, a more significant correlation is given by the winter/spring differences, when the variability of seasonal discharges is neglected.

4.3.3.3 Lead

A similar seasonal trend was observed for lead, which was found to occur at higher particulate concentrations in spring than in winter (Fig. 4.3.9), despite the higher absolute values in the coastal water at that time. During winter, particulate lead was about 50% assuming total lead (ASV) = 100%. The particulate lead concentrations calculated per water volume reflect the distribution of suspended material along a belt from the British coast towards the continental coastal water with an extension off the northwest Danish coast (Fig. 4.3.10).

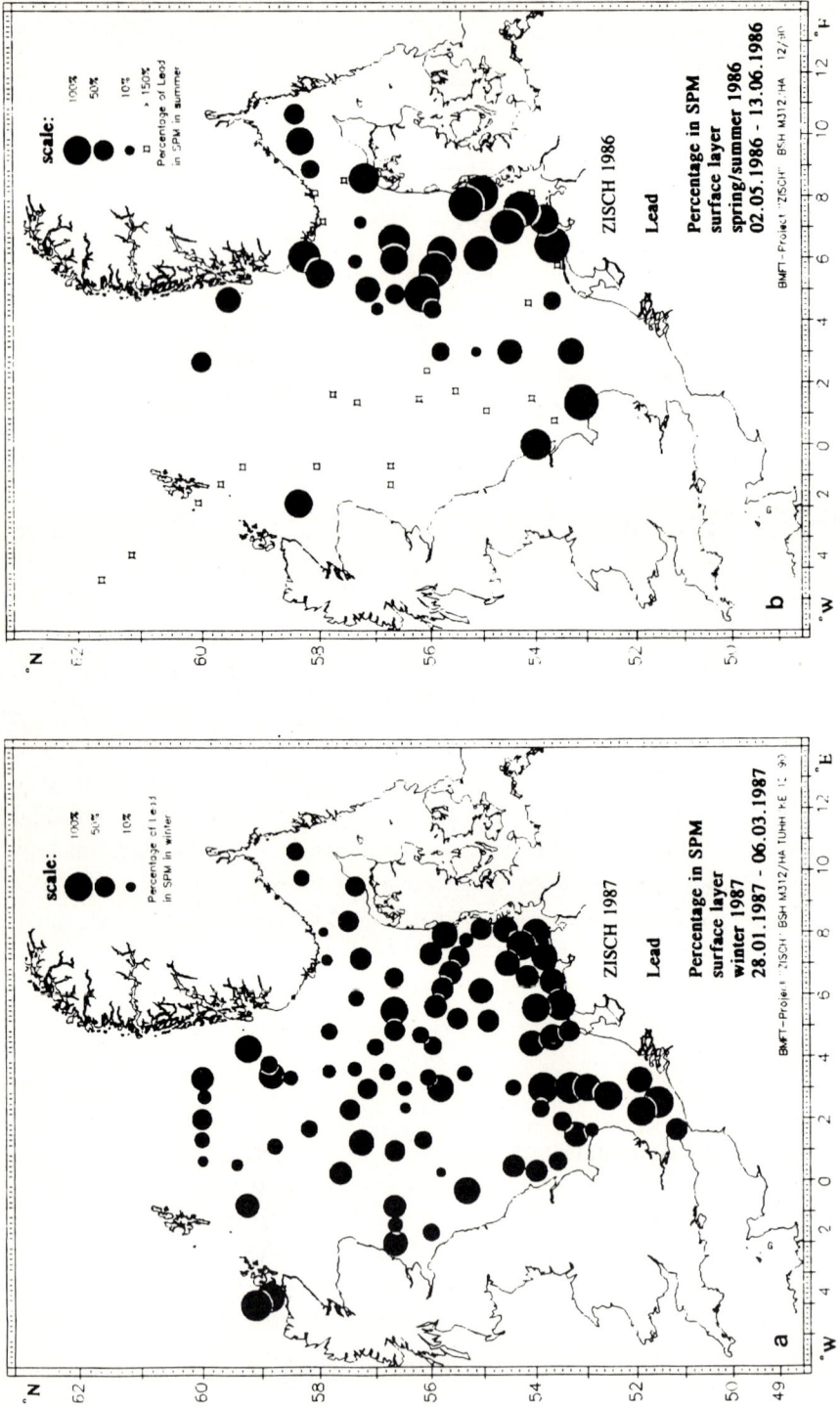

Fig. 4.3.9a,b. Pb concentration as percentage of suspended material in winter (*a*) and spring (*b*), 10 m layer; total Pb (ASV) = 100 %

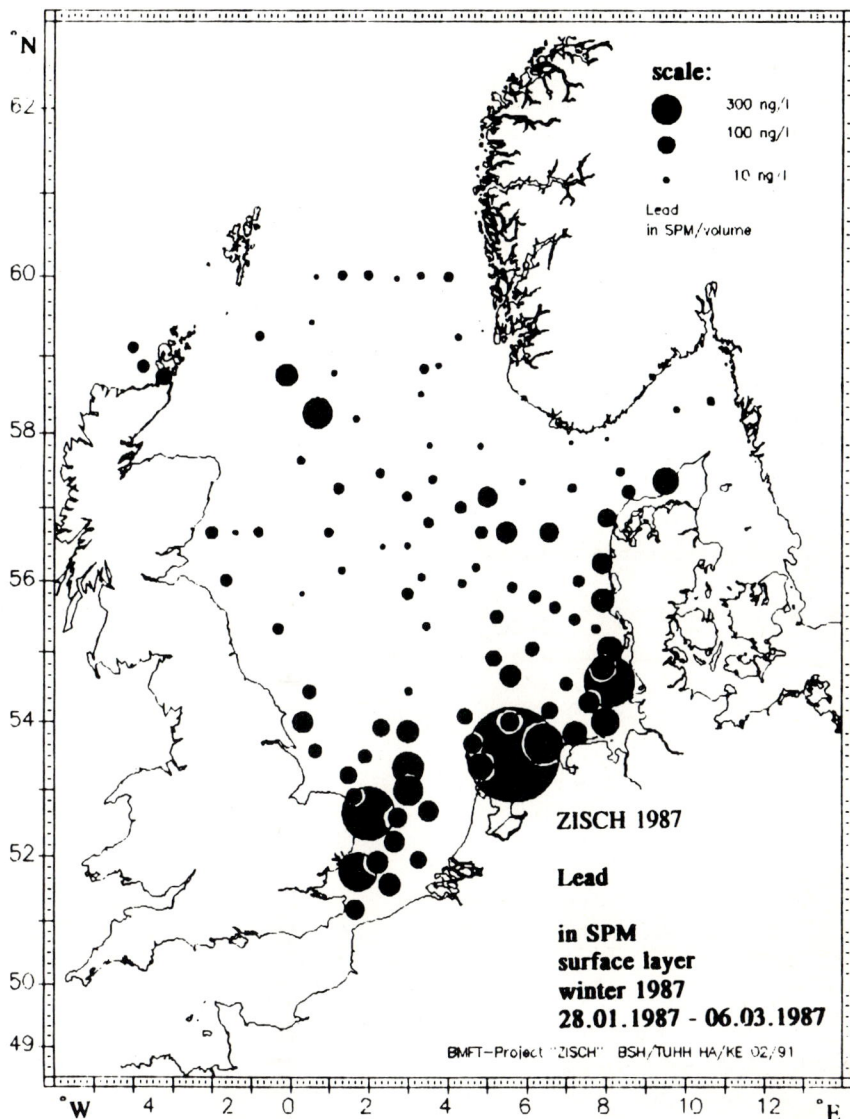

Fig. 4.3.10. Pb concentration of suspended material, referred to extracted (centrifuged) water volume

The content of lead in suspended material during summer was mostly close to 100% of total Pb (ASV). At some stations in the western North Sea, the particulate lead content was higher than the total estimate from smaller samples. These stations are indicated by white circles. The reason

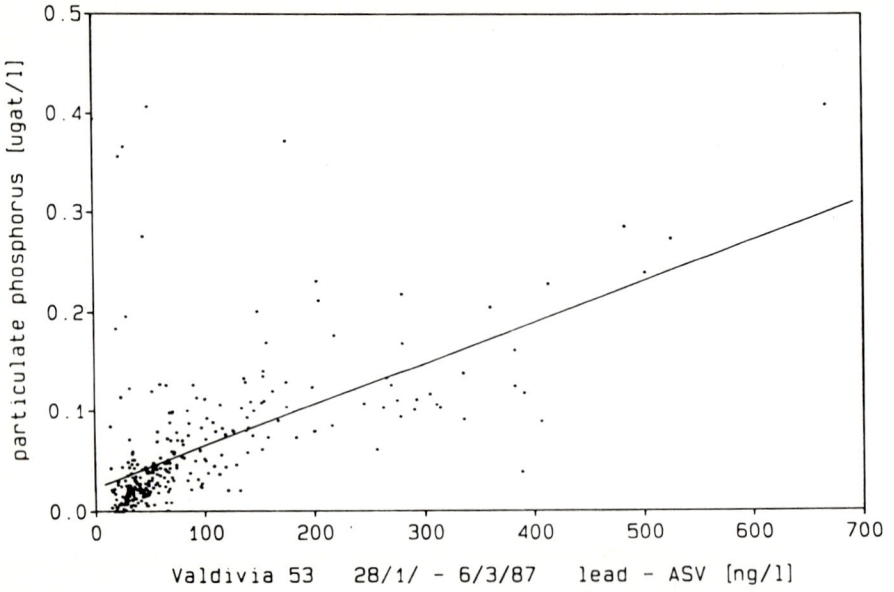

Fig. 4.3.11. Correlation between particulate P and Pb (ASV) during winter

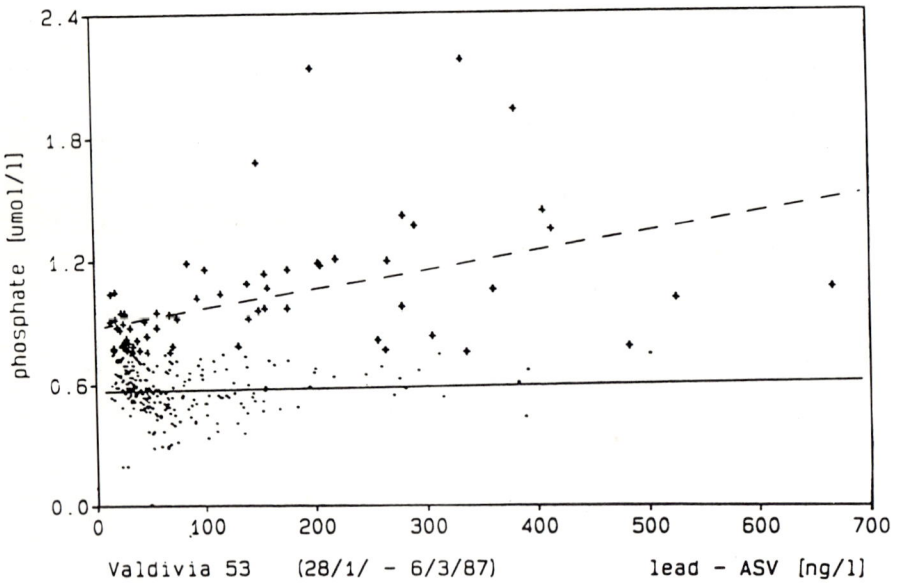

Fig. 4.3.12. Correlation between phosphate and Pb (ASV) during winter. + = continental coastal water (r=0.44); · = central North Sea and British coastal water (r=0.05)

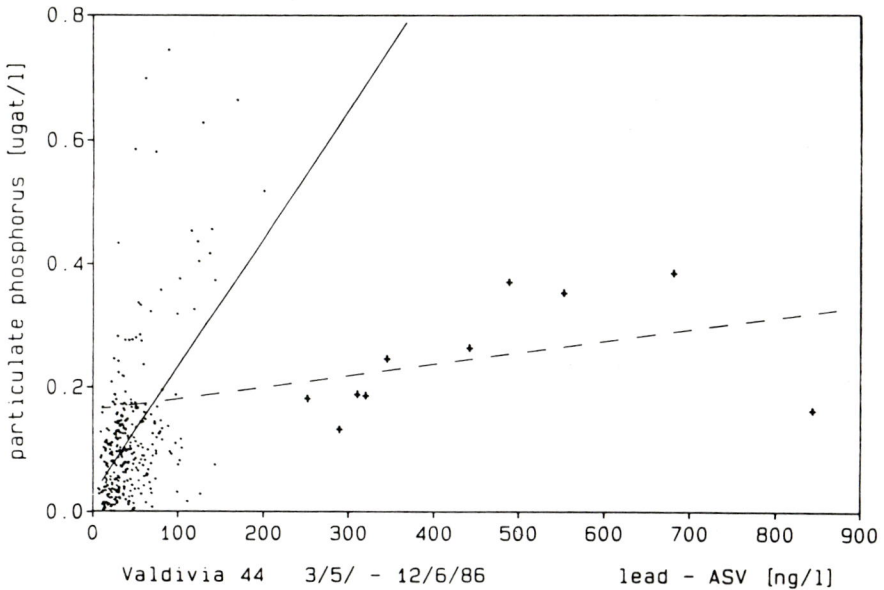

Fig. 4.3.13. Correlation diagram of particulate P and Pb (ASV) during spring. + = British coastal water, r=0.39; · = North Sea and continental coastal water (r=0.51)

for this discrepancy is the insufficient detection of trace metals by ASV in the presence of plankton blooms (see Sect. 4.3.2). These data indicate an overlapping of two processes controlling the lead distribution in the particulate phase during spring: first, hydrodynamic erosion/sedimentation (Fig. 4.3.1) and second, biological particularization.

In winter there was a significant positive linear correlation between particulate phosphorus and total lead based on the analysis of unfiltered samples (Fig. 4.3.11). A couple of samples from the northern part of the German Bight showed lead concentrations as low as 150 ng/l associated with high particulate phosphorus content. Erosion and advection caused by a storm event (20 m/s; 320°) dominate both the high concentrations of particulate phosphorus and total lead in the wedge off the west coast of Denmark (see Chap. 2.2 and 2.3).

In winter there were two overlapping groups of phosphate/total lead relationship covering more than 97% of the data (Fig. 4.3.12). One group characterized by phosphate concentrations between 0.7 and 1.4 µM was located in the continental coastal waters. The other group with phosphate values between 0.3 and 0.9 µM was mainly located off the northern coast of Britain, in the wedge of highly suspended material off Denmark, and in the transient zone between British and continental coastal water.

In spring, no significant linear correlations between particulate phosphorus and total lead were found (Fig. 4.3.13), but there were two separated mixing lines: first, one at low lead concentrations with high particulate phosphorus in samples from the continental coastal water, especially along the Dutch coast, and second, one with low particulate phosphorus and high lead concentrations from the British coastal water, indicating that in winter as well as in spring the loading with lead along the British east coast was much higher than in the Dutch coastal water. No significant linear correlation was found between phosphate and total lead.

4.3.3.4 Copper

Dissolved copper concentrations were calculated as difference between particulate and total measurements (AAS) in spring from the few available stations (stations with parallel measurements of total copper, particulate copper and suspended material). Concentrations occasionally reached 200 ng/l (Fig. 4.3.14b). This was contrary to winter, where dissolved copper was mostly above 200 ng/l, reaching values up to 500 ng/l, especially in the German Bight (Fig. 4.3.14a).

Seasonal differences of dissolved copper were only found at some stations from both cruises (Fig. 4.3.15). Highest differences occurred in the continental coastal water, where the differences of phosphate concentrations between winter and spring were also highest (Fig. 4.3.16). This again is a confirmation of the seasonal scavenging effect for nutrients and some heavy metals, but for a more extended data base the differences between winter and summer of the unfiltered copper showed that, apart from the continental coastal water, these effects also occurred at some spots in the eastern part of the North Sea.

4.3.3.5 Organic Contaminants

Generally, specific relations between organic contaminants and nutrients are difficult to detect in the North Sea. The reason is that the sources for the few well-detectable organic contaminants are spread over the North Sea (oil rigs as for phosphoric esters) or discharges dominate their distribution during both seasons (as for HCH). Another important variable source is the atmosphere (Duce et al. 1990), the estimated inputs of which are in the order of surface concentrations (Van Aalst et al. 1982).

The concentrations of γ-HCH (Chap. 2.3) in the entire North Sea were in spring especially high in areas where the percentage of phosphate to total

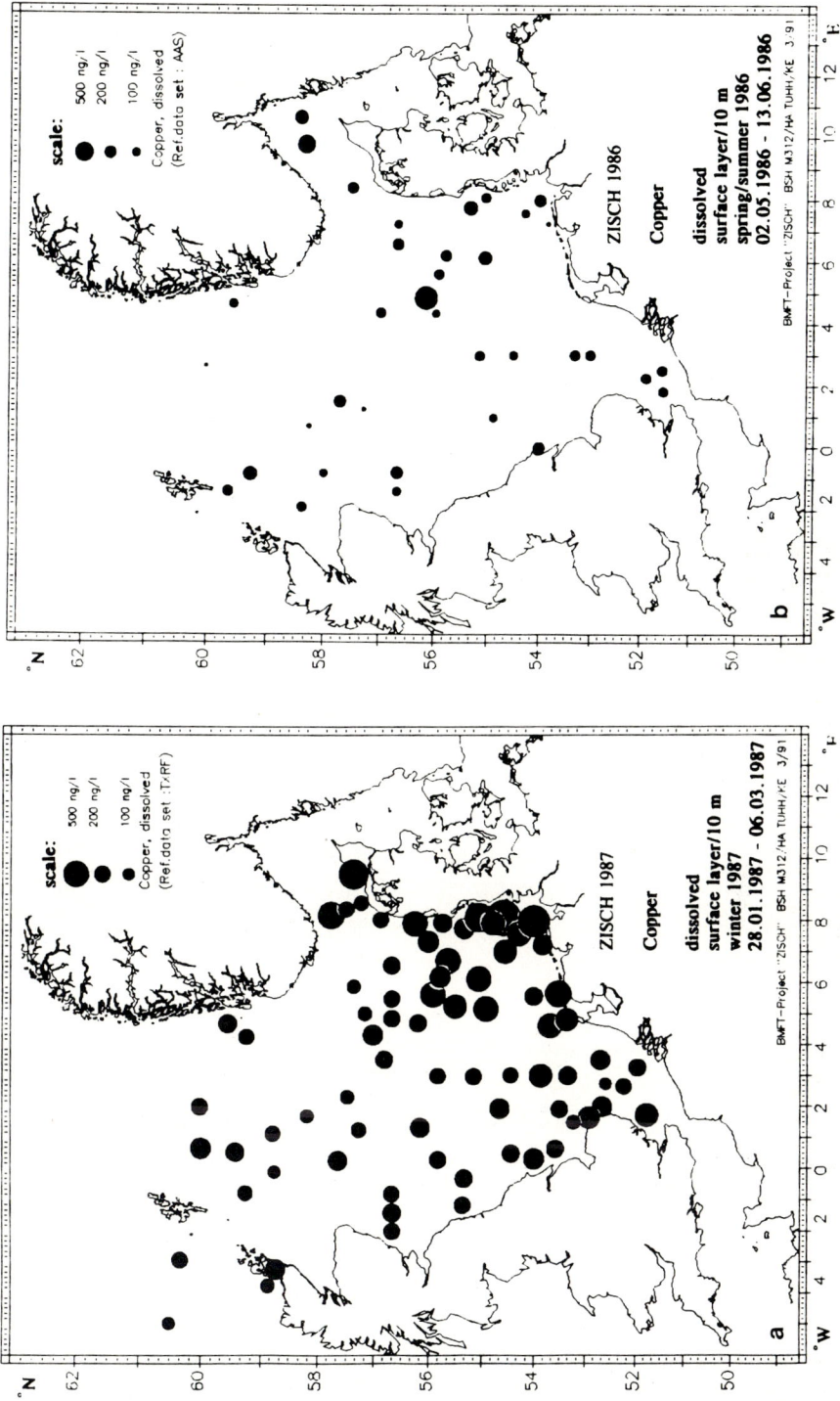

Fig. 4.3.14a,b. Dissolved copper (calculated from total and particulate Cu) during winter (*a*) and spring (*b*)

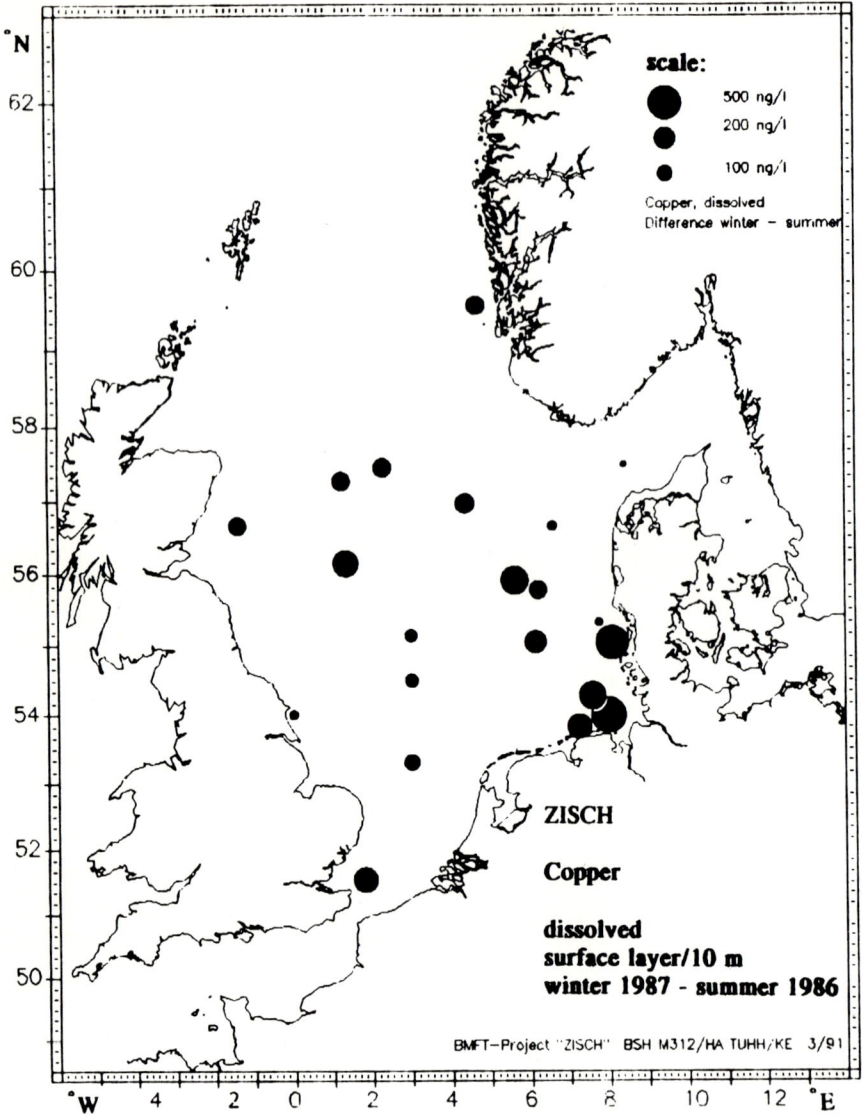

Fig. 4.3.15. Differences of dissolved copper between winter and spring

phosphorus (Fig. 4.3.2) was low. These were the areas in the continental coastal water and the Norwegian coastal waters. This indicated that the more water-soluble γ-HCH (Table 4.3.4) was still abundant where scavenging had already occurred transferring phosphate to the particulate fraction. On the other hand, the concentrations of α-HCH, the degradation- and by-product of γ-HCH which is much less soluble than γ-HCH (Table 4.3.4),

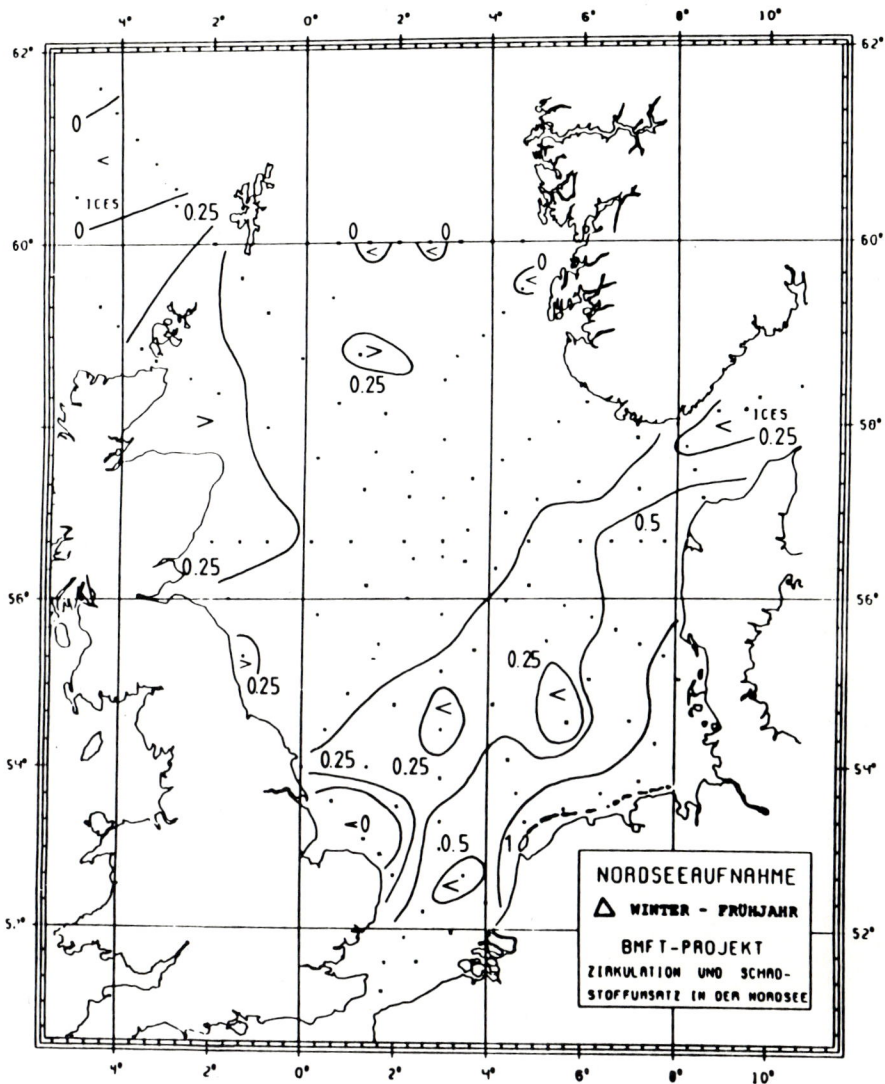

Fig. 4.3.16. Phosphate difference (μM) between spring and winter (integrated values)

were highest around the Norwegian water as well as in the Dogger Bank area. These are regions where production also occurs during winter, enhanced by frequent stratification or shallow water-depths which are not as turbid as the continental coastal zones.

The hypothesis of scavenging α-HCH is supported by the shape of γ/α-HCH ratio near the surface during spring (Fig. 4.3.17) which was similar

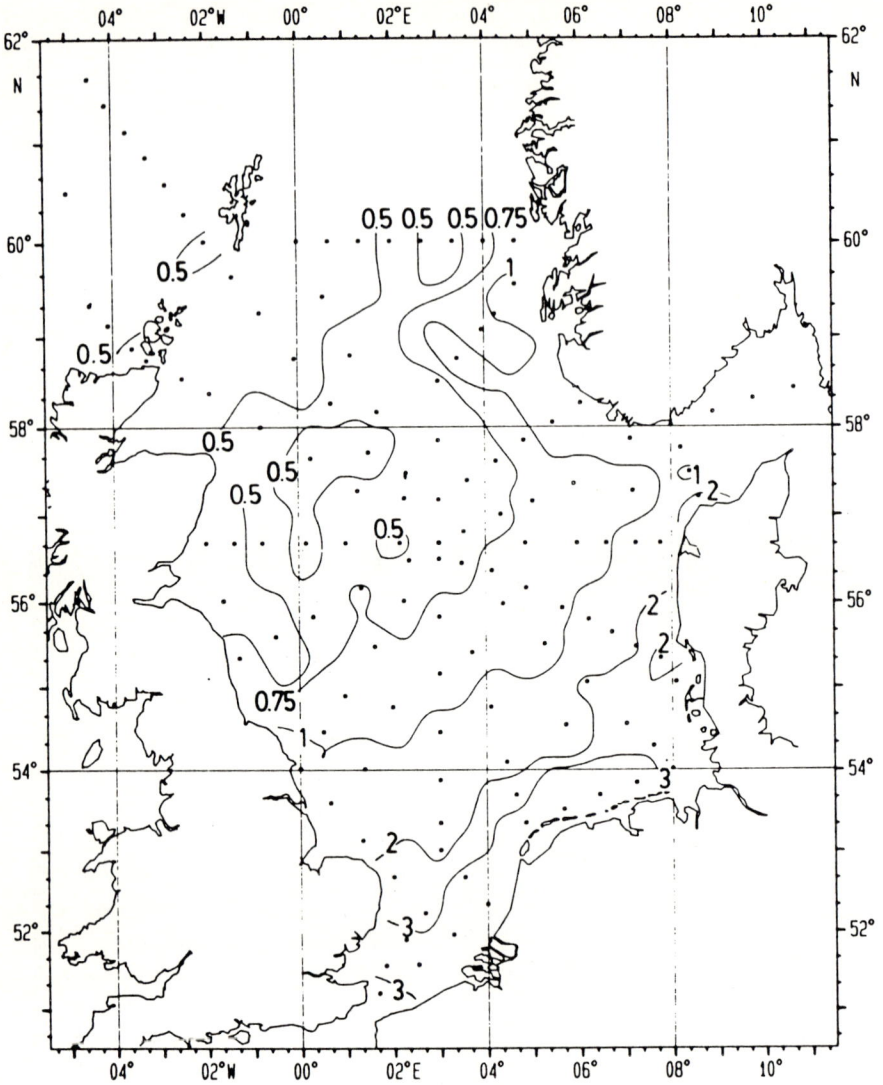

Fig. 4.3.17. γ/α-HCH ratio in the surface layer (5 m) during spring

to the extension of low phosphate percentage. These findings are interpreted such that in the mixed layer the much more water-soluble γ-HCH was not adsorbed by the organic matrices formed during the spring phytoplankton bloom to such an extent as α-HCH which was additionally regionally enriched in areas of increased conversion activity. By these processes the isomerization from γ- to α-HCH may be enhanced by shifting the equilibrium at some spots, removing α-HCH from the water phase.

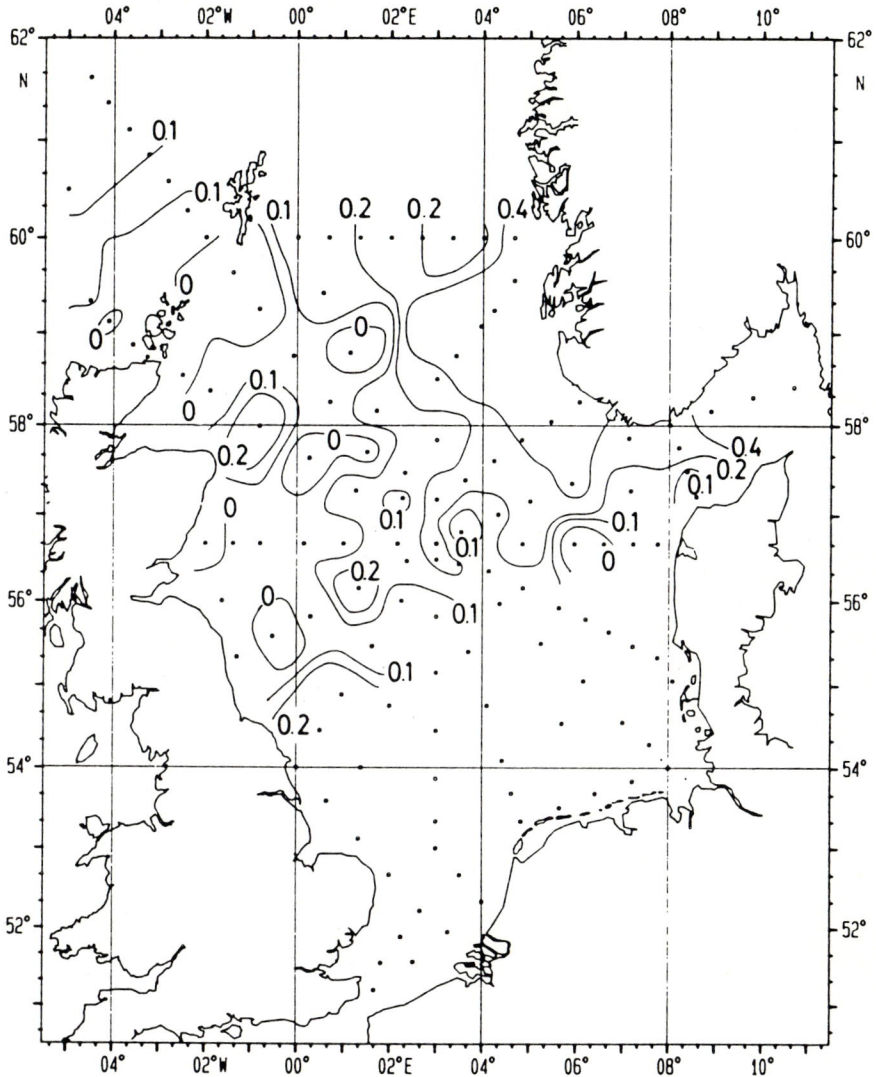

Fig. 4.3.18. γ/α-HCH ratio as difference between mixed layer (5 m) and bottom layer during spring

The differences of γ/α-HCH ratios between upper and lower layer show that in the northwestern North Sea, partly covered by the area of high vertical phosphate differences, α-HCH was reduced in the upper layer (Fig. 4.3.18). Vertical differences of γ/α-ratios of more than 0.2 were also observed during winter close to the Norwegian coast. This area was much less extended than during spring. These results are interpreted such that the

more adsorptive α-HCH is removed from surface water by biogenic scavenging, since the primary production is also occurring during winter in
the stratified waters of the Norwegian coastal current (see Chap. 2.2)
though to a lesser extent, indicated by nutrient minima at the surface.

These relations are, however, masked by topical local sources as for the
γ-HCH and its by-product α-HCH along the south Norwegian coast.

The ratios of γ/α-HCH in the surface sediments are especially low in the
Norwegian trench (Knickmeyer and Steinhart 1988). This corresponds well
to the high concentrations of α-HCH in the water column (Fig. 4.3.18).
Due to year-around production in the region of the Norwegian coastal
current the sedimentation rate is high, causing a "dilution" of ΣPC (see
Sect. 4.3.5), but also an enrichment of less water-soluble organic pollutants, which can be detected well by isomeric ratios.

4.3.4 Correlations Between Nutrients and Contaminants in the Water Phase of the German Bight

Parallel to the decrease of nutrients in the German Bight from January to
May in 1989 (Chaps. 2.4, 2.5), a decrease of heavy metals occurred. This
relationship was most consistent for copper and cadmium (Table 4.3.2).
The complexing capacity is increasing at the same time, which was confirmed by the simultaneous increase of dissolved organic nitrogen compounds,
of which the amino-groups have a high complexing potential (Houghton
1979). This process has been reported also for other coastal areas, like the
Channel (Butler et al. 1979).

The phytoplankton metal contents (Zn, Cd, Pb, and Cu, listed in Table
4.3.3) during April and May were significantly higher than during February. The high lead content (27%) of phytoplankton cells supports the interpretation of the data set from the entire North Sea that, apart from hydrodynamic forces, biological effects also significantly influence the phase
distribution of lead by providing adsorptive matrices and, by this, modify
its transport behaviour.

In May, diatoms were associated with the Elbe river plume in an area of
substantial nitrate input (Chap. 2.4), whereas in the East Frisian coastal
water *Phaeocystis globosa* dominated (Fig. 4.3.19). The slime-producing
Phaeocystis bloom was associated with highly soluble carbohydrates and
the highest content of cadmium was found in the phytoplankton cells (Fig.
4.3.20) due to adsorption by the slime-surrounded colonies, resulting in
low Cd concentrations in the water phase (see Chap. 2.4). Generally, *P.*

Table 4.3.2. Means, minimum and maximum values of metals and nutrients of filtered water in the German Bight (5-10 m depth) from December 1988 to May 1989

	Zinc (ng/l)			Cadmium (ng/l)			Lead (ng/l)			Copper (ng/l)			Speciation (lead) (ng/l)			Nitrate (μmol/l)			Phosphate (μmol/l)			Silicate (μmol/l)		
	x	min	max	x	min	max	x	min	max	x	min	max	x	min	max	x	min	max	x	min	max	x	min	max
Dec.																17.30	1.99	48.23	1.01	0.44	1.87	12.26	4.21	27.90
Jan.	7290	1310	16690	51	11	163	160	4	1060	740	230	1610	-	-	-	37.34	12.09	155.25	1.36	0.26	2.29	19.77	5.44	57.54
Feb.	3700	860	8360	49	0	91	150	0	430	840	200	2280	1900	1450	2300	39.03	18.15	97.88	1.27	0.74	2.05	16.06	5.05	44.40
March	5400	2030	8050	42	16	70	160	0	350	900	280	2010	2500	1830	3430	27.79	12.57	43.35	0.98	0.42	1.40	7.80	1.05	17.68
April	4300	550	8670	43	0	95	30	0	250	530	260	950	3200	2120	3990	32.23	10.60	83.18	0.89	0.47	1.54	7.84	1.15	26.10
May	2060	280	4440	36	2	65	70	0	310	430	170	970	2310	950	4530	19.88	0.09	52.27	0.28	0.02	0.77	1.08	0.12	4.20

Table 4.3.3. Estimation of the metal load of water and phytoplankton in the German Bight on the basis of monthly sampling from January to May 1989. (Basis of estimation: water area = 435 km^2, water volume = 4.35 x 10^{14} l)

	Biomass of phytoplankton (tons carbon)	Zinc			Cadmium		
		Water (tons)	Phytoplankton (tons)	%	Water (tons)	Phytoplankton (tons)	%
January	15,000	3170	--	--	22	--	--
February	34,000	1830	19.4	1.06	20	0.1	0.50
March	--	2260	--	--	17	--	--
April	53,000	1870	42.8	2.24	16	0.2	1.23
May	365,000	953	158.2	14.20	16	1.5	8.57

	Lead			Copper		
	Water (tons)	Phytoplankton (tons)	%	Water (tons)	Phytoplankton (tons)	%
January	70	--	--	323	--	--
February	66	1.6	2.37	395	0.7	0.18
March	66	--	--	447	--	--
April	13	3.5	21.21	218	1.3	0.59
May	29	10.7	26.95	194	5.9	2.95

globosa showed much higher adsorption than the diatom cells (Fig. 4.3.19) relative to their biomass (see Chap. 4.4).

PCP concentrations measured in the German Bight from November 1988 to May 1989 showed highly significant correlations with the nutrients nitrate, silicate and phosphate. These correlations were mainly an effect of dilution, evident by the significant negative correlation between PCP and salinity (r = 0.92). The correlation coefficients were even higher for silicate (r = 0.94) and nitrate (r = 0.93) than for salinity. For PCP the dilution effect is masking any seasonal trend (Hühnerfuss et al. 1990).

4.3.5   Regional Contents of Organic Material and Contaminants in Sediment

The patterns of individual PCB congeners in the sediment differ between depositional and erosional areas. For example, samples from the Skagerrak/Norwegian Trench as well as the inner German Bight show greater amounts of highly chlorinated PCB congeners than samples from the cen-

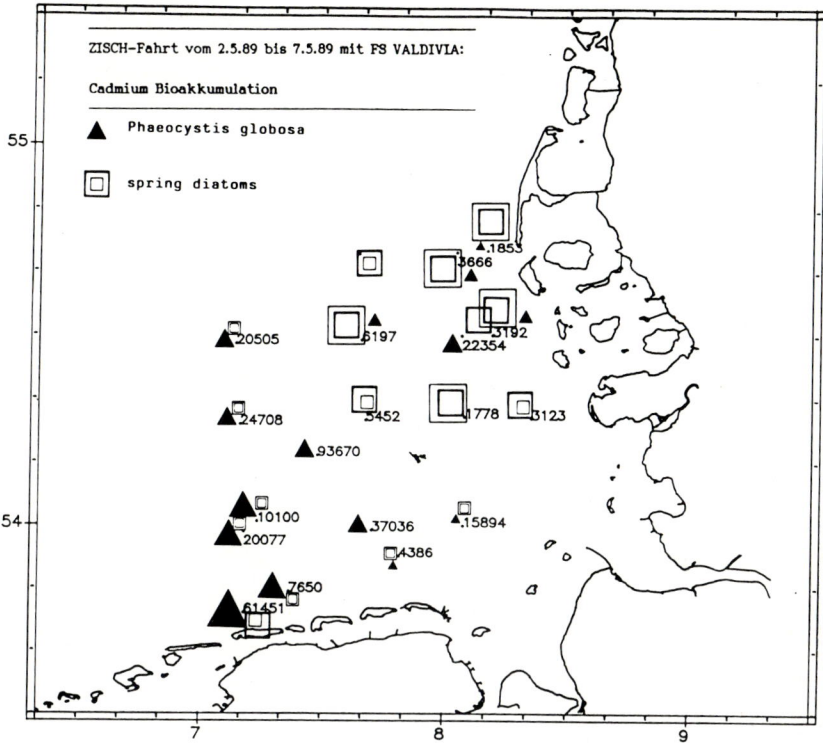

Fig. 4.3.19. Phytoplankton distribution and Cd bioaccumulation in the German Bight

tral North Sea. The geographical differences in organochlorine contaminations were demonstrated relative to the content of total organic matter (TOM) in dried sediments. High concentrations of Σ PCB reflect areas of recent deposition of suspended matter in the North Sea (Fig. 4.3.21). They correlate with the areas of maximum concentration of higher chlorinated PCBs. A correlation of Σ PCB to total organic matter does not show a distinct trend, due to differences in water solubility (Table 4.3.4 and Fig. 4.3.22). Not only in coastal areas with high discharges, also an extended part of the central North Sea is characterized by high Σ PCB concentrations related to organic material. This area is also characterized by ongoing primary production during winter, as indicated by low phosphate and high particulate phosphorous concentrations (Figs. 4.3.2, 4.3.3). On the other hand, in the area of the Norwegian coastal current which is also known for winterly production (Brockmann et al. 1990), high concentrations of organic material in the sediment (Knickmeyer and Steinhart 1989) cause a "dilution effect" of adsorbed Σ PCB. Therefore, it can be assumed that by a

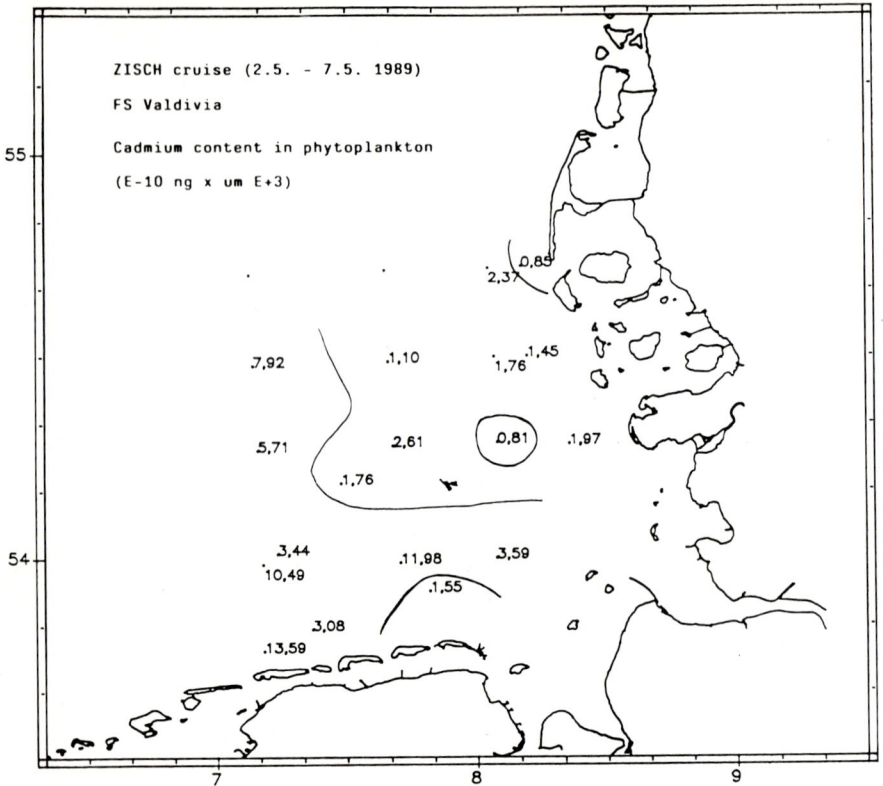

Fig. 4.3.20. Cd content of phytoplankton in the German Bight during May 1989

permanent scavenging, contaminants with a high adsorption affinity will be enriched in the sediment. The reason for this enrichment is that the nutrients in the North Sea are cycled many times during the course of a year (Brockmann et al. 1990; Rick 1990) and, parallel to this process, contaminants are transferred to the particulate phase, and amphipatic compounds are enriched by adsorption.

Hydrophobic sorption of PCBs to extensive surface area, associated with suspended matter coming from the central North Sea, moving on through the continental coastal water northward implies that highly chlorinated molecules can favorably be adsorbed by the particulate phase.

Figure 4.3.23 shows a typical chromatogram with great amounts of highly chlorinated congeners from station 72 in the inner German Bight, which is very similar to sediments from the Norwegian Trench. The reasons for this behaviour in the water phase are in fact (1) the lower water solubility and therefore the tendency of these congeners to hydrophobic sorption on suspended matter (Table 4.3.4) and (2) that in these regions

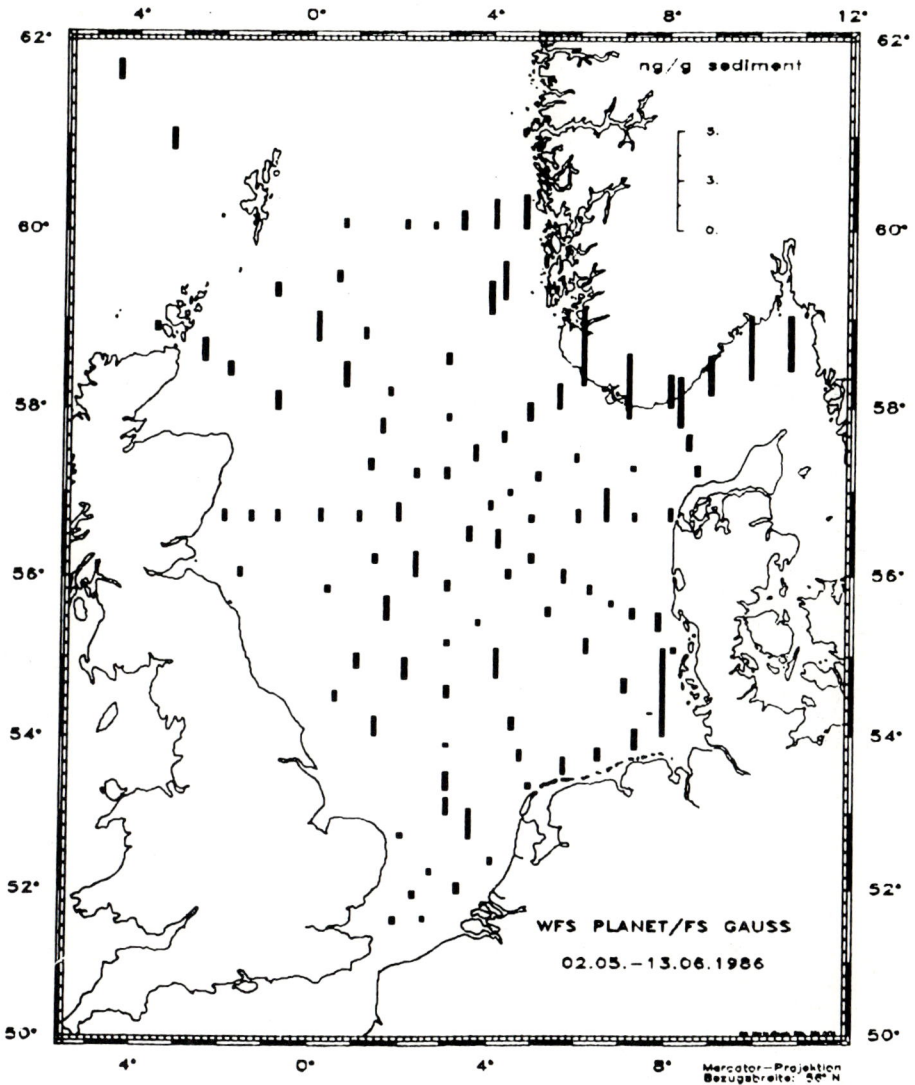

Fig. 4.3.21. Concentrations of ΣPCB (as the sum of 24 components) in sediments throughout the North Sea. Concentration ng PCBs/g sediment

year-round production will offer permanently fresh biogenic adsorptive material for sedimentation.

In contrast, lower chlorinated congeners, which are more soluble in water, exhibit larger dispersion over longer distances, because they remain in solution for a longer time. This is the reason for the greater amounts of lower chlorinated congeners in offshore sediment at station 78 (Fig. 4.3.24).

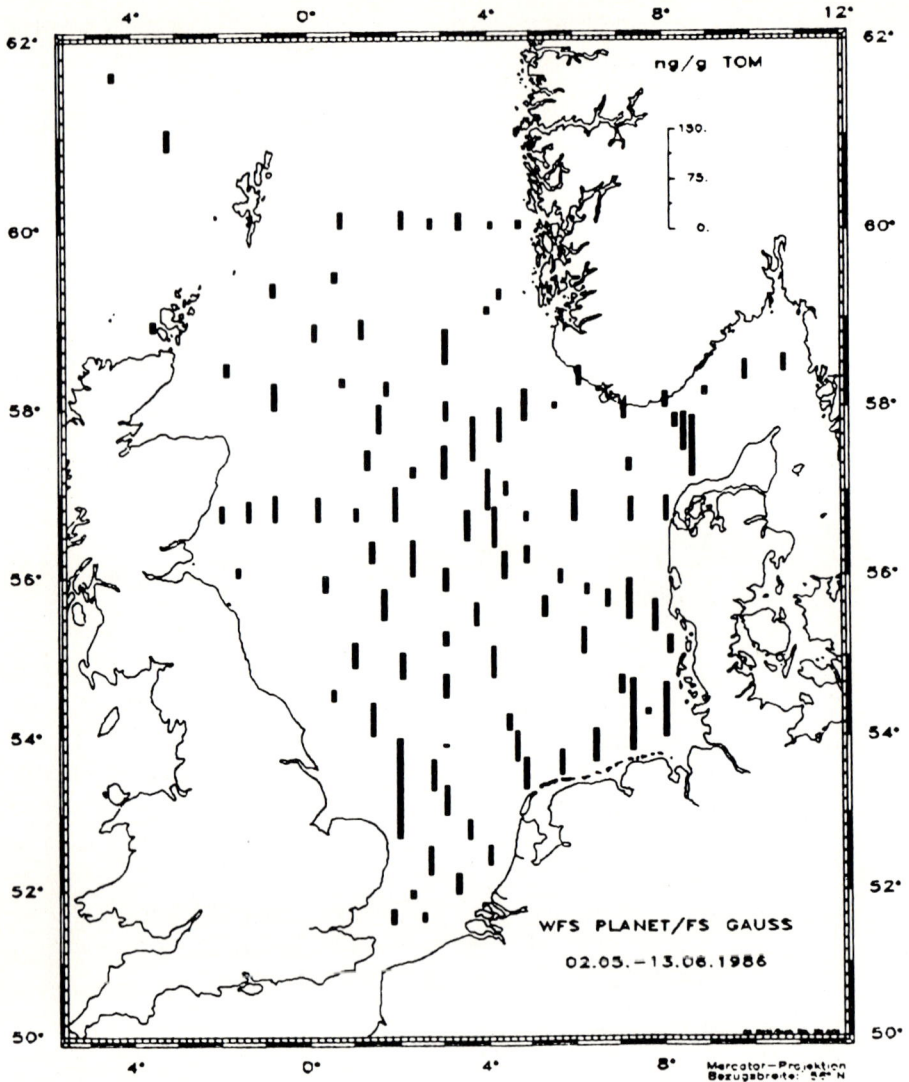

Fig. 4.3.22. Concentrations of ΣPCB (as sum of 24 components) in sediments related to organic matter throughout the North Sea. Concentration ng PCBs/g total organic matter (TOM)

Also, the distributions of p,p'-DDE and HCB reflects areas of recent deposition of suspended matter in the North Sea. The Ems-Dollart estuary is a source of HCB contaminants for the southern North Sea. P,p'-DDE associated with suspended matter from the Baltic Sea presumably accumulates in the Skagerrak/Norwegian Trench. Concentrations of Lindane and

Table 4.3.4. Organochlorines: water solubility correlations with organic material in the sediment.

	p,p'-DDE	PCB 15	PCB 209	HCB	α-HCH	Lindane
Water solubility (25°C;μg/l)	1-10	56	0.02	7.9	2000	7800
Correlation (r) (Sediment/TOM)	0.81	∑PCB 0.68		0.67	0.47	0.32

even α-HCH showed only slight or even no gradients in the investigated area because of their high water solubility (Table 4.3.4).

4.3.6 Conclusions

The effect of scavenging by biogenic material during spring can be postulated for some of the trace metals, especially for cadmium and copper, when data from spring and winter are compared. In order to identify these processes in more detail, it is necessary to measure dissolved and particulate metals from the same sample parallel to nutrients and particulate organic material. The latter gives integrated information on the developmental state of the ecosystem and the scavenging process. The scavenging effect for trace metals and α-HCH was not only shown by comparison of the seasonal distributions, but also by detection of vertical gradients following the spring phytoplankton bloom in stratified areas.

Investigation of the scavenging process in detail would require sampling with a higher resolution in time and space, because the conversion processes in the water column are always interfered with by discharges from the atmosphere, from direct dumping and from coastal sites as well as by remobilization from the sediment. Most of the sources of pesticides are unknown; especially, for the atmospheric input, only rough estimates are available. It can be assumed that the atmosphere is one of the main sources of organic pollutants. Therefore, it is necessary in future investigations to stress the sampling of the air/sea interface and to couple sampling of surface layers and water compartments with measurements of pesticide inputs from the atmosphere.

Despite the complex hydrodynamic situation in the German Bight, the monthly measurements from a station net, including the different water

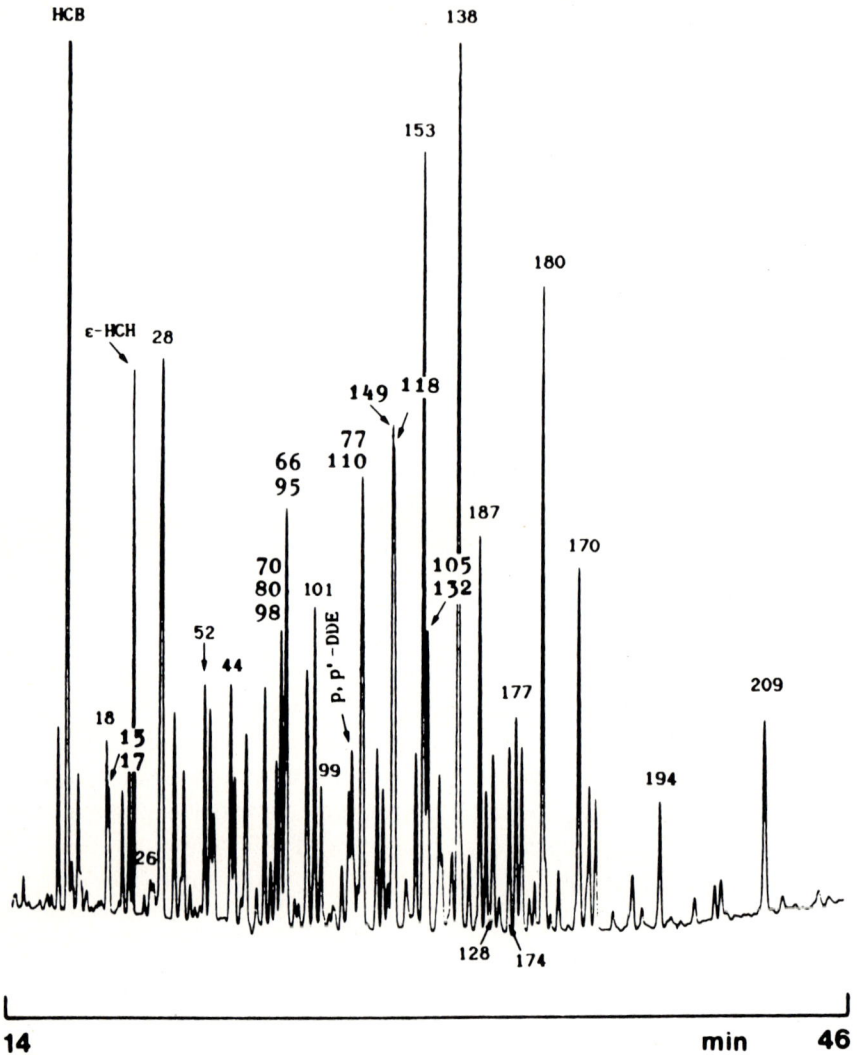

Fig. 4.3.23. Example of a chromatogram of station 72 from the SE 54-CB fused silica column. Components detected in sample (numbering of components according to Ballschmiter and Zell 1980); for co-eluting congeners quantitation is on the basis of the first *(upper)* component given. ε-HCH = internal standard

masses from the coasts, estuaries and central North Sea, indicate parallel trends for nutrients and some dissolved trace elements at the beginning of the spring phytoplankton bloom. Since changing river discharges and variable extension of coastal and river plume water are dominating the dis-

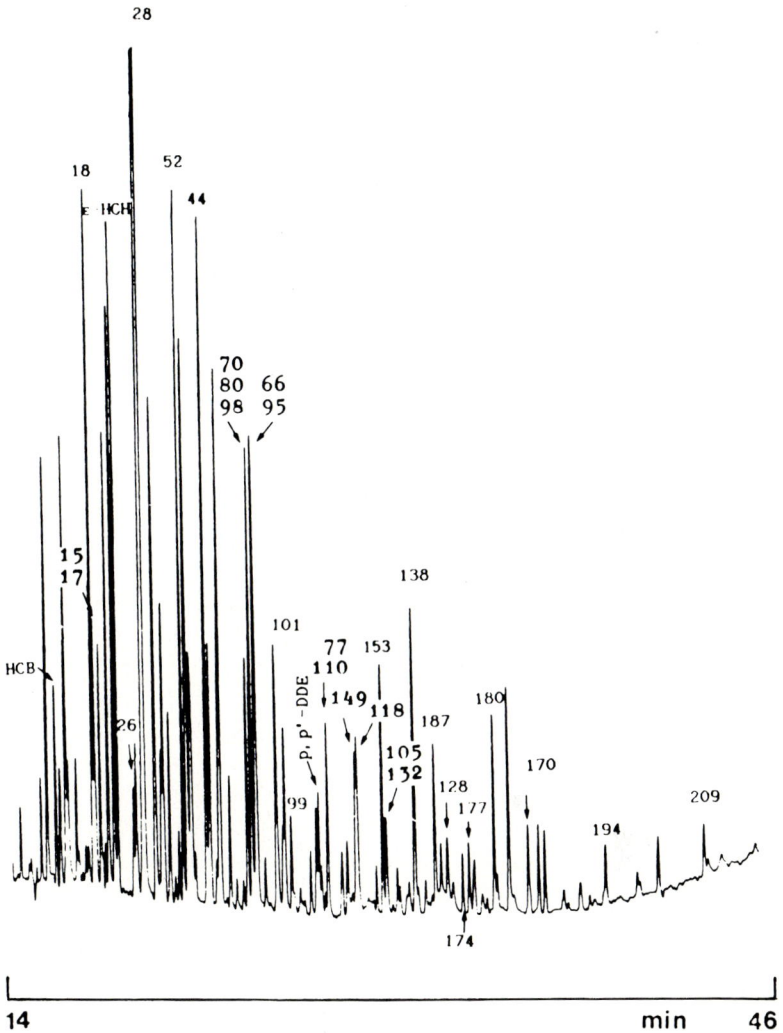

Fig. 4.3.24. Example of a chromatogram of station 78 from the SE 54-CB fused silica column. Components detected in sample (numbering of components according to Ballschmiter and Zell 1980); for co-eluting congeners quantitation is on the basis of the first *(upper)* component given. ϵ-HCH = internal standard

tribution of plankton as well as dissolved compounds, the quantitative relationship will be influenced not only by seasonal biological development, but also by variable dilution processes. Only by extended investigations can results be gained which enable us to distinguish the different influences.

This possibility is already indicated by the relationships found during May, when the biological effects on the trace metal distribution can clearly be evaluated due to the regional dominance of phytoplankton species with different adsorption qualities.

References

Ballschmiter K, Zell M (1980) Analysis of polychlorinated biphenyls (PCB) by glass capillary gas chromatography. Fresenius Z Anal Chem 302:20-31

Beddig S, Sündermann J (1988) Zirkulation und Schadstoffumsatz in der Nordsee. Geowissenschaften 6:207-220

Boyle EA, Huested SS, Jones SP (1981) On the distribution of copper, nickel and cadmium in the surface waters of the North Atlantic and North Pacific Ocean. J Geophys Res 86:8048-8066

Brockmann UH, Kattner G (1993) Nutrient distribution in the North Sea during spring 1986 and winter 1987. Continent Shelf Res(submitted)

Brockmann UH, Laane RWPM, Postma H (1990) Cycling of nutrient elements in the North Sea. Neth J Sea Res 26:239-264

Bruland KW, Franks RP (1983) Mn, Ni, Cu, Zn and Cd in the western North Atlantic. In: Wong CS, Boyle E, Bruland KW, Burton JD, Goldberg ED (eds) Trace metals in the sea water. Plenum, New York, pp 395-414

Butler EJ, Knox S, Liddicoat MJ (1979) The relationship between inorganic and organic nutrients in sea water. J Mar Biol Assoc UK 59:239-250

Calmano W, Ahlf W, Förstner U (1990) Exchange of heavy metals between sediment components and water. NATO ASI Ser 23:503-522

Danielsson LG, Westerlund S (1983) Trace metals in the Arctic Ocean. In: Wong CS, Boyle E, Bruland KW, Burton JD, Goldberg ED (eds) Trace metals in the sea water. Plenum, New York, pp 85-95

Dehairs F, Baeyens W, Van Gansbeke D (1989) Tight coupling between enrichment of iron and manganese in North Sea suspended matter and sedimentary redox processes: evidence for seasonal variability. Estuarine Coastal Shelf Sci 29:457-471

Dicke M, Schmidt D, Michel A (1987) Trace metal distribution in the North Sea. Proceedings, Int Conf Heavy metals in the environment. New Orleans, Sept. 1987, 2:312-314, Edinburgh

Duce RA, Liss PS, Merrill JT, Atlas EL, Buat-Menard P, Hicks BB, Miller JM, Prospero JM, Arimoto R, Church TM, Ellis W, Galloway JN, Hansen L, Jickells TD, Knap AH, Reinhardt KH, Schneider B, Soudine A, Tokos JJ, Tsunogai S, Wollast R, Zhoa M (1990) GESAMP-reports and studies No 38, WMO, Geneva

Duinker JC (1983) Effects of particle size and density on the transport of metals to the oceans. In: Wong CS, Boyle E, Bruland KW, Burton JD, Goldberg ED (eds) Trace metals in the sea water. Plenum, New York, pp 209-226

Freimann P, Schmidt D (1989) Application of total reflection X-ray fluorescence analysis for the determination of trace metals in the North Sea. Spectrochim Acta 44B:505-510

Houghton RP (1979) Metal complexes in organic chemistry. Cambridge University Press, Cambridge, 307 pp

Hühnerfuss H, Dannhauer H, Faller J, Ludwig P (1990) Concentration variations of anthropogenic and biogenic organic substances in the German Bight due to changing meteorological conditions. Dtsch Hydrogr Z 43:253-272

Kersten M, Dicke M, Kriews M, Naumann K, Schmidt D, Schulz M, Schwikowski M, Steiger M (1988) Distribution and fate of heavy metals in the North Sea. In: Salomons W, Bayne BL (eds) Pollution of the North Sea: an assessment. Springer, Berlin Heidelberg New York, pp 300-347

Kersten M, Irion G, Förstner U (1991) Particulate trace metals in surface waters of the North Sea. In: Vernet J-P (ed) Heavy metals in the environment. Elsevier, Amsterdam, pp 137-159

Knickmeyer R, Steinhart H (1988) The distribution of cyclic organochlorines in North Sea sediments. Dtsch Hydrogr Z 41:1-44

Knickmeyer R, Steinhart H (1989) Cyclic organochlorines in North Sea sediments, relation with size and organic matter. Dtsch Hydrogr Z 42:43-59

Kremling K (1985) The distribution of cadmium, copper, nickel, manganese, and aluminium in surface waters of the open Atlantic and European shelf area. Deep-Sea Res 32:531-555

Kremling K, Wenck A, Pohl C (1987) Summer distribution of dissolved Cd, Co, Cu, Mn, and Ni in central North Sea waters. Dtsch Hydrogr Z 40:103-114

Rick H-J (1990) Ein Beitrag zur Abschätzung der Wechselbeziehung zwischen den planktischen Primärproduzenten des Nordseegebietes und den Schwermetallen Kupfer, Zink, Cadmium und Blei auf Grundlage von Untersuchungen an natürlichen Planktongemeinschaften und Laborexperimenten mit bestandsbildenden Arten. Diss RWTH Aachen, 94 pp

Saager PM, De Baar HJW, Howland RJ (1992) Cd, Zn, Ni and Cu in the Indian Ocean. Deep-Sea Res 39:9-35

Santschi PH (1988) Factors controlling the biogeochemical cycles of trace elements in fresh and coastal marine waters as revealed by artificial radioisotopes. Limnol Oceanogr 33:848-866

Schmidt D, Dicke M (1990) Schwermetalle im Wasser. In: Lozán JL, Lenz W, Rachor E, Watermann B, Westernhagen H von (eds) Warnsignale aus der Nordsee. Parey, Hamburg, pp 30-41

Schmidt D, Wendlandt U (1987) Spurenbestimmung von Quecksilber im Meerwasser der Nordsee. In: Welz (ed) 4. Kolloquium Atomspektrometrische Spurenanalytik, Konstanz, April 1987, Überlingen, pp 617-628

Schmidt D, Freimann P, Zehle H (1986) Changes in trace metal levels in the coastal zone of the German Bight. Rapp PV Réun Cons Int Explor Mer 186:321-328

Van Aalst RM, Van Ardenne RAM, de Kreuk JF Jr, Lems T (1982) Pollution of the North Sea from the atmosphere. TNO Rep CL 82/152, Netherlands Organization for Applied Scientific Research, Den Helder, 124 pp

Whitfield M, Turner DR (1987) The role of particles in regulating the composition of seawater. In: Stumm W (ed) Aquatic surface chemistry. Wiley, New York, pp 457-493

4.4 Effects of Abiotic Processes on the Fate of Contaminants

W. Puls, M. Haarich and D. Schmidt

In this Chapter, the abiotic processes relevant for the fate of contaminants are described and their effects are outlined. There will be first a survey of these processes, and thereafter the results of a numerical simulation concerning the fate of lead in the North Sea will be shown. The modelling is based on the results of Chapter 3.2. The geographic locations mentioned here are given in Fig. 3.2.11.

4.4.1 Abiotic Processes

4.4.1.1 Input

The fate and the environmental impact of a contaminant depend (a) on the spatial and (b) on the temporal input distribution.

The spatial distribution concerns the question whether the input comes mainly from the atmosphere (diffusive input) or from rivers or dumping (point sources).

The temporal distribution concerns (1) the atmospheric input: according to Petersen et al. (1989) there is, for instance, a minimum deposition of lead into the North Sea in April, the time of phytoplankton blooms, (2) the input from estuaries with a turbidity maximum: at low fluvial discharge the estuary is a trap for pollutants bound to particles, while at high fluvial discharge the turbidity maximum is pushed out of the estuary, leading to a

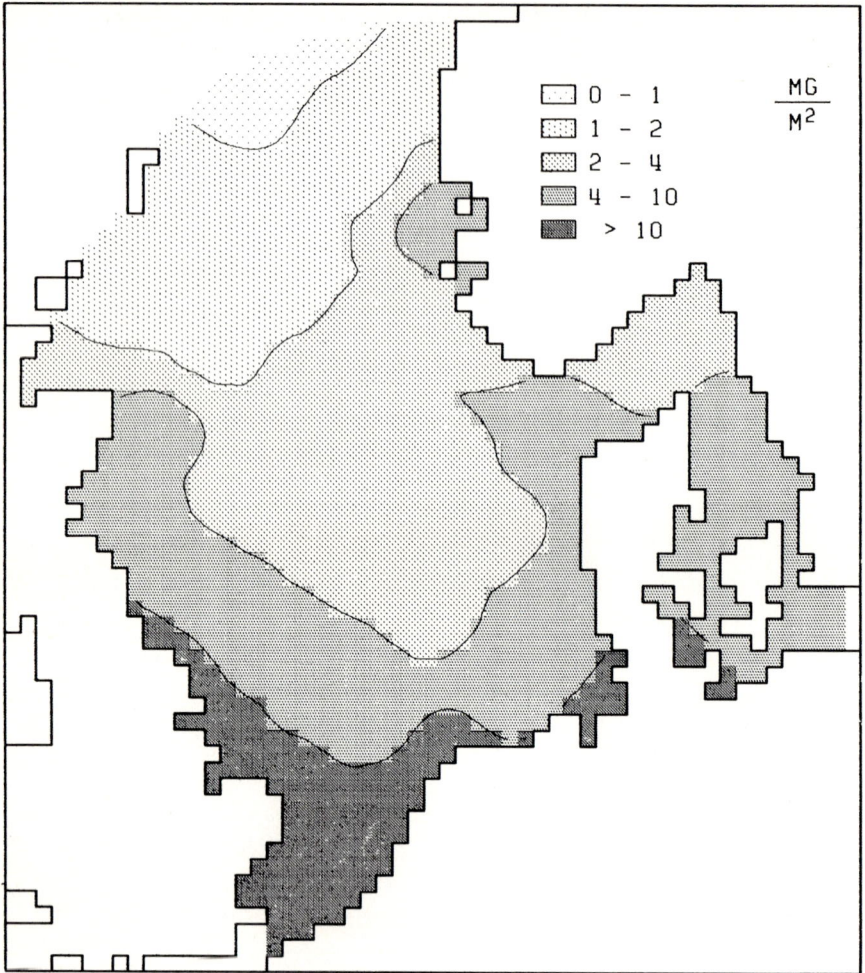

Fig. 4.4.1. Deposition of lead from the atmosphere into the model area, computed for 1985

sudden strong input of contaminants into the sea (Flügge 1988), and (3) the input from Wadden Seas in the stormy season.

A second temporal aspect is the long-term distribution, e.g. the reduction of lead input since 1986. The question arising then is: how much and how fast is the pollution level in the sea decreasing?

In the following, the lead input to the North Sea is given in some detail, also with regard to the model simulations given in the second part of this Chapter.

Input of Lead from the Atmosphere. The input of lead into the North Sea is mainly due to deposition from the atmosphere. A major source of lead in the environment are the emissions of vehicles.

The main supplier of the North Sea with lead from the atmosphere is the UK: 54 % of the total deposition in 1980 (Petersen et al. 1989). Next in the order of rank are France (11.5 %), the Netherlands (10 %) and former West Germany (9 %).

The input of lead into the North Sea has decreased during the last 20 years as a consequence of introducing (1) low lead and (2) unleaded gasoline. This decrease is not uniform: concerning its beginning and its intensity, there are great regional differences within Europe.

The trend towards low-lead gasoline in Europe began in 1972 in the FRG, resulting in a reduction of Pb consumption as leaded gasolines from a maximum value of 10,000 tons in 1971 to 3400 tons in 1984.

The introduction of unleaded petrol started in 1985. The lead concentration in air at Hoek van Holland decreased by a factor of 0.75 from 1985 to 1986 and by a factor of 0.63 from 1985 to 1987 (van Daalen, pers. comm. 1990). The same reductions were obtained from measurements of the annual deposition of lead near Cologne (Schladot et al. 1990).

The North Sea input of lead from the atmosphere used by the model was calculated for the year 1985 by the trajectory model described in Petersen et al. (1989). Figure 4.4.1 shows the distribution of the atmospheric lead input into the model area as it was simulated for the year 1985; the total deposition into the model area is 2870 t. This distribution differs from that given in Fig. 3.2.10a, Chapter 3.2.

The lead input from the atmosphere in 1986 and during the first 3 months of 1987 is taken to be 90 % of the 1985 values. This general value is a rough estimate, taking into account (1) the decrease at Hoek van Holland and near Cologne given above, (2) an (optimistic) decrease by the factor 0.95 for UK and (3) the above given "atmosperic lead input quota" of each nation, computed by Petersen et al. (1989).

Lead Input from Land. The input of lead from rivers, eroded cliffs and by dumping is given by the following numbers (in tons per year):

Elbe: 160	Forth: 30
Thames: 20	Holderness cliffs: 10
Tyne/Tees: 50	Rhine/Meuse/Scheldt: 200
Norfolk cliffs: 8	London sewage sludge: 90
Weser: 40	North England sewage sludge: 10
Humber: 170	North England fly ash: 14

Most of the locations mentioned in this Chapter are shown in Fig. 3.2.11 of Chapter 3.2. The input numbers are taken from different sources: Rat von Sachverständigen für Umweltfragen (1980), Wood and Franklin (1983), Carlson (1986), Bundesminister für Umwelt, Naturschutz und Reaktorsicherheit (1987), Hupkes (1990). If it was deemed necessary, the literature numbers were modified, taking into account harbour dredging and subsequent dumping on land.

Lead input due to dumping of dredged spoils (2000 t/a according to the Bundesminister für Umwelt, Naturschutz und Reaktorsicherheit 1987) is no real input because the most part by far of the dredged material comes from the North Sea itself (e.g. Malle 1987). Of no present environmental importance is the lead input due to dumping of colliery waste off North England (about 200 t/a) because colliery waste consists of rocks with only a small portion of fine material.

The seasonal variation of lead input from land is the same as that of suspended matter input, see Section 3.2.3.7.

Lead Input from Seaward Boundaries. The input of lead through the seaward model boundaries is defined by prescribing at the boundaries either the total (dissolved plus particulate, see Sect. 4.4.1.2) lead concentration Pb_{TOT} (Pentland Firth: 67 ng/l; Atlantic between Orkney and Norway: 55 ng/l) or the lead concentration Pb_{SPM} on SPM (64 mg/kg in Dover Strait, 100 mg/kg for the Baltic Sea). The data were taken from measurements during the ZISCH star cruise 1987, see Chapter 2.3. Lacking Baltic Sea data, a typical lead concentration on the Baltic Sea sediment (Irion 1984) was assumed as a boundary condition.

With the above numbers, there is the following net input of lead (both particulate and dissolved) across the seaward boundaries in 1986: Dover Strait: 1240 t, Baltic Sea: 100 t, Atlantic: 1720 t. Considering seaward boundaries only, there is thus a net output of lead from the North Sea to the Atlantic. Section 4.4.2.4 gives a comprehensive estimation of the North Sea lead budget.

A remark: standing alone, the input numbers given above are not meaningful - they must always be regarded together with the SPM input numbers given in Section 3.2.3.7. For example: The concentration "lead on SPM" is only 20 mg/kg for the input from the cliffs, which is assumed to be the natural background concentration for the sediment fraction < 20 µm. Considering that the lead concentration on SPM and on fine sediment in the southern North Sea is more than twice as large as 20 mg/kg, the input from the cliffs leads to a dilution of lead concentration in the North Sea. An opposite example is the input from the atmosphere: be-

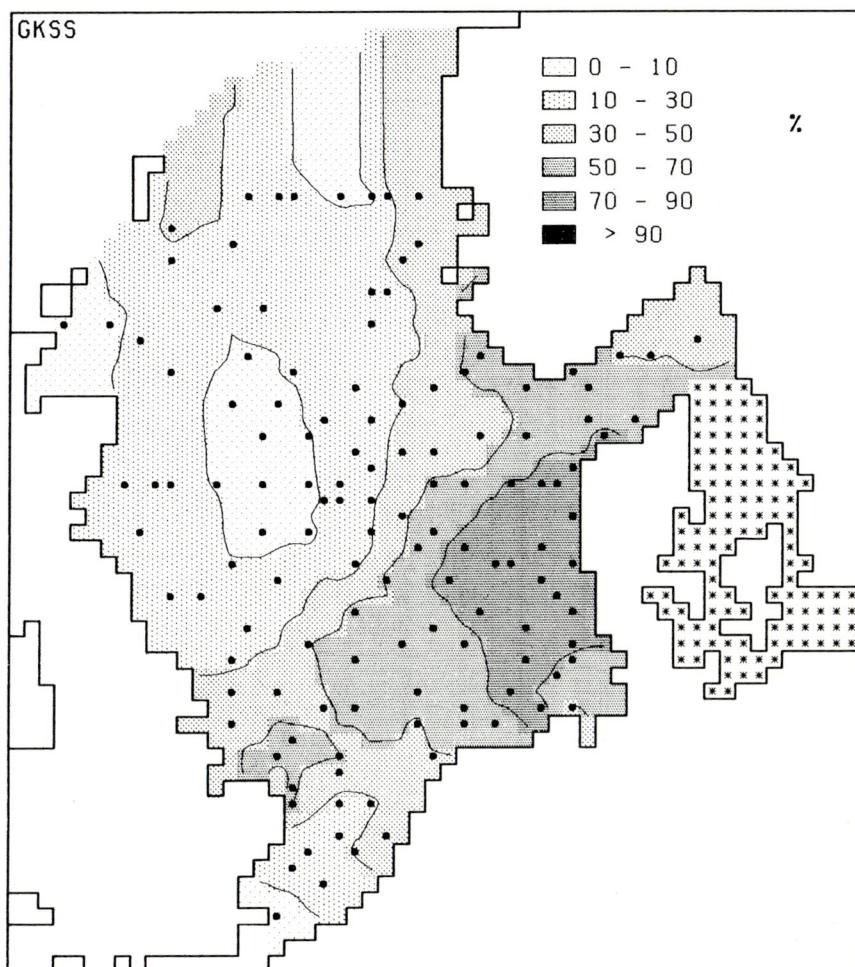

Fig. 4.4.2. Percentage proportion of autochthonous (eroded) Pb for the computed total Pb concentration Pb_{TOT} which is shown in Fig. 4.4.6

cause the model does not consider an atmospheric input of SPM (which is the small amount of about 1.6 million tons per year according to Eisma and Irion 1988), the lead from the atmosphere always means an increase of the lead concentration in sea water.

Some additional numbers for the background concentration of lead: Based on the grain size fraction < 20 μm, Irion (1984) gives a background value for the western Baltic Sea of 25 mg/kg, the same value is given by ARGE Elbe (1988) for "Elbe-SPM". Also, for grain sizes < 20 μm,

Kersten (1988) and Albrecht (pers. comm. 1990) measured Pb concentrations in the Fladen Ground sediment between 10 and 30 mg/kg.

4.4.1.2 Distribution Between Dissolved and Particulate Phase

A characteristic of a contaminant in natural waters is its distribution between dissolved and particulate phases, which depends on the distribution coefficient K_d and on the concentration of suspended particulate matter (SPM) in the water. The relations governing this distribution are given here, taking lead as "the contaminant". $K_d(Pb)$ is defined as (e.g. Balls 1988),

$$K_d(Pb) = \frac{Pb_{SPM} \cdot 10^6}{Pb_{DISS}} \tag{1}$$

where Pb_{SPM} is the lead concentration on the particulate material (mg lead per kg suspended matter) and Pb_{DISS} is the dissolved lead concentration (ng lead per kg sea water). From Pb_{SPM} and the SPM concentration c (mg/l), the particulate phase lead concentration in the sea water $c \cdot Pb_{SPM}$ (ng/l) is determined.

The total lead concentration Pb_{TOT} ($= c \cdot Pb_{SPM} + \rho_w \cdot Pb_{DISS}$, determined from "unfiltered" sea water, see Chap. 2.3) is connected with Pb_{DISS}, $c \cdot Pb_{SPM}$ and Pb_{SPM} via

$$\frac{Pb_{DISS}}{Pb_{TOT}} = \left(\rho_w + \frac{c \cdot K_d(Pb)}{10^6} \right)^{-1} \tag{2}$$

$$\frac{c \cdot Pb_{SPM}}{Pb_{TOT}} = \left(1 + \rho_w \frac{10^6}{c \cdot K_d(Pb)} \right)^{-1} \tag{3}$$

$$\frac{Pb_{SPM}}{Pb_{TOT}} = \left(c + \rho_w \frac{10^6}{K_d(Pb)} \right)^{-1} \tag{4}$$

where ρ_w is the sea water density ($\approx$ 1 kg/l).

From the measurements during the ZISCH star cruise 1987 the value $K_d(Pb) = 1.4 \times 10^6$ was determined.

4.4.1.3 Transport as Dissolved Matter

A contaminant existing totally in the dissolved phase is subject to only one abiotic process (disregarding diffusive exchange with the sediment pore water): the transport with the sea water. The water movement and the concentration patterns of dissolved substances in the North Sea are discussed in Section 3.2.2.

The turnover time of water masses (see Chap. 2.1) is important for the level of contamination. This is best illustrated by an example:

A water mass starts at the Dover Strait with a total lead concentration $Pb_{TOT} = 50$ ng/l. The water mass travels to the German Bight within 60 days, the average water depth is 30 m. Assuming a constant atmospheric Pb input of 10 mg Pb per m^2 and year (see Fig. 4.4.1), the Pb_{TOT} value in the water mass increases continuously, reaching a value of 105 ng/l in the German Bight. If, however, the transport period would not be 60 but 120 days, the total lead concentration in the German Bight would be 160 ng/l.

4.4.1.4 Transport on SPM

Contaminants on SPM are subject to the same transport processes as the SPM itself. Apart from the transport with the water, these processes are erosion, settling and deposition - they are described in Chapter 3.2.3. Their significance for the North Sea is shown later in this Chapter, where the results of the lead transport simulation are presented.

Exchange of Contaminants between the Water Body and the Sediment. The lead concentrations Pb_{SPM} and Pb_{SED} (lead concentration on the sediment with grain sizes < 20 μm, given in mg/kg) control the contaminant's exchange between the water body and the sediment in sandy areas of the southern North Sea, where erosion and deposition of fine sediment are dominating processes.

If Pb_{SPM} and Pb_{SED} in a certain area are different, the processes of (1) erosion of fine sediment, (2) equalization of the Pb_{SPM} values of resident and eroded SPM and (3) re-deposition of the eroded SPM will lead to a mutual adjustment of Pb_{SPM} and Pb_{SED}.

The efficiency of the contaminant exchange between the water body and the sediment is demonstrated by a result of the model simulations. The model distinguishes between (1) allochthonous lead which has entered the North Sea from rivers, cliff erosion, dumping, adjacent seas and the atmosphere, and (2) autochthonous lead which originates from the North Sea sediment. Due to continuous erosion and deposition of fine sediment and due to the exchange of Pb (1) between the dissolved and the particulate phase and (2) within the SPM itself, there is a gradual increase of autochthonous Pb and a gradual reduction of allochthonous Pb in the sea water of the southern North Sea from west to east (= direction of the main residual current). This is shown in Fig. 4.4.2, which gives the percentage proportion of autochthonous Pb for the total Pb concentration Pb_{TOT}. The patch with a percentage < 10 % in the NW central North Sea is due to the local input of atmospheric (i.e. allochthonous) Pb.

Accumulation of SPM in the Coastal Zone. An accumulation of SPM means also an accumulation of particle bound contaminants. Kersten et al. (1990) conclude that the increased SPM concentration in the coastal zones of the southern North Sea is "the last barrier for anthropogenic pollutants between their source and the open sea".

Three processes are outlined that produce an accumulation of SPM in the coastal zone:

1. The starting point of the first process is a low nearshore water salinity. The horizontal difference in water salinity between the nearshore and the offshore water causes horizontal pressure differences which are highest near the bottom. This induces a residual onshore bottom current. For compensation, there is an offshore directed surface current.

SPM concentration is (due to its settling) higher near the bottom than at the surface. The combination of (1) the residual water circulation and (2) the vertical distribution of SPM concentration leads to an enrichment of SPM in the coastal zone. The above-described water circulation is best developed in the vicinity of high fresh water inputs, e.g. the Dutch coast north of the Rhine mouth.

2. Offshore winds push surface water away from the coast - for compensation there is a landward component in the bottom water. Just as with the density-driven water circulation, the wind-driven currents lead to an enrichment of SPM and of particulate contaminants in the coastal zone. Because of the predominance of westerly winds, this mechanism mainly affects the English and the Scottish coast (Lee 1980).

3. Eisma and Irion (1988) also report a net transport of SPM inward into the German Bight, caused by asymmetries of flood and ebb currents near the bottom

The enrichment of SPM in a certain area, however, does not mean that there must also be SPM sedimentation. Whether or not sedimentation takes place depends on the local wave and current conditions.

Accumulation of SPM in the Wadden Sea. The Wadden Sea is a deposition area for SPM and, thus, also for contaminants. Three main mechanisms lead to SPM accumulation in the Wadden Sea:

1. The first mechanism is due to a residual density driven water circulation which is produced by the same factors as the vertical water circulation in the coastal zone.

2. The primary cause for the second mechanism is the asymmetry of tidal currents in shallow water which affects SPM transport as follows:

At the end of the ebb phase, the ebb current decreases quickly, being immediately followed by a rapid increase of the flood current. Due to the short ebb slack water period (assuming that there is no emerging shoal), only a small percentage of SPM settles to the bottom. The developing flood current soon re-suspends the non-consolidated settled mud and transports SPM landward.

At the end of the flood phase the flood current decreases gradually, being followed by a slowly rising ebb current. The flood slack water period is thus rather long, allowing particles to settle to the bottom and to consolidate. The developing ebb current re-suspends the consolidated mud only after some time (if at all): there is a substantial time lag between the begin of the ebb current and the time of re-suspension.

Summing up, the described processes cause SPM to be carried over a longer distance landward with the flood current than seaward with the ebb current. The processes belong to the field of "tidal pumping"; other key words are "settling lag" and "scour lag" (e.g. Postma 1981; Uncles et al. 1985 or Dyer 1988).

3. Van Straaten and Kuenen (1958) point to a third SPM accumulation mechanism in Wadden Seas that complements the second mechanism: around high tide ($\approx$ flood slack time) the water spreads horizontally on the tidal flats, the water depth is small. This means that particles reach the bottom within a short time. Around low tide ($\approx$ ebb slack time) the same

water body has moved to the deeper tidal channels where SPM needs a longer time to settle out.

The deposition of mud (plus particulate contaminants) in the Wadden Sea takes place in the very nearshore parts where waves and currents are weak, or in areas which are sheltered from the prevailing westerly winds. Eisma and Irion (1988) estimate the net annual mud deposition in the Wadden Sea and the Wash to be about 5×10^6 t.

Erosion of mud on the tidal flats takes place essentially by strong wave action during storms.

Accumulation of Marine SPM in Estuaries. SPM in estuarine turbidity maxima is to a large extent of marine origin (e.g. Schwedhelm et al. 1988). The transport effects responsible for this are (1) the vertical density driven current circulation in the brackish water zone and (2) tidal pumping. For the North Sea, Eisma and Irion (1988) specify the annual deposition of marine SPM in estuaries as 1.8×10^6 t. This, of course, means also the deposition of a corresponding amount of contaminants.

4.4.1.5 Processes in the Sediment

As a result of the exchange between the water body and the sediment, contaminants can be enriched in the sediment, which is then a reservoir for particle reactive elements. The average residence time of a contaminant in the sediment is higher than the North Sea water turnover time (maximum: about 3 years) and is thus the determining factor for the long-term existence of contaminants in the North Sea.

The fate of a contaminant within the sediment depends on the processes (1) bioturbation of sediment particles as carriers of contaminants and (2) transport of dissolved contaminants by diffusion or by pore water movement. These processes are treated thoroughly by Kersten (1988), together with the transfer of contaminants from the sediment pore water to the water body (remobilization).

4.4.2 Numerical Simulation of Lead Transport in the North Sea

In this Chapter, only the fate of lead in the North Sea is simulated. Lead was chosen because

1.   Pb has a high affinity to the particulate phase.

2.   According to Sect. 2.3.2, Pb is "predominantly enriched
        by organic matter but bound to the mineral oxide
        components of SPM",
3.   Pb shows the highest enrichment factors in the fine fraction
        of the present-day North Sea sediments compared with the
        mean concentration in pre-industrial time (Kersten and
        Klatt 1988; Irion and Müller 1990).

The modelling of the fate of a contaminant with a high SPM affinity must
start with modelling the fate of SPM itself. This model is described in
Sect. 3.2.3. The time period that is simulated with the model is June 1984
- March 1987.

Generally, the same tracer model as that for simulating the fate of SPM
is used for the particulate and the dissolved phase of lead. The lead that is
adsorbed to SPM undergoes the same transport processes as the SPM itself.
In particular, this means the deposition, the mixing in the sediment by
bioturbation and the erosion of lead. Remobilization is not simulated
because it is insignificant for Pb. It must be stressed that the SPM model
of Sect. 3.2.3 does not include phytoplankton; the consequences will be
discussed in Sect. 4.4.2.4. Further, the model is not able to simulate the
accumulation of SPM in coastal zones, Wadden Sea and estuaries.

The movement of dissolved lead is simulated by separate tracer particles
which have no settling velocity.

A dynamic equilibrium is assumed for the lead concentration between
dissolved and particulate phases, using the distribution coefficient $K_d(Pb)$
$= 1.4 \times 10^6$. The model uses this value for determining ("updating") the
distribution of the lead concentration between the dissolved and the
particulate phase. The updating is done every 2 days for each single box of
the model, using the total lead concentration Pb_{TOT} and the SPM
concentration c in the box.

4.4.2.1 Lead in North Sea Sediment

Figure 4.4.3 shows the measured distribution of lead concentration Pb_{SED}
(in mg/kg) in the fine sediment fraction (< 20 μm) in the upper 0.2 m of
the sea floor. Highest concentrations were measured (1) off North England
between Tees and Humber and (2) in the central North Sea.

Generally, the lead content in the fine sediment fraction (see Fig.
3.2.12, Chap. 3.2) is lower in muddy than in sandy areas.

The distribution shown in Fig. 4.4.3 gives the impression of a complete

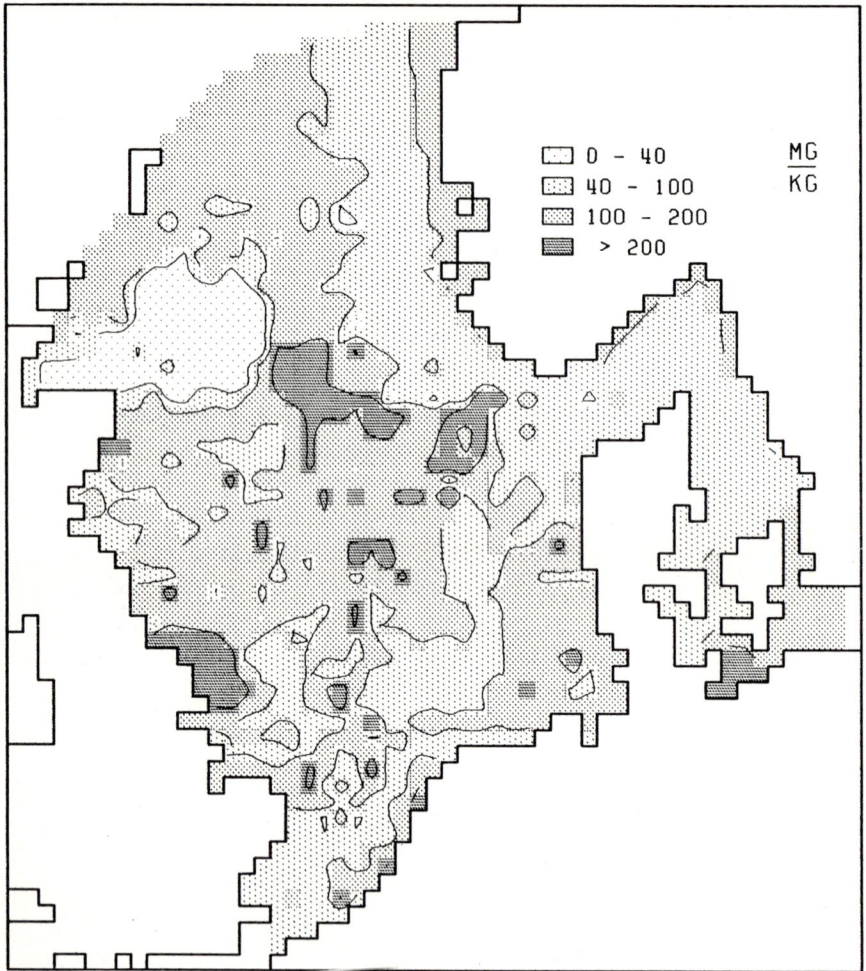

Fig. 4.4.3. Distribution of lead content Pb_{SED} in the fine sediment fraction ($<$ 20 µm) in the upper 2 cm of the bed, based on data of Kersten (Sect. 2.3.3, Chap. 2.3) and Albrecht (pers. comm. 1990); additional information is taken from Irion and Müller (1990)

knowledge of lead in the North Sea sediment. This impression is wrong; knowledge is still poor, especially in the northern North Sea.

The Pb mass distribution given in Fig. 4.4.4 is the product of the factors (1) sediment dry density ρ_{dry}, see Sect. 3.2.3.3, (2) the fine sediment fraction $c_{20\mu m}$ given in Fig. 3.2.12 of Chapter 3.2, (3) Pb_{SED} given in Fig. 4.4.3 and (4) the sediment depth 0.2 m.

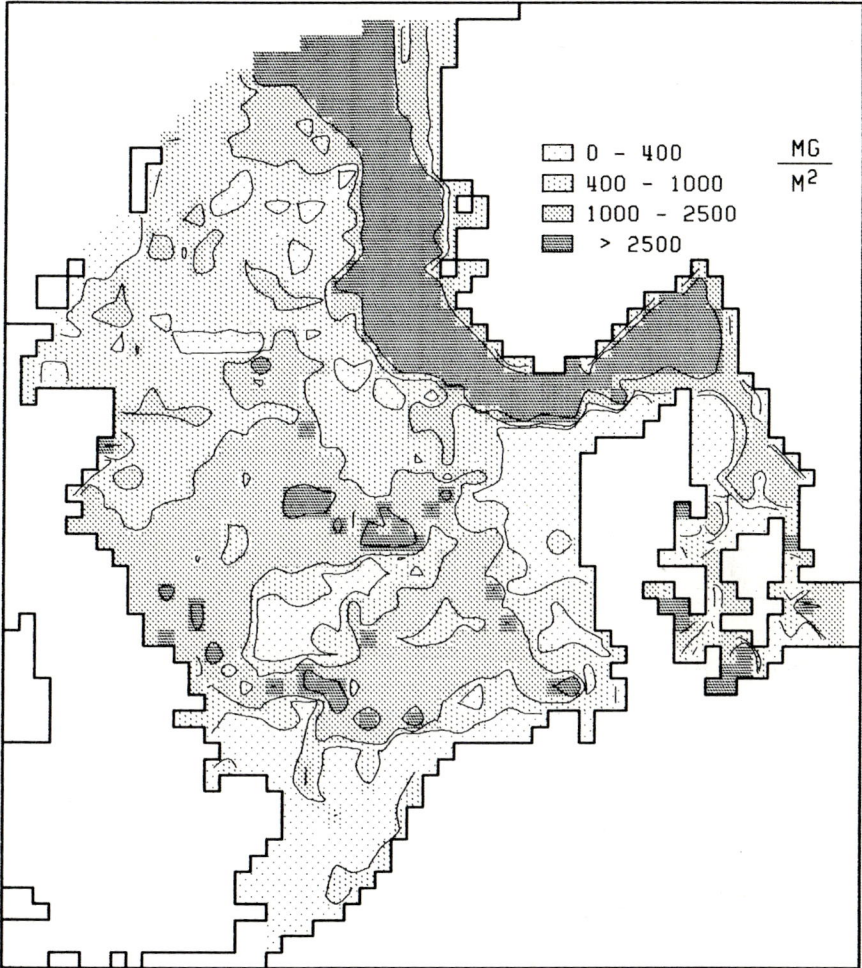

Fig. 4.4.4. Distribution of Pb mass in the sediment in the upper 20 cm of the sea bed

A remark: assuming a natural Pb background concentration of $Pb_{SED} =$ 20 mg/kg (see Chap. 2.8), the Pb mass in the upper 0.2 m of the Skagerrak is about 1000 mg/m^2. Thus, a high Pb mass in the sediment must not automatically mean a hazard for the environment.

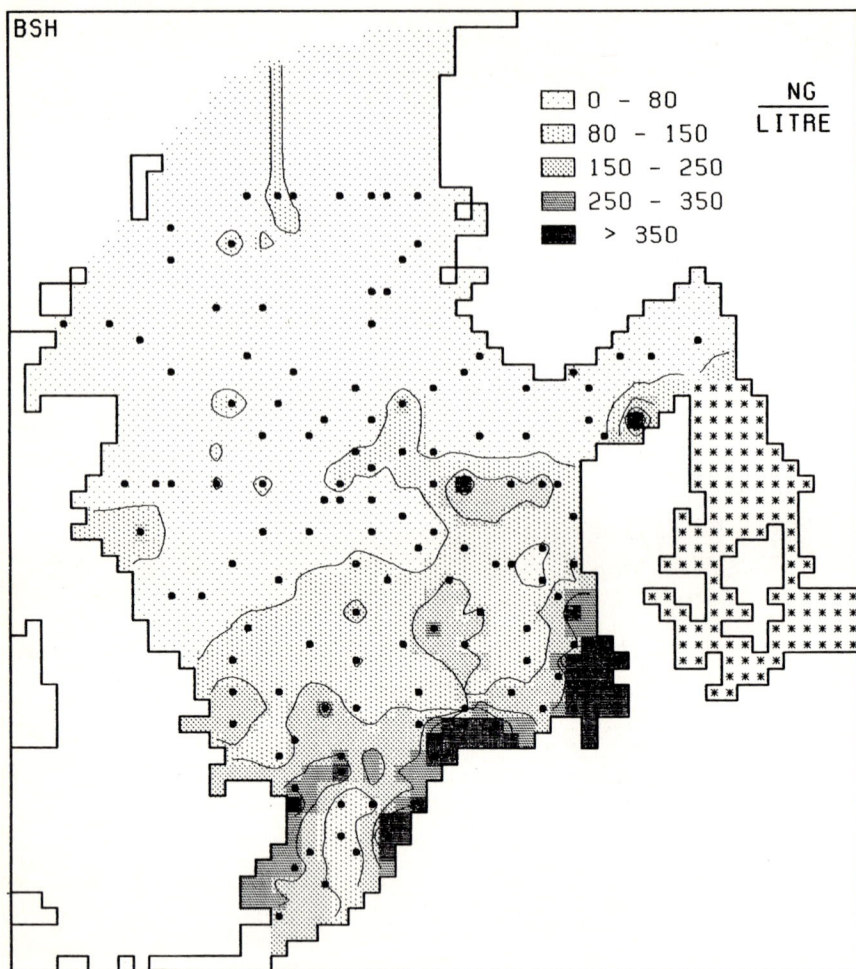

Fig. 4.4.5. Measured distribution of total lead concentration Pb_{TOT}, star cruise in January - March 1987, water depth: 10 m

4.4.2.2 Lead in North Sea Water

In this section, measured distributions of lead concentrations in the North Sea are compared with results computed by the model.

The measured Pb data is (1) the total lead concentration Pb_{TOT} given in Fig. 4.4.5 and (2) the lead concentration on SPM (= particulate lead concentration Pb_{SPM}) given in Fig. 4.4.11. The data was measured in a water depth of 10 metres.

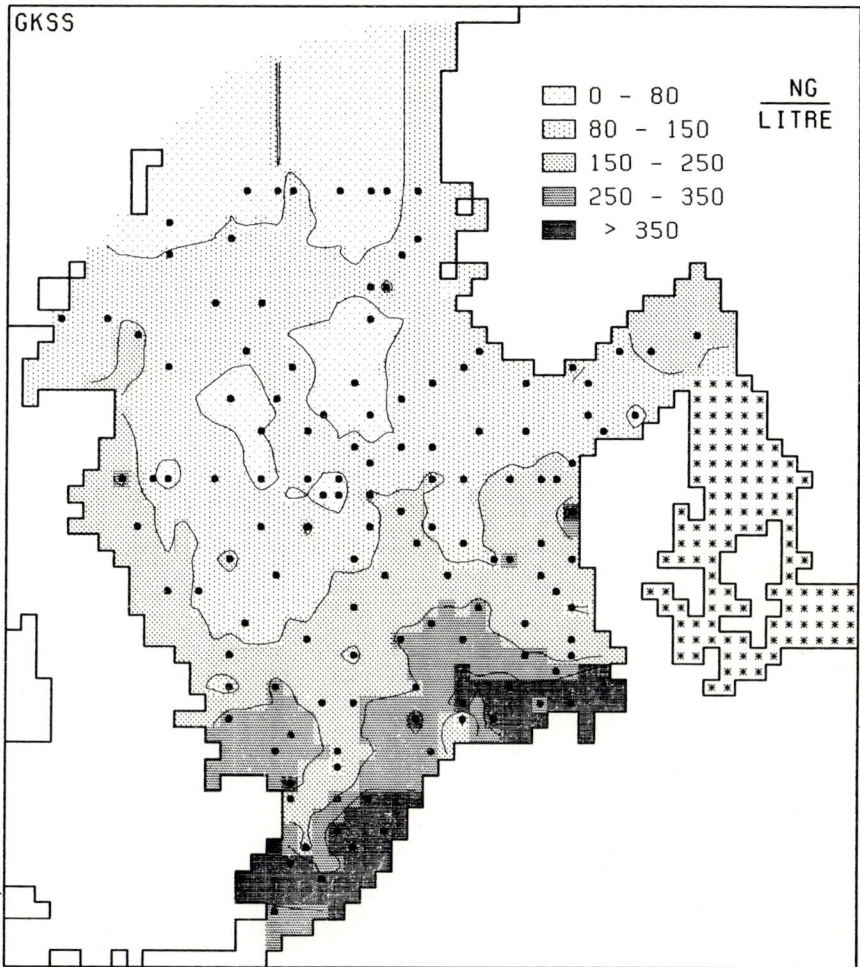

Fig. 4.4.6. Computed distribution of total lead concentration Pb_{TOT}, star cruise in January - March 1987, water depth: 0 - 20 m

The computed lead concentrations are determined by averaging the water column from the surface down to a maximum depth of 20 m.

Figure 4.4.6 shows the computed total lead concentration; for the generation of the computed distributions see the text accompaying Fig. 3.2.17, Chapter 3.2. The black spots in the figures are the measurement stations - the space between the stations is inter- or extrapolated. There are no measurement stations in the Kattegat and the Belt Sea; this area is covered by stars.

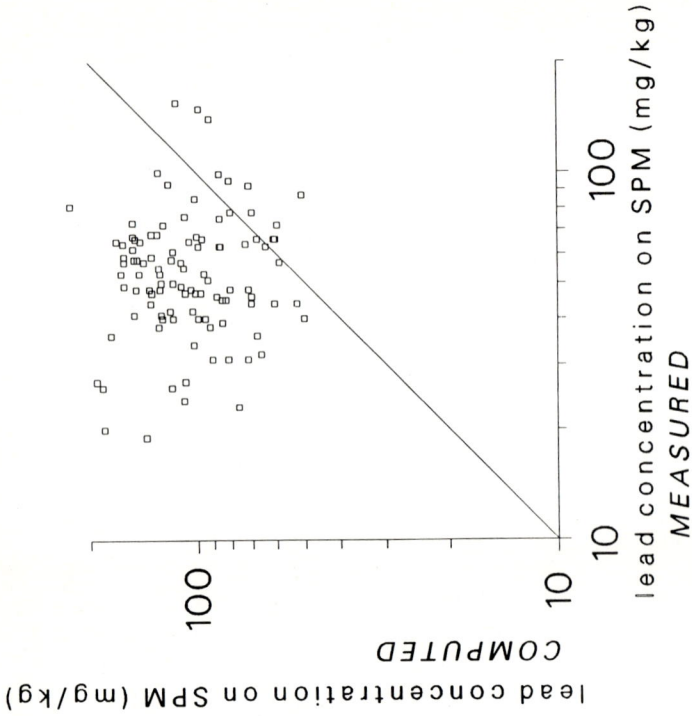

Fig. **4.4.8.** Comparison of measured and computed concentrations Pb$_{SPM}$, star cruise 1987

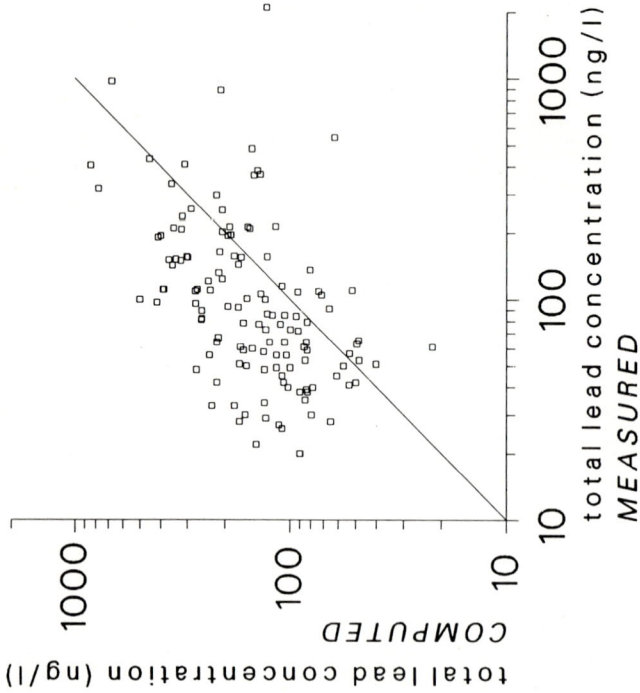

Fig. **4.4.7.** Comparison of measured and computed total lead concentration Pb$_{TOT}$, star cruise 1987

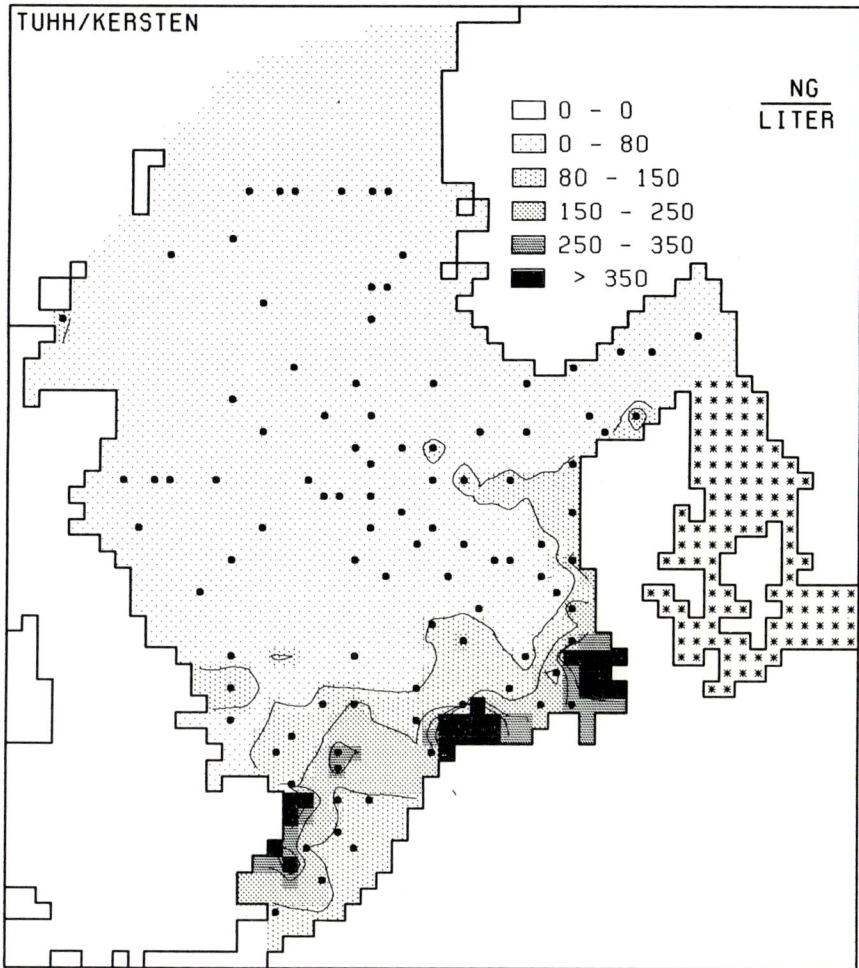

Fig. 4.4.9. Measured distribution of volume concentration $c \cdot Pb_{SPM}$ of particulate lead in January - March 1987, water depth: 10 m

The distribution of total lead is dominated to a great extent by the SPM concentration: a high SPM concentration also means a high total lead concentration (compare Fig. 4.4.5 with Fig. 3.2.18 of Chap. 3.2 or Fig. 4.4.6 with Fig. 3.2.19 of Chap. 3.2).

Comparing now the measured data of Fig. 4.4.5 with the computed data of Fig. 4.4.6, the obvious result is: the computed total lead concentration is mostly higher than the measured concentration. This is best made clear by Fig. 4.4.7, where the computed Pb_{TOT} values are plotted against the measured values. Ideally, all points in Fig. 4.4.7 should lie on the 45°

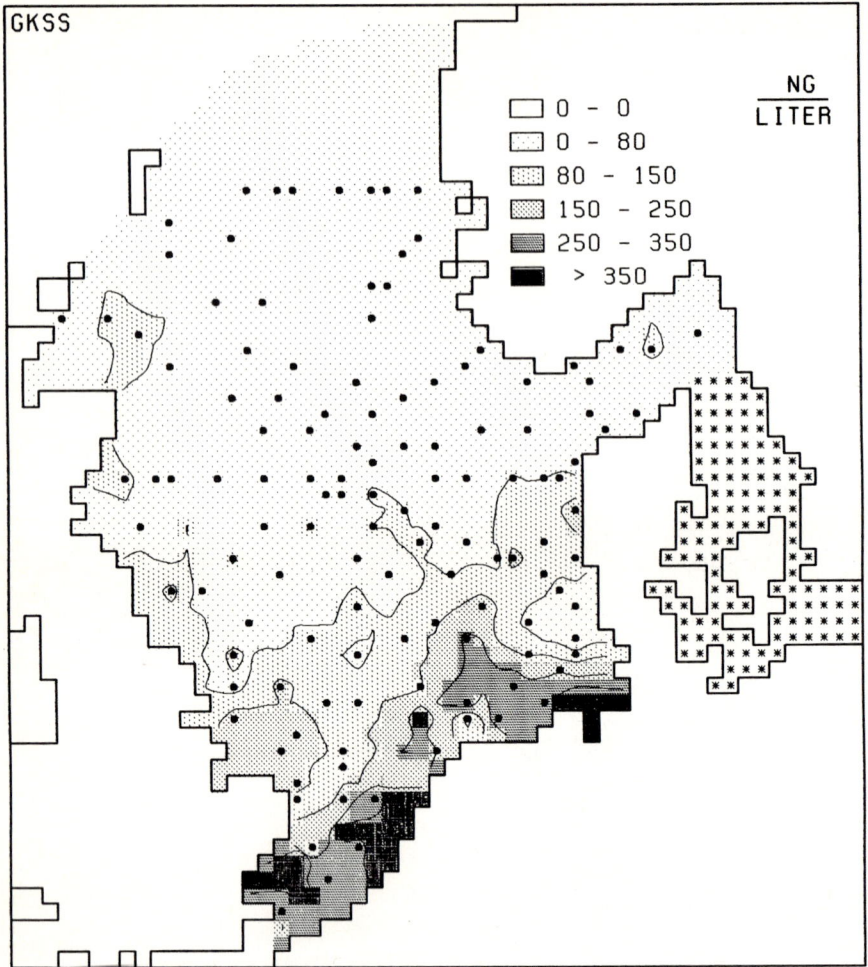

Fig. 4.4.10. Computed distribution of volume concentration $c \cdot Pb_{SPM}$ of articulate lead in January - March 1987, water depth: 10 m

line. The logarithmic mean value of the computed Pb_{TOT} is by about 60 % higher than the log-mean value of the measured Pb_{TOT}.

There is an increase of Pb_{TOT} from N to S both in Figs. 4.4.5 and 4.4.6. The non-ability of the model to accumulate SPM (and thus particulate Pb) in the coastal zone is obvious: the highest measured Pb_{TOT} values are found very near to the coast (see also Brügmann et al. 1985), while high computed Pb_{TOT} values are also found 100 km away from the coast.

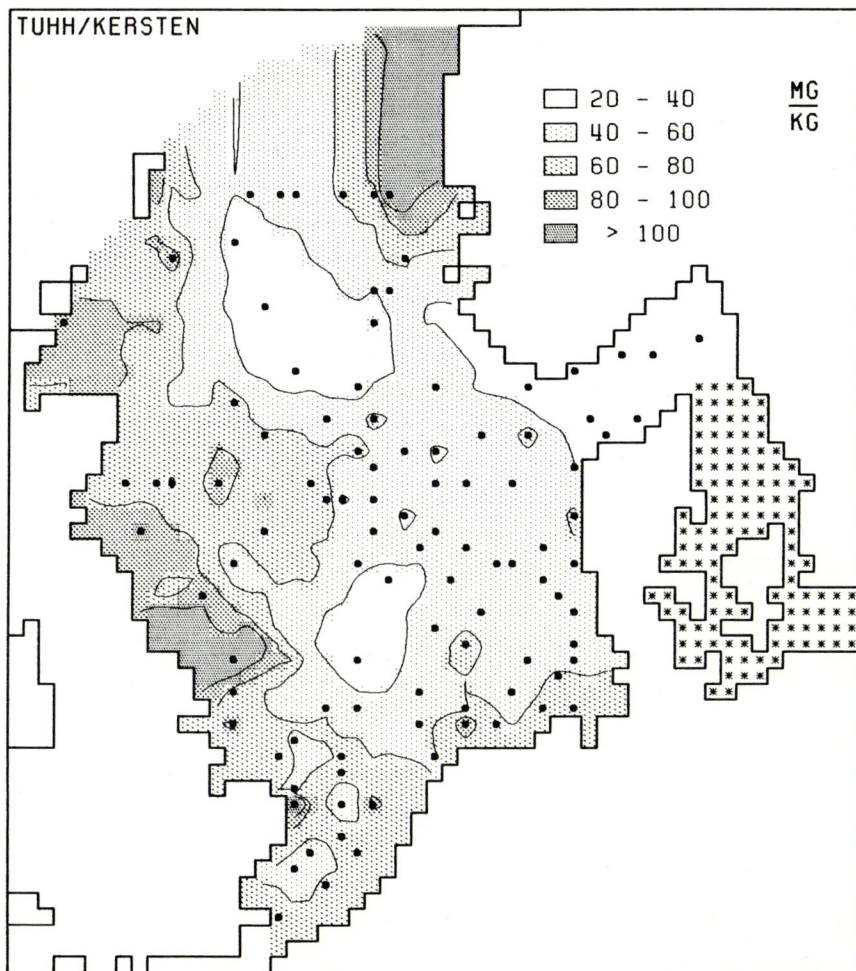

Fig. 4.4.11. Measured distribution of the concentration Pb_{SPM}, star cruise in January - March 1987, water depth: 10 m. (Data from Kersten et al. 1988)

The next comparison is done for the volume concentration $c \cdot Pb_{SPM}$ of particulate lead. The existing measured data are (1) the lead concentration Pb_{SPM}, given in Fig. 4.4.11 and (2) the measured SPM concentration c given in Fig. 3.2.18 of Chapter 3.2. The product of these two quantities is the measured $c \cdot Pb_{SPM}$, shown in Fig. 4.4.9. The corresponding computed $c \cdot Pb_{SPM}$ distribution is given in Fig. 4.4.10. Again, as for the total lead concentration, the computed values are higher than the measured ones. The log-mean of the computed $c \cdot Pb_{SPM}$ is about 45 % higher than the log-mean of the measured $c \cdot Pb_{SPM}$.

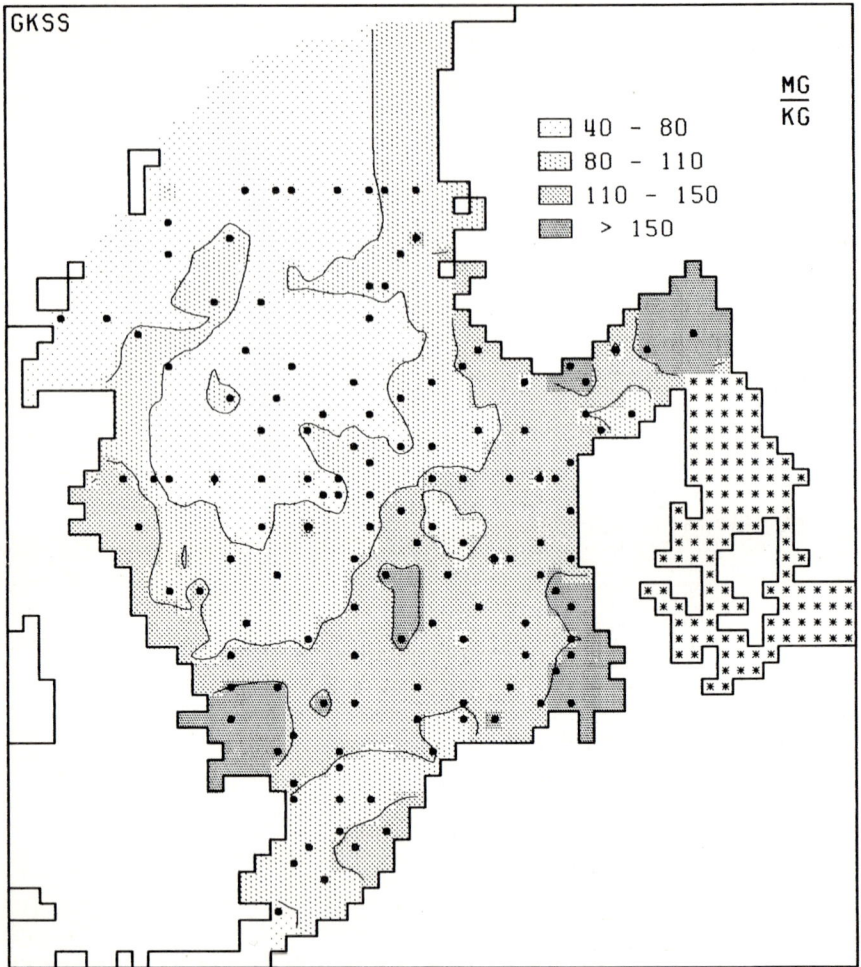

Fig. 4.4.12. Computed distribution of the concentration Pb$_{SPM}$, star cruise in January - March 1987, water depth: 0 to 20 m

The reason for the too high computed lead content in the North Sea is not known. The obvious explanation, of course, is an over-estimation of the seaward lead input numbers. This is probably the case for the northern North Sea, where the inflow from the Atlantic is dominating the lead concentration. For the southern North Sea, the inflow through the Dover Strait might be over-estimated. Other explanations could be:

- The above-mentioned model's inability to keep the particulate Pb concentrated in the nearshore zone.

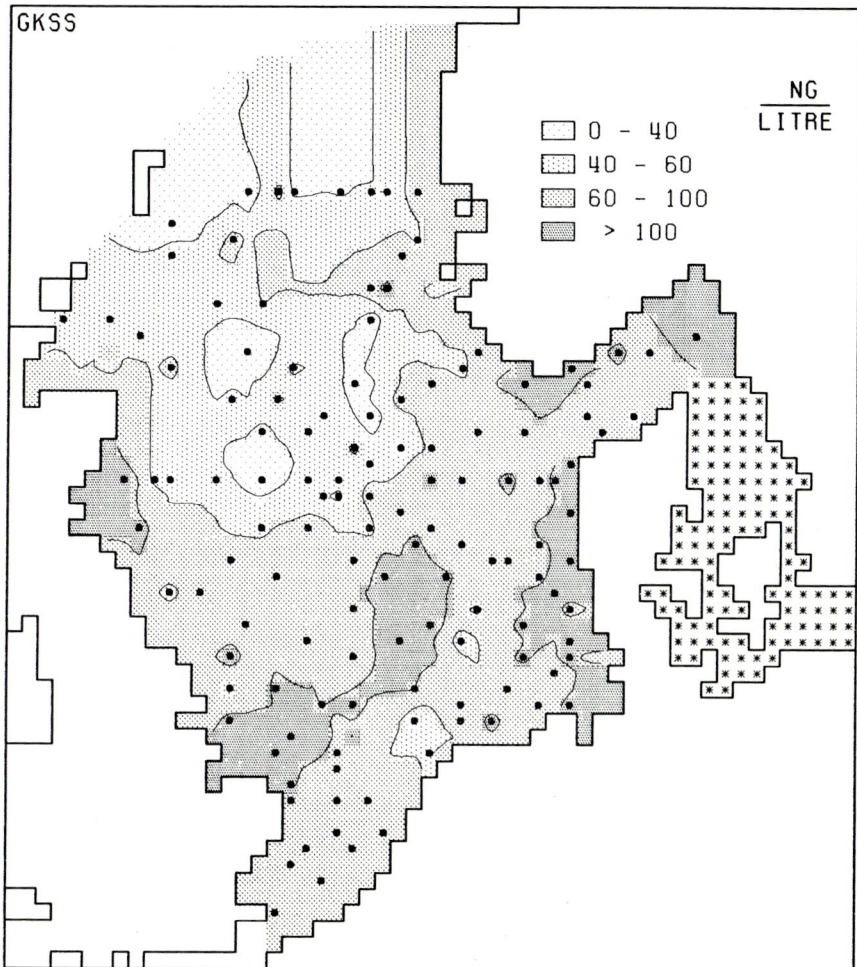

Fig. 4.4.13. Computed distribution of dissolved lead, star cruise 1987

- The absence of phytoplankton simulation; phytoplankton adsorbs contaminants and transfers it to the sediment.
- A too high mobility of particulate Pb, especially the exchange between Pb on newly eroded fine sediment and Pb on resident SPM. This effect increases the Pb content in the water body in the areas with a high Pb_{SED} value.

The computed lead concentration Pb_{SPM} in Fig. 4.4.12 shows the worst agreement with the measured data (Fig. 4.4.11), which is also obvious from Fig. 4.4.8. The log-mean of the computed Pb_{SPM} is 105 mg/kg and thus by about 100 % higher than the log-mean of the measured Pb_{SPM} (52 mg/kg). Note: the shading keys in Figs. 4.4.11 and 4.4.12 are not identical.

The comparison of the distributions of measured and computed Pb_{SPM} values shows too high computed values, particularly in the Skagerrak and in the German Bight, where the computed SPM concentrations (Fig. 3.2.19, Chap. 3.2) are small compared to the measured values (Fig. 3.2.18, Chap. 3.2).

The measured Pb_{SPM} distribution shows highest values in the coastal zone off North England. Here, the computed values are not excessively high, except off the Humber and the Wash.

Regarding the measured Pb_{SPM} distribution in Fig. 4.4.11 alone, the main increase is from E to W, and not, as in Fig. 4.4.5 for Pb_{TOT}, from N to S. This behaviour is explained with the help of the SPM concentrations c which are low along the UK coast north of the Humber, increasing the Pb_{SPM} level automatically, cf. Eq. (4).

The highest Pb_{SPM} value off the British coast appears off York, where Pb_{SED} values are also extremely high, cf. Fig. 4.4.3.

The computed concentration Pb_{DISS} of dissolved lead is shown in Fig. 4.4.13. According to Eq. (1) there is the same distribution as in Fig. 4.4.12 for the lead concentration Pb_{SPM}.

Dissolved lead concentrations measured in June 1984 in the northern North Sea (Balls 1985) lie between 19 and 45 ng/l; the mean value is 31 ng/l. In the same area, the computed Pb_{DISS} values in Fig. 4.4.13 are between 40 and 60 ng/l.

4.4.2.3 Computed Erosion and Deposition of Lead

As a consequence of the deposition of SPM on the sea bed and the erosion of fine sediment (see Figs. 3.2.22 and 3.2.23, Chap. 3.2), there is an increase or a decrease of lead in the sediment.

The net input (atmosphere, rivers, etc.) of lead into the North Sea during 1985 is about 3300 t, which is also the net deposition. A budget of lead in the North Sea is given in Section 4.4.2.4.

Assuming a North Sea bottom area of 575,000 km^2 (Reid et al. 1988), the above given total annual net deposition means an average lead deposition of about 6 mg/m^2/a. The computed spatial distribution of Pb deposition and -erosion is given in Figs. 4.4.14 and 4.4.15.

There is deposition in most parts of the North Sea except the sandy parts SW of the Norwegian Trench (see Fig. 3.2.11, Chap. 3.2), the Dogger Bank and some areas of the southern North Sea.

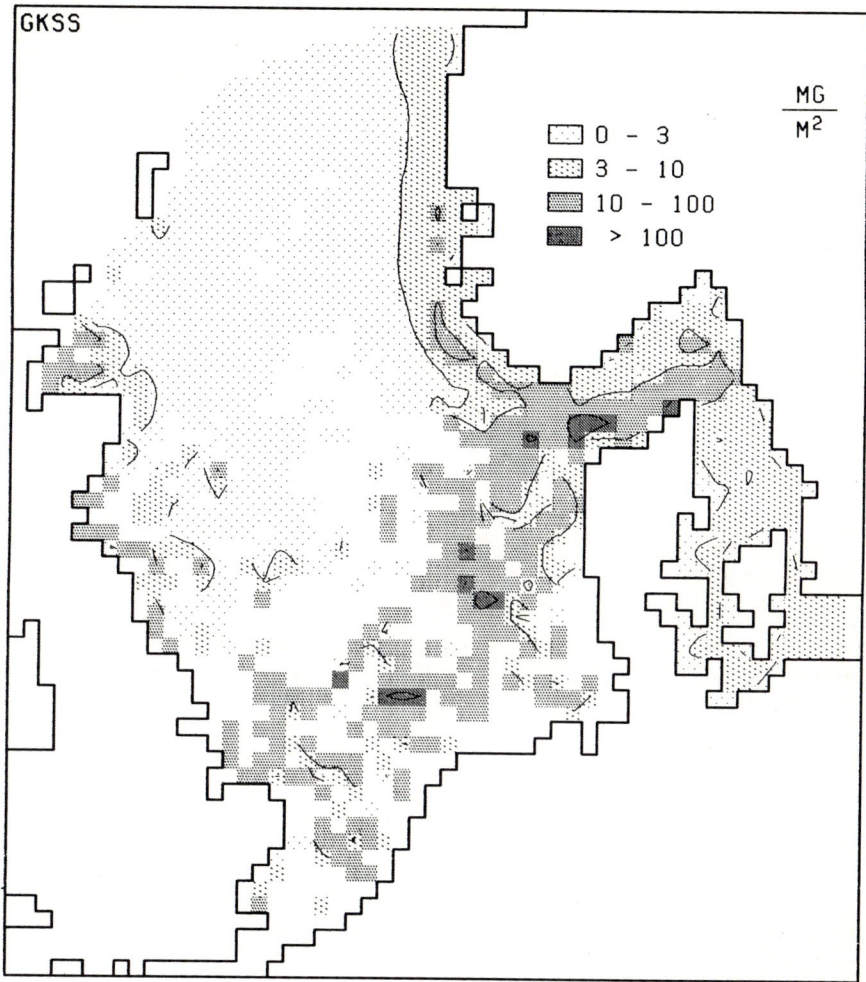

Fig. 4.4.14. Computed distribution of Pb mass deposition during the year 1985

Lead deposition is most intensive in the deeper parts of the southern North Sea, the Skagerrak and the Norwegian Trench. Highest deposition takes place on the southern slope of the Skagerrak, with maximum values up to 200 mg/m^2/a. This is also the area with the highest SPM deposition, see Fig. 3.2.22, Chapter 3.2. The significance of the deposited or eroded Pb masses can be assessed by comparing the numbers of Figs. 4.4.14 and 4.4.15 with the existing Pb masses in the sea bed (Fig. 4.4.4).

The stress on benthic animals is not due to the total Pb mass in the sediment, but due to the lead concentration Pb_{SED} on the fine sediment which is given in Fig. 4.4.3.

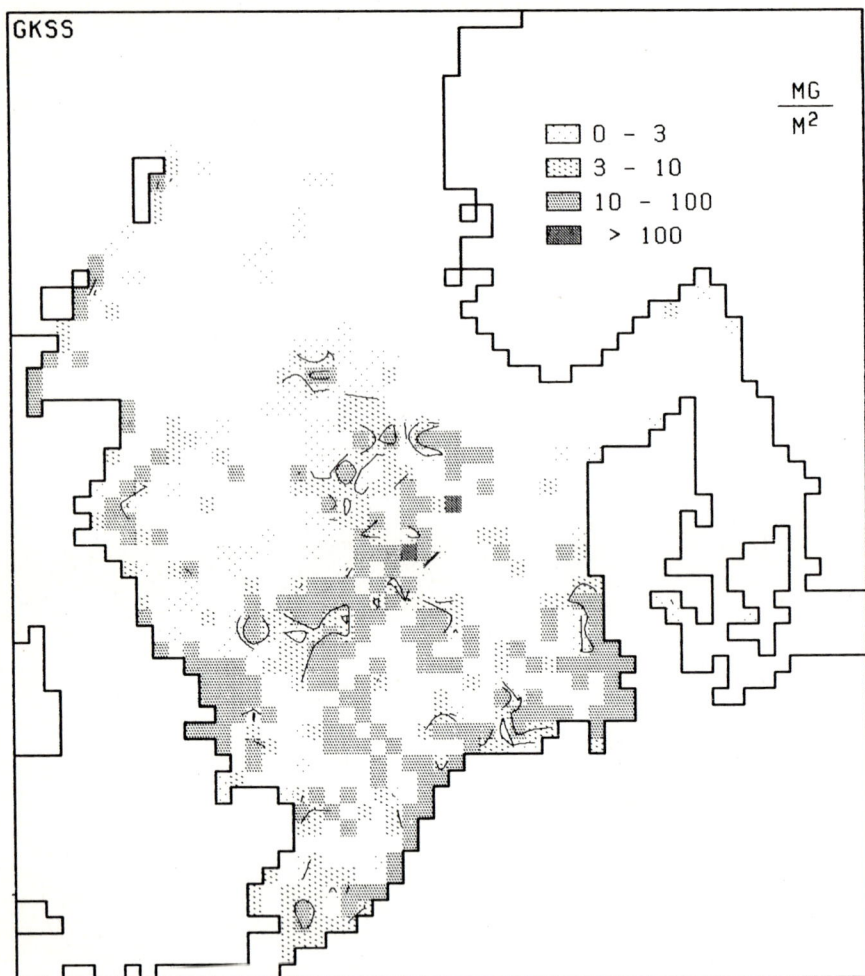

Fig. 4.4.15. Computed distribution of Pb mass erosion during the year 1985

The differences between Pb_{SED} values of December 31 and January 1, 1985, are shown in Figs. 4.4.16 and 4.4.17.

Comparing Fig. 4.4.16 with Fig. 4.4.14, it is obvious that the increase of the Pb mass (Fig. 14) in regions with a high fine sediment content (see Fig. 3.2.12, Chap. 3.2) means only a slight increase of Pb_{SED}. This is mainly true in in the Skagerrak and in parts of the Oyster Grounds. The opposite effect is true in sandy areas (e.g. off Jutland) where a small increase of the Pb mass in the sediment can already mean a large increase of Pb_{SED}. Moreover, Pb_{SED} can increase even if there is a net erosion of Pb mass. This happens when a large amount of fine sediment with low

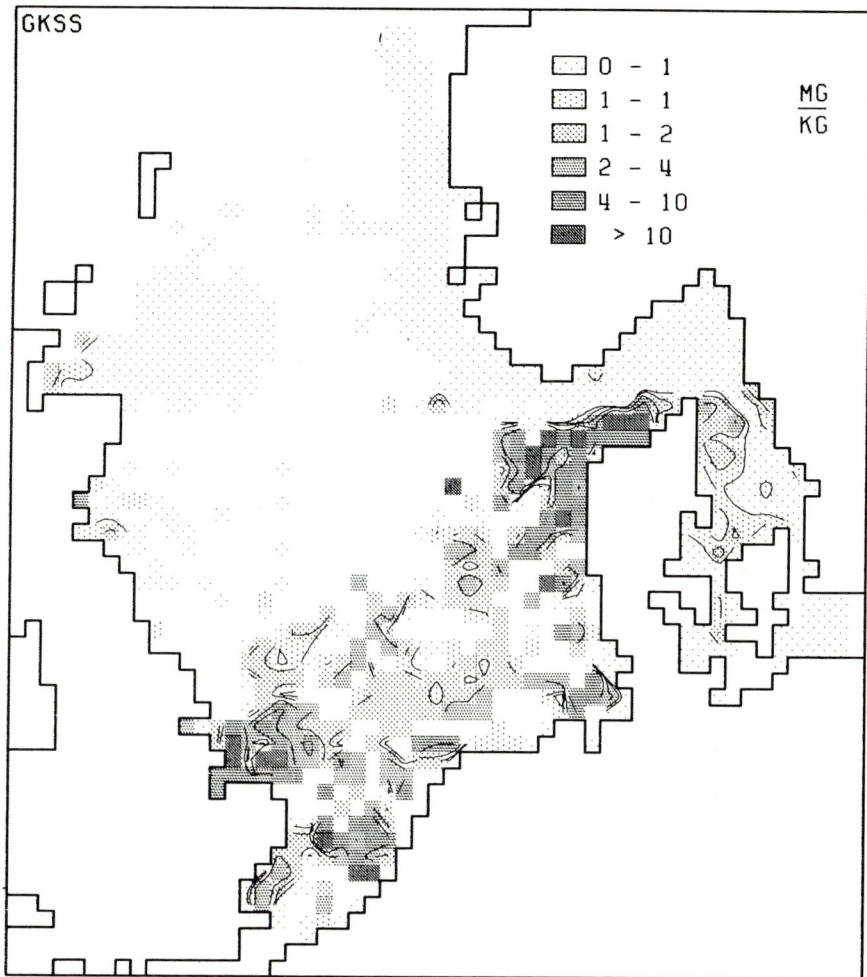

Fig. 4.4.16. Computed distribution of the increase in Pb concentration on fine sediment Pb_{SED} during 1985

Pb_{SED} is eroded, and a small amount of highly Pb-contaminated SPM is deposited. The reciprocal process (Pb_{SED} decreases in spite of a net accumulation of Pb mass) takes place as well, of course.

Comparing now Figs. 4.4.16 and 4.4.17 with Fig. 4.4.3, the general observation is: there is an increase of Pb_{SED} where the existing Pb_{SED} is low and vice versa. The obvious conclusion: The model is not able to simulate the fate of lead in the North Sea as far as the accumulation or the erosion of lead is concerned. This essential defect, however, is not necessarily due to a wrong simulation of processes. It was shown in Fig.

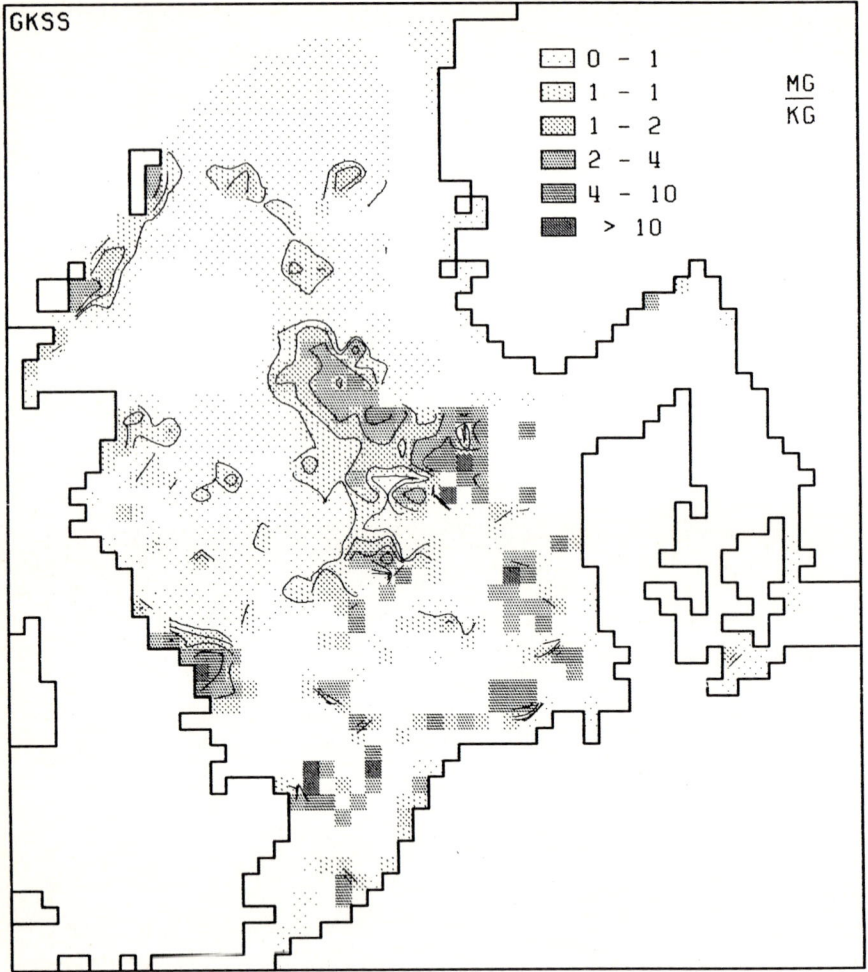

Fig. 4.4.17. Computed distribution of the decrease in Pb concentration on fine sediment Pb_{SED} during 1985

4.4.11 that the measured concentration Pb_{SPM} in the central North Sea does not exceed 100 mg/kg, while the measured concentrations Pb_{SED} on the fine sediment (Fig. 4.4.3) exceed 200 mg/kg. In such a case, as a consequence of erosion and re-deposition, there can only be a decrease of Pb_{SED}, irrespective of a wrong or a correct simulation of the sediment-water interaction.

The big deficit of the model is: in its present stage it does not simulate the effect of phytoplankton on the fate of contaminants. Only when phytoplankton is adequately represented, will a reliable assessment of the model's ability to simulate the fate of contaminants be possible.

4.4.2.4 Lead Budget

Table 4.4.1 shows the computed Pb masses that enter ("input") and leave ("output") the model's water body during the year 1985. The numbers include both dissolved and particulate lead. An analogous table for SPM is Table 3.2.2 of Chapter 3.2.

Table 4.4.1. Sources and sinks of Pb (in tons per year) for the North Sea water body, computed for the year 1985. The "net amount" is the sum of "input" and "output"

Sources/sinks	Input	Output	Net amount
Pentland Firth and Fair Isle Strait	4320	3020	1300
North Atlantic between Shetland and	9080	12100	-3020
Norway	2620	1380	1240
Dover Strait	460	360	100
Baltic Sea			
Rivers, cliff erosion, dumping	810	-	810
Atmosphere	2870	-	2870
Sediment	15,000	18,200	-3200

The sum of net amounts in Table 4.4.1 is "plus 100 tons". This residual input appears as an (accidental) increase of Pb mass in the water body at the end of 1985, compared with the begin of 1985.

The model does not "keep in mind" the origin of the lead in detail; it only distinguishes between allochthonous and autochthonous lead (see Sect. 4.4.1.4). This is because of limited storage space of the computer. It is planned, however, to handle at least the lead from the atmosphere separately.

The particulate Pb mass computed for the year 1985 in the North Sea water body is between 3000 t during calm weather and 8000 t during a severe storm. The mass of dissolved Pb is approximately invariant with 4000 t. This is because the erosion of fine sediment with particulate lead does not alter the Pb_{SPM} value substantially, which is then [according to Eq. (1)] also true for Pb_{DISS}. At the end of the simulated year 1985, the model's water body contained 7040 t of lead with 60 % being allochthonous, i.e. having never been incorporated in the sediment.

The total mass of lead in the upper 0.2 m of the sediment bed shown in Fig. 4.4.4 is about 860,000 t. Assuming a natural background concentration $Pb_{SED} = 20$ mg/kg, the natural Pb mass in the upper 0.2 m bed layer is 200,000 t. The anthropogenic Pb mass is thus 660,000 t, which means an annual net supply of about 6000 t since the beginning of anthropogenic pollution around 1880. The net supply to the sediment in 1985 was calculated to be only 3200 t, see Table 4.4.1.

The biggest numbers in Table 4.4.1 concern the sediment. These high numbers demonstrate the importance of the lead interchange between the water body and the sediment - they are mainly due to erosion and re-deposition of autochthonous lead. The increase of allochthonous lead in the sediment during 1985 was 5400 t, the decrease of allochthonous lead was 2200 t, resulting in the net number of 3200 t in Table 4.4.1.

4.4.3 Discussion

In this section the Pb_{SED} distribution shown in Fig. 4.4.3 will be discussed, also with respect to the model results.

The exchange of fine sediment between the water body and the sediment bed results in a renewal of the fine bottom sediment within a "fine sediment turnover time T_{SED}". It can be defined as follows:

The annually eroded and the annually deposited mass of fine sediment is determined. The greater of the two values is divided by the total mass of fine sediment in the upper 0.2 m of the bed. The result is assumed to be the portion of the renewed fine sediment per year. The turnover time T_{SED} [years] is the reciprocal value of that portion.

Stricly speaking, T_{SED} is well defined only for a (sandy) bottom with an equilibrium between eroded and deposited masses. For a muddy bottom, with deposition only, the definition of T_{SED} is useless: in that case the turnover time is the period of a bed thickness increase of 0.2 m.

Figure 4.4.18 shows the sediment turnover times T_{SED} determined from simulating the year 1985. Values of more than 300 years appear in muddy areas - they are meaningless. Also meaningless are the values for the Oyster Ground and the northern Elbe Rinne, because in most parts there is SPM deposition only and no equilibrium.

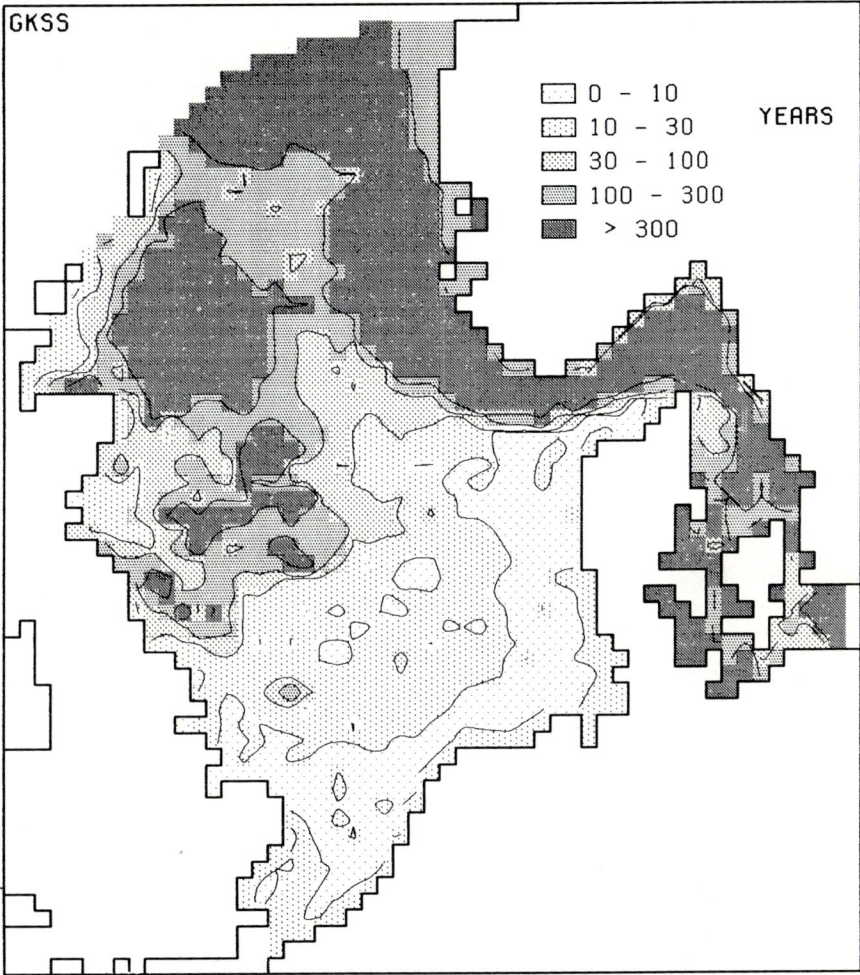

Fig. 4.4.18. Turnover time T_{SED} of fine sediment in the upper 0.2 m of the bed, determined from simulating the year 1985

As was mentioned in Section 3.2.3.9 (text connected with Fig. 3.2.22, Chap. 3.2), there is unrealistic continuous erosion of fine sediment on the Dogger Bank, near the Dover Strait and in the near-coast zone of the southern North Sea. This has resulted in an exhaustion of fine sediment in the near-surface bed layers, which in turn means reduced erosion rates. The reduced erosion rates finally produce unrealistically high turnover times in the mentioned areas, see Fig. 4.4.18.

On the whole, the T_{SED} values in Fig. 4.4.18 bring up more confusion than answers, but nevertheless the use of a fine sediment turnover time can be helpful when trying to explain the pattern of sediment contamination in the North Sea.

Comparing the content $c_{20\mu m}$ of fine sediment in the bed (Fig. 3.2.12, Chap. 3.2) with the concentration Pb_{SED} of lead on fine sediment (Fig. 4.4.3), the muddy areas show comparatively low Pb_{SED} values. This is true for (1) the Skagerrak and the Norwegian Trench, (2) the Oyster Ground and the Northern Elbe Rinne and (3) the Fladen Ground. The reason: there is a high portion of relict (unpolluted) fine sediment in muddy areas, which dilutes the deposited, modern (polluted) fine sediment. This is mostly pronounced in the Fladen Ground, where the deposition of polluted mud since about 1880 (first effects of the industrial age) has only been 0.5 cm, which is small compared to the depth of the bioturbated layer ($\approx$ 20 cm). It is thus not surprising that, concerning Pb, the Fladen Ground sediment is "clean".

There are, however, high deposition rates of modern (polluted) SPM in parts of the Oyster Ground and of Skagerrak/Norwegian Trench. As we have seen in Fig. 4.4.11, the present concentration Pb_{SPM} in the North Sea is mostly below 100 mg/kg. So, starting out from present day conditions, even a total replacement of relict by modern fine sediment will not increase the Pb_{SED} level beyond 100 mg/kg.

In sandy areas, Pb_{SED} varies widely: (1) from 70 mg/kg in the Southern Bight and NW of Jutland (2) over 100-200 mg/kg in the German Bight and on the Dogger Bank (where the Pb_{SED} distribution is very heterogeneous, increasing from the shallow SW part towards the Dogger Tail End) and (3) up to more than 200 mg/kg off York (between Tees and Humber) and in the triangle Ling Bank-Little Fisher Bank-north slope of Dogger Tail End (see Fig. 3.2.11 of Chap. 3.2). Generally, the fine sediment in a sandy bottom is more contaminated than in a muddy bottom. The reason: in sandy areas the relict fine sediment fraction has been replaced by modern fine sediment because of the continuous exchange of sediment between the water body and the bottom, resulting in higher pollution levels than in the muddy parts.

This cannot explain, however, the extremely high Pb_{SED} values off York and in the triangle. A possible explanation: the input of lead into the North Sea was much higher in previous times. This, however, is unlikely, because before 1985 there was no substantial decrease of lead emissions by vehicles in UK. The high Pb_{SED} values can only be explained by phytoplankton-induced pollution-pumping: after the production of phytoplankton (which takes place all over the North Sea), it adsorpts a certain portion of Pb. This scavenging is most effective if the concentration of competitive terrigeneous SPM is low. The depositing phytoplankton transfers Pb to the bottom. In the sediment, the organic matter is re-mineralized, leaving a certain amount of Pb behind. The crucial point is thus: the Pb mass in the sediment is increasing, but not the mass of fine sediment. This gradually increases Pb_{SED} up to values of more than 200 mg/kg.

Summing up, the Pb_{SED} values measured in the southern North Sea can be explained as follows: there is a strong supply of terrigeneous SPM, mainly from the Dover Strait. The lead concentration on the supplied SPM is increased by lead from the atmosphere. The resulting Pb concentration Pb_{SPM} is mostly less than 100 mg/kg. The intensive exchange of terrigeneous SPM and fine sediment (resulting in low sediment turnover times) has led to an equilibrium of Pb_{SPM} and Pb_{SED} values.

There are areas with increased Pb_{SED} values in the German Bight and along the Dutch coast. This is due to an increased local input of lead.

The dilution of modern (polluted) with relict (unpolluted) fine sediment in the Oyster Ground (and in Skagerrak/Norwegian Trench) is distinct, but not dominating, because the deposition of modern (polluted) terrigeneous SPM is "sufficiently" high.

There is, of course, phytoplankton pumping of lead in the southern North Sea. Its scavenging effectiveness, however, is reduced because of the competition with the highly concentrated terrigeneous SPM. Its effect on Pb_{SED} is dominated by the high supply of terrigeneous SPM and the strong exchange between the water body and the sediment.

In the northern North Sea the supply of terrigeneous SPM is small (and if there is any, it has a very small settling velocity). Therefore, phytoplankton pumping is dominant. The mass of relict, unpolluted fine sediment in the bed becomes very important: the effect of phytoplankton pumping on Pb_{SED} is highest on a sandy bottom (as in the triangle) and seems to be negligible in the Fladen Ground.

The highly polluted sediment off York exists in an intermediate position: north of it the bed is muddy, which impedes a high Pb_{SED}, and south of it the concentration of terrigeneous SPM is high, which also impedes a high Pb_{SED}. Only in the gap in between was a $Pb_{SED} > 200$ mg/kg able to develop.

In the above considerations, the total Pb concentration in the water column was not mentioned. What determines the scavenged amount of a contaminant is the contaminant's mass in the water column (i.e. under 1 m^2 surface area). Because of the increased water depth in the northern North Sea, the Pb masses under 1 m^2 are comparable to the Pb masses under 1 m^2 in the southern North Sea.

4.4.4 Summary and Conclusion

Together with storm surges and algae blooms, the effect of contaminants in the North Sea is one of the major interests of administrators and the public. It is thus obvious to attempt the modelling of contaminant transport in the North Sea, see for instance van Pagee et al. (1988). In the present Chapter, the fate of lead is simulated, including (1) realistic input data, (2) transport by currents in the dissolved phase and on suspended matter and (3) the interchange (erosion, deposition) with the sediment.

The results of the simulation runs are compared with measured data of the ZISCH star cruise in winter 1987. The computed distributions of lead erosion and -deposition disclose one weak point of the model: the non-consideration of phytoplankton.

On the whole, the results of the simulation runs are encouraging rather than discouraging. The inclusion of phytoplankton is only a matter of time. A permanent problem, however, is the insufficient knowledge of so many transport-relevant abiotic processes (e.g. sediment erosion, settling and deposition of suspended matter). One could argue now that before continuing the numerical simulation of contaminant transport, one should first investigate all those superficially known processes. This view, however, disregards the fact that only the co-operation and the mutual stimulation of measuring and modelling groups will guarantee substantial progress in simulating the fate of contaminants. Attention must therefore be directed to that co-operation.

References

ARGE Elbe (1988) Schwermetalldaten der Elbe 1984-1988. Bericht der Arbeitsgemeinschaft für die Reinhaltung der Elbe, Wassergütestelle Elbe, Hamburg, 193 pp

Balls PW (1985) Trace metals in the northern North Sea. Mar Pollut Bull 16(5):203-207

Balls PW (1988) The control of trace metal concentrations in coastal seawater through partition onto suspended particulate matter. Neth J Sea Res 22(3):213-218

Brügmann L, Danielsson L-G, Magnusson B, Westerlund S (1985) Lead in the North Sea and the north east Atlantic Ocean. Mar Chem 16:47-60

Bundesminister für Umwelt, Naturschutz und Reaktorsicherheit (1987) Bericht der Bundesregierung an den Deutschen Bundestag zur Vorbereitung der 2. Internationalen Nordseeschutz-Konferenz (2.INK). Drucksache 11/878, Deutscher Bundestag, 11. Wahlperiode

Carlson H (1986) Quality status of the North Sea. Dtsch Hydrogr Z Ergänzungsh Reihe B 16:424

Dyer KR (1988) Fine sediment particle transport in estuaries. In: Dronkers J, van Leussen W (eds) Physical Processes in Estuaries. Springer, Berlin Heidelberg New York, pp 295-310

Eisma D, Irion G (1988) Suspended matter and sediment transport. In: Salomons W, Bayne BL, Duursma EK, Förstner U (eds) Pollution of the North Sea. Springer, Berlin Heidelberg New York, pp 20-35

Flügge G (1988) Keulenschläge ins Seewasser. TAZ Hamburg 30. 1. 1988, p 29

Hupkes R (1990) Pollution of the North Sea imposed by West European rivers (1984-1987). ICWS-rep 90.03, International Center of Water Studies, Amsterdam, 102 pp

Irion G (1984) Schwermetallbelastung in Oberflächensedimenten der westlichen Ostsee. Naturwissenschaften 71:536-537

Irion G, Müller G (1990) Lateral distribution and sources of sediment-associated heavy metals in the North Sea. In: Ittekot V, Kempe S, Michaelis W, Spitzi A (eds) Facets of modern biogeochemistry. Springer, Berlin Heidelberg New York, pp 175-210

Kersten M (1988) Geobiological effects on the mobility of contaminants in marine sediments. In: Salomons W, Bayne BL, Duursma EK, Förstner U (eds) Pollution of the North Sea. Springer, Berlin Heidelberg New York, pp 36-58

Kersten M, Klatt V (1988) Trace Metal Inventory and Geochemistry of the North Sea Shelf Sediments. In: Kempe S, Liebezeit G, Dethlefsen V, Harms U (eds) Biogeochemistry and distribution of suspended matter in the North Sea and implications to fisheries biology. Mitt Geol Paläontol Inst Univ Hamb, 65:289-311

Kersten M, Dicke M, Kriews M, Naumann K, Schmidt D, Schulz M, Schwikowski M, Steiger M (1988) Distribution and fate of heavy metals in the North Sea. In: Salomons W, Bayne BL, Duursma EK, Förstner U (eds) Pollution of the North Sea. Springer, Berlin Heidelberg New York, pp 300-347

Kersten M, Kienz W, Koelling S, Schröder M, Förstner U (1990) Schwermetallbelastung in Schwebstoffen und Sedimenten der Nordsee. Vom Wasser 75:245-272

Lee AJ (1980) North Sea: physical oceanography. In: Banner FT, Collins MB, Massie KS (eds) The north-west European shelf seas: the sea bed and the sea in motion II. Physical and chemical oceanography, and physical resources. Elsevier, Amsterdam, pp 467-493

Malle K-G (1987) Wie schmutzig ist die Nordsee? Chem unserer Zeit 21(1):9-16

Petersen G, Weber H, Graßl H (1989) Modelling the atmospheric transport of trace metals from Europe to the North Sea and the Baltic Sea. In: Pacyna JM, Ottar B (eds) Control and fate of atmospheric trace metals. Kluwer, Dordrecht, pp 57-83

Postma H (1981) Exchange of materials between the North Sea and the Wadden Sea. Mar Geol 40:199-213

554   Puls et al.

Rat von Sachverständigen für Umweltfragen (1980) Umweltprobleme der Nordsee. Sondergutachten. Kohlhammer, Stuttgart, 503 pp

Reid PC, Taylor AH, Stephens JA (1988) The hydrographie and hydrographic balances of the North Sea. In: Salomons W, Bayne BL, Duursma EK, Förstner U (eds) Pollution of the North Sea. Springer, Berlin Heidelberg New York, pp 3-19

Schladot JD, Stoeppler M, Schwuger MJ (1990) Umweltprobenbank Jülich: Ein Projekt für das nächste Jahrhundert. Annu Rep Forschungszentrum Jülich, pp 95-103

Schwedhelm E, Salomons W, Schoer J, Knauth H-D (1988) Provenance of the sediments and the suspended matter of the Elbe estuary. GKSS Rep 88/E/20, GKSS Research Centre, Geesthacht, 76 pp

van Pagee JA, Glas PCG, Markus AA, Postma L (1988) Mathematical modeling as a tool for assessment of North Sea pollution. In: Salomons W, Bayne BL, Duursma EK, Förstner U (eds) Pollution of the North Sea. Springer, Berlin Heidelberg New York, pp 400-422

van Straaten LNJU, Kuenen PH (1958) Tidal action as a cause for clay accumulation. J Sediment Petrol 28:406-413

Uncles RJ, Elliott RCA, Weston SA (1985) Observed fluxes of water and suspended sediment in a partly mixed estuary. Estuarine Coastal Shelf Sci 20:147-167

Wood PC, Franklin A (1983) The bioavailability of metals in fly ash and colliery waste dumped at sea. ICES CM 1983/E:15, International Council for the Exploration of the Sea, Copenhagen, 3 pp

4.5 Bioaccumulation and Effects of Plankton and Benthos on the Fate of Contaminants

L. Karbe, L. Aletsee, C.-D. Dürselen, K. Heyer,
U. Kammann, M. Krause, H.-J. Rick and H. Steinhart

There is no doubt that interactions with biota have a strong impact on the fate of contaminants in the environment. Having this in mind, it is obvious that contaminant circulation, pathways and fluxes in the North Sea can only be understood by considering the regional and seasonal variability of biological structures and functional interdependences. Within this Chapter, it will be attempted to draw some general conclusions on the fate of contaminants in the North Sea as it is influenced by biological processes within the pelagic and benthic environment.

Contrary to the alternative hypothesis of random distribution made in Chapter 4.7, it is assumed that there is a non-random distribution caused by functional interdependences. However, structures are often not obvious at first sight, due to the complexity of multivariate dependences. It is the major intention of this Chapter to identify such structures and to draw some conclusions on cause effect relationships.

When evaluating the data on contaminants distribution, presented in Chapters 2.3 and 2.5, problems arise due to limitations of data available, which are not sufficient in number to calculate statistical significances of multivariate cause-effect interdependences. Data are also lacking for a good evaluation at a coherent scale. However, some structures and trends are obvious and plausible considering synecological interactions. Priority is given to plausibility before (statistical) significance.

The inorganic and organic xenobiotic contaminants investigated are representatives of compounds that are of considerable environmental interest because of their ubiquity, stability and persistence and their potential for bioaccumulation and toxicity. Our studies are focussed on a selection of contaminants known to differ with respect to their physical and chemical characteristics and, as a consequence, with respect to dynamics and kinetics of interactions with biota. These are the α- and γ-isomers of hexachlorocyclohexane (HCH), hexachlorobenzene (HCB), a selection of polychlorinated biphenyls (PCBs) and the metals mercury (Hg), cadmium (Cd), copper (Cu) and lead (Pb).

In order to explain the results of the distributions of the contaminants on large (North Sea - Chap. 2.3) and meso scale (German Bight- Chap. 2.5), it is necessary to understand the mechanisms which influence the uptake and turnover of the contaminants by the biota.

4.5.1   Factors Influencing Uptake and Concentration of Contaminants by Biota

To come to an understanding of the regional and seasonal variability of contaminant concentrations of biota, it has to be considered that there are major differences among the various xenobiotic compounds with respect to their tendency to be accumulated in biota. These are related to the physico-chemical properties and substitution patterns of the individual compounds. Furthermore, there are major differences among the higher taxa of phytoplankton, zooplankton, benthos and fish, and minor interspecific differences in the sorption capacity for contaminants. Last but not least, dynamics and kinetics of contaminants bioaccumulation are influenced by a number of factors highly variable in space and time.

Within the organism, contaminants are bound to various compartments which are different with respect to their sorption capacity. Binding may be of low or of high affinity. Binding may be unspecific, or specific in the case of contaminants which are bound to specific biochemical structures (receptors). The sorption of contaminants may follow short-term or long-term kinetics characterized by half-lives of several hours, days, weeks or months and depending on several factors (Fig. 4.5.1).

4.5.1.1 Differences Among Chemicals Investigated

Contaminants will be accumulated by the biota in different amounts. For a comparison of the rate of contaminant uptake between species and contaminants, accumulation factors of the biota relative to water were cal-

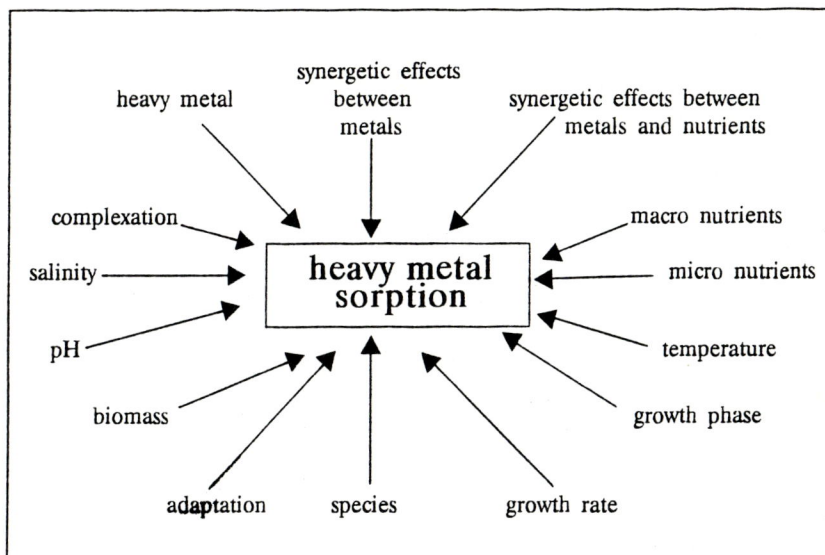

Fig. 4.5.1. Factors influencing the sorption of metals by phytoplankton. (After Rick 1990)

culated. Since the reasons for bioaccumulation are complex (Fig. 4.5.1), these accumulation factors represent only a rough estimate.

Organic Contaminants. Some data to illustrate the situation in the marine ecosystem are presented in Table 4.5.1 and Fig. 4.5.2. Polychlorinated biphenyls (PCBs) are calculated as the sum of the 35 individual main congeners (unless noted otherwise). The organisms listed are roughly ordered according to increasing trophic level in the marine food web. It is also in this order that the maximum levels of PCBs have been found to increase: these levels increase when going from zooplankton to the hermit crab and from dab to whelk. This leads to the conclusion that for the benthic organisms under investigation, a biomagnification of lipophilic chemicals like PCBs may exist.

Cyclic organochlorines and other organic contaminants are enriched from water to sediment and to biota as a consequence of hydrophobic/lipophilic interactions. The tendency to be accumulated is highest in the case of the highly hydrophobic and lipophilic xenobiotics like DDT and its metabolites, as well as in the case of the PCBs and less in the case of the rather polar less hydrophobic xenobiotics like γ-HCH (Lindane). The water solubility

Table 4.5.1. Polychlorinated biphenyls (Σ-PCB determined from 23 or 24 individual PCB congeners) and Lindane in different compartments of the ecosystem

	Σ-PCB	γ-HCH	Reference
Water[a]	1×10^{-7} - 4×10^{-6} μg/ml water[b]	3×10^{-4} - 2.7×10^{-3} ng/ml water	(1),(2)
Sediment[a]	1×10^{-4} - 4.4×10^{-3} μg/g sediment[c]	1×10^{-3} - 6.2×10^{-2} ng/g sediment	(3)
Zooplankton[a]	0.2 - 8.3 μg/g lipid[c]	2 - 130 ng/g lipid	(4)
Hermit crab[d]	1.1 - 12.3 μg/g lipid	15 - 147 ng/g lipid	(5)
Dab[d]	2.5 - 16.2 μg/g lipid	32 - 98 ng/g lipid	(6)
Whelk[d]	1.6 - 25.9 μg/g lipid	25 - 88 ng/g lipid	(7)

[a] Data covering the entire North Sea.
[b] Σ-PCB determined from 23 individual PCB congeners.
[c] Σ-PCB determined from 24 individual PCB congeners.
[d] data covering the German Bight.
References:
(1) Schulz (1990), (2) Faller et al. (1991), (3) Knickmeyer and Steinhart (1988c), (4) Knickmeyer and Steinhart (1989b), (5) Knickmeyer and Steinhart (1990c), (6) Kammann et al. (1992) and (7) Knickmeyer and Steinhart (1990a).

or the water octanol partitioning coefficient may be used to quantify the hydrophobic properties of the chemicals (Fig. 4.5.3).

The concentration of Lindane in water is strikingly higher than the PCB concentrations due to the relatively high water solubility (low log octanol/water partition coefficient) of Lindane. Even the carnivorous whelk exhibits a lindane contamination level of 25-88 ng/g lipid which is within the the range of the other organisms presented in Table 4.5.1. The water solubilities of the hexachlorocyclohexane (HCH) isomeres are about two to four orders of magnitude higher than those of most PCB congeners. Less lipophilic chemicals, such as HCHs, are both taken up and lost by various organisms within a few hours. In contrast, clearance half-lives of hexachlorobiphenyls vary from about 35 to 70 days in some fish species. Therefore, Lindane shows no biomagnification in the marine food web. Lindane contamination of marine organisms is probably due to equilibration processes between organisms and sea water, dependent on lipid polarity of the tissues. It has to be noted that Σ-PCB is a total of a number of PCB congeners very different with respect to physico-chemical properties and their tendency to be accumulated and to interact with biota (Tanabe et al. 1987; Kannan et al. 1989).

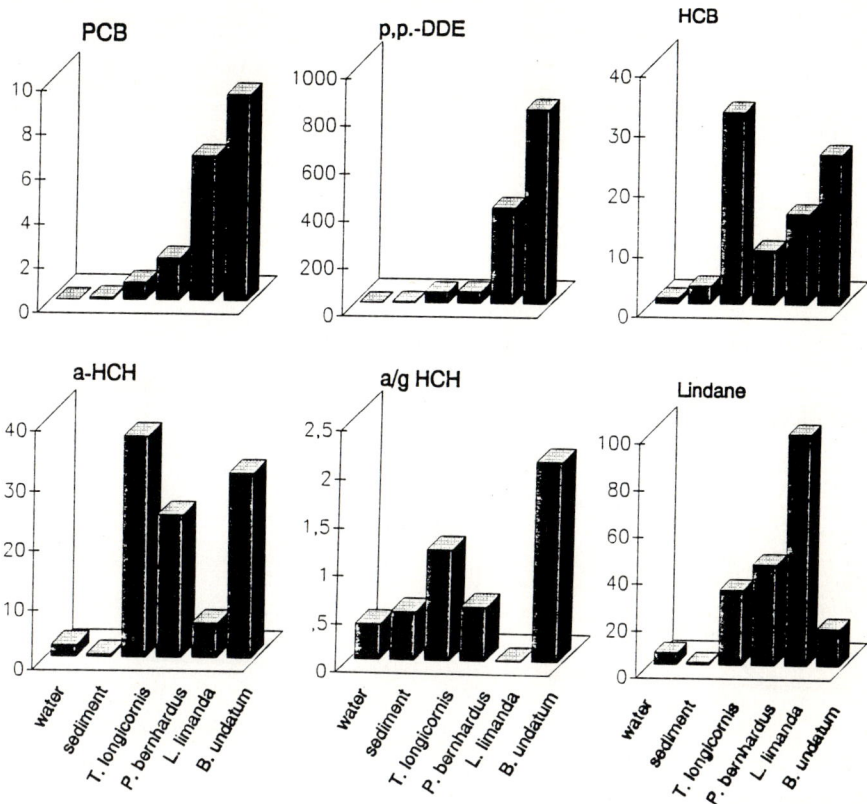

Fig. 4.5.2. Cyclic organochlorines in different compartments of the ecosystem taken in early summer at station 73 (inner German Bight): Σ-PCB, p,p'-DDE, HCB, α-HCH, γ-HCH and the relation α-HCH/γ-HCH in water (ng/dm^3), sediment (ng/kg), calanoid copepod *Temora longicornis*, hermit crab *Pagurus bernhardus*, dab *Limanda limanda* and whelk *Buccinum undatum* (μg/kg lipid)

Metals. As in the case of organic contaminants, the different metals also show significant differences with respect to their tendency to be bio-accumulated. The ranking of tendencies for bioaccumulation estimated from results of laboratory experiments or field studies may be different, depending on the type of organism investigated, and on whether the sorption capacity is dominated by binding to structures at the body surfaces or by incorporation followed by binding to the interior body compartments.

Ranges of accumulation factors for blue mussels (total soft parts), hermit crabs (abdominal part) and dab (liver) estimated from results of the ZISCH field studies are compiled in Table 4.5.2 and compared with accumulation factors for sediment samples (fraction < 20 μm). Comparing accumulation

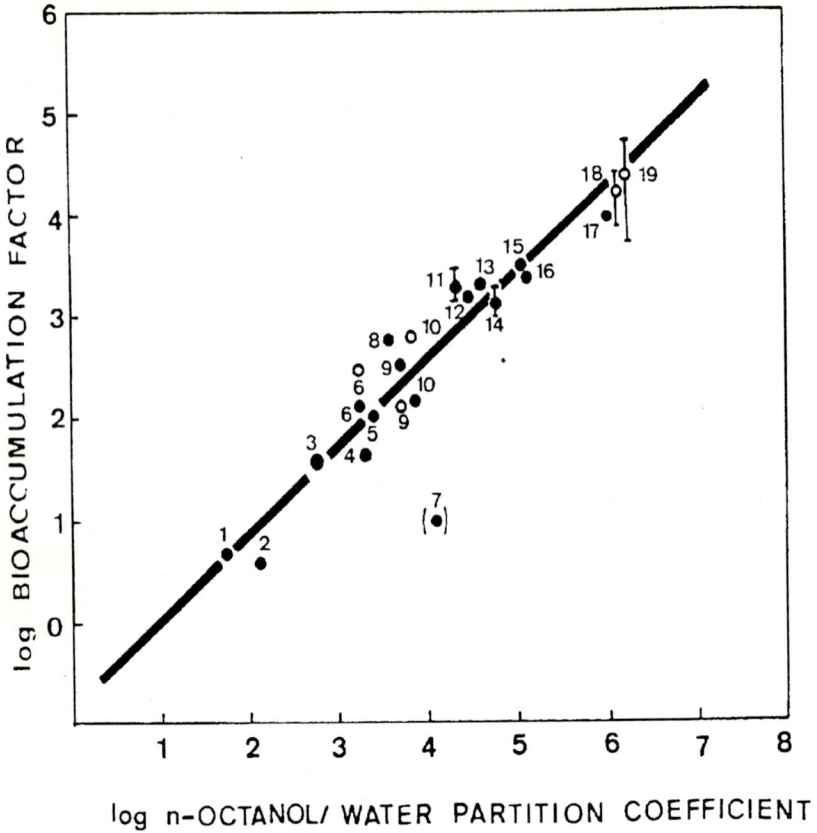

Fig. 4.5.3. Bioaccumulation of organic xenobiotics related to differences in physico-chemical properties: correlation of log bioaccumulation factor in the blue mussel *Mytilus edulis* with log octanol/water partition coefficient. After Geyer et al. (1982). *6* γ-HCH; *10* α-HCH: *18* PCB (Acrolor 1248)

factors for water to the fine-grained fraction of sediments, and those for water to biota, it is obvious that accumulation may be higher in sediments than in the benthic organisms. However, it has to be considered that accumulation in sediments is calculated for the fine grained fraction only (this may be < 0.65 % of the total) but for the total soft parts of the body in the case of the biota.

Accumulation factors are ranked Hg > Cd > Cu > Pb for the blue mussel and Fe > Cu > Mn > Cd > Hg > Pb for the hermit crab, but Mn > Pb > Hg > Cu > Cd > Fe in the case of the sediment samples. Obviously, there is a difference in ranking in the case of cadmium and lead.

Table 4.5.2. Accumulation of heavy metals in blue mussels (*Mytilus edulis*, soft parts dry weight), hermit crabs (*Pagurus bernhardus*, abdominal parts dry weight), dab (*Limanda limanda*, liver fresh weight) and sediments (fraction < 20 μm) from offshore areas of the German Bight. Ranges of accumulation factors estimated from results of Borchardt et al. (1988), the ZISCH cruise in May 1989, Claussen (1988) and Albrecht (pers.comm.)

	Accumulation factors (x 10^6)						
	Hg	Cd	Pb	Cu	Zn	Fe	Mn
M.edulis	0.02-0.10	0.04-0.20	0.02-0.04	0.02-0.03	0.05-0.08		
P.bernhardus	0.08-0.20	0.02-0.07	0.004-0.06	0.35-0.84	0.05-0.09	0.04-0.06	0.007-0.02
L.limanda	13.9-28.4	4.0-11.2	5.0-8.0				
Sediment	0.05-0.26	0.01-0.1	1.2-3.7	0.08-0.17	0.07-0.43		0.61-8.12

This can be understood as there is a great difference among these two metals with respect to their tendency of sorption to inorganic and organic surfaces (concentration to fine-grained particles) and uptake by resorptive processes into organic matrices. The tendency of sorption to surfaces is high in case of lead and rather low for cadmium. Both metals are taken up into the body of the benthic organisms following kinetics with a long half-lives for uptake and elimination. However, the tendency for bioaccumulation as a consequence of high affinity binding to specific biochemical structures is much stronger for cadmium than for lead. In Fig. 4.5.4 accumulation factors calculated on the basis of data obtained in the German Bight and in the central northern North Sea are compared. The figure clearly shows the well-known fact that accumulation factors depend on the concentration in the environment, these being relatively low under conditions of the high metal concentrations in the water of the German Bight, but much higher in the northern North Sea, where metals in the water are at lower concentrations.

Data calculated for accumulation factors for phytoplankton show a figure more comparable to sediments than to fish and crustaceans. This is to be understood as the sorption to phytoplankton is dominated by binding to the body surface. Results on bioaccumulation of heavy metals compiled in Table 4.5.3 show that lead is accumulated by the highest rate followed by copper, zinc and cadmium (Pb > Cu > Zn > Cd).

When evaluating field data on partitioning of metals between water and biota and during the calculation of accumulation factors, it has to be taken into account that metal accumulation in biota is a long term process. Under conditions of changing metal concentrations within the water it may take a

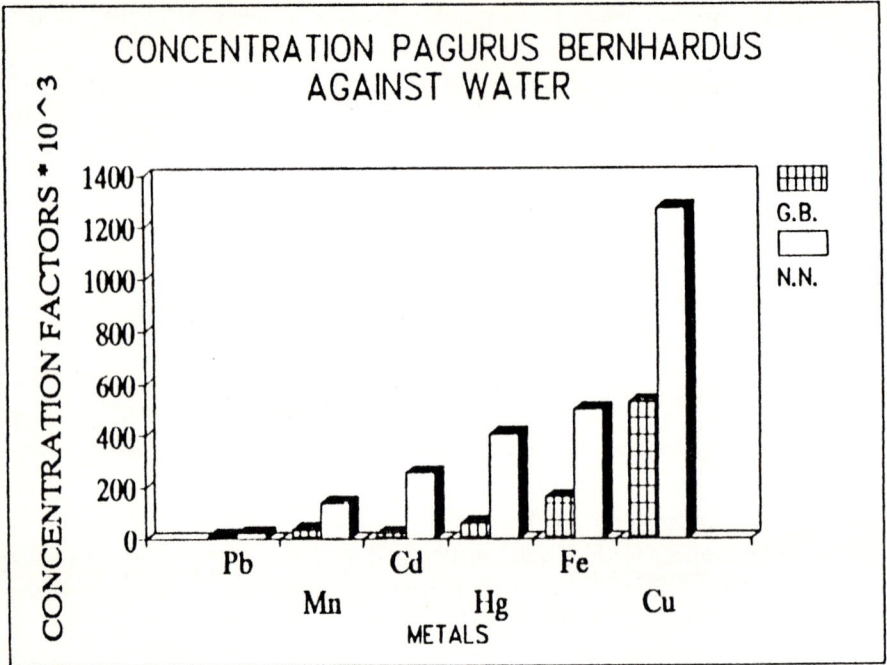

Fig. 4.5.4. *Pagurus bernhardus* Accumulation of some metals. Accumulation factors water to body dry weight comparing results of the German Bight (*GB*) and central northern North Sea (*NN*)

few hours in case of the phytoplankton, but several weeks or even months in the case of mussels, crustaceans or fish until a new steady state between metal uptake and elimination is reached.

4.5.1.2 Interspecific Differences Among Different Biota

Species differ in the bioaccumulation of contaminants. For this reason it is not sufficient to analyze pooled samples of phyto-, zooplankton or benthos but it is necessary to analyze single species.

Heavy Metals in Phytoplankton. Results on the regional and seasonal variability of metals bound to the phytoplankton as described in Chap. 2.5 can be partly explained by regional differences and seasonal changes in the composition of the phytoplankton assemblages.

Table 4.5.3. Bioaccumulation of heavy metals by different phytoplankton species. Accumulation factors determined in laboratory experiments. (Rick 1990)

Organisms	Accumulation factors (x 10^3)			
	Cadmium	Zinc	Copper	Lead
Diatomeae				
Thalassiosira punctigera	0.09-0.24	4.2-5.0	1.1-11.0	42.0-64.0
Th. rotula	0.4-2.2	2.9-22.0	2.2-25.0	41.0-73.0
Coscinodiscus wailesii	0.5-3.2	1.2-2.2	0.19-0.83	
Dinophyceae				
Gyrodinium aureolum	0.8-1.0	8.0	0.75-1.65	80.0-90.0
Prasinophyceae				
Micromonas pusilla	1.0-8.0	1.0-3.4	17.0-35.0	100.0-190.0
Prymnesiophyceae				
Phaeocystis globosa	0.98-6.7	3.0-64.0	4.0-90.0	30.0-93.0

The Figs. 4.3.19 and 4.3.20 in Chapter 4.3 of cadmium in phytoplankton of the German Bight show highest concentrations in regions of phytoplankton blooms of *Phaeocystis globosa*. This is in agreement with results of laboratory experiments summarized in Table 4.5.3, comparing metal accumulation by diatoms and prymnesiophyceans (Rick 1990; Rick et al. 1990). The data show that cadmium is taken up by the prymnesiophycean to concentrations much higher than by the diatoms investigated. The interspecific differences in sorption capacity may be explained by both, differences in nature of the cell surfaces and differences in cell surface areas to cell volume (Myers et al. 1975; Button and Hostetter 1977; Werner 1977). Relating the sorption to cell surface areas, it has been found to be reduced in the smaller types of phytoplankton. The same table also shows that the interspecific differences in sorption capacity vary among the four metals investigated. The highest accumulation of cadmium and lead have been found in the prasinophycean *Micromonas pusilla* but that of zinc and copper was highest in the prymnesiophycean *Phaeocystis globosa*.

Heavy Metals in Zooplankton, Macrobenthos and Fish. There are no data on heavy metal concentrations in zooplankton originating from the cruises covering the North Sea as a total. Some conclusions on metal accumulation can be drawn from analyses made using material obtained from local studies in the German Bight. Detailed results are presented in Chapter 2.5. Results

Table 4.5.4. Zooplankton compared with hermit crab: intra- and interspecific variability of metal accumulation. Concentration factors (CFW: Water to biomass dry weight) calculated from data of monthly sampling winter to spring 1989 in the German Bight

| | | Jan | Feb | Mar | Apr | May | | Jan | Feb | Mar | Apr | May |
		CFW x 10^3						CFW x 10^3				
Bulk zoopl.	Pb	*	76	123	3810	3.7	Cd	*	19	30	12	5.2
Calanus spp.		2.7	3.7	*	13.7	9.3		20	36.7	*	25.2	25.1
Cumacea		17.6	35.3	*	228	40.4		13.1	26.2	*	15	19.3
Mysidacea		2.9	2.9	*	8	2.1		1.3	0.4	*	0.2	0.2
Sagitta spp.		3.6	10.2	2.4	16	2.3		9.3	10.5	5.9	6.2	10.8
Fish larvae		0.9	2	2.7	8	2.3		1.1	3.7	4	2.6	1.6
Pagurus	Pb	2	3	*	17	12	Cd	17	16	*	15	19
	Hg	190	314	*	550	410	Zn	20	39	*	31	66
	Cu	240	238	*	326	488						

show that there are major differences in metal accumulation comparing different types of zooplankton. An example may be the copepod *Calanus* sp. showing accumulation factors for cadmium about four times above those calculated for chaetognaths (Fig. 4.5.9, Table 4.5.4). There are also interspecific differences with respect to the seasonal variability of the metal contamination. Under the conditions of the winter to spring survey of 198, cumaceans and chaetognaths have been found highly contaminated with lead with a maximum in April, whilst *Calanus* was observed to be highly contaminated with cadmium in February.

For an interspecific comparison of metal concentrations in macrobenthic organisms and fish, the data of hermit crabs (*Pagurus bernhardus*), blue mussels (*Mytilus edulis*) and dab (*Limanda limanda*) are shown in Fig. 4.5.5. For the hermit crabs, all data obtained from various locations scattered over the North Sea are combined. Data on blue mussels are those from offshore specimens exposed at specifically designed monitoring stations or sampled from anchor chains of offshore buoys (Borchardt et al.1988). Data for dab are concentrations in samples of liver. There are general differences in the level of metal concentrations between the three types of samples. In the case of copper, silver, zinc and cadmium, the median metal contamination was highest within the abdominal parts of the crabs. The highest median concentration of mercury was calculated for the samples of dab liver, whereas lead looks most enriched in mussels.

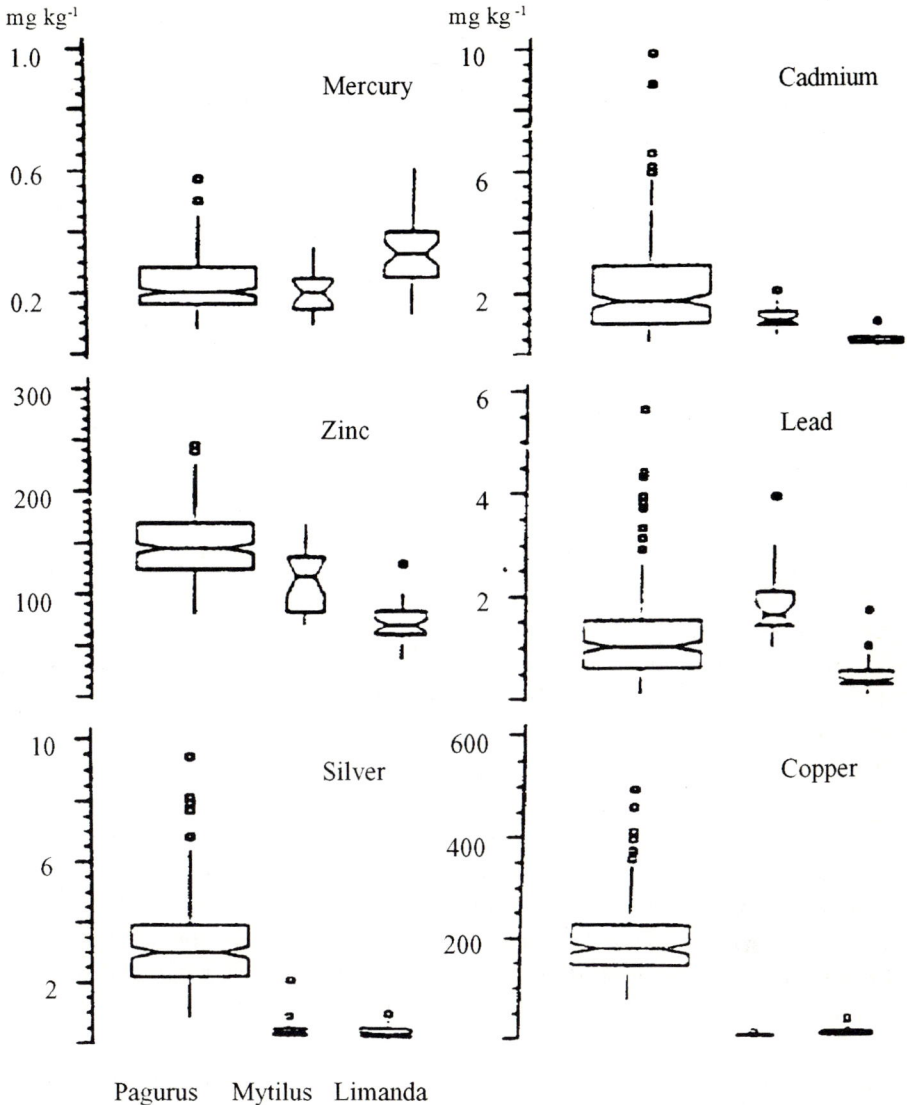

Fig. 4.5.5. Notched box- and whisker plots of metal contents in *Pagurus bernhardus*, *Mytilus edulis* and liver of *Limanda limanda* in offshore regions of the North Sea. (Karbe et al.1988)

Organic Contaminants in Zooplankton, Macrobenthos and Fish. Constantly increasing levels of the particularly lipophilic PCBs and p,p'-DDE have been found in the line from herbivorous zooplankton (*Temora longicornis*) over omnivorous hermit crab (*Pagurus bernhardus*) and carnivorous dab (*Limanda limanda*) to the carnivorous/carcarrivorous whelk (*Buccinum undatum*) (Fig. 4.5.2). The reason for the conspicuously high HCB content of the zooplankton may be seen in the high surface/mass ratio of *T.longicornis* compared to the other animals under investigation. Therefore, surface adsorption may be of major importance in HCB accumulation by the zooplankton crustacean. The relatively low Lindane concentration of the whelk can be seen within the context of its partly carcarrivorous type of feeding. Concentrations of the γ-isomer of HCH are low, whereas concentrations of the α-isomer are high. Ths may be an indication of metabolic isomerization within the food web, leading to an enrichment of the more resistant α-isomer. Comparable conclusions can be drawn with respect to patterns of PCB congeners changing along the food chains.

The group of PCBs is comprised of 209 individual congeners having different toxic and biological responses. All animals investigated here exhibit a distinct PCB-pattern, independent of sampling station and season, when individual congener contributions are expressed as per cent of the sum. Typical PCB patterns of water and various organisms are presented in Fig. 4.5.6. PCB patterns of water and zooplankton are dominated by tri- and tetrachlorobiphenyls. The patterns of dab, hermit crab, whelk and seal, on the other hand, are dominated by penta- and hexachlorobiphenyls with large contributions from the highly persistent congeners 138 and 153. The amounts of individual PCB congeners in the water pattern decrease as the water solubility of the congeners decreases. The similarity of the zooplankton pattern to the water pattern can be explained by the permanent equilibration process with the surrounding sea water, a short life cycle and the absence of any obvious enzymatic metabolizing activity.

Patterns of PCB congeners in marine organisms capable of degrading and metabolizing some congeners are affected by the degree and capacity of biodegradability and the storage capacity of the organisms. Various species, in particular those on a higher trophic level, are known to possess mixed function oxygenase enzyme systems (MFO systems) composed of NADPH and cytochromes, which eliminate lipophilic xenobiotics by converting them to hydrophilic metabolites (Lee 1981). This system is activated in many species by contaminants such as benzo(a)pyrene and PCB mixtures. PCB congeners known to be MFO inducers were also detected in samples of several benthic organisms (Knickmeyer and Steinhart 1988a). The tendency of the MFO system to react with distinct PCB congeners varies between species.

The presence of vicinal hydrogen atoms in meta-para position and only one or no para-chlorine atoms in the biphenyl system appear to be necessary for a rapid enzymatic degradation to an epoxide intermediary. PCB congeners with two para-chlorine atoms in the biphenyl are sterically hindered in interacting with the enzyme receptors. These congeners, partic-

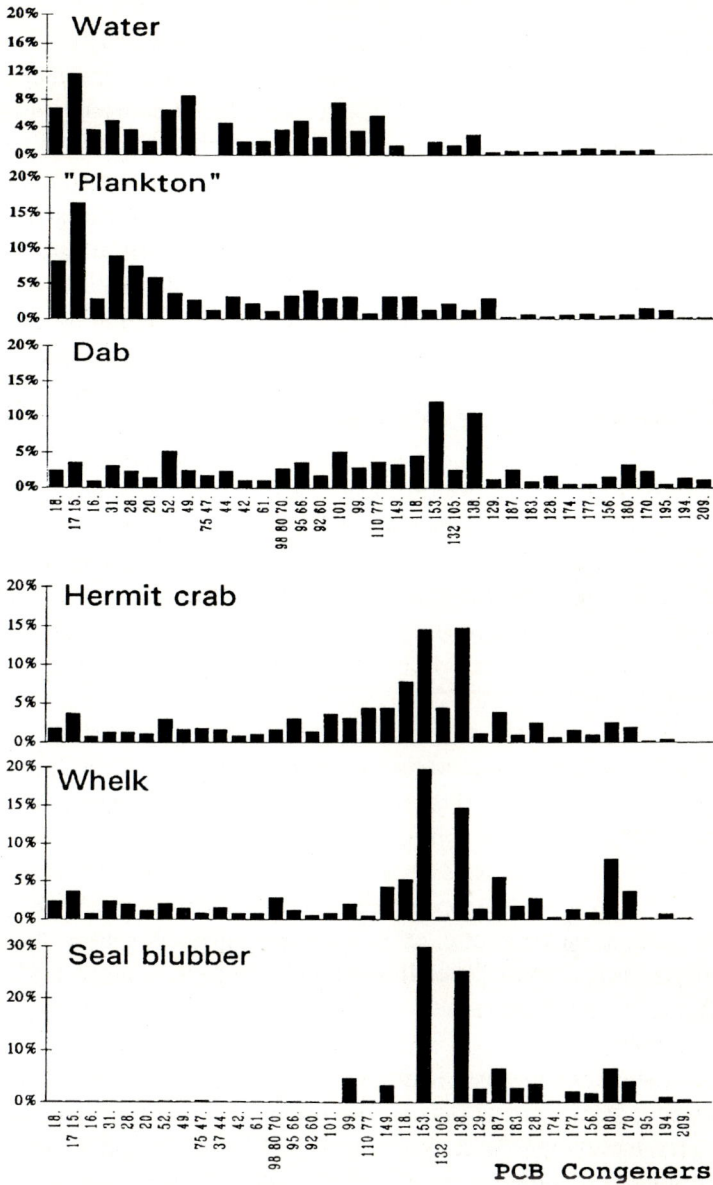

Fig. 4.5.6. Typical PCB patterns for water and different organisms in the German Bight. The PCB pattern of zooplankton was reconstructed using a PCB mixture (Clophen A30/A40/A50/A60, 4.0/1.0/0.8/1.2) (Knickmeyer and Steinhart 1990c). The sample of seal blubber (*Phoca vitulina*) was generously provided by Dr. Heidemann, Institut für Haustierkunde Univ Kiel. The contribution of each congener to the sum is given in order of the elution from the SE-54 GC column (numbering according to Ballschmiter and Zell (1980). For co-eluting components, the quantification was done at the basis of the upper component

ularly PCB 138 and 153, are enriched in the food chain. A comparison of relative PCB pattern in hermit crab, dab, whelk, and zooplankton from the German Bight (Fig. 4.5.6) shows that the higher members of the marine food web accumulate PCB congeners which do not interact easily with enzymes of the MFO system (Knickmeyer and Steinhart 1990b,c). The PCB pattern found in the seal, an animal at the end of many marine food chains, is the result of enzymatic degradation of the less persistent PCB congeners and contrasts with the pattern in organisms at the beginning of the marine food web, represented here by zooplankton.

4.5.1.3 Intraspecific Differences

Differences of the sorption capacity of contaminants not only exist among species but also within single species. These are, e.g. due to different stage of development of the organisms or to seasonal variability. Furthermore, the concentrations of contaminants are dependent on sex and the body condition of the organisms.

Factors Influencing the Sorption of Metals by the Phytoplankton. The sorption of metals and organic xenobiotics by phytoplankton is influenced by various factors (Fig. 4.5.1) modifying the bioavailability of the contaminants and the sorption capacity of the biota. Of major importance are all factors influencing the dynamics of metal complexes formed by different species of heavy metals, synergistic effects among different metals, as well as antagonistic effects among nutrients and contaminants. The sorption capacity is depending on both, physical and biochemical as well as autecological differences among the single species of phytoplankton (examples in Table 4.5.3.) and different stages of the same species within the development (and breakdown) of a plankton bloom as well as by synecological interactions within the plankton assemblages.

Examples for combined effects are given in Figs. 4.5.7 and 4.5.8. Increasing complexation reduces the uptake rate of copper (Fig. 4.5.7). The sorption of zinc, lead and copper by diatoms is a function of the prevailing silicate content (Fig. 4.5.8). For a more detailed discussion, the reader is referred to Rick (1990) and Dürselen (1990).

Effects of Feeding and Reproduction on the Accumulation of Organo-chlorines in the Hermit Crab. There is a seasonal and regional change with respect to the food source of the hermit crab. During times and at locations

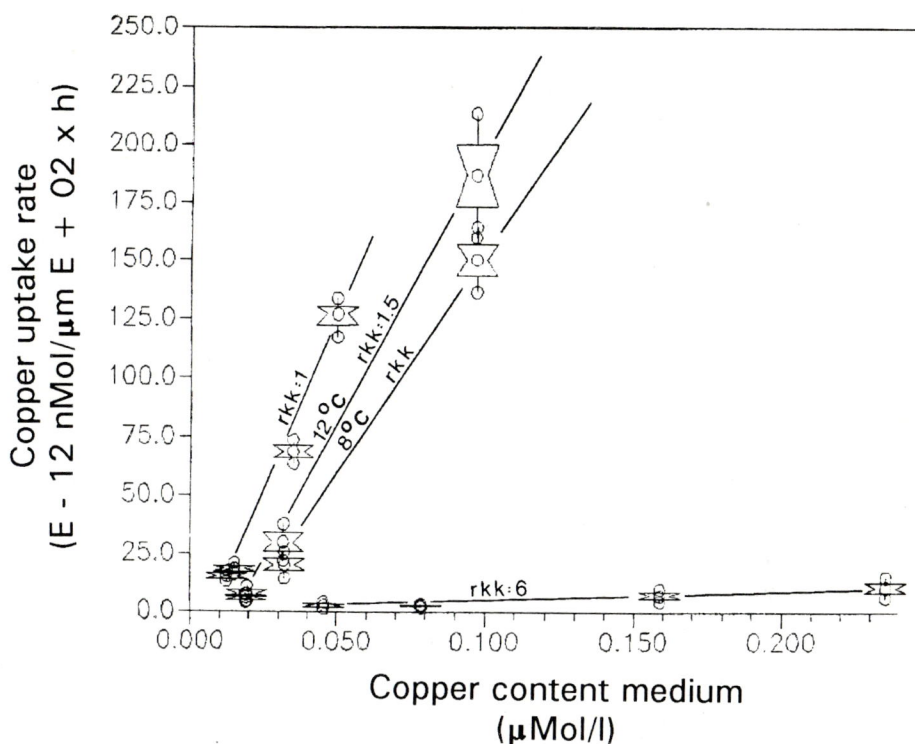

Fig. 4.5.7. Effects of complexation on the uptake of copper by phytoplankton Uptake by the diatom *Thalassiosira punctigara* depending on copper concentrations and complexation. *rkk* relative complexation capacity related to lead; *rkk 1* capacity of complexing 5000 ng Pb/l

of high phytoplankton productivity, the main food source is the freshly deposited phytoplankton material. In times or at locations of lower phytoplankton productivity, the main food sources are sedimentary organic matter and components of the benthic fauna. The changing food composition directly influences the contamination pattern of the animal (Knickmeyer and Steinhart 1988a).

Male whelks and dabs have been discovered to be significantly more highly polluted with PCBs than females because of the transfer of cyclic organochlorines from females to their offspring (Knickmeyer and Steinhart 1989a). The lipid-water partitioning equilibria of organochlorines from ambient water via gills and surface to body lipids of the organisms is generally considered to be dominant in gill-breathing aquatic animals (Falkner and Simonis 1982; Tanabe et al. 1987). The dab *Limanda limanda* however, shows no direct influence of equilibration with the surrounding

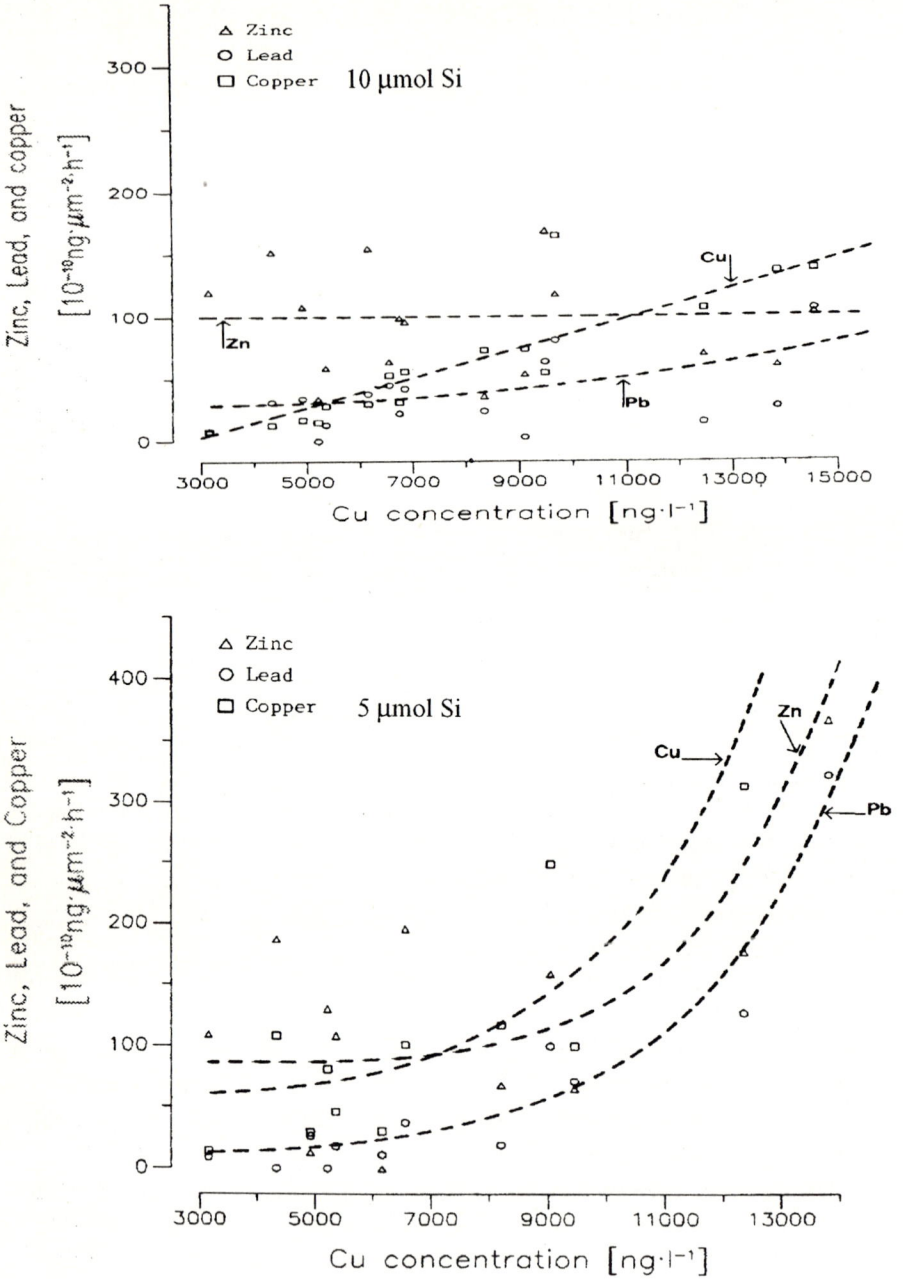

Fig. 4.5.8. Effects of interactions among different metals and silicate Uptake rates of zinc, lead, and copper by the diatom *Thalassiosira punctigera* as dependent on copper concentration in the medium at two different silicate concentrations. (Dürselen 1990)

water: the pattern of the PCB congener composition of the gill exhibited a pattern comparable to the liver. Further, the gill lipids contained a great proportion of neutral lipids, which is typical for liver lipids. These results suggest that lipid composition is more important for determining congener distribution in dab gills than equilibration with sea water (Kammann et al. 1990).

Local Variability of Heavy Metal Accumulation in Zooplankton and Benthos. In Chapter 2.5, results have been given on the local variability of heavy metal concentrations in hermit crabs of the German Bight during the winter to spring 1988/89 period. Within this period there was a significant decrease of metal concentrations dissolved in the water and an increase in the sorption capacity of the benthic organisms with the consequence that concentration factors to be calculated did increase for all metals analyzed from January to May, most for mercury and copper. Comparable results could not be obtained for zooplankton samples taken during the same study (Fig. 4.5.9).

Relation Metal Body Burden to the Somatic Condition of Hermit Crabs. The relation of the dry weight of the abdominal parts to the carapax length was used as a somatic index for a rough characterization of the body condition of *Pagurus bernhardus*. The total fresh weight, total dry weight and the dry weight of the abdominal parts have been found to be highly correlated. For practical reasons, the relationship between abdominal dry weight (g) and the carapax length (cm) was chosen, as corresponding data are available from all crabs analyzed during the ZISCH field studies. The body somatic indices calculated are in a range from 0.04 to 8.9 with a median of 0.5 (0.9). High values indicate crabs with well-developed abdominal parts. The data are not fully evaluated yet with respect to their meaning for the nutritional status and the reproductive cycle of the crabs. However, this index may be used as a practical approach to correlate contaminants concentrations to an easily determined somatic covariate.

Due to the low number of samples taken during the ZISCH winter campaign 1987, calculations had been made only for the data originating from the summer cruise 1986. As a result of calculations made for different sub-regions of the North Sea, it can be concluded that there is a general tendency of somatic indices decreasing from the southern to the northern area (Table 4.5.5). Crabs of similar length are of lower biomass in the north compared to the southern North Sea. Within the northern North Sea the somatic indices proved to be negatively correlated with the water depth

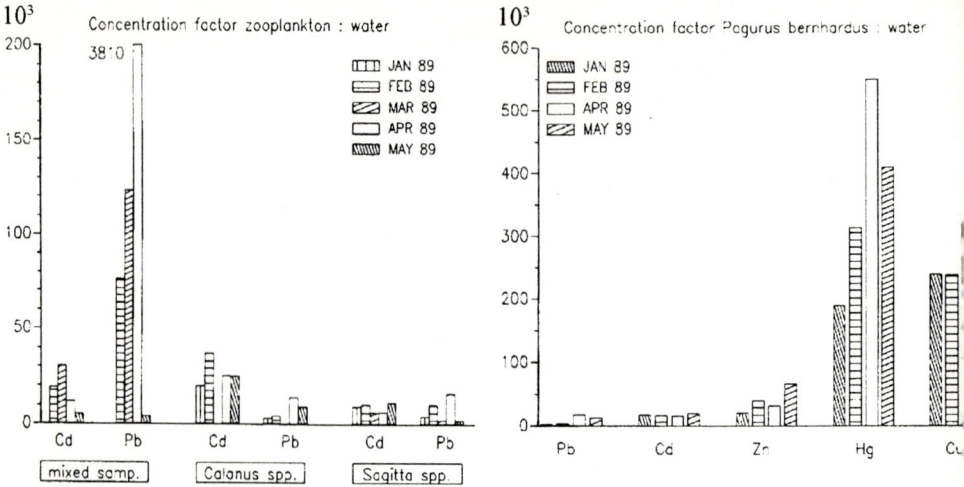

Fig. 4.5.9. Heavy metal accumulation factors (10^3): concentrations in zooplankton (bulk sample, copepod *Calanus* spp., chaetognat *Sagitta* spp.), and hermit crab (dry weight) related to water

of the location from which the crabs had been sampled. This may be taken as an indication of a nutritional status of the crabs decreasing with increasing water depth. Also, the concentrations of mercury and cadmium are correlated positively with water depth and negatively with the somatic condition. Crabs with a low weight to length relationship are more highly contaminated with cadmium and mercury than those in a better condition with a higher weight to length ratio. This conclusion cannot be generalized, as there are local conditions for instance at the Dogger Bank where crabs in better condition proved to be higher in cadmium compared to those in bad condition. It is obvious that there are relations between metal concentrations and the nutritional status of the crabs. This can be shown by analyzing regional differences as well as seasonal differences, as shown by monthly resampling of different stations in the German Bight Nov. 1988 to May 1989 (Chap. 2.5).

Seasonality of Specific and Unspecific Binding of Organochlorines. It is known that there are major differences among the various compartments of an organism with respect to its sorption capacity for xenobiotic contaminants. Kinetics of accumulation as well as of elimination can be described assuming a two-compartment model, discriminating two types of binding: unspecific sorption resulting in a partitioning of concentrations between the surrounding water and the biotic matrix correlated to the log octanol/water partitioning coefficient (see above), and sorption by binding to specific biochemical structures following saturation type sorption kinetics (Fig. 4.5.10). From the toxicological point of view, the latter may be the more important as it stands for the more reactive portion, biochemically.

During the ZISCH 6 monthly samplings, some combined field- and laboratory-experiments were conducted to investigate seasonal differences and interdependencies of lipids and proteins with respect to the sorption capacity of blue mussels for (^{14}C-labelled) γ-HCH (Gonzalez- Valero

Table 4.5.5. *Pagurus bernhardus:* correlation matrix: metal concentrations, somatic index (SI), depth (Dp), salinity (S). Northern North Sea: upper right, southern North Sea: lower left. Data originating from the ZISCH summer cruises. Correlation coefficients and level of significance. Correlations with probability of error less than 0.10 are in boldface type

	Hg	Cd	Ag	Zn	Cu	Pb	Fe	Mn	SI	Dp	S	
Hg		**0.64**	0.06	0.16-	0.00-	**0.42**	0.11	0.00	**0.26-**	**0.57**	**0.27**	N
		0.00	0.69	0.31	0.99	**0.01**	0.50	0.98	**0.10**	**0.00**	**0.10**	o
Cd	0.04		0.10-	0.14-	0.05-	**0.27**	0.07	0.01-	**0.32-**	**0.66**	**0.54**	r
	0.77		0.52	0.39	0.77	**0.09**	0.66	0.97	**0.04**	**0.00**	**0.00**	t
Ag	0.21	0.17-		0.16	**0.30**	0.16-	0.04-	0.02-	0.14-	0.25-	0.12-	h
	0.12	0.23		0.31	**0.06**	0.33	0.83	0.89	0.38	0.12	0.44	e
Zn	**0.31**	0.00	0.06-		**0.37**	0.03-	0.23-	0.17-	**0.30**	**0.30-**	0.14-	r
	0.02	0.98	0.69		**0.02**	0.83	0.15	0.29	**0.06**	**0.06**	0.39	n
Cu	0.25	0.26	0.01	0.09		0.23	0.00	0.23	0.01-	0.24-	**0.26-**	
	0.07	**0.06**	0.93	0.52		0.15	0.99	0.15	0.94	0.14	**0.10**	
Pb	0.03	0.06	0.01	0.17-	0.10		**0.38**	0.25	0.07-	0.16	0.13-	
	0.84	0.65	0.96	0.22	0.49		**0.02**	0.12	0.68	0.32	0.42	
Fe	0.02-	0.09	0.20	**0.33-**	**0.36**	0.44		**0.67**	0.16	**0.30**	0.12-	
	0.91	0.50	0.14	**0.02**	**0.01**	0.00		**0.00**	0.32	**0.06**	0.45	
Mn	0.09-	0.10-	0.05	0.20-	0.14	0.50	0.64		0.01	0.14	0.01-	
	0.51	0.47	0.74	0.14	0.31	0.00	0.00		0.94	0.38	0.94	
SI	0.13	0.13-	0.02-	**0.32**	**0.51-**	**0.28-**	**0.40-**	0.12-		**0.27-**	0.21-	
	0.33	0.35	0.87	**0.02**	**0.00**	**0.04**	**0.00**	0.37		**0.10**	0.19	
Dp	0.02-	**0.34**	0.09	**0.29-**	0.01-	0.02-	0.18	0.15	0.09		**0.56**	
	0.88	**0.01**	0.53	**0.03**	0.95	0.90	0.18	0.26	0.53		**0.00**	
S	**0.28-**	**0.65**	0.12-	**0.35-**	0.16	0.14-	0.11	0.17-	0.18-	**0.56**		
	0.04	**0.00**	0.39	**0.01**	0.23	0.31	0.42	0.20	0.18	**0.00**		

Southern

bioconcentrated Lindane (µg/kg)

Lindane concentration in water (µg/L)

bioconcentrated Lindane (µg/kg)

Lindane concentration in water (µg/L)

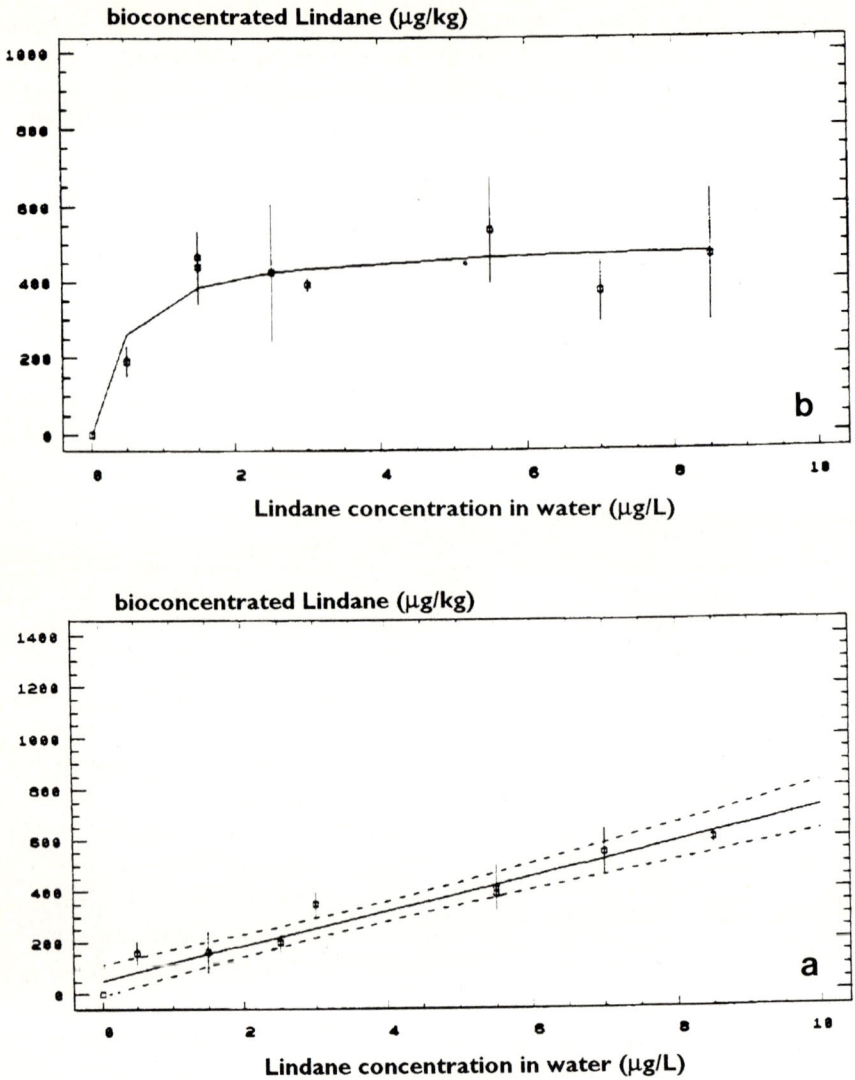

Fig. 4.5.10.a,b *Mytilus edulis* Sorption isotherms of the non saturation type of binding (**a**) and the saturation type of binding (**b**). (Results of laboratory experiments by Gonzalez-Valero 1990)

1990). It could be shown that there was a seasonally anatropous trend of sorption to the lipid and the non-lipid compartments. The high portion of lipid-bound HCH resulted in a trend of concentrations in lipids corresponding to the total body burden. However, there are indications of a negative correlation between lipid concentration and concentrations of γ-HCH within the non-lipid compartments (Fig. 4.5.11). The consequence may be that during times of low lipid contents, γ-HCH concentrations are low related to the total biomass but of increased availability for specific receptors and toxic effects.

4.5.2 Contaminant Fluxes: Biological Transport, Turnover and Biodeposition Related to Trophodynamic Interactions in the Pelagic and Benthic North Sea Environment

4.5.2.1 Biotic Effects on the Fate of Contaminants within the Epipelagic Trophogenic Zone

Sorption of Contaminants by the Phytoplankton. The connection between the biomass of dominating phytoplankton groups and the heavy metal sorption rates of the same plankton species, determined in laboratory experiments, enables us to make an estimate of the integrated heavy metal transfer of the planktonic primary producers for the periods of the ZISCH cruises in May/June 1986 and February/March 1987, knowing the in-situ heavy metal concentration. Obviously, the diatoms seem to play the most important role concerning the copper and zinc transfer (Table 4.5.6), while the cadmium input into the food chain is caused by the other phytoplankton groups.

The results of the sorption data from both research periods are projected for the whole year. Results and the amounts of the annual inputs (natural and anthropogenic input, rivers, atmosphere, direct inputs, dumping etc.) and the existing amounts of heavy metals dissolved in the water of the North Sea (in winter) are listed in Table 4.5.7. The calculations made may be taken as an indication that the projected yearly sorption capacities of copper and zinc by the phytoplankton are both higher than the heavy metal amount in the water of the North Sea and considerably above the annual input of these heavy metals. Therefore, phytoplankton activity arithmetically causes a heavy metal turnover of 1.4-2.3 times for copper and 1.3-1.5 times for zinc. This shows not only the importance of the turnover of heavy

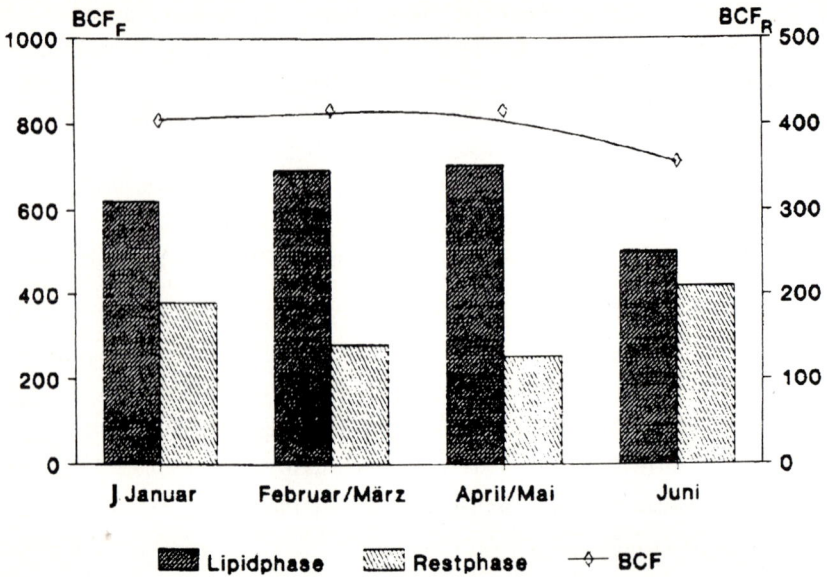

Fig. 4.5.11. *Mytilus edulis* Bioconcentration of γ-HCH to the hepatopancreas as a total (bioconcentration factor: *BCF*) and to the lipid (*BCF*$_L$) and the non-lipid fractions (*BCF*$_{NL}$). Results of laboratory experiments using mussels exposed at an offshore monitoring station within the German Bight by Gonzalez-Valero (1990)

metals by the primary producers, but also that the as yet unquantified process of remineralization of organically bound heavy metals and with that the cycling back into the dissolved phase is an important factor for understanding the cycle of heavy metals in the North Sea ecosystem. The corresponding data for cadmium point to a considerably lower turnover by the phytoplankton. The estimate shows that only 10% of the amount in water is taken up by the phytoplankton and the annual input may exceed the sorption of the primary producers by the factor of 3-4.

In areas of restricted metal availability, the sorption by the phytoplankton may lead to an exhaustion of metals in the most productive euphotic layers and a nutrient type vertical distribution of metals among the water layers. Such phenomena are well documented from oceanic locations (Bruland 1980; Bruland and Franks 1983). Results of investigations at the NE Atlantic continental slope and within the northern North Sea show the same figure.

The major part of the contaminants taken up by the phytoplankton and released again when plankton stocks decay may be recycled by a microbial

Table 4.5.6. Estimated potential heavy metal sorption capacity by different phytoplankton groups in the North Sea (575 10^3 km^2) during the ZISCH campaigns May/June 1986 and February/March 1987. (Rick 1990)

Research period May/June 1986					
Diatomeae			Dinophyceae		
Copper	Zinc	Cadmium	Copper	Zinc	Cadmium
5.3 ± 2.0 t/h	13.8 ± 1.7 t/h	3.2 ± 0.3 kg/h	98.3 ± 36.9 kg/h	3.2 ± 0.4 t/h	1.2 ± 0.1 kg/h
Prymnesiophycea			Prasinophyceae		
Copper	Zinc	Cadmium	Copper	Zinc	Cadmium
2.4 ± 0.8 t/h	10.1 ± 3.8 t/h	15.2 ± 7.1 kg/h	43.8 ± 5.7 kg/h	7.2 ± 3.7 kg/h	0.1 ± 0.1 kg/h
Total					
Copper	Zinc	Cadmium			
7.8 ± 2.8 t/h	27.1 ± 5.9 t/h	19.8 ± 7.5 kg/h			
Research period February/March 1987					
> 20 μm			< 20 μm		
Copper	Zinc	Cadmium	Copper	Zinc	Cadmium
0.9 ± 0.4 t/h	1.6 ± 0.2 t/h	0.5 ± 0.1 kg/h	13.5 ± 1.6 kg/h	6.5 ± 3.4 kg/h	0.1 ± 0.1 kg/h
Total					
Copper	Zinc	Cadmium			
1.0 ± 0.4 t/h	1.6 ± 0.2 t/h	0.5 ± 0.1 kg/h			

loop within the euphotic layer or be taken up by grazing of the herbivorous zooplankton. In any case, there is a substantial net vertical transport of contaminants via the particle flux of detrital and faecal material (biodeposition). It can be concluded that calculations made in Chapter 4.4

Table 4.5.7. Annual uptake of metals by the phytoplankton, input and existing amounts in the water (situation in winter). All data given in 10^3t or 10^3t/a. The data of the annual input are taken from Kersten et al. (1988), the existing amounts in the water of the North Sea are estimated from data by Lee and Ramster (1981), Dicke et al.(1987) and Kersten et al.(1988). (Rick 1990)

	Copper	Zinc	Cadmium
Phytoplankton sorption/year	17-36.8	68-102	0.04-0.08
Annual input	3-4	22-28	0.123-0.35
Existing amount	12-16	50-70	0.5-0.9

on effects of abiotic processes on the fate of contaminants underestimate the real fluxes, as the physical transport of contaminants bound to suspended matter is enhanced by the biotic processes.

4.5.2.2 Biotic Effects on the Fate of Contaminants within the Deeper Pelagic and Benthic Tropholytic Zone

Sorption of Contaminants by the Zooplankton. Heavy metals and organic xenobiotics are taken up by the zooplankton primarily from the water corresponding to its body surface sorption capacity, and, in addition, by food chain carry-over. The major part of contaminants taken up together with the food will not be assimilated but excreted, bound to faecal material. By this route, material originally bound to phytoplankton will settle out to the deeper water layers and to the bottom, depending on the amount of zooplankton present within the water column. Specific attention has to be paid to the diurnal vertical migrations of zooplankton species and ladders of migration that have a substantial contribution to contaminant fluxes from the trophogenic to the deeper layers of the tropholytic zone.

There is a major difference between the southern and northern North Sea with respect to zooplankton biomass and zooplankton species diversity within the water column and, as a consequence, in trophodynamic interactions among phytoplankton and zooplankton. It is clear that contaminant fluxes are strongly influenced by regional differences of both. The length of food chains and the complexity of food webs resulting from the more diversified zooplankton dynamics in the deeper areas of the central and northern North Sea has to be taken into account for an explanation of local differences in the availability of contaminants for benthic organisms and deep-dwelling fish. Among the different components of the zooplankton, the copepod *Calanus finmarchicus* is of major importance, and this species is known to exhibit extended diurnal vertical migrations from its subsurface feeding grounds to deep waters, at least during times of reduced phytoplankton productivity.

Knickmeyer and Steinhart (1989b) determined the concentrations of Σ-PCB, p,p'-DDE, HCB, γ-HCH and α-HCH in mixed zooplankton samples. The *Temora longicornis* dominated samples in the southern North Sea were distinctly more highly contaminated with organic pollutants compared with the northern North Sea, dominated by the copepod *Calanus finmarchicus*. The distribution patterns of the organic contaminants of the zooplankton were similar to the dissolved contaminants in water (Gaul and Ziebarth 1983; Carlson 1986). But when the biomass of the zooplankton was taken into account, the distribution pattern of the load of organic contaminants by the zooplankton shows a different picture. Krause and Knickmeyer (1992) calculated amounts of organic contaminants bound to depth integrated biomass of zooplankton to get an idea on the role of zooplankton in the pollutant transfer. Due to the high zooplankton biomass in the central and northern North Sea, the total of plankton bound organic contaminants was higher in this area (Fig. 4.5.12). In the southern North Sea, only some spots near the sources were visible (Firth of Forth, rivers Thames, Rhine and Elbe).

Krause and Martens (1990) found that the horizontal distribution of the zooplankton biomass mainly imitated the cyclic current pattern of Atlantic water, the zooplankton biomass forming a large left-turning helix. Therefore, the zooplankton was gradually shifted to deeper water layers. It can be assumed that in the northern North Sea a considerable proportion of zooplankton and its load of pollutants will be shifted into deeper water layers and will finally reach the sea floor (Krause and Knickmeyer 1992). The contaminant flux to the benthos via faecal pellets of the zooplankton seemed not to play the anticipated important role because of a quick remineralization of the faecal material, so that no faecal pellets were detectable in water layers deeper than 100 m (Martens and Krause 1990).

Sorption of Contaminants by the Benthos. To obtain a rough idea on the amounts of heavy metals within the North Sea that might be bound to the standing stocks of benthic organisms, some calculations are made for the North Sea as a total and for the different subregions known as the ICES Boxes 1 to 7 (in the north not identical with the NSTF Boxes, see Fig. 2.1.13, Chap. 2.1). Calculations are made assuming the boxes to be homogeneous with respect to biomass of the macrobenthos and the hermit crab as a selected component of the epibenthic megafauna. The estimated

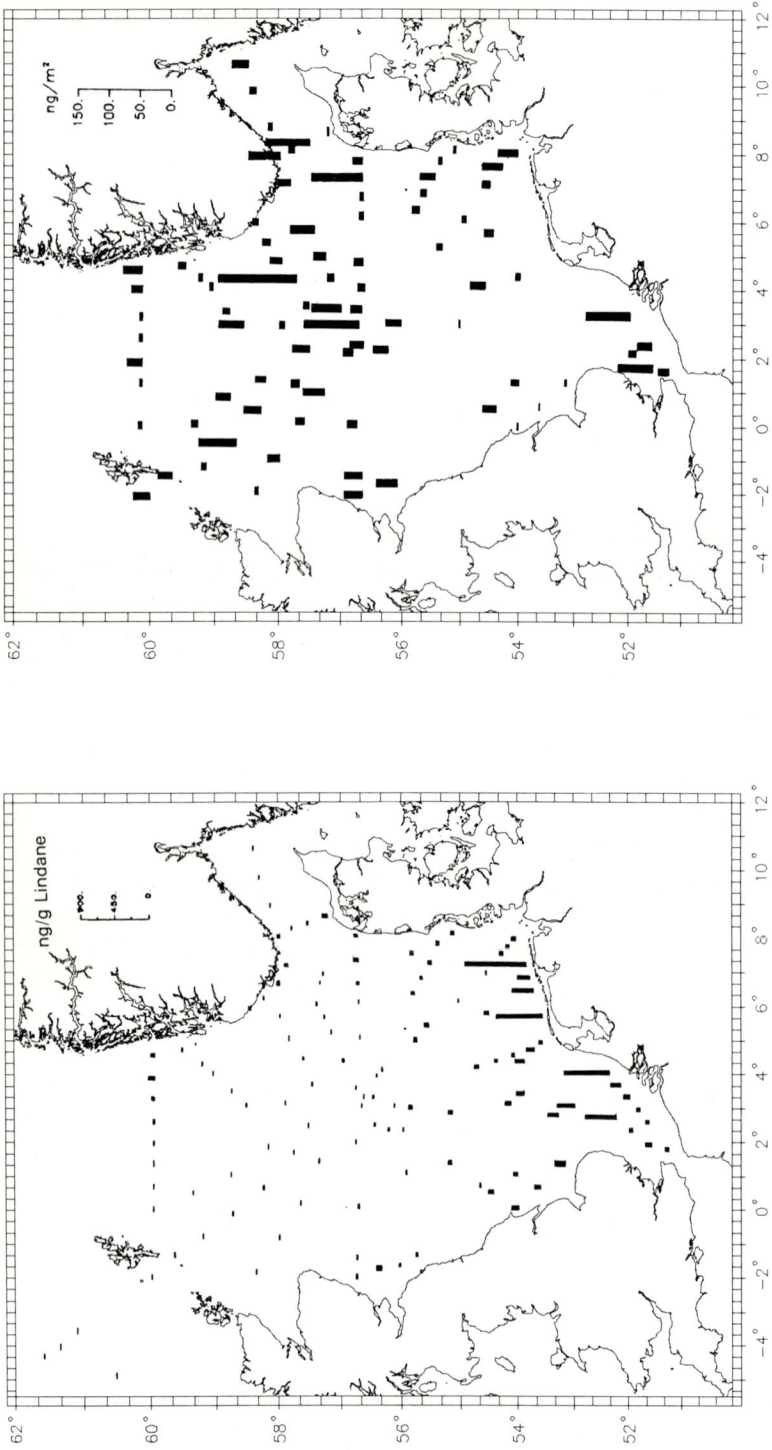

Fig. 4.5.12. γ-HCH in zooplankton Concentrations (ng/g) related to hexane extractable lipids (*left*) and amounts of zooplankton bound contaminants in the water column (ng/m^2) calculated considering regional differences in zooplankton biomass (*right*) during the ZISCH North Sea survey May to June 1986. (Knickmeyer and Steinhart 1989b, Krause and Knickmeyer 1992)

biomass is multiplied by a metal concentration estimated as a rough approximation on the basis of results available for the hermit crab, determined in samples of the respective boxes. There are no adequate data for an estimation of the abundance of hermit crabs in Box 3b (English coastal) and no data for the macrofauna in Box 6 (Norwegian Trench). Results for cadmium, mercury and lead are listed in Table 4.5.8 and Fig. 4.5.13.

There is a distinct difference between the southern and northern North Sea with respect to the macrobenthos biomass. As a consequence, amounts of metals per unit area bound to the hermit crab or the macrobenthos as a total show quite a different regional pattern to metal concentrations of the biota. This difference is most obvious in the case of cadmium.

Mean cadmium concentrations in hermit crabs are in a range of 0.8 to 5.6 mg/kg. Concentrations per unit area have been estimated to be in a range of about 10 to 30 mg/km^2 for the total biomass of hermit crabs and (as a rough estimation) of about 10 to 30 g/km^2 for the total macrobenthos. The highest cadmium concentrations were determined in samples collected in the northernmost offshore North Sea (ICES box 1). Although the macrobenthos biomass is relatively low in this area, the cadmium concentrations per unit area are estimated to be highest (about 28 g/km^2). Next to the northern North Sea, high areal concentrations of cadmium have been estimated also for the offshore southern North Sea (ICES box 7b: 22 g/km^2). The total amount of cadmium bound to the macrobenthos is estimated high in the southern North Sea, irrespective of low cadmium concentration per unit weight. This the case as the low concentration is multiplied by a biomass per unit area about three times the biomass in the deep waters of the Northern North Sea.

Quite another figure is approximated for the benthic areal concentration of mercury. The estimated amount of mercury bound to the benthos as a total is in a range of about 1 to 5 g/km^2. Areal concentrations were estimated highest in the German and Southern Bights (ICES boxes 5 and 4). These are areas of high benthic productivity where mercury concentrations are also high. An obvious difference between concentrations and amounts per unit area results for the northern North Sea. The mercury concentrations determined for crabs of that area are only slightly below the concentration level in the German Bight (0.26 compared to 0.35 mg/kg); but, as a consequence of major differences in macrobenthos biomass, the total amount of mercury bound to the macrobenthos can be estimated as being about twice as much in the German Bight compared with the less productive northern North Sea.

The highest concentrations of lead were determind in samples from the offshore central North Sea (ICES box 7a). As in the case of cadmium, lead

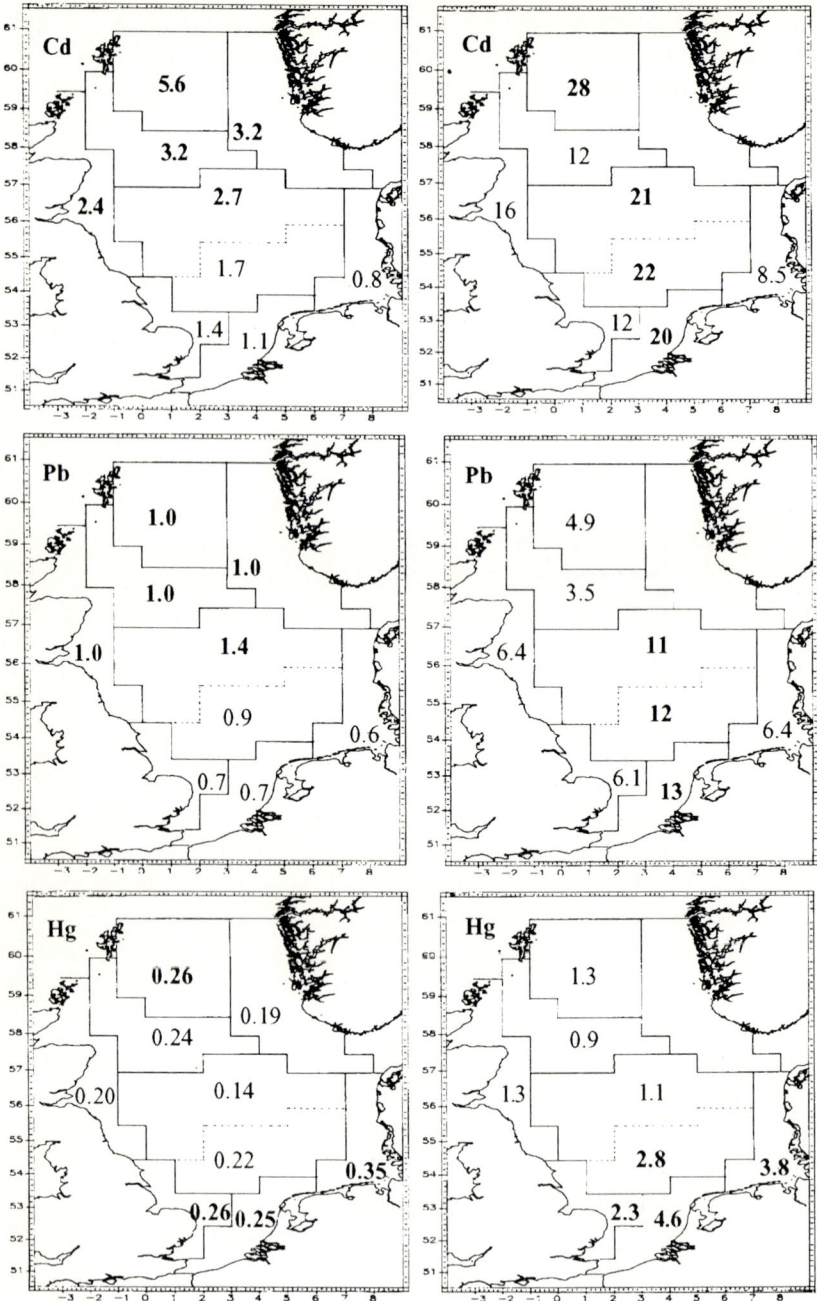

Fig. 4.5.13. Cadmium, lead and mercury bound to the biomass of the macro-benthos: Mean concentration of the hermit crab (mg/kg dry weight), *left side*, and amounts per unit area bound to the macrofauna (g/km), *right side*, calculated as a rough approximation at the basis of data compiled in Table 4.5.8

Table 4.5.8. Contaminants in benthic organisms calculated for the North Sea as a total and for the different subregions: Cd, Hg and Pb in *Pagurus* and in the total macrofauna. Calculations made on the basis of assumptions discussed in the text. Concentrations above the North Sea mean, in bold type

ICES box no.	1	2	3a	3b	4	5	6	7a	7b	NS Total
10^3 km^2	52.2	56.0	48.4	37.0	47.8	43.3	68.7	90.4	69.3	513
Pagurus bernhardus										
Biom. t	193	510		1335	685	1587	149	825	361	5645
Biom kg/km^2	3.7	9.1		15.6	**14.3**	**36.7**	2.2	9.1	5.2	11.0
Cd mg/kg	**5.6**	**3.2**	**2.4**	1.4	1.1	0.8	**3.2**	**2.7**	1.7	1.9
Hg mg/kg	**0.26**	0.24	0.20	**0.26**	**0.25**	**0.35**	0.19	0.14	0.22	0.25
Pb mg/kg	**1.0**	**1.0**	**1.0**	0.7	0.7	0.6	**1.0**	**1.4**	**0.9**	0.9
n	13	17	7	15	26	120	7	70	63	338
Cd kg	1.08	1.63		2.5	0.75	1.27	0.48	2.23	0.62	11
Hg kg	0.05	0.12		0.31	0.17	0.56	0.03	0.12	0.08	1.4
Pb kg	0.19	0.48		1.13	0.48	0.95	0.15	1.16	0.32	5.0
Cd mg/km^2	21	**29**		30	16	**29**	7.0	25	8.8	21
Hg mg/km^2	1.0	2.2		3.6	**3.6**	**13**	0.4	1.3	1.1	3.3
Pb mg/km^2	3.6	8.6		13	10	**22**	2.2	**13**	4.7	9.6
Macrobenthos										
Biom.[a]t/km^2 AFDW	4.05	2.93	5.33	6.94	**14.6**	**8.58**		6.07	**10.3**	7.35
Biom. t/km^2 DW	5.06	3.67	6.66	8.68	**18.3**	**10.7**		7.58	**12.9**	9.19
Biom. 10^3t DW	264	205	322	321	873	464		686	892	>4027
Cd t	1.5	0.66	0.77	0.45	0.96	0.37		1.9	1.5	>8.1
Hg t	0.06	0.05	0.06	0.08	0.22	0.16		0.10	0.20	>0.9
Pb t	0.25	0.19	0.31	0.22	0.61	0.28		0.96	0.80	>3.6
Cd g/km^2	**28**	12	16	12	**20**	8.5		21	22	18
Hg g/km^2	1.3	0.9	1.3	**2.3**	**4.6**	**3.8**		1.1	**2.8**	2.1
Pb g/km^2	4.9	3.5	6.4	6.1	**13**	6.4		**11**	12	8.2

[a]according to ICES Benthos Group (Heip et al. 1990). Biomass dry weight concentrations calculated assuming that dry weight is 1.25 x ash-free dry weight

concentrations show a general tendency to be higher in the northern (ICES boxes 1, 2, 3a, 6, and 7a) compared to the southern North Sea (ICES boxes 3b, 4, 5, and 7b). This regional pattern is turned to the opposite, calculating the total amounts bound to the macrobenthos per unit area, which is estimated highest in the Southern Bight and the adjacent offshore Southern North Sea.

The amounts of cadmium, lead and mercury accumulated by the macrobenthos of the North Sea as a total can be estimated to be more than 8 t Cd, 4 t Pb and 1 t Hg, respectively. Assuming an input of 135-335 t/a Cd and 50-75 t/a Hg, but 6000-11,000 t/a Pb (INC 1987), it can be concluded that significant parts of the annual input of cadmium and mercury are not only taken up by the phytoplankton but are also transferred to the benthic organisms. Quite a different situation can be concluded for lead. As a consequence of its tendency to be adsorbed to surfaces, concentrations of lead are high in the phytoplankton; but due to its low bioavailability for incorporation, the portion of this metal taken up by the benthic organisms (and fish) is very small (less than one part per thousand).

4.5.2.3 Gradients of Contaminants at a South to North Transect

Comparing the results of the distribution of contaminants for the total North Sea on a large scale with those of the German Bight (Chaps. 2.3 and 2.5), it can be seen that several parameters show similar regional patterns and follow a covariant seasonal variability whereas others differ.

The boxes of Fig. 4.5.14 will be used for a rough evaluation of data obtained during the ZISCH campaigns on a south to north transect. Within the southern North Sea two transects are plotted, German Bight to Dogger Bank and Southern Bight to Dogger Bank. Concentrations measured at different stations are combined to a line of boxes to obtain a greater data basis. Boxes are more or less homogeneous with respect to depth and distance to the coast. With the exception of Box 3, they are defined in a way that all stations combined are within the limit of one of the above-mentioned ICES boxes.

In evaluating the data available on contaminants in water, suspended matter, sediments and different types of biota, the problem arises that all the samples are taken from an environment highly variable in space and time and with a patchy distribution of the various compartments. By that, it is to be understood that only the long-lived bigger macrobenthic biota can be used as elements integrating the local variability in contaminant availability. Some stations have been resampled for hermit crabs at different times within the period 1987 to 1990. The fact that results from different

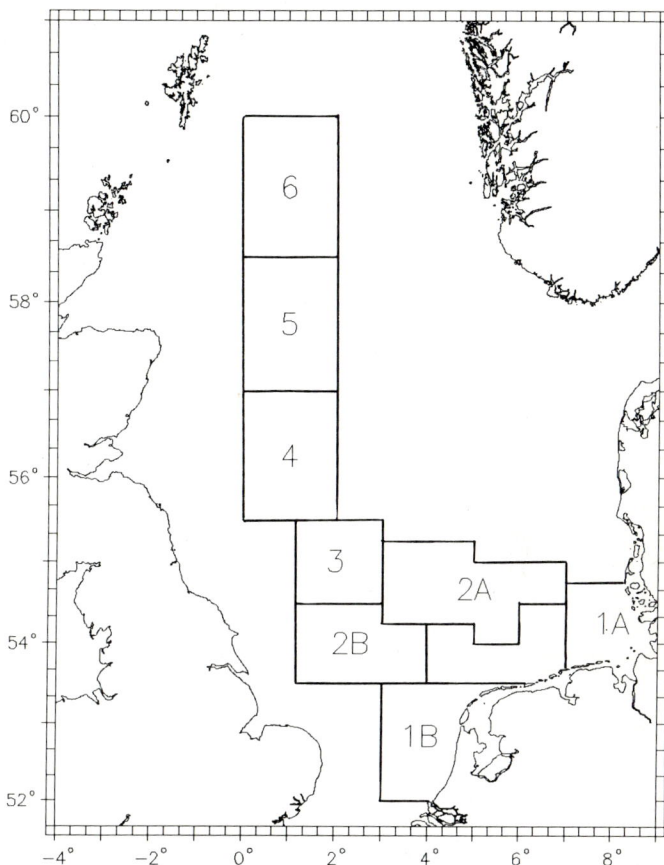

Fig. 4.5.14. Boxes at a transect from southern continental (1A and 1B: ICES 5a and 4) over to southern offshore (2A and 2B: ICES 7a); Dogger Bank (*3*); central offshore (*4* ICES 7b) northern offshore waters (*5* and *6* ICES 2 and 1) used for calculations of some general gradients of contaminants concentrations in hermit crabs

cruises obviously fit into the same pattern of regional distribution may be taken as an argument to accept the hypothetical conclusions made on regional gradients within the North Sea presented in Fig. 4.5.15 and 4.5.16 (heavy metals) and Fig. 4.5.17 and 4.5.18 (chlorinated hydrocarbons). There are only small differences among the two southern gradients (German Bight and Southern Bight). As a result, the following evaluation is made for a single south to north transect only.

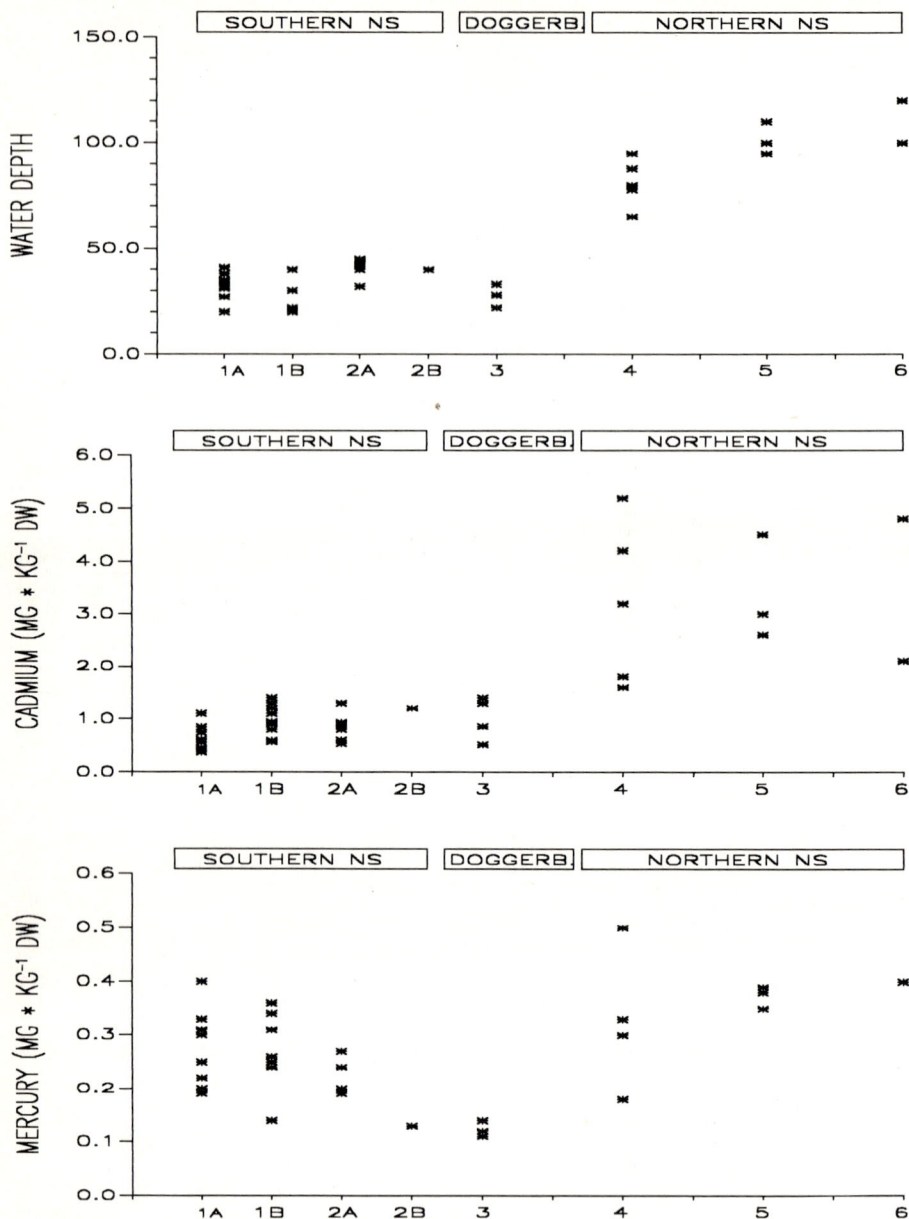

Fig. **4.5.15.** *Pagurus bernhardus* Depth and concentrations of cadmium and mercury along a transect (Fig. 4.5.14) from southern continental over southern offshore, Dogger Bank, central offshore, to northern North Sea waters

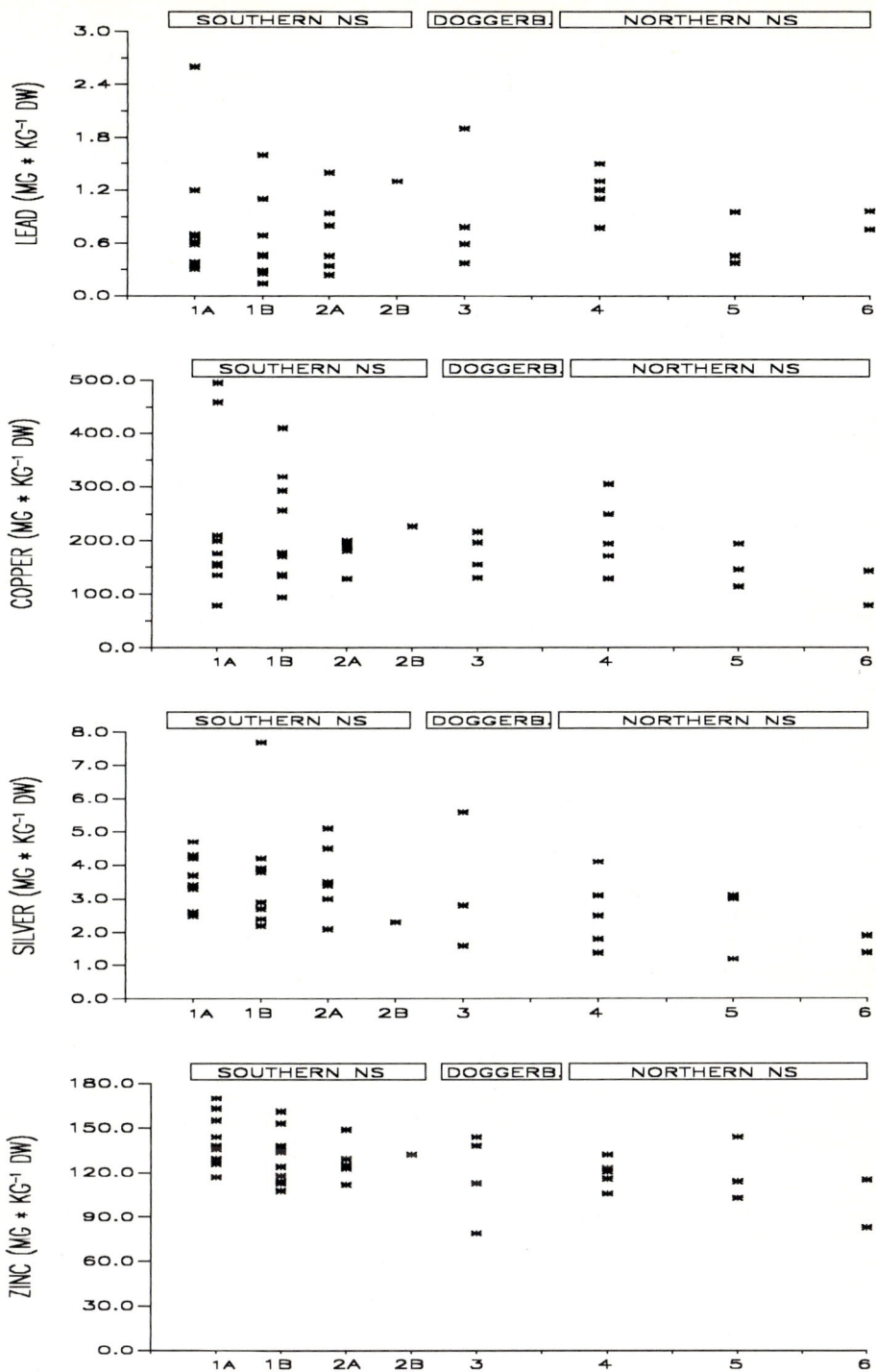

Fig. 4.5.16. *Pagurus bernhardus* Concentrations of copper, lead, silver and zinc along a transect from southern continental to northern North Sea waters

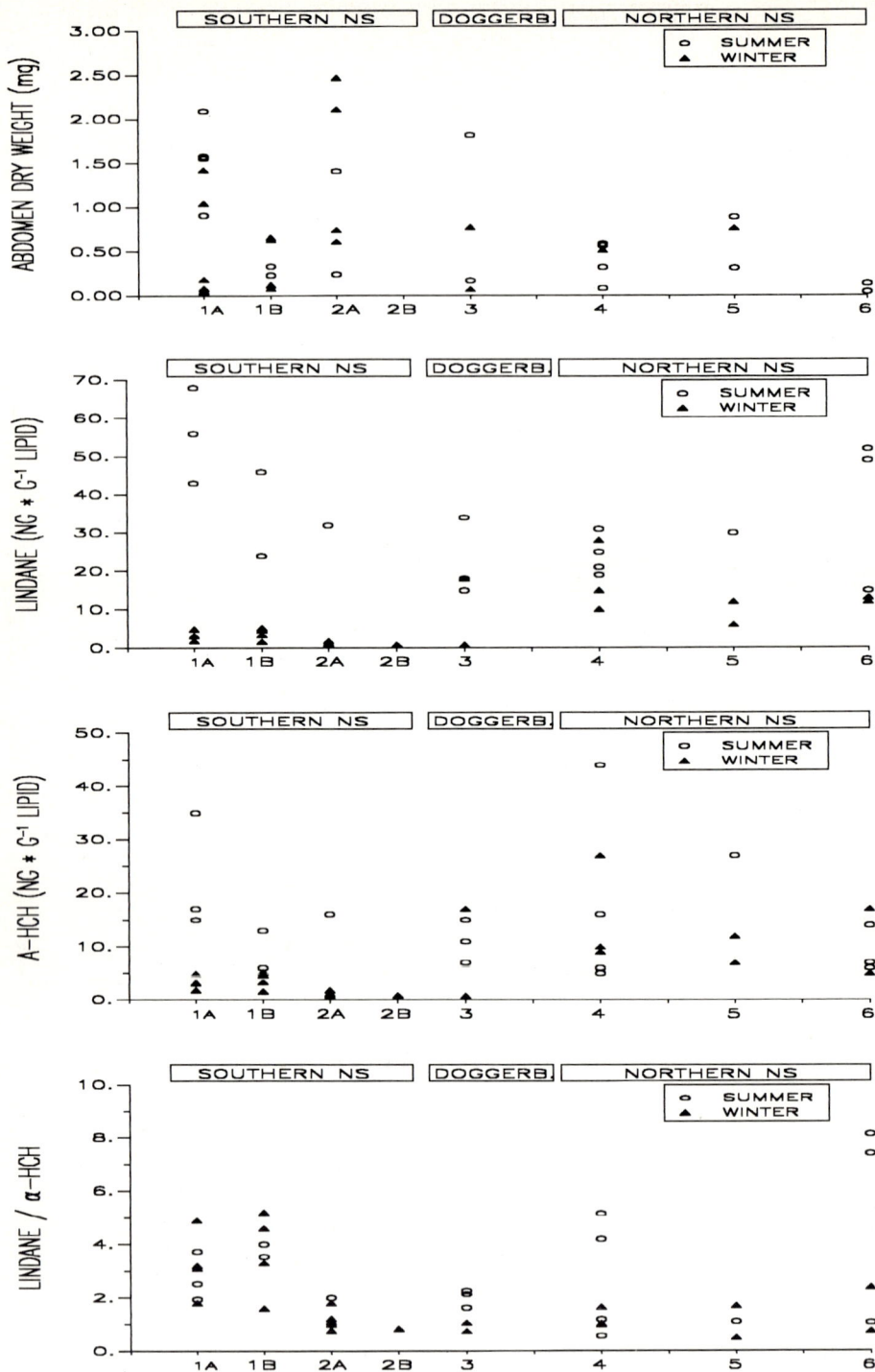

Fig. 4.5.17. *Pagurus bernhardus* Abdomen dry weight and concentrations of γ-HCH, α-HCH and relation γ-HCH to α-HCH along a transect from southern continental to northern North Sea waters

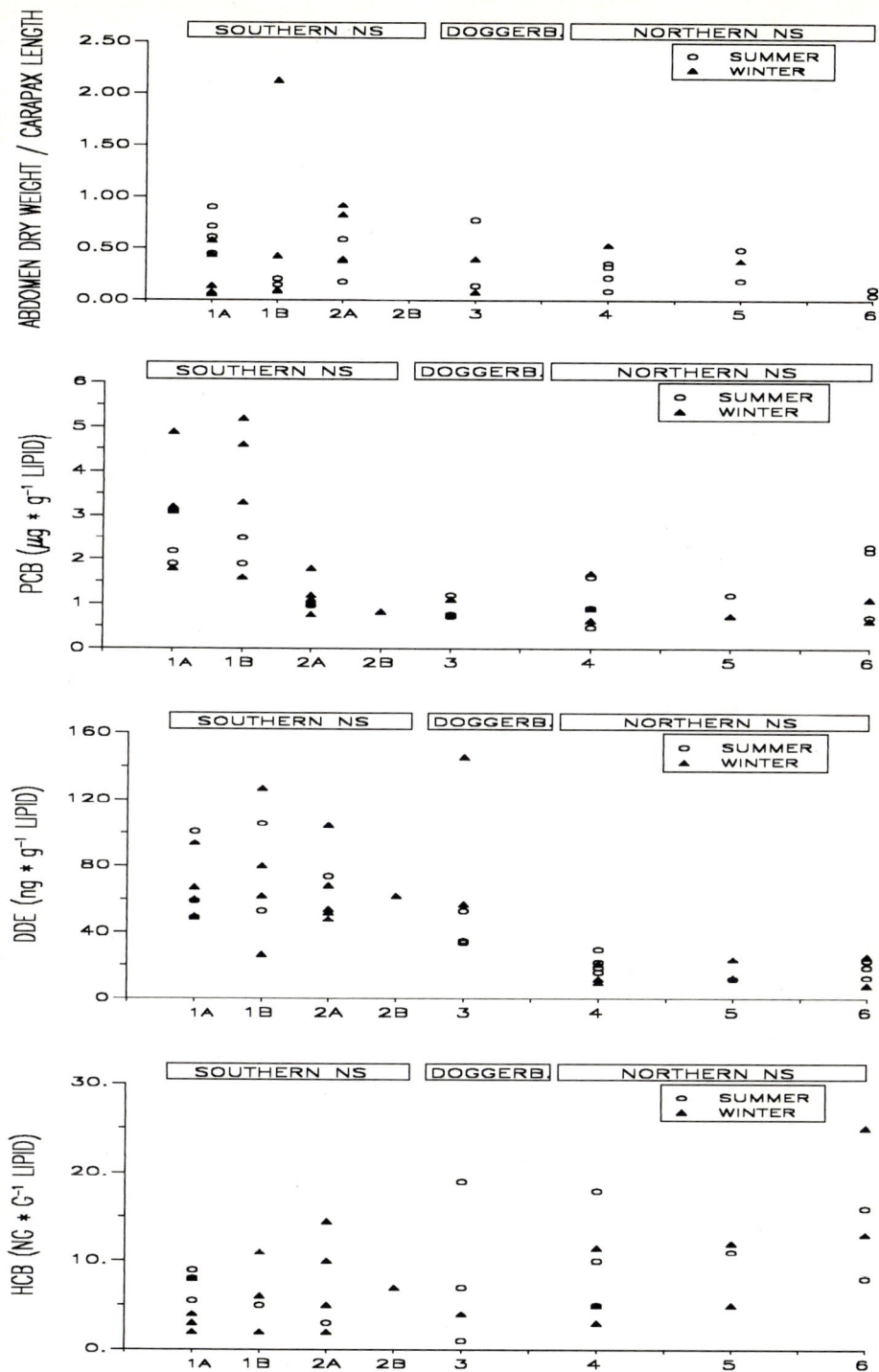

Fig. 4.5.18. *Pagurus bernhardus* Condition index (abdomen dry weight to carapax length) and concentrations of Σ-PCB, DDE and HCB along a transect from southern continental to northern North Sea waters

Some simple questions may be asked and different answers are to be expected for the different contaminants investigated:

- Are there significant gradients of concentrations from the inshore to the far-offshore subregions?
- Are there significant gradients between the different offshore subregions?
- Are there regional gradients corresponding to the estimated advective transports and to the hydrographic frontal systems?
- Are there patterns of regional gradients that can be related to combined effects of plankton and benthos?

Five types of gradients can be concluded from results of contaminants analyzed in hermit crabs. They are listed in Table 4.5.10 and major categories of factors to be considered for an understanding of the regional patterns are summarized in Table 4.5.9. The gradients are those found for contaminant concentrations in hermit crabs during the ZISCH campaigns. As results on contaminants in other biota fit into the same figure, it may be possible to make some general conclusions on contaminant gradients and the bioavailability of contaminants in the North Sea.

It is obvious that there are high concentrations of various contaminants within the nearshore areas of the Southern and the German Bight which are directly influenced by inputs of contaminants from land based continental sources, via discharges of the Meuse/Rhine and the Weser/Elbe river systems as well as via atmospheric transport and mainly nearshore wet and dry deposition of contaminants (Chap. 4.4). Resulting southern nearshore to southern offshore gradients are negatively correlated to the major changes in salinities, indicating decreasing continental and increasing Atlantic influences.

Such a trend, resulting in concentrations along the transect decreasing from the coastal to the southern offshore locations and to the Dogger Bank, has been found to be most significant for a number of chlorinated hydrocarbons such as the various PCB congeners and the two HCH isomers. In the case of γ- and α-HCH, such a gradient of concentrations decreasing with the distance from the coast was found to be most significant during the summer cruise and less developed during the winter. Among the heavy metals, this type of a gradient has been found to be most significant for mercury. Also, results of zinc and copper, although highly variable, may fit into such a figure. The question arises whether this gradient can only be seen as a general difference between coastal and offshore waters, or whether this local gradient may be followed by regional changes of contaminant concentrations also within the far offshore areas of the North Sea. There are no data showing a simple continuous gradient from the southern

Table 4.5.9. General gradients in distribution of contaminants in hermit crabs at a south to north transect from southern continental (Southern and German Bight) to northern waters in summer (S) and winter (W) or equal in both seasons

	Trend	Contaminant	Reasons (see Table 4.5.10)
I	From southern continental to southern offshore decreasing	Hg, γ-HCH (S), PCBs	1a 1b
II	From southern continental to the north continuously decreasing	(Pb, Ag, Cu, Zn)	1a 1b 2a
III	From southern continental to the north continuously increasing	Cd	1e 1f 2a 2c
IV	From southern continental to southern offshore and Dogger Bank decreasing, north of Dogger Bank increasing again	Hg	1a 1c 1e 2b 2c 2d
IVa	Same but less developed	α-HCH (S), PCBs (S)	1a 2a 2b 2c
V	Low in the southern, high in the central and northern North Sea	Cd γ- and α-HCH (W)	1e 1f 2b 2c 2d

to the northernmost waters in a linear way related to decreasing continental and increasing Atlantic influences. To some extent lead, silver, zinc and copper may show such a tendency, taking the means of concentra- tions combined within the boxes along the transect. In contrast, there are examples of contaminant concentrations significantly increasing to the north.

Concentrations of PCBs in hermit crabs from the two southern subregions (1a and 1b) can be concluded to be homogeneous and significantly higher than those from all the other subregions at the south to north transect (2 to 6). Within the northern North Sea, there is a slight tendency of PCB concentrations to increase within the northernmost area of the Shetland Trench. Such a local gradient of concentrations increasing again within the northern North Sea can also be concluded for the data on γ-HCH obtained in summer. It can be assumed that this figure may be a regional phenomenon resulting from local inputs from (diffuse) land based sources during times of the year when this pesticide is used in agriculture and for-

Table 4.5.10. Major factors to be considered for an understanding of the regional pattern of contaminants concentrations in benthic organisms given in Figs. 4.5.15, 4.5.16 and Table 4.5.9

1a	Inputs of contaminants from southern continental sources via the Meuse/Rhine and the Weser/Elbe river systems	South to north gradients of concentrations decreasing related to increasing salinity
1b	Inputs of contaminants from southern continental sources via atmospheric transport and mainly nearshore wet and dry deposition	As above
1c	Advective entrainment of contaminants from western land-based sources via the Scottish and English river systems or by other inputs into the Caledonian and English coastal waters	Increased concentrations of contaminants superimposed onto the south to north gradient resulting from 1a, 1b and 1e
1d	Inputs of contaminants from western land-based sources via atmospheric transport and wet or dry deposition within the central North Sea	As above
1e	Inputs of contaminants from northern sources with Atlantic inflow	Increased concentrations of contaminants in the northern and central North Sea
1f	Sedimentation of particulate matter, sediment resuspension and transport	Final deposition of contaminants bound to suspended matter in deep waters
2a	Regional differences in the bioavailability of contaminants	Concentrations in biota not correlated to concentrations in the abiotic matrices
2b	(Photo decomposition and) biotransformation of the less resistant contaminants	Selective elimination of the less resistant and accumulation of the more resistant contaminants including metabolites
2c	Bioconcentration of contaminants, food chain carry-over and vertical transport from the water column to sediment, deep-dwelling benthic biota	Selective food web enrichment of several highly persistent contaminants in deep waters
2d	Bioaccumulation and food web biomagnification	Selective accumulation of several highly persistant contaminants in the long-lived biota

estry of the adjacent areas. During the winter cruise, γ-HCH concentrations were generally low in the southern continental and southern offshore waters but in a range of higher concentrations in the central and northern North Sea.

This type of general difference between the southern and northern North Sea with low levels of contamination in southern and relatively high contaminant concentrations in the deep waters of the central and northern

North Sea, best documented by the regional distribution of cadmium concentrations in the hermit crab, can only be explained by combined effects of abiotic and biotic factors.

It is known that cadmium is taken up by the phytoplankton, zooplankton and the benthos by high accumulation factors and sorbed by high-affinity binding (Lillelund et al. 1987). As a consequence of food chain carry-over there is a substantial zooplankton-mediated vertical flux of this typical "nutrient-type" metal (see Sect. 4.5.2.2). It is taken up (and eliminated) by the bigger benthic organisms by long-term kinetics characterized by half lives of months, a year or longer. Combining results of calculations made on water masses entering the North Sea with the Atlantic inflow (Chap. 2.1) and cadmium concentrations measured, a substantial inflow of cadmium can be calculated from Atlantic sources. It is to be expected that a major part of this cadmium inflow is effectively sorbed by bioconcentration and deposited within the deep benthic water as a consequence of the very dynamic biological interactions, including transport by the diurnal vertical migration of interzonal zooplankton species and its large biomass typical for the northern North Sea. The same reasons are to be considered as an explanation of the high concentrations of mercury in crabs from the central and northern North Sea.

For the southern North Sea, there is a difference between cadmium and mercury. Whereas cadmium concentrations are generally low within the southern area but slightly increasing from nearshore to offshore, there is a distribution of mercury with a gradient of high concentrations within the coastal waters similar to the deep locations in the central and northern North Sea decreasing offshore to lowest concentrations at the Dogger Bank. This difference can be explained as a combined effect of regional differences in concentrations of the contaminants dissolved in the water and bound to particulate matter and regional differences of the speciation having a major influence on the bioavailability of these two metals. The results of modelling the deposition of contaminant bound to seston particles clearly show that there is a major seasonal variability in amounts of suspended matter deposited in the southern areas and moved away in times of increased sediment resuspension (Chap. 4.4). Contaminants like mercury that can be considered highly bioavailable are taken up effectively by the benthic organisms with relatively low half lives for uptake but a prolonged half-life of elimination. As a consequence, they show a significant response to elevated concentrations in the estuarine and coastal environment. Concentrations in the biota decrease offshore corresponding to decreasing concentrations in water and particulate matter. In the case of cadmium, there is a major part of the total metal dissolved in water and bound to particulate phases of a reduced bioavailability. There are inorganic complexes of lower bioavailability and dissolved and particulate organic species of cadmium that are taken up most effectively by the biota. The portion of cadmiums generally increases from nearshore to offshore and, as a consequence, there is a gradient of the bioavailable species not correlated to the total concentration.

The results presented show clearly that the North Sea on the whole is a well-structured ecosystem. The regional patterns of contaminant concentrations of the different compartments are structured in a way similar to that known for the regional differences within the structure of the planktonic and benthic assemblages. They can be related to inputs from landbased sources and atmospheric and aquatic advective transport. Within the northern North Sea, the Atlantic inflow may be of major importance. However, abiotic processes are enhanced and modified by a variety of biotic processes in such a way that the patterns of contaminant concentrations measured can only be understood as a result of a combined abiotic and biotic processes within a system of complex functional interdependences.

References

Ballschmitter K, Zell M (1980) Analysis of polychlorinated biphenyls (PCB) by glass capillary gas chromatotography. Fresenius Z Anal Chem 302:20-31

Borchardt T, Burchert S, Hablizel H., Karbe L, Zeitner R (1988) Trace metal concentrations in mussels: comparison between estuarine, coastal and offshore regions in the southeastern North Sea from 1983 to 1986. Mar Ecol Progr Ser 42:17-31

Bruland KW (1980) Oceanographic distribution of cadmium, zinc, nickel, and copper in the North Pacific. Earth Planet Sci Lett 47:176

Bruland KW, Franks RP (1983) Mn, Ni, Cu, Zn, and Cd in the Western North Atlantic. In: Wong CS, Boyle E, Bruland KW, Burton JD, Goldberg ED (eds) Trace metals in seawater. Plenum, New York, pp 185-302

Button KS, Hostetter HP (1977) Copper sorption and release by *Cyclotella meneghiana* (Bacill.) and *Chlamydomonas reinhardtii* (Chlorophyta). J Phycol 13:198-20

Carlson H (ed) (1986) Quality status of the North Sea. Dtsch Hydrogr Z Ergänzungsh B 16:424

Claussen T (1988) Levels and spatial distribution of trace metals in dabs (Limanda limanda) of the southern North Sea. Mitt Geol Paläontol Inst Univ Hamb 65:467-496

Dicke M, Schmidt D, Michel A (1987) Trace metal distribution in the North Sea. In: Lindberg SE, Hutchinson TC (eds) Proc Int Conf heavy metals in the environment. New Orleans 1987, vol 2. CEP Consultants, Edinburgh

Dürselen C-D (1990) Untersuchungen zur Schwermetallakkumulation von Phytoplanktongemeinschaften der Deutschen Bucht mit ergänzenden Laborversuchen zur Deutung der Ergebnisse. Dipl Arb RWTH Aachen. 127 (unpublished)

Falkner R, Simonis W (1982) Polychlorierte Biphenyle (PCB) im Lebensraum Wasser (Aufnahme und Anreicherung in Organismen - Probleme der Weitergabe in der Nahrungspyramide). Arch Hydrobiol Beih Ergbn Limnol 17:1-74

Faller J, Hühnerfuss H, König WA, Ludwig P (1991) Gas chromatographic separartion of the enantiomers of marine organic pollutants. Distribution of α-HCH enantiomers in the North Sea Mar Pollut Bull 22:82-86

Gaul H, Ziebarth U (1983) Method for the analysis of lipophilic compounds in water and results about the distribution of different organochlorine compounds in the North Sea. Dtsch Hydrogr Z 36,191-212

Geyer H, Sheehan P, Kotzias D, Freitag D, Korte F (1982) Prediction of ecotoxicological behaviour of chemicals: relationships between physico-chemical properties and bioaccumulation of organic chemicals in the mussel *Mytilus edulis*. Chemosphere 11:1121-1134

Gonzalez-Valero J (1990) Saisonalität toxikokinetischer Beziehungen von γ-HCH in der Miesmuschel *Mytilus edulis* L. Diss Fachber Biologie, Univ Hamburg

Heip C, Herman PM, Craeymeersch, Soetaert K (1990) Statistical analysis and trends in biomass and diversity of North Sea Macrofauna. ICES CM 1990/MINI:10

INC (1987) Second International North Sea Conference. Quality Status of the North Sea. Report of the Scientific Technical Working Group, London

Kammann U, Knickmeyer R, Steinhart H (1990) Distribution of polychlorobiphenyls and hexachlorobenzene in different tissues of the dab (*Limanda limanda* L.) in relation to lipid polarity. Bull Environ Contam Toxicol 45:552-559

Kammann U, Landgraff O, Steinhart H (1992) Cyclic organochlorines in benthic organisms from the North Sea and the German Bight. Analusis 22:M70-M73

Kannan N, Tanabe S, Tatsukawa R (1989) Persistence of highly toxic PCBs in aquatic ecosystems: uptake and release kinetics of coplanar PCBs in green-lipped mussels (*Perna viridis* L.). Environ Pollut 56:65-76

Karbe L, Gonzalez-Valero J, Borchardt T, Dembinski M, Duch A, Hablizel H, Zeitner R (1988) Heavy metals in fish and benthic organisms from the northwestern, central and southern North Sea: Regional patterns comparing dab, blue mussel and hermit crab (*Limanda limanda*, *Mytilus edulis*, *Pagurus bernhardus*). ICES C M 1988/E:22

Kersten M, Dicke M, Kriews M, Naumann K, Schmidt D, Schulz M, Schikowski M, Steiger M (1988) Distribution and fate of heavy metals in the North Sea. In: Salomons W, Bayne BL, Duursma EK, Förstner U (eds) Pollution of the North Sea - an assessment. Springer, Berlin Heidelberg New York

Knickmeyer R, Steinhart H (1988a) Seasonal differences of cyclic organochlorines in eggs of the hermit crab *Pagurus bernhardus* L. from the North Sea. Sarsia 73:291-298

Knickmeyer R, Steinhart H (1988b) Cyclic organochlorines in the hermit crabs *Pagurus bernhardus* and *Pagurus pubescens* from the North Sea. A comparison between winter and early summer situation. Neth J Sea Res 22 (3): 237-251

Knickmeyer R, Steinhart H (1988c) The distribution of cyclic organochlorines in North Sea sediments. Dtsch Hydrorg Z 41:1-21

Knickmeyer R, Steinhart H (1989a) On the distribution of polychlorinated biphenyl congeners and hexachlorobenzene in different tissues of dab (*Limanda limanda*) from the North Sea. Chemosphere 19:1309-1320

Knickmeyer R, Steinhart H (1989b) Cyclic organochlorines in plankton from the North Sea in spring. Estuarine Coastal Shelf Sci 28:117-127

Knickmeyer R, Steinhart H (1990a) Seasonal variations and sex related differences of organochlorines in whelks (*Buccinum undatum*) from the German Bight. Chemosphere 20:109-122

Knickmeyer R, Steinhart H (1990b) Patterns of cyclic organochlorine contamination in livers of male pleuronectiformes from the North Sea, winter 1987. Mar Pollut Bull 21 (4):187-189

Knickmeyer R, Steinhart H (1990c) Concentrations of organochlorine compounds in the hermit crab *Pagurus bernhardus* from the German Bight. December 1988 - May 1989. Neth J Sea Res 25 (3):365-376

Krause M, Knickmeyer R (1992) Estimation of the load of cyclic organochlorines in North Sea zooplankton. Helgol Meeresunters 46:69-91

Krause M, Martens P (1990) Distribution patterns of mesozooplankton biomass in the North Sea. Helgol Meeresunters 44:295-327

Lee RF (1981) Mixed function oxygenase (MFO) in marine invertebrates. Mar Biol Lett 2:87-105

Lee AJ, Ramster JW (1981) Atlas of the seas around the British Isles. Ministry of Agriculture Fisheries and Food, London, 80 pp

Lillelund K, de Haar U, Elster H-J, Karbe L, Schwörbel I, Simonis W (eds) (1987) Bioakkumulation in Nahrungsketten. Zur Problematik der Akkumulation von Umweltchemikalien in aquatischen Systemen. Ergebnisse aus dem Schwerpunktprogramm Nahrungskettenprobleme. DFG und VCH, Weinheim, 327 pp

Martens P, Krause M (1990) The fate of faecal pellets in the North Sea. Helgol Wiss Meeresunters 44:9-19

Myers VB, Iverson RL, Harris RC (1975) The effect of salinity and dissolved organic matter on the surface charge characteristics of some euryhaline phytoplankton. J Exp Mar Biol Ecol 17:59-68

Rick H-J (1990) Ein Beitrag zur Abschätzung der Wechselbeziehung zwischen den planktischen Primärproduzenten des Nordseegebietes und den Schwermetallen Kupfer, Zink, Cadmium und Blei auf Grundlage von Untersuchungen an natürllichen Planktongemeinschaften und Laborexperimenten mit bestandsbildenden Arten. Diss Fachber Biologie Techn Univ Aachen

Rick HJ, Aletsee L, Schmidt D (1990) Wechselbeziehungen zwischen Phytoplankton und Schwermetallen. In: Lozan JL, Lenz W, Rachor E, Watermann B, von Westernhagen H (eds) Warnsignale aus der Nordsee. Parey, Berlin, pp 138-148

Schulz DE (1990) Chlorobiphenyle im Meerwasser des Nordatlantiks und der Nordsee. Ber Inst Meereskunde Univ Kiel, Bd 197

Tanabe S, Tatsukawa R, Phillips DJH (1987) Mussels as bioindicators of PCB pollution: a case study on uptake and release of PCB isomers and congeners in green-lipped mussels (*Perna viridis*) in Hong Kong waters. Environ Pollut 47:41-62

Werner D (1977) Regulation of metabolism by silicate in diatoms. In: Benz G, Lindquist U (eds) Biochemistry of silicon and related problems. Plenum, New York

4.6 Combined Effects of Abiotic and Biotic Factors on Heavy Metal Fluxes

M. KERSTEN, M. KRIEWS, W. KÜHN and H.-J. RICK

4.6.1 Introduction

Heavy metal concentration distributions have been documented recently for a variety of abiotic compartments of the entire North Sea (Kersten et al. 1988, 1990, 1991). It is readily obvious, however, that no evaluation of metal concentration distributions in different ecosystem compartments of the North Sea is possible without an at least rough knowledge of the biogeochemical behaviour and fate of these elements. During its passage and stay in a marine environment, a metal undergoes a series of complex interactions and transformation processes with and within the sediment, the water column, biota and other dissolved or suspended matter, as well as with the atmosphere (Fig. 4.6.1). The air-water and sediment-water interfaces are the most important with respect to trace metal cycling in the marine environment. Biological activity can represent an important transfer mechanism for heavy metals from the aqueous to the particulate phase in the water column. It may interact thereby with geochemical processes at the sea surface and the sea bed to create possibilities of contaminant transport and redistribution (Kersten 1988). However, even in well-studied regions like the coastal areas of the North Sea, knowledge is still insufficient to predict reliably the impact of anthropogenic contaminants. Little is known about net fluxes of metals at the air-water and sediment-water interfaces and even less about their residence times in the water column once having passed these interfaces. It is therefore timely to advance past "descriptive" science and adopt more intensively the "process" approach to the marine

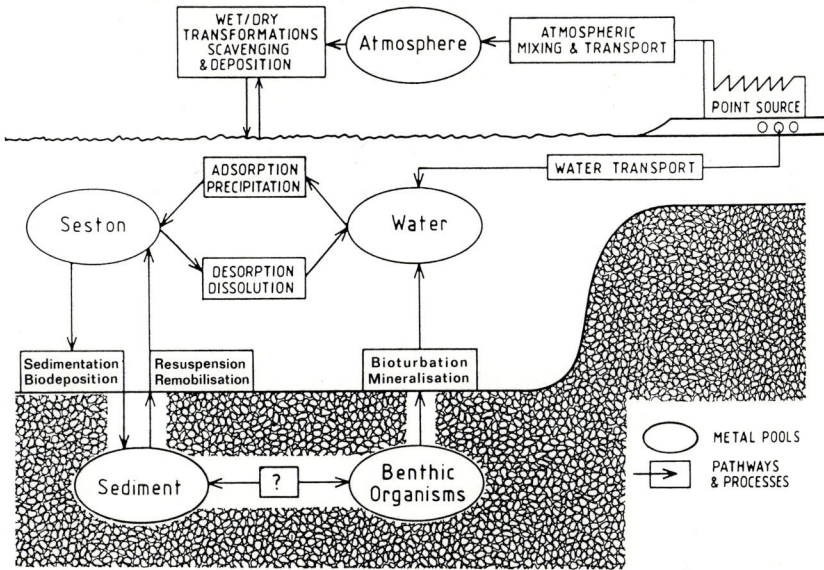

Fig. 4.6.1. Schematic review of the major pathways and pools of heavy metals in a marine ecosystem

system of the North Sea currently fostered by such interdisciplinary efforts as described in this Volume.

A complete process model of the heavy metal distribution and fate in such a marine environment would have to incorporate rate expressions for each process into a general statement of mass conservation, which ultimately yields the temporal and spatial distribution of the contaminant in each phase. In its most general form, the model would be comprised of several non-linear, multi-dimensional partial differential equations, augmented by appropriate equilibrium relations. Because of the complexity of such systems, it has become customary to introduce simplifications of various types. In the simplest approach, the system is partitioned into compartments of uniform concentration which are at instantaneous equilibrium with each other. Rate processes are neglected and the solution merely yields time-dependent equilibrium concentrations. Thus, the model proceeds by development of a time variable mass balance equation for the sources and sinks of a system such as depicted in Fig. 4.6.1. The weakness of this approach, however, is that the system reservoir is considered as a black box where the processes responsible for the fluxes are ignored. Clearly, this assumption is not justified for pollution scenarios of the non-conservatively behaving heavy metals. The next level of analytical sophistication thus consists of deriving steady-state distributions under conditions of constant contaminant influx to the system and introduction of a first order rate formulation of sorption-desorption, which leads to powerful extensions of useful concepts such as response time and the possible instability behaviour of a system (DiToro et al. 1980; Wollast 1986).

The need for such a modelling framework arises from two points: the need to (1) understand what mechanisms are operative and important in the fate of heavy metals, and (2) predict quantitatively how this system would react to perturbations due to man's activities. We will present here such a geochemical compartment model including a first-order rate process formulation to approximate biological processes controlling the dissolved metal concentrations in the water column. To facilitate the evaluation and discussion of the model development presented below, field data is presented for the ecotoxic elements Cd, Pb and Cu from the German Bight. It should be noted, however, that a specific concentration of a heavy metal in water only reflects the fate of this element and not the effect. The effect and consequence of a given potentially toxic element concentration must be part of a separate evaluation.

4.6.2 The Air-Water Interface

Weekly aerosol sampling was performed on the island of Helgoland (54°10'N, 7°53'E) in the context of a long-term monitoring study from June 1986 until May 1990 (Fig. 4.6.2). Atmospheric particulate matter was collected on dried preweighed low-background quartz microfibre filters (Stora MK 360, 472 cm^2, DOP-test 99.98 %) using a high volume sampler with a flow rate of 100 m^3/h. After digestion of the filters and aerosols with $HF/HNO_3/HClO_4$, the element concentrations were analyzed by atomic spectrometric techniques (ICP-AES and GFAAS, Kriews et al. 1989). With these data, estimates of the total dry deposition mode were attempted (see also Chap. 2.4).

The estimates of total atmospheric deposition and the percentage of dry deposition for the German Bight are based on the average mean concentrations and calculated average dry deposition velocities. The deposition velocities, in turn, were estimated from size-separated aerosol samples using the dry deposition model of Slinn and Slinn (1980). The wet deposition flux is based on the average mean concentrations, the yearly precipitation amount and summarized washout factors from the last GESAMP report (Duce et al. 1990). Details of the calculation model are

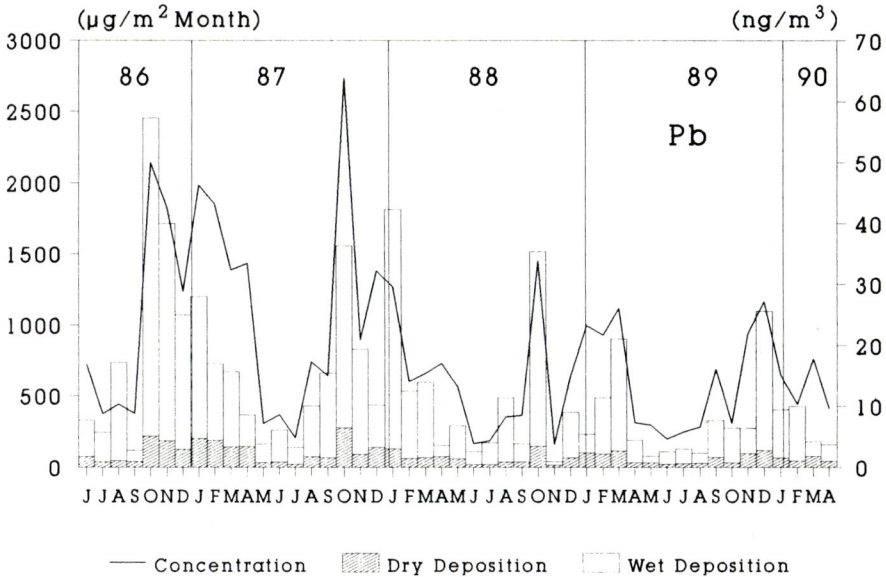

Fig. 4.6.2. Monthly averaged atmospheric concentrations, dry and wet mode depositions of lead at Helgoland during the period June 1986 to April 1990

published elsewhere (Kriews 1992). A brief discussion of the reliability of this approach is presented in Chapter 2.4. The average atmospheric fluxes of the elements Cd, Cu, and Pb (Table 4.6.1) show a high standard deviation which can be explained by the strong variability of atmospheric element concentrations in the atmospheric boundary layer at Helgoland (Schulz et al. 1988). This is exemplarily illustrated for the element Pb in Fig. 4.6.2. The long-term monitoring data reveal that both the concentrations and the total deposition of Pb tend to be higher in the winter than in the summer season. In winter there are often situations with stable stratification in the boundary layer which are associated with southeasterly air masses, while in summer westerly flows predominate. It is interesting to note from this long-term study at Helgoland that there is a general tendency of decreasing atmospheric input to the sea surface which coincides with the introduction of unleaded petrol started in 1985. The relevance and possible reason for this development are discussed in Chapters 2.4 and 4.4.

The percentage of dry deposition (Table 4.6.1) agrees quite well with other estimates of dry deposition fluxes in the coastal environment (Krell and Roeckner 1988; Graßl et al. 1989; cf. Chap. 3.1). The wet deposition of heavy metals is the predominant means for the four elements studied, but at least one-third is introduced via dry deposition. In order to assess the impact that atmospherically transported heavy metals from both natural and

Table 4.6.1. Total atmospheric deposition rates of heavy metals on island of Helgoland based on average air concentration data (n = 154) between 1986 and 1990

Element	Mean $\mu g/m^2 d$	Standard deviation $\mu g/m^2 d$	Percentage dry deposition wt. %
Cd	0.89	0.94	43
Pb	9.1	9.3	30
Cu	1.9	1.8	41

anthropogenic sources have on seawater chemistry, it is necessary to evaluate their subsequent fate in the water column, i.e. to understand how they behave within the marine biogeochemical cycles. One of the critical processes which determines the fate of the metals in aerosol once they have been deposited at the sea surface is the extent to which they undergo solubilization in seawater. In order to study the sea water solubilities of elements, leaching experiments were made with selected samples of the marine aerosol monitoring series, representing different source regions (Kriews 1992). The data listed in Table 4.6.2 indicate in fact considerable seawater solubilities of aerosol heavy metals ranging from 25 to 67 %. Other studies (Hodge et al. 1978; Hardy and Crecelius 1981) indicate similarly high sea water solubilities of these elements in marine aerosols collected at coastal sites. Cations and dissolved organic matter in sea water

Table 4.6.2. Average sea water solubility of the North Sea aerosol samples (n = 20, leaching time 24 h)

Element	Mean	Relative standard deviation %	Min. wt. %	Max. wt. %
Cd	67.0	35.2	19.9	98.4
Pb	28.1	28.8	3.7	63.5
Cu	25.5	12.2	3.9	60.8

might increase metal solubility by displacing them from the aerosol particle surfaces (Förstner 1986). Low pH values most likely also increase aerosol metal solubility (Lindberg and Harriss 1983); however, if metals are solubilized from aerosol by ion displacement, as evidenced by Kersten et al. (1991), then the relatively higher concentration of cations in sea water would most likely displace more aerosol metals than would the smaller hydronium ion concentration difference between acid rainwater and sea water. This was evidenced by Maring and Duce (1989), who found more aerosol copper remaining in particulate form when exposed to artificial, even acidified rainwater (62 $\pm$ 20 %) than when exposed to sea water (26 $\pm$ 10 %). In conclusion, there is strong evidence that whatever the ratio of wet to dry deposition may be, a significant portion of the atmospherically deposited toxic metals enter the marine ecosystem finally in dissolved form and can readily be incorporated into the biogeochemical cycle.

4.6.3 The Sediment-Water Interface

Apart from atmospheric inputs, sediments may also become an important nonpoint source of heavy metals, especially in the case of metals characterized by a nutrient-type remobilization process. The sediment can be viewed only conditionally as a "secondary source" of heavy metals, as the metals emerging at this interface originate in the first instance from the water column via sedimentation of suspended matter. The redistribution from the particulate to dissolved phases of trace metals occur within surficial sediments either from the release of organically bound metals from decomposing organic matter or as a result of redox reactions in the sediments (Shaw et al. 1990). This process will redistribute metals from particles that are settling out of the water column into the dissolved phase or onto finer suspended particles with longer residence times in the water column. Independent evidence for the existence of these particulate to dissolved conversion processes was obtained from an examination of the metal-salinity relationships in coastal waters. These analyses (GESAMP 1987) demonstrate that the intercepts at zero salinity of salinity-metal concentration relationships for various shelf areas can yield effective freshwater end-member concentrations exceeding those of dissolved concentrations in rivers for Co, Cu, Zn and Cd. Clearly, with respect to differentiating anthropogenic metal input from this natural cycling, it is important to quantitatively estimate metal fluxes at the sediment-water interface.

Concentrations of chemical constituents in water overlying marine sediments are controlled by complex interacting influences of both biotic and abiotic mechanisms (Kersten 1988; Santschi et al. 1990). These often result in chemical gradients between sediment pore water and overlying water, with respect to soluble components. Diffusion operates to dissipate the concentration differences. Concentration profiles of the dissolved species which bridge the concentration difference at the sediment-water interface are straight lines, if the diffusion coefficients (factor of proportionality between flux and concentration gradient: Fick's first law of diffusion) are constant, or curved ones, if the diffusion coefficients vary with time or depth of burial. For the particle-reactive trace elements curved profiles are common, which result (1) from reactions within the sediment, (2) from heterogeneous composition and redox regime of surface sediments, and (3) from the existence of diffusive sublayer resistance above the sediment-water interface.

Emerson et al. (1984) compared the contribution of both biological advection and molecular diffusion to the transport of dissolved components across the sediment-water interface using a combination of an in-situ tritium tracer experiment and pore water chemistry. The results indicated that biologically mediated diffusive exchange dominates the flux only for components which do not have strong gradients at the sediment-water interface. This is true for alkalinity, silicate, ammonia, As and Ni, but not for other particle-reactive trace elements. McCaffrey et al. (1980) found in their study of epibenthic element fluxes in Narragansett Bay that even at a site of high benthic activity molecular diffusion is as important as the respiratory activity of macrobenthic organisms in transporting heavy metals to the overlying water column. This result was attributed to the fact that metal release to pore waters by early diagenesis is limited to the depths of oxygen penetration in the topmost centimetres of the sediment profile. At these depths, diffusion to the overlying water column is rapid due to the strong gradient which is dominant compared to the patchy biopumping events. Although these results encourage using pore water gradients for metal flux determinations, the reliability of the results of this approach is limited due to the high sampling resolution (millimetre scale) necessary to resolve the steep gradients. Development of in situ-sampling techniques at sub-millimetre intervals such as diffusive equilibration of pore water in a thin film (DET-technique: Davison et al. 1991) are in progress but yet not available.

A more reliable method for determination of net diffusive exchange fluxes of dissolved metal species are in-situ benthic chamber incubation techniques or tank experiments (Santschi et al. 1990); or, with less but still satisfactory reliability, intact core incubation. As in-situ experiments were not possible in the highly dynamic environment of the German Bight due to excess logistic expense, the low-cost open incubation diffusion technique was used in this work. The basic principle in this laboratory experiment is to expose the surface of a sediment core to a well-stirred water reservoir of fixed volume. Diffusive exchange of solutes occurs between the sediment pore water and over-lying water reservoirs. Because the overlying water reservoir is generally finite, solute concentration in each reservoir usually changes with time to an extent which is determined in part by production rates of sedimentary solutes as well as appropriate diffusion coefficients and geometric scaling. Benthic diffusive exchange fluxes (FS) of solute species can then be calculated from the concentration increase $\Delta C = C_1 - C_0$, over time t, but at small values of $\Delta C/C_0$ (i.e. < 0.1) for at least close approximations to ambient steady-state conditions:

$$(1)$$

$$F_S = (C_1 - C_0)\frac{(V/A)}{(t_1 - t_0)}$$

where V is the chamber volume and A is the entrapped sediment area (in cylindrical core incubation methods this ratio equals the height of the enclosed water volume). If the concentration increases are excessive ($\Delta C/C_0$ > 0.1), then chemical reactions of reactive elements such as metals in the enclosed water volume would complicate flux determination. In order to maintain a pseudo-steady-state flux of the trace elements in the open incubation experiments, a continuous-flow adaptor was used for a steady renewal of the overlying water.

The continuous-flow adaptor was designed by Twinch and Ashton (1984) to enclose small volumes of water which facilitates rapid flushing of the overlying water during incubation experiments. Firstly, this prevents development of excess solute concentrations in the enclosed water and secondary reactions in the water phase. This is especially important with less contaminated North Sea sediments, where a slight overloading of the water column may already reverse metal fluxes back into the sediment. Secondly, the existence of diffusive sublayer resistance above the sediment-water interface requires that turbulence levels inside incubation chambers equal more or less those in nature, especially for the measurement of species that react within the upper few millimetres of sediments. The third but most important aspect is that flushing of a significant amount of water through a cation exchange column facilitates detection of the trace elements. The concentrations in the outflow may be even less than the already very low concentrations in the inflow (which is natural unspiked sea water), when the sediment acts as a sink rather than a source for the dissolved metals. Moreover, contamination problems are avoided as well because there is no necessity to open the sediment-water system during exprimentation for sampling.

The continuous-flow incubation method was used to examine metal release rates from sediment cores collected from an area southeast from the island of Helgoland in the German Bight (muddy sand bottom at $54°1'N/7°49'E$). Details of the experimental setup and evaluation tests will be published elsewhere (Kersten and Kienz 1993). The average sediment metal release rates for Cd, Cu and Pb are listed in Table 4.6.3. Clearly, these release rates are dependent on the inflow concentrations since the direction and rate of metal flux between sediments and overlying water has been shown to be dependent on the metal gradient between sediment pore water and the overlying water. Therefore, the average seawater metal concentrations at which these fluxes have been estimated are listed in the left hand portion of Table 4.6.3. They were on the lower end of the dissolved metal concentration range found in the German Bight (Kersten et al. 1988). The flux results indicate that the German Bight sediment acts as a source for dissolved Cd, Cu, and Pb. This result provides a useful predictive tool for assessing the potential role of sediments in metal cycling. It should be stressed, however, that these results might not hold when fresh plankton detritus from a decaying bloom is deposited which is known to be mineralized within a very short time (Graf 1990). Such biological events are not easy to measure in field but have been recently simulated in mesocosm experiments (Slauenwhite and Wangersky 1991).

Table 4.6.3. Dissolved metal fluxes at the sediment-water interface measured on sediment cores from the German Bight (with bulk metal concentrations as indicated) at ambient dissolved metal concentrations by the continuous-flow incubation technique

Element	Total conc. in top 2 cm of sediment mg/kg	Dissolved metal in natural sea water ng/l	Flux at natural sea water conc. $\mu g/m^2 d$
Cd	0.5 ± 0.1	15 ± 5	10 ± 5
Pb	12.0 ± 1.1	100 ± 30	20 ± 10
Cu	5.3 ± 0.7	200 ± 50	200 ± 50

4.6.4   Heavy Metal Scavenging by Phytoplankton

Once introduced into the water column, heavy metal concentrations are controlled by partitioning between the dissolved phase and both biotic and abiotic seston components. A common finding from both field and mesocosm experiments is that metals have shorter water column residence times when background particulate loadings are augmented by phytoplankton cells (Balls 1989). This is not surprising, bearing in mind that all the constituents of marine phytoplankton are extracted from sea water. In the pelagic system this mode of biotic particulation of both essential and toxic elements seems to be the most dynamic in character. During winter, the seston in the German Bight is built up mainly by abiotic components, together with biogenic detritus, which both also take part in the turnover of heavy metals. During the remaining season (May-October), phytoplankton becomes the dominating substrate in the water column. Due to the standing crop of the planktonic primary producers (summer mean 1.34 Mio. t, winter mean 0.27 Mio. t organic carbon for the entire North Sea: Rick 1990), an enormous amount of reactive surface is available for additional pollutant uptake throughout the year. This corresponds to a specific organic matter surface (S_{tot}) of 10 m^2 (summer) and 2 m^2 (winter), respectively, under 1 m^2 of water surface (Rick 1990). It can be further assumed that the bioaccumulation of metals into higher trophic levels of the pelagic food web seems to be of little importance for pollutant turnover considering the minor biomass transfer ($< 15 \%$) between the linked levels (cf. Chap. 4.5).

Table 4.6.4. Assessment of the mass (in tons) balance between dissolved and phytoplankton-bound heavy metals in the German Bight given a water volume of 4.35 x 10^{11} m^3. (Rick 1990)

Month	Biomass (in dry wt. C_{org})	Cd plankton/diss.	Pb plankton/diss.	Cu plankton/diss.
February	34 x 10^3	0.1/20	1.6/66	0.7/395
April	53 x 10^3	0.2/17	3.5/13	1.3/218
May	365 x 10^3	1.5/16	10.7/29	5.9/194

The seasonal effects on the efficiency of heavy metal scavenging by the seston are clearly a result of the at least fivefold higher summer values in active surface binding sites than in winter. Thus, estimations of the metal sorption capacity (cf. Chap. 4.5) reflect the same seasonal pattern: during summer, 8 times the amount of copper, 17 times the amount of zinc, and 36 times the amount of cadmium can be bound by phytoplankton compared to the winter values. The close link of the extent of metal sorption to the stage of phytoplankton development is verified by data obtained from monthly sampling campaigns in the German Bight in 1989. The data show a reciprocal relationship of Cd, Cu, and Zn dissolved in the surficial waters to the amount of phytoplankton biomass (Table 4.6.4). Both the results from heavy metal concentration of the ambient phytoplankton species and the quantitative phytoplankton community analysis led to an estimation of the time necessary for the prevailing phytoplankton population to incorporate about half of the dissolved heavy metal inventory, under the assumptions that (1) no mineralization, (2) no exponential growth and (3) efficient removal of the biomass by sedimentation within one generation would occur. This estimation can be used to calculate a potential turnover time of a metal in the water column which is defined as the ratio of the total dissolved concentration of the metal to its rate of transfer from the dissolved to particulate phase. The numbers calculated in this way for Cd, Pb and Cu as a rough estimate of the response time of the ecosystem to any metal input are given in Table 4.6.5.

Biological heavy metal turnover times calculated thereby for the winter season (February and March) are in the order of 1 year. In April, planktonic primary producers become an increasing factor for metal turnover, leading to response times in the order of 1 month. In May, finally, they decreased to values in the order of a few days. Similar results were found with an experimental mesocosm system (Slauenwhite and Wangersky 1991), in which a half-life of 50 h was found for the transfer process of Cd to the particulate phase following a bacillariophytes bloom (*Chaetoceros gracilis*). The kinetics of this phase transition were found to be of first order with a rate constant k of 0.33/d. The maximum rate at which Cd was deposited to the sediment in association with the organic flocs was 2 $\mu g/m^2/d$. This benthic flux was obtained 5 days after the biomass and production peak of the bloom in the tank (Slauenwhite and Wangersky 1991).

The incorporation of trace metals in phytoplankton should be primarily dependent on the binding of metals to hydrophilic functional groups (such as -COOH, -NH$_2$, -SH, -OH, etc.) of proteins, which are embedded in the lipid bilayer of cell membranes, and subsequent transport of organometallic complexes into the cell interior through complicated biochemical mecha-

Table 4.6.5. Turnover times for a typical phytoplankton population to remove 63 % of dissolved heavy metal content (wt %) of the German Bight (in days)

Period	Cd	Pb	Cu
2/89	415	85	1180
2/89 C.w.[a]	1690	110	640
3/89 C.w.	150	37	205
4/89	70	30	140
5/89	5	1.51	7
5/89 D.[b]	10	0.6	1.4

[a] C.w. = bloom of *Coscinodiscus wailesii*.
[b] D.   = bloom of Spring diatoms.

nisms. It has been reported that dead and living cells take up metals to a similar extent, suggesting that surface rather than metabolic processes are dominant (Fisher 1986; Xue et al. 1988). For our modelling approach of the biological removal of heavy metals from a sea water reservoir, we may thus consider that the removal is governed by the adsorption processes onto the algal cell surfaces S_{tot}. We will consider here that a metal M is adsorbed on an active surface site. The adsorption of an amount C_M of M on the active plankton surface site S_0 may be described by:

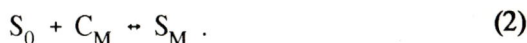

$$S_0 + C_M \leftrightarrow S_M . \tag{2}$$

We will assume that the adsorption is at equilibrium in the system and is of the Langmuir type (i.e. linear at ambient concentrations; Xue et al. 1988). The equilibrium condition is given by:

$$K_{eq} = \frac{S_M}{(S_0 C_M)} = \frac{S_X}{C_M(S_{tot} - S_M)} . \tag{3}$$

We will further assume that the metal transfer from dissolved to particulate phase (F_B) is proportional to the amount S_M of M bound to the active sites. The transfer kinetics tend to be "fast" (i.e. completion times in the order of minutes to hours) compared to the kinetics inherent in any seasonal fluctuations of S_{tot}. The "fast" kinetics of sorption-desorption indicate that for time scales of days to months, there will be a virtually continuous equilibration of the dissolved and particulate forms depending on the local solids concentration. This means simply that the fluctuation of S_{tot} is small over the equilibration time considered and thus may be taken constant as a first approximation. The local equilibrium assumption gives a rate equation which is a classical expression known as the Michaelis-Menten relation used by the biologists to describe the uptake rate F_B of a substrate by a biological population, which, in turn, governs its growth:

$$F_B = kS_M = \frac{kS_{tot}C_M}{\left(\dfrac{1}{K_{eq}} + C_M\right)} = \frac{k_m C_M}{(K_M + C_M)} \tag{4}$$

where k is the rate constant of the first-order phase transfer kinetics. The kinetic constant k_m $(= kS_{tot})$ is the maximum rate that the process can reach at high C_M, and K_M is the Michaelis constant. Typical adsorption rates determined for heavy metals on living phytoplankton cells in cultivation experiments with North Sea water range between 5×10^{-12} and 5×10^{-10} nmol/μm^2/h (cf. Chapter 4.5).

As pointed out above, the residence time t can be estimated at steady-state conditions by simply dividing the amount of the component in the reservoir by its input or removal flux, respectively:

$$\tau = \frac{C_M}{F_B} = \frac{(K_M + C_M)}{k_m} \tag{5}$$

Equation (5) shows clearly that the residence time of a metal is directly proportional to K_M, which is, in turn, inversely proportional to the partitioning coefficient K_{eq} of that metal between sea water and the particulate organic matter (or equal to the ratio S_{tot}/K_d because at low heavy metal concentrations $S_M << S_{tot}$). Partitioning coefficients measured in the German Bight increase with increasing S_{tot}-values in the order Cu < Cd < Pb (Table 4.6.4). Since for heavy metals in a natural sea water ecosystem K_M may be assumed to be much greater than C_M, the residence time becomes:

$$\tau = \frac{K_M}{k_m} \tag{6}$$

An interesting comparison can be drawn between the behaviour of a pure spring diatom bloom and the total phytoplankton population which was built up to a considerable extent by dinoflagellates and prymnesiophytes (*Phaeocystis globosa*) in May (Table 4.6.5). The sorption of cadmium by diatoms is considerably lower and of copper significantly higher, respectively, than the sorption by other members of the planktonic community which do not need silicate as a nutrient. Laboratory mesocosm experiments gave supporting results: copper uptake by diatoms increases due to silicate deficiency, while cadmium showed a higher affinity to the surface of dinoflagellates and prymnesiophytes (Rick 1990). These results indicate a close relationship between nutrient supply and heavy metal residence times in the North Sea.

This kinetic expression (4) may also be used to represent the uptake of a metal M by a phytoplankton population with a specific reactive surface S_{tot}, which is then eliminated from the reservoir by settling. F_B refers to the dissolved metal portion scavenged by the phytoplankton crop from the water column, which comprises a flux from dissolved to particulate phases; it should not be confused with particulate loss by sedimentation. The residence times calculated in Table 4.6.5 are based on biogenic particle settling fluxes which can be calculated directly from distribution coefficients using the following equation after Santschi (1984):

$$\log \tau = -\log K_d - \log (S/h) , \qquad (7)$$

where S is the particle settling flux and h is the mean water column high (24 m). Using the data from Tables 4.6.4 and 4.6.5, particle settling fluxes range from 9 g/d/m^2 in Winter to 390 g/d/m^2 in spring. This particle flux, however, does not include the metal fraction which is still integrally associated with lithogenous minerals even though this will be present in fecal pellets which apparently provide its main mode of transport in the water column. This fraction can be assumed to be inert within the short time scale considered by our model, so that the supplies of particulate metals equal the quantities removed from the reservoir. This distinction is necessary since our simple box model is concerned only with primary sources and sinks for dissolved heavy metals.

The subsequent pathway of heavy metals adsorbed by phytoplankton populations is still unknown and remains to be described quantitatively. A part of the total biomass is settling, but as discussed in the previous section, it serves not only for the input of pollutants to the sediment, but is in part remobilized back into the overlying water column. Moreover, thin-walled or nanoplankton species can already be lysed in the water column by microbial remineralization processes, thereby releasing their metal load before reaching finally the sea bottom even in shallow water environments such as the German Bight. Though all these processes have not yet been quantified, it can be anticipated that they would significantly increase the net heavy metal residence times in the water column. For our first-order approximation, however, we will neglect these constraints.

4.6.5   Discussion and Synthesis

The results presented in the preceding sections enable us to compare the flux of heavy metals at the two main interfaces, i.e. the atmospheric deposition rates with the flux of dissolved heavy metals from the sediment-water interface. Although this estimate is based on a limited number of samples, a simple comparison clearly shows that for the metals Cd and Cu mean atmospheric deposition rates are much lower than the remobilization rates of the respective elements at the sediment-water interface, whereas for Pb they are in the same order of magnitude (Tables 4.6.1 and 4.6.3). However, this result may not be representative for all seasons, since no seasonal fluctuation of the remobilization rates has been measured.

The fluxes which are being compared here are undoubtedly all quite variable, as shown in the preceding sections: the first on account of meteorological conditions, the second due to the variability of the sediment mineralogy and pollution degree, and the last due to variable biological productivity. The last two quantities are likely to show some correlation since the productivity may exert a yet unkown influence on the metal fluxes at the sediment-water interface. Since the atmospheric input of heavy metals into the water column of the German Bight is fluctuating biannually, it is rather easy to treat this case as a perodic fluctuation of input into a reservoir (Holland 1978):

$$F_A = f_a + a \sin \omega t , \tag{8}$$

where F_A represents the atmospheric deposition flux of a metal to the reservoir, f_a its constant fraction, a the amplitude of the fluctuation and ω its circular frequency which is expressed by $\omega = 2\pi/T$, with $T \leq 1$ year. To start with a first-order approximation, we might assume that the amplitude of the atmospheric input function is constant with time and neglect thereby the decrease observed at least for the Pb input during the last few years at Helgoland (Fig. 4.6.2). The periodic fluctuation of F_A, however, must be explicitly included in the geochemical model because its frequency induces significant changes at time scales comparable to those due to other kinetic processes occurring in the system. Such a periodic fluctuation is included by the maximum removal rate k_m in the biological flux term F_B which fluctuates at least seasonally with S_{tot}. In Chapter 3.3 it was evidenced that the mean annual cycle of phytoplankton productivity in the southern North Sea can be simply represented by a maximum culminating in summer, and a minimum during winter season, which tend to be anticyclic to the atmospheric input.

The interactions between these fluxes, biological reaction rates and the hydrodynamic residence time can be studied with the aid of a simple model of the German Bight, or a segment thereof, to be represented by a well-mixed box of fixed volume V. For an estimation of the advective net flux F_M of a metal M across the boundaries of the box (assuming that the heavy metal concentrations in the adjacent boxes are all equal, which is, however, usually not the case as shown by Kühn et al. 1992) one may use the following kinetic expression:

$$F_M = RC_M \, , \tag{9}$$

where R denotes the net portion of the water reservoir V which is exchanged during a certain time $t = V/R$. Summarizing all the processes described above, the following time dependent mass balance equation for the dissolved heavy metal content within the box can be set up:

$$\frac{dC_M}{dt} = F_A + F_S - F_M - F_B \, . \tag{10}$$

Considerable insight can be obtained by simply assuming a steady state by $dC_M/dt = 0$. Considering the above-deduced Eqs. (4), (8) and (9) for the three time-dependent flux terms, Eq. (10) can be transformed:

$$RC_M^2 - XC_M - Y = 0 \tag{11}$$

with

$$X = F_S + f_a + a \sin \omega t - RK_M - k_m \tag{11a}$$

and

$$Y = (f_a + a \sin \omega t + F_S)K_M \, . \tag{11b}$$

The solution for the steady-state amount C_0 becomes:

$$C_0 = [X \pm (X^2 + 4YR)^{0.5}](2R)^{-1} . \tag{12}$$

Since the square root is necessarily greater than X, only the positive root leads to positive values of C_0. This result shows that there is thus only one possible steady state for this system for any finite values of F_S, f_a, R, a, K_M, and k_m which fix the amount C_0 of a given heavy metal in the reservoir.

For problem contexts where the time variable behaviour of the system is important, this equation must be solved numerically. However, it is also possible to analyze the response time of the system to short term perturbations of an assumed steady state condition without integrating this equation. One might assume at time t = 0 a small instantaneous perturbation α_0 (such as an anthropogenic input event) in the box initially at steady state. The evolution of the metal amount in this reservoir then has the following form:

$$C(t) = C_0 + \alpha(t) . \tag{13}$$

The adjustment of the reservoir to a new equilibrium can be calculated by considering Eq. (11). The mass balance of the perturbated system is then given by:

$$\frac{d(C_0 + \alpha)}{dt} = f_a + a \sin \omega t + F_S - R(C_0 + \alpha)$$
$$- k_m(C_0 + \alpha)(K_M + C_0 + \alpha)^{-1} \tag{14}$$

Since for heavy metals K_M may be assumed to be much greater than C_0 and necessarily even much greater than α in sea water, Eq. (14) reduces at steady state (dC/dt = 0) to the classical first-order decay equation:

$$\frac{d\alpha}{dt} = -(R + \frac{k_m}{K_M})\alpha , \tag{15}$$

which can be easily integrated:

$$\alpha = \alpha_0 \exp(-k' t) , \tag{16}$$

where $k' = R + k_m/K_M$. Given the above assumptions, the system may thus be considered to be stable since it goes back to its initial steady state. The imposed small perturbation is decaying exponentially with a response time τ, which has often been chosen by convenience as the time required to reduce the initial perturbation from steady state to $1/e$ ($= 0.37$), or in other words, to reduce the perturbation by 63 %. According to Eq. (16), the response time τ is:

$$\tau = \frac{1}{k'} = \frac{1}{(R + \dfrac{k_m}{K_M})} = \frac{K_M}{(K_M R + k_m)} \ . \tag{17}$$

The response time is thus inversely proportional to both the kinetic constant R for the advective water transport (which is in this case the gross transport) and the biological removal rate k_m/K_M (cf. Sect. 4.6.4).

The evolution of the perturbation α/α_0 of C_0 as a function of time and the effect of these kinetic constants on the response time are illustrated graphically for cadmium in Fig. 4.6.3 with data taken for the German Bight. In this Figure the decay of a small perturbation from the same steady state is shown for the various advective and biological rates determined in February and May 1989. The biological removal rate is lower by nearly two orders of magnitude than the advection rate in winter, which yields a net residence time of 15 days. Although the advection rate is still relatively high for the German Bight in May, biological scavenging now dominates Cd removal at a rate three times higher.

4.6.6 Conclusions and Outlook

In this Chapter we have tried to elaborate the dynamic aspect of the most important biotic and abiotic processes which determine the dissolved heavy metal cycle in the water column of the North Sea. The aim was to show by simple geochemical modelling how these processes may be coupled in a natural, mixed one-reservoir system such as the water column of the German Bight. We have demonstrated thereby that the dissipation of an anthropogenic metal input event is controlled by physical advection in winter and by biological uptake in summer (Fig. 4.6.3). It is evident that a number of processes interact to rapidly reduce the magnitude of changes in sea water concentration arising from anthropogenic inputs. We conclude, therefore, in agreement with Balls (1989) that for a given coastal North Sea

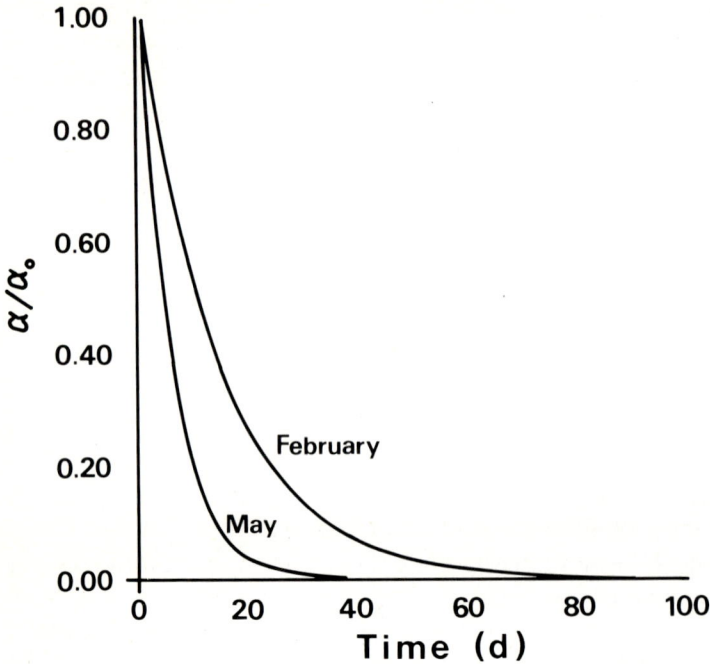

Fig. 4.6.3. Evolution of the pertubation α/α_0 of the Cd steady state concentration in the German Bight described by Eq. (17). Data for biological removal term k_m/K_M were calculated from biological Cd residence time ($t_{1/2} = \ln 2 \times K_M/k_m$) as assessed for the German Bight in February and May 1989 (Table 4.6.5). The respective advection rates R for the gross water exchange across the boundaries of the German Bight calculated by the hydrodynamic model discussed in Chapter 3.2

area such as the German Bight it is extremely difficult to identify an anthropogenic signal merely on the basis of concentration trend monitoring in the water column. An alternative approach is the use of isotopic compositions, although this is generally only possible where there is a source with a well-characterized composition such as in case of lead from gasoline combustion (Kersten et al. 1992).

The next step would be to extend this first model approximation by increasing the number of rate expressions and introducing additional time-dependent variables for the sediment-water exchange and seston concentration before in a third step any quantitative predictions of the heavy metal cycles may be tried. One of these time-dependent processes neglected in our first approach is the relatively rapid regeneration of phytodetritus at the sediment-water interface. Unfortunately, no experimental or field data exists yet on this process, although our preliminary data demonstrate that fluxes at the sediment-water interface might exceed atmospheric deposition rates. Without these data, therefore, it seems unlikely that knowledge of heavy metal cycling in the marine coastal environment can be furthered. Moreover, geochemical cycles usually involve several reservoirs connected by multi-directional fluxes of components. The water column, e.g. should be divided into at least two sub-reservoirs, because of the importance of physically induced stratification for the development of phytoplankton blooms. In conclusion, a lot of work has to be done before a 1-D heavy metal model can be presented similar to that described for nutrients in Chapter 3.3, this Volume.

References

Balls PW (1989) Trend monitoring of dissolved trace metals in coastal sea water - a waste of effort? Mar Pollut Bull 20:546-548

Bender M, Martin W, Hess J, Sayles F, Ball L, Lambert C (1987) A whole-core squeezer for interstitial porewater sampling. Limnol Oceanogr 32:1214-1225

Davison W, Grime GW, Morgan JAW, Clarke K (1991) Distribution of dissolved iron in sediment pore waters at submillimetre resolution. Nature 352:323-325

DiToro DM (1980) Simplified model of the fate of partitioning chemicals in lakes and streams. In: Mackay D (ed) Modelling the fate of chemicals in the aquatic environment. Ann Arbor Sci Publ, Ann Arbor, Mich, pp 165-190

Duce RA, Liss PS, Merrill JT et al. (1990) GESAMP-Reports and studies No 38, WMO, Geneva

Emerson S, Jahnke R, Heggie D (1984) Sediment-water exchange in shallow water estuarine sediments. J Mar Res 42:709-730

Fisher NS (1986) On the reactivity of metals for marine phytoplankton. Limnol Oceanogr 31:443-449

Förstner U (1986) Chemical forms and environmental effects of critical elements in solid-waste materials: Combustion residuals. In: Bernhard M, Brinkmann FE, Sadler PJ (eds) The importance of chemical "speciation" in environmental processes. Springer, Berlin Heidelberg New York, pp 465-491

GESAMP (1987) Land-sea boundary flux of pollutants. United Nations Joint Group of Experts on the Scientific Aspects of Marine Pollution (GESAMP) Rep Studies Ser 32, Unesco, New York, 172 pp

Graf G (1989) Benthic-pelagic coupling in a deep-sea benthic community. Nature (Lond) 341:437-439

Graßl H, Eppel D, Petersen G, Schneider B, Weber H, Gandraß J, Reinhardt KH, Wodarg D, Fließ J (1989) Stoffeintrag in Nord- und Ostsee über die Atmosphäre. GKSS Rep 89/E/8, Geesthacht

Hardy JT, Crecelius EA (1981) Is atmospheric particulate matter inhibiting marine primary productivity? Environ Sci Technol 15:1103-1105

Hodge V, Johnson SR, Goldberg ED (1978) Influence of atmospherically transported aerosols on surface ocean water composition. Geochem J 12:7-20

Holland HD (1978) The chemistry of the atmosphere and oceans. Wiley, New York

Kersten M (1988) Geobiological effects on the mobility of contaminants in marine sediments. In: Salomons W et al. (eds) Pollution of the North Sea - an assessment. Springer, Berlin Heidelberg New York, pp 36-58

Kersten M, Kienz W (1993) A continuous-flow system for the determination of trace metal fluxes using intact sediment cores. Limnol Oceanogr (submitted)

Kersten M, Dicke M, Kriews M, Naumann K, Schmidt D, Schulz M, Schwikowski M, Steiger M (1988) Distribution and fate of heavy metals in the North Sea. In: Salomons W et al. (eds) Pollution of the North Sea - an assessment. Springer, Berlin Heidelberg New York, pp 300-347

Kersten M, Kienz W, Koelling S, Schröder M, Förstner U (1990) Schwermetallbelastung in Schwebstoffen und Sedimenten der Nordsee. Vom Wasser 75:245-272

Kersten M, Kriews M, Förstner U (1991) Partitioning of trace metals released from polluted marine aerosol in North Sea water. Mar Chem 36:165-182

Kersten M, Förstner U, Krause P, Kriews M, Dannecker W, Garbe-Schönberg C-D, Höck M, Terzenbach U, Grassl H (1992) Pollution source reconnaissance using stable lead isotopes (^{206}Pb/^{207}Pb). In: Vernet J-P (ed) Impact of heavy metals in the environment, Elsevier, Amsterdam, pp 311-325

Krell U, Roeckner E (1988) Model simulation of the atmospheric input of lead and cadmium into the North Sea. Atmos Environ 22:375-381

Kriews M (1992) Charakterisierung mariner Aerosole in der Deutschen Bucht sowie Prozeßstudien zum Verhalten von Spurenelementen beim Übergang Atmosphäre-Meerwasser. PhD Thesis, Univ Hamburg

Kriews M, Naumann K, Dannecker W (1989) Einsatz atomspektrischer Methoden zur Multielementbestimmung in marinen Aerosolen. In: Welz B (ed) 5. Colloquium Atomspektrische Spurenanalytik. Perkin-Elmer, Überlingen, pp 633-646

Kühn W, Radach G, Kersten M (1992) Cadmium in the North Sea - a mass balance. J Mar Systems 3:209-224

Lindberg SE, Harriss RC (1983) Water- and acid-soluble trace metals in atmospheric particles. J Geophys Res 88:5091-5100

Maring HB, Duce RA (1989) The impact of atmospheric aerosols on trace metal chemistry in open ocean surface seawater. J Geophys Res 94:1039-1045

McCaffrey RJ, Myers AC, Davey E, Morrison G, Bender M, Luedtke N, Cullen D, Froelich P, Klinkhammer G (1980) The relation between pore water chemistry and benthic fluxes of nutrients and manganese in Narragansett Bay, Rhode Island. Limnol Oceanogr 25:31-44

Rick HJ (1990) Ein Beitrag zur Abschätzung der Wechselbeziehung zwischen den planktischen Primärproduzenten des Nordseegebietes und den Schwermetallen Kupfer, Zink, Cadmium und Blei. PhD Thesis, RWTH Aachen

Santschi PH (1984) Particle flux and trace metal residence time in natural waters. Limnol Oceanogr 29:1100-1108

Santschi PH, Höhener P, Benoit G, Buchholtzten Brink M (1990) Chemical processes at the sediment-water interface. Mar Chem 30:269-315

Schulz M, Steiger M, Schwikowski M, Kriews M, Naumann K., Dannecker W (1988) Variability of ambient trace element concentrations at the North Sea with respect to air mass history. J Aerosol Sci 19:1171-1174

Shaw TJ, Gieskes JM, Jahnke RA (1990) Early diagenesis in differing depositional environments: the response of transition metals in pore water. Geochim Cosmochim Acta 54:1233-1246

Slauenwhite DE, Wangersky PJ (1991) Behavior of copper and cadmium during a phytoplankton bloom: a mesocosm experiment. Mar Chem 32:37-50

Slinn SA, Slinn WGN (1980) Predictions for particle deposition on natural water. Atmos Environ 14:1013-1016

Twinch AJ, Ashton PJ (1984) A simple gravity corer and continuous-flow adaptor for use in sediment/water exchange studies. Water Res 18:1529-1534

Wollast R (1986) Basic concepts in geochemical modelling. In: Buat-Menard P (ed) The role of air-sea exchange in geochemical cycling. Reidel, Dordrecht, pp 1-34

Xue HB, Stumm W, Sigg L (1988) The binding of heavy metals to algal surfaces. Wat Res 22:917-926

4.7 On the Reliability of our North Sea Assessment

M. BOHLE-CARBONELL

The objective of the two ZISCH surveys was to obtain a comprehensive, uniform data set for contaminants and ecosystem parameters in the entire North Sea. Indeed, before ZISCH no comparable data set existed. It is still necessary to explore the limits of this data in order to avoid overinterpretation, in particular as the ZISCH surveys had to be designed without reference to statistical considerations on sampling schemes (Cochran 1977).

The strategy of the ZISCH surveys was to observe all investigated parameters under comparable sampling conditions (sampling interval, coverage etc.). The sampling scheme, the ZISCH star, was the usual compromise between scientific requirements and logistical constraints. This compromise determines "what is observed" and the magnitude of biases due to a particular interplay between sampling scheme and the variability of the observed object (Wunsch 1989). In this Chapter a limited set of descriptive analyses is presented to illustrate some statistical aspects of the representativity of our data.

Assessment of reliability in a positive sense requires a fairly comprehensive knowledge of the investigated subject. Otherwise, reliability only can be assessed in an indirect manner. The method "proof by contradiction" combined with analysis of statistical power (e.g. Peterman 1990) would be most promising, if reasonable assumptions appropriate for falsification could be formulated. Otherwise, it is necessary to limit, as done here, the discussion to basic sampling problems which are obstacles to reliability.

4.7.1   A View Back

Knowing the considerable amount of marine research carried out in the North Sea during the past 100 years, in particular in the German Bight, it seems obvious that the data coverage of these seas is sufficient, at least for many parameters and for the German Bight. Nevertheless, reviewing observations from the German Bight with respect to sampling gives rise to the suspicion that salinities are still undersampled (Bohle-Carbonell 1992), contrary to the impression left by eye inspection. Relevant variance on small scales seems not sufficiently resolved, and large scale structures are therefore aliased[1]. If this is true for salinity, then many environmental parameters are undersampled, if these parameters are conservative with respect to salinity.

Reasons to question the sufficiency of the data coverage for salinity in shelf seas with strong runoff like the North Sea arise from the following line of argument (Bohle-Carbonell 1990): Historically, most data there were collected from ships operating on coarse grids. Modern observation techniques, either by remote sensing or by towed, continuously recording instruments, showed subgrid structures which develop swiftly. These structures cannot be resolved by the usual, coarse grid ship surveys. Large scale structures may be artificially enhanced by aliasing, because variance attributed to non-resolved time or space scales is transferred to the resolved scales.

4.7.1.1   Variability of Mean Fields

Charts of mean salinity and temperature in the North Sea have been prepared by several authors. Well known are, for example, the charts by Goedecke et al. (1967). The isolines on these charts show lines of same sample mean.

In reference to the German Bight, one of the key regions of environmental concern in the North Sea, Goedecke and colleagues describe their analysis as follows: First, the total number of data was about 9600 measurements, unevenly scattered over 12 0.5° x 0.5° fields. Second, the variability was so high that the authors draw only lines for each part per thousand. Third, spatial smoothing was done for three 0.5° fields in each

---

[1]Chambers Science and Technology dictionary (1988): "aliasing": error or image imperfections resulting from limited detail in raster display caused by insufficient number of data points.

direction. The result is that smoothing makes each detail of an isoline plot with scales larger than 3° relevant.

Frey and Becker (1986) analyze differences between surface and bottom densities in the German Bight on a 0.17° (10 min) grid (grid cell area is about 200 km^2) in all data (profiles) from the period 1919-1985. Their monthly charts are not spatially smoothed but are sometimes interpolated if single values are missing (no data for month and grid cell). Information on the scatter is given as charts of the upper bound of density differences. These maxima are roughly double the climatic means. Taking this as an estimate of the range, then the standard deviation can be estimated (Stange 1970) to be about two thirds of the mean value.

Thus, as commonly inferred, the apparent smoothness of mean salinity fields (and fields of other data) is mainly due to averaging, and considerable data scatter is consistent with smoothly varying mean values.

4.7.1.2 Common Sampling

Sampling the German Bight is commonly done in surveys lasting several days. One research vessel samples on a regular grid. It is assumed that the hydrography of the German Bight changes only slightly during the sampling period and that corrections due to advection by tidal currents can be obtained from a numerical model (Brockmann and Dippner 1988).

There is evidence for quasi-synopsis, especially for summer conditions (e.g. Katsaros et al. 1983, plate 1, 2). However, the presence of important small-scale structures of the salinity field in the German Bight has been known since Becker and Prahm-Rodewald (1980) published observations obtained with a continuously recording towed instrument. They discovered that the convergence zone in the German Bight is composed of multiple salinity (and temperature) fronts and a complicated patchwork of water masses of different characteristics. The relevant cross-frontal scales are less than 3' EW, and structures are anisotropic, as has been already observed by Lüneburg (1963) for southern parts of the German Bight.

Evidence is therefore established that the usual procedure for sampling salinity in the German Bight must miss important structures. Salinity differences, for example, are due the varying influence of fresh water runoff. The spatial distribution of any parameter influenced by runoff therefore will be biased if sampled on common grids, and visible structures might be artefacts. However, such sampling will be sufficient for taking a random sample as needed for most monitoring tasks.

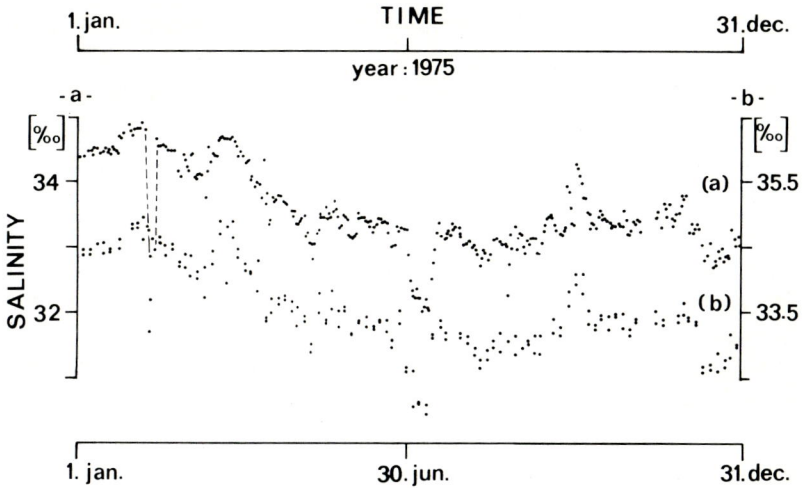

Fig. 4.7.1. Two apparently similar salinity time series from 1975 at Light vessel "TW EMS" (position at 54°10'N, 6°21'E; about 75 km offshore and 45 km off the 10 m depth contour, respectively; water depth 34 m). The scales -*a*- and -*b*- refer to time series (*a*) and (*b*), respectively. The *broken line* is for perception of a short-lasting (about 5 days) salinity excursion. Observations at the light vessel have been done in two modes, first (*a*) once a day, at 0800 h, and second (*b*) twice a day at slack water every fourth day. Each pair of measurements at slack water shows up as a "*double point*". The *vertical spacing* of the double point indicates salinity differences due to tidal advection of horizontal salinity gradients. These variations are not resolved by regular daily observations, and therefore cause aliasing in the record of daily salinity observations which is not detectable by eye inspection

4.7.1.3 Aliasing in Time

Prahm (1961) described salinity fluctuations at the southern fringes of the Dogger Bank as recorded by daily observations over a 5 year period. Diurnal and weekly variations are typically 0.06. Diurnal and weekly variations above 0.1 have a relative frequency of 13 and 26%, respectively. Most of these variations are due to non-resolved tidal advection of the outer fringes of the Rhine plume (Damm 1989). Some are due to aliasing. Aliasing of salinity time series (Fig. 4.7.1) due to non-resolved tides is common for the North Sea, because of its strong lateral salinity gradients (Damm 1989).

4.7.2   A Simulation of the ZISCH Sampling Scheme

The ZISCH surveys will not be repeated. Thus, structures observed once in 1986 and 1987 cannot be checked for persistence. It is therefore difficult to estimate whether the data revealed typical or particular situations. Furthermore, it is less than evident how efficient the star-shape sampling pattern was or whether it caused biased results. In order to consider these and similar questions, effects of sampling schemes should be simulated (Bohle-Carbonell 1992). Results of such an exercise will be reported in the following.

Part of the ZISCH program was to simulate the current field of the North Sea driven by real wind fields. Results of this simulation exercise will be used to do the proposed simulation of effects of sampling schemes. In a first step, the variability which would be encountered by repeated ZISCH surveys will be simulated.

4.7.2.1   Material and Methods

In order to make the problem as simple as possible, only one source of variability will be considered - the advection due to variable wind forcing, seasonally varying density fields and tides (Backhaus and Hainbucher 1987). To simplify further, currents will be averaged over 1 day and over the vertical. These current fields will be used to calculate Lagrangian advection of passive particles in order to analyze their distribution in time and space for certain particle sources.

Hainbucher et al. (1987) have carried out such a simulation for particle sources in those regions where water masses enter the North Sea, either from the Atlantic or through runoff. The source strength was set to one particle a day. The model spins up/down in 2 years and the model run covers 14 years. Positions of particles are calculated for 2 years. Thus, after spin up, positions of about 730 particles from any source are known for about 9 years (see Chap. 3.2).

These results will be used. Here, particles are understood as labels indicating the origin of water masses. The simulation experiment consists simply in counting the number of labels found near certain positions at certain times (year, date, hour). The positions are given by the stations of the ZISCH survey. All labels are counted which are in a rectangular area of 40 to 40 km around the ZISCH stations. That accounts for non-simulated tidal dispersion over a period of some days (estimated after Nihoul 1980, for observations, e.g. Watson et al. 1991). The particle count is for ZISCH station times (date and hour) plus a given set off (year, day). Since the model has a daily time step, index counts are interpolated to station times.

The variable parameters of this simulation are the source, the year (1971-1979) and the time (ZISCH station time +/- 0, 1, 3, 10 or 30 days). Thus, different label counts for different years and times will measure the interannual and synoptic variability for each source, due to different advection. These label counts therefore will be named "advective influence index" (AII).

The results of Hainbucher and colleagues showed that the particles are not smoothly distributed but form varying clusters in coastal regions of the North Sea. It will be quantified now, how strongly this heterogeneity effects the AII at ZISCH positions. Two cases will be discussed in some detail, first, the interannual winter variability 1971-1979 (with two subcases), and, second, the synoptic winter variability of the years 1975 and 1979. Two sources are considered, first, the Atlantic inflow between Orkneys and Shetlands, and, second, the runoff from the Rhine. AII's for Tyne, Humber, Thames and Elbe are similar and summer and winter seasons show no qualitative difference.

Results are given as minimum, 25%-quartile, median, 75%-quartile and maximum of AII's. These statistics are used to derive features of the AII which are plotted as a function of the station number (Fig. 4.7.2) or the distance covered by the ship on its way along the star (Fig. 4.7.3a).

The sample path of the ZISCH survey cuts several times the coastal band of higher numbers of labels. The plot of AII along the ships' track therefore shows several spikes which can be identified clearly with certain regions (Fig. 4.7.2). These spikes are particularly well defined, as a uniform distribution of tracing particles over the North Sea would lead to a mean AII of 0 to 1 per box. Any non-uniform distribution therefore will lead to a pattern of high and small values.

The interannual variability is estimated from either nine simulations, one each year at ZISCH dates (ZISCH date = station time and day of the ZISCH survey but unspecified year), or 36 simulations covering the 9 years. In the latter case, the ZISCH station times are shifted by +30, +10, -10 and -30 days. It seems reasonable to attribute 9 degrees of freedom to the AII's in the first case, but only 18 degrees of freedom (instead of 36) in the second case due to the serial correlation of current events.

The winter synoptic variability is estimated from 18 simulations for the years 1971 and 1979 (station dates shifted by +30, +10, +3, +1, 0, -1, -3, -10, -30 days within 1 year). These simulations cannot be considered as independent cases, but it seems reasonable that this sample has 9 degrees of freedom. The 95% confidence interval of the median is either the range

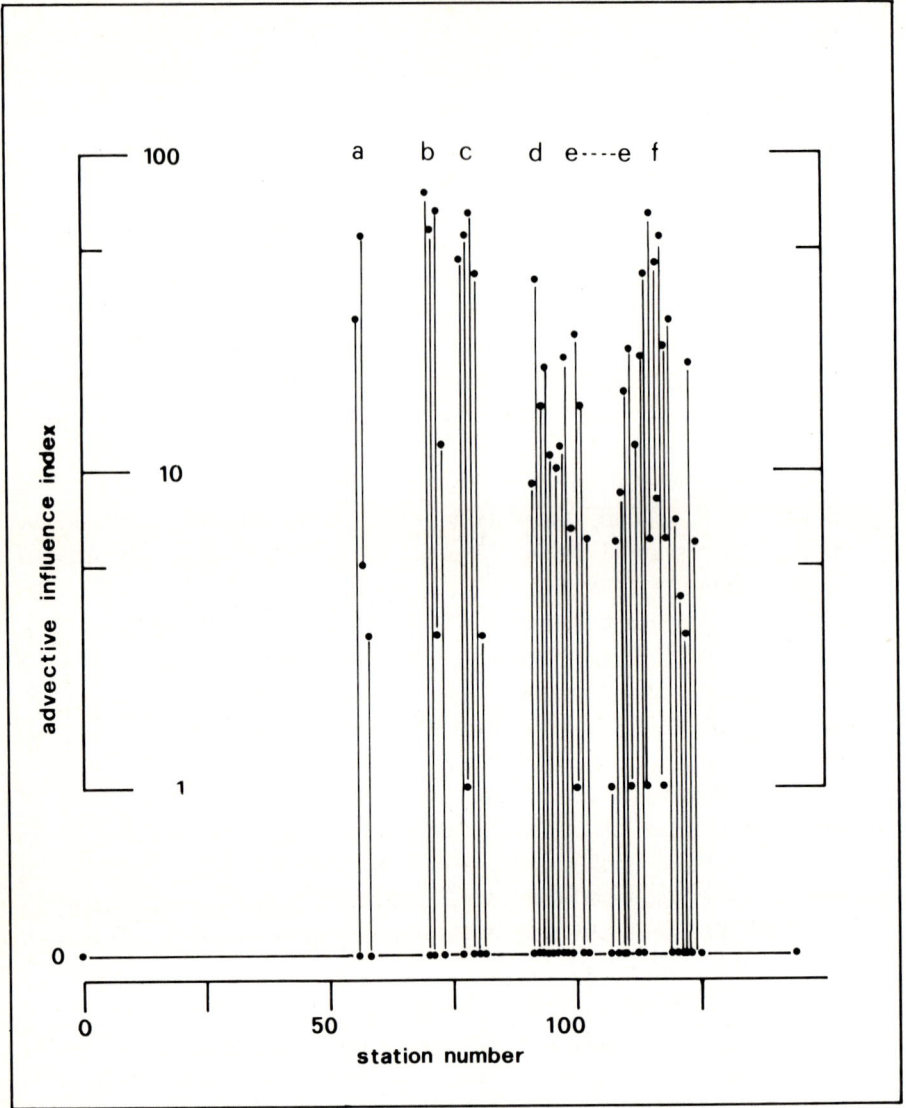

Fig. 4.7.2. Interannual variability of the advective influence index (AII). The simulated range of advective influence index (AII) is shown as function of the station number for the case: source Rhine, years 1971 to 1979 and dates and stations of the winter ZISCH survey. These ranges equal the 95% confidence limits of the median AII. The peaks labelled *a* to *f* refer to positions off the Rhine estuary, off island Texel, at western and northern limits of the German Bight, off Jutland and in the Skagerak

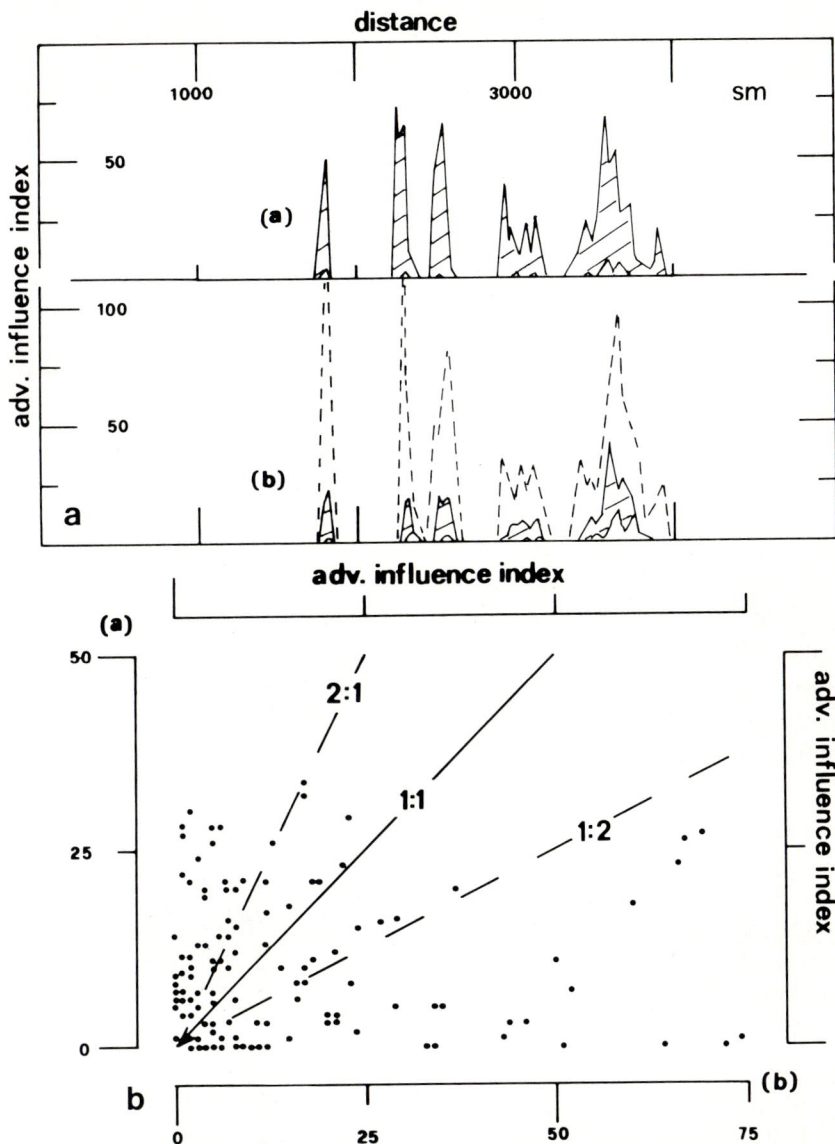

Fig. 4.7.3. Comparison of synoptic and interannual variability of the advective influence index (AII). The *hatched surface* in 4.7.3a gives the simulated 95% confidence interval of advective influence index (AII) as a function of the distance steamed (along the ZISCH star) for the cases: source Rhine, years 1971 to 1979 and stations of the winter ZISCH-survey. *Upper part, a* Dates are as during the ZISCH-survey (compared Fig. 4.7.2). *Lower part, b* Dates shifted +30, +10, -10 and -30 days (*broken lines* refer to maximum values); The quartiles (25-25%, 75-75%) of the AII's of years 1975 and 1979 are shown as scatter plot (4.7.3b). For similar conditions a scatter around the line (1:1) would be expected

of the sample or the distance between the 75%-quantile and 25%-quantile, if the degrees of freedom are 9 or 18, respectively.

4.7.2.2 Results

Source: Rhine. First, the interannual variability is considered and, in particular, the AII's found for the ZISCH dates of the years 1971 to 1979. AII's for the source Rhine show five well-defined spikes (Figs. 4.7.2 and 4.7.3.a, upper part). Their geographical locations are: off the Rhine estuary (maximum at station 57), off the island Texel (72), western German Bight (79), northern German Bight (92) and off southern Jutland and off northern Jutland and Skagerak (115). The corresponding ranges of the AII's are 5 to 52, 3 to 66, 0 to 66, 0 to 40 and 6 to 67, respectively. The ranges of the AII's cover an order of magnitude and are similar for all regions. Particularly low AII's (0) can be found for the German Bight. Thus, the AII's at ZISCH stations are very different from year to year.

Next, the interannual variability is considered for shifted ZISCH dates covering a two month period (Fig. 4.7.3.a, lower part) compared to the ZISCH dates (Fig. 4.7.3.a, upper part). The result is nearly the same. The spikes of AII's are somewhat broader and higher (maximum at Texel, 121), and the confidence interval for the medians are narrower, e.g. 8 to 23 off the Rhine estuary. However, the scatter of the AII's has not considerably decreased in spite of averaging over more cases.

Now, the synoptic variability is considered. Again, the five spike structure is found (not shown). The data scatter within each year is somewhat reduced, in particular for the year 1975. Maximum values in 1975 are lower than in 1979. The range of AII's found in 1979 is comparable to the interannual variability found for non-shifted ZISCH dates. The scatter plot of corresponding quantiles of years 1975 and 1979 shows how strongly both years differ (Fig. 4.7.3.b). Nevertheless, the median AII's of the years 1975 and 1979 are not much different if their confidence intervals are considered. Thus, considering the ZISCH survey, different station times imply different AII's - differences of half an order of magnitude, at least. Thus, the pattern found once is a strong function of the actual station times.

Source: Shetlands. The basic structure is the same as found for the source Rhine. The spikes are less clear cut, due to the stronger dispersion of AII's from this source. Maximum values and ranges are smaller. Maximum AII's of zero (meaning no influence from this source in that region) are found, e.g. upstream of the source and in the eastern Channel. Interannual variability is different than winter synoptic variability. In the latter case, the AII spikes are more clear cut because AII is spatially distributed in a manner which is characteristic for the simulated year. The interannual distribution of AII's is therefore more dispersed.

This simulation exercise illustrates how the advective influence arising from one particular source is a strong function of the year and the date. Individual AII's are typical for a certain year and month. However, the result of this simulation exercise may not be applied directly to nature, because diffusive and dispersive processes counteract the heterogeneity caused by advection, but the strength of this "smoothing operation" is undetermined and might be counteracted by different sources of variability.

4.7.3 A Statistical Analysis

The ZISCH experiment was conceived on the basis of the precautionary principle of environmental politics. Accordingly, the emphasis was on identification of plausible structures within the patterns of, for example, pollutants in the North Sea. For this reason, ZISCH's data interpretations are done using a non-specified, vague null hypothesis "the observed patterns are structured". The corresponding alternative hypothesis, "the observed patterns are random", is of minor importance.

Putting these considerations on needs and aims, which are extra-statistical or "meta-methodological" (Hofstadter 1988), into statistical terms means: scientists in ZISCH study their data sample in a way such that the probability of "not identifying a pattern as structured has to be small if the pattern is structured". It is of minor importance if random features are erroneously identified as structured.

Defining one's approach in this way means attributing particular values for the statistical errors of the first and second kind - "not identifying a true hypothesis" and "not rejecting a false hypothesis", respectively. These particular values might be expressed as different risks or levels of significance for both kinds of statistical errors.

Inherent to any data analysis presented in the preceding chapters is an argument on risks (or significances) of erroneous statements, even in cases where there is no handy means or no practical need to quantify these statistical parameters. Knowledge and understanding generally handle these risks in a qualitative way. However, it would be illustrative to check how the quantification of risks of both statistical errors would qualify our assessment of the North Sea environment. To put this in methodological terms means to suggest a statistical analysis of power to distinguish between two hypotheses. A systematic power analysis of ZISCH data would be a tremendous piece of work. Therefore, I will limit myself in the following to giving some illustrative examples which show that the ZISCH results are significant up to a certain degree, but the alternative hypothesis "random patterns" is consistent with the data, if modestly high levels of significance are required.

In the following sub section I present an analysis under three constraints. First, compared to the standard ZISCH hypothesis, I interchange null and alternative hypotheses, also "random patterns" is my null hypothesis. Second, the risk should be low for the case "not identifying a pattern as random" if "random" would be true, or I accept only a small statistical error of the first kind. Third, I take any risk of identifying falsely a pattern as random which is structured, or I do not specify the statistical error of the second kind.

Thus, I will take voluntarily the opposite of the standard ZISCH approach, but do it with similar methodological drawbacks. This means that I do not quantify my statistical error of the second kind. Note that, due to the interchanging of hypotheses, my statistical error of the first kind corresponds to the statistical error of the second kind in the standard ZISCH approach and vice versa. Results obtained under the standard and the opposite approach might be compared. However, due to the voluntarily taken methodological drawbacks, this cannot be done in quantified statistical terms, but has to be done in more general terms. However, such comparison might be considered as a qualitative equivalent to a statistical power analysis and will illustrate how approximate is our quantified description of the North Sea environment[2].

---

[2]Before proceeding, the reader is kindly asked to remember:

- Testing null hypotheses (H0) and alternative hypotheses (Ha) based on the same data sample has four possible outcomes: rejecting H0 and Ha, accepting H0 and Ha, rejecting H0 and accepting Ha, or accepting H0 and rejecting Ha. Any of these cases has a different probability, depending on the significance level of the tests for H0 and Ha.
- The power to distinguish between hypotheses H0 and Ha will depend on the sample size, the applied tests, the detailed way in which the hypotheses H0 and Ha are expressed in a testable statistic.
- An analysis of statistical power (the ability to distinguish between two hypotheses) might well lead to the result that both hypotheses are consistent with a given data sample. Anyhow, the argument for accepting a particular hypothesis is related to the required levels of significance of both hypotheses,

4.7.3.1 Material and Methods

For this analysis, only data for water samples from the upper 5 m along several transections through the ZISCH star were considered, which is only a small subset of the complete ZISCH data set but taken from a regime showing an outstanding variability (see Fig. 4.7.4a,b):

1. from English coast over the centre of the star into the Skagerak (mainly SW-NE across the North Sea),
2. from NW off the Shetlands down into the German Bight (mainly NW-SE across the North Sea,
3. from just SW of the Shetlands along the British, Dutch, German, Danish and Norwegian coasts,
4. as in the preceding case but further offshore,
5. as in the preceding case but around the centre of the ZISCH star.

Transection (5) is less a section in space than a time series of observations within the centre of the survey region. For synoptic conditions, therefore, it is to be expected that the data sample along this transection exhibits only structures which are consistent with random fluctuations.

The analyzed parameters are concentrations of lead in suspended matter, phytoplankton carbon, phosphate and nitrate (summer survey) or the sum of nitrite and nitrate (winter survey). The observed values were taken from the ZISCH data report (Moll and Radach 1990, codes G2-7, G4-2, G5-15, G6-2, 31). The minimum sample size was set to the modest value of 14 observations, so these were the only parameters we could use for these tests. Three simple non-parametric tests were applied to the data (Schönwiese 1985):

- The iteration test after Wallis and Moore for testing whether the sample consists of independent data. Rejection of data independence indicates that the data sequence along the transection is not random.

- The trend test after Cox and Stuart for testing whether a trend-like change is consistent with a sequence of randomly changing data - failure of this test indicates that the data sequence changes systematically.

---

which might be fairly different. Furthermore, the required levels of significance are set according to non-statistical requirements.

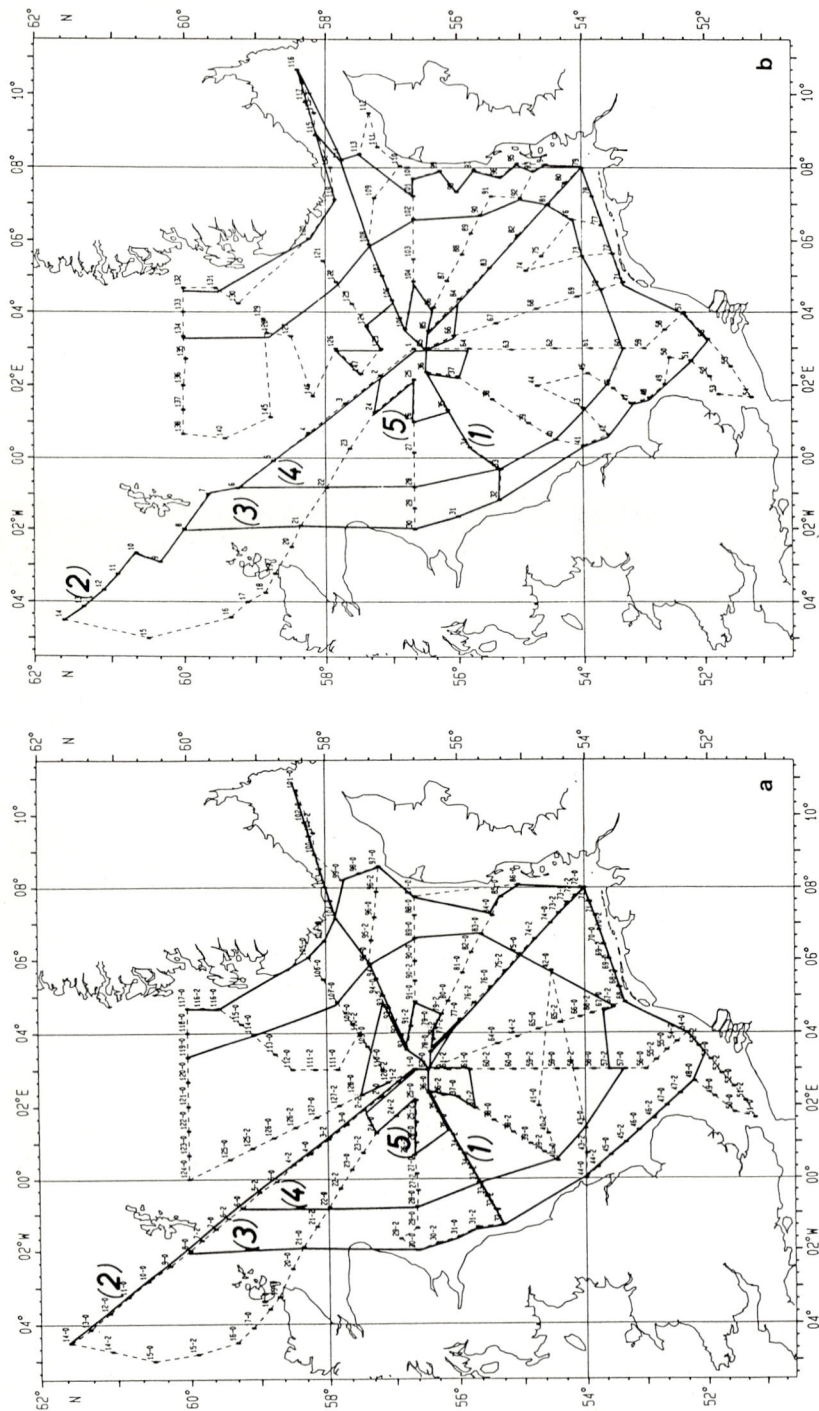

Fig. 4.7.4a,b. Positions of the transections through the ZISCH star along which data sequences have been analyzed. *a* summer *b* winter

- The rank test after Wilcoxon, Mann and Whitney for testing whether the data sampled along transection (2) and (3) are from the same distribution - failure of this test indicates that near-coast and cross-North Sea data sets differ systematically.

All three tests express the test statistic (number of ups and downs along the data sequence, comparison of values in the first and last third of the data sequence by pairs, sum of ranks of each original sample in the combined sample) as a standardized, normally distributed random variable (z). Evaluation is now simple.

If z exceeds a predefined threshold, then my null hypothesis, "random", is rejected. The threshold is defined with respect to rejecting falsely the null hypothesis. Setting the threshold to $z = 1.3$ means that in one of ten cases the rejection of the null hypothesis will be done erroneously. This seems to be a good compromise between significance of the result and anticipated variability of the system. The results of this analysis are given as a dot chart (Fig. 4.7.5), which should be read keeping both methodological problems in mind, first, those discussed above and, second, those mentioned in the preceeding chapters which concern technical matters of measuring.

4.7.3.2 Results

The amount of data regarding lead in suspended matter and phytoplankton carbon is either too small to do the tests or the statistics are mainly below or marginally above the 10% threshold but below the 5% threshold. As a consequence, in most cases the null hypothesis could not be tested or must be accepted. At the chosen confidence levels there is no reason to distinguish the observed patterns from random. There are, however, two exceptions to this: the summer NW-SE transection (no. 2) and the winter nearshore transection (no. 3). Thus, for example, winter lead concentrations apparently decrease from the British to the Danish coasts. However, no strong reason is given to qualify the main pattern of lead and phytoplankton carbon as different from random.

The estimates for the nutrients indicate that the null hypothesis of random fluctuations of these parameters is less consistent with the data sample.

Summer nutrient samples deviate most from the hypothesis that all fluctuations are random. Most data samples of both nutrients are probably not a sequence of independent observations. The data samples from transection (5) are the least consistent (below 5%) with the hypothesis of independent data and, furthermore, show a trend-like decrease of nutrient

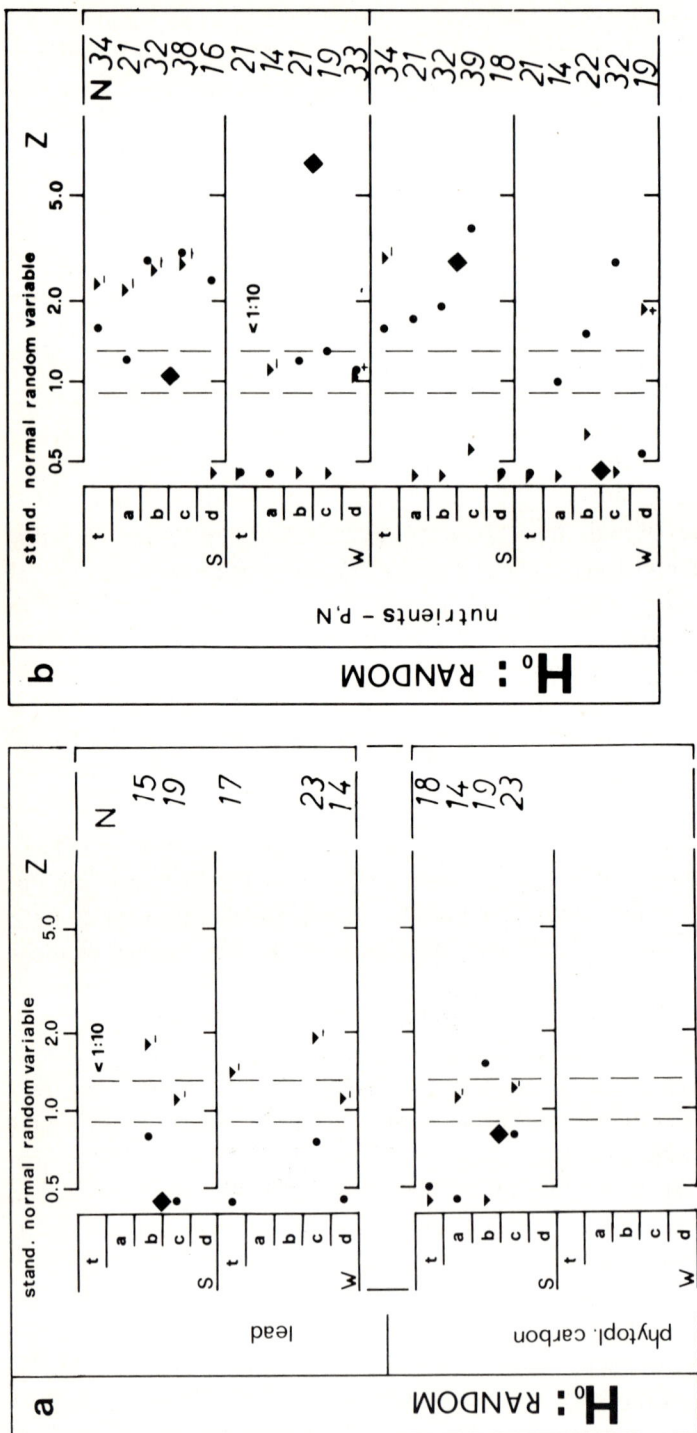

Fig. 4.7.5a,b. This dot chart shows estimates of standardized normal random variable for the null hypothesis "deviations are random" (see text) for three tests and four environmental parameters observed along five transections through the ZISCH star (summer and winter survey S, W). *N* gives the size of the sample. The transections are: *t* round the centre (understood as time series); *a* from SWW to NEE (from the British coast into the Skagerak); *b* NW to SE (Shetlands to German Bight); *c* and *d* counterclockwise along the coast (from Shetlands to Norwegian coast without entering in the Southern Bight and Skagerak) with *c* stations next to the coast and *d* further offshore. The test where (*dot*) sequence of independent data points, (*triangle*) trend-like changes (*minus sign* refers to decrease) and (*lozenge*) data of transection *b,c* are from the same distribution

concentrations. Such trend-like temporal changes are to be expected for the season. Generally, phosphate shows trend-like changes along most transections, but not nitrate. Phosphate and nitrate concentrations therefore form a pattern which probably has not been created by random fluctuations. However, the non-random fluctuations found for phosphate might be mainly due to seasonal changes. This conjecture is supported the high and low values, respectively, of the z-statistic of the range tests of phosphate and nitrate samples across the North Sea and along the coasts. It is accepted that the nitrate samples are from the same distribution, but not the phosphate distributions. On the other hand, these seem mainly to be structured along the coast, but without a trend-like change. Thus, concentrations along the British coast are not distinguishable from concentrations along the Danish-Norwegian coast.

The summer structure of nutrient concentrations does not imply only random fluctuations in their space and time dependency. As with other nutrients, patches of high nitrate concentrations were found along the coast. The observed pattern is unlikely for a purely random distribution. Less conspicuous features are consistent with the hypothesis "they are produced by random fluctuations".

Beyond the general statement that mean winter nutrient concentrations are comparatively higher, our analysis of winter data along transection (5) does not support trends or patterns. This is to be expected for the winter season. The along-coast transection shows a pattern which is improbable for random fluctuations. Samples collected along the coast and across the North Sea are accepted to be from the same distribution. The difference between the samples of phosphate concentrations collected across the North Sea and along the coast is the strongest indication for structure within the winter nutrient distribution. The probability of producing such a difference by chance is less than one in a million. However, the patterns found along the transections are mainly consistent with the hypothesis of random fluctuations.

Although in detail data along the transections are distributed at random, there are two exceptions: (1) there is a clear difference between the near-coast and cross-North Sea sections for the phosphate concentrations in winter and (2) the winter nitrate concentrations along the coast vary in a non-random way.

In summary: although predominantly consistent with the hypothesis of random changes, the analyzed data reveal significant structures with respect to nutrient concentration in a near-coastal transection around the North Sea. The result "predominantly consistent with hypothesis of random changes" is not as discouraging as it sounds, considering the fact that our tests were done in a way to take only a small risk of not identifying a feature as "random" if "random" is true. Furthermore, both the data samples were relatively small, and the dynamics of the North Sea are of partially random nature. Anyhow, accepting (or rejecting) both null hypotheses and alternative hypotheses is not inconsistent, but indicates that the sample is too small for the required levels of significance.

4.7.4   A Supplementary Comment

The issues discussed above have now to be brought into perspective with regard to the overall sampling scheme of the ZISCH survey. It should be recalled that the surveys had a duration of 39 and 45 days in summer and winter, respectively. Measurements were carried out at 205 and 142 stations. Distance and time lag between consecutive stations were between 10 and 50 sea miles and 1 and 7.5 h, respectively. Several parameters were observed on an even coarser grid. Lower bounds of resolvable scales are about 50 nm and one day.

Salinity, as the best-known environmental parameter, and the simulation of the advective influence index illustrate the high variability of the North Sea environment, and the ZISCH surveys thus in a certain sense were of snapshot quality. With this in mind, the results have to be interpreted with caution.

Many of the observed features are consistent with the hypothesis of random fluctuations. This result is not in contradiction to conclusions of the preceding chapters, where structures within the data are identified. It is inherent to a statistical reasoning that apparently contradictory statements are obtained - opposite hypotheses are accepted with respect to the data sample, but with different probability. The decision, whether any, none, or one of both hypotheses is accepted follows from the levels of significance which are required due to the extra-statistical understanding of the analyzed features (Hofstadter 1988).

In conclusion, the ZISCH survey provided a reliable set of data from a highly variable marine environment. This data set defines a restricted sample of the North Sea environment. The risk of over-interpreting this data set in terms of structures is therefore relatively high. For positive examples of interpreting this data set, the reader is referred to the other chapters of this Volume.

One topic stressed above was the basic problem of aliasing, due to undersampling of fluctuations, which are organized on short scales. No estimate of the degree of aliasing within the ZISCH sampling scheme can be given due to the missing knowledge on small-scale variability. However, it seems to be misleading to interpret the existing knowledge in a way that aliasing might cause only small distortions. Irregular sampling, interpolating or smoothing are no remedies to aliasing but causes more particular aliases (Shapiro and Silverman 1960; Thompson 1971; Jones 1972; Moore et al. 1988; Wunsch 1989). Thus, particular features observed once might be artefacts. However, a well-defined cause-effect relationship identified from the observations may counter this suspicion. In that case, a particular critical experiment might be designable (Bohle-Carbonell 1992), either as simulation exercise or field study.

References

Backhaus JO, Hainbucher D (1987) A finite difference general circulation model for shelf seas and its application to low frequency variability on the North European Shelf. In: Nihoul JCJ, Jamart BM (eds) Three-dimensional models of marine and estuarine dynamics. Elsevier, Amsterdam, pp 221-244

Becker GA, Prahm-Rodewald G (1980) Fronten im Meer, Salzgehaltsfronten in der Deutschen Bucht. Seewart 41:12-21

Bohle-Carbonell M (1990) On random sampling and adapted data interpretation - Case: Inner German Bight. In: Yap HT, Bohle- Carbonell M, Gomez ED (eds) Oceanography and marine pollution: an ASEAN-EC perspective. Marine Science Institute, Univ Philippines, Quezon City, Philippines, pp 310-333

Bohle-Carbonell M (1992) Pitfalls in sampling, Comments on variability and suggestions for simulation. Continent Shelf Res 12:3-24

Brockmann CW, Dippner JW (1987) Tidal correction of hydrographic measurements. Dtsch Hydrogr Z 40:241-260

Cochran WG (1977) Sampling techniques. Wiley, New York

Damm P (1989) Klimatologischer Atlas des Salzgehaltes, der Temperatur und der Dichte in der Nordsee, 1968-1985, Technischer Report. Inst Meereskunde, Univ Hamburg, Troplowitzstraße 7, D-2000 Hamburg 54

Frey H, Becker GA (1986) Untersuchung der langzeitigen Variation der hydrographischen Schichtung in der Deutschen Bucht. Umweltforschungsplan des Bundesministeriums für Umwelt, Naturschutz und Reaktorsicherheit Wasserforschungsbericht 102 04 215/20, 173 pp

Goedecke E, Smed J, Tomczak G (1967) Monatskarten des Salzgehaltes der Nordsee dargestellt für verschiedene Tiefenhorizonte. Dtsch Hydrogr Z Erg-H B9:13

Hainbucher D, Pohlmann T, Backhaus J (1987) Transport of conservative passive tracers in the North Sea: first results of a circulation and transport model. Continent Shelf Res 7:1161-1179

Hofstadter DR (1988) Gödel, Escher, Bach - ein endlosgeflochtenes Band. Klett-Cotta, Stuttgart, 844 pp

Jones RH (1972) Aliasing with unequally spaced observations. J Appl Meteorol 11:245-254

Katsaros KB, Fiuza A, Sousa F (1983) Sea surface temperature patterns and air-sea fluxes in the German Bight during MARSEN 1979, Phase 1. J Geophys Res C88:9871-9882

Lüneburg H (1963) Wassermischvorgänge vor der Weser- und Elbmündung. Veröff Inst Meeresforsch Bremerhaven 8:111-141

Moll A, Radach G (1990) ZISCH Parameter Report, Compilation of measurements from two interdisciplinary STAR-shaped surveys in the North Sea. Inst Meereskunde, Univ Hamburg, 15 pp

Moore MI, Thompson PJ, Shirtcliffe TGL (1988) Spectral analysis of ocean profiles from unequally spaced data. J Geophys Res C93:655-664

Nihoul JCJ (1980) Residual circulation long waves and mesoscale eddies in the North Sea. Oceanol Acta 3:309-316

Peterman RM (1990) Statistical power analysis can improve fisheries research and management. Can J Fish Aquat Sci 47:2-15

Prahm G (1961) Die Beobachtung auf dem Feuerschiff "S2". Ein Beitrag zur Hydrographie des Grenzbereichs zwischen Deutscher Bucht und südwestlicher Nordsee. Dtsch Hydrogr Z 14:218-239

Schönwiese CD (1985) Praktische Statistik für Meteorologen und Geowissenschaftler. Bornträger, Berlin, 270 pp

Shapiro HS, Silverman R (1960) Alias-free sampling of random noise. J Soc Ind Appl Math 8:225-248

Stange K (1970) Angewandte Statistik, Erster Teil, Eindimensionale Probleme. Springer, Berlin Heidelberg New York, 592 pp

Thompson R (1971) Spectral estimation from irregulary spaced data. IEEE Trans Geoscie Electr 9:107-110

Watson AJ, Upstill-Goddard RC, Liss PS (1991) Air-sea gas exchange in rough and stormy seas measured by a dual-tracer technique. Nature 349:145-147

Wunsch C (1989) Sampling characteristics of satellite orbits. J Atmos Ocean Technol 6:891-907

5 A Composite View of the North Sea Ecosystem and Future Research Needs

J. SÜNDERMANN

At the end of a book on environmental research in the North Sea, scientists and policymakers naturally want to know what conclusions can be drawn from the results with regard to the present situation and future developments. In this section, I will make an attempt to balance the necessary scientific rigour with the need for political action, well aware of the thin ice upon which I am treading. Considering the high natural variability of the North Sea system, considering our insufficient data base (even after ZISCH), and considering the deficiencies in our understanding of the processes underlying the system, all non-trivial statements regarding the North Sea ecosystem must remain simplifications which are uncertain, even subjective. The probability of making a false statement which will be refuted by further research is relatively large.

For this reason, we included a critical evaluation of our results in this book, Chapter 4.7: *On the Reliability of Our North Sea Assessment*. To my knowledge, this type of evaluation is an innovative addition to a book on the North Sea, and, admittedly, it was controversial within the ZISCH group. In this connection it is important to note that the selection of the analyzed data sets and the application of the tests were - as the author acknowledges - subjective and one-sided. Neither the - hardly quantifiable - "pre-knowledge" of an experienced researcher nor four-dimensional space-time relations, both of which would have appreciably reduced the probability of false statements, were considered. Nevertheless, the article demonstrates clearly the high risk of falsely interpreting results, the care which must be taken when making general statements or prognoses and the necessity for improved data bases. Hence, further North Sea research is essential.

Still, society and politics demand immediate steps to alleviate the "obvious" damage to the North Sea ecosystem - even if it means stretching the limits of reliable scientific knowledge.

ZISCH wishes to accept this challenge, since, either with or without the constructive cooperation of scientists, action will be taken, in the latter case based on a much weaker foundation of knowledge. In this conflict, the socially conscious scientist is compelled to abbreviate and to improvise in order to give policymakers (for example in connection with the International North Sea Conferences) a reliable basis for the decisions that must be made. He does this according to his best knowledge, conscious that this knowledge will be corrected by future research. The present Chapter (and this whole Volume) should be understood in this spirit.

Our use of the term "ecosystem" in the ZISCH project corresponds to the definition by K. Buchwald in his book *Nordsee - Ein Lebensraum ohne Zukunft?* (1990, p. 476), which translates:

"Entirety of biological community and environment, i.e. the network of relationships and interactions between organisms of a community and the natural and/or anthropogenic abiotic elements and structures in the environment, including the fluxes of energy, matter and information."

Not an isolated biological system is referred to, but rather the complete marine milieu, consisting of the dynamics of air and water (meteorology and physical oceanography), the sediment (geology), the transformation of matter in water and sediment (chemistry) and the organisms in water and the ocean bed (biology) as well as the interactions between the individual compartments. ZISCH concentrated, with few exceptions, on the lower trophic levels. Moreover, no work was carried out on biological effects; only the contaminant fluxes were investigated. This constitutes a further limitation on the statements regarding the North Sea ecosystem.

Our analysis and assessment of the state of the North Sea are based on a project-specific data set which we complied in the course of ZISCH. This data set is described below.

As emphasized in the introduction, ZISCH is distinguished by its holistic approach to analysis of the North Sea system. In recognition of the close interactions between the individual compartments - atmosphere, water body, dissolved and suspended matter, organisms and sediment - the result of which represents the "quality status" of the North Sea, contributions from the fields of meteorology, physical oceanography, marine chemistry, sedimentology, botany, zoology and ecology have been brought together for an overall interpretation. For example, primary production, a key influence on the flux of matter in the sea, is dependent on atmospheric forcing (wind, temperature, light) as well as on the hydrodynamic milieu (advection, turbulence, stratification) and the supply of nutrients by rivers, atmosphere and remineralization. It is also influenced by organic and anorganic contaminants (Chap. 4.3).

This and other considerations made it clear at the beginning of the project in 1985 that it would be necessary to acquire a new, comprehensive data set for the North Sea ecosystem. We were conscious of the fact that extensive, qualified research has been going on in the countries surrounding the North Sea, making it one of the most thoroughly investigated parts of the world ocean. Our analysis of available data showed, however, that in spite of the many cruises that have been carried out, these were almost never coordinated with each other and

-     neither blanket coverage of all relevant (physical, chemical and biological) ecological parameters including contaminants existed,

-     nor were time series available for analysis of statistical trends below, say, a 2-year signal - not because these trends do not exist, but rather because the scattered data do not reveal them.

Conception of an appropriate measurement strategy had to be preceded by estimation of the space and time scales of dominant biological processes (such as the phytoplankton spring bloom) and mass transports in a region characterized by extreme variability. Analysis of high-resolution satellite images of temperature, seston and chlorophyll distributions at the surface of the North Sea reveal coherency scales, i.e. the intervals within which structures and events are directly related, of around 5 km and several hours. More highly resolved measurements are appropriate for process studies and formulations. Above these scales, events are statistically independent and probability steps in place of deterministic laws. If, notwithstanding, causal models are applied on these scales, then an appropriate parameterization of sub-scale processes is necessary.

Since both source strength and advection affect the concentrations of contaminants, the variability of the latter will have a substantial effect on the contaminant load. Advection is forced by air and water currents, and these are highly variable even at the roughest approximation due to the stochastic character of atmospheric forcing. There are no mean conditions in nature and, consequently, no mean concentrations. Strong stochastic fluctuations (around a theoretical mean) are typical for a northern, temperate marine region. Analogously, fluctuations in emissions will also greatly affect the variance of contaminant concentrations.

This means that any survey will acquire only a random ensemble of statistically distributed parameters. Without an objective evaluation of the representativity of such data sets and their relation to the natural variance at each measurement site, their scientific merit is limited, to say nothing of their value for political decisions regarding protective measures.

Then how can one obtain information on the natural variability of the ecosystem? Naturally, every well-conceived and properly scaled survey is also a gain in statistics. But reliable statements of probabilities must be based on very large data sets, and cruises are costly and time-consuming. Model calculations, which are inexpensive in comparison to fieldwork, thus constitute a valuable tool for objective interpolation between data points and for simulating the stochastic variance of natural phenomena. This coupling of fieldwork and model calculations was a specialty of ZISCH.

As a consequence of the considerations mentioned above, the following double strategy was developed:

- *Phase 1*, consisting of two comprehensive surveys in space and time of the main parameters at as many stations and depths as possible in order to characterize the North Sea ecosystem above the coherence scales, supplemented by model calculations for the atmosphere and the water. This phase would produce large-scale information on the general environmental state of the North Sea and reveal statistical correlations between the individual ecosystem components. At its conclusion, it would permit a description of the present status of the North Sea without, however, providing information on causal relationships. That means that the questions posed in the introduction regarding future developments and the possibility of influencing them could not be answered. This requires further analyses.

- *Phase 2*, consisting of field and model investigations in a limited area (the German Bight) and with monthly resolution in order to study processes, formulate causal relationships between individual ecosystem compartments and work out parameterizations for large-scale models. At the end of this phase the interactions mentioned at the beginning of the introduction and the contaminant fluxes within the ecosystem should be formulated, quantified and incorporated into a (simple) causal model.

During the ZISCH project, the results of which are presented in this volume, the objectives of Phase 1 were realized. A star-shaped cruise track was chosen for the surveys (see Fig. 2.2.1), which were carried out by two or three ships over periods of approximately 6 weeks: 2 May - 13 June 1986 and 26 January - 9 March 1987. As mentioned in the Introduction, the observations were thus not synoptic. The star-shaped cruise track, however, meant that the central area of the North Sea was sampled several times during the cruises, so that an impression could be gained regarding the temporal variability of this area during the measurement period. More important, a Lagrangian back-calculation of the station positions was carried out, the results of which are shown in Fig. 2.2.2.

As a transition to Phase 2, but still during the ZISCH project, 6-monthly cruises were carried out in the German Bight from November 1988 to May 1989 (see Chap. 2.5). The frequent measurements during the half-year period were aimed at obtaining an overall view of the developments in one part of the North Sea by a simultaneous survey of as many relevant ecosystem parameters as possible. The objective was to gain an information basis for later process-oriented ecosystem studies in the German Bight (Phase 2). The cruises were carried out parallel to those of the British NERC (Natural Environment Research Council) North Sea Project in the western and southern North Sea.

The specific ZISCH data base upon which the descriptions and analyses in this book are grounded include not only these "classical" data obtained by cruises but also satellite images (Chap. 4.1), radar measurements (Chap. 2.7) and numerical data from calculations of the three-dimensional circulation in atmosphere and ocean together with the dispersion of contaminants. Further numerical data are obtained from a model of primary production (Chap. 3.3). In accordance with our recognition of the high variability of North Sea dynamics, the latter are forced by the actual data on meteorological conditions (Chaps. 2.1, 3.1) and on the known inputs for the periods of concern and rather than by theoretical, mean values.

What, then, does this data set tell us about the status of the North Sea environment? Let us first examine the results of the large-scale surveys. They show that the North Sea on the whole is burdened by nitrogen, heavy metals and halogenated hydrocarbons, but that the various regions are affected differently (Chaps. 2.2 and 2.3). Generally, it can be stated that the contamination in the coastal regions is greater than in the central North Sea. The German Bight, in particular, is threatened by inputs from the Rhine and the Elbe. This is compounded by the counterclockwise direction of the circulation in the North Sea, which causes inputs from Great Britain, Belgium and the Netherlands to eventually reach the German and Danish coasts. This situation is especially unfortunate in view of the fact that this most polluted zone contains the biologically extremely active Wadden Sea, which is so important for recruitment of many species in the North Sea.

The high level of contamination in coastal regions was to be expected due to the pollution of major rivers and the atmosphere. However, the ZISCH observations revealed concentration maxima in some compartments of the ecosystem in areas far away from the coasts. For example, unexpectedly high cadmium concentrations were found in sediment and in benthic organisms (hermit crabs) from the northwestern North Sea, far removed from presumed sources (Chaps. 4.5 and 4.6). This implies that, along with the coasts, the open sea is also endangered. The processes leading to these accumulations of contaminants are still not fully understood. Contributing factors are selective physico-chemical reactions on suspended particles and in organisms. These lead to completely different concentration patterns for water, suspended sediment and benthos.

The measurements of inputs from the atmosphere over the open North Sea (Chap. 2.4) have shown that this is an important transport medium for contaminants. Depending on meteorological conditions, movement over large distances is possible. For many heavy metals (e.g. cadmium, lead, zinc) most of the input enters via the atmosphere, and this is the case for one third of the total nitrogen input (see Table 1.1).

Results of model calculations of tracer dispersion for inputs from rivers (Chap. 3.2) and from the atmosphere (Chap. 3.1) generally showed a good qualitative agreement with survey data. Moreover, the models proved to be a helpful tool for revealing the high degree of variability in the contaminant load due to the stochastic fluctuations in the weather (time scales from weeks to years)(see Chap. 4.2). Systematic investigations have shown that one and the same contaminant load (assuming constant inputs) can lead to vastly different local concentrations, depending on the weather situation. The critical factor is the residence time of water bodies, and this depends mainly on the wind. Westerly winds transport water (including contaminants) relatively rapidly out of the North Sea and into the Atlantic. There are, however, unfavorable meteorological conditions, e.g. easterly winds lasting several weeks, which impede water exchange with the ocean and lead to transient peaks in contaminant concentrations. Although these situations are relatively seldom, they can have a limiting effect on the ecosystem. Calculations have shown that the range of variation is so large that - under constant input rates - the concentrations can range from very low to extremely high values (see Fig. 3.2.8). Data which have served as the basis for previous policy decisions could not reflect this sufficiently: they only describe a momentary state and necessarily underestimate worst cases.

Dissolved contaminants with sufficiently conservative properties leave the North Sea and enter the North Atlantic, depending on the advection and diffusion fields (see Chap. 3.2), after a certain residence time. For the western and southern North Sea this is - depending on type of input - from 2 to 4 years. These dissolved substances are responsible for a slow but steady contamination of the open ocean.

A large proportion of the contaminants, however, is transported in association with suspended particles (e.g. lead) (see Chap. 4.4). They are deposited in a few characteristic areas of the North Sea, leaving these (after resuspension) only after decades, if at all. Model calculations of the transport of suspended matter together with field measurements (Chap. 2.6) have yielded a clear picture regarding deposition of sediments and its seasonal variability (Chap. 3.2). They have also shown that the influence of the surface wave field (thus, of the wind and, with it, the climate and its changes) cannot be neglected for water depths up to 100 m. The most important "final resting place" in the North Sea is the Norwegian Trench, including the Skagerrak; the Wadden Sea must be regarded as an "intermediate dump" with exchange times on the order of 10 years.

There can be no doubt that the North Sea ecosystem has changed significantly in the past decades or that its "health" has declined. This is particularly evident in phenomena such as excessive algae blooms and oxygen deficiencies as well as fish diseases and the spectacular seal mortality in 1988/89 (Lozan et al. 1990). A reduction in species diversity has been observed (Kempe 1988). The simultaneous increase in contaminant concentrations in all compartments of the ecosystem leads to the conclusion that the observed environmental degradation has been caused by human activity.

Admittedly, however, this causal chain has not been stringently substantiated. Basically, environmental activists and scientists are in a dilemma on several planes. On the one hand, the "health" of the North Sea, the biological-pathological evaluation of a constantly changing extraordinarily complex system cannot be clearly defined. Moreover, the changes are in part evolutionary - thus intrinsic - and in part caused by external forces, the latter also partially natural (through global changes in the system Earth) and partially anthropogenic. These changes occur on vastly different time scales, which can extend from diurnal fluctuations up to geological periods, and encompass a wide diversity, e.g chemical compounds, physical forces or overfishing. It requires a certain courage of simplification and dedication to the natural environment to take a definite stand regarding potential and actual harm to the North Sea ecosystem by human activities. In view of the irreversibility of these processes and their significance for the entire biosphere - consequently, also for the existence of humanity - the only reaction to this dilemma can be the principle of precautionary action, whereby all anthropogenic discharges must be regarded as potentially hazardous unless proven otherwise.

Therefore, contaminant concentrations in all compartments of the North Sea ecosystem must be reduced. The results of ZISCH show clearly that effective protection of the North Sea necessitates that those responsible for inputs substantially reduce or eliminate them. In contrast, a policy based only on monitoring of local immissions (degree of dilution) is of little practical value due to the difficulty in defining the true potential for harm. In addition, considering the high stochastic variability of the system, a "quality status approach" can hardly be realized. Fluctuations in contaminant concentrations of orders of magnitude and on scales of hours and kilometres are typical and intrinsic to the North Sea system.

Pollution of the rivers and the atmosphere reach far inland. The catastrophe at Chernobyl and the chemical accident in Basle both demonstrated how severely the North Sea can be affected by faraway events. Protection of the North Sea is therefore a responsibility for all elements of society, and must emanate from all types of industry, agriculture, transportation, municipal disposal and, last but not least, from the environmental behaviour of each individual.

Convincing and encouraging examples for improvement are the reduction of phosphate inputs (due to the introduction of phosphate-free laundry soaps and to phosphate elimination in sewage plants) and lead emissions (due to the introduction of lead-free gasoline). The measurements of atmospheric lead deposition over the North Sea carried out in the ZISCH project indicate a slight reduction in this element (Chap. 2.4). These are signs that the chance of reversing the anthropogenic harm to the North Sea does exist, if the attitudes of society change sufficiently and if the necessary funds (e.g. for building sewage plants, developing new technology) are invested. These examples are convincing enough to make this path worth following.

At present, we are still at the very beginning, though. A number of persistent xenobiotics such as polychlorinated biphenyls (PCB) and hexachlorobenzenes (e.g. Lindane) are still entering the North Sea. New synthetic compounds in sea water are constantly being identified, and research on their effects on the ecosystem (and that of their degradation products) is just beginning. Some basic results are elaborated within ZISCH. A major problem is posed by nitrogen, inputs of which are increasing. Its role in primary production and eutrophication makes it a controlling factor for the entire food web in those regions where it was the limiting nutrient up to now, for example the central North Sea. In this way it could influence the whole carbon/nitrogen cycle and, hence, long-term climate change.

The long residence times for contaminants transported into the North Sea mean that even drastic protective measures will not be immediately effective. We have seen that the residence times for contaminants dissolved in water are in the order of years; for particle-bound contaminants they are in the order 10-100 years. At first these substances, originating from rivers or wet deposition from the atmosphere, are deposited near the coasts (primarily in the Wadden Sea or in the southern North Sea). Here, e.g. through bioturbation or fishing activity, they may be worked into deeper sediment layers. This means that surface concentrations may be decreased through dilution by older, less contaminated sediments, but that, on the other hand, a thicker layer of sediment becomes contaminated. Waves caused by extreme wind events resuspend this material so that the contaminants repeatedly return to the biogenic cycle. The floor of the southern North Sea must thus be seen as a long-term source for contaminants even after inputs are reduced. Our results (Chap. 3.2) show that this holds true for water depths of less than 50 m, for example in a sensitive area such as the Dogger Bank.

The general mechanisms of advection and deposition of suspended matter in the North Sea lead to a kind of "final dump" in the Norwegian Trench. The currents at the depths of 300 m and more are generally so small that the sediment remains in place. Accordingly, cores from this region document the course of the industrialization of Europe and the attendant contamination of the environment. It can be presumed that this material will be reactivated only on geological time scales, but radical climate change could accelerate this process.

A word will be said in the following about ecological quality objectives. In the same sense that the present state of the North Sea must be regarded relative to some largely "undisturbed" reference level (see Chap. 2.8), the question arises as to what kind of "ideal condition" should be striven for by environmental policymakers.

This is a point which scientists have debated heatedly in recent times. Politicians and administrators tend to have a more practical definition of quality objectives, the gradual fulfillment of which should be measurable (e.g. with statistics such as threshold values). Environmental protection costs a tremendous amount of money and requires radical changes in government and society. Its chances of success and acceptance increase if it can describe present and future states as a numerical matrix and connect both in an optimal way (with respect to cost-benefit relations). The necessary investments thus become more transparent, and it is possible to determine whether fulfillment of the quality objectives is actually progressing.

Many ecologists, physicists and system theorists, on the other hand, dispute the usefulness of ecological quality objectives on the grounds that our environment, in particular the North Sea, represents an extraordinarily complex, highly variable dynamic system which is in a state of constant evolution. Any fixation of quality objectives formulated by humans would automatically achieve an artificial result. Measures for realizing the objectives would - since they could not allow for the myriad unknowns in the system - necessarily constitute an impediment to the natural course of evolution. Arguing along these lines one would have to reject practically all inputs on the grounds that they are a potential hazard to the environment. At the same time, the environment would have to be constantly monitored for previously undetected contamination sources.

In the ZISCH project, we have tried to mediate between these two extreme viewpoints. Scientific reasoning would tend to question the validity of defining ecological quality objectives for a highly complex, non-linear system. The contribution on the reliability of our data (Chap. 4.7) reveals how fragile our description of the present state of the North Sea is. How then, can a reliable way of achieving them be found? On the other hand, scientists are members of society and know that protective measures must be accepted politically and be financed. We cannot suppose that all inputs of potentially harmful substances will be reduced to zero immediately. Financial reasons alone will prolong this process over a period of decades. Priorities must therefore be set, and this is where a contribution by science is crucial. Important questions regard which measures are the most pressing and which regulations must be made first in order to allow the North Sea ecosystem the most natural course of development - even without an exact idea of a desired "final state". Answering these necessitates instruments for reliable prediction of the effects of specific emission reductions. The development of such prediction instruments has been one of the main objectives of ZISCH. Work was done on development of comprehensive models for simulating the causal chain input - circulation and transport - chemical/biologicaltransformation - environmentalload(seeFig. 1.1 in Chap. 1). This program system includes separate models for the atmosphere, the water body, suspended matter, primary production etc., which must be verified by appropriate data sets from nature. Formulation of the governing physical, sedimentological, chemical and biological processes is decisive for this, and represents one of the main accents of ZISCH and the ensuing PRISMA (Processes Influencing Contaminant Fluxes in the North Sea) project. Eventually, it should be possible to simulate specific protective measures and predict the probable consequences of environmental actions such as the building of sewage plants and the reduction of NO_x emissions for the regional distribution of contaminants in the North Sea. We are of the opinion that clarifying the relationship between financial investment on land and improvement of the marine environment (in the sense of a reduction in contaminant concentrations) would convince the public and policymakers of the need and the benefits of long-term investments in environmental protection.

However, we are well aware of the fact that we are far from a complete understanding of the causal relationships in the complex North Sea ecosystem and from hind- and forecasts of them. Perhaps this will always be the case. The stochastic influences of weather and climate are too strong, the chaotic structures too prevalent and the processes involved too little known. Moreover, the ZISCH/PRISMA projects have excluded almost the entire aspect of contaminant effects on organisms. This means that we cannot draw conclusions from the contaminant loads of the individual compartments regarding the damage to higher organisms. This is a task for a comprehensive research program in its own right. Nevertheless, the overall goal of ZISCH/PRISMA to describe the causal relationships input-contamination under the atmospheric-hydrographic conditions in the North Sea is a necessary, fundamental step for improvement of the present state. At least trends and time scales can be correctly predicted using our methods.

At the end of this book questions remain concerning the main research deficits after ZISCH. I would divide these into the following six theme groups, some of which have been subject to investigation in the PRISMA project.

1.   Smaller spatial and temporal time scales must be resolved.

The characteristic resolution of the ZISCH surveys was several tens of kilometres and from several hours to days. The North Sea models with grid-spacing of around 20 km and time steps of 20 min to 12 h (as must be the case for a comparison between observations and model simulations) lie in approximately the same order of magnitude.

With these scales it is not possible to resolve important processes in North Sea dynamics. Some pronounced orographic and topographic features of the southern North Sea have sizes in the range of several kilometres. This is particularly the case for the extended Wadden Sea areas which - as we have seen - are an important internal source/sink for the system. An over-riding factor for the scaling of a stratified water body on the rotating earth is the baroclinic Rossby radius

$$R = \frac{Nh}{f}$$

where

$$N = \left(-\frac{g}{\rho} \frac{d\rho}{dz} \right)^{1/2}$$

is the Brunt-Väisälä frequency of the internal waves, h the water depth, f the Coriolis parameter, g the earth's acceleration, ρ the density and z the vertical coordinate (positive upwards). For the southern North Sea one obtains values of R = 2-5 km, as can actually be seen for the eddy structures recognizable in satellite images. The corresponding periods are around 20-60 min.

Only by going below these scales is it possible to detect coherent structures for which deterministic, causal relations can be formulated. Process studies are also only possible below this boundary. The subsequent PRISMA project therefore applied itself to the smaller scales. However, this does not mean that there is no need for future work on these scales or that past work on larger scales is meaningless. Financial and logistical expenses alone dictate further large-scale work. In the corresponding models, sub scale processes then need to be appropriately parameterized in order to describe the relationship between them and the global field properly. This presupposes representative process studies in limited areas.

2.   Dominant processes for circulation and contaminant cycling must be investigated more thoroughly.

(Small-scale) processes such as turbulence, mixing, erosion, adsorption, assimilation, respiration and grazing determine the dynamics of the physical-biological system. They appear in the basic equations of the appropriate simulation models, but they have by no means been adequately studied, especially considering the problems posed by contaminants to the marine ecosystem. Here it is necessary to gain new knowledge through coordinated field, laboratory and model experiments. For example, in the field of experimental and theoretical hydrodynamics much is known about turbulent flow, but the application to actual marine conditions with complicated topography, stratification, tides and atmospheric momentum and energy input has still hardly been accomplished. The same is true for the determination of shear velocities at the bottom or sinking velocities for flocculating suspended matter. Also, cycles of matter in marine water bodies as influenced by real oceanographic and meteorological peripheral conditions are not well enough understood; in situ experiments with closed mesocosms are well suited for this.

After ZISCH, attention must be paid to identifying and observing as well as quantitatively and qualitatively describing processes. With regard to the larger-scale models, parameterizations of processes and examination of their performance in test runs - compared with appropriate data sets from nature - are the most important tasks at hand.

3.   Internal sources and sinks of the system must be studied in more detail.

Along with the processes, internal sources and sinks represent a group of important terms in the budget equations for the system. They take into consideration whether a contaminant behaves in a conservative or non-conservative manner, whether it remains in the system or leaves, and whether it accumulates or decomposes. This category includes deposition and resuspension of sediment, bioturbation, biological decomposition by bacteria, contaminant uptake by phytoplankton etc. The magnitudes, mechanisms and processes determine the mass and energy budget of the ecosystem to a decisive degree. Thus, ZISCH was able to show that the sediment distribution in the North Sea can only be simulated correctly by considering the influence of the wave field on resuspension (Chap. 3.2). It also elucidated the significant role of phytoplankton in the temporary withdrawal of copper from the water body. The role of microbiota may be likewise important and we are well aware of this deficiency in ZISCH. The sources and sinks mentioned above can be so important that mass budgets are completely false to the first approximation. This results in a corresponding research need which, as with the processes, also should concentrate on the small scales. The step to global models must then include formulation of appropriate parameterizations which have been verified by observational data.

4.   More detailed information must be made available on contaminant inputs, particularly from rivers and the atmosphere.

This refers to the external, usually anthropogenic sources in the system. Despite environmental education and regulations, the continental inputs are insufficiently known. The dynamical variability of the system dictates - as we have seen above - time scales on the order of hours. At the moment we are far from having any such highly resolved information about inputs by the major European rivers into the North Sea, not to mention differentiation between contaminants.

    The unification of the European community and the disintegration of the communist block may have eliminated the political barriers to this kind of information; however, the legal, administrative and logistical difficulties are still so great that these necessary input data will remain a primary deficiency of model calculations. This is no task for research. It is an environmental debt of European dimensions which can only be settled by means of coordinated monitoring concepts and central data administration.

The same applies to the atmosphere. Since it will hardly ever be possible to measure the inputs at the North Sea margins on a routine basis, this contribution must be made by model calculations. These models, which reach far inland, need terrestrial input data in the required resolution. The available emission data are still too coarse - in spite of the effort and progress that has gone into improving them. It is an anachronism that modern transport and dispersion models in the atmosphere and the ocean can resolve the internal dynamics of the system for periods of time far shorter than days, while the input data for contaminants are still in the form of monthly or even yearly means (and then only made available after long delays).

5.   Lateral interactions with the bordering marine areas have not been investigated sufficiently.

As is the case with any marginal sea, the North Sea is not a closed system but rather a heterogeneous body which interacts with the bordering marine areas. Important fluxes of momentum, energy and matter cross these (imaginary) borders. Corresponding models are influenced decisively by the formulation of the boundary conditions. Their determination usually occurs - more or less schematically - on the basis of historical data, model assumptions or is provided for by a higher ranking model. The situation is unsatisfactory on the whole; for some state variables not even the sign of the flux between the Atlantic and the North Sea is known.

Here, research deficiencies at three open borders are referred to. The first of these is the interaction between the North Atlantic and the North Sea along the shelf break. As with the other borders, transfers in both directions are relevant here. The Atlantic may have an immensely greater momentum and energy budget and thus dominate the the North Sea in this respect, but for the global ecosystem and climate system important mass inputs occur from the land, eventually affecting the ocean. This research deficit is to be taken up by the international programs LOICZ (Land-Ocean Interaction in the Coastal Zone) and OMEX (Ocean Margin Experiment).

The North Sea and the Baltic also influence each other. The Baltic is a humid sea from which a net efflux of freshwater ensues, resulting in a high input of contaminants due to a high level of emissions. On the other hand, the Baltic is periodically supplied with oxygenated deep water from the North Sea. The interactions between these two areas have not been investigated sufficiently, and the present international programs will not alleviate this deficit.

Finally, there are the interactions between the open North Sea and the bordering Wadden Sea. The latter are a characteristic particularity which, in spite of their relatively small area, exert a decisive influence on the flow of matter and the regeneration of organisms in the entire North Sea. As in all North Sea research programs, ZISCH/PRISMA have practically neglected this region; the spatial scales resolved are too coarse. We were aware of this deficiency when the projects were conceived, but inclusion of the Wadden Sea was not feasible for both financial and logistical reasons. Moreover, currently there are several specific Wadden Sea research projects in the Netherlands and in Germany. It must be noted, however, that these are aimed more at clarifying internal dynamics of an amphibian ecosystem than at investigating the interrelationships with the bordering open sea. Since shelf sea models do not regard the Wadden Sea, its processes and its internal sources and sinks, they must be false in this respect. Setting river input at a single grid point of the North Sea model neglects the enormous transforming and retarding effects of the Wadden Sea.

The experience from ZISCH/PRISMA reveals a considerable research deficit with respect to the fluxes between the Wadden Sea and the open sea and to the coupling of the corresponding models.

6.   Causal relationships must be formulated and scenarios simulated.

ZISCH had the objective of describing the causal relationships in the system shown schematically in Fig. 1.2 (Chap. 1). This has been accomplished to a certain degree; however, it must be extended and continued corresponding to the progress which has been made in the areas (1) to (5). Specific investigations with verified models must simulate causal correlations within the entire physical-biological system. This should include, for example, such obvious or presumed relationships as

- Topography causes eddies and fronts; these affect deposition of suspended matter.
- Long periods of easterly winds block water exchange in the North Sea and lead to an increase in contaminant concentrations.
- A warm winter with profuse precipitation is conducive to energy and nutrient input and, thus, algae blooms.

Another aspect which also belongs to this theme group is investigation of the relative significance of mean meteorological-hydrographical conditions as opposed to extreme events for shaping the ecosystem. Answering this question for the North Sea as a whole and in its parts is crucial for the conception of measurement strategies, monitoring schemes and protective measures.

Finally, after sufficient verification of the models and comprehensive analysis of causal relationships in the ecosystem, suitable scenario simulations should be conceived and carried out. On the one hand, they should aim at distinguishing "natural" changes (those induced by climatological influences or the inner evolution of the system) from "artificial" ones (those caused by human activity). On the other, they should at least qualitatively reproduce and predict the effects of particular protective measures on the contaminant load in the North Sea. This would make the models available for political decisions as an objective, scientific intrument.

Common sense tells us, however, that they will never be able to completely describe the manifold interrelationships in nature and must therefore remain tools for achieving something between the ecological optimum and that which is politically realistic. They do not release us from the obligation of the precautionary principle.

References

Buchwald K (1990) Nordsee - ein Lebensraum ohne Zukunft? Die Werkstatt, 552 pp

Kempe S, Liebezeit G, Dethlefsen V, Harms U (eds) (1988) Biogeochemistry and distribution of suspended matter in the North Sea and implications to fisheries biology. Mitt Geol-Paläontol Inst Univ Hamb 65:547 pp

Lozan JL, Lenz W, Rachor E, Watermann B, von Westernhagen H (eds) (1990) Warnsignale aus der Nordsee. Parey, Hamburg, 428 pp

Printing: Mercedesdruck, Berlin
Binding: Buchbinderei Lüderitz & Bauer, Berlin